# *Embedded Systems Architecture*

*To the engineer and man I respect and admire the most,*
*my father,*

*Dr Al M. Zied*

# Embedded Systems Architecture

*A Comprehensive Guide for Engineers and Programmers*

Tammy Noergaard

AMSTERDAM • BOSTON • HEIDELBERG • LONDON • NEWYORK • OXFORD • PARIS
SAN DIEGO • SAN FRANCISCO • SINGAPORE • SYDNEY • TOKYO

Newnes is an imprint of Elsevier

Newnes is an imprint of Elsevier
The Boulevard, Langford Lane, Kidlington, Oxford, OX5 1GB
225 Wyman Street, Waltham, MA 02451, USA

First edition 2005
Second edition 2013



**Notices**

Knowledge and best practice in this field are constantly changing. As new research and experience broaden our understanding, changes in research methods, professional practices, or medical treatment may become necessary.

Practitioners and researchers must always rely on their own experience and knowledge in evaluating and using any information, methods, compounds, or experiments described herein. In using such information or methods they should be mindful of their own safety and the safety of others, including parties for whom they have a professional responsibility.

To the fullest extent of the law, neither the Publisher nor the authors, contributors, or editors, assume any liability for any injury and/or damage to persons or property as a matter of products liability, negligence or otherwise, or from any use or operation of any methods, products, instructions, or ideas contained in the material herein.

**British Library Cataloguing-in-Publication Data**
A catalogue record for this book is available from the British Library

**Library of Congress Cataloguing-in-Publication Data**
A catalog record for this book is available from the Library of Congress

ISBN: 978-0-12-382196-6

For information on all Newnes publications
visit our website at store.elsevier.com

Typeset by MPS Limited, Chennai, India
www.adi-mps.com

Printed and bound in the United States

13 14 15 16    10 9 8 7 6 5 4 3 2 1

# Contents

# Foreword to the First Edition

When Tammy Noergaard first told me she wanted to write a soup-to-nuts book about building embedded systems I tried to dissuade her. This field is so vast, requiring insight into electronics, logic circuits, computer design, software engineering, C, assembly, and far more. But as we talked she showed me how the industry's literature lacks a definitive work on the subject. I warned her of the immensity of the project.

A year and many discussions later Fedex arrived with the review copy of this book. At approximately 700 pages it's appropriately twice the size of almost any other opus on the subject. The book you're holding truly is "A Comprehensive Guide for Engineers and Programmers." Sure, the minutiae of programming a PIC's timer might have been left out, but the scope is vast and important.

Tammy starts with the first principles of electronics and advances through software to the expensive end-phase of maintenance. She treats hardware and software as an integrated whole, which sort of defines the nature of embedded systems. Ironically, though, developers are increasingly specialized. More than a few software folks haven't a clue about transistors while too many EEs can't accurately define middleware. I fear readers may skip those chapters that don't immediately pertain to the project at hand.

Resist any such temptation, gentle reader! Become a true master, an embedded sage, by broadening your horizons to cover all aspects of this fascinating field. We engineers are professionals; you and I know this in our hearts. Yet true professionals are those who learn new things, who apply newly evolving technologies to solve problems. Consider doctors: the discovery and production of penicillin in the 1940s changed the profession of medicine forever. Any doc who ignored this new technology, who continued to practice using only the skills learned in college, was suddenly rendered a butcher. Software and hardware developers are faced with the same situation. C wasn't taught when I went to school. The FPGA hadn't been invented. GOTOs were still just fine, thank you. We learned to program microprocessors in machine code using primitive toolchains. Today—well, we know how much has changed.

The rate of change is increasing; change's first derivative is an ever-escalating positive number. Professional developers will read this book from cover to cover, and will constantly seek out other sources of information. If you're not at least surfing through a half dozen technical magazines a month and reading a handful of books like this per year, then it won't take a Cretaceous asteroid to make you a dinosaur.

Some of this book might surprise you. Ten pages about reading datasheets? Fact is, datasheets are dense formal compilations of contractual material. The vendor promises the part will do "X" as long as we use it in an agreed-on manner. Violate any of perhaps thousands of specifications and the part will either not work or will be unreliable. With some parts dissipating 100 watts or more, even such arcana as thermal characteristics are as important as the device's instruction set.

Tammy's generous use of examples elucidates the more obscure points. Engineering—whether hardware or software—is the art of building things and solving problems. The academics can work with dry theory; we practicing developers often learn best by seeing how something works. So the chapter on device drivers does explain the intricacies of building these often-complex bits of code, but couples the explanation to a wealth of real-world examples.

Finally, Tammy's words about the Architecture Business Cycle of embedded systems resonate strongly with me. We don't build these things just to have a good time (though we sure hope to have one along the way), but to solve important business problems. Every decision we make has business implications. Use too little horsepower and development costs skyrocket—sometimes to the point of making the project unviable. A poor analysis of the problem that leads you to toss in an excess of Flash might drive costs unacceptably high. Select a component (hardware or software) from a failing company and your outfit may share in the vendor's demise.

Enjoy this book, and futureproof your career at the same time.

**Jack Ganssle**

# Acknowledgments

My greatest debt in creating this second edition goes to my readers and reviewers, who I hope will be pleasantly surprised to see how many of their suggestions have been incorporated into the book. They include Dr Al M. Zied, both of my brothers (especially my younger brother who also provided me the inspiration to write this book in the first place), Jack Ganssle, and Steve Bailey.

Thank you to my publisher Elsevier, specifically to the Elsevier team for their hard work and dedication in making this book a reality.

I would also like to acknowledge my Terma team members as some of the most talented I have been lucky enough to work with, as well as my mentor when I was with Sony Electronics, Kazuhisa Maruoka. Maruoka-san patiently trained me to design televisions and gave me such a strong foundation upon which to grow, as well as my manager at Sony Electronics, Satoshi Ishiguro, who took a chance and hired me. My journey in the embedded systems field that has led me to writing this book began with the great people I worked with at Sony in Japan and in San Diego.

Thank you to my family for their support. For my beautiful girls Mia and Sarah, thank you for always gifting me with endless smiles, hugs, and kisses. Thank you, Mom and my brothers and sister in Southern California for their support and for encouraging me every step of the way.

Finally, a very special thanks to Kenneth Knudsen and his welcoming, loving family. Also, the US Embassy for all their support, other fellow Americans here in Denmark, and to the lovely Danes I have been lucky enough to have in my life—shining the light on the hardest days while I was working to finish this edition in Denmark. To all the other friends who have touched Mia's, Sarah's, and my life here in Denmark. I will always be grateful.

# *About the Author*

Tammy Noergaard is uniquely qualified to write about all aspects of embedded systems. Since beginning her career, she has gained wide experience in product development, system design and integration, operations, sales, marketing, and training. She has design experience using many hardware platforms, operating systems, middleware, and languages. Tammy worked for Sony as a lead software engineer developing and testing embedded software for analog TVs, and also managed and trained new embedded engineers and programmers. The televisions she helped to develop in Japan and California were critically acclaimed and rated #1 in Consumer Reports magazines. She has consulted internationally for many years, for companies including Esmertec and WindRiver, and has been a guest lecturer in engineering classes at the University of California at Berkeley, Stanford University, as well as giving technical talks at the invitation of Aarhus University for professionals and students in Denmark. Tammy has also given professional talks at the Embedded Internet Conference and the Java User's Group in San Jose over the years. Most recently, her experience has been utilized in Denmark to help ensure the success of fellow team members and organizations in building best-in-class embedded systems.

SECTION I

# Introduction to Embedded Systems

## Introduction to Embedded Systems

The field of embedded systems is wide and varied, and it is difficult to pin down exact definitions or descriptions. However, Chapter 1 introduces a useful model that can be applied to any embedded system. This model is introduced as a means for the reader to understand the major components that make up different types of electronic devices, regardless of their complexity or differences. Chapter 2 introduces and defines the common standards adhered to when building an embedded system. Because this book is an overview of embedded systems architecture, covering every possible standards-based component that could be implemented is beyond its scope. Therefore, significant examples of current standards-based components have been selected, such as networking and Java, to demonstrate how standards define major components in an embedded system. The intention is for the reader to be able to use the methodology behind the model, standards, and real-world examples to understand any embedded system, and to be able to apply any other standard to an embedded system's design.

CHAPTER 1

# A Systems Approach to Embedded Systems Design

**In This Chapter**

- Defining embedded system
- Introducing the design process
- Defining an embedded systems architecture
- Discussing the impact of architecture
- Summarizing the remaining sections of the book

## 1.1 What Is an Embedded System?

An embedded system is an applied computer system, as distinguished from other types of computer systems such as personal computers (PCs) or supercomputers. However, you will find that the definition of "embedded system" is fluid and difficult to pin down, as it constantly evolves with advances in technology and dramatic decreases in the cost of implementing various hardware and software components. Internationally, the field has outgrown many of its traditional descriptions. Because the reader will likely encounter some of these descriptions and definitions, it is important to understand the reasoning behind them and why they may or may not be accurate today, and to be able to discuss them knowledgeably. The following are a few of the more common descriptions of an embedded system:

- *Embedded systems are more limited in hardware and/or software functionality than a PC.* This holds true for a significant subset of the embedded systems family of computer systems. In terms of hardware limitations, this can mean limitations in processing performance, power consumption, memory, hardware functionality, etc. In software, this typically means limitations relative to a PC—fewer applications, scaled-down applications, no operating system (OS) or a limited OS, or less abstraction-level code. However, this definition is only partially true today as boards and software typically found in PCs of the past and present have been repackaged into more complex embedded system designs.
- *An embedded system is designed to perform a dedicated function.* Most embedded devices are primarily designed for one specific function. However, we now see devices such as personal data assistant (PDA)/cell phone hybrids, which are embedded systems designed to be able to do a variety of primary functions. Also, the latest digital TVs

include interactive applications that perform a wide variety of general functions unrelated to the "TV" function but just as important, such as e-mail, web browsing, and games.

- *An embedded system is a computer system with higher quality and reliability requirements than other types of computer systems.* Some families of embedded devices have a very high threshold of quality and reliability requirements. For example, if a car's engine controller crashes while driving on a busy freeway or a critical medical device malfunctions during surgery, very serious problems result. However, there are also embedded devices, such as TVs, games, and cell phones, in which a malfunction is an inconvenience but not usually a life-threatening situation.
- *Some devices that are called embedded systems, such as PDAs or web pads, are not really embedded systems.* There is some discussion as to whether or not computer systems that meet some, but not all, of the traditional embedded system definitions are actually embedded systems or something else. Some feel that the designation of these more complex designs, such as PDAs, as embedded systems is driven by non-technical marketing and sales professionals, rather than engineers. In reality, embedded engineers are divided as to whether these designs are or are not embedded systems, even though currently these systems are often discussed as such among these same designers. Whether or not the traditional embedded definitions should continue to evolve or a new field of computer systems be designated to include these more complex systems will ultimately be determined by others in the industry. For now, since there is no new industry-supported field of computer systems designated for designs that fall in between the traditional embedded system and the general-purpose PC systems, this book supports the evolutionary view of embedded systems that encompasses these types of computer system designs.
- Electronic devices in just about every engineering market segment are classified as embedded systems (see Table 1-1). In short, outside of being "types of computer systems," the only specific characterization that continues to hold true for the wide spectrum of embedded system devices is that *there is no single definition reflecting them all.*

## 1.2 An Introduction to Embedded Systems Architecture

In order to have a strong technical foundation, all team members must first start with understanding the **architecture** of the device they are trying to build. The *architecture* of an embedded system is an *abstraction* of the embedded device, meaning that it is a generalization of the system that typically doesn't show detailed implementation information such as software source code or hardware circuit design.[3] At the architectural level, the hardware and software components in an embedded system are instead represented as some composition of interacting *elements*. Elements are representations of hardware and/or software whose implementation details have been abstracted out, leaving only behavioral and inter-relationship information. Architectural elements can

**Table 1-1: Examples of embedded systems and their markets[1]**

| Market | Embedded Device |
|---|---|
| Automotive | Ignition system<br>Engine control<br>Brake system (i.e., antilock braking system) |
| Consumer electronics | Digital and analog televisions<br>Set-top boxes (DVDs, VCRs, cable boxes, etc.)<br>Personal data assistants (PDAs)<br>Kitchen appliances (refrigerators, toasters, microwave ovens)<br>Automobiles<br>Toys/games<br>Telephones/cell phones/pagers<br>Cameras<br>Global Positioning Systems (GPS) |
| Industrial control | Robotics and control systems (manufacturing) |
| Medical | Infusion pumps<br>Dialysis machines<br>Prosthetic devices<br>Cardiac monitors |
| Networking | Routers<br>Hubs<br>Gateways |
| Office automation | Fax machines<br>Photocopiers<br>Printers<br>Monitors<br>Scanners |

be internally integrated within the embedded device or exist externally to the embedded system and interact with internal elements. In short, an embedded architecture includes elements of the embedded system, elements interacting with an embedded system, the properties of each of the individual elements, and the interactive relationships between the elements.

Architecture-level information is physically represented in the form of *structures*. A structure is one possible representation of the architecture, containing its own set of represented elements, properties, and inter-relationship information. A structure is therefore a "snapshot" of the system's hardware and software at design time and/or at runtime, given a particular environment and a given set of elements. Since it is very difficult for one "snapshot" to capture all the complexities of a system, an architecture is typically made up of more than one structure. All structures within an architecture are inherently related to each other, and it is the *sum* of all these *structures* that is the embedded *architecture* of a device. Table 1-2

**Table 1-2: Examples of architectural structures[4]**

| Structure Types[1] | | | Definition |
|---|---|---|---|
| Module | | | Elements (referred to as modules) are defined as the different functional components (the essential hardware and/or software that the system needs to function correctly) within an embedded device. Marketing and sales architectural diagrams are typically represented as modular structures, since software or hardware is typically packaged for sale as modules (an OS, a processor, a Java Virtual Machine (JVM), etc.). |
| | Uses (also referred to as subsystem and component) | | A type of modular structure representing system at runtime in which modules are inter-related by their usages (e.g., what module uses what other module). |
| | | Layers | A type of uses structure in which modules are organized in layers (i.e., hierarchical) in which modules in higher layers use (require) modules of lower layers. |
| | | Kernel | Structure presents modules that use modules (services) of an operating system kernel or are manipulated by the kernel. |
| | | Channel architecture | Structure presents modules sequentially, showing the module transformations through their usages. |
| | | Virtual machine | Structure presents modules that use modules of a virtual machine. |
| | Decomposition | | A type of modular structure in which some modules are actually subunits (decomposed units) of other modules and inter-relations are indicated as such. Typically used to determine resource allocation, project management (planning), data management (e.g., encapsulation, privatization), etc. |
| | Class (also referred to as generalization) | | This is a type of modular structure representing software and in which modules are referred to as classes, and inter-relationships are defined according to the object-oriented approach in which classes are inheriting from other classes or are actual instances of a parent class, for example. Useful in designing systems with similar foundations. |
| Component and connector | | | These structures are composed of elements that are either components (main hardware/software processing units, e.g., processors, a JVM) or connectors (communication mechanism that interconnects components, such as a hardware bus, or software OS messages, etc.). |

*(Continued)*

**Table 1-2: (Continued)**

| Structure Types[1] | | | Definition |
|---|---|---|---|
| | Client/server (also referred to as distribution) | | Structure of system at runtime where components are clients or servers (or objects) and connectors are the mechanisms used (protocols, messages, packets, etc.) used to intercommunicate between clients and servers (or objects). |
| | Process (also referred to as communicating processes) | | This structure is a software structure of a system containing an operating system. Components are processes and/or threads (see Chapter 9 on OSs), and their connecters are the interprocess communication mechanisms (shared data, pipes, etc.) Useful for analyzing scheduling and performance. |
| | Concurrency and resource | | This structure is a runtime snap shot of a system containing an OS and in which components are connected via threads running in parallel (see Chapter 9 on OSs). Essentially, this structure is used for resource management and to determine if there are any problems with shared resources, as well as to determine what software can be executed in parallel. |
| | | Interrupt | Structure represents the interrupt handling mechanisms in system. |
| | | Scheduling (Earliest Deadline First (EDF), priority, round-robin) | Structure represents the task scheduling mechanism of threads demonstrating the fairness of the OS scheduler. |
| | Memory | | This runtime representation is of memory and data components with the memory allocation and deallocation (connector) schemes—essentially the memory management scheme of the system. |
| | | Garbage collection | This structure represents the garbage allocation scheme (more in Chapter 2). |
| | | Allocation | This structure represents the memory allocation scheme of the system (static or dynamic, size, etc.). |
| | Safety and reliability | | This structure is of the system at runtime in which redundant components (hardware and software elements) and their intercommunication mechanisms demonstrate the reliability and safety of a system in the event of problems (its ability to recover from a variety of problems). |

(*Continued*)

**Table 1-2: (Continued)**

| Structure Types[1] | | Definition |
|---|---|---|
| Allocation | | This structure represents relationships between software and/or hardware elements, and external elements in various environments. |
| | Work assignment | This structure assigns module responsibility to various development and design teams. Typically used in project management. |
| | Implementation | This is a software structure indicating where the software is located on the development system's file system. |
| | Deployment | This structure is of the system at runtime where elements in this structure are hardware and software, and the relationship between elements are where the software maps to in the hardware (resides, migrates to, etc). |

[1]Note that in many cases the terms "architecture" and "structure" (one snapshot) are sometimes used interchangeably, and this will be the case in this book.

summarizes some of the most common structures that can make up embedded architectures, and shows generally what the elements of a particular structure represent and how these elements inter-relate. While Table 1-2 introduces concepts to be defined and discussed later, it also demonstrates the wide variety of architectural structures available to represent an embedded system. Architectures and their structures—how they inter-relate, how to create an architecture, etc.—will be discussed in more detail in Chapter 11.

In short, an embedded system's architecture can be used to resolve these types of challenges early in a project. Without defining or knowing any of the internal implementation details, the architecture of an embedded device can be the first tool to be analyzed and used as a high-level blueprint defining the infrastructure of a design, possible design options, and design constraints. What makes the architectural approach so powerful is its ability to informally and quickly communicate a design to a variety of people with or without technical backgrounds, even acting as a foundation in planning the project or actually designing a device. Because it clearly outlines the requirements of the system, an architecture can act as a solid basis for analyzing and testing the quality of a device and its performance under various circumstances. Furthermore, if understood, created, and leveraged correctly, an architecture can be used to accurately estimate and reduce costs through its demonstration of the risks involved in implementing the various elements, allowing for the mitigation of these risks. Finally, the various structures of an architecture can then be leveraged for designing future products with similar characteristics, thus allowing design knowledge to be reused, and leading to a decrease of future design and development costs.

By using the architectural approach in this book, I hope to relay to the reader that **defining and understanding the architecture of an embedded system is an essential component of good system design**. This is because, in addition to the benefits listed above:

1. Every embedded system has an architecture, whether it is or is not documented, because every embedded system is composed of interacting elements (whether hardware or software). An architecture by definition is a set of representations of those elements and their relationships. Rather than having a faulty and costly architecture forced on you by *not* taking the time to define an architecture before starting development, take control of the design by defining the architecture first.
2. Because an embedded architecture captures various views, which are representations of the system, it is a useful tool in understanding *all* of the major elements, why each component is there, and why the elements behave the way they do. None of the elements within an embedded system works in a vacuum. Every element within a device interacts with some other element in some fashion. Furthermore, externally visible characteristics of elements may differ given a different set of other elements to work with. Without understanding the "whys" behind an element's provided functionality, performance, etc., it would be difficult to determine how the system would behave under a variety of circumstances in the real world.

Even if the architectural structures are rough and informal, *it is still better than nothing*. As long as the architecture conveys in some way the critical components of a design and their relationships to each other, it can provide project members with key information about whether the device can meet its requirements, and how such a system can be constructed successfully.

## 1.3 The Embedded Systems Model

Within the scope of this book, a variety of architectural structures are used to introduce technical concepts and fundamentals of an embedded system. I also introduce emerging architectural tools (i.e., reference models) used as the foundation for these architectural structures. At the highest level, the primary architectural tool used to introduce the major elements located within an embedded system design is what I will simply refer to as the Embedded Systems Model, shown in Figure 1-1.

What the Embedded Systems Model implies is that all embedded systems share one similarity at the highest level; that is, they all have at least one layer (hardware) or all layers (hardware, system software and application software) into which all components fall. The hardware layer contains all the major physical components located on an embedded board, whereas the system and application software layers contain all of the software located on and being processed by the embedded system.

This reference model is essentially a layered (modular) representation of an embedded systems architecture from which a modular architectural structure can be derived. Regardless of the differences between the devices shown in Table 1-1, it is possible to understand the

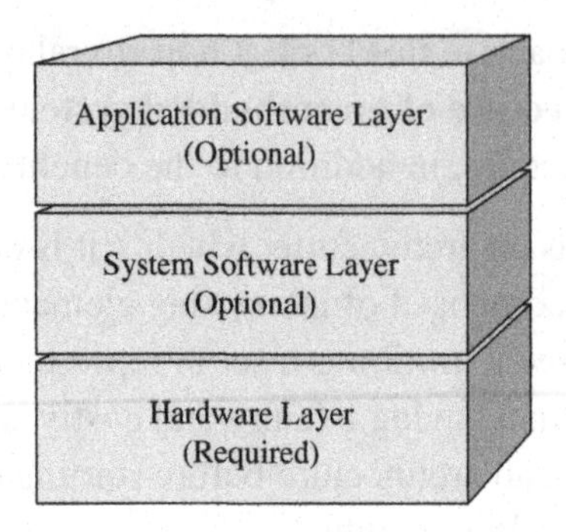


**Figure 1-1**
Embedded Systems Model.

architecture of all of these systems by visualizing and grouping the components within these devices as *layers*. While the concept of layering isn't unique to embedded system design (architectures are relevant to all computer systems, and an embedded system is a type of computer system), it is a useful tool in visualizing the possible combinations of hundreds, if not thousands, of hardware and software components that can be used in designing an embedded system. In general, I selected this modular representation of embedded systems architecture as the primary structure for this book for two main reasons:

1. *The visual representation of the main elements and their associated functions.* The layered approach allows readers to visualize the various components of an embedded system and their inter-relationship.
2. *Modular architectural representations are typically the structures leveraged to structure the entire embedded project.* This is mainly because the various modules (elements) within this type of structure are usually functionally independent. These elements also have a higher degree of interaction, thus separating these types of elements into layers improves the structural organization of the system without the risk of oversimplifying complex interactions or overlooking required functionality.

Sections II and III of this book define the major modules that fall into the layers of the Embedded Systems Model, essentially outlining the major components that can be found in most embedded systems. Section IV then puts these layers together from a design and development viewpoint, demonstrating to the reader how to apply the technical concepts covered in previous chapters along with the architectural process introduced in this chapter. Throughout this book, real-world suggestions and examples are provided to present a pragmatic view of the technical theories, and as the key teaching tool of embedded concepts. As you read these various examples, in order to gain the maximum benefits from this text and to be able to apply the information provided to future embedded projects, I recommend that the reader note:

- *The patterns that all these various examples follow*, by mapping them not only to the technical concepts introduced in the section, but ultimately to the higher-level architectural

representations. These patterns are what can be universally applied to understand or design any embedded system, regardless of the embedded system design being analyzed.
- *Where the information came from.* This is because valuable information on embedded systems design can be gathered by the reader from a variety of sources, including the internet, articles from embedded magazines, the Embedded Systems Conference, data sheets, user manuals, programming manuals, and schematics, to name just a few.

## 1.4 Why the Holistic, Architectural Systems Engineering Approach? The Golden Rules …

This book uses a holistic, architectural systems engineering approach to embedded systems in demystifying embedded systems and the different types of components that can make up their internal designs. This is because one of the most powerful methods of insuring the success of an engineering team is by taking the systems approach to defining the architecture and implementing the design.

A systems engineering approach addresses the reality that more than the pure embedded systems technology, alone, will impact the successful engineering of a product. In other words, ***Rule 1*** to remember is that building an embedded system and taking it to production successfully **requires more than just technology**!

Many *different* influences will impact the process of architecting an embedded design and taking it to production. This can include influences from financial, technical, business-oriented, political, and/or social sources, to name a few. These different *types of influences* generate the requirements, the requirements in turn generate the embedded system's architecture, this architecture then is the basis for producing the device, and the resulting embedded system design in turn provides feedback for requirements and capabilities back to the team. So, it is important for embedded designers to understand and plan for the technical as well as the non-technical aspects of the project, be they social, political, legal, and/or financial influences in nature. This is because the most common mistakes which kill embedded systems projects are typically unrelated to one specific factor in itself, for example:

- The process of defining and capturing the design of a system.
- Cost limitations.
- Determining a system's integrity, such as reliability and safety.
- Working within the confines of available elemental functionality (processing power, memory, battery life, etc.).
- Marketability and salability.
- Deterministic requirements.

The key is for the team to identify, understand, and engage these different project influences from the start and throughout the life cycle of the project. The core challenges that real-world development teams building any embedded system face are handling these influences while balancing quality versus schedule versus features. Team members who recognize this wisdom from day 1 are more likely to insure project success within quality standards, deadlines, and costs.

The next rule to succeeding at embedded systems design is team members having the discipline in following **development processes** and **best practices** (***Rule 2***). Best practices can be incorporated into any development team's agreed upon process model and can include everything from focusing on programming language-specific guidelines to doing code inspections to having a hard-core testing strategy, for example. In the industry, there are several different process models used today, with newer software process schemes and improvements being introduced, constantly. However, most of these approaches used by embedded design teams are typically based upon one or some hybrid combination of the following general schemes:

- *Big-bang*: projects with essentially no planning, specific requirements, or processes in place before and during the development of a system.
- *Code-and-fix*: projects with no formal processes in place before the start of development, but in which product requirements are defined.
- *XP (extreme programming)* and *TDD (test-driven development)*: projects driven by re-engineering and ad-hoc testing of code over-and-over until the team gets it right, or the project runs out of money and/or time.
- *Waterfall*: projects where the process for developing a system occurs in steps and where results of one step flow into the next step.
- *Hybrid spiral*: projects in which the system is developed in steps, and throughout the various steps, feedback is obtained and incorporated back into the project at each process step.
- *Hybrid iterative models*: such as the Rational Unified Process (RUP), which is a framework that allows for adapting different processes for different stages of projects.
- *Scrum*: another framework for adapting different processes for different stages of projects, as well as allowing team members to hold various roles throughout the project. Scrum incorporates shorter-term, more stringent deadlines and continual communication between team members.

Whether an embedded design team is following a similar process to what is shown in Figure 1-2 or some other development model, the team members need to objectively evaluate how well a development process model is working for them. For instance, the team can begin by doing practical and efficient assessments by first outlining the development goals the team wishes to achieve, as well as what challenges team members are facing. Then team members focus on objectively investigating and documenting what existing development processes team members are following, including:

- One-shot project activities.
- Reoccurring project activities.
- Functional roles of team members at various stages of a project.
- Measuring and metering of development efforts (what is working versus what is crippling development efforts).

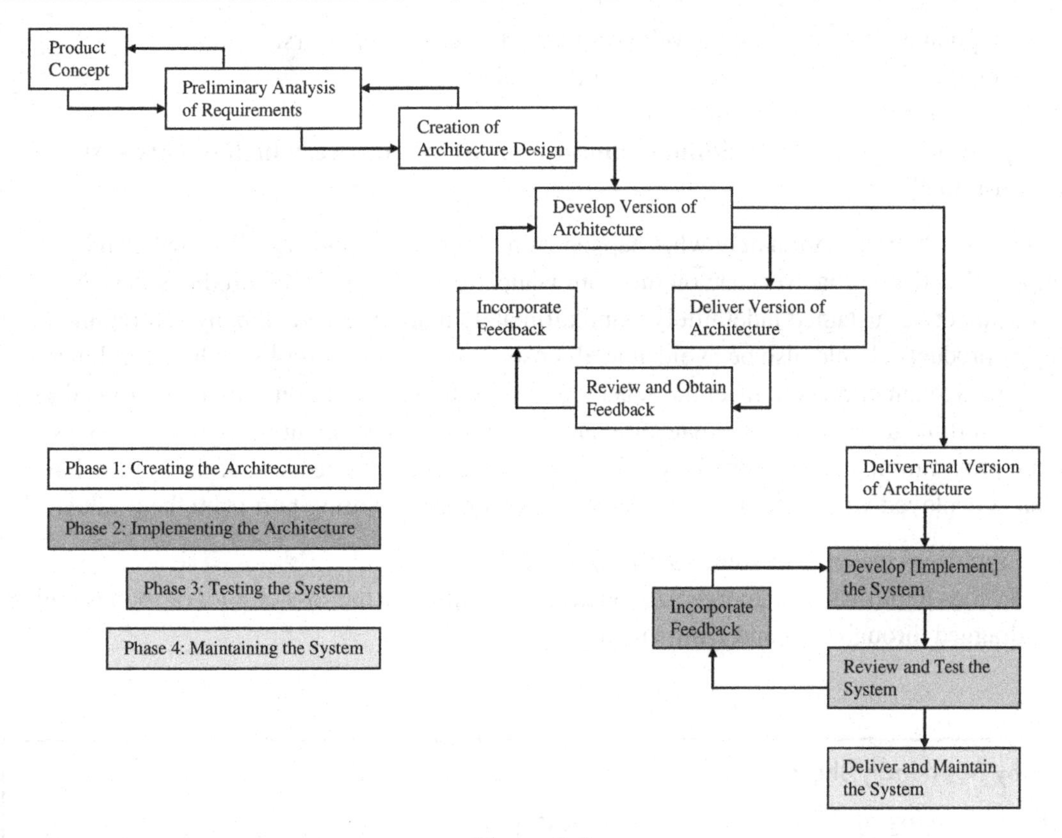


**Figure 1-2**
Embedded Systems Design and Development Lifecycle Model.[2]

- Project management, release management, and configuration management efforts.
- Testing and verification efforts.
- Infrastructure and training (for getting team members up and running efficiently).

Finally, follow through with defining improvements to these existing processes that all team members must adhere to. This means looking into the different possibilities of what team members are ready to implement relative to development efforts, in terms of more disciplined measures. There are also standard industry approaches, such as via CMMI (Capability Maturity Model Integration), that a team can use to introduce improvements and increase discipline in order to save money and save time, as well as improve the quality of the product.

The next step is then for the team to transform these processes into pragmatic tools for solving the everyday challenges, and finding the *best* solution. Project teams typically face some combination of the following solutions when facing the challenge to successfully build an embedded system:

× Option 1 Don't ship.
× Option 2 Blindly ship on time, with buggy features.

× Option 3 Pressure tired developers to work even longer hours.
× Option 4 Throw more resources at the project.
× Option 5 Let the schedule slip.
**√ Option 6 Healthy shipping philosophy: "Shipping a very high quality system on-time."**

Solutions 1–5 are unfortunately what happens too often in the industry. Obviously, "not shipping" is the option everyone on the team wants to avoid. With "no" products to sell, a team cannot be sustained indefinitely—and ultimately neither can a company. "Shipping a buggy product" should also be avoided at all costs. It is a serious problem when developers are forced to cut corners to meet the schedule relative to design options, are being forced to work overtime to the point of exhaustion, and are undisciplined about using best practices when programming, doing code inspections, testing, etc. This is because of the high risk that what is deployed will contain serious defects or of someone getting hurt from the product:[5]

- Employees can end up going to prison
- Serious liabilities for the organization arise—resulting in the loss of a lot of money, being dragged through civil and criminal courts

**Why not blindly ship?**

**Programming and Engineering Ethics Matter**

*Breach of Contract*

- if bug fixes stated in contract are not forthcoming in timely manner

*Breach of Warranty and Implied Warranty*

- delivering system without promised features

*Strict and Negligence Liability*

- bug causes damage to property
- bug causes injury
- bug causes death

*Malpractice, i.e.*

- customer purchases defective product

*Misrepresentation and Fraud*

- product released and sold that doesn't meet advertised claims

*Based on the chapter "Legal Consequences of Defective Software," in* Testing Computer Software, *C. Kaner, J. Falk, and H. Q. Nguyen, 2nd edn, Wiley, 1996*

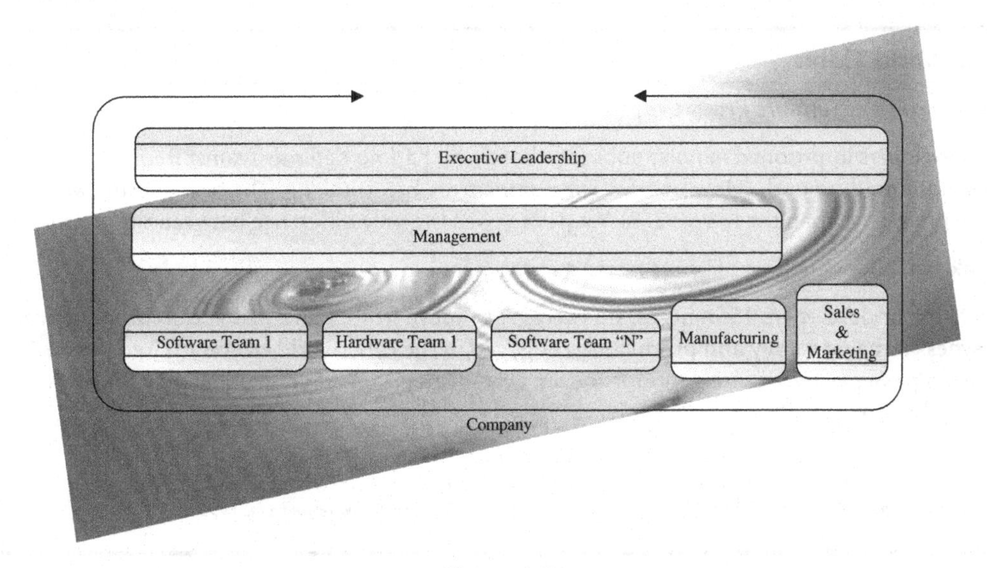


**Figure 1-3**
Problems radiate and impact the environment.[6]

The key is to "not" remove calm from an engineering team and to "not" panic. Pushing exhausted developers to work even longer overtime hours will only result in more serious problems. Tired, afraid, and/or stressed out engineers and developers will result in mistakes being made during development, which in turn translates to additional costs and delays. Negative influences on a project, whether financial, political, technical, and/or social in nature, have the unfortunate ability to negatively harm the cohesiveness of an ordinarily healthy team within a company—eventually leading to sustaining these stressed teams as unprofitable in themselves. Within any organization, even a single weak link, such as a team of exhausted and stressed out engineers, will be debilitating for an entire project and even an entire company. This is because these types of problems radiate outwards influencing the entire environment, like waves (see Figure 1-3).

*Decrease* the stress and the interruptions for a development team during their most productive programming hours within a normal work week, so that there is more focus and fewer mistakes.

Another approach in the industry to avoid a schedule from slipping has been to throw more and more resources at a project. Throwing more resources ad-hoc at project tasks without proper planning, training, and team building is the surest way to hurt a team and guarantee a missed deadline. As indicated in Figure 1-4, productivity crashes with the more people there are on a project. A limit in the number of communication channels can happen through more than one smaller sub-teams, as long as:

- It makes sense for the embedded systems product being designed, i.e.:
  - not dozens of developers and several line/project managers for a few MB of code;

**Real-World Tidbit**

*Underpinnings of Software Productivity*[7]

"... developers imprisoned in noisy cubicles, those who had no defense against frequent interruptions, did poorly. How poorly? The numbers are breathtaking. The best quartile was 300% more productive than the lowest 25%. Yet privacy was the only difference between the groups.

Think about it—would you like 3× faster development?

It takes your developers 15 minutes, on average, to move from active perception of the office busyness to being totally and productively engaged in the cyberworld of coding. Yet a mere 11 minutes passes between interruptions for the average developer. Ever wonder why firmware costs so much? ..."

---

*"A Boss's Quick-Start to Firmware Engineering," J. Ganssle, http://www.ganssle.com/articles/abossguidepi.htm;* Peopleware: Productive Projects & Teams, *T. DeMarco and T. Lister, Dorset House Publishing, 2nd revised edn, 1999.*

  - not when few have embedded systems experience and/or experience building the product;
  - not for corporate empire-building! This results in costly project problems and delays = bad for business!
- In a healthy team environment:
  - no secretiveness;
  - no hackers;
  - best practices and processes not ignored.
- Team members have sense of professional responsibility, alignment, and trust with each other, leadership, and the organization.

So, ultimately what is recommended is ***Rule 3*—teamwork! ... teamwork! ... teamwork!** Get together with fellow team members to discuss the various process models and determine via consensus together what is what is the best "fit" for your particular team. Meaning, there is not yet "one" particular process that has been invented that is the right approach for "all" teams in the industry, or even "all" projects for one particular team. In fact, most likely what works for the team is some hybrid combination of a few models, and this model will need to be tuned according to the types of team members and how they function best, the project's goals, and system requirements.

Then all team members, from junior to the most senior technical members of the team, as well as leadership, align together to come to an agreed consensus for a process model that will achieve business results (***Rule 4*—alignment behind leadership**). Each team member then understands the big picture, the part each plays in it, and commits to the discipline to follow through. If along the way, it is discovered the process isn't optimally working as

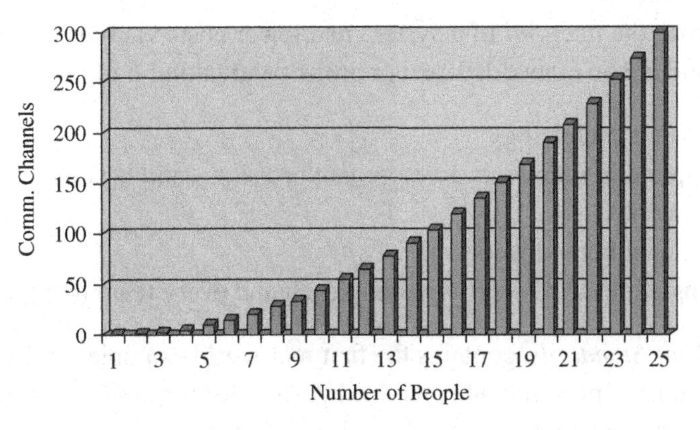


**Figure 1-4**
Too many people.[7]

expected, team members get together again. Openly and respectfully discuss the challenges and frustrations together in a constructive manner; then immediately tune and adjust the process, with team members each doing their part to improve software development efforts. Finally, do not forget ***Rule 5*—strong ethics and integrity among each and every team member**, to continue moving forward as agreed upon together towards success.

While more related to this discussion will be covered in the last chapter of this book, ultimately the most powerful way to meet project schedules, costs, and successfully take an embedded system solution to production is:

√ **by shipping a very high quality product on-time**

- Have a strong technical foundation (i.e., most of this text focuses on providing a strong technical foundation for understanding the major components of an embedded system design).
- Sacrifice fewer essential features in the first release.
- Start with a skeleton, then hang code off the skeleton.
- Do not overcomplicate the design!
- Systems integration, testing, and verification from day 1.

## 1.5 Summary

This chapter began by defining what an embedded system is, including in the definition the most complex and recent innovations in the market. It then defined what an embedded systems architecture is in terms of the sum of the various representations (structures) of a system. This chapter also introduced why the architectural approach is used as the approach to introducing embedded concepts in this book: because it presents a clear visual of what the system is, or could be, composed of and how these elements function. In addition, this approach can provide early indicators into what may and may not work in a system, and possibly improve the integrity of a system and lower costs via reusability. Finally, successfully completing an embedded design project and taking it to production requires:

**Rule 1** More than technology
**Rule 2** Discipline in following development processes and best practices
**Rule 3** Teamwork
**Rule 4** Alignment behind leadership
**Rule 5** Strong ethics and integrity among each and every team member

Chapter 2, *Know Your Standards*, contains the first real-world examples of the book in reference to how industry standards play into an embedded design. Its purpose is to show the importance of knowing and understanding the standards associated with a particular device, and leveraging these standards to understand or create an embedded system's architectural design.

## Chapter 1: Problems

1. Name three traditional or not-so-traditional definitions of embedded systems.
2. In what ways do traditional assumptions apply and not apply to more recent complex embedded designs? Give four examples.
3. [T/F] Embedded systems are all:
   A. Medical devices.
   B. Computer systems.
   C. Very reliable.
   D. All of the above.
   E. None of the above.
4. [a] Name and describe five different markets under which embedded systems commonly fall.
   [b] Provide examples of four devices in each market.
5. Name and describe the four development models that most embedded projects are based upon.
6. [a] What is the Embedded Systems Design and Development Lifecycle Model [draw it]?
   [b] What development models is this model based upon?
   [c] How many phases are in this model?
   [d] Name and describe each of its phases.
7. Which of the stages below is not part of creating an architecture, phase 1 of the Embedded Systems Design and Development Lifecycle Model?
   A. Understanding the architecture business cycle.
   B. Documenting the architecture.
   C. Maintaining the embedded system.
   D. Having a strong technical foundation.
   E. None of the above.

8. Name five challenges commonly faced when designing an embedded system.
9. What is the architecture of an embedded system?
10. [T/F] Every embedded system has an architecture.
11. [a] What is an element of the embedded system architecture?
    [b] Give four examples of architectural elements.
12. What is an architectural structure?
13. Name and define five types of structures.
14. [a] Name at least three challenges in designing embedded systems.
    [b] How can an architecture resolve these challenges?
15. [a] What is the Embedded Systems Model?
    [b] What structural approach does the Embedded Systems Model take?
    [c] Draw and define the layers of this model.
    [d] Why is this model introduced?
16. Why is a modular architectural representation useful?
17. All of the major elements within an embedded system fall under:
    A. The hardware layer.
    B. The system software layer.
    C. The application software layer.
    D. The hardware, system software, and application software layers.
    E. A or D, depending on the device.
18. Name six sources that can be used to gather embedded systems design information.

## Endnotes

[1] *Embedded Microcomputer Systems*, p. 3, J. W. Valvano, CL Engineering, 2nd edn, 2006; *Embedded Systems Building Blocks*, p. 61, J. J. Labrosse, Cmp Books, 1995.

[2] The Embedded Systems Design and Development Lifecycle Model is specifically derived from Software Engineering Institute (SEI)'s Evolutionary Delivery Lifecycle Model and the Software Development Stages Model.

[3] The six stages of creating an architecture outlined and applied to embedded systems in this book are inspired by the Architecture Business Cycle developed by SEI. For more on this brainchild of SEI, read *Software Architecture in Practice*, L. Bass, P. Clements, and R. Kazman, Addison-Wesley, 2nd edn, 2003, which does a great job in capturing and articulating the process that so many of us have taken for granted over the years or have not even bothered to think about. While SEI focus on software in particular, their work is very much applicable to the entire arena of embedded systems, and I felt it was important to introduce and recognize the immense value of their contributions as such.

[4] *Software Architecture in Practice*, L. Bass, P. Clements, and R. Kazman, Addison-Wesley, 2nd edn, 2003; *Real-Time Design Patterns*, B. P. Douglass, Addison-Wesley, 2002.

[5] *Software Testing*, pp. 31–36, R. Patton, Sams, 2003.

[6] *Demystifying Embedded Systems Middleware*, T. Noergaard, Elsevier, 2010.

[7] "Better Firmware, Faster," Seminar, J. Ganssle, 2007.

CHAPTER 2

# Know Your Standards

**In This Chapter**

- Defining the meaning of standards
- Listing examples of different types of standards
- Discussing the impact of programming language standards on the architecture
- Discussing the OSI model and examples of networking protocols
- Using digital TV as an example that implements many standards

Some of the most important components within an embedded system are derived from specific methodologies, commonly referred to as *standards*. Standards dictate how these components should be designed, and what additional components are required in the system to allow for their successful integration and function. As shown in Figure 2-1, standards can define functionality that is specific to each of the layers of the embedded systems model, and can be classified as *market-specific* standards, *general-purpose* standards, or standards that are applicable to *both* categories.

Standards that are strictly market-specific define functionality that is relative to a particular group of related embedded systems that share similar technical or end-user characteristics, including:

- *Consumer Electronics.* Typically includes devices used by consumers in their personal lives, such as personal data assistants (PDAs), TVs (analog and digital), games, toys, home appliances (i.e., microwave ovens, dishwashers, washing machines), and internet appliances.[1]

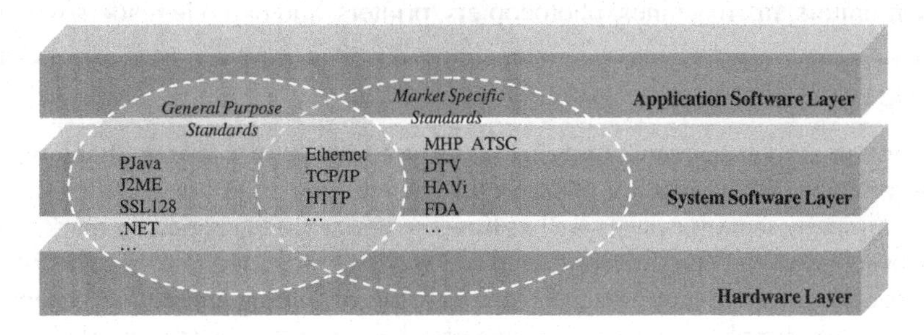


**Figure 2-1**
Standards diagram.

- *Medical*. Defined as "... any instrument, apparatus, appliance, material or other article, whether used alone or in combination, including the software necessary for its proper application intended by the manufacturer to be used for human beings for the purpose of:
  - diagnosis, prevention, monitoring, treatment or alleviation of disease,
  - diagnosis, monitoring, treatment, alleviation of or compensation for an injury or handicap,
  - investigation, replacement or modification of the anatomy or of a physiological process,
  - control of conception, and which does not achieve its principal intended action in or on the human body by pharmacological, immunological or metabolic means, but which may be assisted in its function by such means ...."

*European Medical Device Directive (93/42/EEC).*[14]

This includes dialysis machines, infusion pumps, cardiac monitors, drug delivery, prosthetics, etc.[1]

- *Industrial Automation and Control.* "Smart" robotic devices (smart sensors, motion controllers, man/machine interface devices, industrial switches, etc.) used mainly in manufacturing industries to execute a cyclic automated process.[1]
- *Networking and Communications.* Intermediary devices connecting networked end systems, devices such as hubs, gateways, routers, and switches. This market segment also includes devices used for audio/video communication, such as cell phones (includes cell phone/PDA hybrids), pagers, video phones, and ATM machines.[1]
- *Automotive.* Subsystems implemented within automobiles, such as entertainment centers, engine controls, security, antilock brake controls, and instrumentation.[1]
- *Emergency Services, Police, and Defense.* Systems that are implemented within embedded systems used by the police or military, such as within "smart" weapons, police patrol, ambulances, and radar systems, to name a few.[1]
- *Aerospace and Space.* Systems implemented within aircraft or used by the military, such as flight management, "smart" weaponry, and jet engine control. This category also includes embedded systems that must function in space, such as on a space station or within an orbiting satellite.[1]
- *Energy and Oil.* Embedded systems that are used in the power and energy industries, such as control systems within power plant ecosystems for wind turbine generators and solar.[1]
- *Commercial Office/Home Office Automation.* Devices used in an office setting, such as printers, scanners, monitors, fax machines, photocopiers, printers, and barcode readers/writers.[1]

**Practical Tip**

Embedded system market segments and their associated devices are always changing as new devices emerge and other devices are phased out. The market definitions can also vary from company to company semantically as well as how the devices are grouped by market segment. For a quick overview of the current terms used to describe embedded markets and how devices are being vertically grouped, go to three or four websites of leading embedded system software vendors. Alternately, simply use a search engine with keywords "embedded market segments" and take a look at the latest developments in device grouping.

Most market-specific standards, excluding networking and some TV standards, are typically implemented into specific types of embedded systems, because by definition they are intended for specific groups of embedded devices. General-purpose standards, on the other hand, are typically not intended for just one specific market of embedded devices; some are adopted (and in some cases originated) in non-embedded devices as well. Programming language-based standards are examples of general-purpose standards that can be implemented in a variety of embedded systems as well as non-embedded systems. Standards that can be considered both market-specific as well as general-purpose include networking standards and some television standards. Networking functionality can be implemented in devices that fall under the networking market space, such as hubs and routers; in devices across various markets, such as wireless communication in networking devices, consumer electronics, etc.; and also in non-embedded devices. Television standards have been implemented in PCs, as well as in traditional TVs and set-top boxes (STBs).

Table 2-1 lists some current real-world standards and some of the purposes behind their implementations.

The next three sections of this chapter contain real-world examples showing how specific standards define some of the most critical components of an embedded system. Section 2.1 presents general-purpose programming language standards that can affect the architecture of an embedded system. Section 2.2 presents networking protocols that can be implemented in a specific family of devices, across markets, and in stand-alone applications. Finally, Section 2.3 presents an example of a consumer appliance that implements functionality from

**Warning!**

While Table 2-1 lists market-specific standards in the context of a single market, some market-specific standards listed in this table have been implemented or adopted by other device market segments. Table 2-1 simply shows "some" real-world examples. Furthermore, different countries, and even different regions of one country, may have unique standards for particular families of devices (i.e., DTV or cell phone standards; see Table 2-1). Also, in most industries, competing standards can exist for the same device, supported by competing interests. Find out who has adopted which standards and how these competing standards differ by using the Internet to look up published data sheets or manuals of the particular device and the documentation provided by the vendors of the components integrated within the device, or by attending the various tradeshows, seminars, and conferences associated with that particular industry or vendor, such as the Embedded Systems Conference (ESC), Java One, Real-Time Embedded and Computing Conference.

This warning note is especially important for hardware engineers, who may have come from an environment where certain standards bodies, such as the IEEE, have a strong influence on what is adopted. In the embedded software field there is currently no single standards body that has the level of influence that the IEEE has in the hardware arena.

**Table 2-1: Examples of Standards Implemented in Embedded Systems**

| Standard Type | | Standard | Purpose |
|---|---|---|---|
| Market specific | *Consumer electronics* | Java TV | The Java TV API is an extension of the Java platform that provides access to functionality unique to a DTV receiver, such as audio video streaming, conditional access, access to in-band and out-of-band data channels, access to service information data, tuner control for channel changing, on-screen graphics control, media synchronization (allows interactive television content to be synchronized with the underlying video and background audio of a television program), and application lifecycle control. (Enables content to gracefully coexist with television programming content such as commercials.)[3] |
| | | DVB (Digital Video Broadcasting)–MHP (Multimedia Home Platform) | The Java-based standard used in DTV designs introduces components in the system software layer, as well as provides recommendations for hardware and the types of applications that would be compatible with MHP. Basically, it defines a generic interface between interactive digital applications and the terminals ranging from low-end to high-end STBs, integrated DTV sets, and multimedia PCs on which those applications execute. This interface decouples different providers' applications from the specific hardware and software details of different MHP terminal implementations, enabling digital content providers to address all types of terminals. The MHP extends the existing DVB open standards for broadcast and interactive services in all transmission networks including satellite, cable, terrestrial, and microwave systems.[2] |
| | | ISO/IEC 16500 DAVIC (Digital Audio Visual Council) | DAVIC is an industry standard for end-to-end interoperability of broadcast and interactive digital audio-visual information, and of multimedia communication.[4] |
| | | ATSC (Advanced Television Standards Committee)–DASE (Digital TV Applications Software Environment) | The DASE standard defines a system software layer that allows programming content and applications to run on a "common receiver." Interactive and enhanced applications need access to common receiver features in a platform-independent manner. This environment provides enhanced and interactive content creators with the specifications necessary to ensure that their applications and data will run uniformly on all brands and models of receivers. Manufacturers will thus be able to choose hardware platforms and OSs for receivers, but provide the commonality necessary to support applications made by many content creators.[5] |

(*Continued*)

Table 2-1: (Continued)

| Standard Type | | Standard | Purpose |
|---|---|---|---|
| | | ATVEF (Advanced Television Enhancement Forum)—SMPTE (Society of Motion Picture and Television Engineers) DDE-1 | The ATVEF Enhanced Content Specification defines fundamentals necessary to enable creation of HTML-enhanced television content that can be reliably broadcast across any network to any compliant receiver. ATVEF is a standard for creating enhanced, interactive television content and delivering that content to a range of television, STB, and PC-based receivers. ATVEF [SMPTE DDE-1] defines the standards used to create enhanced content that can be delivered over a variety of mediums—including analog (NTSC) and digital (ATSC) television broadcasts—and a variety of networks, including terrestrial broadcast, cable, and satellite.[6] |
| | | DTVIA (Digital Television Industrial Alliance of China) | DTVIA is an organization made up of leading TV manufacturers, research institutes, and broadcasting academies working on the key technologies and specifications for the China TV industry to transfer from analog to digital. DTVIA and Sun are working together to define the standard for next-generation interactive DTV leveraging Sun's Java TV API specification.[7] |
| | | ARIB-BML (Association of Radio Industries and Business of Japan) | ARIB in 1999 established their standard titled "Data Coding and Transmission Specification for Digital Broadcasting" in Japan—an XML-based specification. The ARIB B24 specification derives BML (broadcast markup language) from an early working draft of the XHTML 1.0 Strict document type, which it extends and alters.[7] |
| | | OCAP (OpenCable Application Forum) | The OpenCable Application Platform (OCAP) is a system software layer that provides an interface enabling application portability (applications written for OpenCable must be capable of running on any network and on any hardware platform, without recompilation). The OCAP specification is built on the DVB MHP specification with modifications for the North American Cable environment that includes a full-time return channel. A major modification to the MHP is the addition of a Presentation Engine (PE), that supports HTML, XML, and ECMAScript. A bridge between the PE and the Java Execution Engine (EE), enables PE applications to obtain privileges and directly manipulate privileged operations.[8] |

(Continued)

**Table 2-1: (Continued)**

| Standard Type | Standard | Purpose |
|---|---|---|
| | OSGi (Open Services Gateway Initiative) | The OSGi specification is designed to enhance all residential networking standards, such as Bluetooth™, CAL, CEBus, Convergence, emNET, HAVi™, HomePNA™, HomePlug™, HomeRF™, Jini™ technology, LonWorks, UPnP, 802.11B, and VESA. The OSGi Framework and Specifications facilitate the installation and operation of multiple services on a single Open Services Gateway (STB, cable or DSL modem, PC, web phone, automotive, multimedia gateway, or dedicated residential gateway).[9] |
| | OpenTV | OpenTV has a proprietary DVB-compliant system software layer, called EN2, for interactive television digital STBs. It complements MHP functionality and provides functionality that is beyond the scope of the current MHP specification, such as HTML rendering and web browsing.[10] |
| | MicrosoftTV | MicrosoftTV is a proprietary interactive TV system software layer that combines both analog and DTV technologies with Internet functionality. MicrosoftTV Technologies support current broadcast formats and standards, including NTSC, PAL, SECAM, ATSC, OpenCable, DVB, and SMPTE 363M (ATVEF specification) as well as Internet standards such as HTML and XML.[11] |
| | HAVi (Home Audio Video Initiative) | HAVi provides a home networking standard for seamless interoperability between digital audio and video consumer devices, allowing all audio and video appliances within the network to interact with each other, and allow functions on one or more appliances to be controlled from another appliance, regardless of the network configuration and appliance manufacturer.[12] |
| | CEA (Consumer Electronics Association) | The CEA fosters consumer electronics industry growth by developing industry standards and technical specifications that enable new products to come to market and encourage interoperability with existing devices. Standards include ANSI-EIA-639 Consumer Camcorder or Video Camera Low Light Performance and CEA-CEB4 Recommended Practice for VCR Specifications.[17] |

(*Continued*)

Table 2-1: (Continued)

| Standard Type | | Standard | Purpose |
|---|---|---|---|
| | *Medical devices* | Food and Drug Administration (US) | US government standards for medical devices relating to the aspects of safety and/or effectiveness of the device. Class I devices are defined as non-life-sustaining. These products are the least complicated and their failure poses little risk. Class II devices are more complicated and present more risk than Class I, though are also non-life-sustaining. They are also subject to any specific performance standards. Class III devices sustain or support life, so that their failure is life threatening. Standards include areas of anesthesia (Standard Specification for Minimum Performance and Safety Requirements for Resuscitators Intended for Use with Humans, Standard Specification for Ventilators Intended for Use in Critical Care, etc.), cardiovascular/neurology (intracranial pressure monitoring devices, etc.), dental/ENT (Medical Electrical Equipment—Part 2: Particular Requirements for the Safety of Endoscope Equipment, etc.), plastic surgery (Standard Performance and Safety Specification for Cryosurgical Medical Instrumentation, etc.), ObGyn/Gastroenterology (Medical electrical equipment—Part 2: Particular requirements for the safety of hemodialysis, hemodiafiltration and hemofiltration equipment, etc.)[13] |
| | | Medical Devices Directive (EU) | European Medical Devices Directives are standards for medical devices for EU member states relating to the aspects of safety and/or effectiveness of these devices. The lowest risk devices fall into Class I (Internal Control of Production and compilation of a Technical File compliance), whereas devices that exchange energy with the patient in a therapeutic manner or are used to diagnose or monitor medical conditions, are in Class IIa (i.e., ISO 9002 + EN 46002 compliance). If this is done in manner that could be hazardous for the patient, then the device falls into Class lib (i.e., ISO 9001 + EN 46001). For a device that connects directly with the central circulatory system or the central nervous system or contains a medicinal product, the device falls into Class III (i.e., ISO 9001 + EN 46001 compliance, compilation of a Design Dossier).[14] |

(*Continued*)

**Table 2-1: (Continued)**

| Standard Type | | Standard | Purpose |
|---|---|---|---|
| | | IEEE1073 Medical Device Communications | IEEE 1073 standards for medical device communication provide plug-and-play interoperability at the point-of-care, optimized for the acute care environment. The IEEE 1073 General Committee is chartered under the IEEE Engineering in Medicine and Biology Society and works closely with other national and international organizations, including HL7, NCCLS, ISO TC215, CEN TC251, and ANSI HISB.[15] |
| | *Industrial control* | DICOM (Digital Imaging and Communications in Medicine) | The American College of Radiology (ACR) and the National Electrical Manufacturers Association (NEMA) formed a joint committee in 1983 to develop the DICOM standard for transferring images and associated information between devices manufactured by various vendors, specifically to:<br>• Promote communication of digital image information, regardless of device manufacturer.<br>• Facilitate the development and expansion of picture archiving and communication systems (PACS) that can also interface with other systems of hospital information.<br>• Allow the creation of diagnostic information databases that can be interrogated by a wide variety of devices distributed geographically.[16]<br>Maintains a website that contains the global medical device regulatory requirements on a per country basis. |
| | | Department of Commerce (US)—Office of Microelectronics, Medical Equipment, and Instrumentation<br>(EU) The Machinery Directive 98/37/EC | EU directive for all machinery, moving machines, machine installations, and machines for lifting and transporting people, as well as safety components. In general, machinery being sold or used in the European Union must comply with applicable mandatory Essential Health and Safety Requirements (EHSRs) from a long list given in the directive, and must undertake the correct conformity assessment procedure. Most machinery considered less dangerous can be self-assessed by the supplier, and being able to assemble a Technical File. The 98/37/EC applies to an assembly of linked parts or components with at least one movable part—actuators, controls, and power circuits, processing, treating, moving, or packaging a material—several machines acting in combination, etc.[18] |

(*Continued*)

**Table 2-1: (Continued)**

| Standard Type | | Standard | Purpose |
|---|---|---|---|
| | | IEC (International Electrotechnical Commission 60204-1) | Applies to the electrical and electronic equipment of industrial machines. Promotes the safety of persons who come into contact with industrial machines, not only from hazards associated with electricity (such as electrical shock and fire), but also those resulting from the malfunction of the electrical equipment itself. Addresses hazards associated with the machine and its environment. Replaces the second edition of IEC 60204-1 as well as parts of IEC 60550 and ISO 4336.[19] |
| | | ISO (International Standards Organization) Standards | Many standards in the manufacturing engineering segment, such as ISO/TR 10450: Industrial automation systems and integration—Operating conditions for discrete part manufacturing-Equipment in industrial environments; and ISO/TR 13283: Industrial automation-Time-critical communications architectures-User requirements and network management for time-critical communications systems.[20] (See www.iso.ch.) |
| | *Networking and communications* | TCP (Transmission Control Protocol)/IP (Internet Protocol) | Protocol stack based on RFCs (Request for Comments) 791 (IP) & 793 (TCP) that define system software components (more information in Chapter 10). |
| | | PPP (Point-to-Point Protocol) | System software component based on RFCs 1661, 1332, and 1334 (more information in Chapter 10). |
| | | IEEE (Institute of Electronics and Electrical Engineers) 802.3 Ethernet | Networking protocol that defines hardware and system software components for LANs (more information in Chapters 6 and 8). |
| | | Cellular | Networking protocols implemented within cellular phones, such as CDMA (Code Division Multiple Access) and TDMA (Time Division Multiple Access) typically used in the United States. TDMA is the basis of GSM (Global System for Mobile telecommunications) European international standard, UMTS (Universal Mobile Telecommunications System) broadband digital standard (third generation). |
| | *Automotive* | GM Global | GM standards are used in the design, manufacture, quality control, and assembly of automotive components and materials related to General Motors, specifically: adhesives, electrical, fuels and lubricants, general, paints, plastics, procedures, textiles, metals, metric, and design.[27] |
| | | Ford Standards | The Ford standards are from the Engineering Material Specifications and Laboratory Test Methods volumes, the Approved Source List Collection, Global Manufacturing Standards, Non-Production Material Specifications, and the Engineering Material Specs & Lab Test Methods Handbook.[27] |

*(Continued)*

**Table 2-1: (Continued)**

| Standard Type | | Standard | Purpose |
|---|---|---|---|
| | | FMVSS (Federal Motor Vehicle Safety Standards) | The Code of Federal Regulations (CFR) contains the text of public regulations issued by the agencies of the US Federal government. The CFR is divided into several titles which represent broad areas subject to Federal Regulation.[27] |
| | | OPEL Engineering Material Specifications | OPEL's standards are available in sections, such as: Metals, Miscellaneous, Plastics and Elastomers, Materials of Body—Equipment, Systems and Component Test Specifications, Test Methods, Laboratory Test Procedures (GME/GMI), body and electric, chassis, powertrain, road test procedures (GME/GMI), body and electric, chassis, powertrain, process, paint & environmental engineering materials.[27] |
| | | Jaguar Procedures and Standards Collection | The Jaguar standards are available as a complete collection or as individual standards collections such as: Test Procedures Collection, Engine & Fastener Standards Collection, Non-Metallic/ Metallic Material Standards Collection, Laboratory Test Standards Collection.[27] |
| | | ISO/TS 16949—The Harmonized Standard for the Automotive Supply Chain | Jointly developed by the IATF (International Automotive Task Force) members, and forms the requirements for automotive production and relevant service part organizations. Based on ISO 9001:2000, AVSQ (Italian), EAQF (French), QS-9000 (United States), and VDA6.1 (German) automotive catalogs.[30] |
| | *Aerospace and defense* | SAE (Society of Automotive Engineers)—The Engineering Society For Advancing Mobility in Land Sea Air and Space | SAE Aerospace Material Specifications, SAE Aerospace Standards (includes Aerospace Standards (AS), Aerospace Information Reports (AIR), and Aerospace Recommended Practices (ARP)).[27] |
| | | AIA/NAS—Aerospace Industries Association of America, Inc. | This standards service includes National Aerospace Standards (NAS) and metric standards (NA Series). It is an extensive collection that provides standards for components, design, and process specifications for aircraft, spacecraft, major weapons systems, and all types of ground and airborne electronic systems. It also contains procurement documents for parts and components of high-technology systems, including fasteners, high-pressure hoses, fittings, high-density electrical connectors, bearings.[27] |

*(Continued)*

**Table 2-1: (Continued)**

| Standard Type | | Standard | Purpose |
|---|---|---|---|
| | | Department of Defense (DOD)—JTA (Joint Technical Architecture) | DOD initiatives such as the Joint Technical Architecture (JTA) permit the smooth flow of information necessary to achieve interoperability, resulting in optimal readiness. The JTA was established by the US Department of Defense to specify a minimal set of information technology standards, including web standards, to achieve military interoperability.[27] |
| | *Office automation* | IEEE 1284.1-1997: IEEE Standard for Information Technology Transport Independent Printer/ System Interface (TIP/SI) | A protocol and methodology for software developers, computer vendors, and printer manufacturers to facilitate the orderly exchange of information between printers and host computers are defined in this standard. A minimum set of functions that permit meaningful data exchange is provided. Thus a foundation is established upon which compatible applications, computers, and printers can be developed, without compromising an individual organization's desire for design innovation.[28] |
| | | Postscript | A programming language from Adobe that describes the appearance of a printed page that is an industry standard for printing and imaging. All major printer manufacturers make printers that contain or can be loaded with Postscript software (.ps file extensions). |
| | | ANSI/AIM BC2-1995: Uniform Symbology Specification for Bar Codes | For encoding general purpose all-numeric data. Reference symbology for UCC/EAN Shipping Container Symbol. Character encoding, reference decode algorithm, and optional check character calculation are included in this document. This specification is intended to be significantly identical to corresponding Commission for European Normalization (CEN) specification.[29] |
| General purpose | *Networking* | HTTP (Hypertext Transfer Protocol) | A WWW protocol defined by a number of different RFCs, including RFC2616, 2016, 2069, 2109. Application layer networking protocol implemented within browsers on any device, for example. |
| | | TCP (Transmission Control Protocol)/IP (Internet Protocol) | Protocol stack based on RFCs (Request for Comments) 791 (IP) & 793 (TCP) that define system software components (more information in Chapter 10). |
| | | IEEE (Institute of Electronics and Electrical Engineers) 802.3 Ethernet | Networking protocol that defines hardware and system software components for LANs (more information in Chapters 6 and 8). |

(*Continued*)

**Table 2-1: (Continued)**

| Standard Type | | Standard | Purpose |
|---|---|---|---|
| | | Bluetooth | Bluetooth specifications are developed by the Bluetooth Special Interest Group (SIG), which allows for developing interactive services and applications over interoperable radio modules and data communication protocols (more information on Bluetooth in Chapter 10).[21] |
| | *Programming languages* | pJava (Personal Java) | Embedded Java standard from Sun Microsystems targeted and larger embedded systems (more information in Section 2.1). |
| | | J2ME (Java 2 Micro Edition) | Set of embedded standards from Sun Microsystems targeting the entire range of embedded systems, both in size and vertical markets (more information in Section 2.1). |
| | | .NET Compact Framework | Microsoft-based system that allows an embedded system to support applications written in several different languages, including C# and Visual Basic (more information in Section 2.1). |
| | | HTML (Hyper Text Markup Language) | Scripting language whose interpreter typically is implemented in a browser, WWW protocol (more information in Section 2.1). |
| | *Security* | Netscape IETF (Internet Engineering Task Force) SSL (Secure Socket Layer) 128-bit Encryption | The SSL is a security protocol that provides data encryption, server authentication, message integrity, and optional client authentication for a TCP/IP connection, and is typically integrated into browsers and web servers. There are different versions of SSL (40-bit, 128-bit, etc.), with "128-bit" referring to the length of the "session key" generated by every encrypted transaction (the longer the key, the more difficult it is to break the encryption code). SSL relies on session keys, as well as digital certificates (digital identification cards) for the authentication algorithm. |
| | | IEEE 802.10 Standards for Interoperable LAN/MAN Security (SILS) | Provides a group of specifications at the hardware and system software layer to implement security in networks. |
| | *Quality assurance* | ISO 9000 Standards | A set of quality management process standards when developing products (not product standards in themselves) or providing a service, including ISO 9000:2000, ISO 9001:2000, ISO 9004:2000 [ISO 9004L:2000]. ISO 9001:2000 presents requirements, while ISO 9000:2000 and ISO 9004:2000 present guidelines. |

a number of different standards. These examples demonstrate that a good starting point in demystifying the design of an embedded system is to simply derive from industry standards what the system requirements are and then determine where in the overall system these derived components belong.

## 2.1 An Overview of Programming Languages and Examples of their Standards

**Why Use Programming Languages as a Standards Example?**

In embedded systems design, there is no single language that is the perfect solution for every system. Programming language standards, and what they introduce into an embedded systems architecture, are used as an example in this section, because a programming language can introduce an additional component into an embedded architecture. In addition, embedded systems software is inherently based on one or some combination of multiple languages. The examples discussed in-depth in this section, such as Java and the .NET Compact Framework, are based upon specifications that add additional elements to an embedded architecture. Other languages that can be based upon a variety of standards, such as ANSI C versus Kernighan and Ritchie C, are not discussed in depth because using these languages in an embedded design does not usually require introducing an additional component into the architecture.

*Note: Details of when to use what programming language and the pros and cons of such usage are covered in Chapter 11. It is important for the reader to first understand the various components of an embedded system before trying to understand the reasoning behind using certain components over others at the design and development level. Language choice decisions are not based on the features of the language alone and are often dependent on the other components within the system.*

The hardware components within an embedded system can only directly transmit, store, and execute *machine code*—a basic language consisting of 1 s and 0 s. Machine code was used in earlier days to program computer systems, which made creating any complex application a long and tedious ordeal. In order to make programming more efficient, machine code was made visible to programmers through the creation of a hardware-specific set of instructions, where each instruction corresponded to one or more machine code operations. These hardware-specific sets of instructions were referred to as *assembly language.* Over time, other programming languages, such as C, C++, and Java evolved with instruction sets that were (among other things) more hardware-independent. These are commonly referred to as *high-level* languages because they are semantically further away from machine code, they more closely resemble human languages, and they are typically independent of the hardware. This is in contrast to a *low-level* language, such as assembly language, which more closely resembles machine code. Unlike high-level languages, low-level languages are hardware-dependent, meaning there is a unique instruction set for processors with different architectures. Table 2-2 outlines this evolution of programming languages.

**Table 2-2: Evolution of Programming Languages**

| | Language | Details |
|---|---|---|
| First generation | Machine code | Binary (0,1) and hardware-dependent. |
| Second generation | Assembly language | Hardware-dependent, representing corresponding binary machine code. |
| Third generation | HOL (high-order languages)/ procedural languages | High-level languages with more English-like phrases and more transportable, such as C and Pascal. |
| Fourth generation | VHLL (very-high-level languages)/non-procedural languages | "Very" high-level languages: object-oriented languages (C++, Java, etc.), database query languages (SQL), etc. |
| Fifth generation | Natural languages | Programming similar to conversational languages, typically used in artificial intelligence (AI). Still in the research and development phases in most cases—not yet applicable in mainstream embedded systems. |

Even in systems that implement some higher-level languages, some portions of embedded systems software are implemented in assembly language for architecture-specific or optimized-performance code.

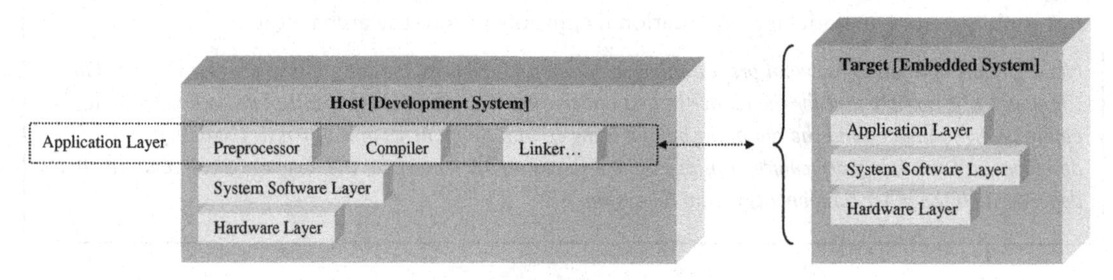


**Figure 2-2**
Host and target system diagram.

Because machine code is the only language the hardware can directly execute, all other languages need some type of mechanism to generate the corresponding machine code. This mechanism usually includes one or some combination of *preprocessing*, *translation*, and *interpretation*. Depending on the language, these mechanisms exist on the programmer's *host* system (typically a non-embedded development system, such as a PC or Sparc station) or the *target* system (the embedded system being developed). See Figure 2-2.

Preprocessing is an optional step that occurs before either the translation or interpretation of source code and whose functionality is commonly implemented by a *preprocessor*. The preprocessor's role is to organize and restructure the source code to make translation or interpretation of this code easier. As an example, in languages like C and C++, it is a preprocessor that allows the use of named code fragments, such as *macros*, that simplify code development by allowing the use of the macro's name in the code to replace fragments of

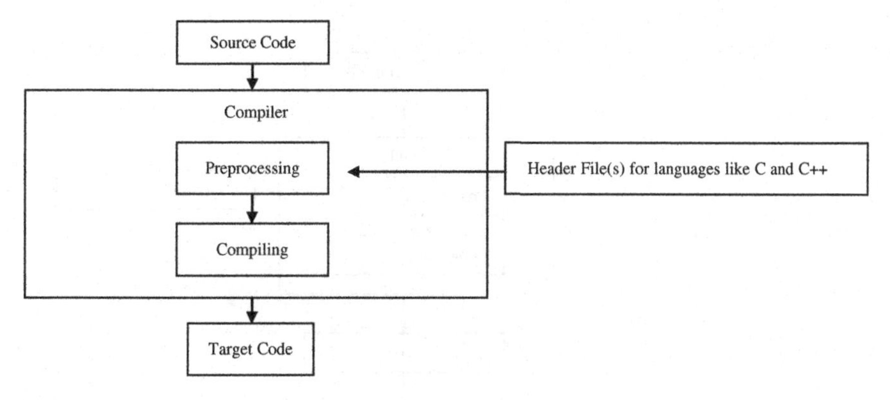


**Figure 2-3**
Compilation diagram.

code. The preprocessor then replaces the macro name with the contents of the macro during preprocessing. The preprocessor can exist as a separate entity, or can be integrated within the translation or interpretation unit.

Many languages convert source code, either directly or after having been preprocessed through use of a *compiler*—a program that generates a particular target language, such as machine code and Java byte code, from the source language (see Figure 2-3).

A compiler typically "translates" all of the source code to some target code at one time. As is usually the case in embedded systems, compilers are located on the programmer's host machine and generate target code for hardware platforms that differ from the platform the compiler is actually running on. These compilers are commonly referred to as *cross-compilers*. In the case of assembly language, the compiler is simply a specialized cross-compiler referred to as an *assembler* and it always generates machine code. Other high-level language compilers are commonly referred to by the language name plus the term "compiler," such as "Java compiler" and "C compiler." High-level language compilers vary widely in terms of what is generated. Some generate machine code, while others generate other high-level code, which then requires what is produced to be run through at least one more compiler or interpreter, as discussed later in this section. Other compilers generate assembly code, which then must be run through an assembler.

After all the compilation on the programmer's host machine is completed, the remaining target code file is commonly referred to as an *object file* and can contain anything from machine code to Java byte code (discussed later in this section), depending on the programming language used. As shown in Figure 2-4, after linking this object file to any system libraries required, the object file, commonly referred to as an *executable*, is then ready to be transferred to the target embedded system's memory.

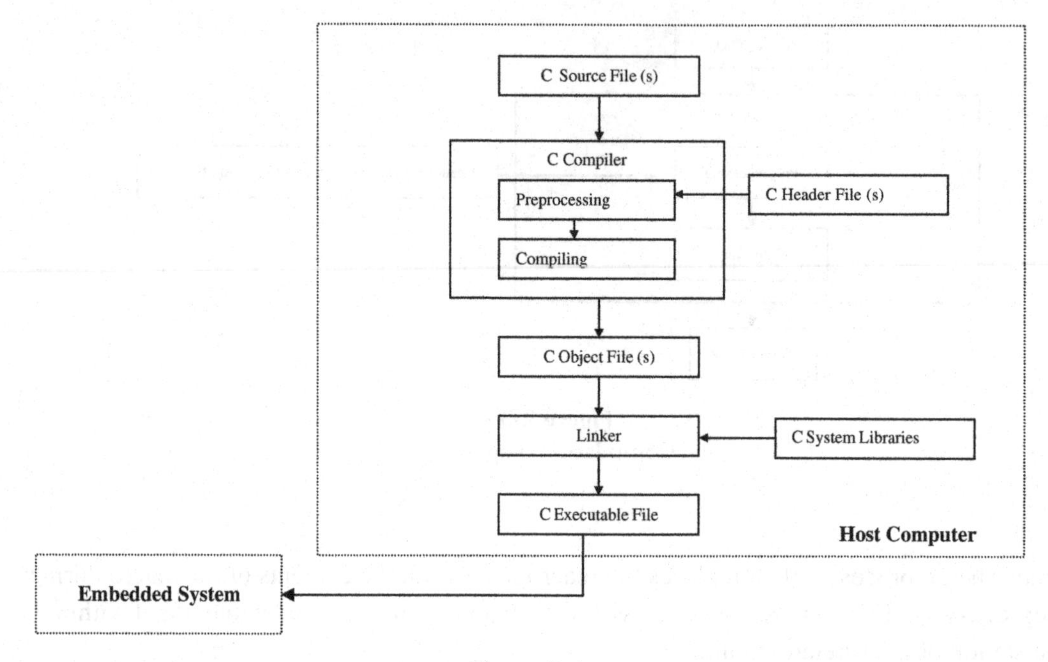


**Figure 2-4**
C Example compilation/linking steps and object file results.

**How Does the Executable Get from the Host to the Target?**

A combination of mechanisms are used to accomplish this. Details on memory and how files are executed from it will be discussed in more detail in Section II, while the different transmission mediums available for transmitting the executable file from a host system to an embedded system will be discussed in more detail in the next section of this chapter (Section 2.2). Finally, the common development tools used will be discussed in Chapter 12.

## 2.1.1 Examples of Programming Languages that Affect an Embedded Architecture: Scripting Languages, Java, and .NET

Where a compiler usually translates all of the given source code at one time, an *interpreter* generates (interprets) machine code one source code line at a time (see Figure 2-5).

One of the most common subclasses of interpreted programming languages are *scripting languages*, which include PERL, JavaScript, and HTML. Scripting languages are high-level programming languages with enhanced features, including:

- More platform independence than their compiled high-level language counterparts.[23]
- Late binding, which is the resolution of data types on-the-fly (rather than at compile time) to allow for greater flexibility in their resolution.[23]

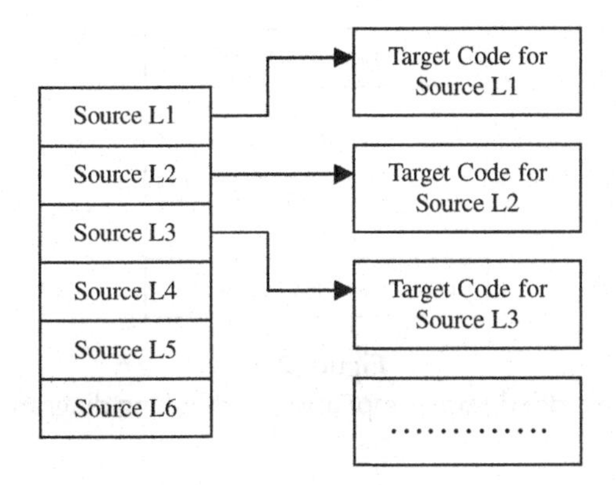


**Figure 2-5**
Interpretation diagram.

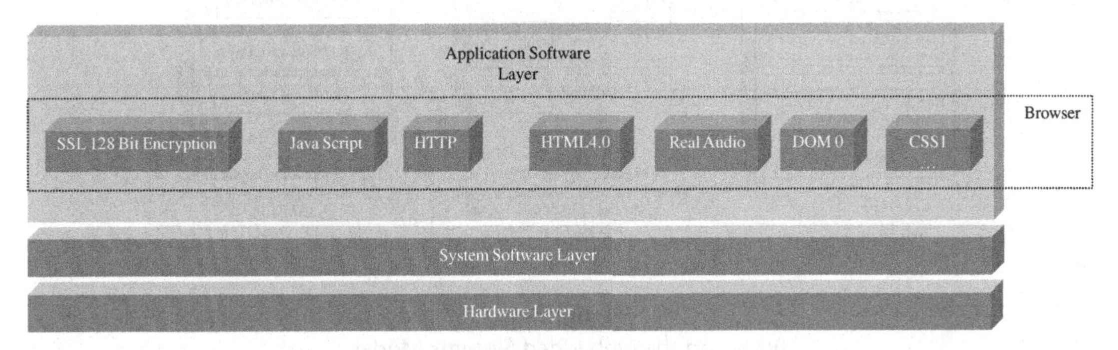


**Figure 2-6**
HTML and Javascript in the application layer.

- Importation and generation of source code at runtime, which is then executed immediately.[23]
- Optimizations for efficient programming and rapid prototyping of certain types of applications, such as internet applications and graphical user interfaces (GUIs).[23]
- With embedded platforms that support programs written in a scripting language, an additional component—an interpreter—must be included in the embedded system's architecture to allow for "on-the-fly" processing of code. Such is the case with the embedded system architectural software stack shown in Figure 2-6, where an internet browser can contain both an HTML and JavaScript interpreter to process downloaded web pages.

While all scripting languages are interpreted, not all interpreted languages are scripting languages. For example, one popular embedded programming language that incorporates both compiling and interpreting machine code generation methods is *Java*. On the programmer's

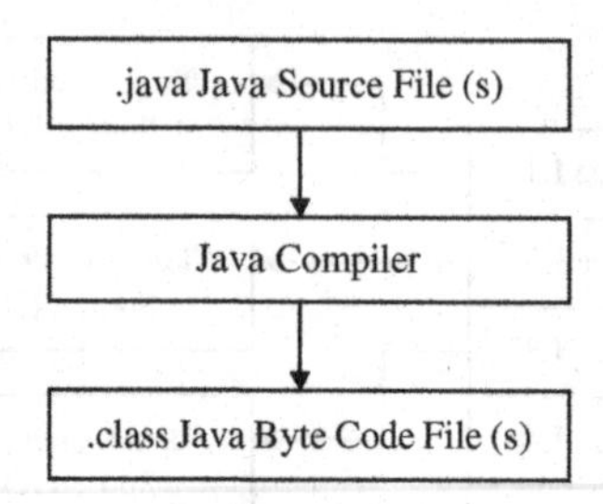


**Figure 2-7**
Embedded Java compilation and linking diagram.

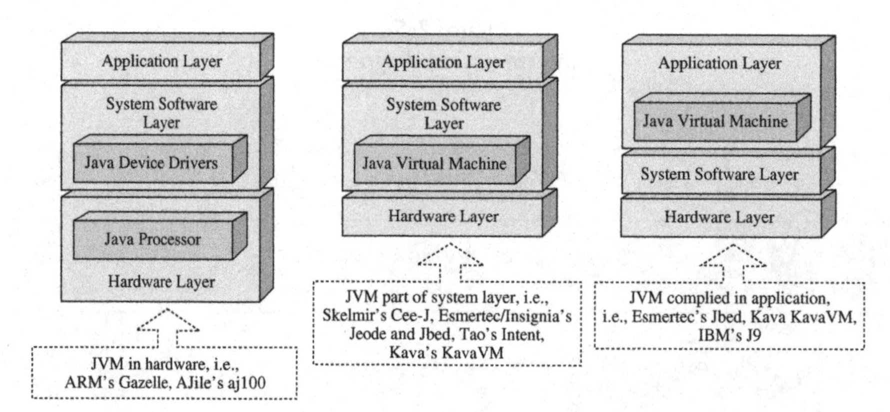


**Figure 2-8**
JVMs and the Embedded Systems Model.

host machine, Java must go through a compilation procedure that generates Java byte code from Java source code (see Figure 2-7).

Java byte code is target code intended to be platform independent. In order for the Java byte code to run on an embedded system, a *Java Virtual Machine (JVM)* must exist on that system. Real-world JVMs are currently implemented in an embedded system in one of three ways: in the hardware, in the system software layer, or in the application layer (see Figure 2-8).

Size, speed, and functionality are the technical characteristics of a JVM that most impact an embedded system design, and two JVM components are the primary differentiators between embedded JVMs: the JVM classes included within the JVM and the execution engine that contains components needed to successfully process Java code (see Figure 2-9).

The JVM classes shown in Figure 2-9 are compiled libraries of Java byte code, commonly referred to as *Java APIs (application program interfaces)*. Java APIs are application-independent libraries provided by the JVM to, among other things, allow programmers to

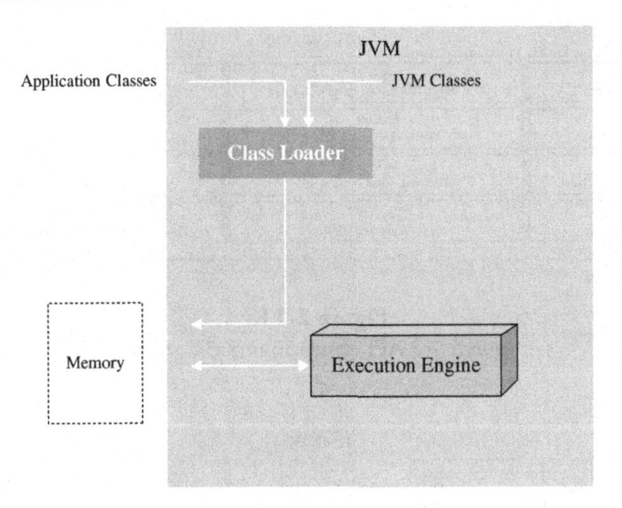


**Figure 2-9**
Internal JVM components.

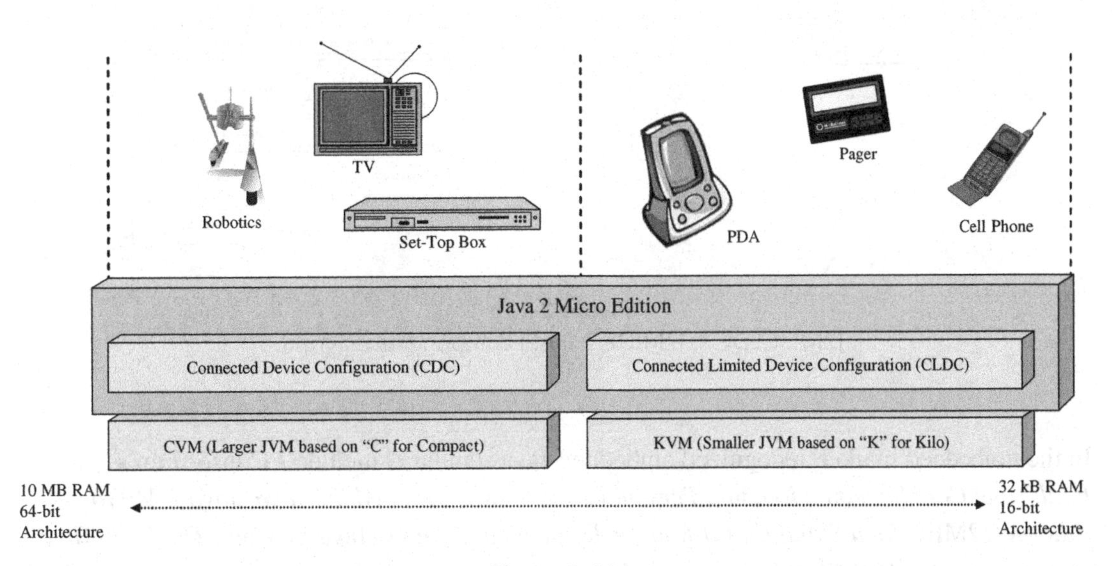


**Figure 2-10**
J2ME family of devices.

execute system functions and reuse code. Java applications require the Java API classes, in addition to their own code, to successfully execute. The size, functionality, and constraints provided by these APIs differ according to the Java specification they adhere to, but can include memory management features, graphics support, networking support, etc. Different standards with their corresponding APIs are intended for different families of embedded devices (see Figure 2-10).

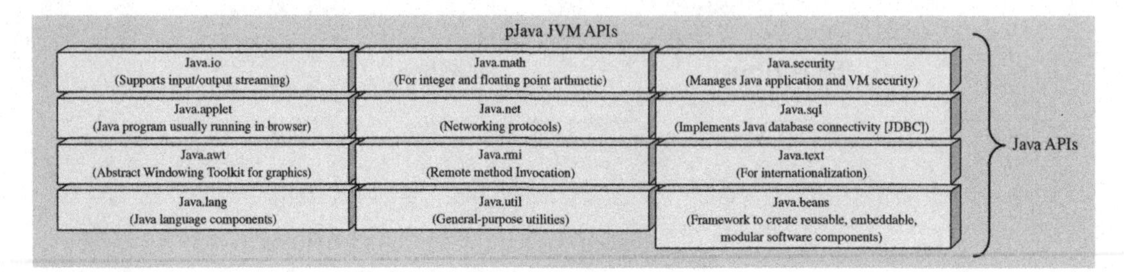


**Figure 2-11a**
pJava 1.2 API components diagram.

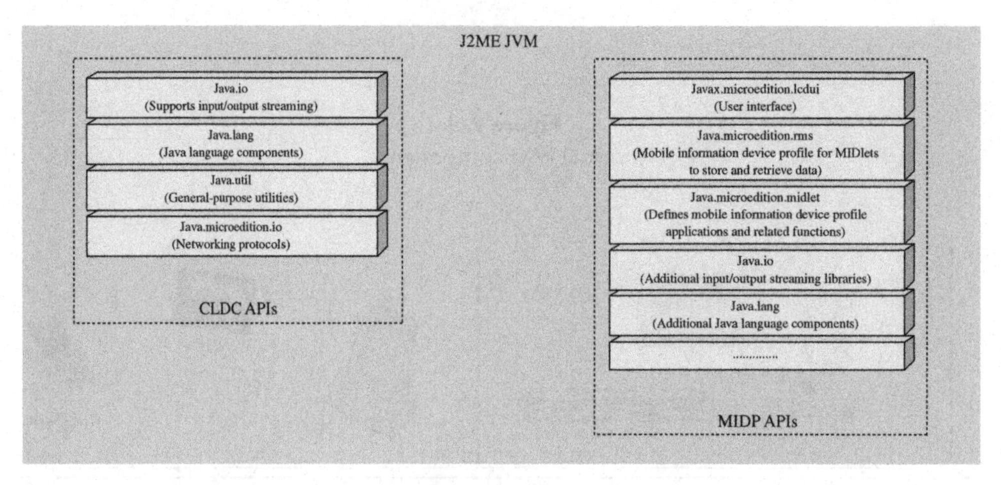


**Figure 2-11b**
J2ME CLDC 1.1/MIDP 2.0 API components diagram.

In the embedded market, recognized embedded Java standards include J Consortium's *Real-Time Core Specification*, and *Personal Java* (pJava), *Embedded Java, Java 2 Micro Edition* (J2ME), *Java Standard Edition for Embedded Systems* (Java SE), and *The Real-Time Specification for Java* from Oracle/Sun Microsystems.

Figures 2-11a and b show the differences between the APIs of two different embedded Java standards.

Table 2-3 shows several real-world JVMs and the standards they adhere to.

Within the execution engine (shown in Figure 2-12), the main differentiators that impact the design and performance of JVMs that support the same specification are:

- The *garbage collector (GC)*, which is responsible for deallocating any memory no longer needed by the Java application.

**Table 2-3: Real-World Examples of JVMs Based on Embedded Java Standards**

| Embedded Java Standard | JVM |
|---|---|
| Personal Java (pJava) | Tao Group's Intent |
| | Insignia's pJava Jeode |
| | NSICom CrE-ME |
| | Skelmir's pJava Cee-J |
| Embedded Java | Esmertec Embedded Java Jeode |
| Java 2 Micro Edition (J2ME) | Esmertec's Jbed for CLDC/MIDP and Insignia's CDC Jeode |
| | Skelmir's Cee-J CLDC/MIDP and CDC |
| | Tao Group's Intent CLDC and MIDP |

Information in the table was gathered at the time this book was written and is subject to change; check with specific vendors for latest information.

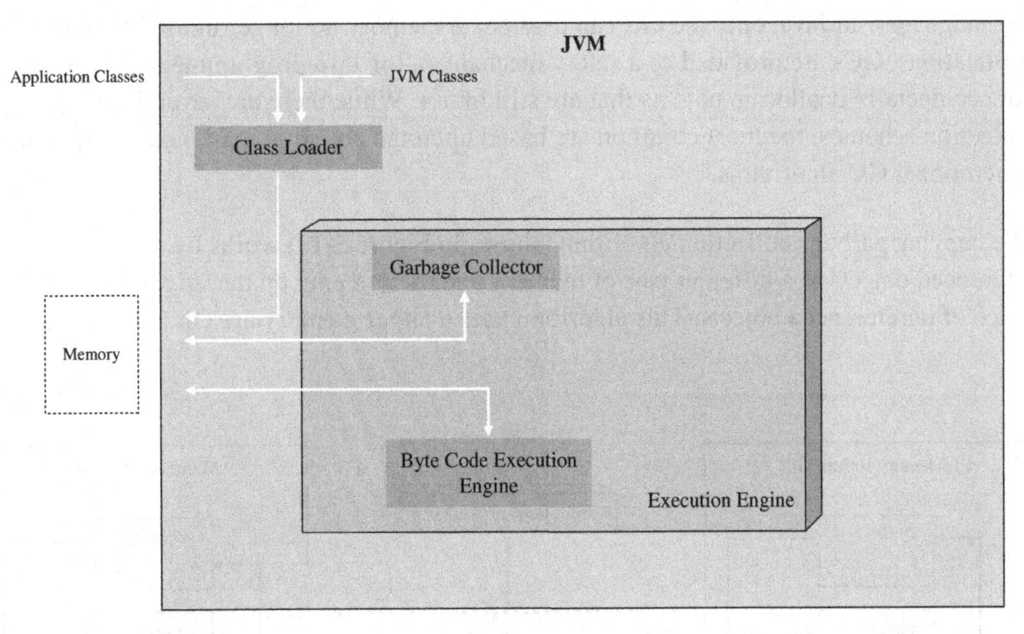


**Figure 2-12**
Internal execution engine components.

- The unit that *processes byte codes*, which is responsible for converting Java byte codes into machine code. A JVM can implement one or more byte code processing algorithms within its execution engine. The most common algorithms implemented are some combination of:
  - interpretation,
  - ahead-of-time (AOT) compilation, such as dynamic adaptive compilers (DAC), ahead-of-time, and way-ahead-of-time (WAT) algorithms,
  - just-in-time (JIT)—an algorithm that combines both compiling and interpreting.

### 2.1.2 Garbage Collection

**Why Talk About Garbage Collection?**

While this section discusses garbage collection within the context of Java, I use it as a separate example because garbage collection isn't unique to the Java language. A GC can be implemented in support of other languages, such as C and C++,[24] that don't typically add an additional component to the system. When creating a GC to support any language, it becomes part of an embedded system's architecture.

An application written in a language such as Java cannot deallocate memory that has been allocated for previous use (as can be done in native languages, such as using "free" in the C language, though as mentioned above, a GC can be implemented to support any language). In Java, only the GC can deallocate memory no longer in use by Java applications. GCs are provided as a safety mechanism for Java programmers so they do not accidentally deallocate objects that are still in use. While there are several garbage collection schemes, the most common are based upon the copying, mark and sweep, and generational GC algorithms.

The copying garbage collection algorithm (shown in Figure 2-13) works by copying referenced objects to a different part of memory and then freeing up the original memory space of unreferenced objects. This algorithm uses a larger memory area in order to work

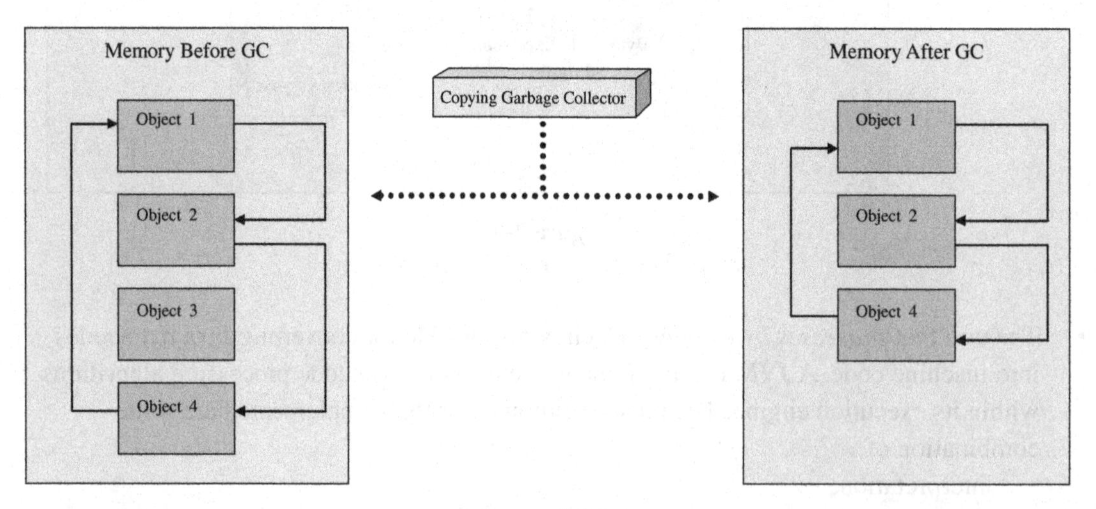


**Figure 2-13**
Copying GC.

and usually cannot be interrupted during the copy (it blocks the system). However, it does ensure that what memory is used is used efficiently by compacting objects in the new memory space.

The mark and sweep garbage collection algorithm (shown in Figure 2-14) works by "marking" all objects that are used and then "sweeping" (deallocating) objects that are unmarked. This algorithm is usually non-blocking, meaning the system can interrupt the GC to execute other functions when necessary. However, it doesn't compact memory the way a copying GC does, leading to *memory fragmentation*—the existence of small, unusable holes where deallocated objects used to exist. With a mark and sweep GC, an additional memory compacting algorithm can be implemented, making it a mark (sweep) and compact algorithm.

Finally, the generational garbage collection algorithm (shown in Figure 2-15) separates objects into groups, called *generations*, according to when they were allocated in memory. This algorithm assumes that most objects that are allocated by a Java program are short-lived; thus copying or compacting the remaining objects with longer lifetimes is a waste of time. So, it is objects in the younger generation group that are cleaned up more frequently than objects in the older generation groups. Objects can also be moved from a younger generation to an older generation group. Different generational GCs also may employ different algorithms to deallocate objects within each generational group, such as the copying algorithm or mark and sweep algorithms described previously.

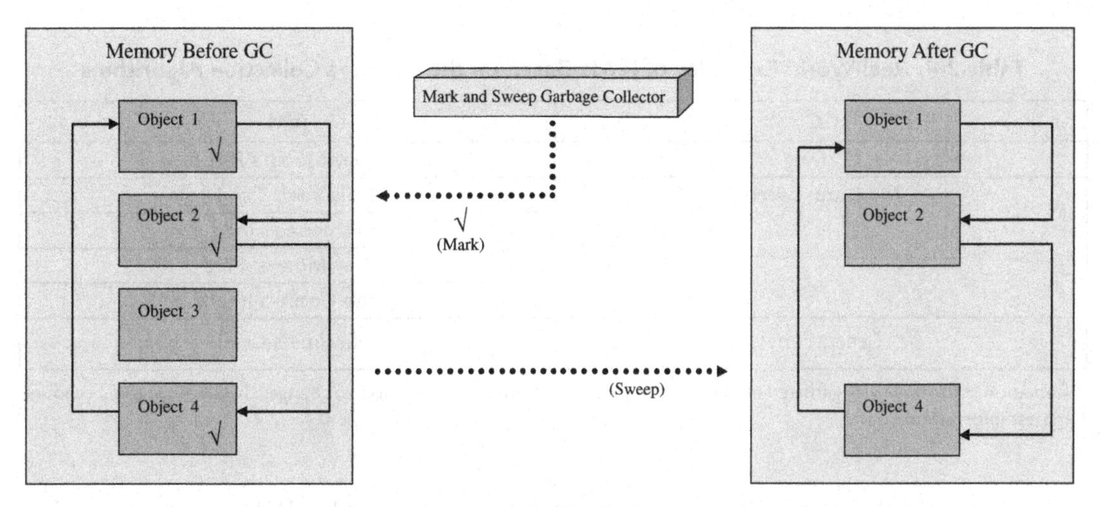


**Figure 2-14**
Mark and sweep (no compaction) GC diagram.

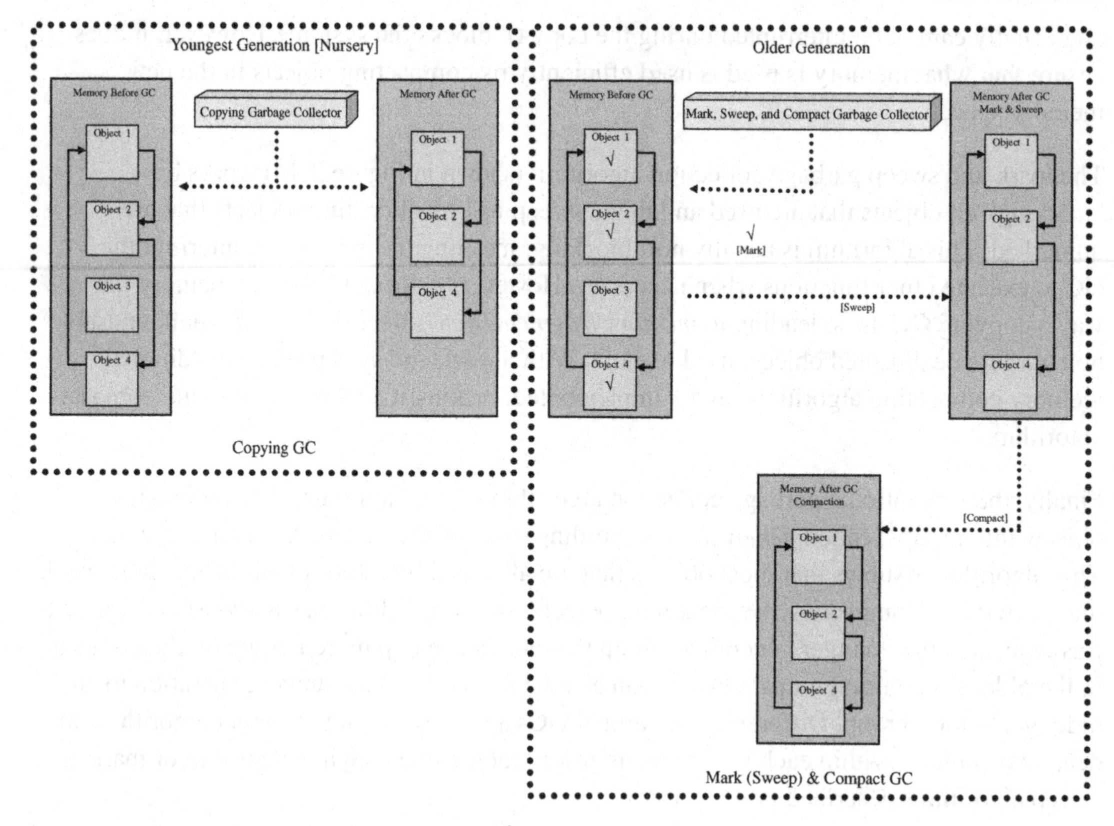


**Figure 2-15**
Generational GC diagram.

**Table 2-4: Real-World Examples of JVMs Based on the Garbage Collection Algorithms**

| GC | JVM |
|---|---|
| Copying | NewMonic's Perc |
| Mark and sweep | Skelmir's Cee-J |
| | Esmertec's Jbed |
| | NewMonics' Perc |
| | Tao Group's Intent |
| Generational | Skelmir's Cee-J |

Information in the table was gathered at the time this book was written and is subject to change; check with specific vendors for latest information.

As mentioned at the start of this section, most real-world embedded JVMs implement some form of the copying, mark and sweep, or generational algorithms (see Table 2-4).

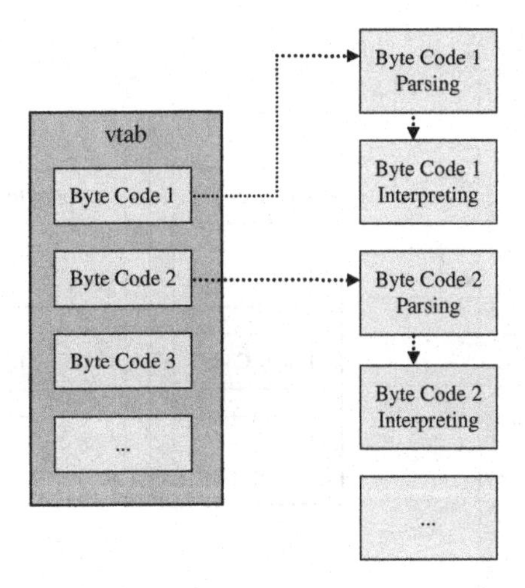


**Figure 2-16**
Interpreter diagram.

### *2.1.3 Processing Java Byte Code*

> **Why Talk About How Java Processes Byte Code?**
>
> This section is included because Java is an illustration of many different real-world techniques that are used to translate source code into machine code in a variety of other languages. For example, in assembly, C, and C++, the compilation mechanisms exist on the host machine, whereas HTML scripting language source is interpreted directly on the target (with no compilation needed). In the case of Java, Java source code is compiled on the host into Java byte code, which then can be interpreted or compiled into machine code, depending on the JVM's internal design. Any mechanism that resides on the target, which translates Java byte code into machine code, is part of an embedded system's architecture. In short, Java's translation mechanisms can exist both on the host and on the target, and so act as examples of various real-world techniques that can be used to understand how programming languages in general impact an embedded design.

The JVM's primary purpose in an embedded system is to process platform-independent Java byte code into platform-dependent code. This processing is handled in the execution engine of the JVM. The three most common byte code processing algorithms implemented in an execution engine to date are interpretation, JIT compiling, and WAT/AOT compiling.

With interpretation, each time the Java program is loaded to be executed, every byte code instruction is parsed and converted to native code, one byte code at a time, by the JVM's interpreter (see Figure 2-16). Moreover, with interpretation, redundant portions of the code are reinterpreted every time they are run. Interpretation tends to have the lowest performance of the three algorithms, but it is typically the simplest algorithm to implement and to port to different types of hardware.

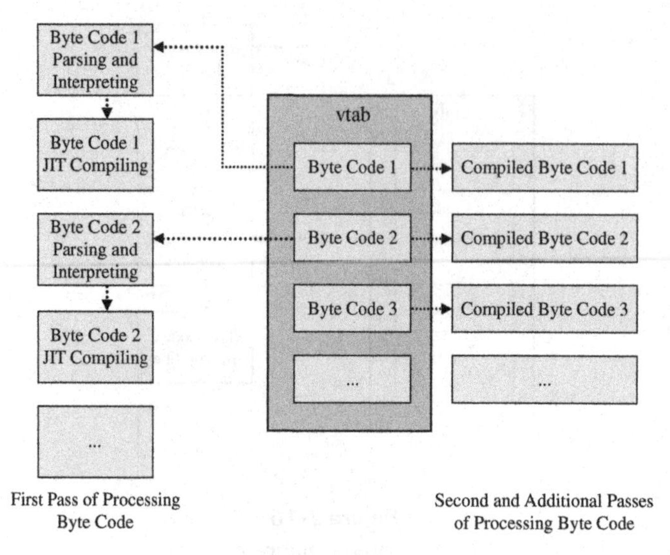


**Figure 2-17**
JIT diagram.

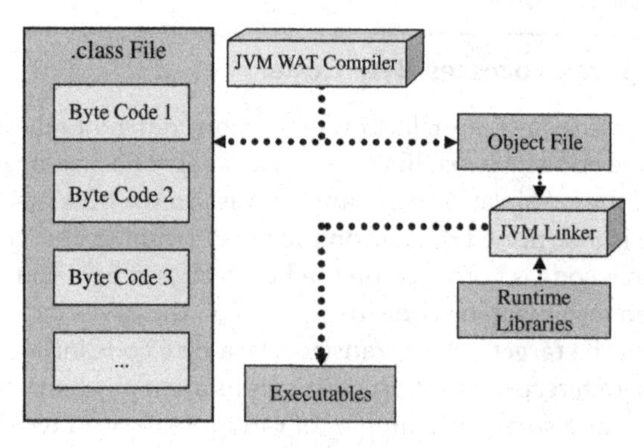


**Figure 2-18**
WAT/AOT diagram.

A JIT compiler, on the other hand, interprets the program once, and then compiles and stores the native form of the byte code at runtime, thus allowing redundant code to be executed without having to reinterpret (see Figure 2-17). The JIT algorithm performs better for redundant code, but it can have additional runtime overhead while converting the byte code into native code. Additional memory is also used for storing both the Java byte codes and the native compiled code. Variations on the JIT algorithm in real-world JVMs are also referred to as translators or DAC.

Finally, in WAT/AOT compiling, all Java byte code is compiled into the native code at compile time, as with native languages, and no interpretation is done (see Figure 2-18). This algorithm performs at least as well as the JIT for redundant code and better than

**Table 2-5: Real-World Examples of JVMs Based on the Various Byte Code Processing Algorithms**

| Byte Code Processing | JVM |
|---|---|
| Interpretation | Skelmir Cee-J |
| | NewMonics Perc |
| | Insignia's Jeode |
| JIT | Skelmir Cee-J (two types of JITS) |
| | Tao Group's Intent–translation |
| | NewMonics Perc |
| | Insignia's Jeode DAC |
| WAT/AOT | NewMonics Perc |
| | Esmertec's Jbed |

Information in the table was gathered at the time this book was written and is subject to change; check with specific vendors for latest information.

a JIT for non-redundant code, but, as with the JIT, there is additional runtime overhead when additional Java classes dynamically downloaded at runtime have to be compiled and introduced into the system. WAT/AOT can also be a more complex algorithm to implement.

As seen in Table 2-5, there are real-world JVM execution engines that implement each of these algorithms, as well as execution engine hybrids that implement some or all of these algorithms.

Scripting languages and Java aren't the only high-level languages that can automatically introduce an additional component within an embedded system. The *.NET Compact Framework* from Microsoft allows applications written in almost any high-level programming language (such as C#, Visual Basic, and Javascript) to run on any embedded device, independent of hardware or system software design. Applications that fall under the .NET Compact Framework must go through a compilation and linking procedure that generates a CPU-independent intermediate language file, called MSIL (Microsoft Intermediate Language), from the original source code file (see Figure 2-19). For a high-level language to be compatible with the .NET Compact Framework, it must adhere to Microsoft's *Common Language Specification*—a publicly available standard that anyone can use to create a compiler that is .NET compatible.

The .NET Compact Framework is made up of a *common language runtime (CLR)*, a class loader, and platform extension libraries. The CLR is made up of an execution engine that processes the intermediate MSIL code into machine code and a GC. The platform extension libraries are within the base class library (BCL), which provides additional functionality to applications (such as graphics, networking, and diagnostics). As shown in Figure 2-20, in order to run the intermediate MSIL file on an embedded system, the .NET Compact Framework must exist on that embedded system. At the current time, the .NET Compact Framework resides in the system software layer.

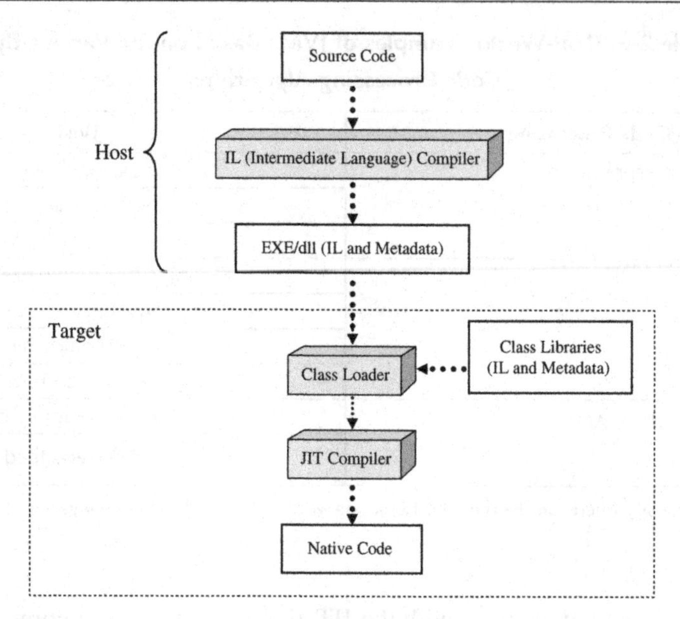


**Figure 2-19**
.NET Compact Framework execution model.

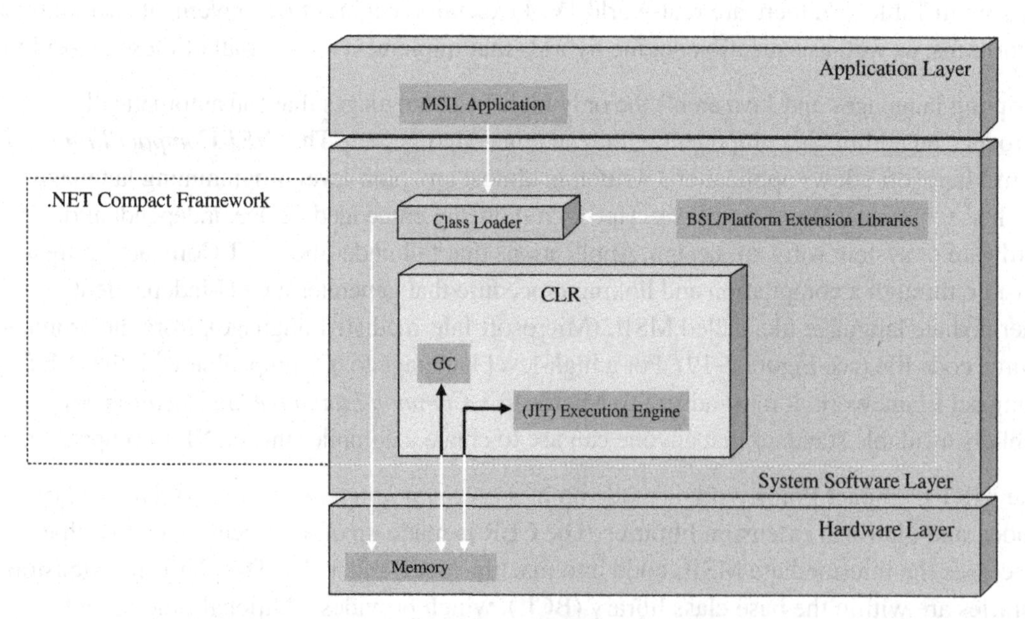


**Figure 2-20**
.NET Compact Framework and the Embedded Systems Model.

## 2.2 Standards and Networking

**Why Use Networking as a Standards Example?**

A network, by definition, is two or more connected devices that can send and/or receive data. If an embedded system needs to communicate with any other system, whether a development host machine, a server, or another embedded device, it must implement some type of connection (networking) scheme. In order for communication to be successful, there needs to be a scheme that interconnecting systems agree upon, and so networking protocols (standards) have been put in place to allow for interoperability. As shown in Table 2-1, networking standards can be implemented in embedded devices specifically for the networking market, as well as in devices from other market segments that require networking connectivity, even if just to debug a system during the development phase of a project.

Understanding what the required networking components are for an embedded device requires two steps:

- Understanding the entire network into which the device will connect.
- Using this understanding, along with a networking model, such as the OSI (Open Systems Interconnection) model, discussed later in this section, to determine the device's networking components.

Understanding the entire network is important, because key features of the network will dictate the standards that need to be implemented within an embedded system. Initially, an embedded engineer should, at the very least, understand three features about the entire network the device will plug into: the distance between connected devices, the physical medium connecting the embedded device to the rest of the network, and the network's overall structure (see Figure 2-21).

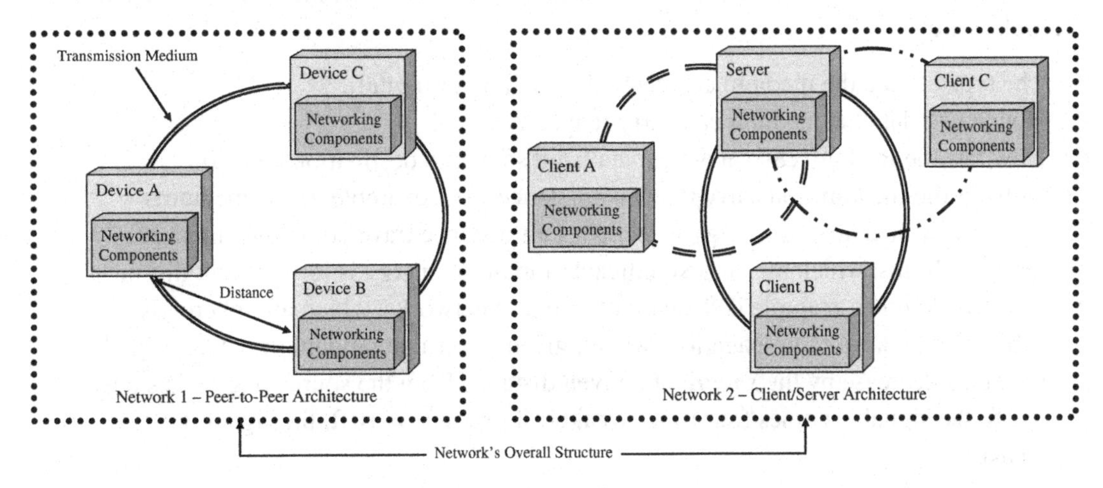


**Figure 2-21**
Network block diagram.

### 2.2.1 Distance between Connected Devices

Networks can be broadly defined as either local area networks (LANs) or wide area networks (WANs). A LAN is a network in which all devices are within close proximity to each other, such as in the same building or room. A WAN is a network that connects devices and/or LANs that are scattered over a wider geographical area, such as in multiple buildings or across the globe. While there are other variations of WANs (such as Metropolitan Area Networks (MANs) for intercity networks, Campus Area Networks (CANs) for school-based networks, etc.) and LANs (i.e., short-distance wireless Personal Area Networks (PANs)) that exist, networks are all essentially either WANs or LANs.

**Warning!**

Watch those acronyms! Many look similar but in fact can mean very different things. For example, a WAN (wide area network) should not be confused with WLAN (wireless local area network).

### 2.2.2 Physical Medium

In a network, devices are connected with transmission mediums that are either bound or unbound. Bound transmission mediums are cables or wires and are considered "guided" mediums since electromagnetic waves are guided along a physical path (the wires). Unbound transmission mediums are wireless connections and they are considered unguided mediums because the electromagnetic waves that are transmitted are not guided via a physical path, but are transmitted through a vacuum, air, and/or water.

In general, the key characteristics that differentiate all transmission mediums whether wired or wireless are:

- The type of data the medium can carry (i.e., analog or digital).
- How much data the medium can carry (capacity).
- How fast the medium can carry the data from source to destination (speed).
- How far the medium can carry the data (distance). For example, some mediums are lossless, meaning no energy is lost per unit distance traveled, while others are lossy mediums, which means a significant amount of energy is lost per unit distance traveled. Another example is the case of wireless networks, which are inherently subject to the laws of propagation, where, given a constant amount of power, signal strengths decrease by the *square* of a given distance from the source (e.g., if distance = 2 ft, the signal becomes four times weaker; if distance = 10 ft, the signal is 100 times weaker).
- How susceptible the medium is to external forces (interference such as electromagnetic interference (EMI), radiofrequency interference (RFI), weather).

**Author Note**

The direction in which a transmission medium can transmit data (i.e., data being able to travel in only one direction versus bidirectional transmission) is dependent on the hardware and software components implemented within the device, and is typically not dependent on the transmission medium alone. This will be discussed later in this section.

Understanding the features of the transmission medium is important, because these impact the overall network's performance, affecting such variables as the network's bandwidth (data rate in bits per second) and latency (the amount of time it takes data to travel between two given points, including delays). Tables 2-6a and b summarize a few examples of wired (bound) and wireless (unbound) transmission mediums, as well as some of their features.

### *2.2.3 The Network's Architecture*

The relationship between connected devices in a network determines the network's overall architecture. The most common architecture types for networks are *peer-to-peer*, *client/server*, and *hybrid* architectures.

*Peer-to-peer* architectures are network implementations in which there is no centralized area of control. Every device on the network must manage its own resources and requirements. Devices all communicate as equals, and can utilize each other's resources. Peer-to-peer networks are usually implemented as LANs because, while simpler to implement, this architecture creates security and performance issues related to the visibility and accessibility of each device's resources to the rest of the network.

*Client/server* architectures are network implementations in which there is a centralized device, called the *server*, in control that manages most of the network's requirements and resources. The other devices on the network, called *clients*, contain fewer resources and must utilize the server's resources. The client/server architecture is more complex than the peer-to-peer architecture and has the single critical point of failure (the server). However, it is more secure than the peer-to-peer model, since only the server has visibility into other devices. Client/server architectures are also usually more reliable, since only the server has to be responsible for providing redundancy for the network's resources in case of failures. Client/server architectures also have better performance, since the server device in this type of network usually needs to be more powerful in order to provide the network's resources. This architecture is implemented in either LANs or WANs.

*Hybrid* architectures are a combination of the peer-to-peer and client/server architecture models. This architecture is also implemented in both LANs and WANs.

**Table 2-6a: Wired Transmission Mediums**

| Medium | Features |
|---|---|
| Unshielded twisted pair (UTP) | Copper wires are twisted into pairs and used to transmit analog or digital signals. Limits in length (distance) depending on the desired bandwidth. UTP used in telephone/telex networks; can support both analog and digital. Different categories of cables (3, 4, and 5) where CAT3 supports a data rate of up to 16 Mbps, CAT4 up to 20 MHz, and CAT5 up to 100 Mbps. Requires amplifiers every 5–6 km for analog signals and repeaters every 2–3 km for digital signals (over long distances signals lose strength and are mistimed). Relatively easy and cheap to install, but with a security risk (can be tapped into). Subject to external EMI. Can act as antennas receiving EMI/RFI from sources such as electric motors, high-voltage transmission lines, vehicle engines, and radio or TV broadcast equipment. These signals, when superimposed on a data stream, may make it difficult for the receiver to differentiate between valid data and EMI/RFI-induced noise (especially true for long spans that incorporate components from multiple vendors). Crosstalk occurs when unwanted signals are coupled between "transmit" and "receive" copper pairs, creating corrupted data and making it difficult for the receiver to differentiate between normal and coupled signals. Lightning can be a problem when it strikes unprotected copper cable and attached equipment (energy may be coupled into the conductor and can propagate in both directions). |
| Coaxial | Baseband and broadband coaxial cables differ in features. Generally, coaxial cables are copper-wire and aluminum-wire connections that are used to transmit both analog and digital signals. Baseband coaxial commonly used for digital—cable TV/cable modems. Broadband used in analog (telephone) communication. Coaxial cable cannot carry signals beyond several thousand feet before amplification is required (i.e., by repeaters or boosters); higher data rates than twisted-pair cabling (several hundred Mbps and up to several km). Coaxial cables not secure (can be tapped into), but are shielded to reduce interference and therefore allow higher analog transmissions. |
| Fiber optic | Clear, flexible tubing that allows laser beams to be transmitted along the cable for digital transmissions. Fiber-optic mediums have a GHz (bandwidth) transmission capability up to 100 km. Because of their dielectric nature, optical fibers are immune to both EMI and RFI and crosstalk rarely occurs. Optical-fiber cables do not use metallic conductors for transmission, so all-dielectric, optical-fiber communications are less susceptible to electrical surges even if struck directly by lightning. More secure—difficult to tap into, but typically more costly than other terrestrial solutions. |

**Author Note**

A network's architecture is *not* the same as its topology. A network's topology is the physical arrangement of the connected devices, which is ultimately determined by the architecture, the transmission medium (wired versus wireless), and the distance between the connected devices of the particular network.

**Table 2-6b: Wireless Transmission Mediums**

| Medium | Features |
|---|---|
| Terrestrial microwave | Classified as SHF (super high frequency). Transmission signal must be line of sight, meaning high-frequency radio (analog or digital) signals are transmitted via a number of ground stations and transmission between stations must be in a straight line that is unobstructed ground—often used in conjunction with satellite transmission. The distance between stations is typically 25–30 miles, with the transmission dishes on top of high buildings or on high points like hill tops. The use of the low GHz frequency range 2–40 GHz provides higher bandwidths (i.e., 2 GHz band has approximately 7 MHz bandwidth, whereas 18 GHz band has approximately 220 MHz bandwidth) than those available using lower frequency radio waves. |
| Satellite microwave | Satellites orbit above the Earth and act as relay stations between different ground stations and embedded devices (their antennas) that are within the satellite's line-of-sight and area covered (footprint), where the size and shape of the footprint on the earth's surface vary with the design of the satellite. The ground station receives analog or digital data from some source (internet service provider, broadcaster, etc.) and modulates it onto a radio signal that it transmits to the satellite, as well as controls the position and monitors the operations of the satellite. At the satellite, a transponder receives the radio signal, amplifies it and relays it to a device's antenna inside its footprint. Varying the footprint changes the transmission speeds, where focusing the signal on a smaller footprint increases transmission speeds. The large distances covered by a signal can also result in propagation delays of several seconds. A typical GEO (geosynchronous earth orbit) satellite (a satellite that orbits about 36,000 km above the equator—the speed of the satellite is matched to the rotation of the Earth at the equator) contains between 20 and 80 transponders, each capable of transmitting digital information of up to about 30–40 Mbps. |
| Broadcast radio | Uses a transmitter and a receiver (of the embedded device) tuned to a specific frequency for the transfer of signals. Broadcast communication occurs within a local area, where multiple sources receive one transmission. Subject to frequency limitations (managed by local communications companies and government) to ensure no two transmissions on the same frequency. Transmitter requires large antennas; frequency range of 10 kHz–1 GHz subdivided into LF (low frequency), MF (medium frequency), HF (high frequency), UHF (ultra high frequency), or VHF (very high frequency) bands. Higher frequency radio waves provide larger bandwidths (in the Mbps) for transmission, but they have less penetrating power than lower frequency radio waves (with bandwidths as low as in the kbps). |
| IR (Infrared) | Point-to-point link of two IR lasers lined up; blinking of laser beam reflects bit representations. THz (1000 GHz–$2 \times 10^{11}$ Hz–$2 \times 10^{14}$ Hz) range of frequencies, with up to 20 Mbps bandwidth. Cannot have obstructions, more expensive, susceptible to bad weather (clouds, rain) and diluted by sunlight, difficult to "tap" (more secure). Used in small, open areas with a typical transmission distance of up to 200 m. |
| Cellular microwave | Works in UHF band—variable—depends on whether there are barriers. Signals can penetrate buildings/barriers, but degradation occurs and reduces the required distance the device must be from the cell site. |

### 2.2.4 OSI Model

To demonstrate the dependencies between the internal networking components of an embedded system and the network's architecture, the distance between connected devices, and the transmission medium connecting the devices, this section associates networking components with a universal networking model, in this case the Open Systems Interconnection (OSI) Reference model. All the required networking components in a device can be grouped together into the OSI model, which was created in the early 1980s by the International Organization for Standardization (ISO). As shown in Figure 2-22, the OSI

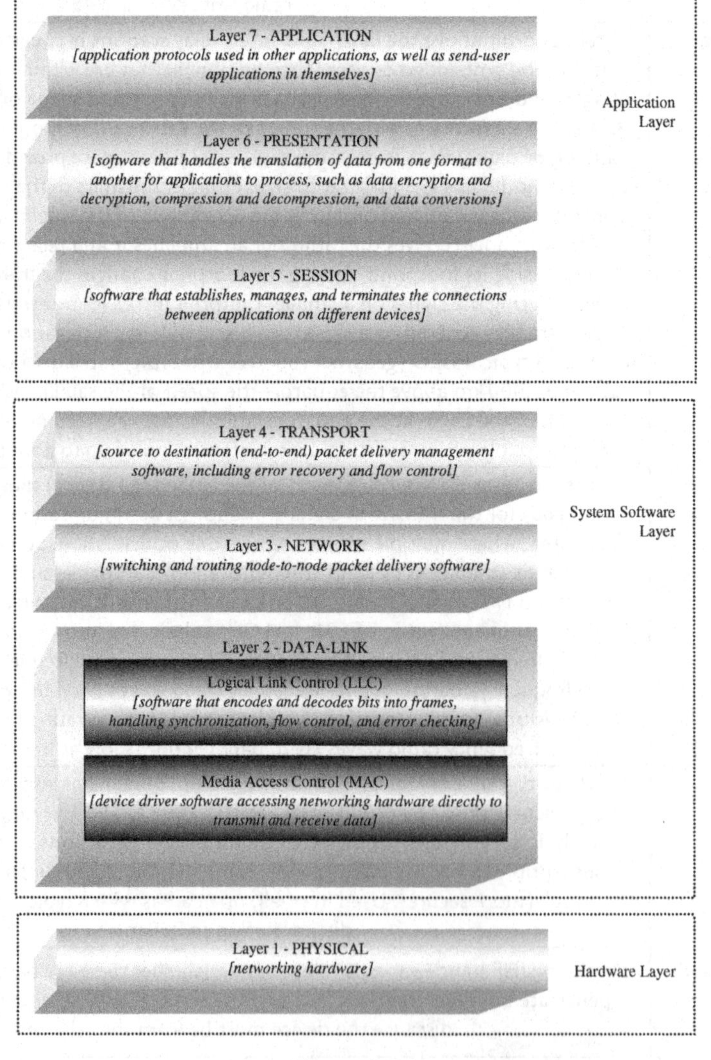


**Figure 2-22**
OSI and Embedded Systems Model block diagram.

model represents the required hardware and software components within a networked device in the form of seven layers: physical, data-link, network, transport, session, presentation, and application layers. In relation to the Embedded Systems Model (see Figure 1-1), the physical layer of the OSI model maps to the hardware layer of the Embedded Systems Model, the application, presentation, and session layers of the OSI model map to the application software layer of the Embedded Systems Model, and the remaining layers of the OSI model typically map to the system software layer of the embedded systems model.

The key to understanding the purpose of each layer in the OSI model is to grasp that networking is not simply about connecting one device to another device. Instead, networking is primarily about the data being transmitted between devices or, as shown in Figure 2-23, between the different layers of each device.

In short, a networking connection starts with data originating at the application layer of one device and flowing downward through all seven layers, with each layer adding a new bit of information to the data being sent across the network. Information, called the *header* (shown

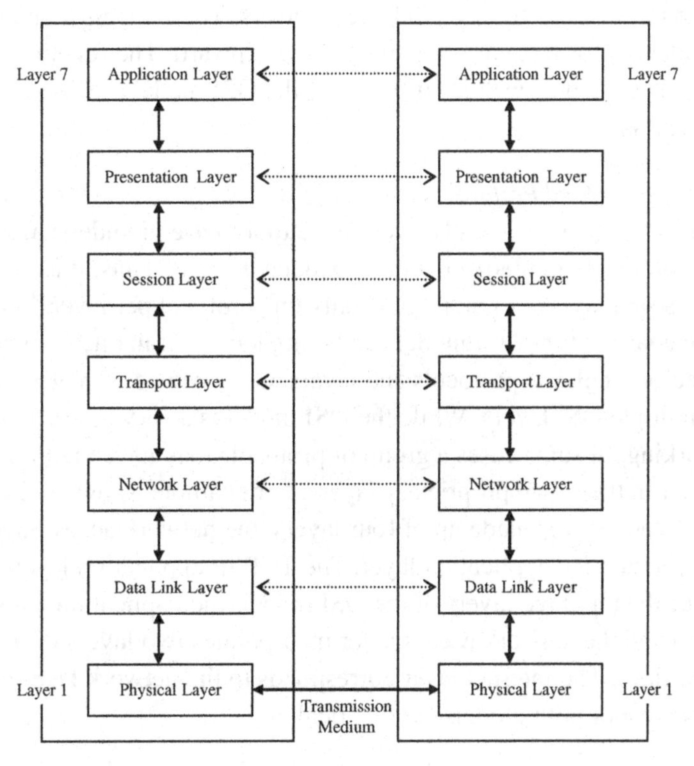


**Figure 2-23**
OSI model data flow diagram.

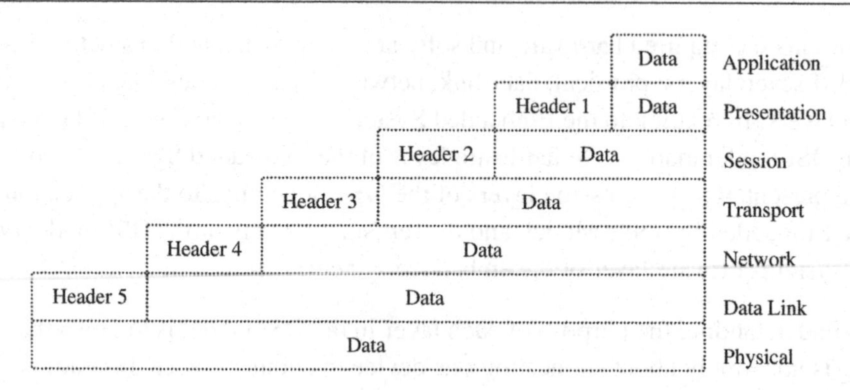


**Figure 2-24**
Header diagram.

in Figure 2-24), is appended to the data at every layer (except for the physical and application layers) for peer layers in connected devices to process. In other words, the data is wrapped with information for other devices to unwrap and process.

The data is then sent over the transmission medium to the physical layer of a connected device and then up through the connected device's layers. These layers then process the data (strip the headers, reformat, etc.) as the data flows upward. The functionality and methodologies implemented at each layer based on the OSI model are also commonly referred to as *networking protocols*.

*The OSI Model and Real-World Protocol Stacks*

Remember that the OSI model is simply a reference tool to use in understanding real-world networking protocols implemented in embedded devices. Thus, it isn't always the case that there are seven layers or that there is only one protocol per layer. In reality, the functionality of one layer of the OSI model can be implemented in one protocol or it can be implemented across multiple protocols and layers. One protocol can also implement the functionality of multiple OSI layers. While the OSI model is a very powerful tool to use to understand networking, in some cases a group of protocols may have their own name and be grouped together in their own proprietary layers. For example, shown in Figure 2-25 is a *TCP/IP* protocol stack that is made up of four layers: the network access layer, internet layer, transport layer, and the application layer. The TCP/IP application layer incorporates the functionality of the top three layers of the OSI model (the application, presentation, and session layers) and the network access layer incorporates two layers of the OSI model (physical and data-link). The internet layer corresponds to the network layer in the OSI model and the transport layers of both models are identical.

As another example, the wireless application protocol (WAP) stack (shown in Figure 2-26) provides five layers of upper layer protocols. The WAP application layer maps to the application layer of the OSI model, as do the transport layers of both models. The session and transaction

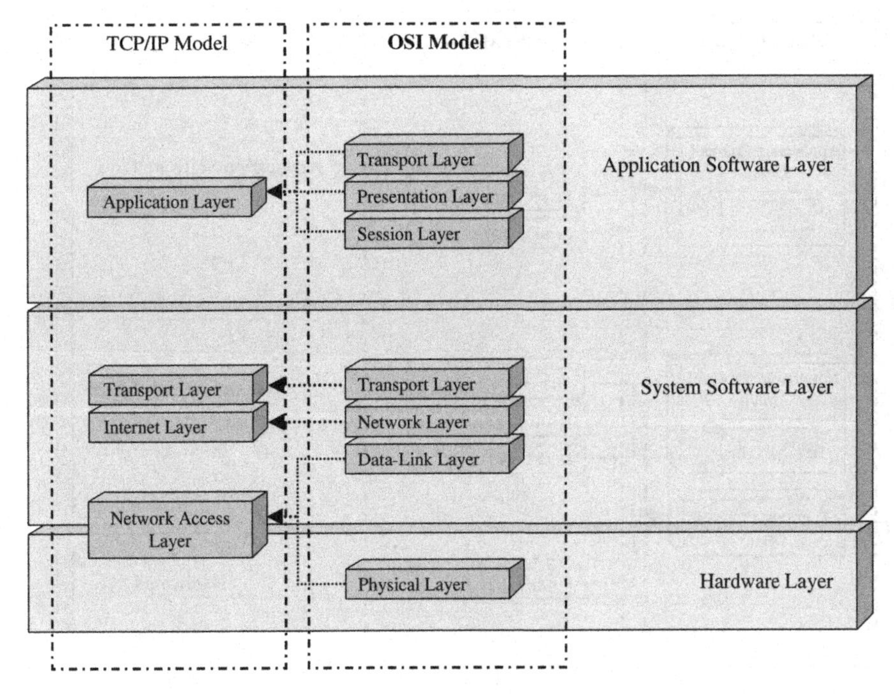


**Figure 2-25**
TCP/IP, OSI models, and Embedded Systems Model block diagram.

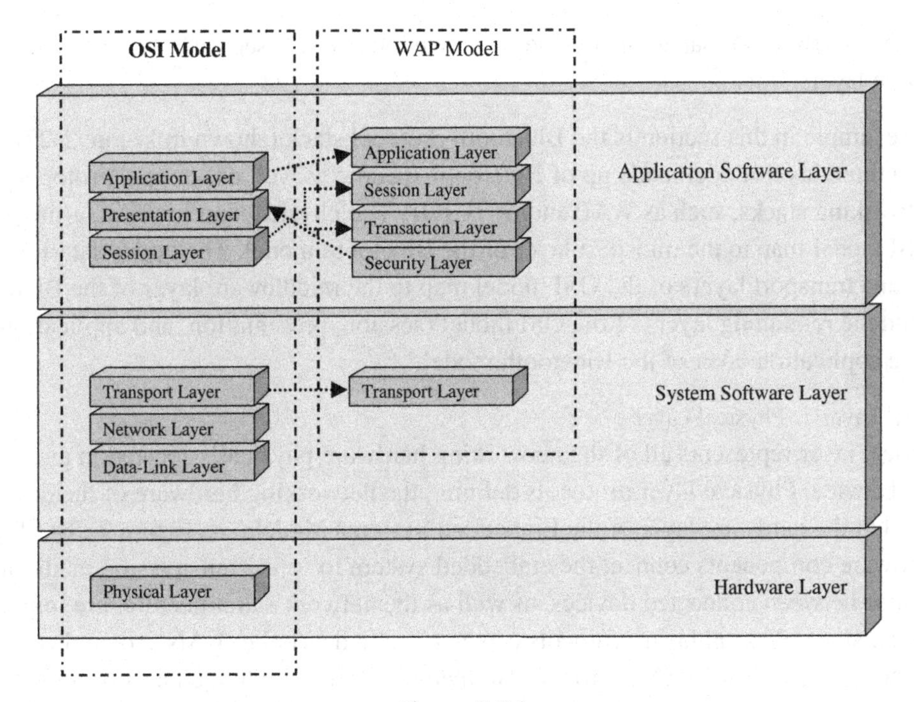


**Figure 2-26**
WAP, OSI, and Embedded Systems Model block diagram.

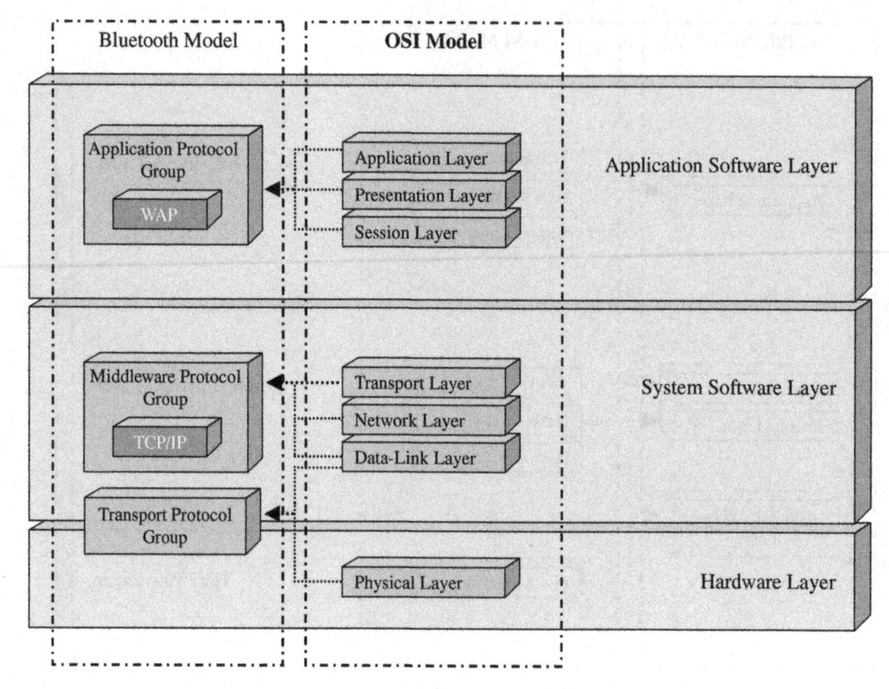


**Figure 2-27**
Bluetooth, OSI, and Embedded Systems Model block diagram.

layers of the WAP model map to the OSI session layer and WAP's security layer maps to OSI's presentation layer.

The final example in this section is the Bluetooth protocol stack (shown in Figure 2-27), which is a three-layer model made up of Bluetooth-specific as well as adopted protocols from other networking stacks, such as WAP and/or TCP/IP. The physical and lower data-link layers of the OSI model map to the transport layer of the Bluetooth model. The upper data-link, network, and transport layers of the OSI model map to the middleware layer of the Bluetooth model, and the remaining layers of the OSI model (session, presentation, and application) map to the application layer of the Bluetooth model.

### OSI Model Layer 1: Physical Layer

The physical layer represents all of the networking hardware physically located in an embedded device. Physical layer protocols defining the networking hardware of the device are located in the hardware layer of the Embedded Systems Model (see Figure 2-28). Physical layer hardware components connect the embedded system to some transmission medium. The distance between connected devices, as well as the network's architecture, are important at this layer, since physical layer protocols can be classified as either LAN protocols or WAN protocols. LAN and WAN protocols can then be further subdivided according to the transmission medium connecting the device to the network (wired or wireless).

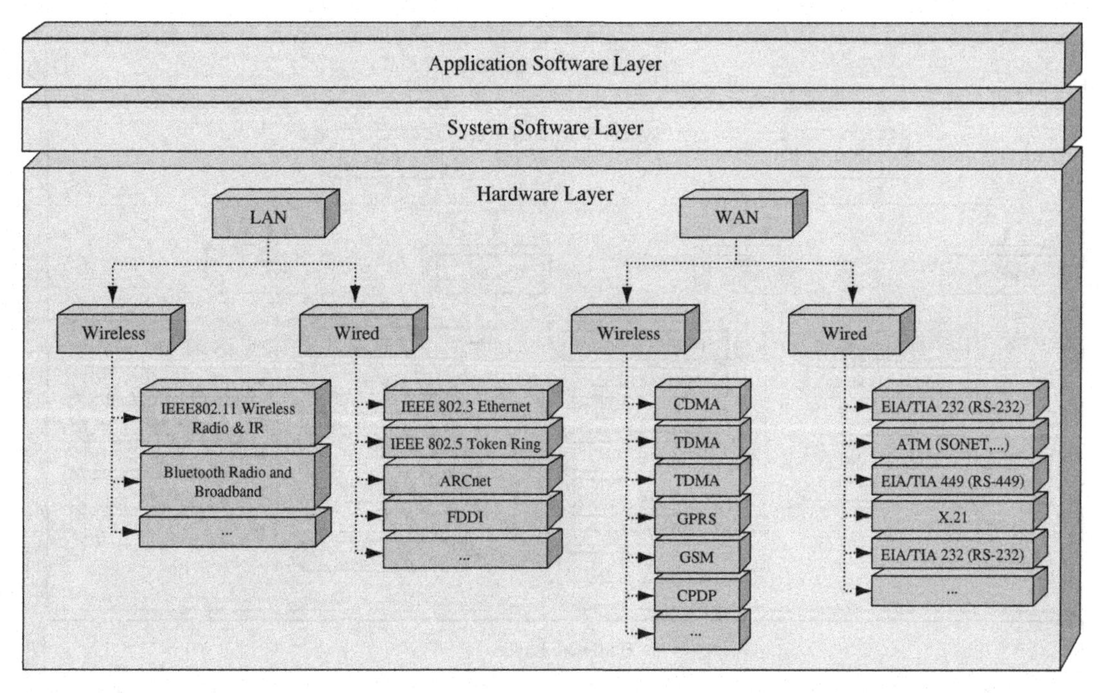


**Figure 2-28**
Physical layer protocols in the Embedded Systems Model.

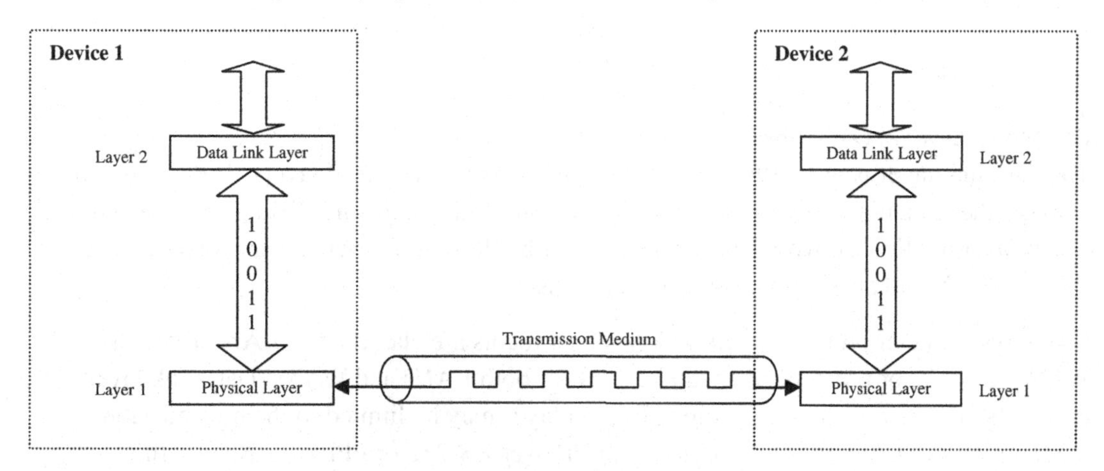


**Figure 2-29**
Physical layer data flow block diagram.

The physical layer defines, manages, and processes, via hardware, the data signals—the actual voltage representations of 1 s and 0 s—coming over the communication medium. The physical layer is responsible for physically transferring the data bits over the medium received from higher layers within the embedded system, as well as for reassembling bits received over the medium for higher layers in the embedded system to process (see Figure 2-29).

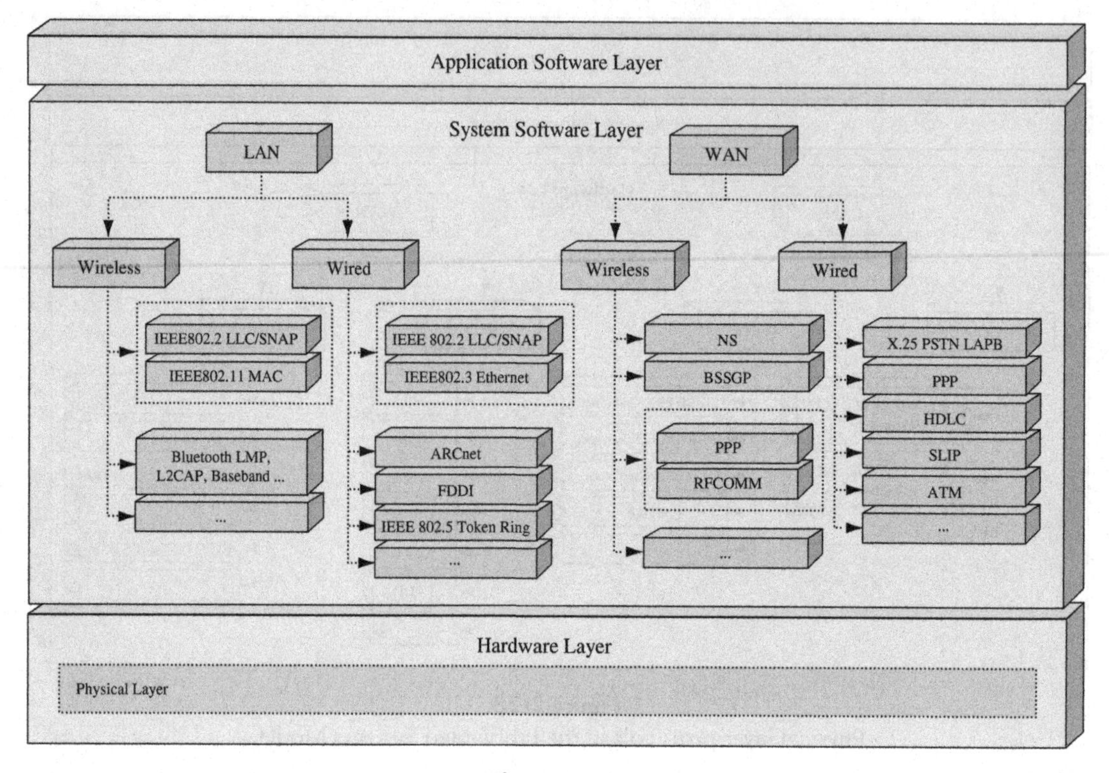


**Figure 2-30**
Data-link layer protocols.

### OSI Model Layer 2: Data-Link Layer

The data-link layer is the software closest to the hardware (physical layer). Thus, it includes, among other functions, any software needed to control the hardware. Bridging also occurs at this layer to allow networks interconnected with different physical layer protocols (e.g., Ethernet LAN and an 802.11 LAN) to interconnect.

Like physical layer protocols, data-link layer protocols are classified as LAN protocols, WAN protocols, or protocols that can be used for both LANs and WANs. Data-link layer protocols that are reliant on a specific physical layer may be limited to the transmission medium involved, but, in some cases (e.g., PPP over RS-232 or PPP over Bluetooth's RF-COMM), data-link layer protocols can be ported to very different mediums if there is a layer that simulates the original medium the protocol was intended for, or if the protocol supports hardware-independent upper-data-link functionality. Data-link layer protocols are implemented in the system software layer, as shown in Figure 2-30.

The data-link layer is responsible for receiving data bits from the physical layer and formatting these bits into groups, called data-link frames. Different data-link standards have varying data-link frame formats and definitions, but in general this layer reads the bit

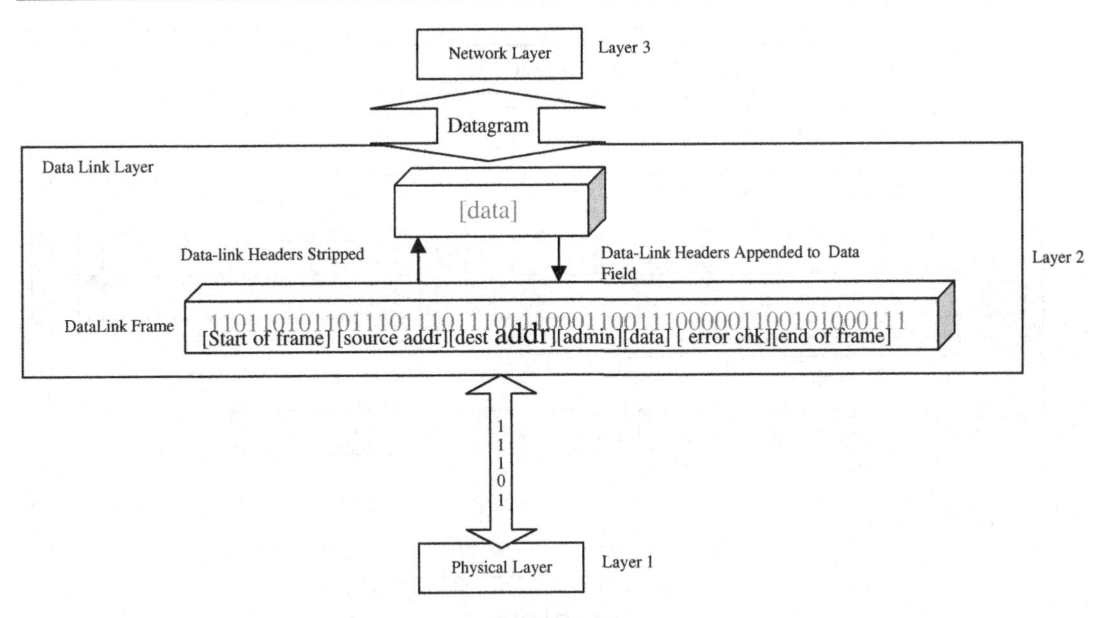


**Figure 2-31**
Data-link layer data flow block diagram.

fields of these frames to ensure that entire frames are received, that these frames are error free, that the frame is meant for this device by using the physical address retrieved from the networking hardware on the device, and where this frame came from. If the data is meant for the device, then all data-link layer headers are stripped from the frame and the remaining data field, called a *datagram*, is passed up to the networking layer. These same header fields are appended to data coming down from upper layers by the data-link layer, and then the full data-link frame is passed to the physical layer for transmission (see Figure 2-31).

### OSI Model Layer 3: Network Layer

Network layer protocols, like data-link layer protocols, are implemented in the system software layer, but, unlike the lower data-link layer protocols, the network layer is typically hardware independent and only dependent on the data-link layer implementations (see Figure 2-32).

At the OSI network layer, networks can be divided into smaller sub-networks, called *segments*. Devices within a segment can communicate via their physical addresses. Devices in different segments, however, communicate through an additional address, called the *network address*. While the conversion between physical addresses and network addresses can occur in data-link layer protocols implemented in the device (ARP, RARP, etc.), network layer protocols can also convert between physical and networking addresses, as well as assign networking addresses. Through the network address scheme, the network layer manages datagram traffic and any routing of datagrams from the current device to another.

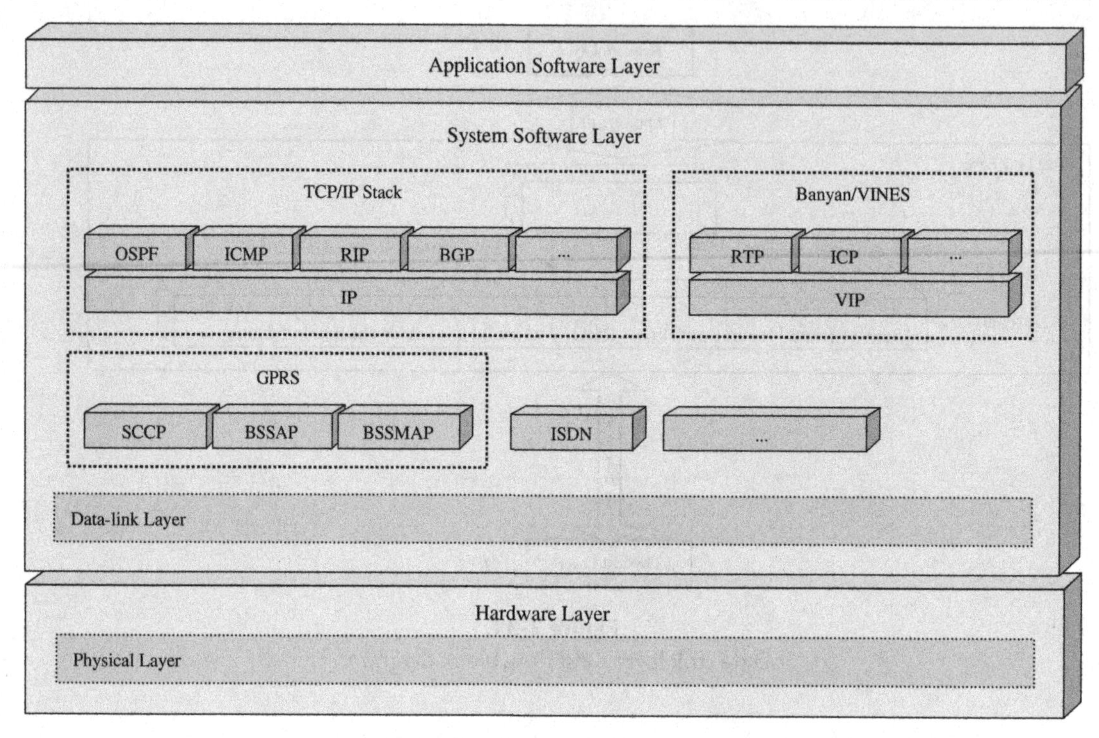


**Figure 2-32**
Network layer protocols in the Embedded Systems Model.

Like the data-link layer, if the data is meant for the device, then all network layer headers are stripped from the datagrams, and the remaining data field, called a *packet*, is passed up to the transport layer. These same header fields are appended to data coming down from upper layers by the network layer and then the full network layer datagram is passed to the data-link layer for further processing (see Figure 2-33). Note that the term "packet" is sometimes used to discuss data transmitted over a network, in general, in addition to data processed at the transport layer.

### OSI Model Layer 4: Transport Layer

Transport layer protocols (shown in Figure 2-34) sit on top of and are specific to the network layer protocols. They are typically responsible for establishing and dissolving communication between two specific devices. This type of communication is referred to as *point-to-point* communication. Protocols at this layer allow for multiple higher-layer applications running on the device to connect point-to-point to other devices. Some transport layer protocols can also ensure reliable point-to-point data transmission by ensuring that packets are received and transmitted in the correct order, are transmitted at a reasonable rate (flow control), and the data within the packets is not corrupted. Transport layer protocols can provide the acknowledgments to other devices upon receiving packets and request packets to be retransmitted if an error is detected.

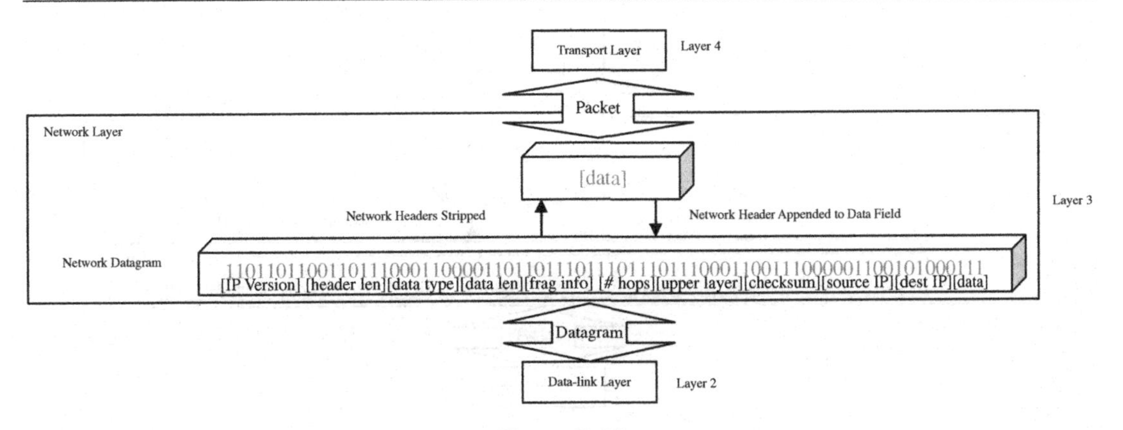


**Figure 2-33**
Network layer data flow block diagram.

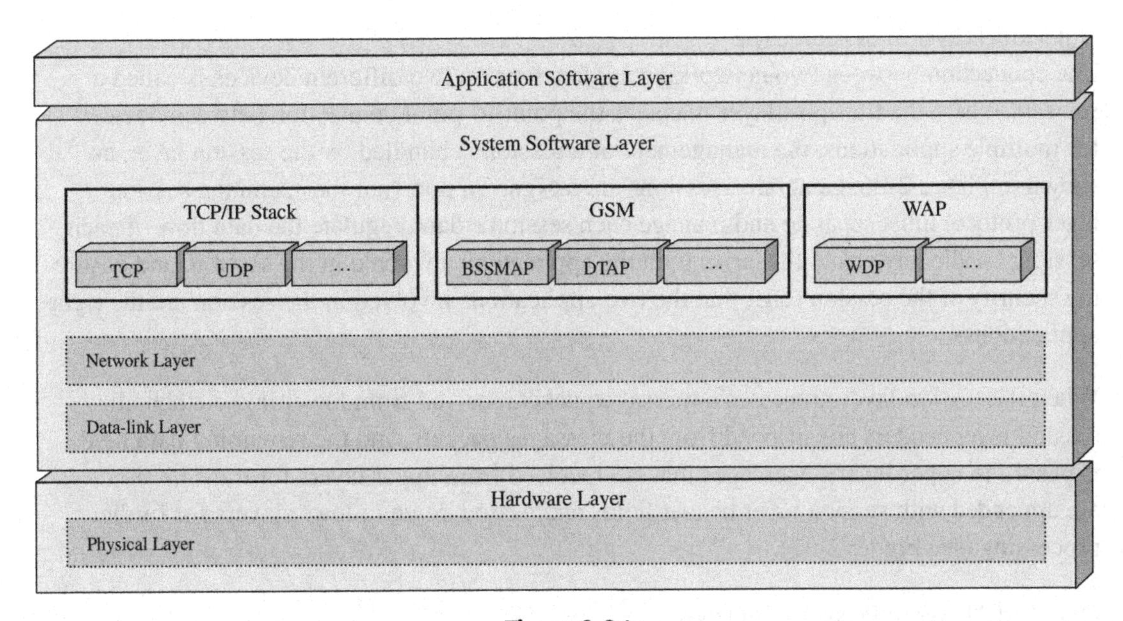


**Figure 2-34**
Transport layer protocols in the Embedded Systems Model.

In general, when the transport layer processes a packet received from lower layers, then all transport layer headers are stripped from the packets, and the remaining data fields from one or multiple packets are reassembled into other packets, also referred to as *messages*, and passed to upper layers. Messages/packets are received from upper layers for transmission, and are divided into separate packets if too long. The transport layer header fields are then appended to the packets, and passed down to lower layers for further processing (see Figure 2-35).

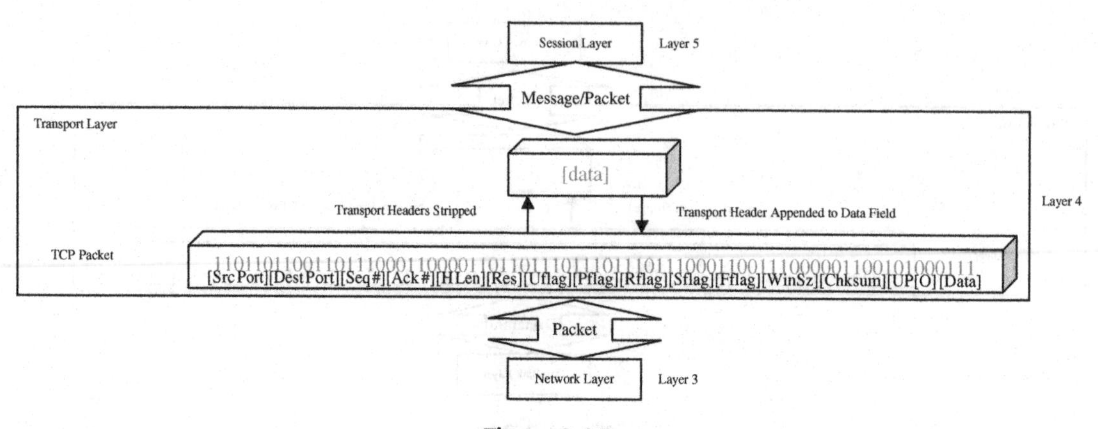


**Figure 2-35**
Transport layer data flow block diagram.

### OSI Model Layer 5: Session Layer

The connection between two networking applications on two different devices is called a *session*. Where the transport layer manages the point-to-point connection between devices for multiple applications, the management of a session is handled by the session layer, as shown in Figure 2-36. Generally, sessions are assigned a port (number), and the session layer protocol must separate and manage each session's data, regulate the data flow of each session, handle any errors that arise with the applications involved in the session, and ensure the security of the session (e.g., that the two applications involved in the session are the right applications).

When the session layer processes a message/packet received from lower layers, then all session layer headers are stripped from the messages/packets, and the remaining data field is passed to upper layers. Messages that are received from upper layers for transmission are appended with session layer header fields and passed down to lower layers for further processing (see Figure 2-37).

### OSI Model Layer 6: Presentation Layer

Protocols at the presentation layer are responsible for translating data into formats that higher applications can process or translating data going to other devices into a generic format for transmission. Generally, data compression/decompression, data encryption/decryption, and data protocol/character conversions are implemented in presentation layer protocols. Relative to the Embedded Systems Model, presentation layer protocols are usually implemented in networking applications found in the application layer as shown in Figure 2-38.

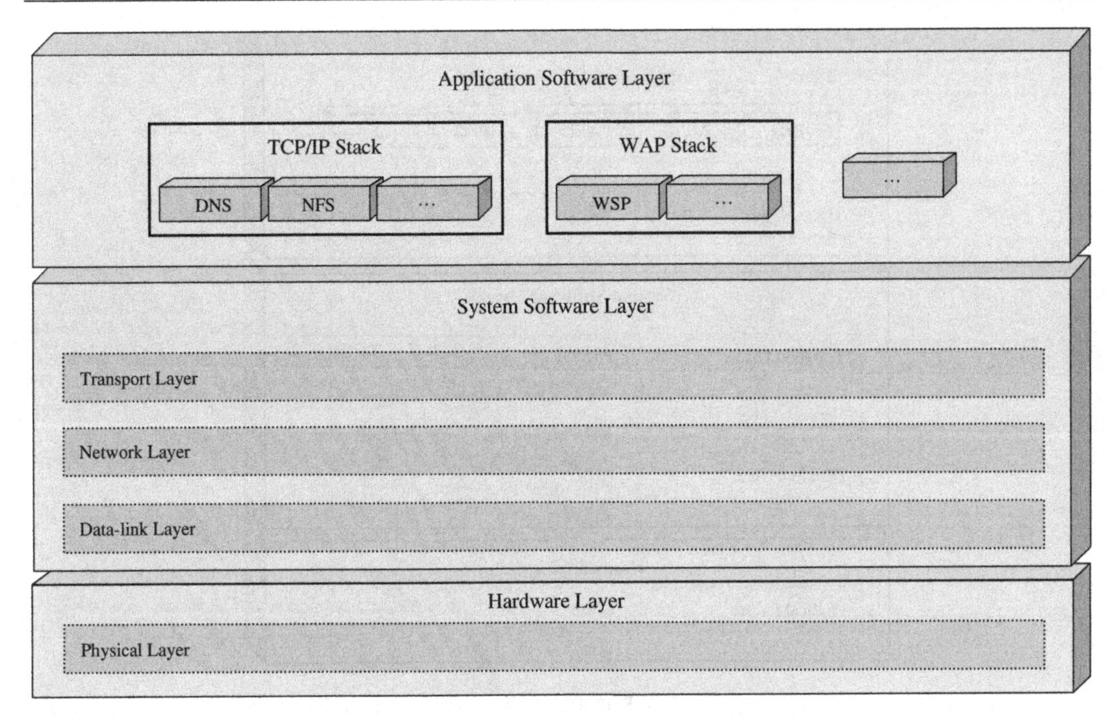


**Figure 2-36**
Session layer protocols in the Embedded Systems Model.

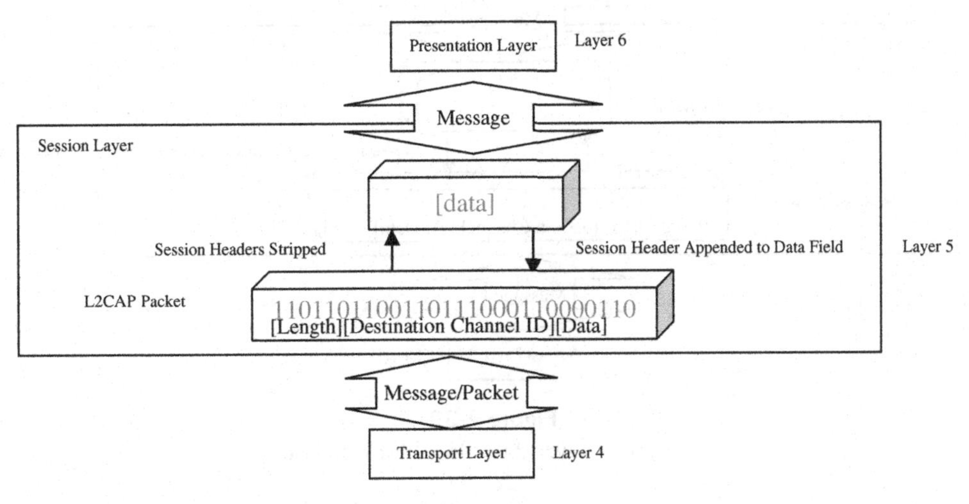


**Figure 2-37**
Session layer data flow block diagram.

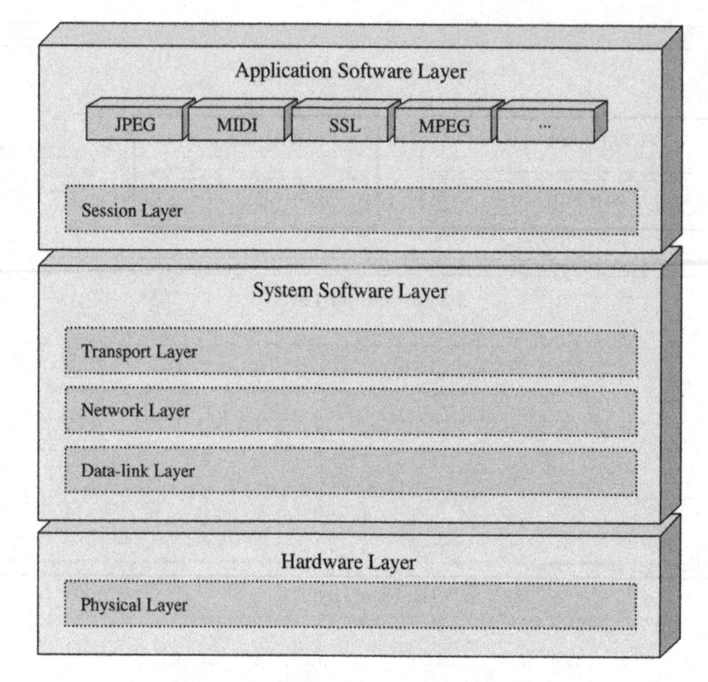


**Figure 2-38**
Presentation layer protocols in the Embedded Systems Model.

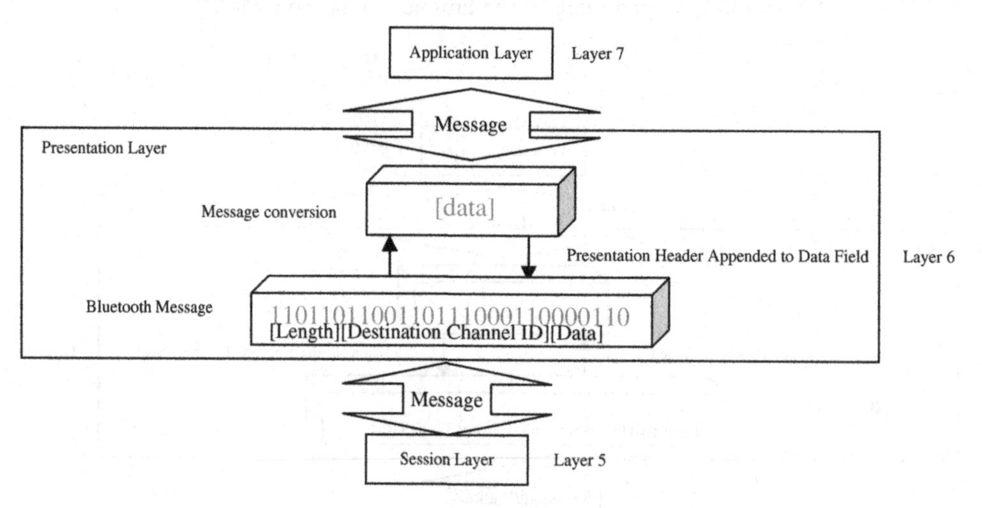


**Figure 2-39**
Presentation layer data flow block diagram.

Basically, a presentation layer processes a message received from lower layers, and then all presentation layer headers are stripped from the messages and the remaining data field is passed to upper layers. Messages that are received from upper layers for transmission are appended with presentation layer header fields and passed down to lower layers for further processing (see Figure 2-39).

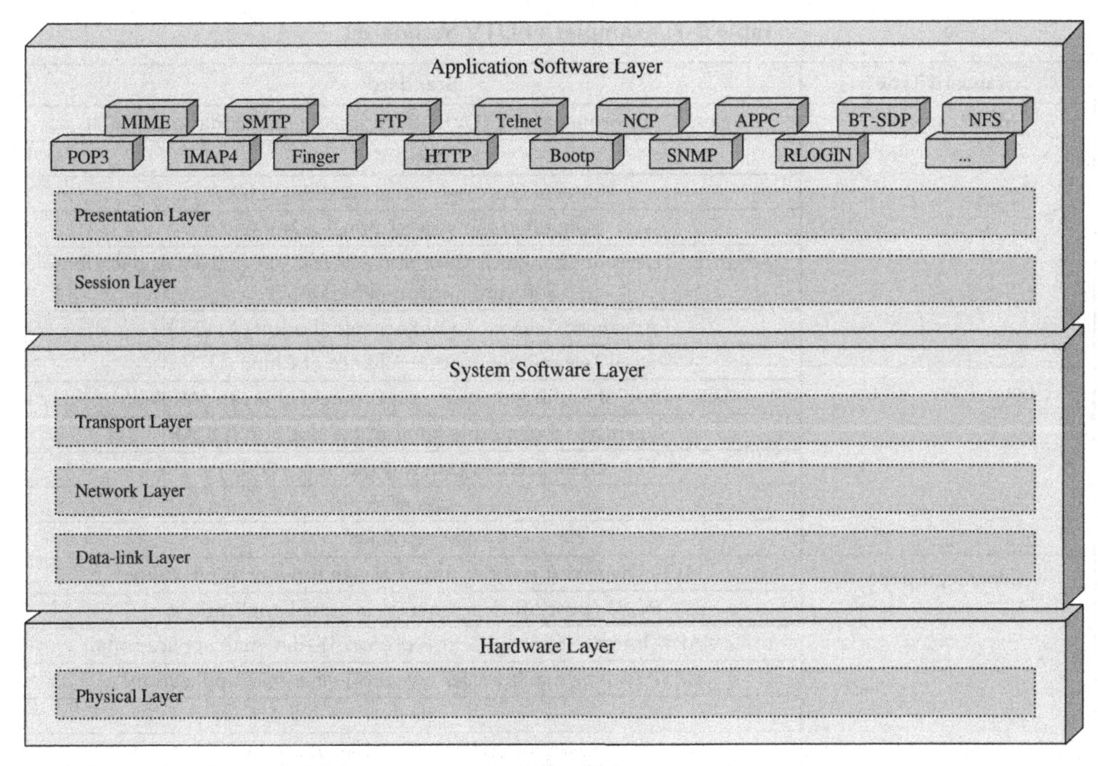


**Figure 2-40**
Application layer protocols in the Embedded Systems Model.

OSI Model Layer 7: Application Layer

A device initiates a network connection to another device at the application layer (shown in Figure 2-40). In other words, application layer protocols are either used directly as network applications by end-users or the protocols are implemented in end-user network applications (see Chapter 10). These applications "virtually" connect to applications on other devices.

## 2.3 Multiple Standards-Based Device Example: Digital Television (DTV)[23]

**Why Use DTV as a Standards Example?**

"There's a frenzy about portals on the internet, but the average person spends about an hour online. The average consumer spends seven hours a day watching TV, and TV is in 99% of US homes."

*Forrester Research*

**Table 2-7: Examples of DTV Standards**

| Standard Type | Standard |
|---|---|
| Market specific | Digital Video Broadcasting (DVB)—Multimedia Home Platform (MHP) |
| | Java TV |
| | Home audio/video interoperability (HAVi) |
| | Digital Audio Video Council (DAVIC) |
| | Advanced Television Standards Committee (ATSC)/Digital TV Applications Software Environment (DASE) |
| | Advanced Television Enhancement Forum (ATVEF) |
| | Digital Television Industrial Alliance of China (DTVIA) |
| | Association of Radio Industries and Business of Japan (ARIB-BML) |
| | OpenLabs OpenCable application platform (OCAP) |
| | Open Services Gateway Initiative (OSGi) |
| | OpenTV |
| | MicrosoftTV |
| General purpose | HTTP (hypertext transfer protocol)—in browser applications |
| | POP3 (post office protocol)—in e-mail applications |
| | IMAP4 (Internet message access protocol)—in e-mail applications |
| | SMTP (simple mail transfer protocol)—in e-mail applications |
| | Java |
| | Networking (terrestrial, cable, and satellite) |
| | POSIX |

Analog TVs process incoming analog signals of traditional TV video and audio content, whereas DTVs process both incoming analog and digital signals of TV video/audio content, as well as application data content that is embedded within the entire digital data stream (a process called data broadcasting or data casting). This application data can either be unrelated to the video/audio TV content (non-coupled), related to video/audio TV content in terms of content but not in time (loosely coupled), or entirely synchronized with TV audio/video (tightly coupled).

The type of application data embedded is dependent on the capabilities of the DTV receiver itself. While there are a wide variety of DTV receivers, most fall under one of three categories: enhanced broadcast receivers, which provide traditional broadcast TV enhanced with graphics controlled by the broadcast programming; interactive broadcast receivers, capable of providing e-commerce, video-on-demand, e-mail, etc., through a return channel on top of "enhanced" broadcasting; and multinetwork receivers that include internet and local telephony functionality on top of interactive broadcast functionality. Depending on the type of receiver, DTVs can implement general-purpose, market-specific, and/or application-specific standards all into one DTV/STB system architecture design (shown in Table 2-7).

These standards then can define several of the major components that are implemented in all layers of the DTV Embedded Systems Model, as shown in Figure 2-41.

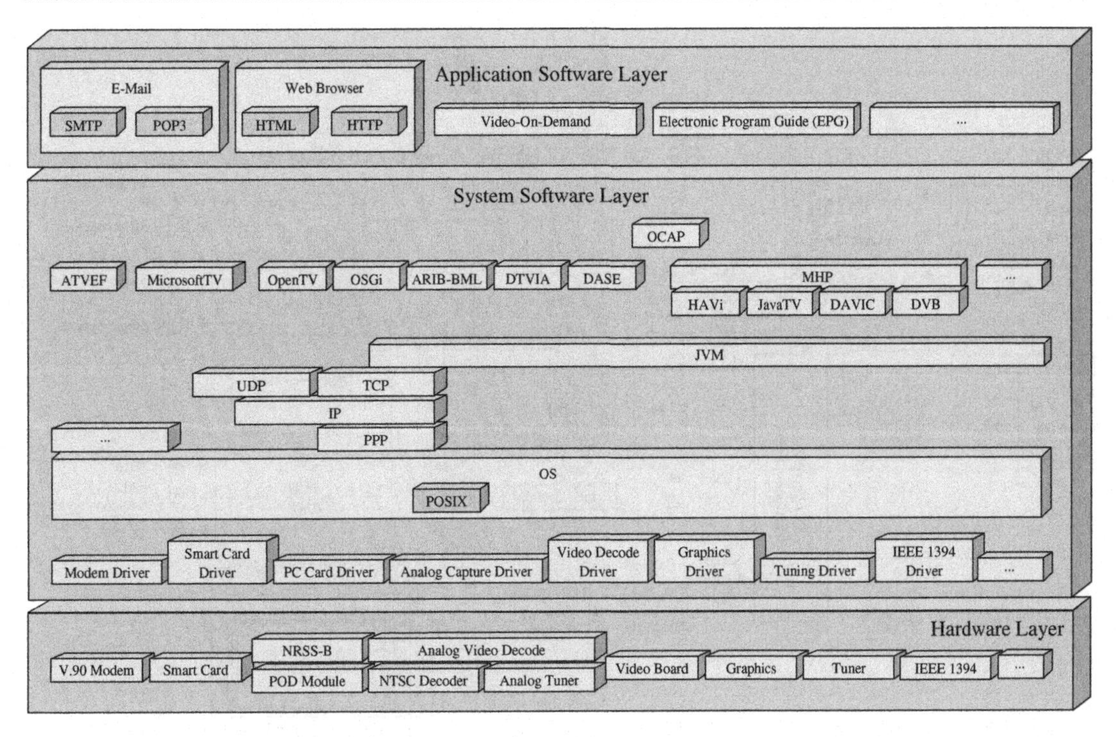


**Figure 2-41**
DTV standards in the Embedded Systems Model.

MHP is Java-based middleware solution based upon the Digital Video Broadcasting (DVB)—Multimedia Home Platform (MHP) specification. MHP implementations in DTV is a powerful example to learn from when designing or using just about any market-specific middleware solution, because it incorporates many complex concepts and challenges that must be addressed in its approach.

In general, as shown in Figure 2-42, hardware board that support MHP include:

- Master processor.
- Memory subsystem.
- System buses.
- I/O (input/output) subsystem:
  - tuner/demodulator
  - demultiplexer
  - decoders/encoders
  - graphics processor
  - communication interface/modem
  - conditional access (CA) module
  - a remote control receiver module.

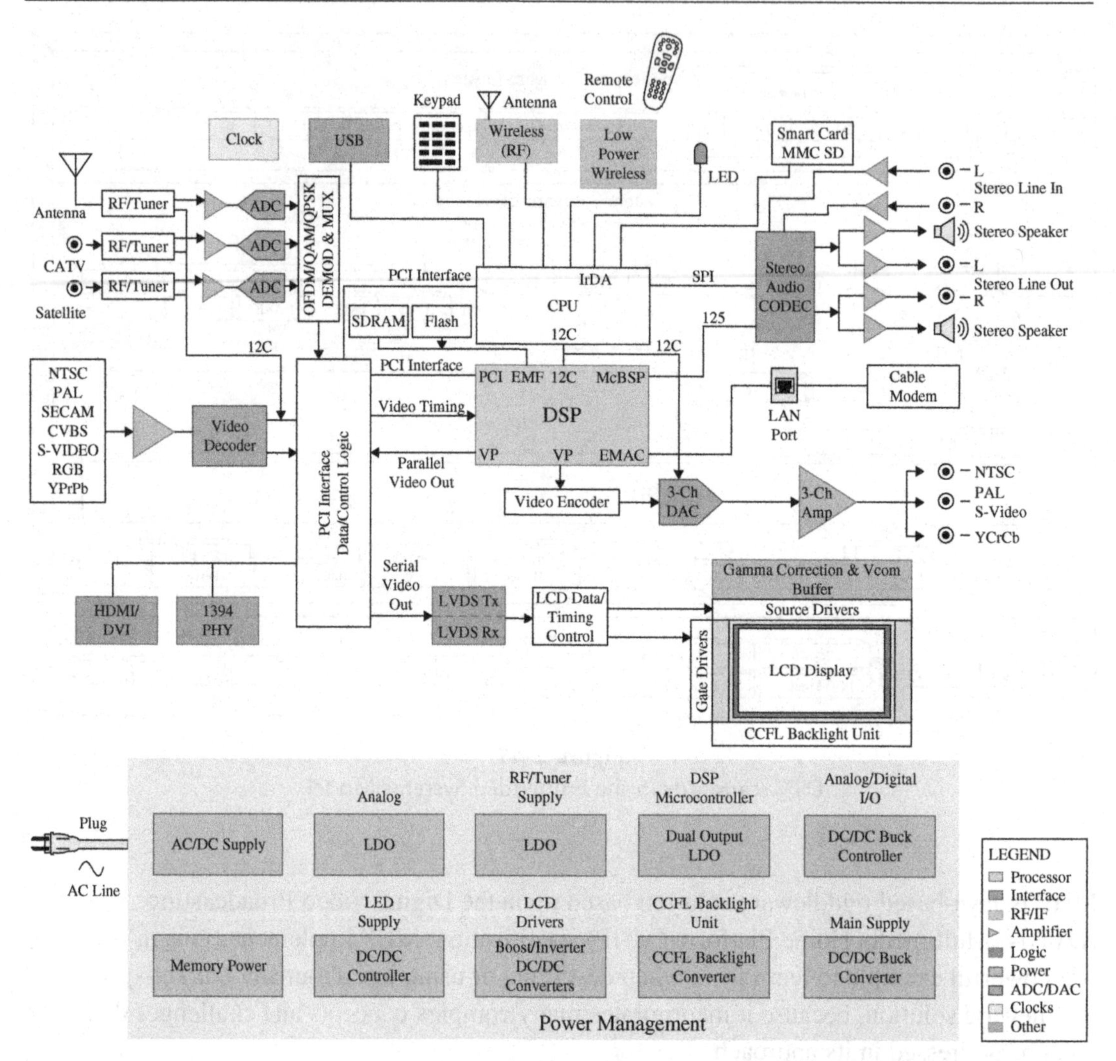


**Figure 2-42**

Texas Instruments DTV block diagram.[34]

Of course, there can be additional components, and these components will differ in design from board to board, but these elements are generally what are found on most boards targeted for this market. MHP and associated system software APIs typically require a minimum of 16 MB of RAM (random access memory), 8–16 MB of Flash, and, depending on how the JVM and operating system (OS) are implemented and integrated, can require a 150–250+ MHz CPU to run at in a practical manner. Keep in mind that depending on the type of applications that will be run over the system software stack, memory and processing power requirements for these applications need to be taken in consideration, and thus may require a change to these "minimum" baseline memory and processing power requirements for running MHP.

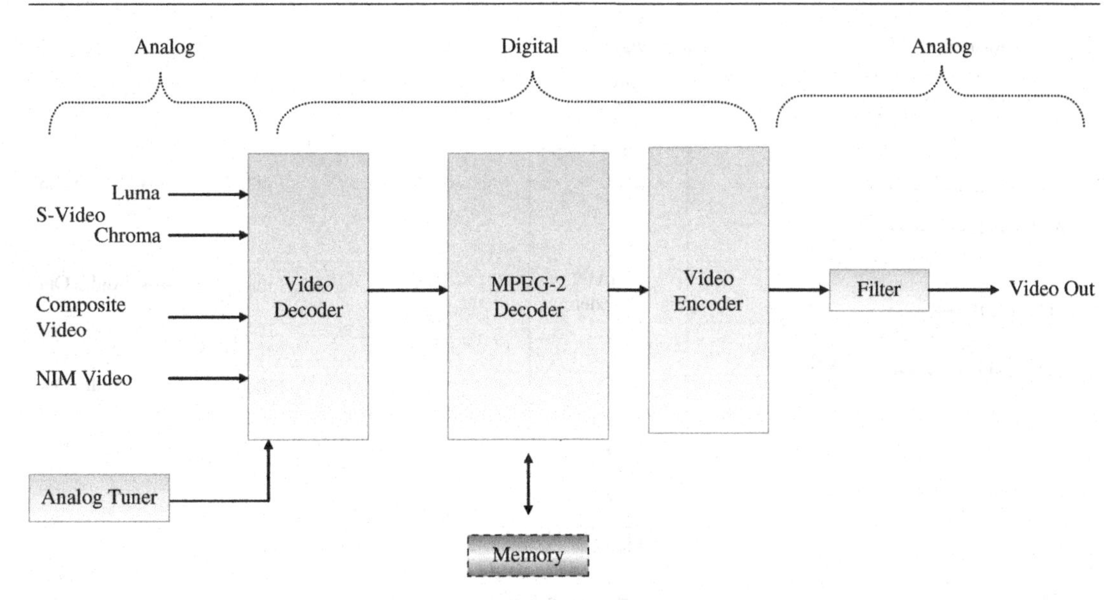


**Figure 2-43**
Example of video data path in DTV.[33]

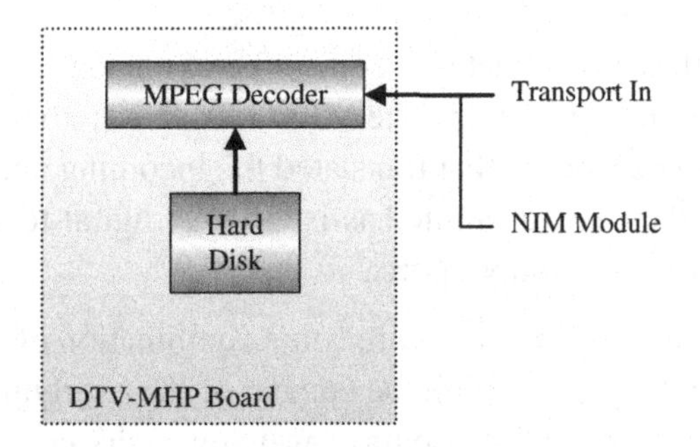


**Figure 2-44**
Example of transport data path in DTV.[33]

The flow of video data originates with some type of input source. As shown in Figure 2-43, in the case of an analog video input source, for example, each is routed to the analog video decoder. The decoder then selects one of three active inputs and quantizes the video signal, which is then sent to some type of MPEG-2 subsystem. An MPEG-2 decoder is responsible for processing the video data received to allow for either standard-definition or high-definition output. In the case of standard-definition video output, it is encoded as either S-video or composite video using an external video encoder. No further encoding or decoding is typically done to the high-definition output coming directly from the MPEG-2 subsystem.

The flow of transport data originating from some type of input source is passed to the MPEG-2 decoder subsystem (see Figure 2-44). The output information from this can be processed and displayed.

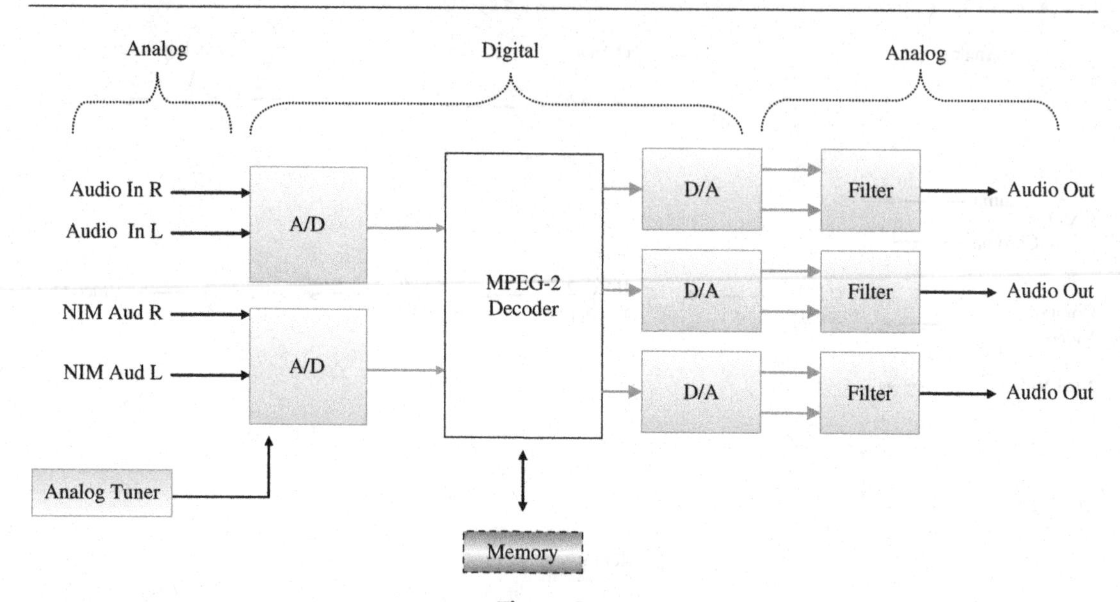


**Figure 2-45**
Example of audio data path in DTV.[33]

In the case of audio data flow, it originates at some type of analog source such as the analog audio input sources shown in Figure 2-45. The MPEG-2 subsystem receives analog data from the A/D (analog-to-digital) converters that translated the incoming analog sources. Audio data can be merged with other data, or transmitted as-is to D/A (digital-to-analog) converters to be then routed to some type of audio output ports.

An MHP hardware subsystem will then require some combination of device driver libraries to be developed, tested, and verified within the context of the overlying MHP compliant software platform. Like the hardware, these low-level device drivers generally will fall under general master processor-related device drivers (see Figure 2-46), memory and bus device drivers (see Figure 2-47), and I/O subsystem drivers.

The I/O subsystem drivers include Ethernet, keyboard/mouse, video subsystem, and audio subsystem drivers to name a few. Figures 2-48a–c show a few examples of MHP I/O subsystem device drivers.

Because MHP is Java-based, as indicated in the previous section and shown in Figure 2-49, a JVM and ported OS must then reside on the embedded system that implements an MHP-stack and underlie this MHP stack. This JVM must meet the Java API specification required by the particular MHP implementation, meaning the underlying Java functions that the MHP implementation calls down for must reside in some form in the JVM that the platform supports.

The open-source example, openMHP, shows how some JVM APIs in its implementation, such as the org.havi.ui library translate, into source code in this particular package (see Figure 2-50).

*Examples of Clock/Timer Drivers*

**enableClock**, for enabling system clock interrupts
**disableClock**, for disabling system clock interrupts
**returnClockRate**, for returning the interrupt rate of the system clock (number of ticks per second)
**connectClock**, for specifying the interrupt handler called at each system clock interrupt
**handleClockInterrupt**, for calling routines that handle the system clock interrupt
**enableTimestampEnable**, for reseting the counter and enabling timestamp timer interrupts
**disableTimestamp**, for disabling timestamp timer interrupts
**connectTimestamp**, for specifying which user interrupt routine is called with each timestamp timer interrupt
**timestampPeriod**, for returning the timestamp timer period (ticks)
**timestampFrequency**, for returning the timer's frequency (ticks per second)
**currentTimestamp**, for returning the current value of the timestamp timer tick counter
**lockTimestamp**, for stopping/reading the timer tick counter and locking interrupts
...

*Examples of Interrupt Controller Drivers*

**initiliazeInterruptController**, interrupt controller initialization
**endOfInterrupt**, for sending an EOI(end of interrupt) signal at the end of the interrupt handler
**disableInterruptController**, for disabling a specified interrupt level
**enableInterruptController**, for enabling a specified interrupt level
**returnInterruptLevel**, for returning an interrupt level in service from some type of interrupt service register
**lockInterrupt**, for saving the mask and locking out interrupt controller interrupts
**unlockInterrupt**, for restoring the mask and unlocking interrupt controller interrupts
...

....

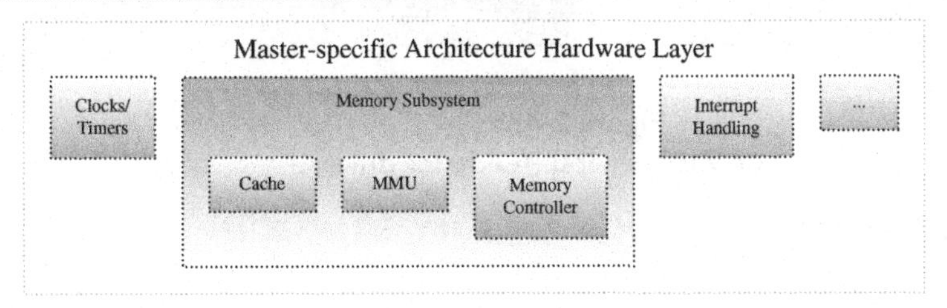


**Figure 2-46**
Examples of general architecture device drivers on MHP platform.[33]

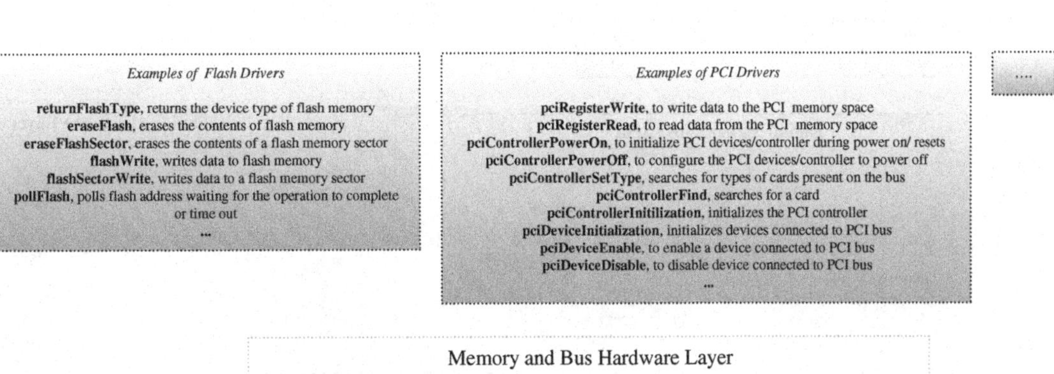

Examples of Flash Drivers
returnFlashType, returns the device type of flash memory
eraseFlash, erases the contents of flash memory
eraseFlashSector, erases the contents of a flash memory sector
flashWrite, writes data to flash memory
flashSectorWrite, writes data to a flash memory sector
pollFlash, polls flash address waiting for the operation to complete or time out
...
Examples of PCI Drivers
pciRegisterWrite, to write data to the PCI memory space
pciRegisterRead, to read data from the PCI memory space
pciControllerPowerOn, to initialize PCI devices/controller during power on/ resets
pciControllerPowerOff, to configure the PCI devices/controller to power off
pciControllerSetType, searches for types of cards present on the bus
pciControllerFind, searches for a card
pciControllerInitilization, initializes the PCI controller
pciDeviceInitialization, initializes devices connected to PCI bus
pciDeviceEnable, to enable a device connected to PCI bus
pciDeviceDisable, to disable device connected to PCI bus
...
....


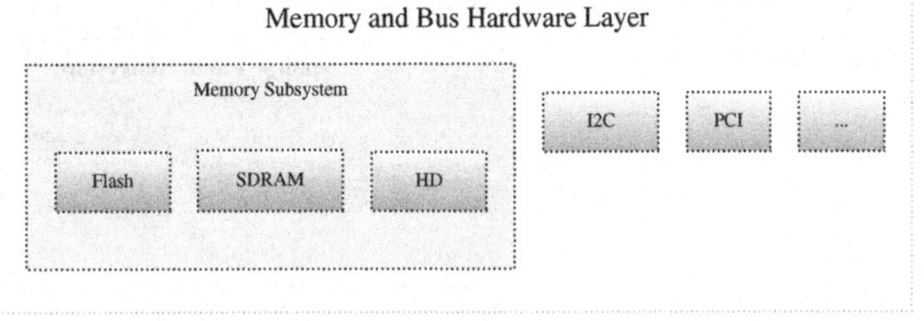


**Figure 2-47**
Examples of memory and bus device drivers on MHP Platform.[33]

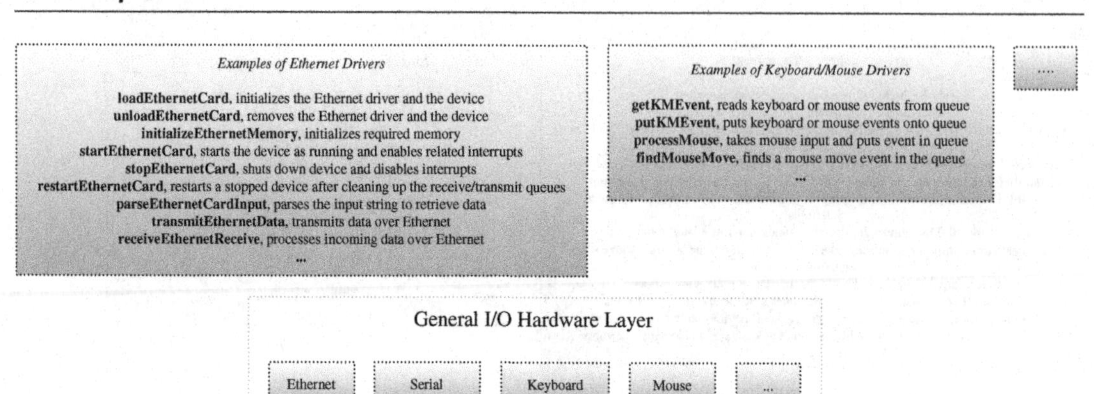


**Figure 2-48a**
Examples of MHP general I/O device drivers.[33]

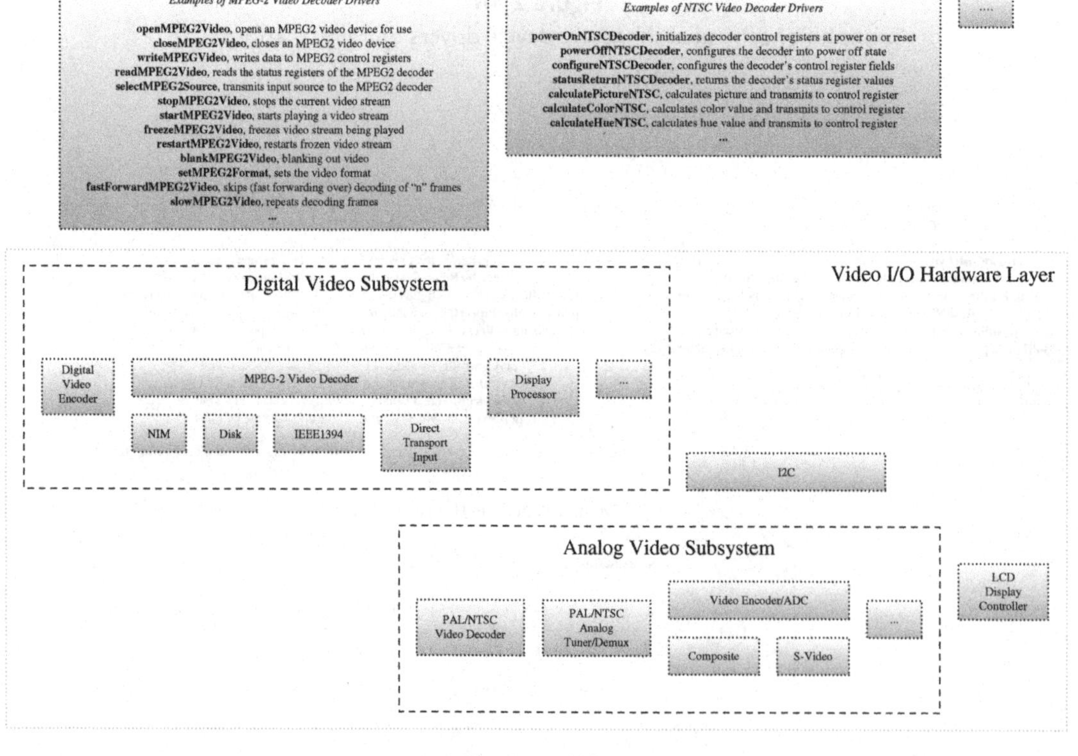


**Figure 2-48b**
Examples of MHP video I/O device drivers.[33]

*Examples of MPEG-2 Audio Drivers*

**openMPEG2audio**, opens an MPEG2 audio device for use
**closeMPEG2Audio**, closes an MPEG2 audio device
**sourceMPEG2Audio**, transmits audio input data source to MPEG2 audio decoder
**writeMPEG2Audio**, transmits data to MPEG2 audio decoder's control registers
**readMPEG2Audio**, reads the values of the status registers of the MPEG2 audio decoder
**stopMPEG2Audio**, instructs decoder to stop playing the current audio stream
**playMPEG2Audio**, instructs decoder to start playing an audio stream from an input source
**pauseMPEG2Audio**, instructs decoder to pause (freeze) in playing the current audio stream
**muteMPEG2Audio**, mutes the sound of an audio stream currently playing
**setChannelMPEG2Audio**, selects a specific audio channel
**setVSyncMPEG2Audio**, instructs the audio decoder how to set A/V synchronization
...

*Examples of Analog Audio Controller Drivers*

**powerONAudio**, initializes an audio processor's control registers at power on/reset
**powerOFFAudio**, configures the audio processor's control registers to power off state
**sendDataAudio**, transmits data to control register fields in audio processor
**calculateVolumeAudio**, calculates volume using the audio volume equation and user "volume" configuration and transmits the value to the audio processor's "volume" control register field
**calculateToneAudio**, calculates bass and treble tone using the audio tone equations and user "bass" or "treble" configuration and transmits the value to the audio processor's control register bass and treble fields
**surroundONAudio**, enables the stereo surround using user configuration and transmits the value to the processor's control register related stereo surround field
**surroundOFFAudio**, disables the stereo surround using user configuration and transmits the value to the processor's control register related stereo surround field
...

....

**Figure 2-48c**
Examples of MHP audio I/O device drivers.[33]

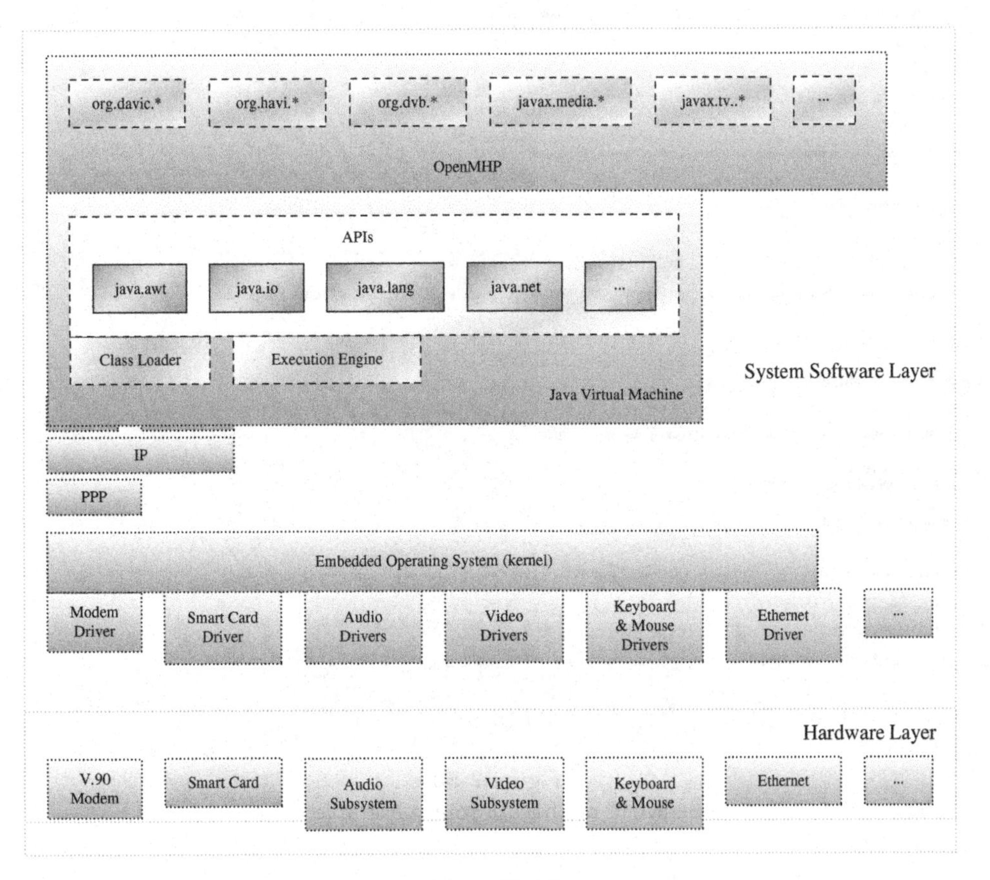


**Figure 2-49**
MHP-based system architecture.[33]

```
package org.dvb.ui;

/* Copyright 2000–2003 by HAVi, Inc. Java is a trademark of Sun Microsystems, Inc. All rights reserved. This program is free software; you can redistribute it and/or modify
 * it under the terms of the GNU General Public License as published by the Free Software Foundation, either version 2 of the License, or (at your option) any later version.
 * This program is distributed in the hope that it will be useful, but WITHOUT ANY WARRANTY, without even the implied warranty of MERCHANTABILITY or FITNESS
 * FOR A PARTICULAR PURPOSE. See the GNU General Public License for more details.
 * You should have received a copy of the GNU General Public License along with this program; if not, write to the Free Software Foundation, Inc., 59 Temple Place, Suite 330,
 * Boston, MA 02111-1307 USA */

import java.awt.Graphics2D;
import java.awt.Graphics;
import java.awt.Dimension;
import javax.media.Clock;
import javax.media.Time;
import javax.media.IncompatibleTimeBaseException;

/** A <code>BufferedAnimation</code> is an AWT component that maintains a queue of one or more image buffers. This permits efficient flicker-free animation by allowing a
 * caller to draw to an off-screen buffer, which the system then copies to the framebuffer in coordination with the video output subsystem. This class also allows an application
 * to request a series of buffers, so that it can get a small number of frames ahead in an animation. This allows an application to be robust in the presence of short delays, e.g. from
 * garbage collection. A relatively small number of buffers is recommended, perhaps three or four. A BufferedAnimation with one buffer provides little or no protection
 * from pauses, but does provide double-buffered animation. .... **/
....

public class BufferedAnimation extends java.awt.Component {

   /**
    * Constant representing a common video framerate, approximately 23.98 frames per second, and equal to <code>24000f/1001f</code>.
    *
    * @see #getFramerate()
    * @see #setFramerate(float)
    **/
   static public float FRAME_23_98 = 24000f/1001f;

   /**
    * Constant representing a common video framerate, equal to <code>24f</code>.
    *
    * @see #getFramerate()
    * @see #setFramerate(float)
    **/
   static public float FRAME_24 = 24f;

   /**
    * Constant representing a common video framerate, equal to <code>25f</code>.
    *
    * @see #getFramerate()
    * @see #setFramerate(float)
    **/
   static public float FRAME_25 = 25f;

   /**
    * Constant representing a common video framerate, approximately 29.97 frames per second, and equal to <code>30000f/1001f</code>.
    *
    * @see #getFramerate()
    * @see #setFramerate(float)
    **/
   static public float FRAME_29_97 = 30000f/1001f;

   /**
    * Constant representing a common video framerate, equal to <code>50f</code>.
    *
    * @see #getFramerate()
    * @see #setFramerate(float)
    **/
   static public float FRAME_50 = 50f;

   /**
    * Constant representing a common video framerate, approximately 59.94 frames per second, and equal to <code>60000f/1001f</code>.
    *
    * @see #getFramerate()
    * @see #setFramerate(float)
    **/
   static public float FRAME_59_94 = 59.94f;

   ....
}
```

**Figure 2-50**

openMHP org.havi.ui source example.[35]

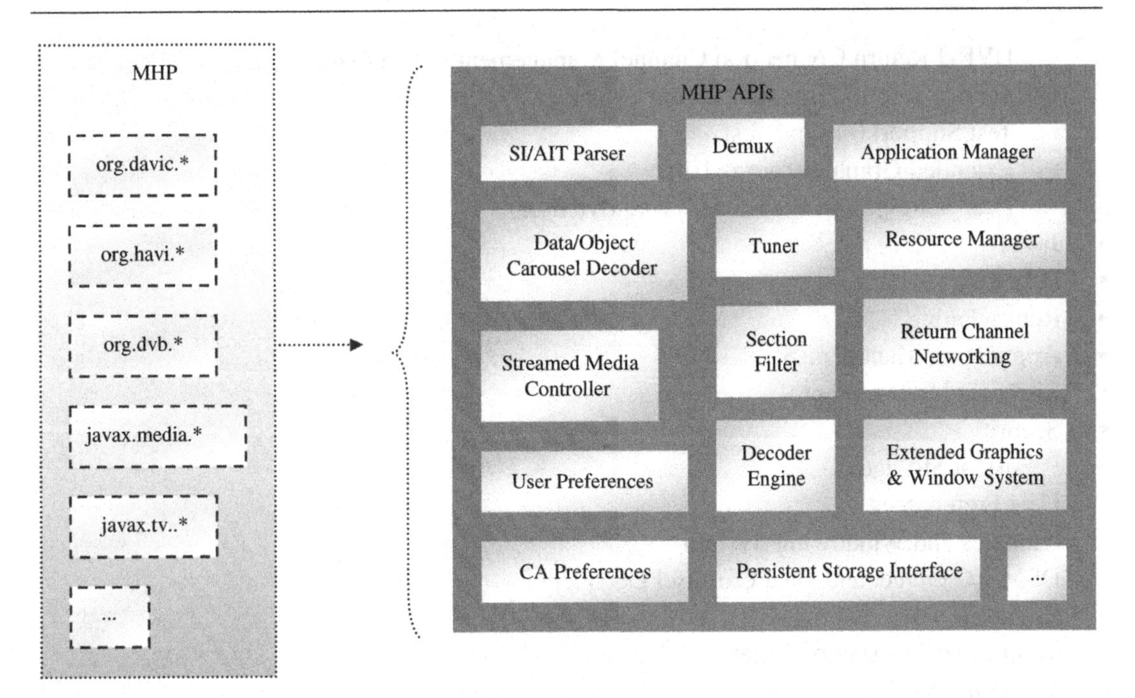


**Figure 2-51**
MHP APIs.[33]

Then, in order to understand MHP components, services, and how to build applications, it is important to understand (as shown in Figure 2-51) that the MHP standard is made up of a number of different substandards which contribute to the APIs, including:

- Core MHP (varies between implementations)
  - DSMCC
  - BIOP
  - Security
- HAVi UI
  - HAVi Level 2 User Interface (org.havi.ui)
  - HAVi Level 2 User Interface Event (org.havi.ui.event)
- DVB
  - Application Listing and Launching (org.dvb.application)
  - Broadcast Transport Protocol Access (org.dvb.dsmcc)
  - DVB-J Events (org.dvb.event)
  - Inter-Application Communication (org.dvb.io.ixc)
  - DVB-J Persistent Storage (org.dvb.io.persistent)
  - DVB-J Fundamental (org.dvb.lang)
  - Streamed Media API Extensions (org.dvb.media)
  - Datagram Socket Buffer Control (org.dvb.net)
  - Permissions (org.dvb.net.ca and org.dvb.net.tuning)

- DVB-J Return Connection Channel Management (org.dvb.net.rc)
- Service Information Access (org.dvb.si)
- Test Support (org.dvb.test)
- Extended Graphics (org.dvb.ui)
- User Settings and Preferences (org.dvb.user)

- JavaTV
- DAVIC
- Return Path
- Application Management
- Resource Management
- Security
- Persistent Storage
- User Preferences
- Graphics and Windowing System
- DSM-CC Object and Data Carousel Decoder
- SI Parser
- Tuning, MPEG Section Filter
- Streaming Media Control
- Return Channel Networking
- Application Manager and Resource Manager Implementation
- Persistent Storage Control
- Conditional Access support and Security Policy Management
- User Preference Implementations

Within the MHP world, the content the end-user of the system interacts with is grouped and managed as services. Content that makes up a service can fall under several different types, such as applications, service information, and data/audio/video streams, to name a few. In addition to platform-specific requirements and end-user preferences, the different types of content in services are used to manage data. For example, when a DTV allows support for more than one type of different video stream, service information can be used to determine which stream actually gets displayed.

MHP applications can range from browsers to e-mail to games to EPGs (electronic program guides) to advertisements, to name a few. At the general level, all these different types of MHP applications will typically fall under one of three general types of profiles:

- *Enhanced broadcasting*: the digital broadcast contains a combination of audio services, video services, and executable applications to allow end-users to interact with the system locally.
- *Interactive broadcasting*: the digital broadcast contains a combination of audio services, video services, executable applications, as well as interactive services and channel that allow end-users to interact with residing applications remotely to their DTV device.
- *Internet access*: the system implements functionality that allows access to the Internet.

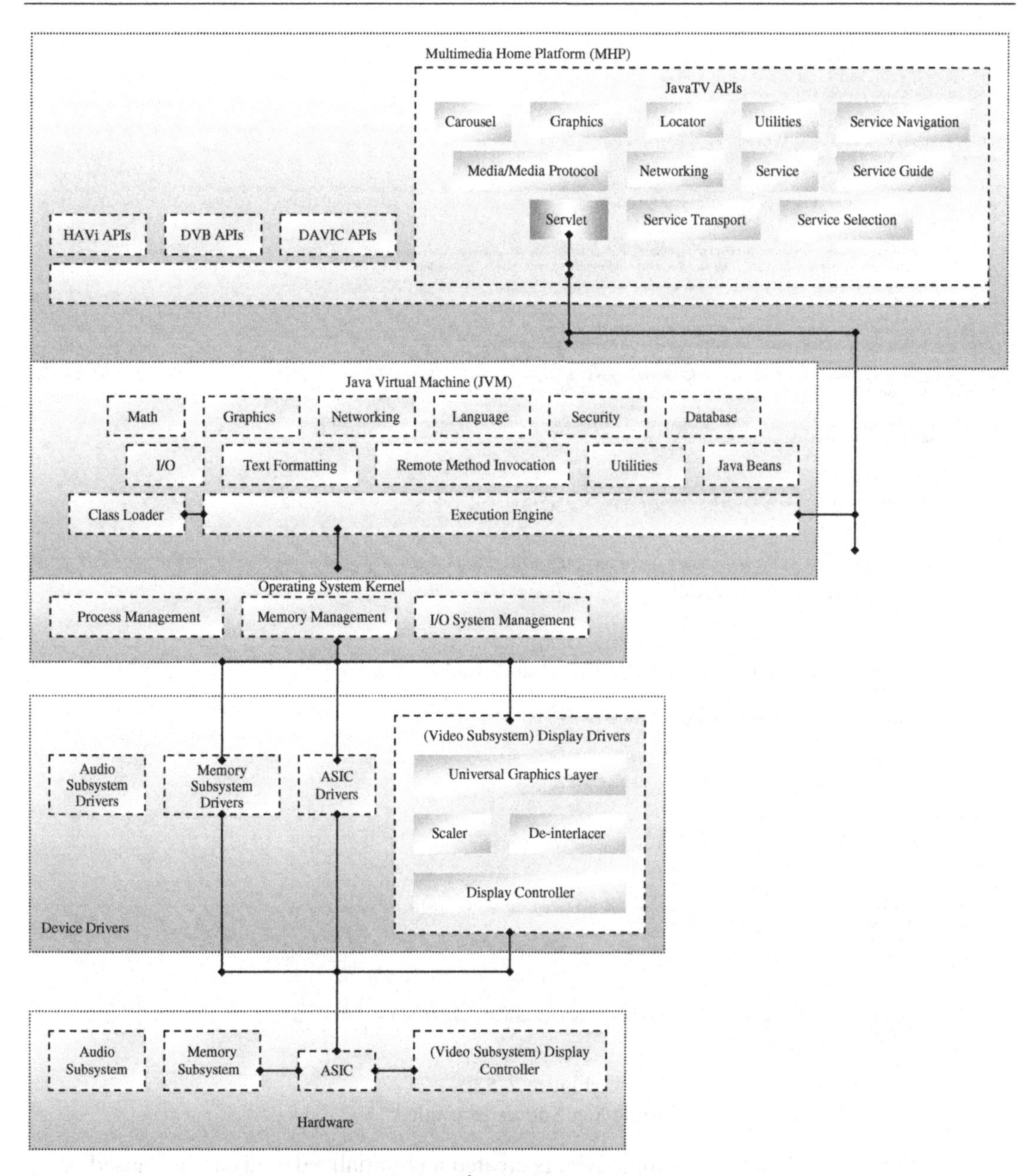


**Figure 2-52a**
Simple Xlet flow example.[36]

An important note is that while MHP is Java-based, MHP DVB-J type of applications are not regular Java applications, but are executing within the context of a Java servlet (Xlet) similar to the concept behind the Java applet. MHP applications communicate and interact with their external environment via the Xlet context. For example, Figures 2-52a and b show

```
import javax.tv.xlet.*;

// The main class of every MHP Xlet must implement this interface
public class XletExample implements javax.tv.xlet.Xlet
{

// Every Xlet has an Xlet context, created by the MHP middleware and passed in to the Xlet as a parameter to the initXlet() method.
private javax.tv.xlet.XletContext context;

// A private field to hold the current state. This is needed because the startXlet() method is called both start the Xlet for the first time and also to make the Xlet resume from // the paused state. This field lets us keep track of whether we're starting for the first time.
private boolean has Been Started;

// Every Xlet should have a default constructor that takes no arguments. The constructor should contain nothing. Any initialization should be done in the initXlet() method, // or in the startXlet method if it's time- or resource-intensive. That way, the MHP middleware can control when the initialization happens in a much more predictable way
// Initializing the Xlet.
public XletExample()

// store a reference to the Xlet context that the Xlet is executing in this .context = context;
 public void initXlet(javax.tv.xlet.XletContext context) throws javax.tv.xlet.XletStateChangeException

// The Xlet has not yet been started for the first time, so set this variable to false.
hasBeenStarted = false;

// Start the Xlet. At this point the Xlet can display itself on the screen and start interacting with the user, or do any resource-intensive tasks.
public void startXlet() throws javax.tv.xlet.XletStateChangeException
{
 if (hasBeenStarted)
  {
    System.out.println("The startXlet() method has been called to resume the Xlet after it's been paused. Hello again, world!");
  }
  else
  {
    System.out.println("The startXlet() method has been called to start the Xlet for the first time. Hello, world!");

    // set the variable that tells us we have actually been started
    hasBeenStarted = true;
  }
}

// Pause the Xlet and free any scarce resources that it's using, stop any unnecessary threads and remove itself from the screen.
 public void pauseXlet()
{
  System.out.println("The pauseXlet() method has been called. to pause the Xlet...");
}

// Stop the Xlet.
public void destroyXlet(boolean unconditional) throws javax.tv.xlet.XletStateChangeException
{
  if (unconditional)
  {
    System.out.println("The destroyXlet() method has been called to unconditionally destroy the Xlet." ");
  }
  else
  {
    System.out.println("The destroyXlet() method has been called requesting that the Xlet stop, but giving it the choice. Not Stopping.");
    throw new XletStateChangeException("Xlet Not Stopped");
  }
}

}
**application example based upon MHP open source by Steven Morris available for download at www.interactivetvweb.org
```

**Figure 2-52b**
Simple Xlet Source Example.[37]

an application example where a simple Xlet is created and initialized, and can be paused or destroyed via an MHP Java TV API package javax.tv.xlet.

The next example shown in Figures 2-53a and b is a sample application that uses the

- JVM packages java.io, java.awt, and java.awt.event.
- MHP Java TV API package javax.tv.xlet.
- MHP HAVi packages org.havi.ui and org.havi.ui.event.
- MHP DVB package org.dvb.ui.

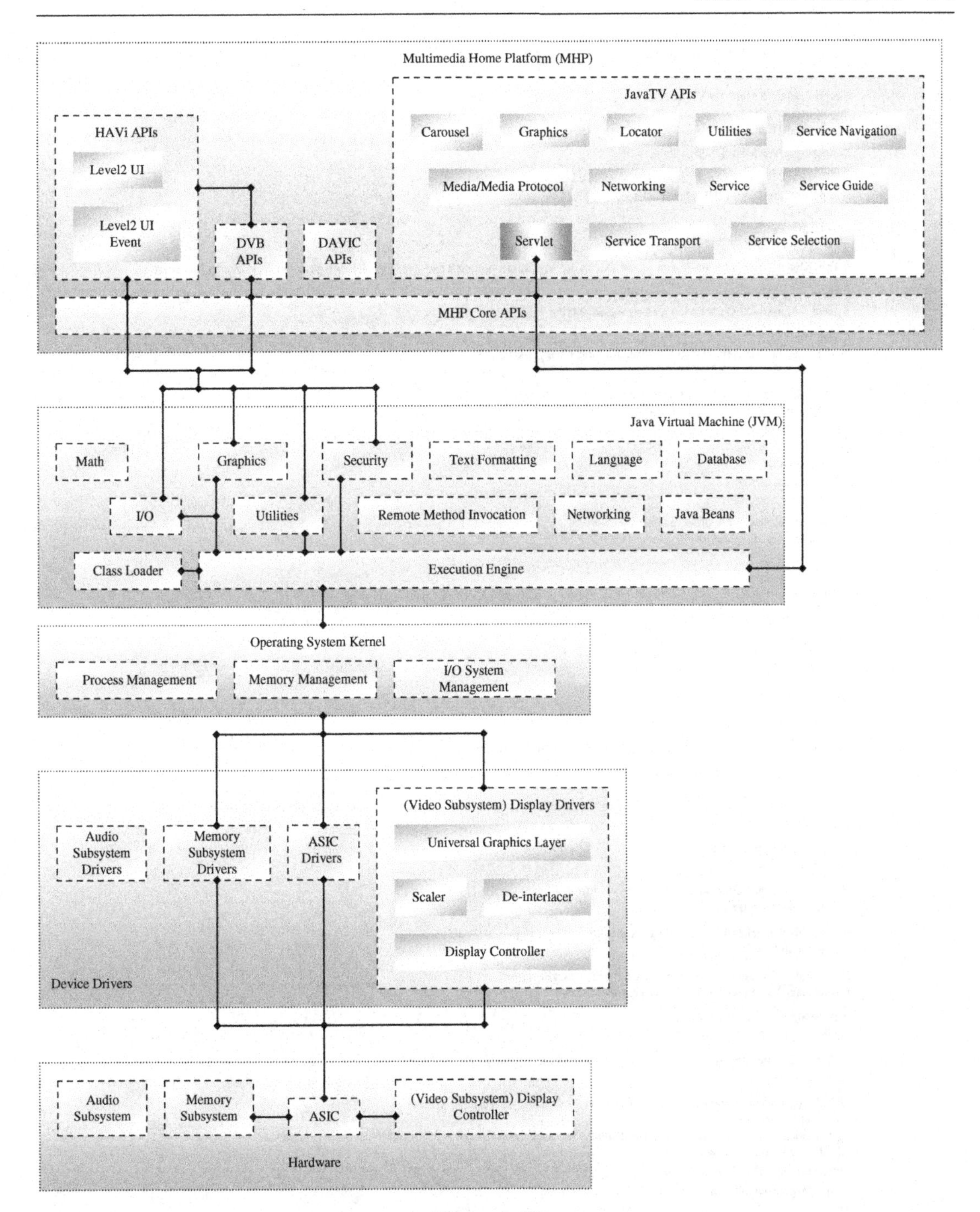


**Figure 2-53a**
Simple MHP HAVi Xlet flow example.[36]

```
// Import required MHP JavaTV package
import javax.tv.xlet.*;

//import required MHP HAVi packages
import org.havi.ui.*;
import org.havi.ui.event.*;

//import required MHP DVB package
import org.dvb.ui.*;

// import required non-MHP pJava packages
import java.io.*;
import java.awt.*;
import java.awt.event.*;
// This Xlet will be visible on-screen, so we extend org.havi.ui.Hcomponent and it also implements java.awt.KeyListener to receive

// Something went wrong reading the message file.
System.out.println("I/O exception reading message.txt"); } }

// Start the Xlet.
public void startXlet() throws javax.tv.xlet.XletStateChangeException
{
  // startXlet() should not block for too long
  myWorkerThread = new Thread(this);
  myWorkerThread.start();
}

// Pause the Xlet.
public void pauseXlet()
{
  // do what we need to in order to pause the Xlet.
  doPause();
}

// Destroy the Xlet.
public void destroyXlet(boolean unconditional) throws javax.tv.xlet.XletStateChangeException
{
  if (unconditional)
  {
   // Unconditional Termination
   doDestroy();
   }
   else
   {
   // Conditional Termination
   throw new XletStateChangeException("Termination Not Required");
   }
 }

// Before we can draw on the screen, we need an HScene to draw into. This variable will hold a reference to our Hscene
private HScene scene;

// The image that we will show
private Image image;

// The message that will get printed. This is read from a file in initXlet()
private String message;

// this holds the alpha (transparency) level that we will be using
private int alpha = 0;

// this object is responsible for displaying the background I-frame
private HaviBackgroundController backgroundManager;

// The main method for the worker thread.
public void run()
{
 // We need quite a few resources before we can start doing anything.
 getResources();

 // This component should listen for AWT key events so that it can respond to them.
 addKeyListener(this);
 // This adds the background image to the display. The background image is displayed in the background plane.
 displayBackgroundImage();

 // The bitmap image is shown in the graphics plane.
 displayForegroundBitmap();
}

........
```

**Figure 2-53b**

Simple MHP HAVi Xlet source example.[37]

Finally, an application manager within an MHP system manages all MHP applications residing on the device from information input from the end-user, as well as via the AIT (application information table) data within the MHP broadcast stream transmitted to the system. AIT data simply instructs the application manager as to what applications are actually available to the end-user of the device and the technical details of controlling the running of the application.

## 2.4 Summary

The purpose of this chapter was to show the importance of industry-supported standards when trying to understand and implement embedded system designs and concepts. The programming language, networking, and DTV examples provided in this chapter demonstrated how standards can define major elements within an embedded architecture. The programming language example provided an illustration of general-purpose standards that can be implemented in a wide variety of embedded devices. These examples specifically included Java, showing how a JVM was needed at the application, system, or hardware layer, and the .NET Compact Framework, for languages such as C# and Visual Basic, which demonstrated a programming language element that must be integrated into the systems software layer. Networking provided an example of standards that can be general purpose, specific to a family of devices (market-driven) or specific to an application (e.g., HTTP in a browser). In the case of networking, both the embedded systems and the OSI models were referenced to demonstrate where certain networking protocols fit into an embedded architecture. Finally, the DTV/STB example illustrated how one device implemented several standards that defined embedded components in all layers.

Chapter 3, *Embedded Hardware Building Blocks and the Embedded Board*, is the first chapter of Section II, *Embedded Hardware.* Chapter 3 introduces the fundamental elements found on an embedded board and some of the most common basic electrical components that make up these elements, as well as those that can be found independently on a board.

## Chapter 2: Problems

1. How can embedded system standards typically be classified?
2. [a] Name and define four groups of market-specific standards.
   [b] Give three examples of standards that fall under each of the four market-specific groups.
3. [a] Name and define four classes of general-purpose standards.
   [b] Give three examples of standards that fall under each of the four general-purpose groups.
4. Which standard below is neither a market-specific nor a general-purpose embedded systems standard?
   A. HTTP—Hypertext Transfer Protocol.
   B. MHP—Multimedia Home Platform.
   C. J2EE—Java 2 Enterprise Edition.

D. All of the above.
E. None of the above.

5. [a] What is the difference between a high-level language and a low-level language?
   [b] Give an example of each.
6. A compiler can be located on:
   A. A target.
   B. A host.
   C. On a target and/or on a host.
   D. None of the above.
7. [a] What is the difference between a cross-compiler and a compiler?
   [b] What is the difference between a compiler and an assembler?
8. [a] What is an interpreter?
   [b] Give two examples of interpreted languages.
9. [T/F] All interpreted languages are scripting languages, but not all scripting languages are interpreted.
10. [a] In order to run Java, what is required on the target?
    [b] How can the JVM be implemented in an embedded system?
11. Which standards below are embedded Java standards?
    A. pJava—Personal Java.
    B. RTSC—Real Time Core Specification.
    C. HTML—Hypertext Markup Language.
    D. A and B only.
    E. A and C only.
12. What are the two main differences between all embedded JVMs?
13. Name and describe three of the most common byte processing schemes.
14. [a] What is the purpose of a GC?
    [b] Name and describe two common GC schemes.
15. [a] Name three qualities that Java and scripting languages have in common.
    [b] Name two ways that they differ.
16. [a] What is the .NET Compact Framework?
    [b] How is it similar to Java?
    [c] How is it different?
17. What is the difference between LANs and WANs?
18. What are the two types of transmission mediums that can connect devices?
19. [a] What is the OSI model?
    [b] What are the layers of the OSI model?
    [c] Give examples of two protocols under each layer.
    [d] Where in the Embedded Systems Model does each layer of the OSI model fall? Draw it.
20. [a] How does the OSI model compare to the TCP/IP model?
    [b] How does the OSI model compare to Bluetooth?

## Endnotes


[1] *Embedded System Building Blocks,* p. 61, J. J. Labrosse, Cmp Books, 1995; *Embedded Microcomputer Systems*, p. 3, J. W. Valvano, CL Engineering, 2nd edn, 2006.
[2] http://www.mhp.org/what_is_mhp/index.html.
[3] http://java.sun.com/products/javatv/overview.html.
[4] www.davic.org.
[5] http://www.atsc.org/standards.html.
[6] www.atvef.com.
[7] http://java.sun.com/pr/2000/05/pr000508-02.html and http://netvision.qianlong.com/8737/2003-6-4/39@878954.htm.
[8] www.arib.or.jp.
[9] http://www.osgi.org/resources/spec_overview.asp.
[10] www.opentv.com.
[11] http://www.microsoft.com/tv/default.mspx.
[12] "HAVi, the A/V digital network revolution," Whitepaper, p. 2, HAVi Organization, 2003.
[13] http://www.accessdata.fda.gov/scripts/cdrh/cfdocs/cfStandards/search.cfm.
[14] http://europa.eu.int/smartapi/cgi/sga_doc?smartapi!celexapi!prod!CELEXnumdoc&lg=EN&numdoc=31993L0042&model=guichett.
[15] www.ieee1073.org.
[16] Digital Imaging and Communications in Medicine (DICOM): Part 1: Introduction and Overview, p.5, http://medical.nema.org/Dicom/2011/11_01pu.pdf
[17] http://www.ce.org/standards/default.asp
[18] http://europa.eu.int/comm/enterprise/mechan_equipment/machinery/
[19] https://domino.iec.ch/webstore/webstore.nsf/artnum/025140
[20] http://www.iso.ch/iso/en/CatalogueListPage.CatalogueList?ICS1=25&ICS2=40&ICS3=1
[21] "Bluetooth Protocol Architecture," Whitepaper, p. 4, R. Mettala, http://www.bluetooth.org/Technical/Specifications/whitepapers.htm.
[22] *Systems Analysis and Design*, p. 17, D. Harris, Course Technology, 3rd revised edn, 2003.
[23] "I/Opener," R. Morin and V. Brown, *Sun Expert Magazine*, 1998.
[24] "Boehm–Demers–Weiser Conservative Garbage Collector: A Garbage Collector for C and C++", H. Boehm, http://www.hpl.hp.com/personal/Hans_Boehm/gc/.
[25] "Selecting the Right Transmission Medium Optical Fiber is Moving to the Forefront of Transmission Media Choices Leaving Twisted Copper And Coaxial Cable Behind," C. Weinstein, **URL?**.
[26] "This Is Microwave," Whitepaper, Stratex Networks, **URL?**. http://www.stratexnet.com/about_us/our_technology/tutorial/This_is_Microwave_expanded.pdf; "Satellite, Bandwidth Without Borders," Whitepaper, S. Beardsley, P. Edin, and A. Miles, **URL?**.
[27] http://www.ihs.com/standards/vis/collections.html and "IHS: Government Information Solutions," p. 4.
[28] http://standards.ieee.org/reading/ieee/std_public/description/busarch/1284.1-1997_desc.html
[29] http://www.aimglobal.org/aimstore/linearsymbologies.htm
[30] http://www.iaob.org/iso_ts.html
[31] http://www.praxiom.com/iso-intro.htm and the "Fourth-Generation Languages," Whitepaper, S. Cummings. P.1.
[32] "Spread Spectrum: Regulation in Light of Changing Technologies," Whitepaper, p. 7, D. Carter, A. Garcia, and D. Pearah, http://groups.csail.mit.edu/mac/classes/6.805/student-papers/fall98-papers/spectrum/whitepaper.html.
[33] *Demystifying Embedded Systems Middleware*, T. Noergaard, Elsevier, 2010.
[34] http://focus.ti.com/docs/solution/folders/print/327.html
[35] openMHP API Documentation and Source Code.
[36] Digital Video Broadcasting (DVB); Multimedia Home Platform (MHP) Specification 1.1.2. European Broadcasting Union.
[37] Application examples based upon MHP open source by S. Morris available for download at www.interactivetvweb.org.

SECTION II

# Embedded Hardware

## Embedded Hardware

Section II consists of five chapters that introduce the fundamental hardware components of an embedded board and show how these components function together. The information reflected in this chapter is not intended to prepare the reader to create a detailed board design, but it will provide an architectural overview of some of the most important elements of an embedded board and information as to the function of these components. Chapter 3 introduces the major hardware components of an embedded board using the von Neumann model, the Embedded Systems Model, and real-world boards as references. Chapters 4–7 discuss the major hardware components of an embedded board in detail.

Wherever possible, the theoretical information introduced is directly related to actual embedded hardware, because it is the underlying physics of embedded hardware that directly impacts board design. Understanding the major hardware elements of an embedded board is critical to understanding the entire system's architecture, because ultimately the capabilities of an embedded device are limited or enhanced by what the hardware is capable of.

CHAPTER 3

# Embedded Hardware Building Blocks and the Embedded Board

**In This Chapter**

- Introducing the importance of being able to read a schematic diagram
- Discussing the major components of an embedded board
- Introducing the factors that allow an embedded device to work
- Discussing the fundamental elements of electronic components

## 3.1 Lesson One on Hardware: Learn to Read a Schematic!

This section is especially important for embedded software engineers and programmers. Before diving into the details, note that it is important for *all* embedded designers to be able to understand the diagrams and symbols that hardware engineers create and use to describe their hardware designs to the outside world. These diagrams and symbols are the keys to quickly and efficiently understanding even the most complex hardware design, regardless of how much or little practical experience one has in designing real hardware. They also contain the information an embedded programmer needs to design any software that requires compatibility with the hardware and they provide insight to a programmer as to how to successfully communicate the hardware requirements of the software to a hardware engineer.

There are several different types of engineering hardware drawings, including:

- *Block diagrams*. Block diagrams typically depict the major components of a board (processors, buses, I/O (input/output), memory) or a single component (e.g., a processor) at a systems architecture or higher level. In short, a block diagram is a basic overview of the hardware, with implementation details abstracted out. While a block diagram can reflect the actual physical layout of a board containing these major components, it mainly depicts how different components or units within a component function together at a systems architecture level. Block diagrams are used extensively throughout this book (in fact, Figures 3-5a–e later in this chapter are examples of block diagrams) because they are the simplest method by which to depict and describe the components within a system. The symbols used within a block diagram are simple, such as squares or rectangles for chips, and straight lines for buses. Block diagrams are typically not detailed enough for a software designer to be able to write all of the low-level software accurately enough to control the hardware (without a lot of headaches, trial and error, and even some burned-out hardware!). However, they are very useful in communicating a basic overview of the hardware, as well as providing a basis for creating more detailed hardware diagrams.

- *Schematics.* Schematics are electronic circuit diagrams that provide a more detailed view of all of the devices within a circuit or within a single component—everything from processors down to resistors. A schematic diagram is not meant to depict the physical layout of the board or component, but provides information on the flow of data in the system, defining what signals are assigned where—which signals travel on the various lines of a bus, appear on the pins of a processor, etc. In schematic diagrams, *schematic symbols* are used to depict all of the components within the system. They typically do not look anything like the physical components they represent, but are a type of "shorthand" representation based on some type of schematic symbol standard. A schematic diagram is the most useful diagram to both hardware and software designers when trying to determine how a system actually operates, to debug hardware, or to write and debug the software managing the hardware. See Appendix B for a list of commonly used schematic symbols.
- *Wiring diagrams.* These diagrams represent the *bus* connections between the major and minor components on a board or within a chip. In wiring diagrams, vertical and horizontal lines are used to represent the lines of a bus, and either schematic symbols or more simplified symbols (that physically resemble the other components on the board or elements within a component) are used. These diagrams may represent an approximate depiction of the physical layout of a component or board.
- *Logic diagrams/prints.* Logic diagrams/prints are used to show a wide variety of circuit information using logical symbols (AND, OR, NOT, XOR, etc.), and logical inputs and outputs (the 1 s and 0 s). These diagrams do not replace schematics, but they can be useful in simplifying certain types of circuits in order to understand how they function.
- *Timing diagrams.* Timing diagrams display timing graphs of various input and output signals of a circuit, as well as the relationships between the various signals. They are the most common diagrams (after block diagrams) in hardware user manuals and data sheets.

Regardless of the type, in order to understand how to read and interpret these diagrams, it is first important to *learn* the standard symbols, conventions, and rules used. Examples of the symbols used in timing diagrams are shown in Table 3-1, along with the conventions for input/output signals associated with each of the symbols.

An example of a timing diagram is shown in Figure 3-1, where each row represents a different signal. In the case of the signal rising and falling symbols within the diagram, the *rise time* or *fall time* is indicated by the time it takes for the signal to move from LOW to HIGH or vice-versa (the entire length of the diagonal line of the symbol). When comparing two signals, a delay is measured at the center of the rising or falling symbols of each signal

**Table 3-1: Timing diagrams symbol table[9]**

| Symbol | Input Signals | Output Signals |
|---|---|---|
| | Input signal must be valid | Output signal will be valid |
| | Input signal doesn't affect system, will work regardless | Indeterminate output signal |
| | Garbage signal (nonsense) | Output signal not driven (floating), tristate, HiZ, high impedance |
| | If the input signal rises | Output signal will rise |
| | If the input signal falls | Output signal will fall |

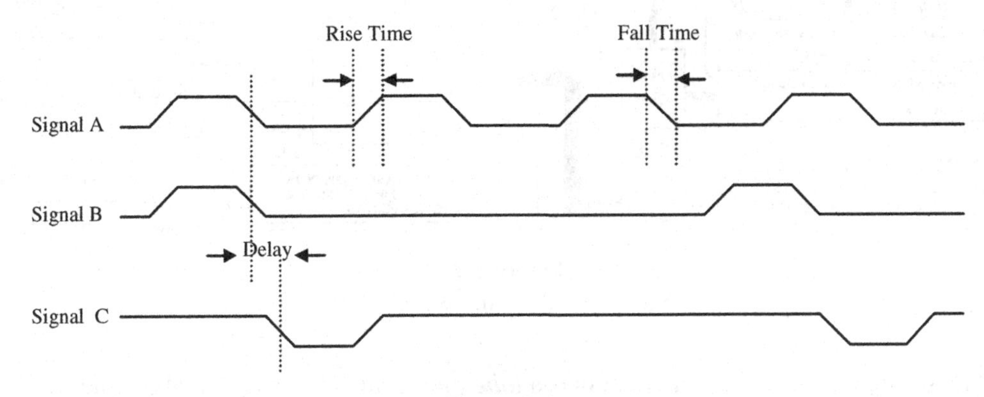


**Figure 3-1**
Timing diagram example.

being compared. In Figure 3-1, there is a fall time delay between signals B and C and signals A and C in the first falling symbol. When comparing the first falling symbol of signals A and B in Figure 3-1, no delay is indicated by the timing diagram.

Schematic diagrams are much more complex than their timing diagram counterparts. As introduced earlier this chapter, schematics provide a more detailed view of all of the devices within a circuit or within a single component. Figure 3-2 shows an example of a schematic diagram.

In the case of schematic diagrams, some of the conventions and rules include:

- A *title section* located at the bottom of each schematic page, listing information that includes, but is not limited to, the name of the circuit, the name of the hardware engineer responsible for the design, the date, and a list of revisions made to the design since its conception.
- The use of *schematic symbols* indicating the various components of a circuit (see Appendix B).

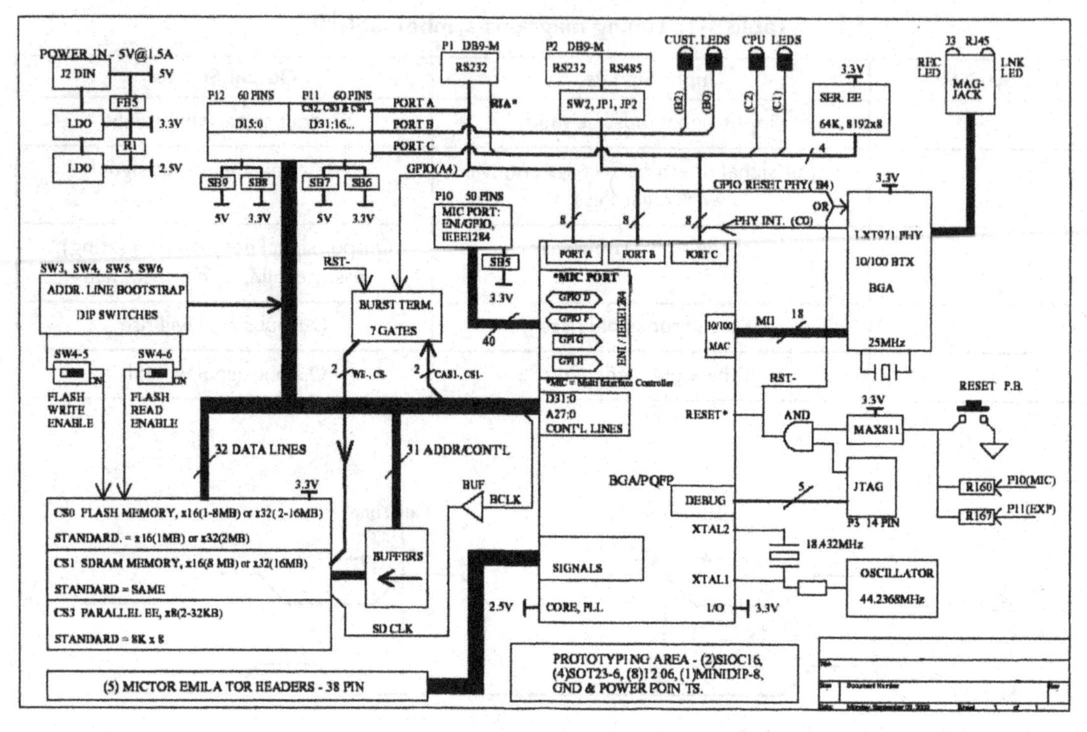

**Figure 3-2**
Schematic diagram example.[7]

- Along with the assigned symbol comes a *label* that details information about the component (size, type, power ratings, etc.). Labels for components of a symbol, such as the pin numbers of an integrated circuit (IC), signal names associated with wires, etc., are usually located outside of the schematic symbol.
- Abbreviations and *prefixes* are used for common units of measurement (i.e., k for kilo or $10^3$, M for mega or $10^6$) and these prefixes replace writing out the units and larger numbers.
- *Functional groups* and subgroups of components are typically separated onto different pages.
- *I/O* and *voltage source/ground terminals.* In general, positive voltage supply terminals are located at the top of the page, and negative supply/ground at the bottom. Input components are usually on the left and output components are on the right.

Finally, while this book provides an introduction into understanding the various diagrams and recognizing schematic symbols, and the devices they represent, it does not replace researching more specifics on the particular diagrams used by your organization, whether through additional reading or purchasing software, or asking the hardware engineers responsible for creating the diagrams what conventions and rules are followed.

(For instance, indicating the voltage source and ground terminals on a schematic isn't required and may not be part of the convention scheme followed by those responsible for creating the schematics. However, a voltage source and a ground are required for any circuit to work, so don't be afraid to ask.) At the very least, the block and schematic diagrams should contain nothing unfamiliar to anyone working on the embedded project, whether they are coding software or prototyping the hardware. This means becoming familiar with everything from where the name of the diagram is located to how the states of the components shown within the diagrams are represented.

One of the most efficient ways of learning how to learn to read and/or create a hardware diagram is via the *Traister and Lisk* method,[10] which involves:

*Step 1* Learning the basic symbols that can make up the type of diagram, such as timing or schematic symbols. To aid in the learning of these symbols, rotate between this step and Steps 2 and/or 3.

*Step 2* Reading as many diagrams as possible, until reading them becomes boring (in that case rotate between this step and Steps 1 and/or 3) or comfortable (so there is no longer the need to look up every other symbol while reading).

*Step 3* Drawing a diagram to practice simulating what has been read, again until it either becomes boring (which means rotating back through Steps 1 and/or 2) or comfortable.

## 3.2 The Embedded Board and the von Neumann Model

In embedded devices, all the electronics hardware resides on a board, also referred to as a printed wiring board (PW) or printed circuit board (PCB). PCBs are often made of thin sheets of fiberglass. The electrical path of the circuit is printed in copper, which carries the electrical signals between the various components connected on the board. All electronic components that make up the circuit are connected to this board, either by soldering, plugging in to a socket, or some other connection mechanism. All of the hardware on an embedded board is located in the hardware layer of the Embedded Systems Model (see Figure 3-3).

At the highest level, the major hardware components of most boards can be classified into five major categories:

- *Central processing unit (CPU): the master processor.*
- *Memory*: where the system's software is stored.
- *Input device(s)*: input slave processors and relative electrical components.
- *Output device(s)*: output slave processors and relative electrical components.
- *Data pathway(s)/bus(es)*: interconnects the other components, providing a "highway" for data to travel on from one component to another, including any wires, bus bridges, and/or bus controllers.

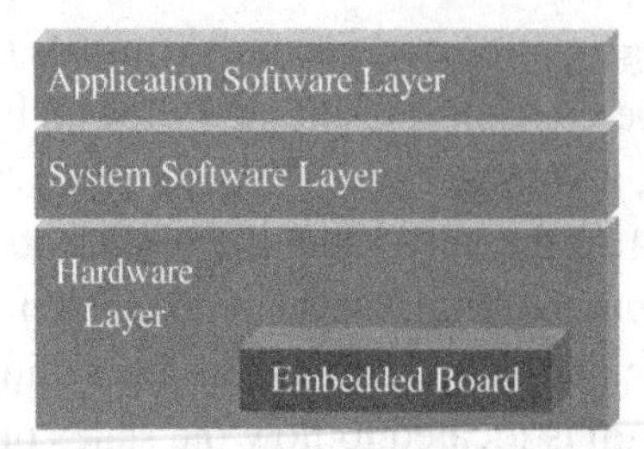


**Figure 3-3**
Embedded board and the Embedded Systems Model.

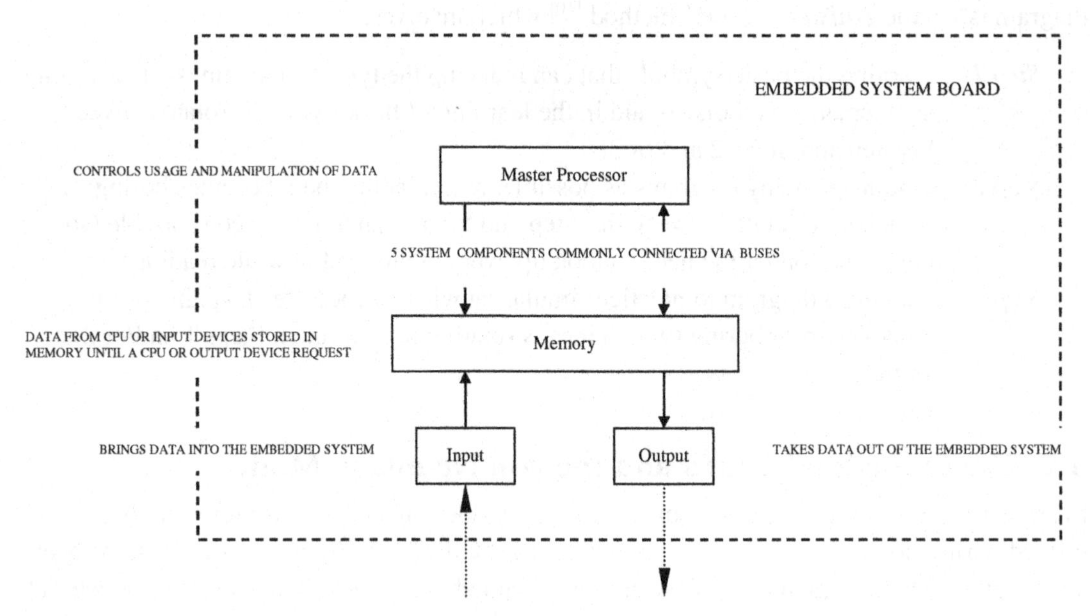


**Figure 3-4**
Embedded system board organization.[11]

These five categories are based upon the major elements defined by the von Neumann model (see Figure 3-4)—a tool that can be used to understand any electronic device's hardware architecture. The von Neumann model is a result of the published work of John von Neumann in 1945, which defined the requirements of a general-purpose electronic computer. Because embedded systems are a type of computer system, this model can be applied as a means of understanding embedded systems hardware.

While board designs can vary widely as demonstrated in the examples of Figures 3-5a–d, all of the major elements on these embedded boards—*and on just about any embedded board*—can be classified as either the master CPU(s), memory, I/O, or bus components.

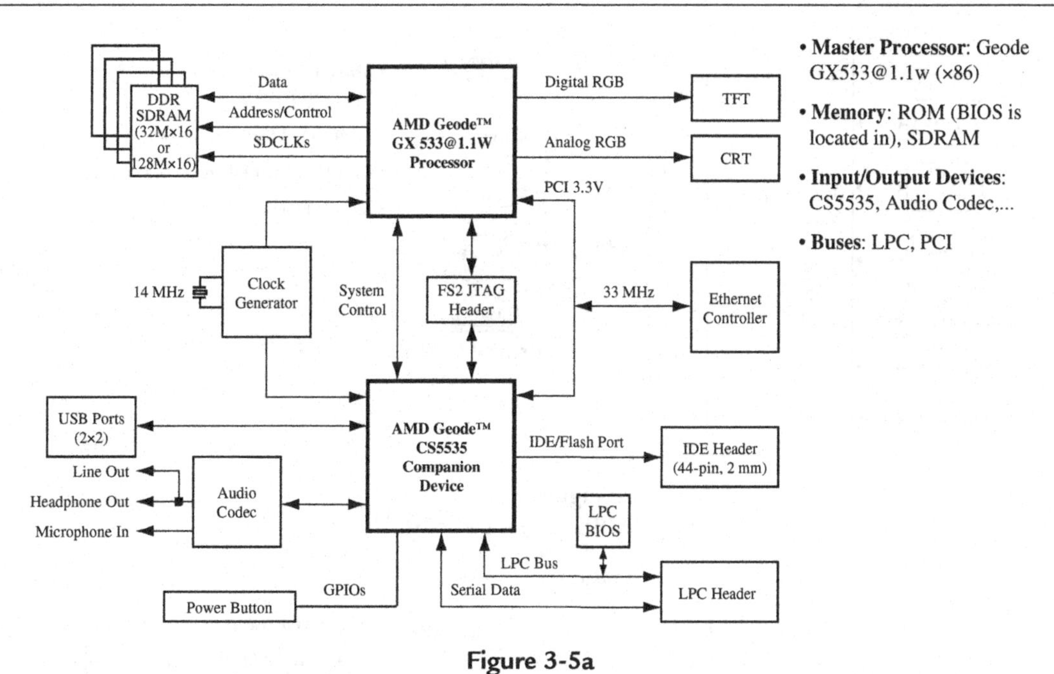


- **Master Processor**: Geode GX533@1.1w (×86)
- **Memory**: ROM (BIOS is located in), SDRAM
- **Input/Output Devices**: CS5535, Audio Codec,...
- **Buses**: LPC, PCI

**Figure 3-5a**
AMD/National Semiconductor ×86 reference board.[1] *© 2004 Advanced Micro Devices, Inc. Reprinted with permission.*

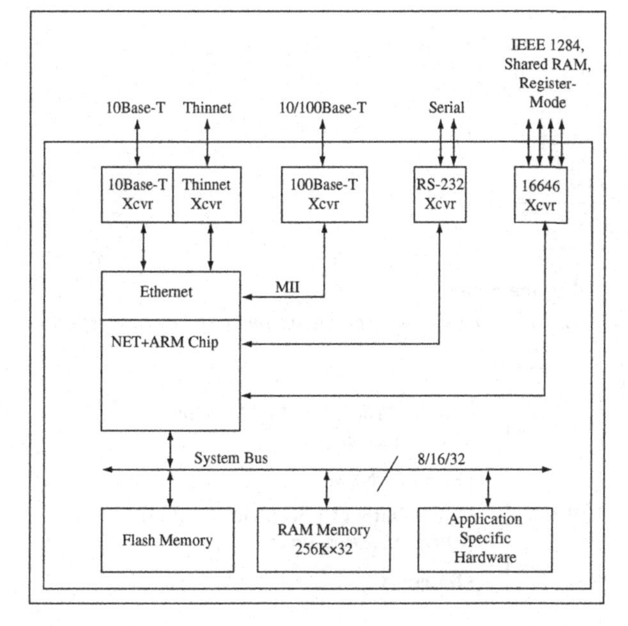


- **Master Processor**: Net+ARM ARM7
- **Memory**: Flash, RAM
- **Input/Output Devices**: 10Base-T transceiver, Thinnet transceiver, 100Base-T transceiver, RS-232 transceiver,16646 transceiver,...
- **Buses**: System Bus, MII,...

**Figure 3-5b**
Net Silicon ARM7 reference board.[2]

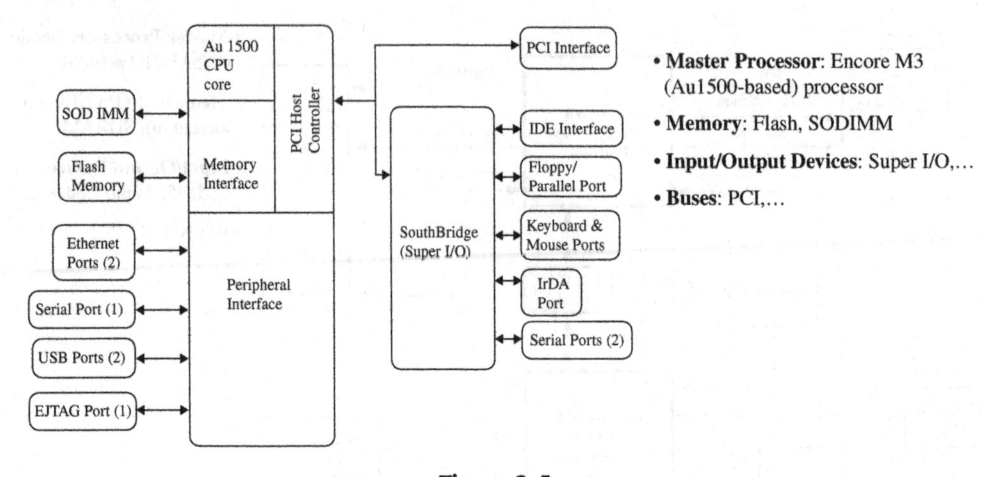


**Figure 3-5c**
Ampro MIPS reference board.[3]

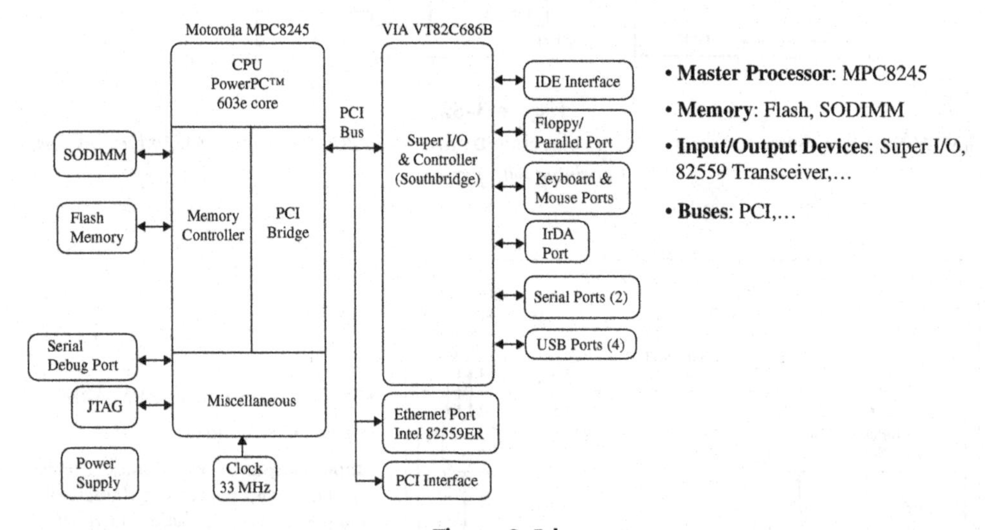


**Figure 3-5d**
Ampro PowerPC reference board.[4] *© 2004 Freescale Semiconductor, Inc. Used by permission.*

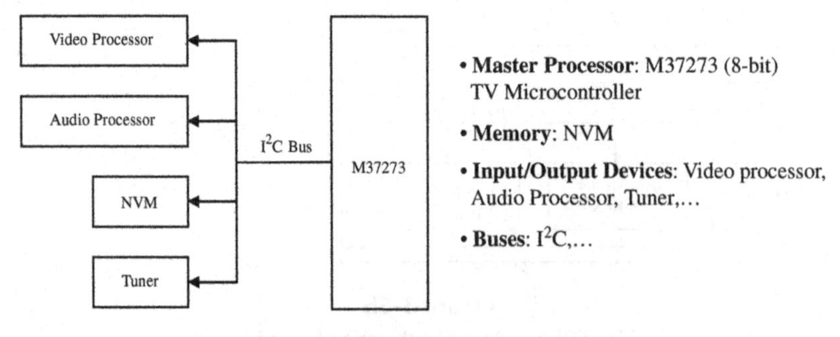


**Figure 3-5e**
Mitsubishi analog TV reference board.

In order to understand how the major components on an embedded board function, it is useful to first understand what these components consist of and why. All of the components on an embedded board, including the major components introduced in the von Neumann model, are made up of one or some combination of interconnected basic electronic devices, such as wires, resistors, capacitors, inductors, and diodes. These devices also can act to connect the major components of a board together. At the highest level, these devices are typically classified as either *passive* or *active* components. In short, passive components include devices such as wires, resistors, capacitors and inductors that can only receive or store power. Active components, on the other hand, include devices such as transistors, diodes, and ICs that are capable of delivering as well as receiving and storing power. In some cases, active components themselves can be made up of passive components. Within the passive and active families of components, these circuit devices essentially differ according to how they respond to *voltage* and *current.*

## 3.3 Powering the Hardware

Power is the rate that energy is expended or work is performed. This means that in alternating current (AC) and direct current (DC) circuits, the power associated with each element on the board equals the current through the element multiplied by the voltage across the element ($P = VI$). Accurate power and energy calculations must be done for all elements on an embedded board to determine the power consumption requirements of that particular board. This is because each element can only handle a certain type of power, so AC/DC converters, DC/AC converters, direct AC/AC converters, etc., may be required. Also, each element has a limited amount of power that it requires to function, that it can handle, or that it dissipates. These calculations determine what type of voltage source can be used on a board, and how powerful the voltage source needs to be.

In embedded systems, both AC and DC voltage sources are used, because each current generation technique has its pros and cons. AC is easier to generate in large amounts using generators driven by turbines turned by everything from wind to water. Producing large amounts of DC from electrochemical cells (batteries) is not as practical. Also, because transmitting current over long transmission lines results in a significant loss of energy due to the resistance of the wire, most modern electric company facilities transmit electricity to outlets in AC current since AC can be transformed to lower or higher voltages much more easily than DC. With AC, a device called a *transformer*, located at the service provider, is used to efficiently transmit current over long distances with lower losses. The transformer is a device that transfers electrical energy from one circuit to another, and can make changes to the current and voltage during the transfer. The service provider transmits lower levels of current at a higher voltage rate from the power plant, and then a transformer at the customer site decreases the voltage to the value required. On the flip-side, at very high voltages, wires offer less resistance to DC than to AC, thus making DC more efficient to transmit than AC over very long distances.

Some embedded boards integrate or plug into *power supplies.* Power supplies can be either AC or DC. To use an AC power supply to supply power to components using only DC, an AC/DC converter can be used to convert AC to the lower DC voltages required by the various components on an embedded board, which typically require 3.3, 5, or 12 V.

*Note: Other types of converters, such as DC/DC, DC/AC, or direct AC/AC can be used to handle the required power conversions for devices that have other requirements.*

Other embedded boards or components on a board (such as non-volatile memory, discussed in more detail in Chapter 5) rely on *batteries* as voltage sources, which can be more practical for providing power because of their size. Battery-powered boards don't rely on a power plant for energy and they allow portability of embedded devices that don't need to be plugged into an outlet. Also, because batteries supply DC current, no mechanism is needed to convert AC to DC for components that require DC, as is needed with boards that rely on a power supply and outlet supplying AC. Batteries, however, have a limited life and must be either recharged or replaced.

### 3.3.1 A Quick Comment on Analog versus Digital Signals

A digital system processes only digital data, which is data represented by only 0 s and 1 s. On most boards, two voltages represent "0" and "1", since all data is represented as some combination of 1 s and 0 s. No voltage (0 V) is referred to as ground, $V_{SS}$, or LOW, and 3, 5, or 12 V are commonly referred to as $V_{CC}$, $V_{DD}$, or HIGH. All signals within the system are one of the two voltages or are transitioning to one of the two voltages. Systems can define "0" as LOW and "1" as HIGH or some range of 0–1 V as LOW and 4–5 V as HIGH, for instance. Other signals can base the definition of a "1" or "0" on edges (LOW-to-HIGH or HIGH-to-LOW).

Because most major components on an embedded board, such as processors, inherently process the 1 s and 0 s of digital signals, a lot of embedded hardware is digital by nature. However, an embedded system can still process analog signals, which are continuous (i.e., not only 1 s and 0 s, but values in between as well). Obviously, a mechanism is needed on the board to convert analog signals to digital signals. An analog signal is digitized by a sampling process and the resulting digital data can be translated back into a voltage "wave" that mirrors the original analog waveform.

**Real-World Advice**

*Inaccurate Signals: Problems with Noise in Analog and Digital Signals*

One of the most serious problems in both the analog and digital signal realm involves noise distorting incoming signals, thus corrupting and affecting the accuracy of data. Noise is generally any unwanted signal alteration from an input source, any part of the input signal generated from something other than a sensor, or even noise generated from the sensor itself. Noise is a common problem with analog signals. Digital signals, on the other hand, are at greater risk if the signals are not generated locally to the embedded processor, so any digital signals coming across a longer transmission medium are the most susceptible to noise problems.

Analog noise can come from a wide variety of sources—radio signals, lightning, power lines, the microprocessor, the analog sensing electronics themselves, etc. The same is true for digital noise, which can come from mechanical contacts used as computer inputs, dirty slip rings that transmit power/data, limits in accuracy/dependability of input source, etc.

The key to reducing either analog or digital noise is:

1. To follow basic design guidelines to avoid problems with noise. In the case of analog noise this includes not mixing analog and digital grounds, keeping sensitive electronic elements on the board a sufficient distance from elements switching current, and limiting length of wires with low signal levels/high impedance. With digital signals, this means routing signal wires away from noise-inducing high-current cables, shielding wires, transmitting signals using correct techniques, etc.
2. To clearly identify the root cause of the problem, which means exactly what is causing the noise.

With Point 2, once the root cause of the noise has been identified, a hardware or software fix can be implemented. Techniques for reducing analog noise include filtering out frequencies not needed and averaging the signal inputs, whereas digital noise is commonly addressed via transmitting correction codes/parity bits, and/or adding additional hardware to the board to correct any problems with received data.

---

*Based on the articles by J. Ganssle "Minimizing Analog Noise" (May 1997), "Taming Analog Noise" (November 1992), and "Smoothing Digital Inputs" (October 1992),* Embedded Systems Programming Magazine.

## 3.4 Basic Hardware Materials: Conductors, Insulators, and Semiconductors

All electronic devices used on a board or that come into contact with an embedded board (such as networking transmission mediums) are made up of materials that are generally classified as conductors, insulators, or semiconductors. These categories differ according to the ability of the materials to conduct an electric current. While conductors, insulators, and semiconductors will all conduct given the right environment, *conductors* are materials that

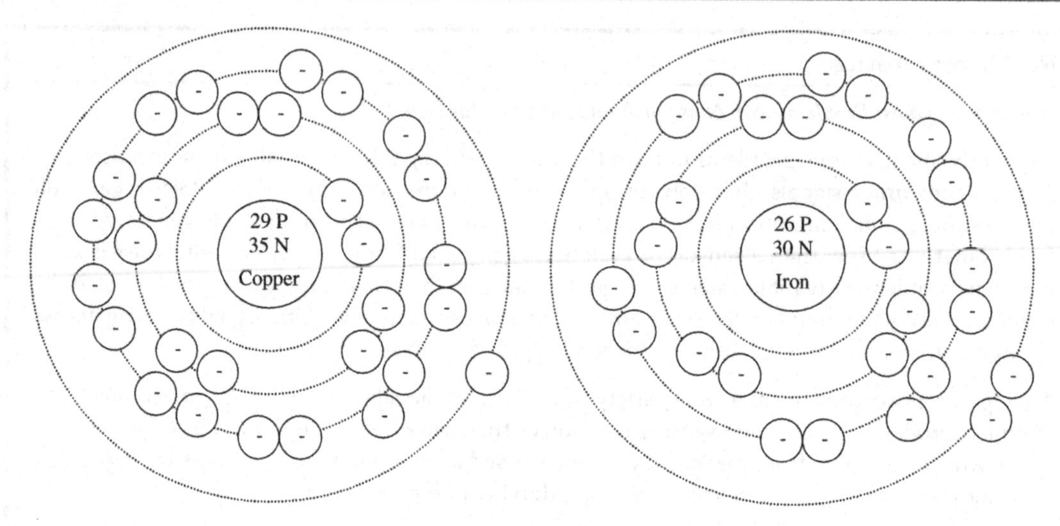


**Figure 3-6**
Conductors.

have fewer impediments to an electric current (meaning they more easily lose/gain valence electrons) and they (coincidentally) have three or fewer valence electrons (see Figure 3-6).

Most metals are conductors, because most metallic elements have a crystalline makeup that doesn't require a lot of energy to free the valence electrons of their atoms. The atomic lattice (structure) of these metals is composed of atoms that are bound so tightly together that valence electrons aren't closely associated with any individual atom. This means that valence electrons are equally attached to surrounding atoms and the force attaching them to an individual nucleus is practically nil. Thus, the amount of energy at room temperature to free these electrons is relatively small. Buses and wired transmission mediums are examples of one or more *wires* made up of conductive metallic material. *A wire, in a schematic diagram, is typically represented by a straight line: "——" (see Appendix B); in other electronic diagrams (i.e., block) wires can also be represented as arrows "←→."*

*Insulators* typically have five or more valence electrons (see Figure 3-7) and impede an electric current. This means that they are less likely to lose or gain valence electrons without a great deal of applied energy to the material. For this reason, insulators are typically not the main materials used in buses. Note that some of the best insulators, like conductive metals, are very regular in their crystal lattice and their atoms do tightly bond. The main difference between a conductor and insulator lies in whether the energy of the valence electrons is enough to overcome any barriers between atoms. If this is the case, these electrons are free floating in the lattice. With an insulator, such as NaCl (sodium chloride, table salt), the valence electrons would have to overcome a tremendous electric field. In short, insulators require greater amounts of energy at room temperature to free their valence electrons in comparison to conductors. Non-metals, such as air, paper, oil, plastic, glass, and rubber, are usually considered insulators.

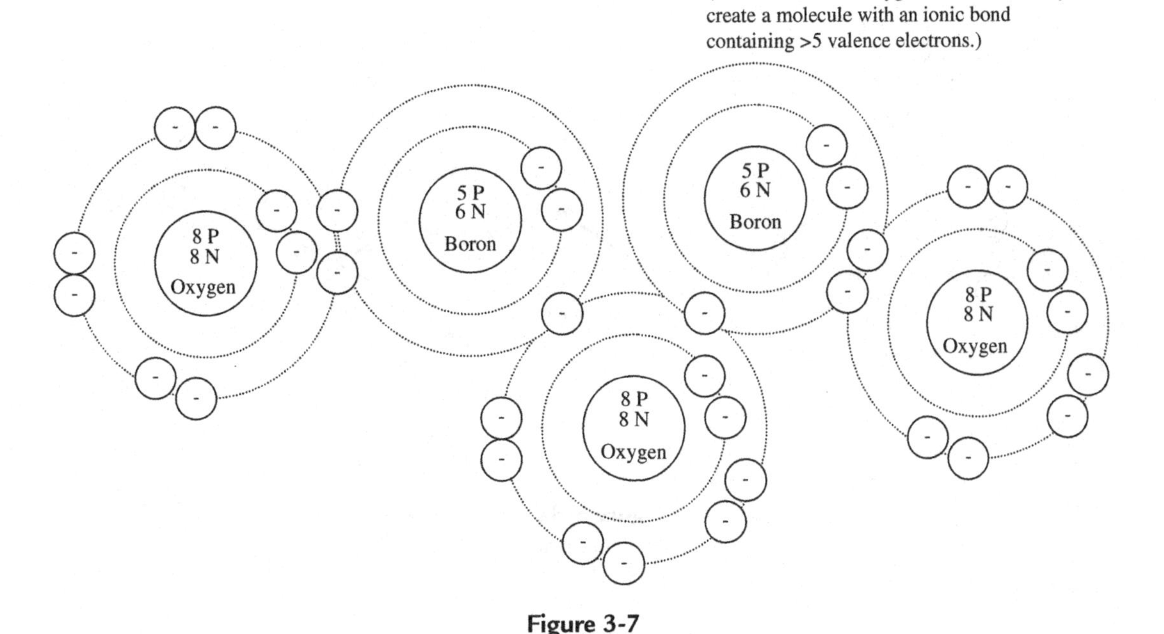


**Figure 3-7**
Insulators.

### Air Transmissions via Electromagnetic Waves

The ability of an insulator, air, to transmit data is the basis of wireless communication. Data is transmitted specifically through electromagnetic waves that have the ability to induce a current in a receiving antenna. An antenna is basically a conducting wire that contains a vibrating current that radiates electromagnetic energy into its surroundings. In short, electromagnetic waves are created when an electric charge oscillates at the speed of light, such as within an antenna. The oscillation of an electric charge can be caused by many things (heat, AC circuitry, etc.), but in essence all elements above the temperature of absolute zero emit some electromagnetic radiation. So, heat (for example) can generate electromagnetic radiation, because the higher the temperature, the faster the electrons oscillate per unit of time, and thus the more electromagnetic energy emitted.

When the electromagnetic wave is emitted, it travels through the empty space, if any, between atoms (of air, of materials, etc.). The electromagnetic radiation is absorbed by that atom, causing its own electrons to vibrate and, after a time, emit a new electromagnetic wave of the same frequency as the wave it absorbed. It is, of course, usually intended at some point for some type of receiver to intercept one of these waves, but the remaining electromagnetic waves will continue to travel indefinitely at the speed of light (though they do weaken the further they travel from their original source–the amplitude/strength of a wave is inversely proportional to the square of the distance). It is for this reason that the different types of wireless mediums (e.g., satellite versus Bluetooth, discussed in Chapter 2) have their limitations in terms of the types of devices and networks they are used in and where their receivers need to be located.

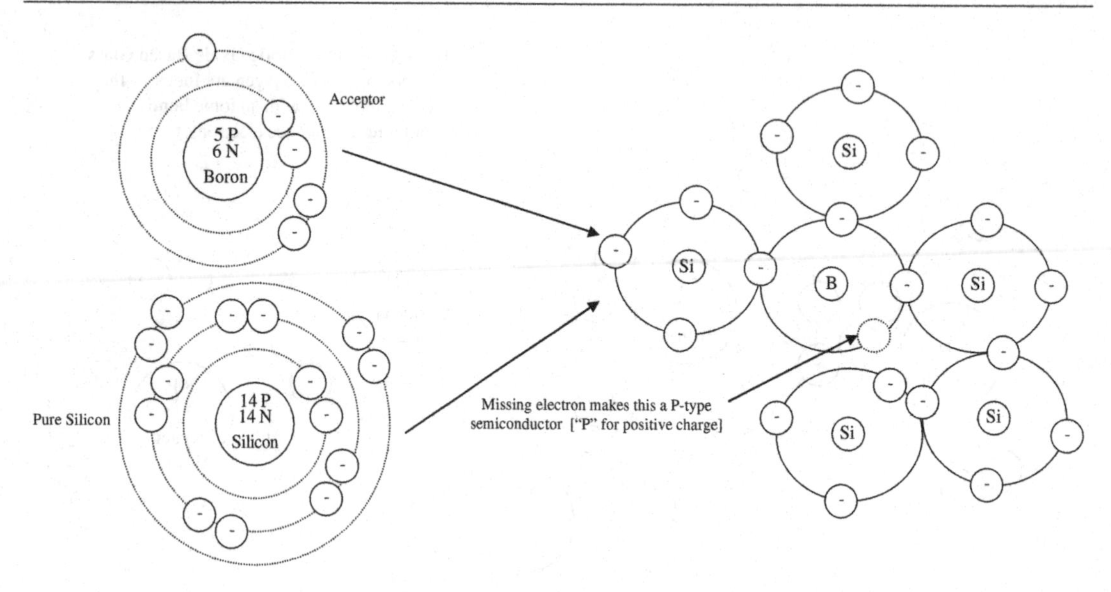


**Figure 3-8a**
P-type semiconductor.

*Semiconductors* usually have four valence electrons and are classified as materials whose base elements have a conductive nature that can be altered by introducing other elements into their structure. This means that semiconductive material has the ability to behave both as a conductor and as an insulator. Elements such as silicon and germanium can be modified in such a manner that they have a resistance about halfway between insulators and conductors. The process of turning these base elements into semiconductors starts with the purification of these elements. After purification, these elements have a crystalline structure in which atoms are rigidly locked together in a lattice with the electrons unable to move, making them strong insulators. These materials are then doped to enhance their abilities to conduct electrons. Doping is the process of introducing impurities, which in turn interweaves the silicon or germanium insulator structure with the conductive features of the donor. Certain impurities (arsenic, phosphorus, antimony, etc.), called donors, create a surplus of electrons creating an N-type semiconductor, while other impurities called acceptors, such as boron, produce a shortage of electrons, creating a P-type semiconductor material (see Figures 3-8a and b).

Note that the fact that semiconductors usually have four valence electrons is a coincidence (e.g., silicon and germanium both have four valence electrons). A semiconductor is defined by the energy of the valence electron with respect to the barriers between lattice atoms.

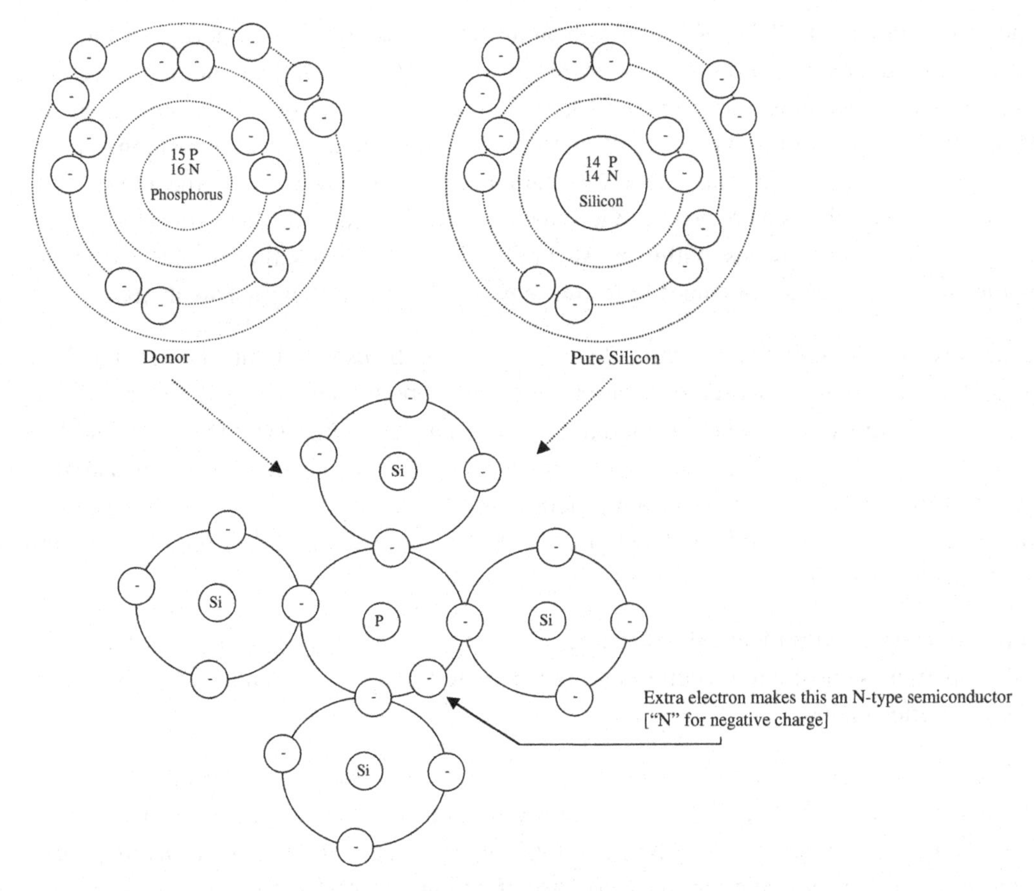


**Figure 3-8b**
N-type semiconductor.

## 3.5 Common Passive Components on Boards and in Chips: Resistors, Capacitors, and Inductors

Passive electronic components, including wires, can be integrated (along with semiconductive devices, discussed later in this chapter) to form processors and memory chips. These components can also be a part of the circuitry (input circuitry, output circuitry, etc.) found on the board. The next several subsections introduce passive components commonly found on an embedded board, mainly the resistor, the capacitor, and the inductor.

### *3.5.1 Resistors*

Even the best of conductors will offer some resistance to current flow. Resistors are devices made up of conductive materials that have had their conductivity altered in some fashion to allow for an increase in resistance. For example, carbon-composition resistors are created by the mixing of carbon (the conductor) with an insulating material (the impurity). Another technique used in creating resistors is to change the physical shape of the material to alter its resistance, such as winding a wire into a coil, as is the case in wire-wound resistors. There are several different types of resistors in addition to wire-wound and carbon-composition, including current-limiting, carbon-film, foil-filament-wound, fuse, and metal-film, to name a few. Regardless of type, all resistors provide the same inherent function: to create a resistive force in a circuit. Resistors are a means, within an AC or DC circuit, to control the current or voltage by providing some amount of resistance to the current or voltage that flows across them.

Because resistors, as reflected in Ohm's law ($V = IR$), can be used to control current and voltage, they are commonly used in a variety of circuitry both on boards and integrated into processor or memory chips when needed to achieve a particular bias (voltage or current level) for some type of circuitry the resistors are connected to. This means that a set of resistors networked properly to perform a certain function (as attenuators, voltage dividers, fuses, heaters, etc.) provides a specific voltage or current value adjustment that is required for some type of attached circuitry.

Given two resistors with identical resistances, depending on how the resistor was made, a set of properties is considered when selecting between the two for use in a particular circuit. These properties include:

- *Tolerance* (%), which represents how much more or less precise the resistance of the resistor is at any given time, given its labeled resistance value. The actual value of resistance should not exceed "plus or minus" (±) the labeled tolerance. Typically, the more sensitive a particular circuit is to error, the tighter (smaller) the tolerances that are used.
- *Power rating.* When a current encounters resistance, heat, along with some other forms of energy at times, such as light, is generated. The power rating indicates how much power a resistor can safely dissipate. Using a low-powered resistor in a higher-powered circuit can cause a melt-down of that resistor, as it is not able to release the generated heat from the current it carries as effectively as a higher-powered resistor can.
- *Reliability level rating* (%), meaning how much change in resistance might occur in the resistor for every 1000 hours of resistor use.
- *Temperature coefficient of resistance (TCR).* The resistivity of materials that make up the resistor can vary with changes in temperature. The value representing a change in resistance relative to changes in temperature is referred to as the *temperature coefficient*. If a resistor's resistivity doesn't change in response to a temperature change, it has a "0" temperature coefficient. If a resistor's resistivity increases when the temperature increases and decreases when the temperature decreases, then that resistor has a "positive" temperature coefficient. If a resistor's resistivity decreases when the temperature increases, and increases when the temperature decreases, then that resistor has a

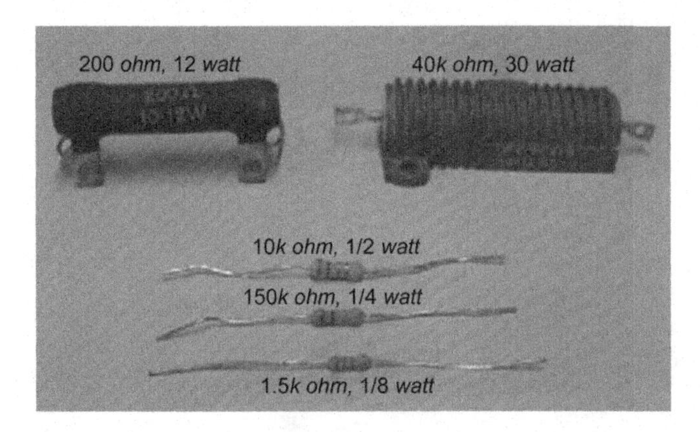


**Figure 3-9a**
Fixed resistors.

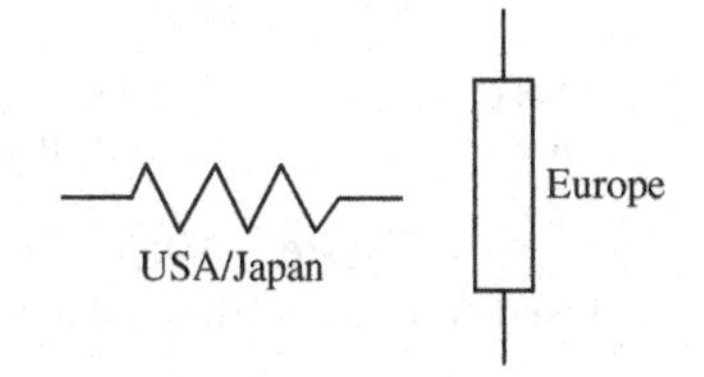


**Figure 3-9b**
Fixed resistor schematic symbols.

"negative" temperature coefficient. For example, conductors typically have a "positive" temperature coefficient and are usually most conductive (have the least resistance) at room temperature, while insulators typically have fewer freed valence electrons at room temperature. Thus, resistors made up of particular materials that display some characteristic at "room temperature," and a measurably different one at warmer or cooler temperatures, impact what types of systems they ultimately may be used in (e.g., mobile embedded devices versus indoor embedded devices).

While there are many different ways to make resistors, each with their own properties, at the highest level there are only two types of resistors: fixed and variable. *Fixed resistors* are resistors that are manufactured to have only one resistance value. Fixed resistors come in many different types and sizes depending on how they are made (see Figure 3-9a), though in spite of the differences in physical appearances, the schematic symbol representing fixed resistors remains the same depending on the schematic standard supported (see Figure 3-9b).

For fixed resistors with bodies that are too small to contain their printed property values, the values are calculated from color coded bands located physically on the resistor's body. These color coded bands appear as either vertical stripes, used on fixed resistors with axial leads as shown in Figure 3-10a, or in various locations on the body, used on fixed resistors with radial leads as shown in Figure 3-10b.

A resistor may also include additional color coded bands representing its various properties, such as reliability level ratings, temperature coefficient, and tolerance. While different types of fixed resistors have different numbers and types of bands, the color definitions are typically the same. Tables 3-2a–d show the various types of bands that can be found on the body of a fixed resistor along with their meanings.

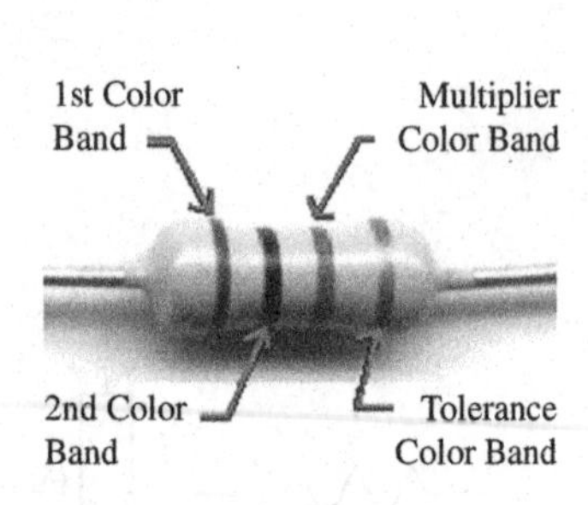


**Figure 3-10a**
Fixed resistors with axial leads.

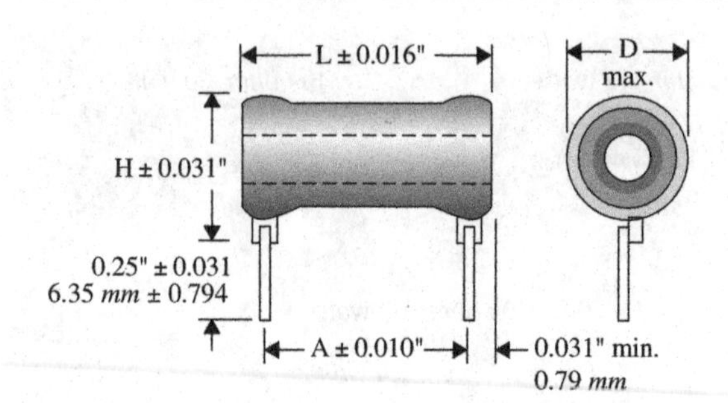


**Figure 3-10b**
Fixed resistors with radial leads.

**Table 3-2a: Resistor color code digits and multiplier table[6]**

| Color of Band | Digits | Multiplier |
|---|---|---|
| Black | 0 | ×1 |
| Brown | 1 | ×10 |
| Red | 2 | ×100 |
| Orange | 3 | ×1 K |
| Yellow | 4 | ×10 K |
| Green | 5 | ×100 K |
| Blue | 6 | ×1 M |
| Purple | 7 | ×10 M |
| Grey | 8 | ×100 M |
| White | 9 | ×1000 M |
| Silver | – | ×0.01 |
| Gold | – | ×0.1 |

**Table 3-2b: Temperature coefficient[6]**

| Color of Band | Temperature Coefficient (p.p.m.) |
|---|---|
| Brown | 100 |
| Red | 50 |
| Orange | 15 |
| Yellow | 25 |

**Table 3-2c: Reliability level[6]**

| Color of Band | Reliability Level (per 1000 hours) |
|---|---|
| Brown | 1 |
| Red | 0.1 |
| Orange | 0.01 |
| Yellow | 0.001 |

**Table 3-2d: Tolerance[6]**

| Color of Band | Tolerance (%) |
|---|---|
| Silver | ±10 |
| Gold | ±5 |
| Brown | ±1 |
| Red | ±2 |
| Green | ±0.5 |
| Blue | ±0.25 |
| Purple | ±0.1 |

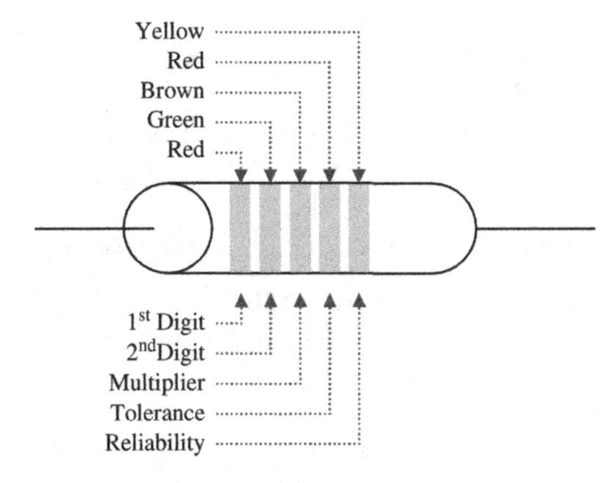


**Figure 3-11**
Five-band fixed resistor with axial leads example.

To understand how the color coding works, let's take an example of a five-band carbon composition resistor with axial leads, in which the bands are arranged as vertical stripes on the resistor's body, with associated colors bands as shown in Figure 3-11. Bands 1 and 2 are digits, band 3 is the multiplier, band 4 is tolerance, and band 5 is reliability. Note that resistors can vary widely in the number and meanings of the bands, and that this is one specific example in which we're given the information and are told how to use the tables to determine resistance and other various properties. This resistor's first three bands are red = 2, green = 5, and brown = ×10. Thus, this resistor has a resistance of 250 Ω (2 and 5 of the red and green bands are the first and second digits; the third brown band "×10" value is the multiplier which is used to multiply "25" by 10, resulting in the value of 250). Taking into account the resistor's tolerance reflected by the red band or ±2%, this resistor has a resistance value of 250 Ω ± 2%. The fifth band in this example is a yellow band, reflecting a reliability of 0.001%. This means that the resistance of this resistor might change by 0.001% from the labeled value (250 Ω ± 2% in this case) for every 1000 h of use. *Note: The amount of resistance provided by a resistor is measured in ohms (Ω).*

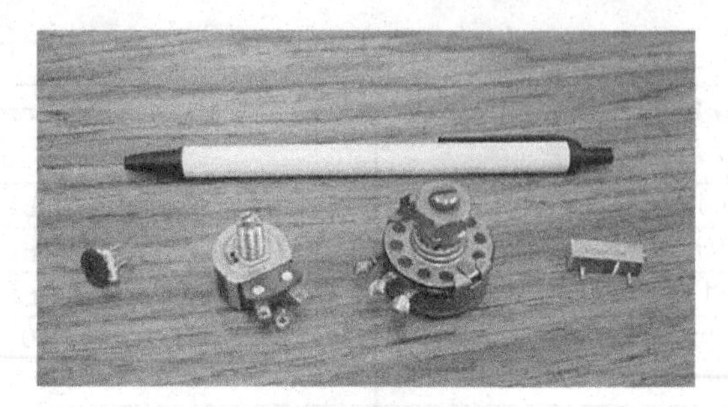

**Figure 3-12a**
Variable resistor's appearance.

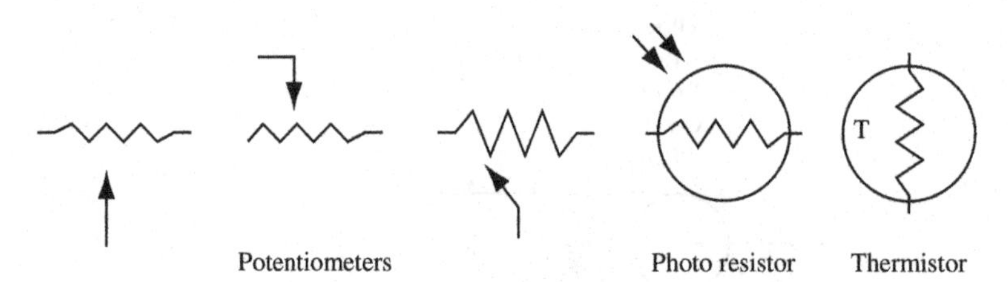


**Figure 3-12b**
Variable resistor's schematic symbols.

*Variable resistors* vary their resistance on-the-fly, as opposed to manufactured fixed resistors. Resistance can be varied manually (potentiometers), by changes in light (photosensitive/photo resistor), by changes in temperature (thermally sensitive/thermistor), etc. Figures 3-12a and b show what some variable resistors physically look like, as well as how they are symbolized in schematics.

### 3.5.2 Capacitors

Capacitors are made up of conductors typically in the form of two parallel metal plates separated by an insulator, which is a dielectric such as air, ceramic, polyester, or mica, as shown in Figure 3-13a.

When each of the plates is connected to an AC voltage source (see Figure 3-13b), the plates accumulate opposing charges—positive in one plate and negative in the other. Electrons are surrounded by electric fields produced by that charge. An electric field emanates outwardly and downwardly from the source, in this case the charged plate, diminishing in field strength as it gets further from the source. The electric field created between the two plates acts to temporarily store the energy and keep the plates from discharging. If a wire were to connect the two plates, current would flow until both plates were no longer charged—or as is the case with AC voltage sources, when the polarity changes, the plates then discharge.

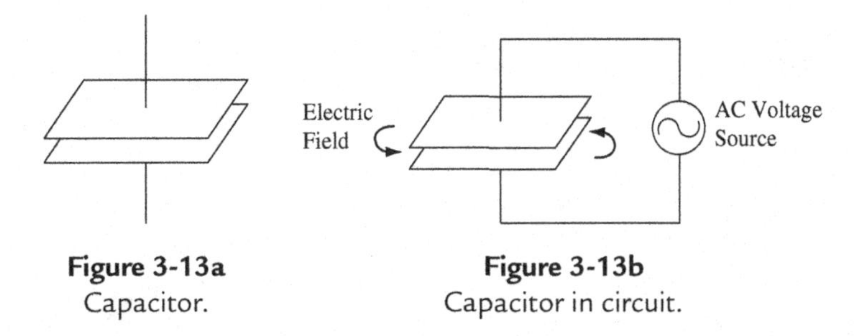


**Figure 3-13a**
Capacitor.

**Figure 3-13b**
Capacitor in circuit.

In short, capacitors store energy in electric fields. Like the resistor, they impede the flow of energy, but unlike the resistor, which dissipates some of this energy intentionally and is typically used in both AC and DC circuits, the capacitor is more commonly used in AC circuits and gives this same energy back to the circuit in its original form (electrically) when the plates are discharged. Note that, depending on how the capacitor is made, manufacturing imperfections may result in a capacitor not functioning perfectly, causing some unintentional loss of energy in the form of heat.

> Any two conductors located in close proximity can act as capacitors (with air being the dielectric). This phenomenon is called interelectrode capacitance. It is for this reason that in some devices (involving radiofrequencies) this phenomenon is minimized by enclosing some electronic components.

A set of properties is considered when selecting capacitors for use in a particular circuit, namely:

- *Temperature coefficient of capacitance.* Similar in meaning to TCR. If a capacitor's conductance doesn't change in response to a temperature change, it has a "0" temperature coefficient. If a capacitor's capacitance increases when the temperature increases, and decreases when the temperature decreases, then that capacitor has a "positive" temperature coefficient. If a capacitor's capacitance decreases when the temperature increases, and increases when the temperature decreases, then that capacitor has a "negative" temperature coefficient.
- *Tolerance* (%), which represents at any one time how much more or less precise the capacitance of a capacitor is at any given time given its labeled capacitance value (the actual value of capacitance should not exceed "plus or minus" (±) the labeled tolerance).

As with resistors, capacitors can be integrated into a chip, and depending on the capacitor, used in everything from DC power supplies to radio receivers and transmitters. Many different types of capacitors exist (variable, ceramic, electrolytic, epoxy, etc.), differing by the material of the plates and dielectric and, like resistors, by whether they can be adjusted on-the-fly (see Figures 3-14a and b).

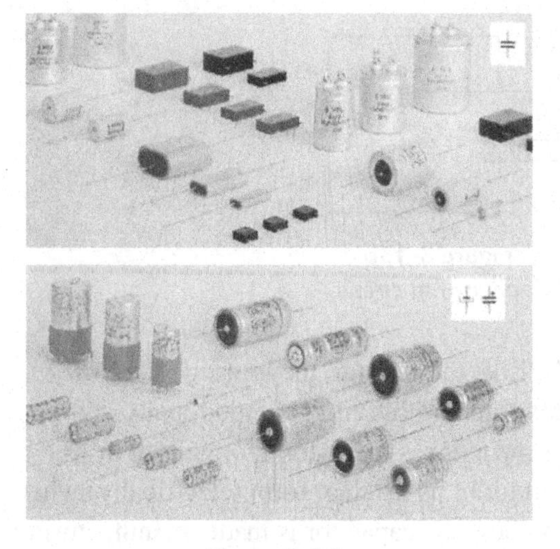

**Figure 3-14a**
Capacitors.

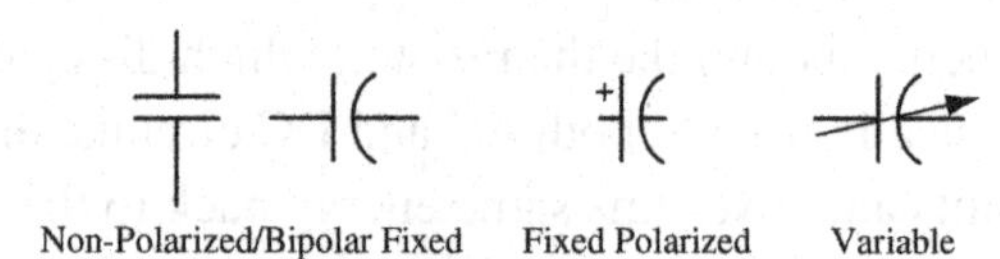


**Figure 3-14b**
Schematic symbols for capacitors.

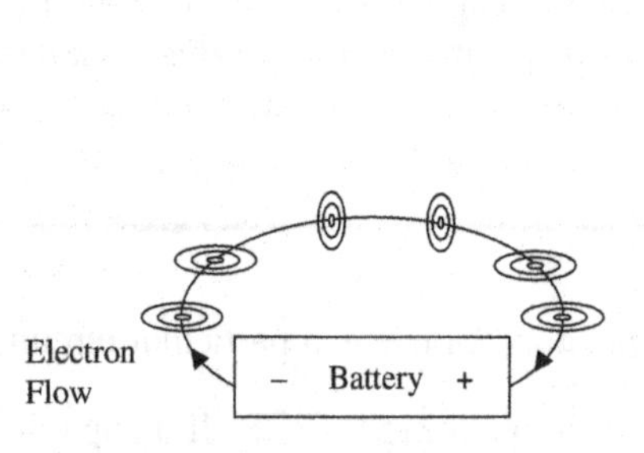


**Figure 3-15a**
Magnetic fields.

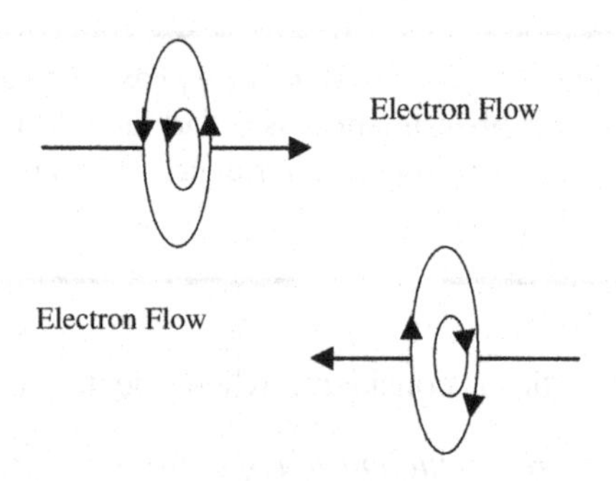


**Figure 3-15b**
Direction of magnetic fields.

### 3.5.3 Inductors

Inductors, like capacitors, store electrical energy in AC circuits. With capacitors, however, energy is temporarily stored in an electric field, whereas inductors temporarily store energy in a *magnetic field.* These magnetic fields are produced by the movement of electrons, and can be visualized as rings surrounding an electric current (see Figure 3-15a). The direction of electron flow determines the direction of the magnetic field (see Figure 3-15b).

All materials, even conductors, have some resistance and thus give off some energy. Some of this energy is stored within the magnetic fields surrounding the wire. Inductance is the storage of energy within the magnetic field surrounding a wire with a current flowing through it (and like capacitance, can occur unintentionally). When a change occurs in the current stream, as

**Figure 3-16a**
Inductor's appearance.

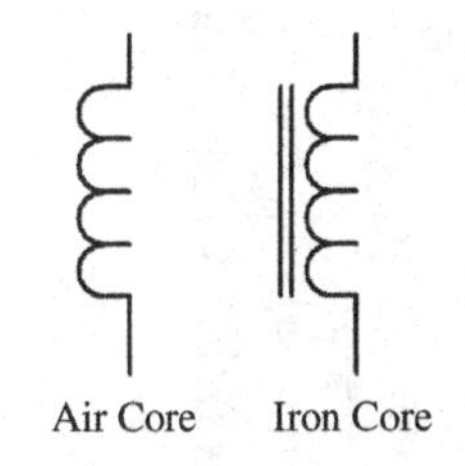


**Figure 3-16b**
Inductor's schematic symbols.

happens in an AC circuit, the magnetic field changes and thus "induces a force on a charged object" (Faraday's Law of Induction). Any expansion, due to a current increase, means an increase in the energy stored by the inductor, whereas a collapse of a magnetic field, due to a lack of current, will release that energy back into the circuit. Changes in current are reflected in how inductance is measured. Measured in units of henries (H), inductance is the ratio between the rate of current change and the voltage across the inductor.

As mentioned, all wires with some current have some sort of inductance, however minimal. Because magnetic flux is much higher for a coiled wire than for a straighter wire, most common inductors are made up of a coiled wire, although, again, inductors can be made up of a single wire or set of wires. Adding some type of core other than air, such as ferrite or powdered iron within the coiled-up wire, increases the magnetic flux density many times over. Figures 3-16a and b show some common inductors and their schematic symbol counterparts.

The properties that define inductance include the number of individual coils (the more coils, the larger the inductance), the diameter of the coils (which is directly proportional to inductance), the overall shape of the coil (cylindrical/solenoidal, doughnut-shaped/toroidal, etc.), and the overall length of the coiled wire (the longer it is, the smaller the inductance).

## 3.6 Semiconductors and the Active Building Blocks of Processors and Memory

While P-type and N-type semiconductors are the basic types of semiconductors, as discussed in Section 3.4, they are not usually very useful on their own. These two types must be combined in order to be able to do anything practical. When P-type and N-type semiconductors are combined, the contact point, called the *P–N Junction*, acts as a one-way gate to allow electrons to flow within the device in a direction dependent on the polarity of the materials. P-type and N-type semiconductive materials form some of the most common basic electronic devices that act as the main building blocks in processor and memory chips: diodes and transistors.

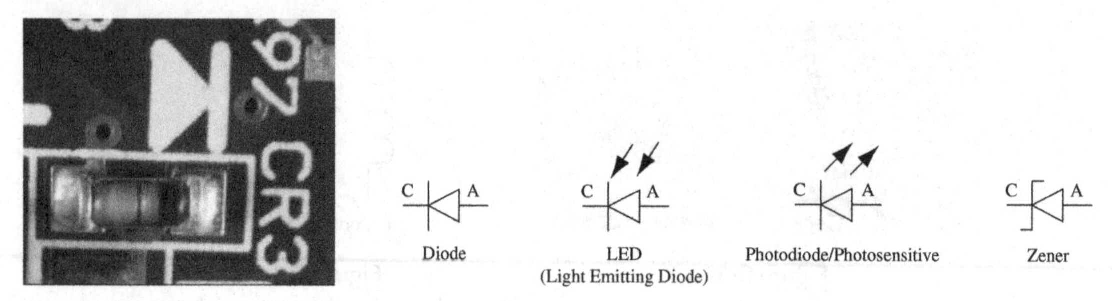


**Figure 3-17a**
Diode and LED.

**Figure 3-17b**
Diode schematic symbols.

### 3.6.1 Diodes

A diode is a semiconductor device made up of two materials, one P-type and one N-type, joined together. A terminal is connected to each of the materials, called an *anode*, labeled "A" in the schematic symbol in Figure 3-17b, and a *cathode*, labeled "C" in the schematic in Figure 3-17b.

These materials work together to allow current to flow in only one direction. Current flows through a diode from the anode to cathode as long as the anode has a higher (positive) voltage; this phenomenon is called forward biasing. Current flows in this condition because the electrons flowing from the voltage source are attracted to the P-type material of the diode through the N-type material (see Figure 3-18a).

When current will not flow through the diode because the cathode has a higher (positive) voltage than the anode, the diode acts like a variable capacitor, whose capacitance changes depending on the amount of reverse voltage. This is called reverse biasing. In this case (as shown in Figure 3-18b), the electrons are pulled away from the P-type material in the diode, creating a *depletion region*, a section surrounding the P–N junction that has no charge and acts as an insulator, resisting current flow.

There are several different types of diodes, each with their own common uses, such as rectifier diodes that convert AC to DC by keeping the polarity constant, PIN diodes as switches, and zener diodes for voltage regulation. Some of the most recognizable diodes on a board are the *light-emitting diodes* (*LEDs*), shown in Figure 3-19. LEDs are the blinking or steady lights that can indicate anything from PowerON, to problems with the system, to remote-control signals, depending on how they are designed. LEDs are designed to emit visible or infrared (IR) light when forward biased in a circuit.

As a final note, keep in mind that higher forms of semiconductor logic are based upon the diode depletion effect. This effect generates a region where the barrier is higher than the average valence electron energy, and the barrier can be influenced by voltage.

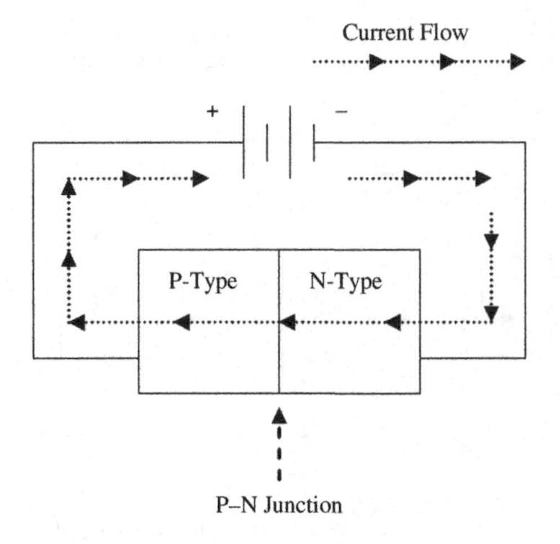


**Figure 3-18a**
Diode in forward bias.

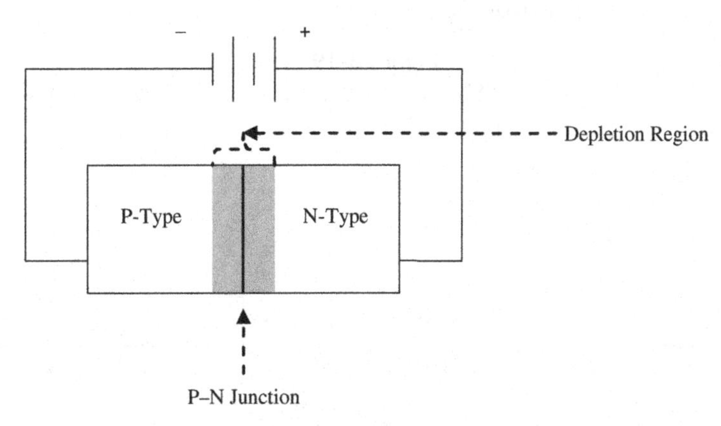


**Figure 3-18b**
Diode in reverse bias.

### *3.6.2 Transistors*

"Transistor" is the contraction for *current-**trans**ferring **resistor***.[5] Transistors are made up of some combination of P-type and N-type semiconductor material, with three terminals connecting to each of the three materials (see Figure 3-20a). It is the combination and versatility of these materials that, depending on the type of transistor, allow them to be used for a variety of purposes, such as current amplifiers (amplification), in oscillators (oscillation), in high-speed integrated circuits (ICs, to be discussed later this chapter), and/or in switching circuits (DIP switches, push buttons, etc., such as commonly found on off-the-shelf reference boards). While there are several different types of transistors, the two main types are the *bipolar junction transistor (BJT)* and the *field-effect transistor (FET)*.

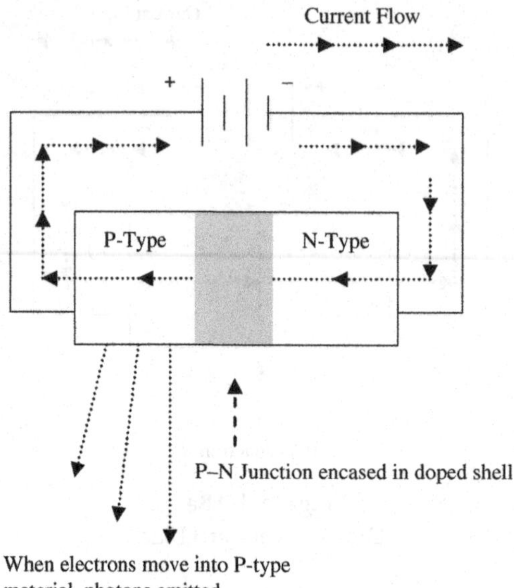


**Figure 3-19**
LED in forward bias.

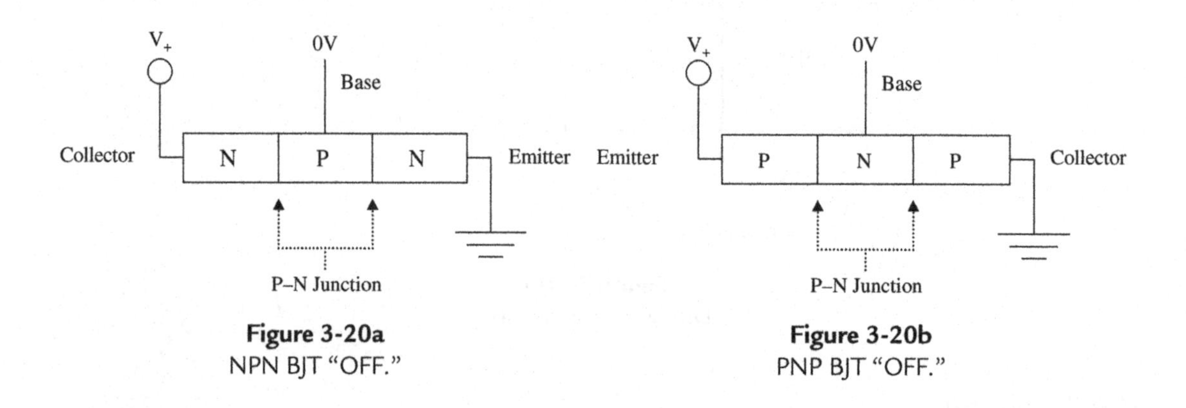


**Figure 3-20a**
NPN BJT "OFF."

**Figure 3-20b**
PNP BJT "OFF."

The BJT, also referred to as the bipolar transistor, is made up of three alternating types of P-type and N-type material, and are subclassed based on the combination of these materials. There are two main subclasses of bipolar transistors, PNP and NPN. As implied by their names, a PNP BJT is made up of two sections of P-type materials, separated by a thin section of N-type material, whereas the NPN bipolar transistor is made up of two sections of N-type material, separated by a thin section of P-type material. As shown in Figures 3-20a and b, each of these sections has an associated terminal (electrode): an *emitter*, a *base*, and a *collector.*

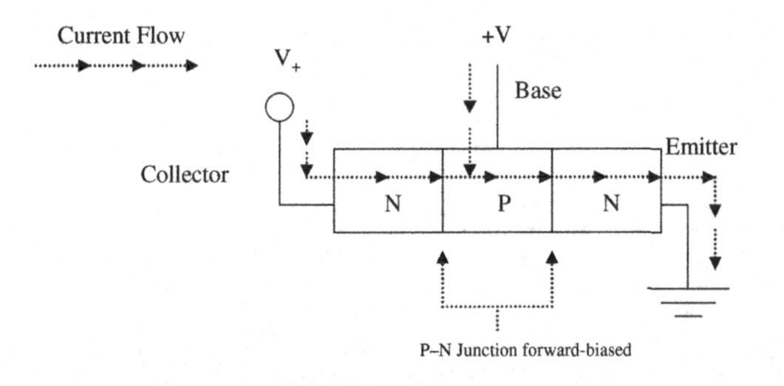


**Figure 3-21a**
NPN BJT "ON."

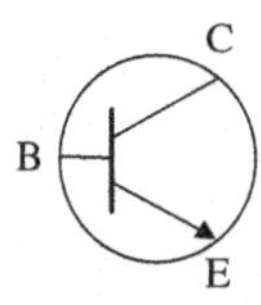


**Figure 3-21b**
NPN BJT schematic symbol.

When the NPN BJT is OFF (as shown in Figure 3-20a), electrons in the emitter cannot bypass the P–N junction to flow to the collector, because there is no biasing voltage (0V) at the base to pressure electrons over the junctions.

To turn the NPN BJT ON (as shown in Figure 3-21a), a positive voltage and input current must be applied at the base so escaping electrons from the emitter are attracted to the P-type base, and, because of the thinness of the P-type material, these electrons then flow to the collector. This then creates a (positive) current flow from the collector to the emitter. This current flow is a combination of the base current and collector current, and so, the larger the base voltage, the greater the emitter current flow. Figure 3-21b shows the NPN BJT schematic symbol, which includes an arrow indicating the direction of output current flow *from* the emitter when the transistor is ON.

When the PNP BJT is OFF (as shown in Figure 3-20b), electrons in the collector cannot bypass the P–N junction to flow to the emitter, because the 0V at the base is placing just enough pressure to keep electrons from flowing. To turn the PNP BJT ON (as shown in Figure 3-22a), a negative base voltage is used to decrease pressure and allow a positive current flow out of the collector, with a small output current flowing out of the base, as well. Figure 3-22b shows the PNP BJT schematic symbol, which includes an arrow indicating the direction of current flow *into* the emitter and out of the collector terminal when the transistor is ON.

In short, PNP and NPN BJTs work in the same manner, given the opposite directions of current flow, the P-type and N-type material makeup, and the voltage polarities applied at the base.

Like the BJT, the FET is made up of some combination of P-type and N-type semiconductor material. Like the BJT, FETs have three terminals, but in FETs these terminals are called a *source*, a *drain/sink*, and a *gate* (see Figure 3-22). In order to function, FETs do not require a biasing current and are controlled via voltage alone. Beyond this, there are several subtypes of

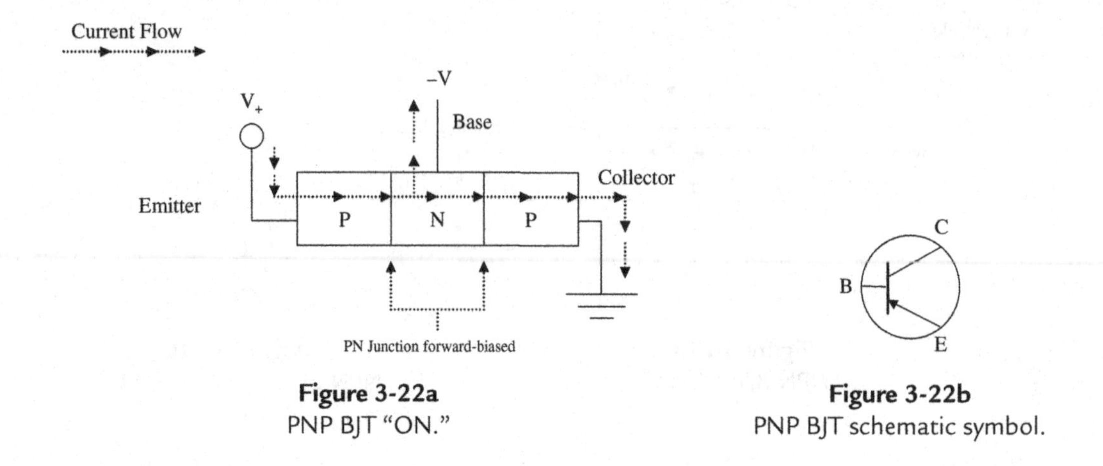


**Figure 3-22a**
PNP BJT "ON."

**Figure 3-22b**
PNP BJT schematic symbol.

FETs that function and are designed differently, the most common falling under the families of the *metal-oxide-semiconductor field-effect transistor (MOSFET)* and the *junction field-effect transistor (JFET)*.

There are several types of MOSFETs, the main two subclasses of which are *enhancement* MOSFETs and *depletion* MOSFETs. Like BJTs, enhancement-type MOSFETs become less resistant to current flow when voltage is applied to the gate. Depletion-type MOSFETs have the opposite reaction to voltage applied to the gate: they become more resistant to current flow. These MOSFET subclasses can then be further divided according to whether they are P-channel or N-channel transistors (see Figures 3-23a–d).

In N-channel enhancement MOSFETs, the source and drains are N-type (negative charge) semiconductor material and sit on top of P-type material (positive charge). In P-channel enhancement MOSFETs, the source and drains are P-type (positive charge) semiconductor material and sit on top of N-type material (negative charge). When no voltage is applied to the gate, these transistors are in the OFF state (see Figures 3-23a and c), because there is no way for current to flow from the source to the drain (for N-channel enhancement MOSFETs) or from drain to source for P-channel enhancement MOSFETs.

N-channel depletion MOSFETs are in the OFF state when a negative voltage is applied to the gate (as shown in Figure 3-23b) to create a *depletion region*—an area in which no current can flow—making it more difficult for electrons to flow through the transistor because of a smaller available channel for current to flow through. The more negative the voltage applied to the gate, the larger the depletion region and the smaller the channel available for electron flow. As seen in Figure 3-23d, the same holds true for a P-channel depletion MOSFET, except because of the reversed type (polarity) of materials, the voltage applied at the gate to turn the transistor OFF is positive instead of negative.

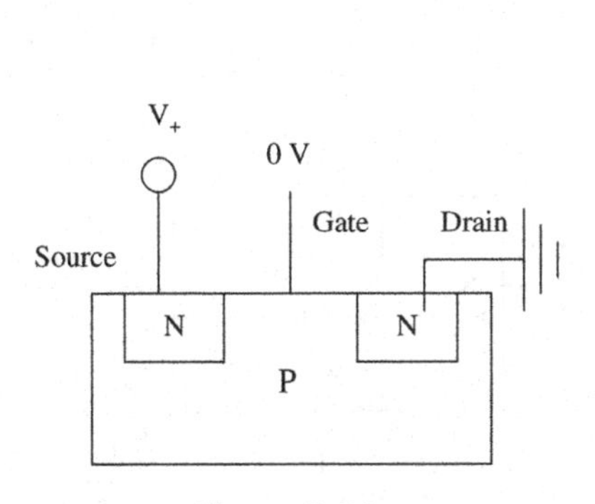


Figure 3-23a
N-channel enhancement MOSFET "OFF."

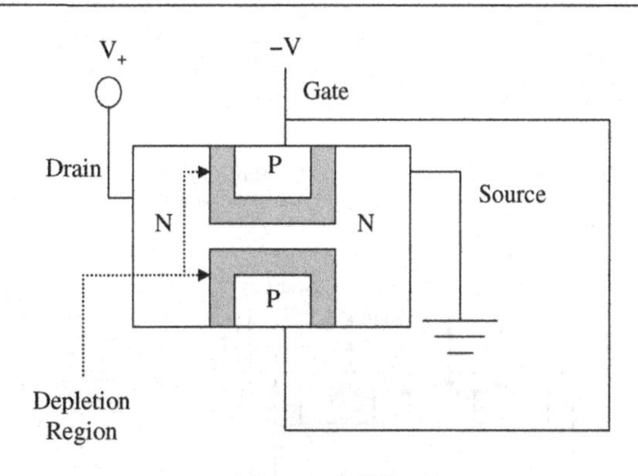


Figure 3-23b
N-channel depletion MOSFET "OFF."

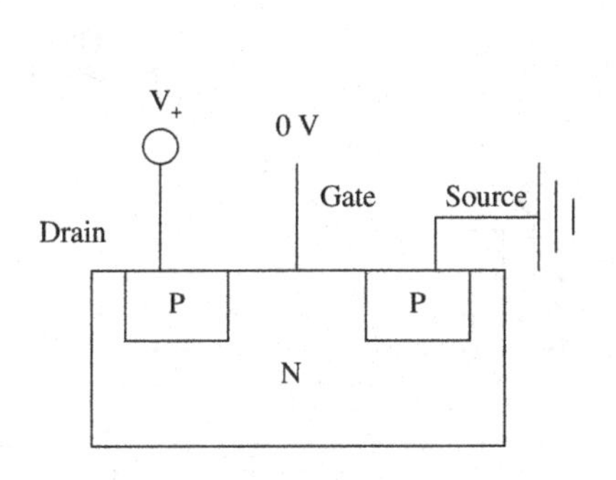


Figure 3-23c
P-channel enhancement MOSFET "OFF."

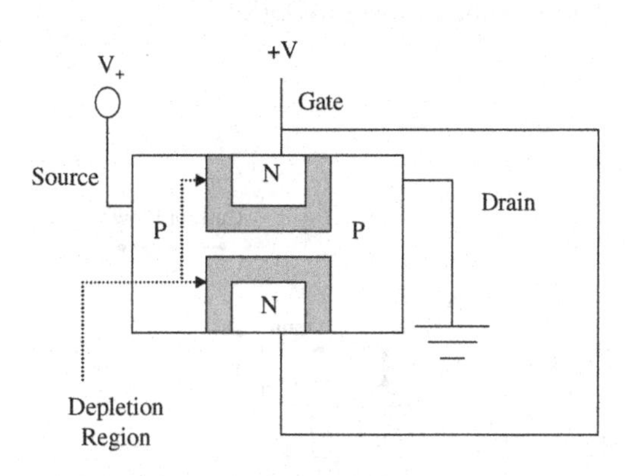


Figure 3-23d
P-channel depletion MOSFET "OFF."

The N-channel enhancement MOSFET is in the ON state when positive voltage is applied to the gate of the transistor. This is because electrons in the P-type material are attracted to the area under the gate when the voltage is applied, creating an electron channel between the drain and source. So, with the positive voltage on the other side of the drain, a current flows from the drain (and gate) to the source over this electron channel. P-channel enhancement MOSFETs, on the other hand, are in the ON state when negative voltage is applied to the gate of the transistor. This is because electrons from the negative voltage source are attracted to the area under the gate when the voltage is applied, creating an electron channel between the source and drain. So, with the positive voltage on the other side of the source, current flows from the source to the drain (and gate) over this electron channel (see Figures 3-24a and c).

Because depletion MOSFETs are inherently conductive, when there is no voltage applied to the gates of an N-channel or P-channel depletion MOSFET, there is a wider channel in which electrons are free to flow through the transistor from, in the case of an N-channel depletion

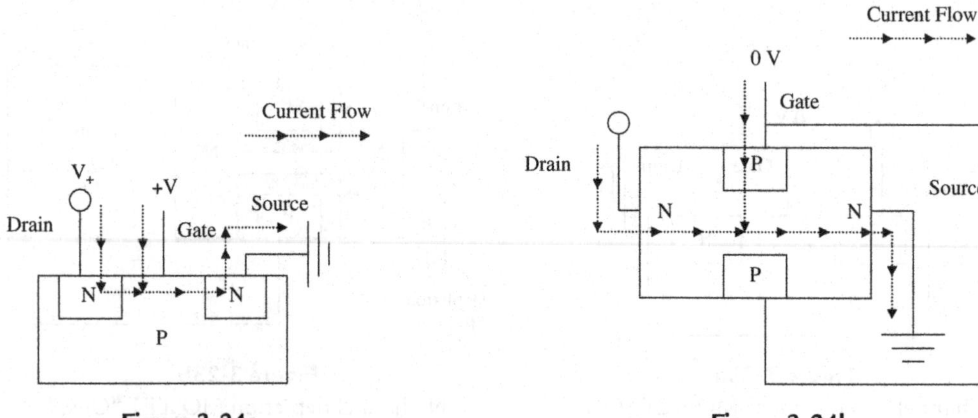


**Figure 3-24a**
N-channel enhancement MOSFET "ON."

**Figure 3-24b**
N-channel depletion MOSFET "ON."

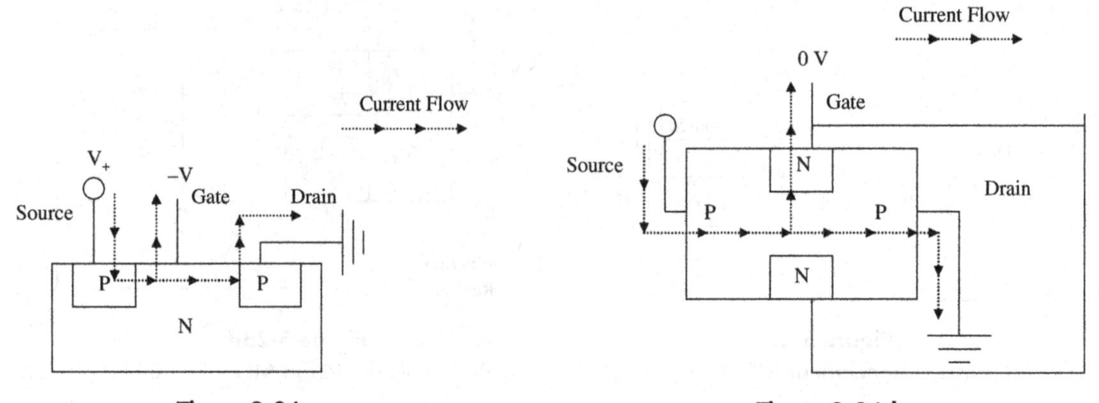


**Figure 3-24c**
P-channel enhancement MOSFET "ON."

**Figure 3-24d**
P-channel depletion MOSFET "ON."

MOSFET, the source to the drain and, in the case of the P-channel depletion MOSFET, the drain to the source. In these cases, the MOSFET depletion transistors are in the ON state (see Figures 3-24b and d).

As seen in Figure 3-25, the schematic symbols for the MOSFET enhancement and depletion N-channel and P-channel transistors contain an arrow that indicates the direction of current flow for N-channel MOSFET depletion and enhancement transistors (into the gate, and with what is coming into the drain, output to the source), and P-channel MOSFET depletion and enhancement transistors (into the source, and out of the gate and drain) when these transistors are ON.

The JFET transistors are subclassed as either N-channel or P-channel JFETs, and like depletion-type MOSFETs, become more resistive to current flow when voltage is applied to

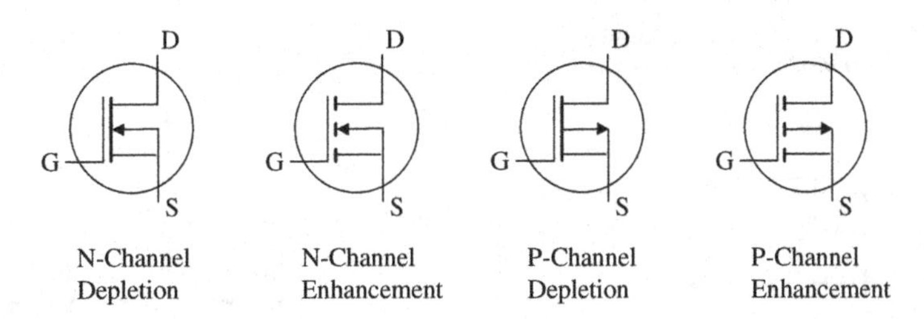


**Figure 3-25**
MOSFET schematic symbols.

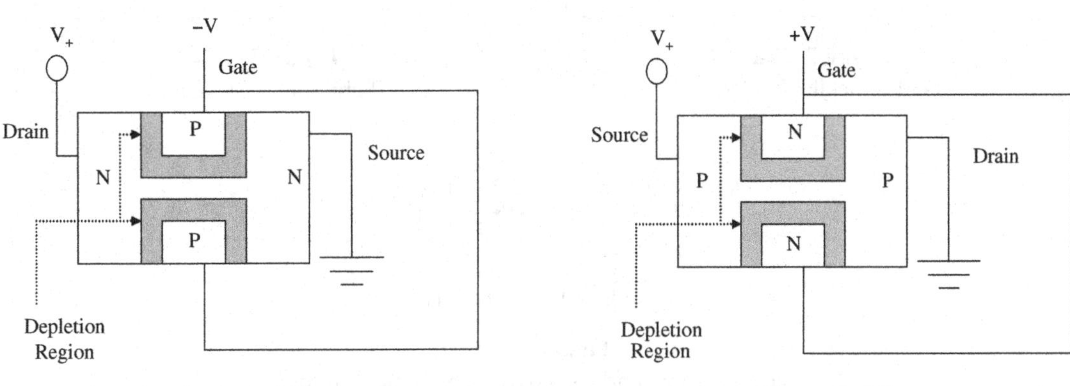


**Figure 3-26a**
N-channel JFET "OFF."

**Figure 3-26b**
P-channel JFET "OFF."

their gates. As shown in Figure 3-26a, an N-channel JFET is made up of the drain and source connecting to N-type material, with the gate connecting to two P-type sections on either side of the N-type material. A P-channel JFET has the opposite configuration, with the drain and source connecting to P-type material, and the gate connecting to two N-type sections on either side of the P-type material (see Figure 3-26b).

In order to turn the N-channel JFET transistor OFF, a negative voltage must be applied to the gate (as shown in Figure 3-26a) to create a depletion region, an area in which no current can flow, making it more difficult for electrons to flow through the transistor because of a smaller available channel for current to flow through. The more negative the voltage applied to the gate, the larger the depletion region, and the smaller the channel available for electron flow. As seen in Figure 3-26b, the same holds true for a P-channel JFET, except because of the reversed type of materials, the voltage applied at the gate to turn the transistor OFF is positive instead of negative.

When there is no voltage applied to the gates of an N-channel or P-channel JFET, there is a wider channel in which electrons are free to flow through the transistor from, in the case of an N-channel JFET, the source to the drain and, in the case of the P-channel JFET, the drain to the source. In this case, the JFET transistors are in the ON state (see Figures 3-27a and b).

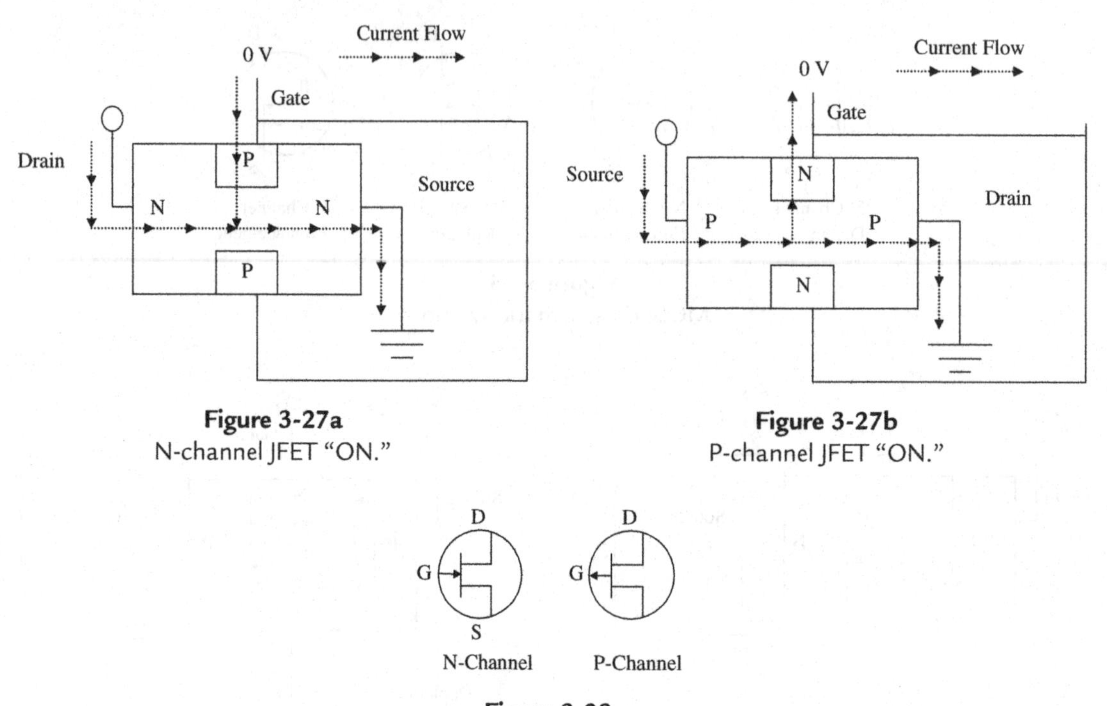


**Figure 3-27a**
N-channel JFET "ON."

**Figure 3-27b**
P-channel JFET "ON."

**Figure 3-28**
JFET N-channel and P-channel schematic symbols.

As seen in Figure 3-28, the schematic symbols for the JFET N-channel and P-channel transistors contain an arrow that indicates the direction of current flow for N-channel (into the gate, and with what is coming into the drain, output to the source) and P-channel (into the source, and out of the gate and drain) when these transistors are ON.

Again, there are other types of transistors (such as unijunction) but essentially the major differences between all transistors include size (FETs can typically be designed to take up less space than BJTs, for instance), price (FETs can be cheaper and simpler to manufacture than BJTs, because they are only controlled via voltage), and usage (FETs and unijunctions are typically used as switches, BJTs in amplification circuits). In short, transistors are some of the most critical elements in the design of more complex circuitry on an embedded board. The following subsections will indicate how they are used.

### *3.6.3 Building More Complex Circuitry from the Basics: Gates*

Transistors that can operate as switches, such as MOSFETs, are operating in one of two positions at any one time: ON (1) or OFF (0). MOSFETs are implemented in a switched electronic circuit in which the switch (the transistor) controls the flow of electrons over the wire by (if an nMOS) being ON (completing the circuit path) or OFF (breaking the circuit path), or vice-versa if a pMOS. It is because embedded hardware communicates

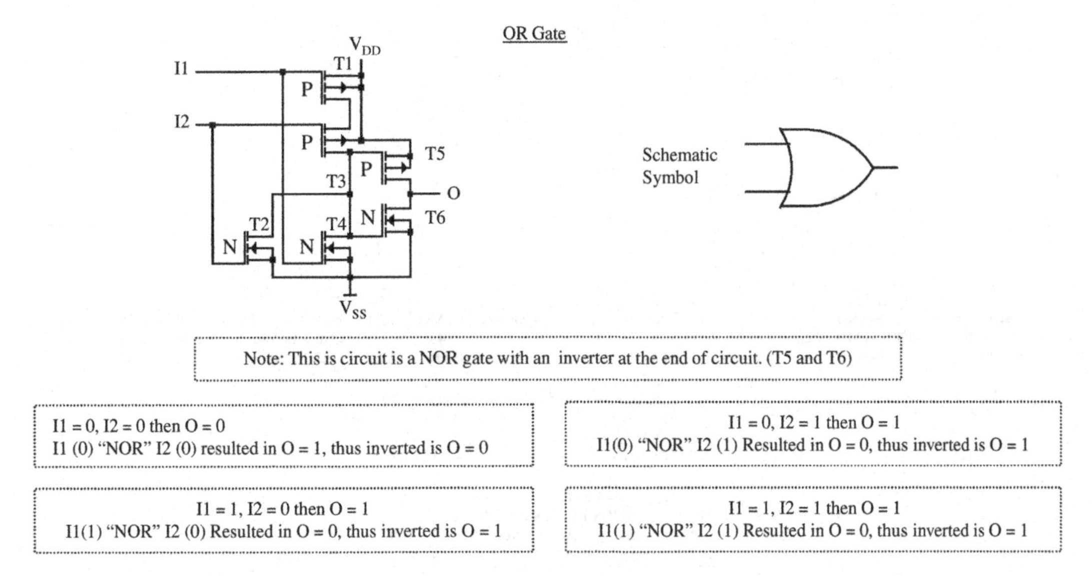
OR Gate
V DD
I1
I2
T1
P
P
T3
P
T5
O
T2
T4
N
T6
N
N
V SS
Schematic Symbol
Note: This is circuit is a NOR gate with an inverter at the end of circuit. (T5 and T6)
I1 = 0, I2 = 0 then O = 0
I1 (0) "NOR" I2 (0) resulted in O = 1, thus inverted is O = 0
I1 = 0, I2 = 1 then O = 1
I1(0) "NOR" I2 (1) Resulted in O = 0, thus inverted is O = 1
I1 = 1, I2 = 0 then O = 1
I1(1) "NOR" I2 (0) Resulted in O = 0, thus inverted is O = 1
I1 = 1, I2 = 1 then O = 1
I1(1) "NOR" I2 (1) Resulted in O = 0, thus inverted is O = 1

**Figure 3-29a**
Truth tables of logical binary operations.

via various combinations of bits (0 s and 1 s) that transistors like the MOSFET are used in circuits that can store or process bits, since these types of transistors can function as a switch that is either a value of "0" or "1." In fact, transistors, along with other electronic components such as diodes and resistors, are the main "building blocks" of more complex types of electronic switching circuits, called logical circuits or *gates*. Gates are designed to perform *logical* binary operations, such as AND, OR, NOT, NOR, NAND, XOR. Being able to create logic circuits is important, because these operations are the basis of all mathematical and logical functions used by the programmer and processed by the hardware. Reflecting logical operations, gates are designed to have one or more input(s) and one output, supporting the requirements to perform logical binary operations. Figures 3-29a and b outline some examples of the truth tables of some logical binary operations, as well as one of the many possible ways transistors (MOSFETs are again used here as an example) can build such gates.

### 3.6.4 Sequential Logic and the Clock

Logic gates can be combined in many different ways to perform more useful and complex logic circuits (called *sequential logic*), such as circuits that have some type of memory. In order to accomplish this, there must be a *sequential* series of procedures to be followed to store and retrieve data at any moment in time. Sequential logic is typically based upon one of two models: a *sequential* or *combinational* circuit design. These models differ in what triggers their gate(s) into changing state, as well as what the results are of a changed

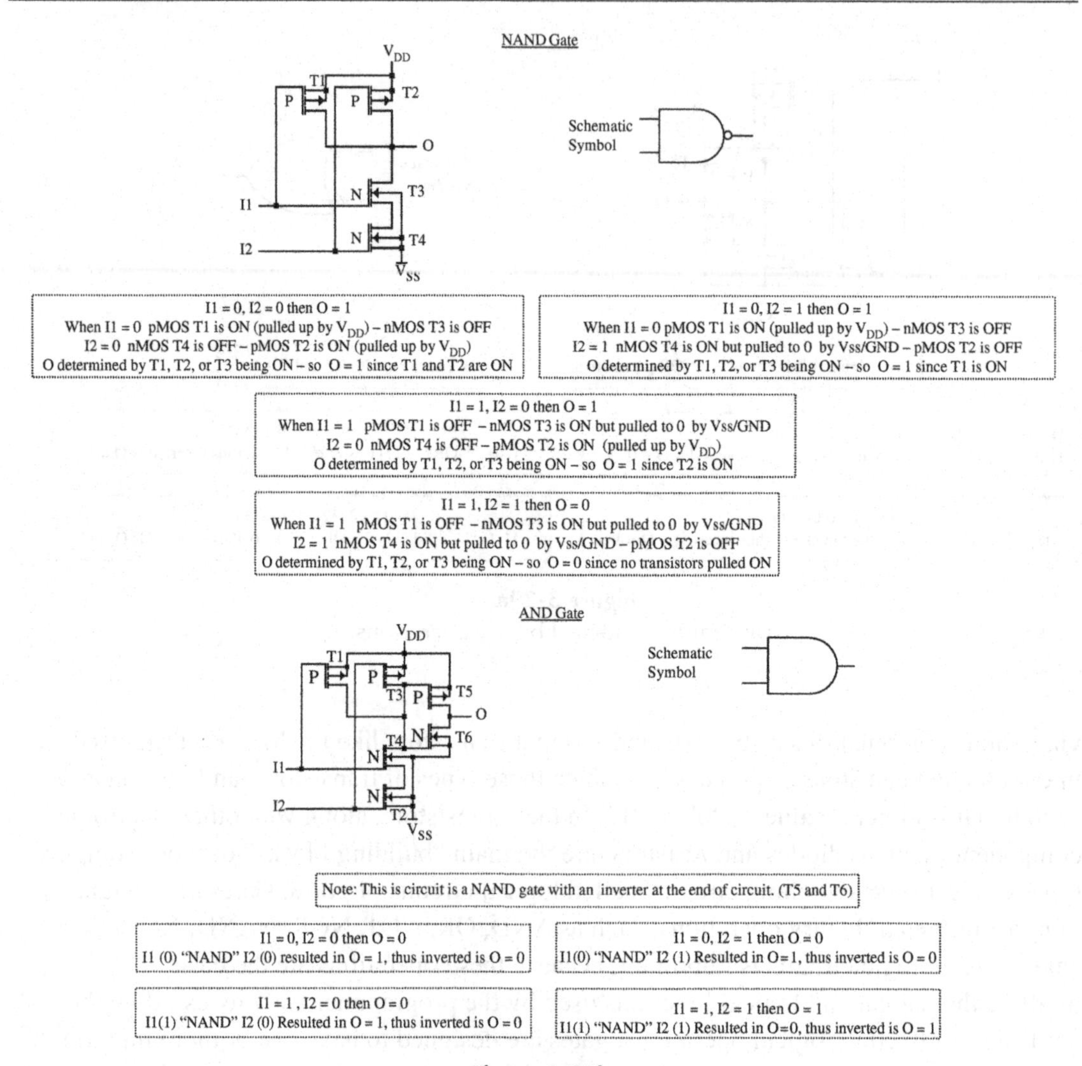

NAND Gate
VDD
T1
P
P
T2
O
N
T3
I1
N
T4
I2
VSS
Schematic Symbol
I1 = 0, I2 = 0 then O = 1
When I1 = 0 pMOS T1 is ON (pulled up by VDD) – nMOS T3 is OFF
I2 = 0 nMOS T4 is OFF – pMOS T2 is ON (pulled up by VDD)
O determined by T1, T2, or T3 being ON – so O = 1 since T1 and T2 are ON
I1 = 0, I2 = 1 then O = 1
When I1 = 0 pMOS T1 is ON (pulled up by VDD) – nMOS T3 is OFF
I2 = 1 nMOS T4 is ON but pulled to 0 by Vss/GND – pMOS T2 is OFF
O determined by T1, T2, or T3 being ON – so O = 1 since T1 is ON
I1 = 1, I2 = 0 then O = 1
When I1 = 1 pMOS T1 is OFF – nMOS T3 is ON but pulled to 0 by Vss/GND
I2 = 0 nMOS T4 is OFF – pMOS T2 is ON (pulled up by VDD)
O determined by T1, T2, or T3 being ON – so O = 1 since T2 is ON
I1 = 1, I2 = 1 then O = 0
When I1 = 1 pMOS T1 is OFF – nMOS T3 is ON but pulled to 0 by Vss/GND
I2 = 1 nMOS T4 is ON but pulled to 0 by Vss/GND – pMOS T2 is OFF
O determined by T1, T2, or T3 being ON – so O = 0 since no transistors pulled ON
AND Gate
VDD
T1
P
P
T3
P
T5
O
T4
N
T6
N
I1
N
I2
T2
VSS
Schematic Symbol
Note: This is circuit is a NAND gate with an inverter at the end of circuit. (T5 and T6)
I1 = 0, I2 = 0 then O = 0
I1 (0) "NAND" I2 (0) resulted in O = 1, thus inverted is O = 0
I1 = 0, I2 = 1 then O = 0
I1(0) "NAND" I2 (1) Resulted in O=1, thus inverted is O = 0
I1 = 1, I2 = 0 then O = 0
I1(1) "NAND" I2 (0) Resulted in O = 1, thus inverted is O = 0
I1 = 1, I2 = 1 then O = 1
I1(1) "NAND" I2 (1) Resulted in O=0, thus inverted is O = 1

**Figure 3-29b**
CMOS (MOSFET) gate transistor design examples.[12].

state (output). All gates exist in some defined "state," which is defined as the current values associated with the gate, as well as any behavior associated with the gate when the values change.

As shown in Figure 3-30, sequential circuits provide output that can be based upon current input values, as well as previous input and output values in a feedback loop. Sequential circuits can change states *synchronously* or *asynchronously* depending on the circuit. Asynchronous sequential circuits change states only when the inputs change. Synchronous sequential circuits change states based upon a *clock signal* generated by a *clock generator* connected to the circuit.

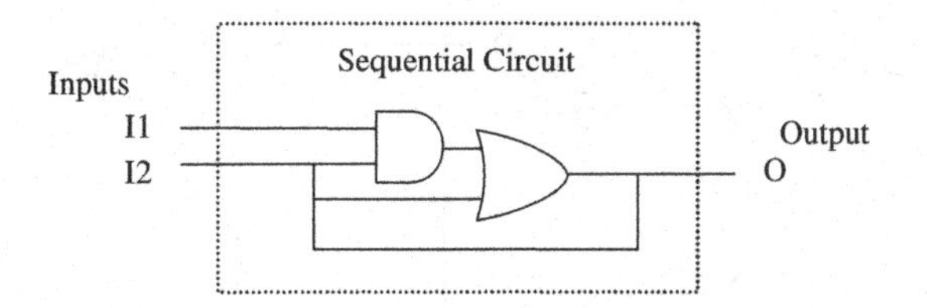


**Figure 3-30**
Sequential circuit diagram.

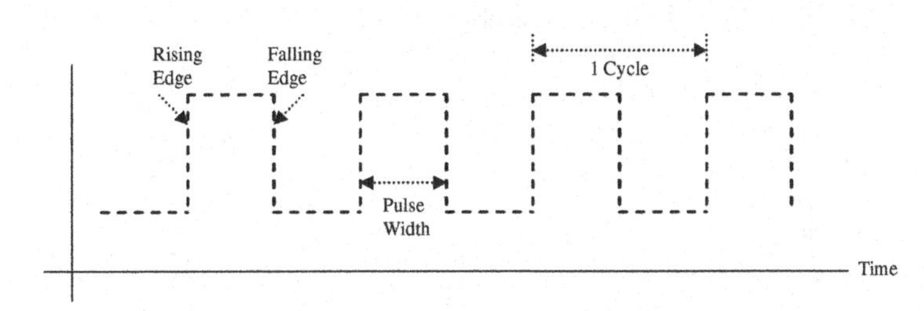


**Figure 3-31**
Clock signal of synchronous sequential circuits.

Almost every embedded board has an oscillator, a circuit whose sole purpose is generating a repetitive signal of some type. Digital clock generators, or simply clocks, are oscillators that generate signals with a square waveform (see Figure 3-31). Different components may require oscillators that generate signals of various waveforms (sinusoidal, pulsed, saw tooth, etc.) to drive them. In the case of components driven by a digital clock, it is the square waveform. The waveform forms a square, because the clock signal is a logical signal that continuously changes from 0 to 1 or 1 to 0. The output of the synchronous sequential circuit is synchronized with that clock.

Commonly used sequential circuits (synchronous and asynchronous) are multivibrators, logic circuits designed so that one or more of its outputs are fed back as input. The subtypes of multivibrators—astable, monostable, or bistable—are based upon the *states* in which they hold stable. Monostable (or one-shot) multivibrators are circuits that have only one stable state, and produce one output in response to some input. The bistable multivibrator has two stable states (0 or 1) and can remain in either state indefinitely, whereas the astable multivibrator has no state in which it can hold stable. *Latches* are examples of bistable multivibrators. Latches are multivibrators, because signals from the output are fed back into inputs, and they are bistable because they have only one of two possible output states they can hold stable at: 0 or 1. Latches come in several different subtypes (S–R, Gated S–R, D Latch, etc.). Figure 3-32 demonstrates how the basic logical gates are combined to make different types of latches.

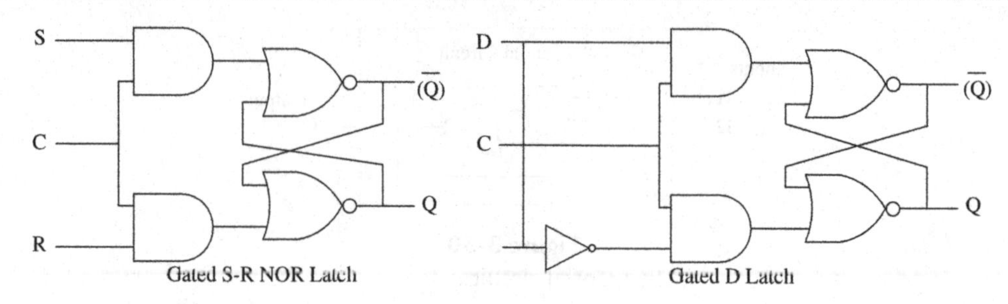


**Figure 3-32**
Latches.[8]

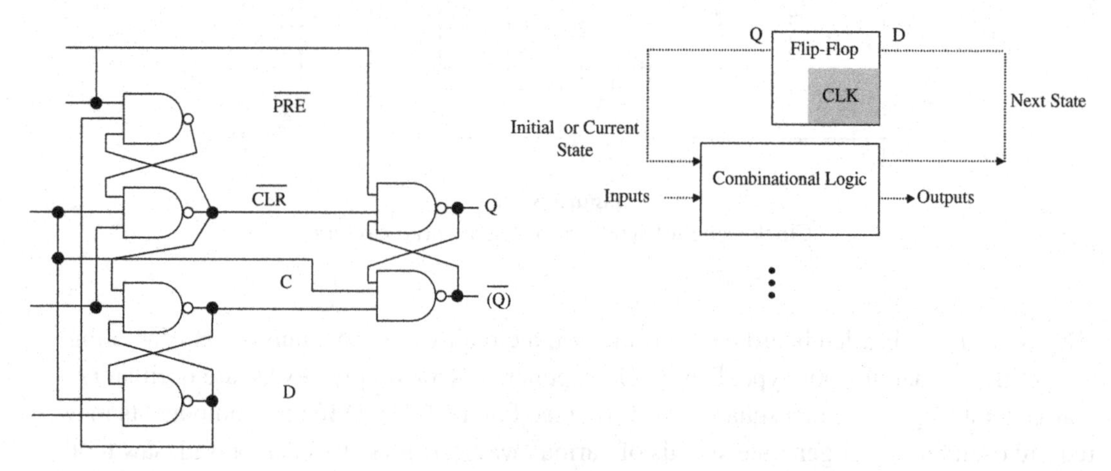


**Figure 3-33**
D flip-flop diagram.[8]

One of the most commonly used types of latches in processors and memory circuitry is the *flip-flop.* Flip-flops are sequential circuits that derived their name because they function by alternating (flip-flopping) between both states (0 and 1), and the output is then switched (e.g., from 0-to-1 or from 1-to-0). There are several types of flip-flops, but all essentially fall under either the asynchronous or synchronous categories. Flip-flops, and most sequential logic, can be made from a variety of different gate combinations, all achieving the same type of results. Figure 3-33 is an example of a synchronous flip-flop, specifically an edge-triggered D flip-flop. This type of flip-flop changes state on the rising edge or falling edge of a square-wave enable signal—in other words, it only changes states, thus changing the output, when it receives a *trigger* from a clock.

Like the sequential circuit, combinational circuits can have one or more input(s) and only one output. However, both models primarily differ in that a combinatorial circuit's output

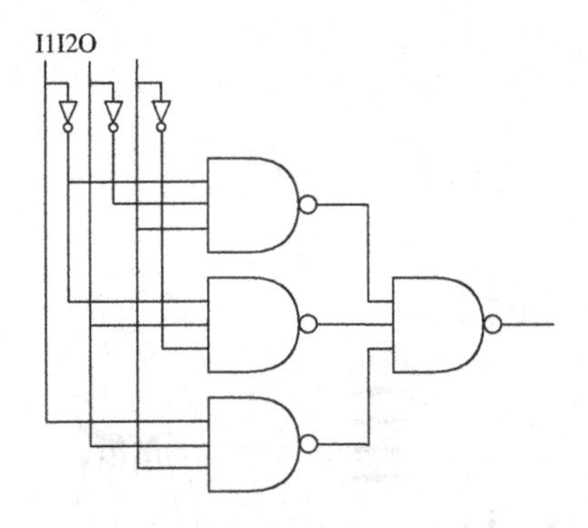


**Figure 3-34**
Combinational circuit (no feedback loop).[9]

is dependent only on inputs applied at that instant, as a function of time, and "no" past conditions. A sequential circuit's output, on the other hand, can be based upon previous outputs being fed back into the input, for instance. Figure 3-34 shows an example of a combinational circuit, which is essentially a circuit with no feedback loop.

All of the various logic gates introduced in the last few sections, along with the other electronic devices discussed in this chapter so far, are the building blocks of more complex circuits that implement everything from the storage of data in memory to the mathematical computations performed on data within a processor. Memory and processors are all inherently complex circuits, explicitly *integrated circuits* (*ICs*).

## 3.7 Putting It All Together: The IC

Gates, along with the other electronic devices that can be located on a circuit, can be compacted to form a single device, called an *IC*. ICs, also referred to as *chips*, are usually classified into groups according to the number of transistors and other electronic components they contain, as follows:

- *SSI (small-scale integration)*: containing up to 100 electronic components per chip.
- *MSI (medium-scale integration)*: containing between 100 and 3000 electronic components per chip.
- *LSI (large-scale integration)*: containing between 3000 and 100 000 electronic components per chip.
- *VLSI (very-large-scale integration)*: containing between 100 000 and 1 000 000 electronic components per chip.
- *ULSI (ultra-large-scale integration)*: containing over 1 000 000 electronic components per chip.

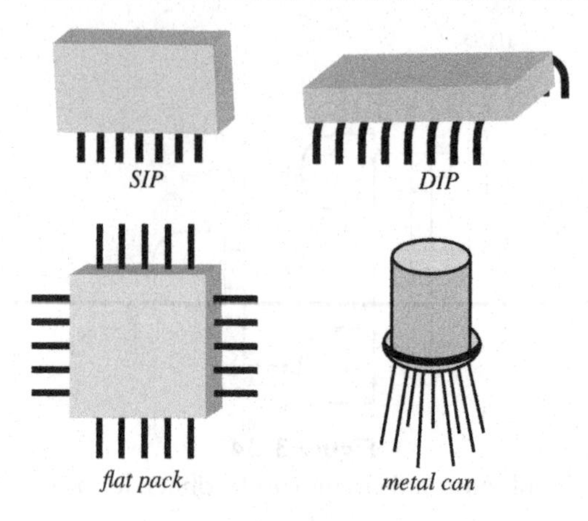


**Figure 3-35**
IC packages.

ICs are physically enclosed in a variety of packages that includes SIP, DIP, flat pack, and others (see Figure 3-35). They basically appear as boxes with pins protruding from the body of the box. The pins connect the IC to the rest of the board.

Physically packaging so many electronic components in an IC has its advantages as well as drawbacks. These include:

- *Size.* ICs are much more compact than their discrete counterparts, allowing for smaller and more advanced designs.
- *Speed.* The buses interconnecting the various IC components are much, much smaller (and thus faster) than on a circuit with the equivalent discrete parts.
- *Power.* ICs typically consume much less power than their discrete counterparts.
- *Reliability.* Packaging typically protects IC components from interference (dirt, heat, corrosion, etc.) far better than if these components were located discretely on a board.
- *Debugging.* It is usually simpler to replace one IC than try to track down one component that failed among 100000 (for example) components.
- *Usability.* Not all components can be put into an IC, especially those components that generate a large amount of heat, such as higher value inductors or high-powered amplifiers.

In short, ICs are the master processors, slave processors, and memory chips located on embedded boards (see Figures 3-36a–e).

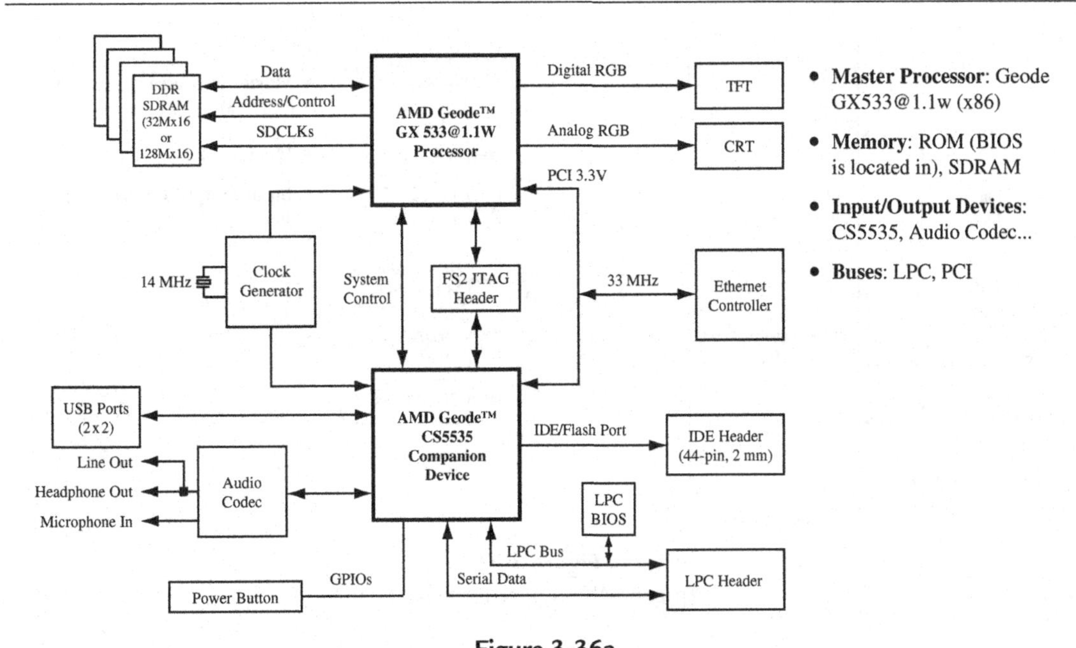


- **Master Processor**: Geode GX533@1.1w (x86)
- **Memory**: ROM (BIOS is located in), SDRAM
- **Input/Output Devices**: CS5535, Audio Codec...
- **Buses**: LPC, PCI

**Figure 3-36a**

AMD/National Semiconductor x86 reference board.[1] *© 2004 Advanced Micro Devices, Inc. Reprinted with permission.*

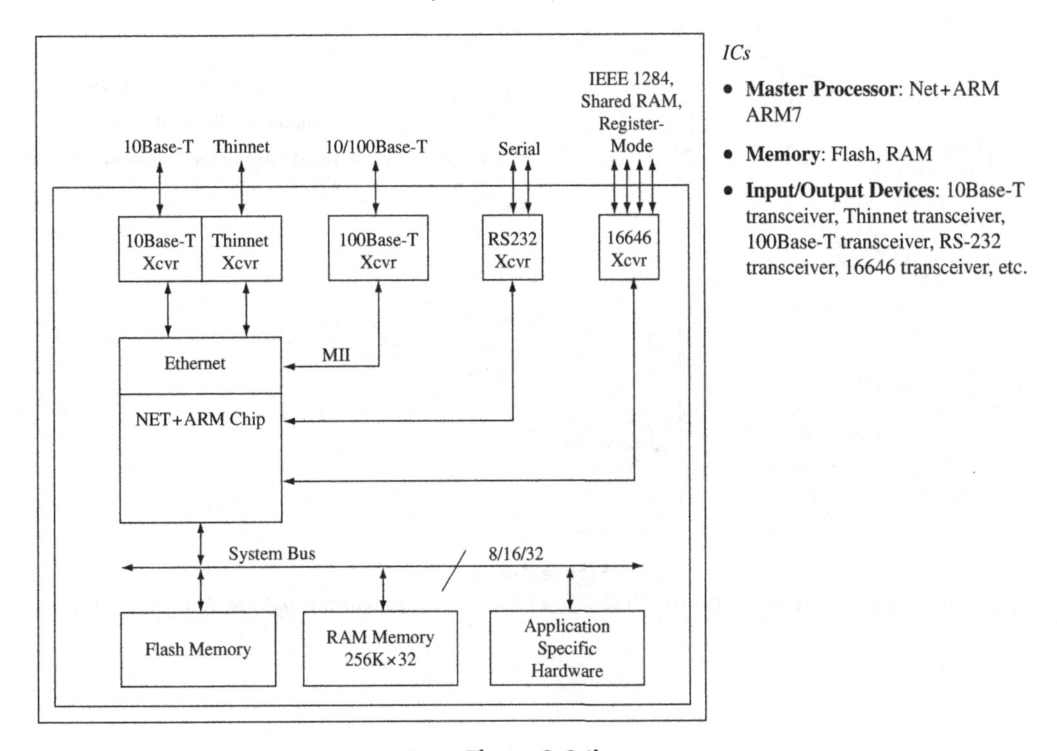


*ICs*

- **Master Processor**: Net+ARM ARM7
- **Memory**: Flash, RAM
- **Input/Output Devices**: 10Base-T transceiver, Thinnet transceiver, 100Base-T transceiver, RS-232 transceiver, 16646 transceiver, etc.

**Figure 3-36b**

Net Silicon ARM7 reference board.[2]

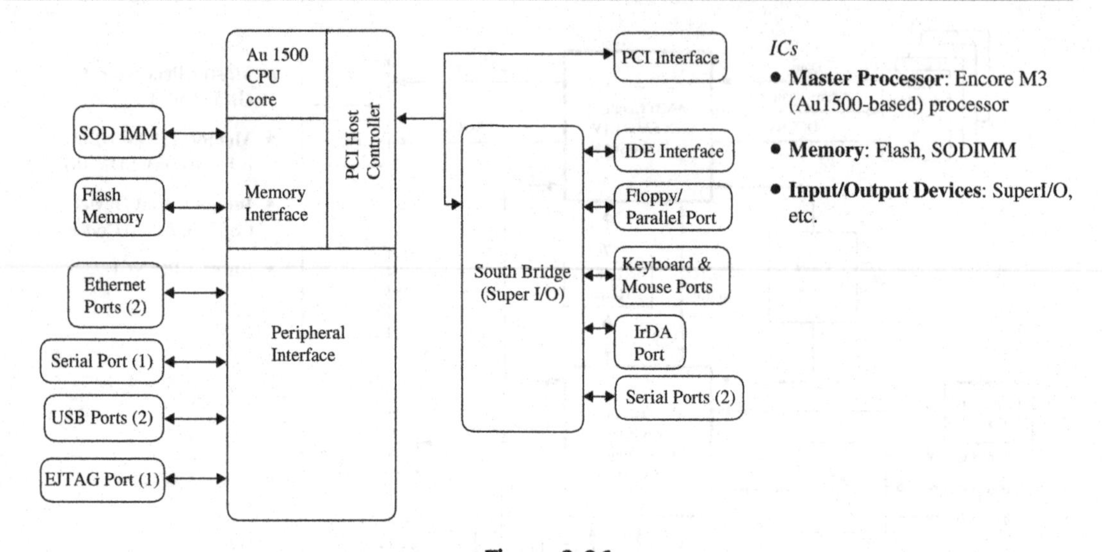


**Figure 3-36c**
Ampro MIPS reference board.[3]

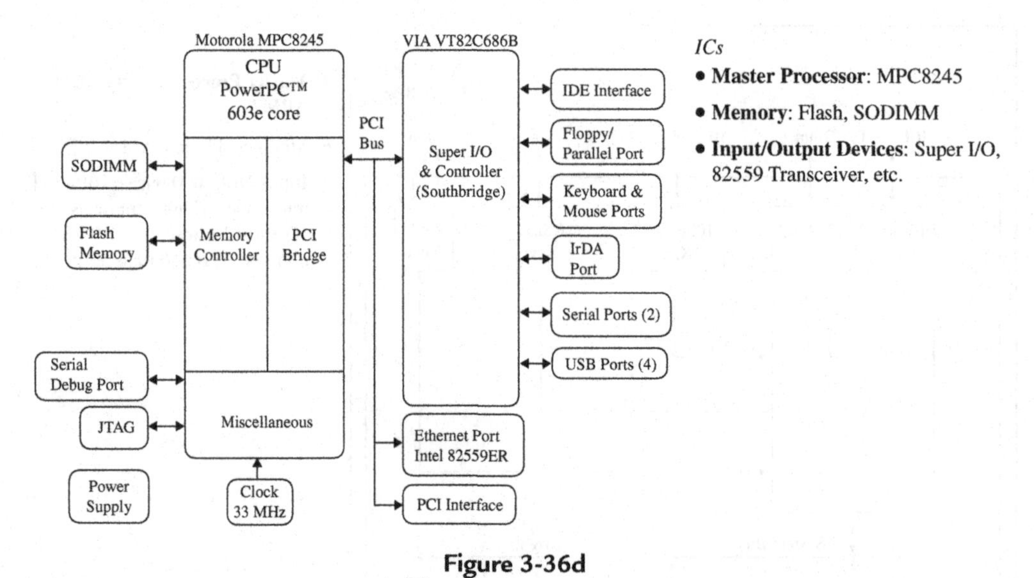


**Figure 3-36d**
Ampro PowerPC reference board.[4] *© 2004 Freescale Semiconductor, Inc. Used by permission.*

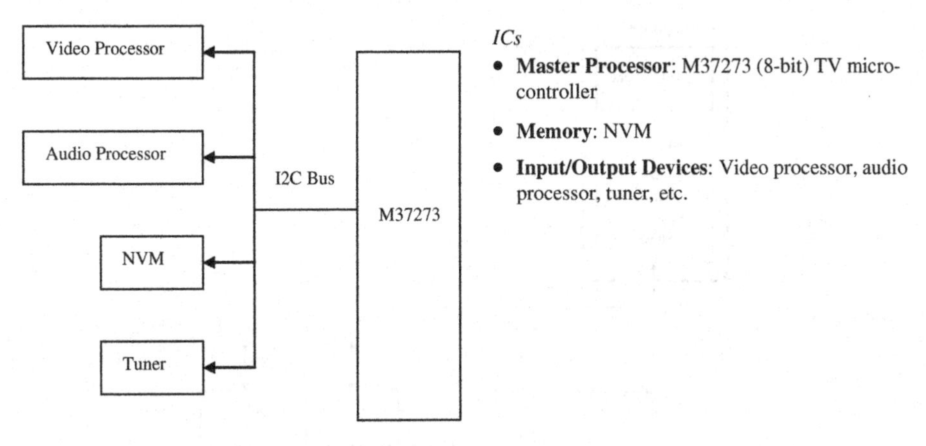


**Figure 3-36e**
Mitsubishi analog TV reference board.

## 3.8 Summary

The purpose of this chapter was to discuss the major functional hardware components of an embedded board. These components were defined as the master processor, memory, I/O, and buses—the basic components that make up the von Neumann model. The passive and active electrical elements that make up the von Neumann components, such as resistors, capacitors, diodes, and transistors, were also discussed in this chapter. It was demonstrated how these basic components can be used to build more complex circuitry, such as gates, flip-flops, and ICs, that can be integrated onto an embedded board. Finally, the importance of and how to read hardware technical documentation, such as timing diagrams and schematics, was introduced and discussed.

The next chapter, Chapter 4, *Embedded Processors*, covers the design details of embedded processors by introducing the different *Instruction Set Architecture (ISA)* models, as well as how the von Neumann model is applicable to implementing an ISA in the internal design of a processor.

## Chapter 3: Problems

1. [a] What is the von Neumann model?
   [b] What are the main elements defined by the von Neumann model?
   [c] Given the block diagrams in Figures 3-37a and b, and data sheet information on the accompanying CD under Chapter 3, files "ePMC-PPC" and "sbcARM7," identify where the major elements in this diagram would fall relative to the von Neumann model.

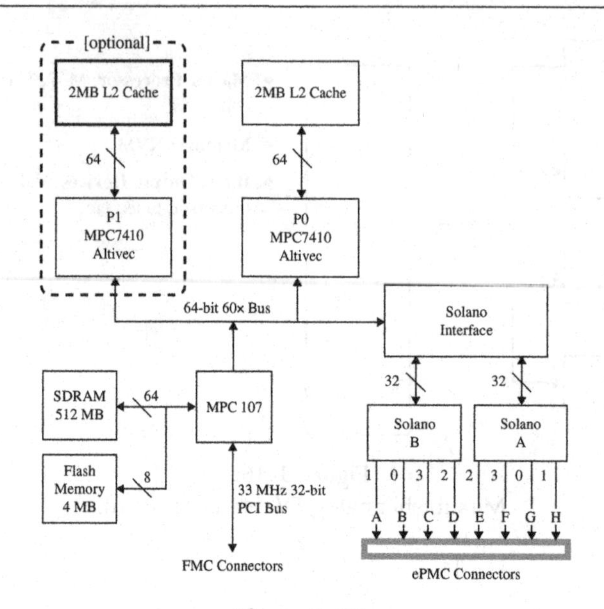


**Figure 3-37a**
PowerPC board block diagram.[13] *© 2004 Freescale Semiconductor, Inc. Used by permission.*

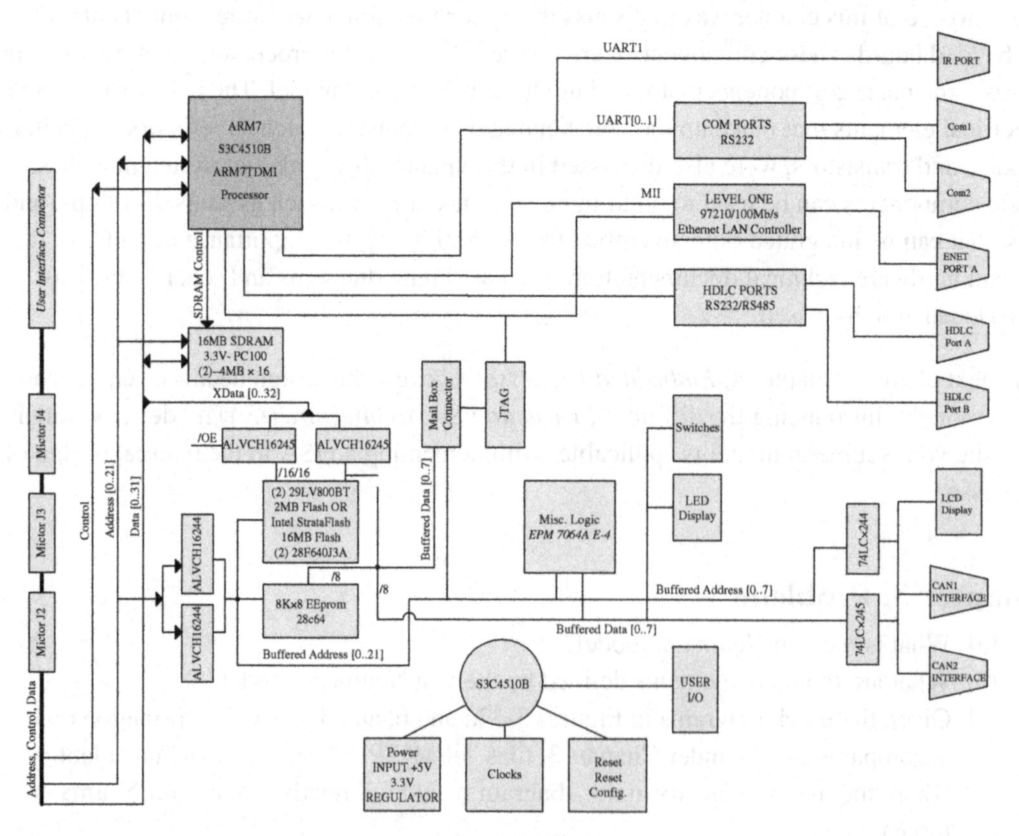


**Figure 3-37b**
ARM board block diagram.[14]

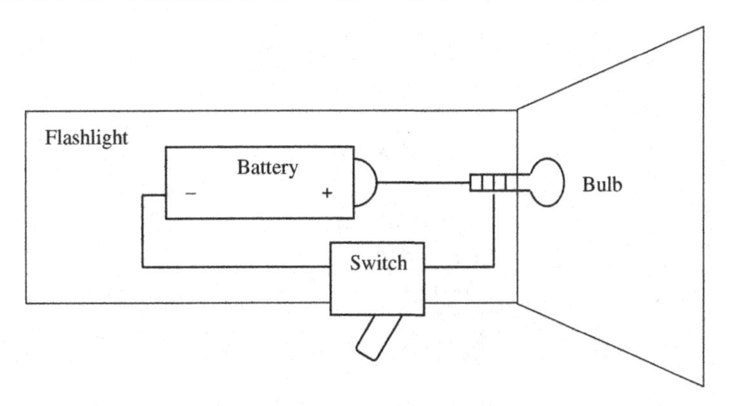


**Figure 3-38**
Simple flashlight.[15]

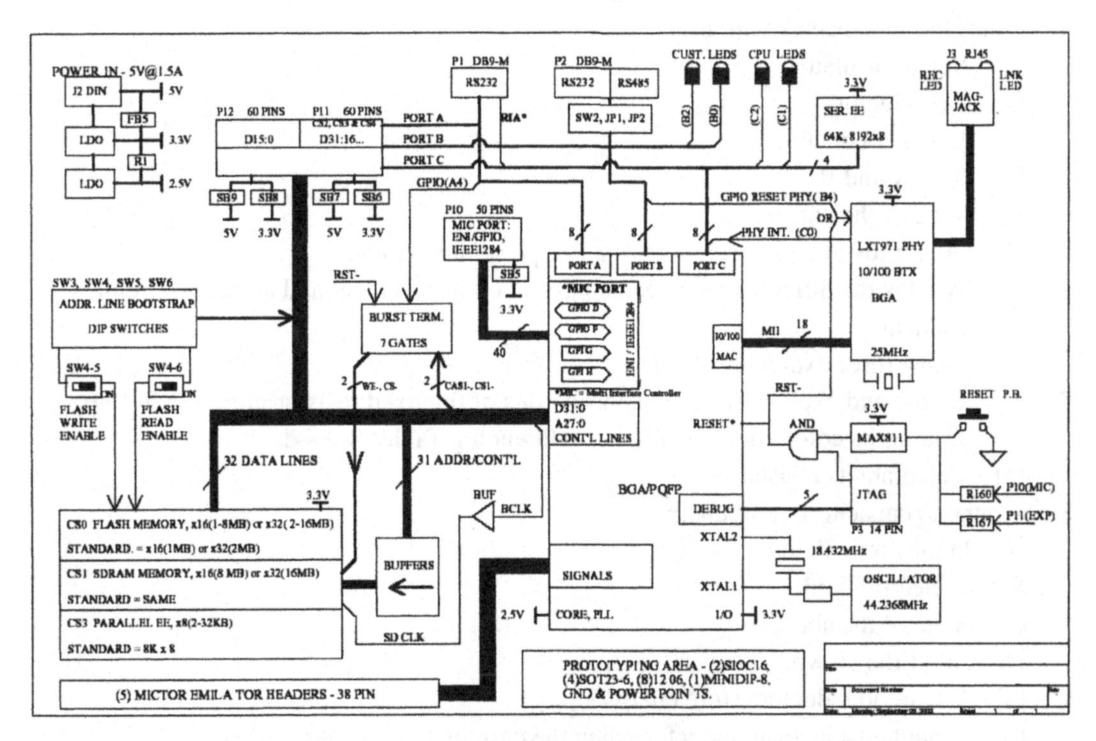


**Figure 3-39**
Schematic diagram example.[7]

2. [a] Given the simple flashlight shown in Figure 3-38, draw a corresponding schematic diagram.
   [b] Read the schematic diagram in Figure 3-39 and identify the symbols therein.
3. [a] What are the basic materials that all components on an embedded board are composed of?
   [b] What are the major differences between these materials?
   [c] Give two examples of each type of material.

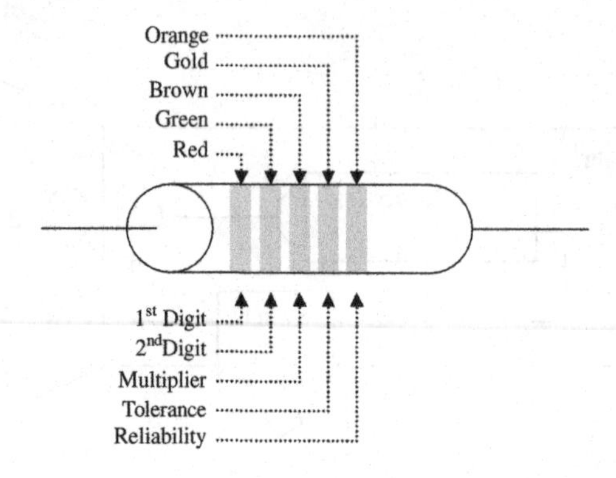


**Figure 3-40**
Fixed resistor.

4. Finish the sentence: A wire is:
   A. Not an insulator.
   B. A conductor.
   C. A semiconductor.
   D. Both A and B.
   E. None of the above.
5. [T/F] A surplus of electrons exists in a P-type semiconductor.
6. [a] What is the difference between a passive circuit element and an active circuit element?
   [b] Name three examples of each.
7. [a] Define and explain the various properties of the fixed resistor in Figure 3-40 by reading its color-coded bands and referencing Tables 3-3a–d.
   [b] Calculate its resistance.
8. Where do capacitors store energy?
   A. In magnetic fields.
   B. In electric fields.
   C. None of the above.
   D. All of the above.
9. [a] Where do inductors store energy?
   [b] What happens to an inductor when the current stream changes?
10. What feature does not affect the inductance of a wire?
    A. The diameter of the wire.
    B. The diameter of the coils.
    C. The number of individual coils.
    D. The type of material the wire is made of.
    E. The overall length of the coiled wire.
    F. None of the above.

**Table 3-3a: Resistor color code digits and multiplier table[6]**

| Color of Band | Digits | Multiplier |
|---|---|---|
| Black | 0 | ×1 |
| Brown | 1 | ×10 |
| Red | 2 | ×100 |
| Orange | 3 | ×1 K |
| Yellow | 4 | ×10 K |
| Green | 5 | ×100 K |
| Blue | 6 | ×1 M |
| Purple | 7 | ×10 M |
| Grey | 8 | ×100 M |
| White | 9 | ×1000 M |
| Silver | – | ×0.01 |
| Gold | – | ×0.1 |

**Table 3-3b: Temperature coefficient[6]**

| Color of Band | Temperature coefficient (p.p.m.) |
|---|---|
| Brown | 100 |
| Red | 50 |
| Orange | 15 |
| Yellow | 25 |

**Table 3-3c: Reliability level[6]**

| Color of Band | Reliability Level (per 1000 hours) |
|---|---|
| Brown | 1 |
| Red | 0.1 |
| Orange | 0.01 |
| Yellow | 0.001 |

**Table 3-3d: Tolerance[6]**

| Color of Band | Tolerance (%) |
|---|---|
| Silver | ±10 |
| Gold | ±5 |
| Brown | ±1 |
| Red | ±2 |
| Green | ±0.5 |
| Blue | ±0.25 |
| Purple | ±0.1 |

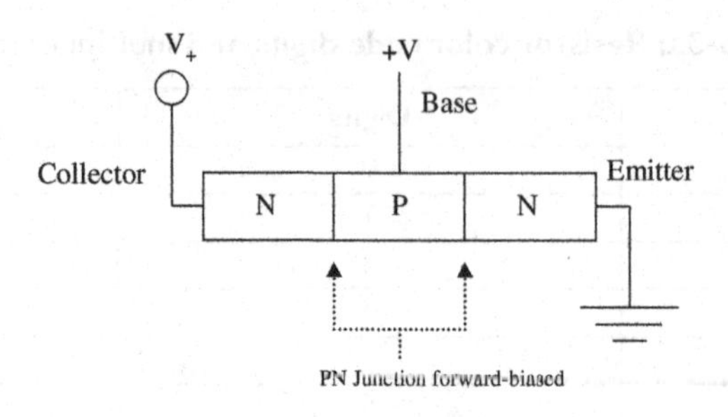


**Figure 3-41**
NPN BJT transistor.

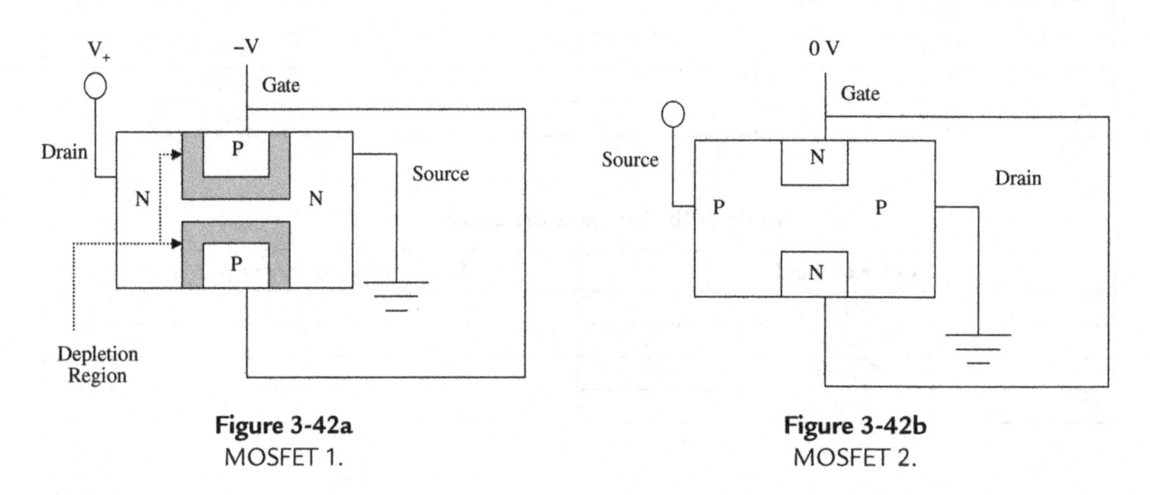


**Figure 3-42a**
MOSFET 1.

**Figure 3-42b**
MOSFET 2.

11. What is the P–N junction?
12. [a] What is an LED?
    [b] How does an LED work?
13. [a] What is a transistor?
    [b] What is a transistor made of?
14. [T/F] The NPN BJT transistor shown in Figure 3-41 is OFF.
15. Which figure, of Figures 3-42a–d, shows a P-channel depletion MOSFET that is ON?
16. [a] What are gates?
    [b] What are gates typically designed to perform?
    [c] Draw the truth tables for the logical binary operations NOT, NAND, and AND.
17. [a] Draw and describe a NOT gate built from CMOS (MOSFET) transistors.
    [b] Draw and describe a NAND gate built from CMOS (MOSFET) transistors.
    [c] Draw and describe an AND gate built from CMOS (MOSFET) transistors. [Hint: this circuit is a NAND gate with an inverter at the end of the circuit.]
18. What is a flip-flop?

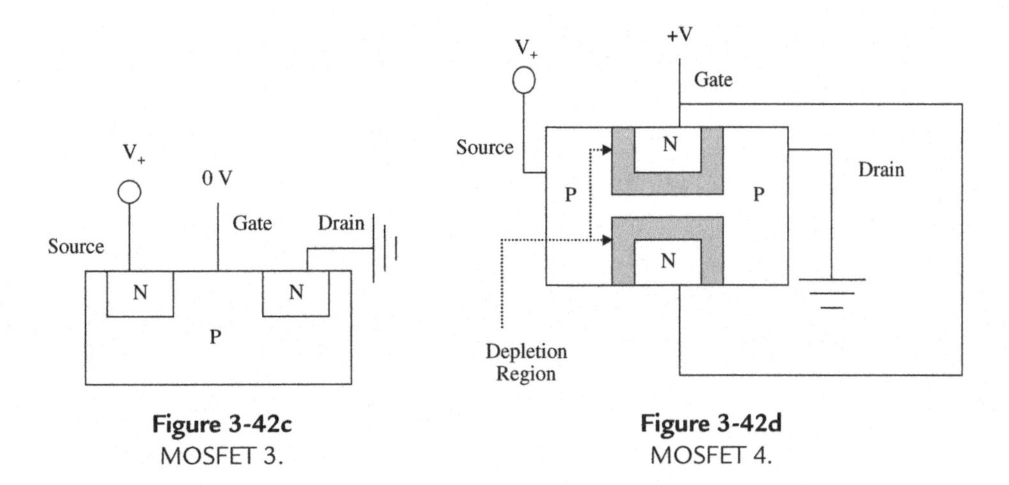


**Figure 3-42c**
MOSFET 3.

**Figure 3-42d**
MOSFET 4.

19. [a] What is an IC?
    [b] Name and describe the classes of ICs according to the number of electronic components they contain.
20. Identify at least five ICs in Figures 3-37a and b under Problem 1 of this section.

## Endnotes


[1] National Semiconductor, Geode User Manual, Revision 1, p. 13.

[2] Net Silicon, Net + ARM40 Hardware Reference Guide, pp. 1–5.

[3] EnCore M3 Embedded Processor Reference Manual, Revision A, p. 8.

[4] EnCore PP1 Embedded Processor Reference Manual, Revision A, p. 9.

[5] *Teach Yourself Electricity and Electronics*, p. 400, S. Gibilisco, McGraw-Hill/TAB Electronics, 3rd edn, 2001.

[6] *Practical Electronics for Inventors*, p. 97, P. Scherz, McGraw-Hill/TAB Electronic, 2000.

[7] Net Silicon, Net50BlockDiagram.

[8] *Electrical Engineering Handbook*, pp. 1637–1638, R. C. Dorf, IEEE Computer Society Press, 2nd edn, 1998.

[9] *Embedded Microcomputer Systems*, p. 509, J. W. Valvano, CL Engineering, 2nd edn, 2006.

[10] *Beginner's Guide to Reading Schematics*, p. 49, R. J. Traister and A. L. Lisk, TAB Books, 2nd edn, 1991.

[11] *Foundations of Computer Architecture*, H. Malcolm. Additional references include: *Computer Organization and Architecture*, W. Stallings, Prentice-Hall, 4th edn, 1995; *Structured Computer Organization*, A. S. Tanenbaum, Prentice-Hall, 3rd edn, 1990; *Computer Architecture*, R. J. Baron and L. Higbie, Addison-Wesley, 1992; *MIPS RISC Architecture*, G. Kane and J. Heinrich, Prentice-Hall, 1992; *Computer Organization and Design: The Hardware/Software Interface*, D. A. Patterson and J. L. Hennessy, Morgan Kaufmann, 3rd edn, 2005.

[12] Intro to electrical engineering website, http://ece-web.vse.gmu.edu/people/full-time-faculty/bernd-peter-paris.

[13] Spectrum signal processing, http://www.spectrumsignal.com/Products/_Datasheets/ePMC-PPC_datasheet.asp.

[14] Wind River, Hardware Reference Designs for ARM7, Datasheet.

[15] *Beginner's Guide to Reading Schematics*, p. 52, R. J. Traister and A. L. Lisk, TAB Books, 2nd edn, 1991.

Figure 3-42c [illegible]

Figure 3-42d [illegible]

9. [illegible] What is an IC?
10. [illegible] and describe the [illegible] of ICs [illegible] in [illegible] components they contain.
10. Identify at least [illegible] ICs in Figures [illegible]

## Endnotes


[1] [illegible] Semiconductor, [illegible] User Manual, Revision [illegible]
[2] [illegible]
[3] [illegible] Reference [illegible] Revision [illegible]
[4] [illegible]
[5] [illegible]
[6] [illegible]
[7] [illegible]
[8] [illegible]
[9] [illegible]
[10] [illegible]
[11] [illegible]
[12] [illegible]
[13] [illegible]
[14] [illegible] Hardware Reference Designs for [illegible]
[15] [illegible]

CHAPTER 4

# Embedded Processors

## In This Chapter

- Discussing what an ISA is and what it defines
- Discussing the internal processor design as related to the von Neumann model
- Introducing processor performance

Processors are the main functional units of an embedded board, and are primarily responsible for processing instructions and data. An electronic device contains at least one *master* processor, acting as the central controlling device, and can have additional *slave* processors that work with and are controlled by the master processor. These slave processors may either extend the instruction set of the master processor or act to manage memory, buses, and I/O (input/output) devices. In the block diagram of an x86 reference board, shown in Figure 4-1, the Atlas STPC is the master processor, and the super I/O and Ethernet controllers are slave processors.

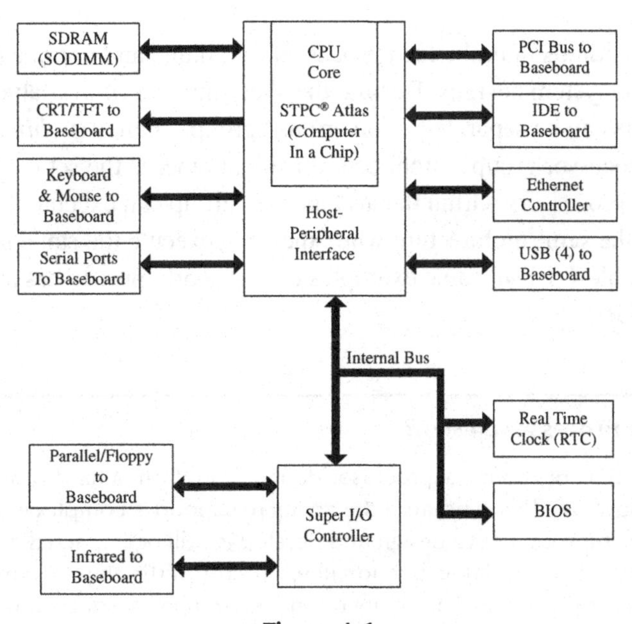


**Figure 4-1**
Ampro's Encore 400 board.[1]

As shown in Figure 4-1, embedded boards are designed around the master processor. The complexity of the master processor usually determines whether it is classified as a *microprocessor* or a *microcontroller*. Traditionally, microprocessors contain a minimal set of integrated memory and I/O components, whereas microcontrollers have most of the system memory and I/O components integrated on the chip. However, keep in mind that these traditional definitions may not strictly apply to recent processor designs. For example, microprocessors are increasingly becoming more integrated.

**Why Use an Integrated Processor?**

While *some* components, like I/O, may show a decrease in performance when integrated into a master processor as opposed to remaining a dedicated slave chip, many others show an increase in performance because they no longer have to deal with the latencies involved with transmitting data over buses between processors. An integrated processor also simplifies the entire board design since there are fewer board components, resulting in a board that is simpler to debug (fewer points of failure at the board level). The power requirements of components integrated into a chip are typically a lot less than for those same components implemented at the board level. With fewer components and lower power requirements, an integrated processor may result in a smaller and cheaper board. On the flip side, there is less flexibility in adding, changing, or removing functionality since components integrated into a processor cannot be changed as easily as if they had been implemented at the board level.

There are literally hundreds of embedded processors available and not one of them currently dominates embedded system designs. Despite the sheer number of available designs, embedded processors can be separated into various "groups" called *architectures*. What differentiates one processor group's architecture from another is the set of machine code instructions that the processors within the architecture group can execute. Processors are considered to be of the same architecture when they can execute the same set of machine code instructions. Table 4-1 lists some examples of real-world processors and the architecture families they fall under.

**Why Care About Processer Design?**

It is very important to care about the processor design, both from a hardware engineer's *and* a programmer's perspective. This is because the ability to support a complex embedded systems design as well as the time it takes to design and develop it, will be impacted by the *Instruction Set Architecture (ISA)* in terms of available functionality, the cost of the chip, and most importantly the performance of the processor. The ability to understand processor performance, and what to look for in a processor's design according to what needs to be accomplished via software,

**Table 4-1: Real-World Architectures and Processors**

| Architecture | Processor | Manufacturer |
|---|---|---|
| AMD | Au1xxx | Advanced Micro Devices, ... |
| ARM | ARM7, ARM9, ... | ARM, ... |
| C16X | C167CS, C165H, C164CI, ... | Infineon, ... |
| ColdFire | 5282, 5272, 5307, 5407, ... | Motorola/Freescale, ... |
| I960 | I960 | Vmetro, ... |
| M32/R | 32170, 32180, 32182, 32192, ... | Renesas/Mitsubishi, ... |
| M Core | MMC2113, MMC2114, ... | Motorola/Freescale |
| MIPS32 | R3K, R4K, 5K, 16, ... | MTI4kx, IDT, MIPS Technologies, ... |
| NEC | Vr55xx, Vr54xx, Vr41xx | NEC Corporation, ... |
| PowerPC | 82xx, 74xx, 8xx, 7xx, 6xx, 5xx, 4xx | IBM, Motorola/Freescale, ... |
| 68k | 680x0 (68K, 68030, 68040, 68060, ...), 683xx | Motorola/Freescale, ... |
| SuperH (SH) | SH3 (7702, 7707, 7708, 7709), SH4 (7750) | Hitachi, ... |
| SHARC | SHARC | Analog Devices, Transtech DSP, Radstone, ... |
| strongARM | strongARM | Intel, ... |
| SPARC | UltraSPARC II | Sun Microsystems, ... |
| TMS320C6xxx | TMS320C6xxx | Texas Instruments, ... |
| x86 | X86 [386, 486, Pentium (II, III, IV)...] | Intel, Transmeta, National Semiconductor, Atlas, ... |
| TriCore | TriCore1, TriCore2, ... | Infineon, ... |

is the key to successfully taking an embedded system to production. This means accepting that processor performance is inherently a combination of:

- *Availability*: length of time a processor runs in normal mode without a failure.
- *Recoverability*: the mean time to recover (MTTR), which is the average time the processor takes to recover from failure.
- *Reliability*: the mean time between failures (MTBF), which is the average time between failures.
- *Responsiveness*: the latency of the processor, which is the length of elapsed time that the processor takes to respond to some event.
- *Throughput*: the processor's average rate of execution, which is the amount of work completed by the processor within a given time period. For example, CPU throughput (in bytes/s or MB/s) = 1/(CPU execution time):
  - CPU execution time (in seconds per total number of bytes) = (total number of instructions)*(CPI)*(clock period) = ((instruction count)*(CPI))/(clock rate). Performance (Processor "1")/performance (Processor "2") = execution time

(Processor "1")/execution time (Processor "2") = "X," meaning that Processor "1" is "X" times faster than Processor "2;"
- CPI = number of cycle cycles/instruction;
- clock period = seconds per cycle;
- clock rate = MHz.

An increase of the overall performance of a processor can be accomplished via an internal processor design feature, such as pipelining within the processor's *ALU (Arithmetic Logic Unit)* or a processor based on the instruction-level parallelism ISA mode. These types of features are what would allow for either an increase in the clock rate or a decrease in the CPI relative to processor performance.

## 4.1 ISA Architecture Models

The *features* that are built into an architecture's instruction set are commonly referred to as the *ISA*. The ISA defines such features as the *operations* that can be used by programmers to create programs for that architecture, the *operands* (data) that are accepted and processed by an architecture, *storage*, *addressing modes* used to gain access to and process operands, and the handling of *interrupts*. These features are described in more detail below, because an ISA implementation is a determining factor in defining important characteristics of an embedded design, such as performance, design time, available functionality, and cost.

### 4.1.1 Features

*Operations*

*Operations* are made up of one or more instructions that execute certain commands. *(Note: Operations are commonly referred to simply as instructions.)* Different processors can execute the exact same operations using a different number and different types of instructions. An ISA typically defines the types and formats of operations.

- *Types of operations*. Operations are the functions that can be performed on the data, and they typically include computations (math operations), movement (moving data from one memory location/register to another), branches (conditional/unconditional moves to another area of code to process), I/O operations (data transmitted between I/O components and master processor), and context switching operations (where location register information is temporarily stored when switching to some routine to be executed and after execution, by the recovery of the temporarily stored information, there is a switch back to executing the original instruction stream). The instruction set on a popular lower-end processor, the 8051, includes just over 100 instructions for math, data transfer, bit variable manipulation, logical operations, branch flow and control, etc. In comparison, a higher-end MPC823 (Motorola/Freescale PowerPC) has an instruction

| *Math and Logical* | *Shift/Rotate* | *Load/Store* | *Compare Instructions ...*<br>*Move Instructions ...*<br>*Branch Instructions ...*<br>..... |
|---|---|---|---|
| Add<br>Subtract<br>Multiply<br>Divide<br>AND<br>OR<br>XOR<br>..... | Logical Shift Right<br>Logical Shift Left<br>Rotate Right<br>Rotate Left<br>..... | Stack PUSH<br>Stack POP<br>Load<br>Store<br>..... | |

**Figure 4-2a**
Sample ISA operations

*CMP crfD, L, rA, rB ...*

```
a ← EXTS(rA)
b ← EXTS(rB)
if a<b then c ← 0b100
else if a>b then  c ← 0b010
else c ← 0b001
CR[4 * crfD-4 *crfD +3] ← c || XER[SO}
```

**Figure 4-2b**
MPC823 compare operation.[2] 

set a little larger than that of the 8051, but with many of the same types of operations contained in the 8051 set along with an additional handful, including integer operations/floating-point (math) operations, load and store operations, branch and flow control operations, processor control operations, memory synchronization operations, and PowerPC VEA operations. Figure 4-2a lists examples of common operations defined in an ISA. In short, different processors can have similar types of operations, but usually have different overall instruction sets. As mentioned above, what is also important to note is that different architectures can have operations with the same purpose (add, subtract, compare, etc.), but the operations may have different names or internally operate very differently, as seen in Figures 4-2b and c.

- *Operation formats.* The format of an operation is the actual number and combination of bits (1s and 0s) that represent the operation, and is commonly referred to as the operation code or *opcode*. MPC823 opcodes, for instance, are structured the same and are all 6 bits long (0-63 decimal) (see Figure 4-3a). MIPS32/MIPS I opcodes are also 6 bits long, but the opcode can vary as to where it is located, as shown in Figure 4-3b. An architecture, like the SA-1100 which is based upon the ARM v4 Instruction Set, can have several instruction set formats depending on the type of operation being performed (see Figure 4-3c).

### *Operands*

Operands are the data that operations manipulate. An ISA defines the types and formats of operands for a particular architecture. For example, in the case of the MPC823

*C.cond.S fs, ft*
*C.cond.D fs, ft ...*

```
if SNaN(ValueFPR(fs, fmt)) or SNaN(ValueFPR(ft, fmt)) or
QNaN(ValueFPR(fs, fmt)) or QNaN(ValueFPR(ft, fmt)) then
less ← false
equal ← false
unordered ← true
if (SNaN(ValueFPR(fs,fmt)) or SNaN(ValueFPR(ft,fmt))) or
(cond3 and (QNaN(ValueFPR(fs,fmt)) or QNaN(ValueFPR(ft,fmt)))) then
SignalException(InvalidOperation)
endif
else
less ←ValueFPR(fs, fmt) <fmt ValueFPR(ft, fmt)
equal ←ValueFPR(fs, fmt) =fmt ValueFPR(ft, fmt)
unordered ← false
endif
condition ← (cond2 and less) or (cond1 and equal)
or (cond0 and unordered)
SetFPConditionCode(cc, condition)
```

**Figure 4-2c**
MIPS32/MIPSI compare operation.[3]

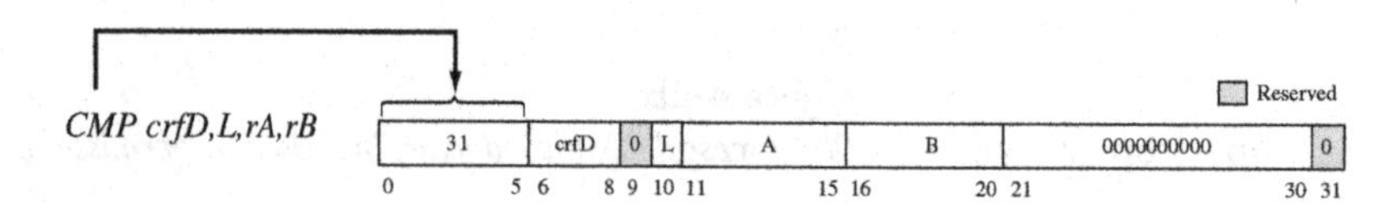


**Figure 4-3a**
MPC823 "CMP" operation size.[2] *© 2004 Freescale Semiconductor, Inc. Used by permission.*

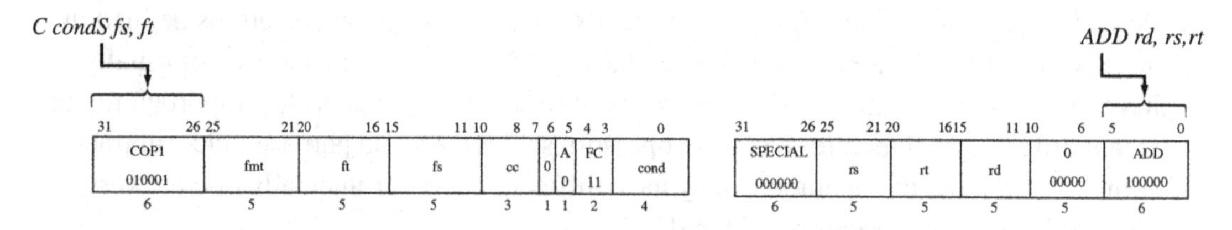


**Figure 4-3b**
MIPS32/MIPSI "CMP" and "ADD" operation sizes and locations.[4]

(Motorola/Freescale PowerPC), SA-1110 (Intel StrongARM), and many other architectures, the ISA defines simple operand types of bytes (8 bits), half words (16 bits), and words (32 bits). More complex data types such as integers, characters, or floating point are based on the simple types shown in Figure 4-4.

An ISA also defines the operand formats (how the data looks) that a particular architecture can support, such as binary, decimal, and hexadecimal. See Figure 4-5 for an example showing how an architecture can support various operand formats.

| Instruction Type | 31 | 2827 | 1615 | | 87 | | 0 |
|---|---|---|---|---|---|---|---|
| Data Processing1/PSR Transfer | Cond | 0 0 I Opcode S | Rn | Rd | Operand2 | | |
| Multiply | Cond | 0 0 0 0 0 0 A S | Rd | Rn | Rs | 1 0 0 1 | Rm |
| Long Multiply | Cond | 0 0 0 0 1 U A S | RdHi | RdLo | Rs | 1 0 0 1 | Rm |
| Swap | Cond | 0 0 0 1 0 B 0 0 | Rn | Rd | 0 0 0 0 | 1 0 0 1 | Rm |
| Load & Store Byte/Word | Cond | 0 1 I P U B W L | Rn | Rd | Offset | | |
| Halfword Transfer: Immediate Offset | Cond | 1 0 0 P U S W L | Rn | Register List | | | |
| Halfword Transfer: Register Offset | Cond | 0 0 0 P U 1 W L | Rn | Rd | Offset 1 | 1 S H 1 | Offset2 |
| Branch | Cond | 0 0 0 P U 0 W L | Rn | Rd | 0 0 0 0 | 1 S H 1 | Rm |
| Branch Exchange | Cond | 1 0 1 L Offset | | | | | |
| Coprocessor Data Transfer | Cond | 0 0 0 1 0 0 1 0 | 1 1 1 1 | 1 1 1 1 | 1 1 1 1 | 0 0 0 1 | Rn |
| Coprocessor Data Operation | Cond | 1 1 0 P U N W L | Rn | CRd | CPNum | Offset | |
| Coprocessor Register Transfer | Cond | 1 1 1 0 Op1 | CRn | CRd | CPNum | Op2 0 | CRm |
| Software Interrupt | Cond | 1 1 1 0 Op1 L | CRn | Rd | CPNum | Op2 1 | CRm |
| ... | Cond | 1 1 1 1 SWI Number | | | | | |

***1 - Data Processing OpCodes***

0000 = AND – Rd: = Op1 AND Op2
0001 = EOR – Rd: = Op1 EOR Op2
0010 = SUR – Rd: = Op1 – Op2
0011 = RSB – Rd: = Op2 – Op1
0100 = ADD – Rd: = Op1 + Op2
0101 = ADC – Rd: = Op1 + Op2 + C
0110 = SEC – Rd: = Op2 – Op1 + C –1
0111 = RSC – Rd: = Op2 – Op1 + C – 1
1000 = TST – set condition codes on Op1 AND Op2
1001 = TEQ – set condition codes on Op1 EOR Op2
1010 = CMP – set condition codes on Op1 – Op2
1011 = CMN – set condition codes on Op1 + Op2
1100 = ORR – Rd: = Op1 OR Op2
1101 = MOV – Rd: = Op2
1110 = BIC – Rd: = Op1 AND NOT Op2
1111 = MVN – Rd: = NOT Op2

**Figure 4-3c**
SA-1100 instruction.[5]

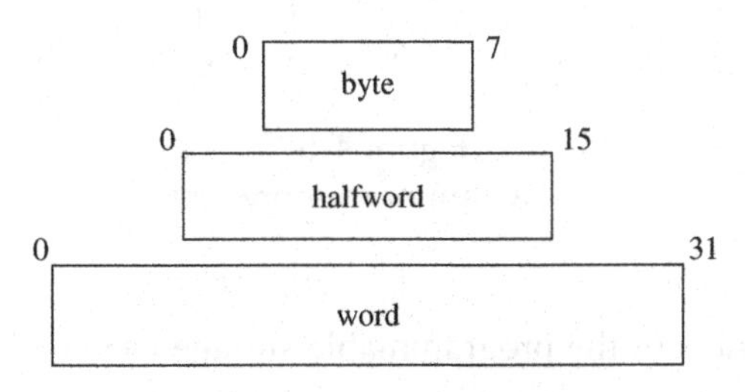


**Figure 4-4**
Simple operand types.

```
MOV     registerX, 10d         ; Move decimal value 10 into register X
MOV     registerX, $0Ah        ; Move hexadecimal value A (decimal 10) to registerX
MOV     registerX, 00001010b   ; Move binary value 00001010 (decimal 10)  to registerX
.....
```

**Figure 4-5**
Operand formats pseudocode example.

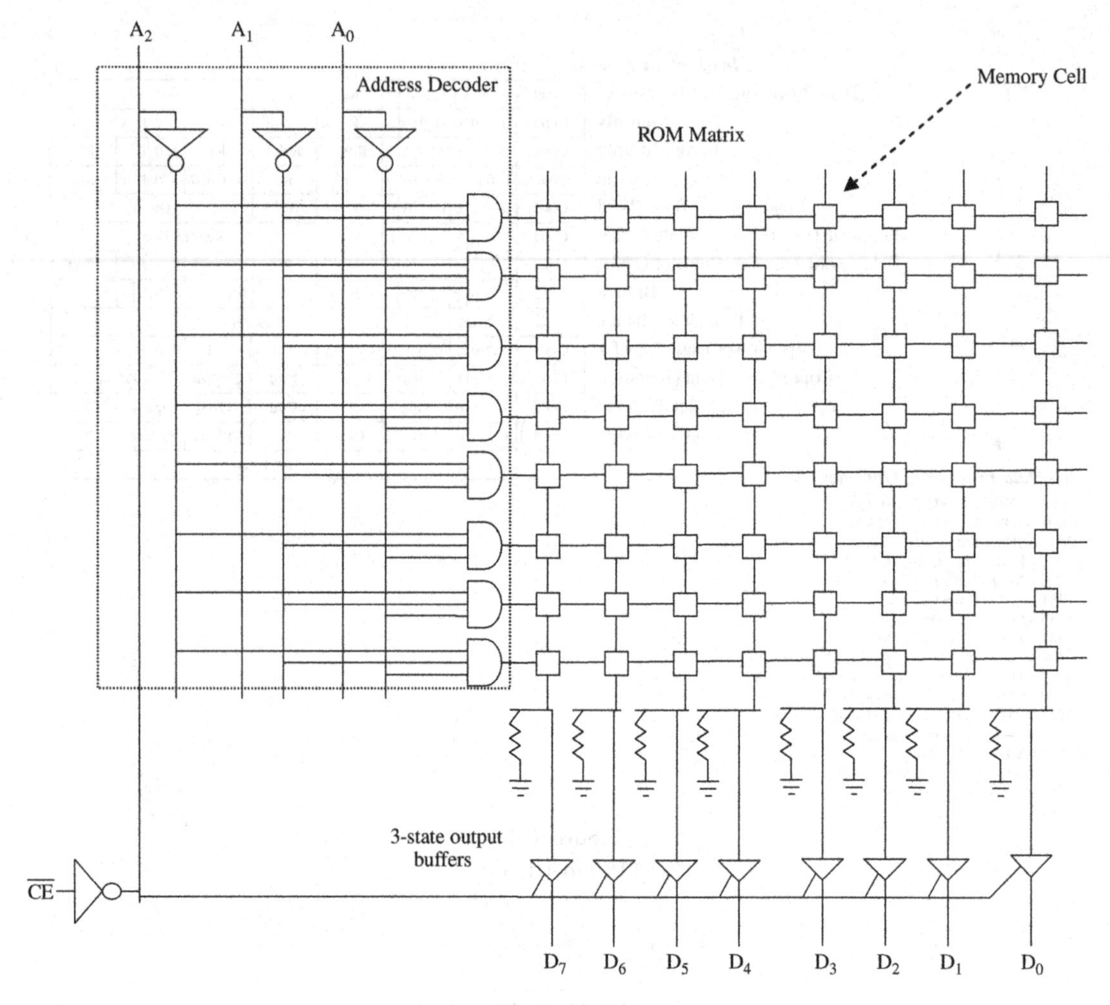


**Figure 4-6**
Block diagram of memory array.[6]

*Storage*

The ISA specifies the features of the programmable storage used to store the data being operated on, primarily the organization of memory used to store operands and register set, and how registers are used.

The Organization of Memory Used to Store Operands

*Memory* is simply an array of programmable storage, like that shown in Figure 4-6, that stores data, including operations and operands.

The indices of this array are locations referred to as *memory addresses*, where each location is a unit of memory that can be addressed separately. The actual physical or virtual range of addresses available to a processor is referred to as the *address space*.

An ISA defines specific characteristics of the address space, such as whether it is:

- *Linear.* A linear address space is one in which specific memory locations are represented incrementally, typically starting at "0" thru $2^{N-1}$, where $N$ is the address width in bits.
- *Segmented.* A segmented address space is a portion of memory that is divided into sections called segments. Specific memory locations can only be accessed by specifying a segment identifier, a segment number that can be explicitly defined or implicitly obtained from a register, and specifying the offset within a specific segment within the segmented address space. The offset within the segment contains a base address and a limit, which map to another portion of memory that is set up as a linear address space. If the offset is less than or equal to the limit, the offset is added to the base address, giving the unsegmented address within the linear address space.
- Containing any *special address regions.*
- *Limited* in any way.

An important note regarding ISAs and memory is that different ISAs not only define where data is stored, but also how data is stored in memory—specifically in what order the bits (or bytes) that make up the data is stored, or *byte ordering.* The two byte-ordering approaches are big-endian, in which the most significant byte or bit is stored first, and little-endian, in which the least significant bit or byte is stored first. For example:

- 68000 and SPARC are big-endian.
- x86 is little-endian.
- ARM, MIPS, and PowerPC can be configured as either big-endian or little-endian using a bit in their machine state registers.

#### Register Set

A register is simply fast programmable memory normally used to store operands that are immediately or frequently used. A processor's set of registers is commonly referred to as the *register set* or the *register file.* Different processors have different register sets and the number of registers in their sets vary between very few to several hundred (even over 1000). For example, the SA-1110 register set has 37 32-bit registers, whereas the MPC823, on the other hand, has about a few hundred registers (general-purpose, special-purpose, floating-point registers, etc.).

#### How Registers Are Used

An ISA defines which registers can be used for what transactions, such as special-purpose, floating point, and which can be used by the programmer in a general fashion (general-purpose registers).

As a final note on registers, one of many ways processors can be referenced is according to the *size* (in bits) of *data* that can be processed and the *size* (in bits) of the *memory space* that can be addressed in a single instruction by that processor. This specifically relates back to

**Table 4-2: "x-Bit" Architecture Examples**

| "x"-Bit | Architecture |
|---|---|
| 4 | Intel 4004, ... |
| 8 | Mitsubishi M37273, 8051, 68HC08, Intel 8008/8080/8086, ... |
| 16 | ST ST10, TI MSP430, Intel 8086/286, ... |
| 32 | 68K, PowerPC, ARM, x86 (386+), MIPS32, ... |

the basic building block of registers, the flip-flop, but this will be discussed in more detail in Section 4.2.

Commonly used embedded processors support 4-, 8-, 16-, 32-, and/or 64-bit processing, as shown in Table 4-2. Some processors can process larger amounts of data and can access larger memory spaces in a single instruction, such as 128-bit architectures, but they are not commonly used in embedded designs.

*Addressing Modes*

Addressing modes define how the processor can access operand storage. In fact, the usage of registers is partly determined by the ISA's *Memory Addressing Modes*. The two most common types of addressing mode models are the:

- *Load-store architecture*: only allows operations to process data in registers, not anywhere else in memory. For example, the PowerPC architecture has only one addressing mode for load and store instructions: register plus displacement (supporting register indirect with immediate index, register indirect with index, etc.).
- *Register-memory architecture*: allows operations to be processed both within registers and within other types of memory. Intel's i960 Jx processor is an example of an addressing mode architecture that is based upon the register-memory model (supporting absolute, register indirect, etc.).

*Interrupts and Exception Handling*

*Interrupts* (also referred to as exceptions or traps depending on the type) are mechanisms that stop the standard flow of the program in order to execute another set of code in response to some event, such as problems with the hardware, resets, and so forth. The ISA defines what, if any, type of hardware support a processor has for interrupts.

*Note: Because of their complexity, interrupts will be discussed in more detail in Section 4.2 later in this chapter.*

### 4.1.2 ISA Models

There are several different ISA models that architectures are based upon, each with their own definitions for the various features. The most commonly implemented ISA models are

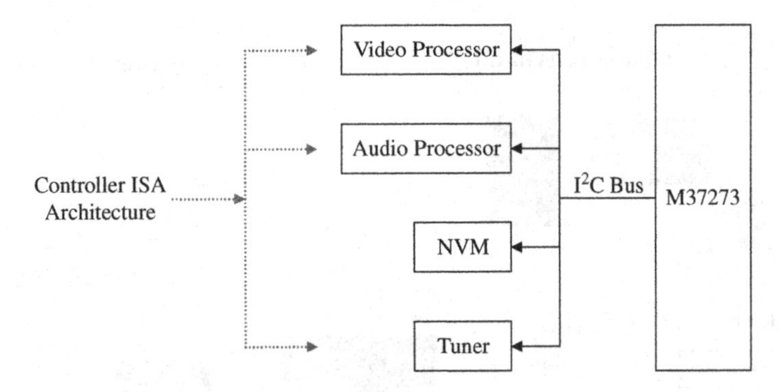


**Figure 4-7**
Analog TV board example with controller ISA implementations.

application-specific, general-purpose, instruction-level parallel, or some hybrid combination of these three ISAs.

*Application-Specific ISA Models*

Application-specific ISA models define processors that are intended for specific embedded applications, such as processors made only for TVs. There are several types of application-specific ISA models implemented in embedded processors, the most common models being:

- *Controller Model.* The Controller ISA is implemented in processors that are not required to perform complex data manipulation, such as video and audio processors that are used as slave processors on a TV board, for example (see Figure 4-7).
- *Datapath Model.* The Datapath ISA is implemented in processors whose purpose is to repeatedly perform fixed computations on different sets of data, a common example being digital signal processors (DSPs), shown in Figure 4-8a.
- *Finite-State Machine with Datapath (FSMD) Model.* The FSMD ISA is an implementation based upon a combination of the Datapath ISA and the Controller ISA for processors that are not required to perform complex data manipulation and must repeatedly perform fixed computations on different sets of data. Common examples of an FSMD implementation are application-specific integrated circuits (ASICs) shown in Figure 4-9, programmable logic devices (PLDs), and field-programmable gate-arrays (FPGAs, which are essentially more complex PLDs). FPGAs have emerged as a powerful and sustaining force in embedded systems designs. At the highest level as seen in Figure 4-8b, FPGAs are some combination of interconnected CLBs (configurable logic blocks) to form a more a complex digital circuit. A wide range of tools are available today to create FPGA custom circuits based on everything from VHDL (type of hardware description language) to Verilog to schematic design techniques.
- *Java Virtual Machine (JVM) Model.* The JVM ISA is based upon one of the JVM standards discussed in Chapter 2, *Sun Microsystem's Java Language.* As described in

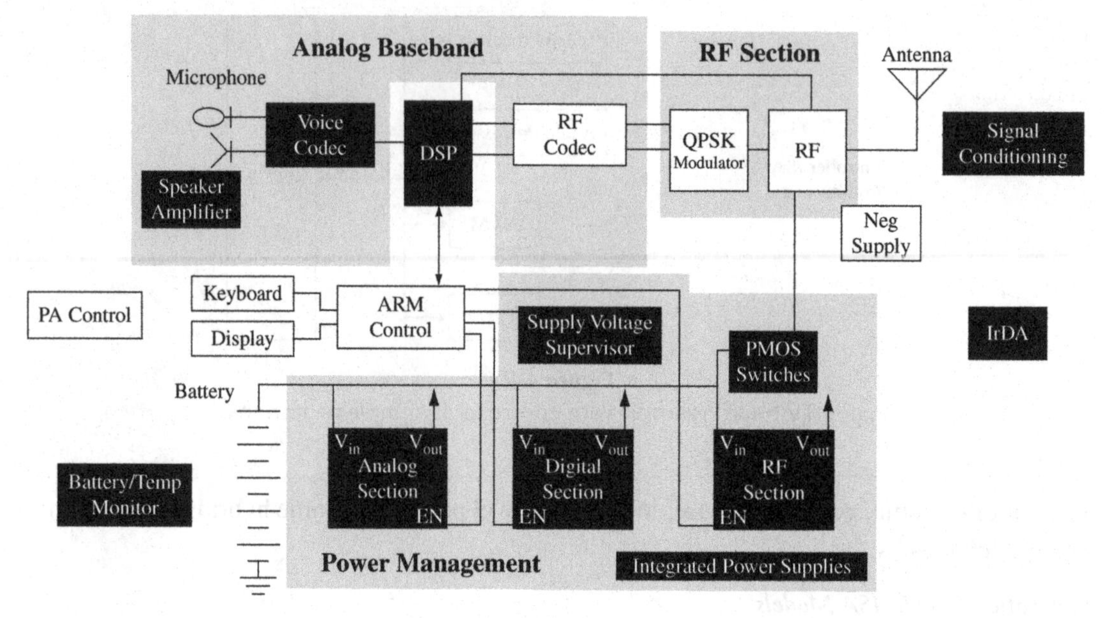


**Figure 4-8a**
Board example with datapath ISA implementation—digital cell phone.[7]

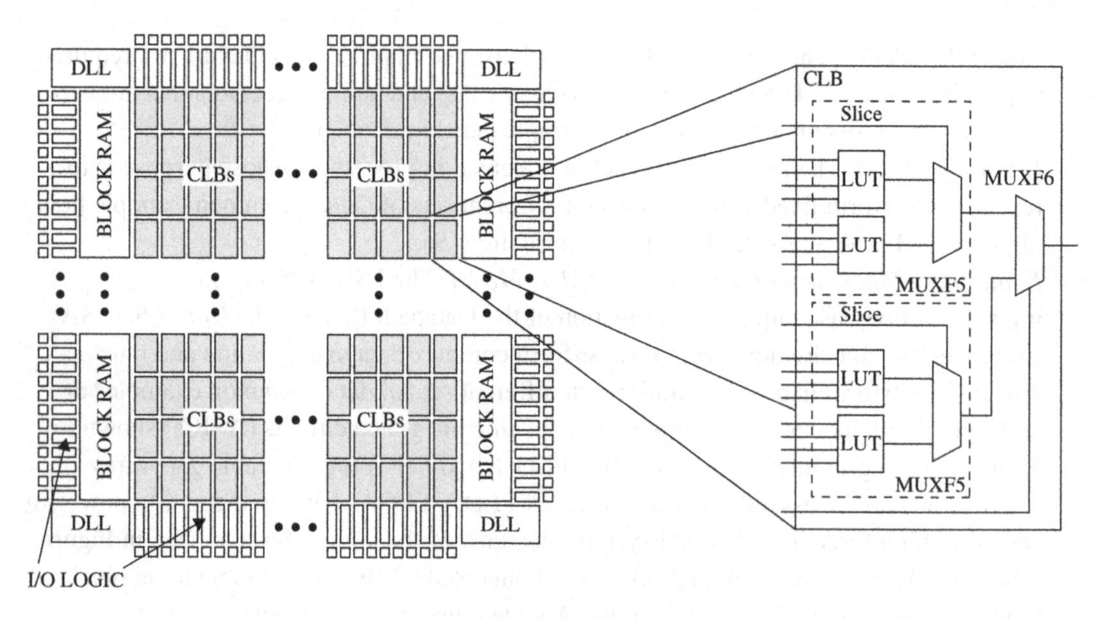


**Figure 4-8b**
Altium's FPGA high-level design diagram.[30]

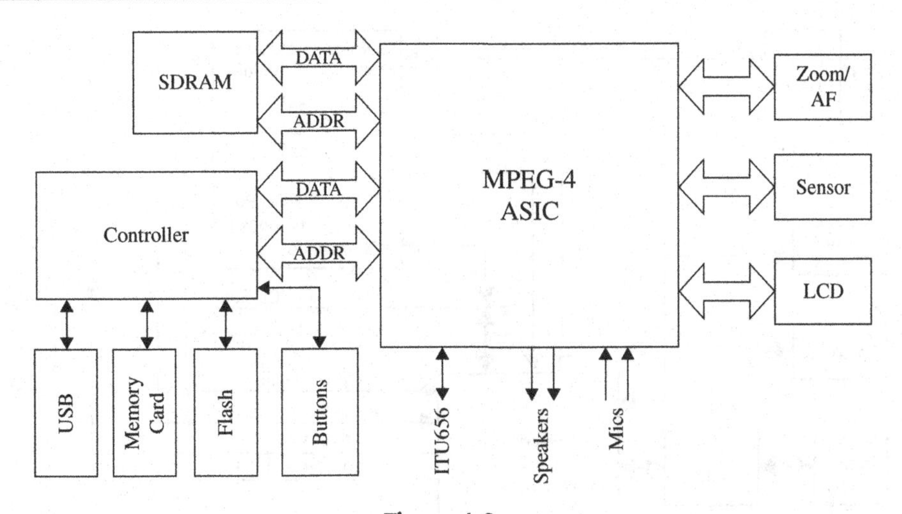


**Figure 4-9**
Board example with FSMD ISA implementation—solid-state digital camcorder.[8]

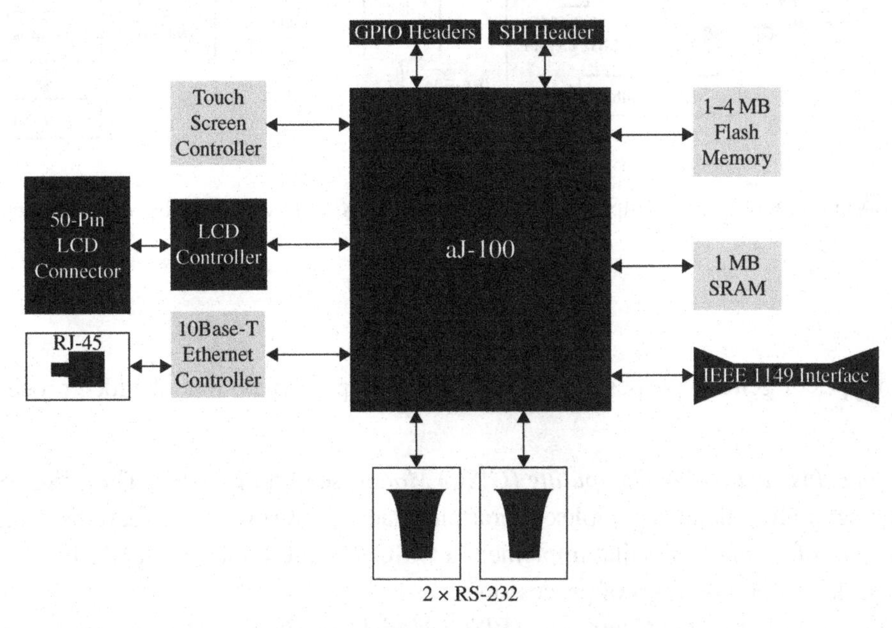


**Figure 4-10**
JVM ISA implementation example.[9]

Chapter 2, real-world JVMs can be implemented in an embedded system via hardware, such as in aJile's aj-80 and aj-100 processors (see Figure 4-10).

### *General-Purpose ISA Models*

General-purpose ISA models are typically implemented in processors targeted to be used in a wide variety of systems, rather than only in specific types of embedded systems. The most

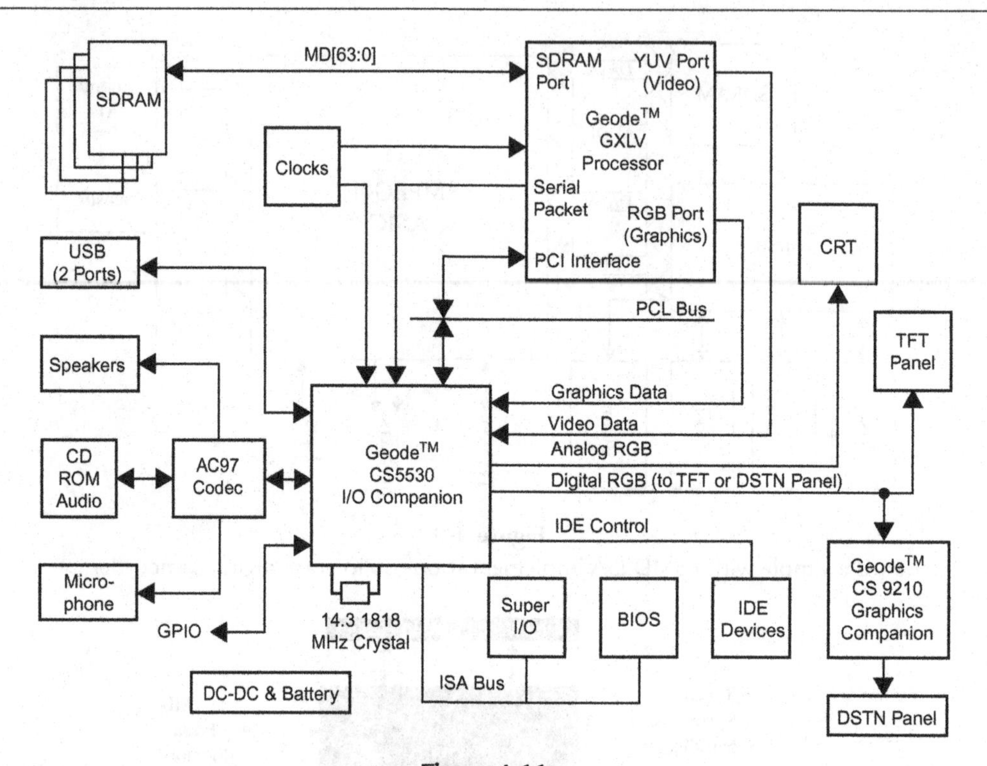


**Figure 4-11**
CISC ISA implementation example.[10] *© 2004 Advanced Micro Devices, Inc. Reprinted with permission.*

common types of general-purpose ISA architectures implemented in embedded processors are:

- *Complex Instruction Set Computing (CISC) Model* (see Figure 4-11). The CISC ISA, as its name implies, defines complex operations made up of several instructions. Common examples of architectures that implement a CISC ISA are Intel's x86 and Motorola/ Freescale's 68000 families of processors.
- *Reduced Instruction Set Computing (RISC) Model* (see Figure 4-12). In contrast to CISC, the RISC ISA usually defines:
  - an architecture with simpler and/or fewer operations made up of fewer instructions;
  - an architecture that has a reduced number of cycles per available operation.

Many RISC processors have only one-cycle operations, whereas CISCs typically have multiple cycle operations. ARM, PowerPC, SPARC, and MIPS are just a few examples of RISC-based architectures.

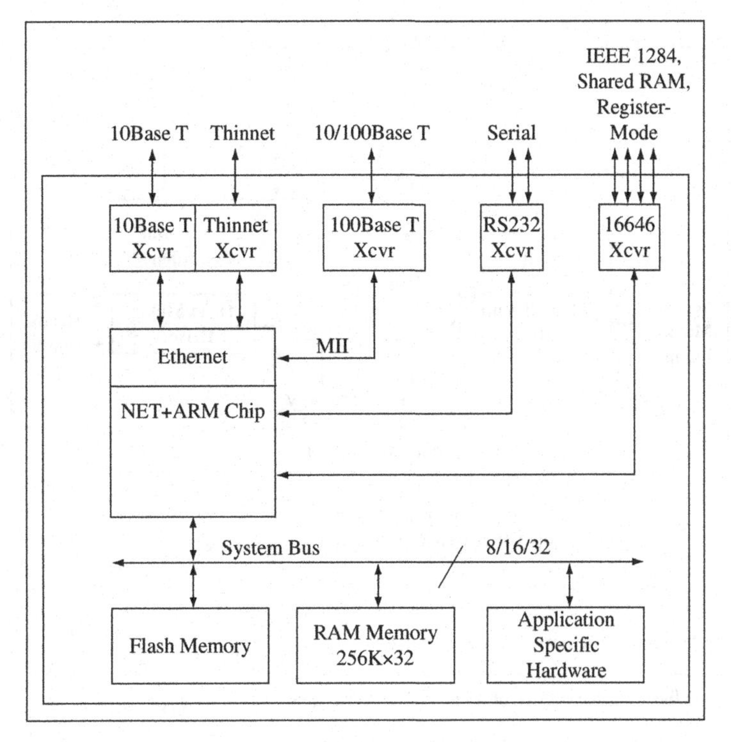


**Figure 4-12**
RISC ISA implementation example.[11]

> **Final Note on CISC versus RISC**
>
> In the area of general-purpose computing, note that many current processor designs fall under the CISC or RISC category primarily because of their heritage. RISC processors have become more complex, while CISC processors have become more efficient to compete with their RISC counterparts, thus blurring the lines between the definition of a RISC versus a CISC architecture. Technically, these processors have both RISC and CISC attributes, regardless of their definitions.

### *Instruction-Level Parallelism ISA Models*

Instruction-level parallelism ISA architectures are similar to general-purpose ISAs, except that they execute multiple instructions in parallel, as the name implies. In fact, instruction-level parallelism ISAs are considered higher evolutions of the RISC ISA, which typically has one-cycle operations, one of the main reasons why RISCs are the basis for parallelism. Examples of instruction-level parallelism ISAs include:

- *Single Instruction Multiple Data (SIMD) Model* (see Figure 4-13). The SIMD Machine ISA is designed to process an instruction simultaneously on multiple data components that require action to be performed on them.

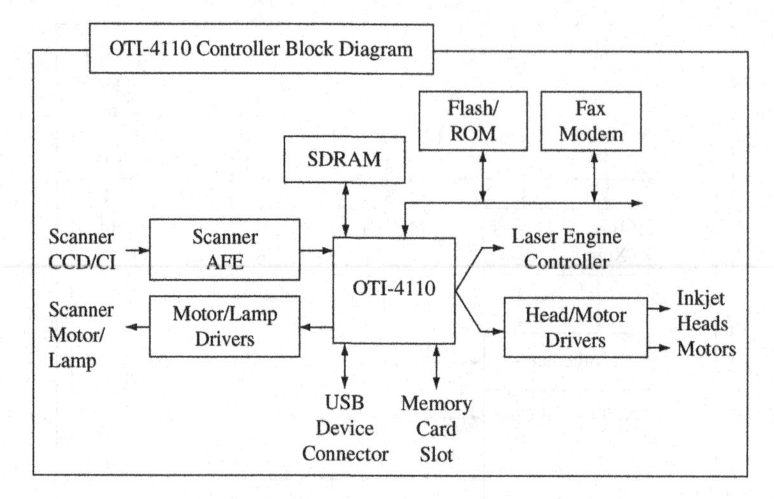


**Figure 4-13**
SIMD ISA implementation example.[12]

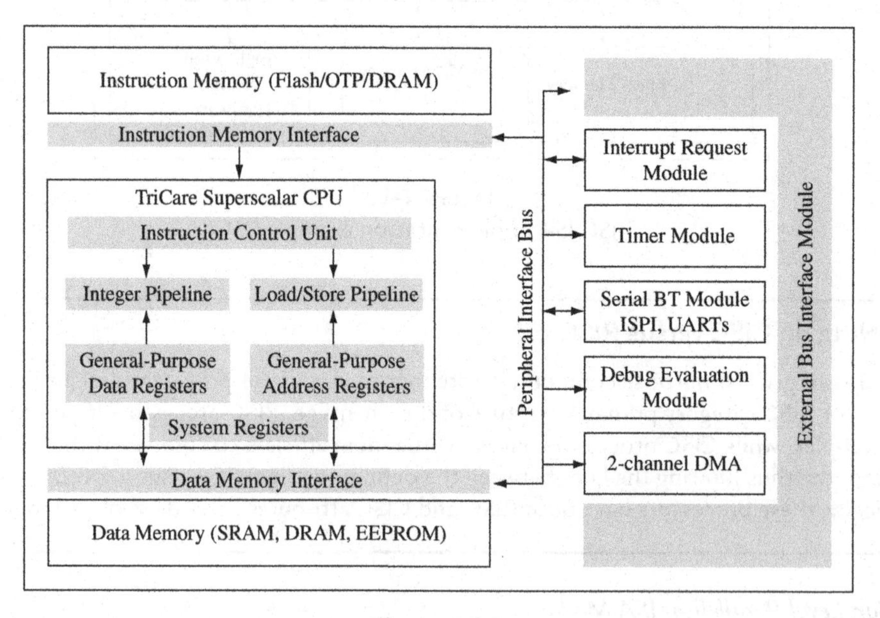


**Figure 4-14**
Superscalar ISA implementation example.[13]

- *Superscalar Machine Model* (see Figure 4-14). The superscalar ISA is able to process multiple instructions simultaneously within one clock cycle through the implementation of multiple functional components within the processor.
- *Very Long Instruction Word Computing (VLIW) Model* (see Figure 4-15). The VLIW ISA defines an architecture in which a very long instruction word is made up of multiple operations. These operations are then broken down and processed in parallel by multiple execution units within the processor.

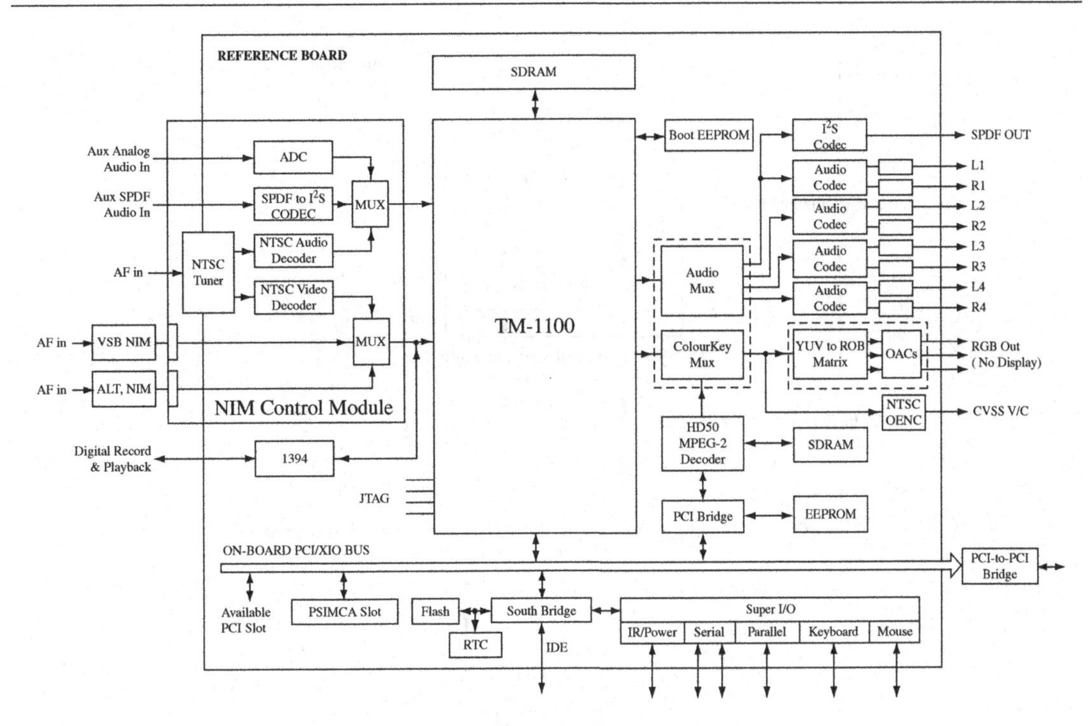


**Figure 4-15**
VLIW ISA implementation example–(VLIW) Trimedia-based DTV board.[14]

## 4.2 Internal Processor Design

The ISA defines *what* a processor can do and it is the processor's internal interconnected hardware components that physically implement the ISA's features. Interestingly, the fundamental components that make up an embedded board are the same as those that implement an ISA's features in a processor: a CPU, memory, input components, output components, and buses. As mentioned in Figure 4-16, these components are the basis of the von Neumann model.

Of course, many current real-world processors are more complex in design than the von Neumann model has defined. However, most of these processors' hardware designs are still based upon von Neumann components or a version of the von Neumann model called the Harvard architecture model (see Figure 4-17). These two models primarily differ in one area—memory. A von Neumann architecture defines a single memory space to store instructions and data. A Harvard architecture defines separate memory spaces for instructions and data; separate data and instruction buses allow for simultaneous fetches and transfers to occur. The main reasoning behind using von Neumann versus a Harvard-based model for an architecture design is *performance.* Given certain types of ISAs, like Datapath model ISAs in

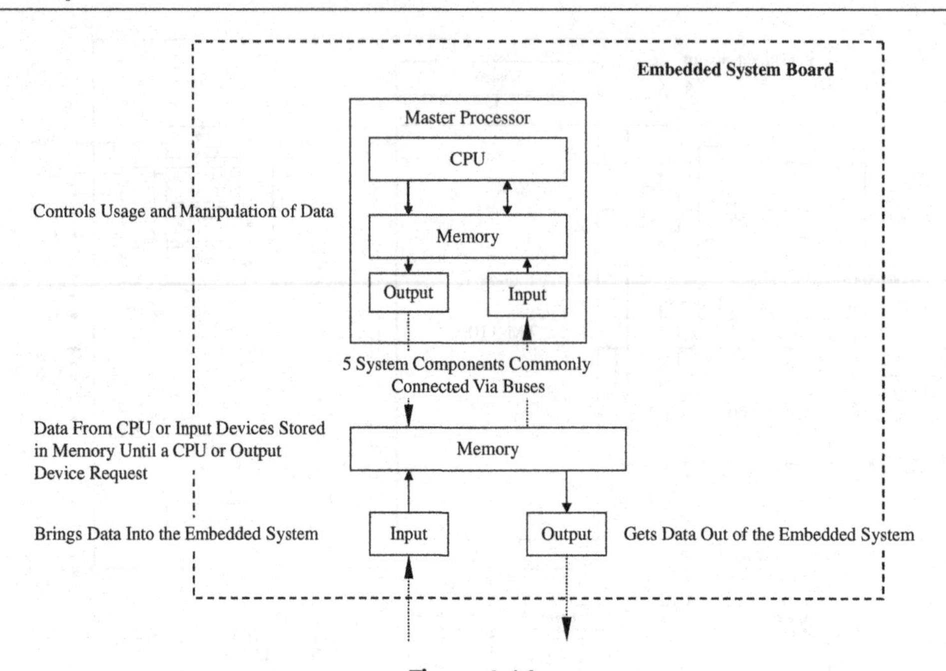


**Figure 4-16**
Von Neumann-based processor diagram.

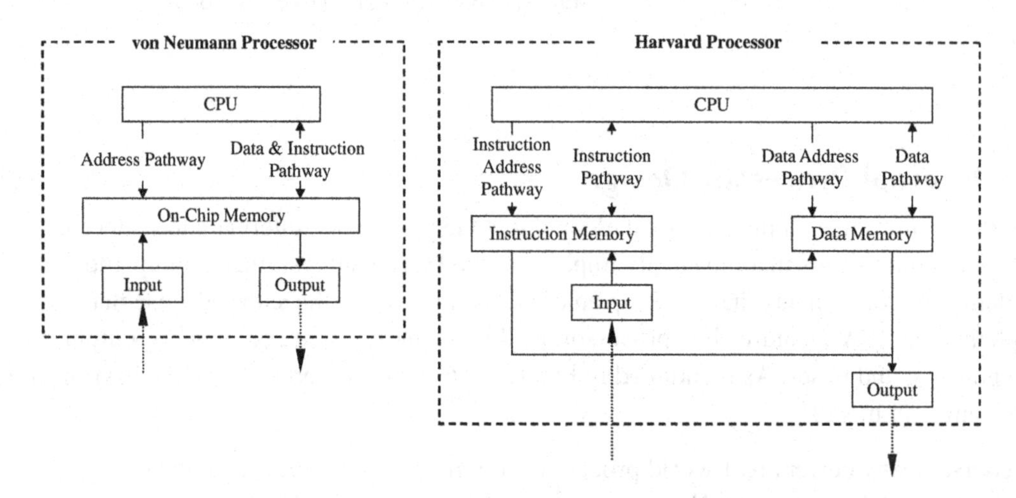


**Figure 4-17**
Von Neumann versus Harvard architectures.

DSPs, and their functions, such as continuously performing fixed computations on different sets of data, the separate data and instruction memory allow for an increase in the amount of data per unit of time that can be processed, given the lack of competing interests of space and bus accesses for transmissions of data and instructions.

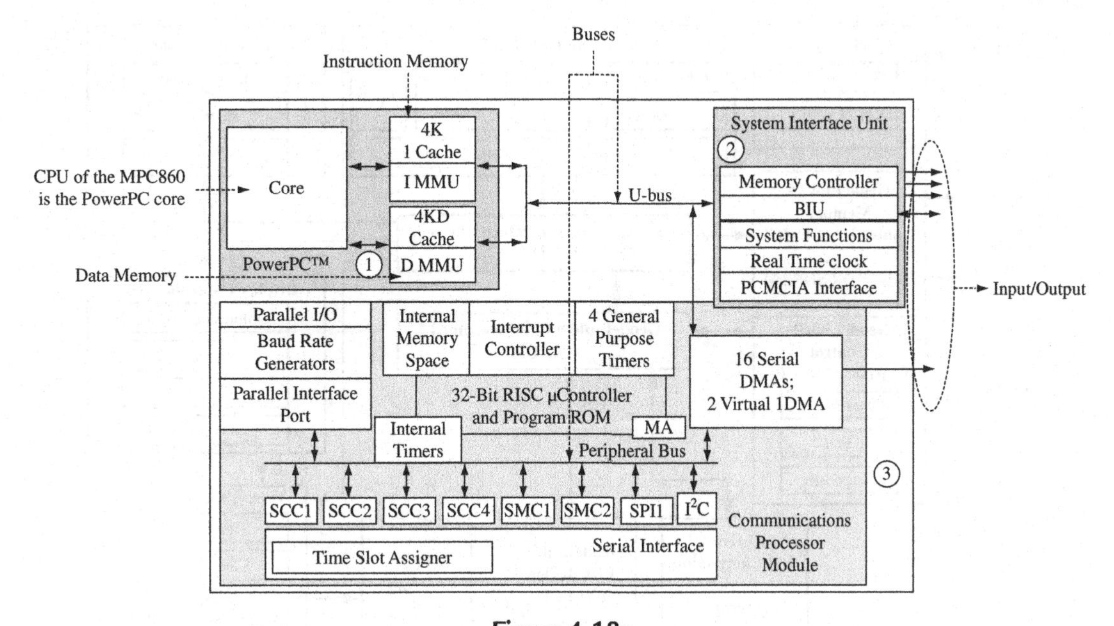


**Figure 4-18a**

Harvard architecture example—MPC860.[15]

While the MPC860 is a complex processor design, it is still based upon the fundamental components of the Harvard model: the CPU, instruction memory, data memory, I/O, and buses.



As mentioned previously, most processors are based upon some variation of the von Neumann model (in fact, the Harvard model itself is a variation of the von Neumann model). Real-world examples of Harvard-based processor designs include ARM's ARM9/ARM10, MPC860, MPC8031, and DSPs (see Figure 4-18a), while ARM's ARM7 and x86 are von Neumann-based designs (see Figure 4-18b).

### Why Talk About the von Neumann Model?

The von Neumann model not only impacts the internals of a processor (what you don't see), but it shapes what you do see and what you can access within a processor. As discussed in Chapter 3, integrated circuits (ICs)—and a processor is an IC—have protruding pins that connect them to the board. While different processors vary widely in the number of pins and their associated signals, the components of the von Neumann model, both at the board and at the internal processor level, also define the signals that all processors have. As shown in Figure 4-19, to accommodate board memory, processors typically have address and data signals to read and write data to and from memory. In order to communicate to memory or I/O, a processor usually has some type of READ and WRITE pins to indicate it wants to retrieve or transmit data.

Of course there are other pins not explicitly defined by von Neumann that are required for practical purposes, such as a synchronizing mechanism like a clock signal to drive a processor, and some method of powering and grounding of the processor. However, regardless of the differences between processors, the von Neumann model essentially drives what external pins all processors have.

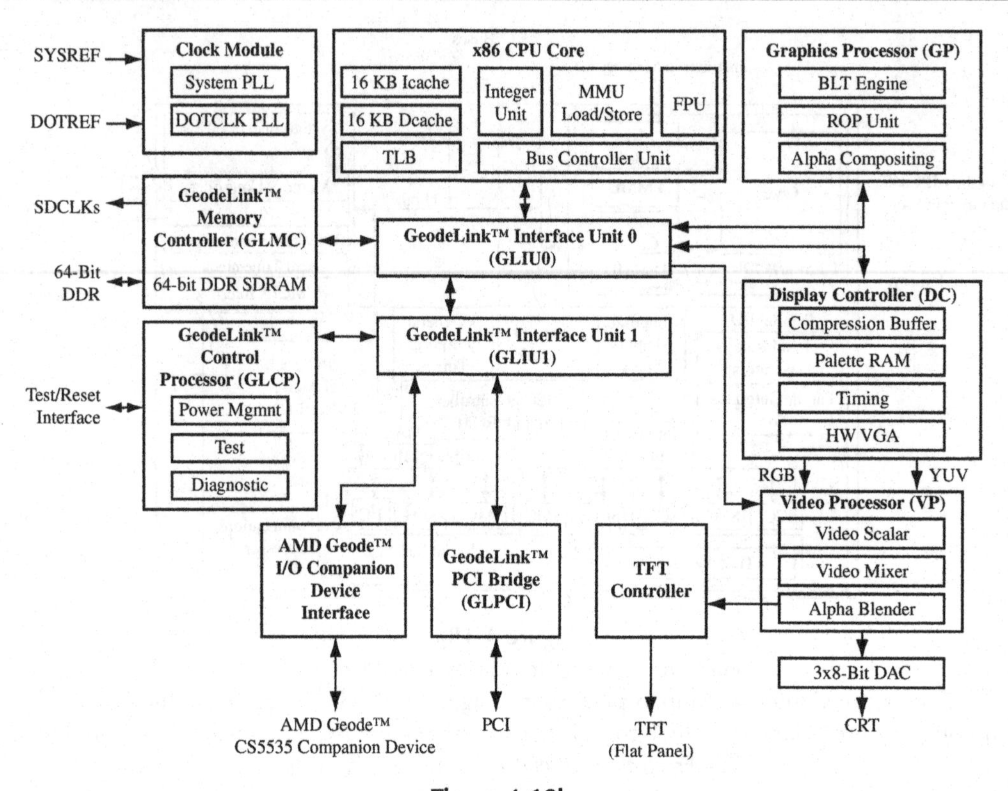


**Figure 4-18b**
Von Neumann architecture example–x86.[16]
x86 is a complex processor design based upon the von Neumann model where, unlike the MPC860 processor, instructions and data share the same memory space.

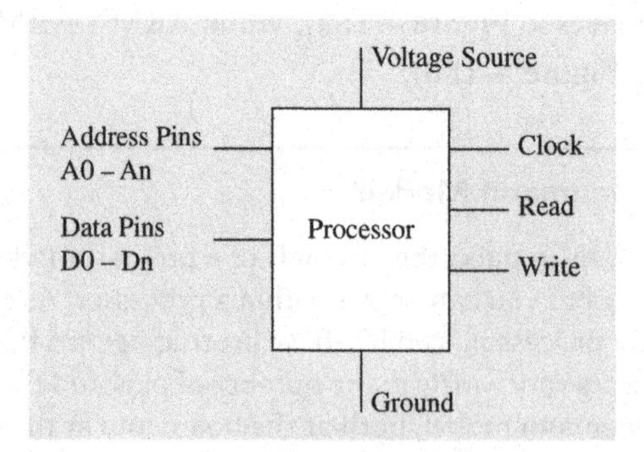


**Figure 4-19**
Von Neumann and processor pins.

### 4.2.1 Central Processing Unit (CPU)

The semantics of this section can be a little confusing, because processors themselves are commonly referred to as CPUs, but it is actually the *processing unit* within a processor that is the CPU. The CPU is responsible for executing the cycle of fetching, decoding, and executing

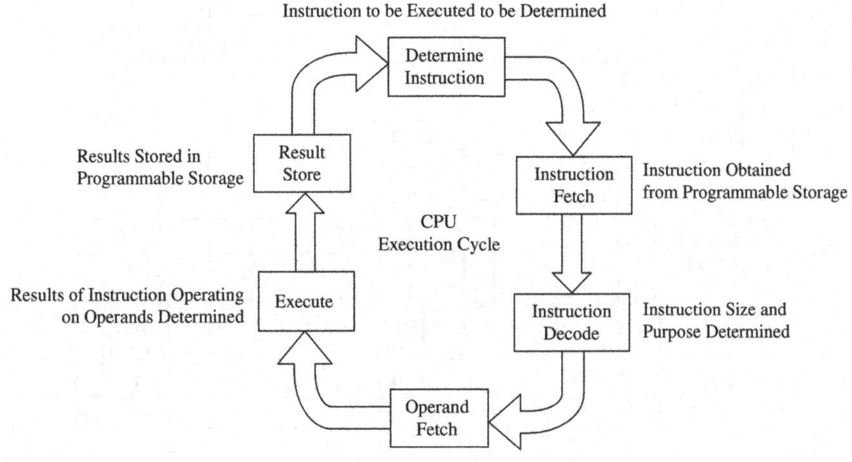


**Figure 4-20**
Fetch, decode, and execution cycle of CPU.

instructions (see Figure 4-20). This three-step process is commonly referred to as a three-stage *pipeline* and most recent CPUs are pipelined designs.

While CPU designs can widely differ, understanding the basic components of a CPU will make it easier to understand processor design and the cycle shown in Figure 4-20. As defined by the von Neumann model, this cycle is implemented through some combination of four major CPU components:

- *The arithmetic logic unit (ALU)*: implements the ISA's operations.
- *Registers*: a type of fast memory.
- *The control unit (CU)*: manages the entire fetching and execution cycle.
- *The internal CPU buses*: interconnect the ALU, registers, and the CU.

Looking at a real-world processor, these four fundamental elements defined by the von Neumann model can be seen within the CPU of the MPC860 (see Figure 4-21).

#### *Internal CPU Buses*

The CPU buses are the mechanisms that interconnect the CPU's other components: the ALU, the CU, and registers (see Figure 4-22). Buses are simply wires that interconnect the various other components within the CPU. Each bus's wire is typically divided into logical functions, such as data (which carries data, bidirectionally, between registers and the ALU), address (which carries the locations of the registers that contain the data to be transferred), control (which carries control signal information, such as timing and control signals, between the registers, the ALU, and the CU), etc.

*Note: To avoid redundancy, buses will be discussed in more detail in Chapter 7.*

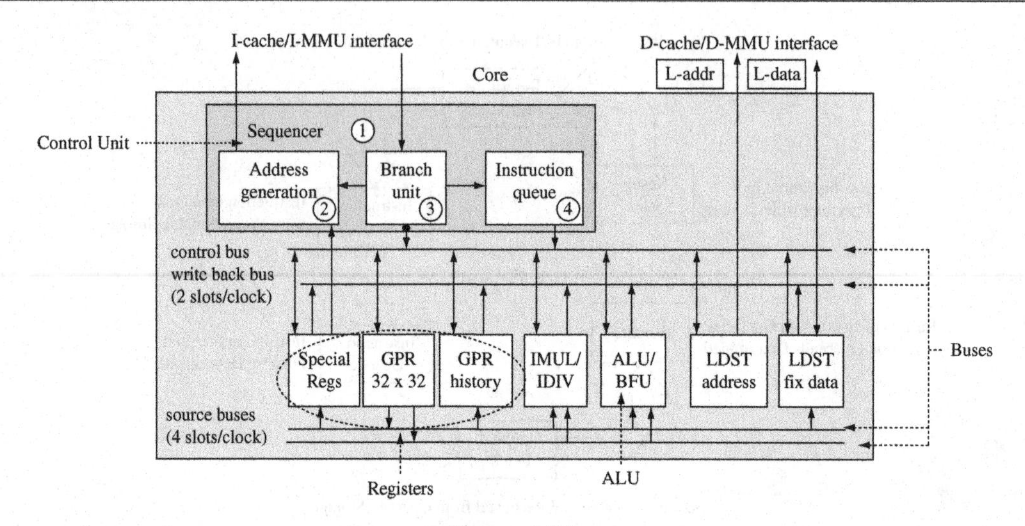


**Figure 4-21**

The MPC860 CPU—the PowerPC core.[15]

Remember: Not all processors have these components as strictly defined by the von Neumann model, but will have some combination of these components under various aliases somewhere on the processor. Remember that this model is a reference tool the reader can use to understand the major components of a CPU design. © *2004 Freescale Semiconductor, Inc. Used by permission.*

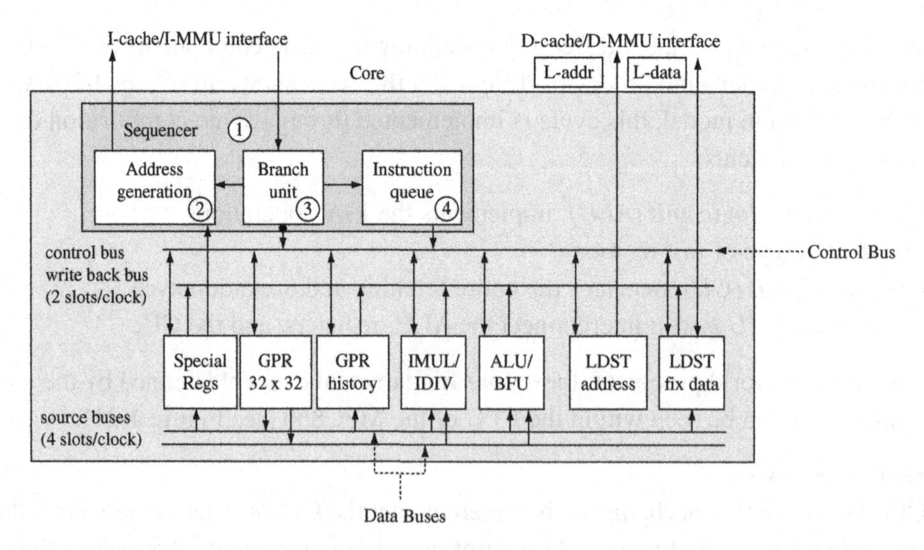


**Figure 4-22**

PowerPC core and buses.[15]

In the PowerPC Core, there is a control bus that carries the control signals between the ALU, CU, and registers. What the PowerPC calls "source buses" are the data buses that carry the data between registers and the ALU. There is an additional bus called the write-back which is dedicated to writing back data received from a source bus directly back from the load/store unit to the fixed or floating-point registers.

### *ALU*

The ALU implements the comparison, mathematical, and logical operations defined by the ISA. The format and types of operations implemented in the CPU's ALU can vary depending on the ISA. Considered the core of any processor, the ALU is responsible for accepting multiple *n*-bit binary operands, and performing any logical (AND, OR, NOT, etc.), mathematical (+, –, *, etc.), and comparison (=, <, >, etc.) operations on these operands.

The ALU is a combinational logic circuit that can have one or more inputs and only one output. An ALU's output is dependent only on inputs applied at that instant, as a function of time, and "no" past conditions (see Chapter 3 on gates). The basic building block of most ALUs (from the simplest to the multifunctional) is considered the *full adder*—a logic circuit that takes three 1-bit numbers as inputs and produces two 1-bit numbers. How this actually works will be discussed in more detail later this section.

To understand how a full adder works, let us first examine the mechanics of adding binary numbers (0s and 1s) together:

Starting with two 1-bit numbers, adding them will produce, at most, a 2-bit number:

| $X_0$ | $Y_0$ | $S_0$ | $C_{out}$ | |
|---|---|---|---|---|
| 0 | 0 | 0 | 0 | ⇒ 0 b + 0 b = 0 b |
| 0 | 1 | 1 | 0 | ⇒ 0 b + 1 b = 0 b |
| 1 | 0 | 1 | 0 | ⇒ 1 b + 0 b = 1 b |
| 1 | 1 | 0 | 1 | ⇒ 1 b + 1 b = 10 b (or 2d). In binary addition of two 1-bit numbers, when the count exceeds 10 (the binary of 2 decimal), the 1 ($C_{out}$) is carried and added to the next row of numbers thus resulting in a 2-bit number |

This simple addition of two 1-bit numbers can be executed via a *half-adder* circuit—a logical circuit that takes two 1-bit numbers as inputs and produces a 2-bit output. Half-adder circuits, like all logical circuits, can be designed using several possible combinations of gates, such as the possible combinations shown in Figure 4-23a.

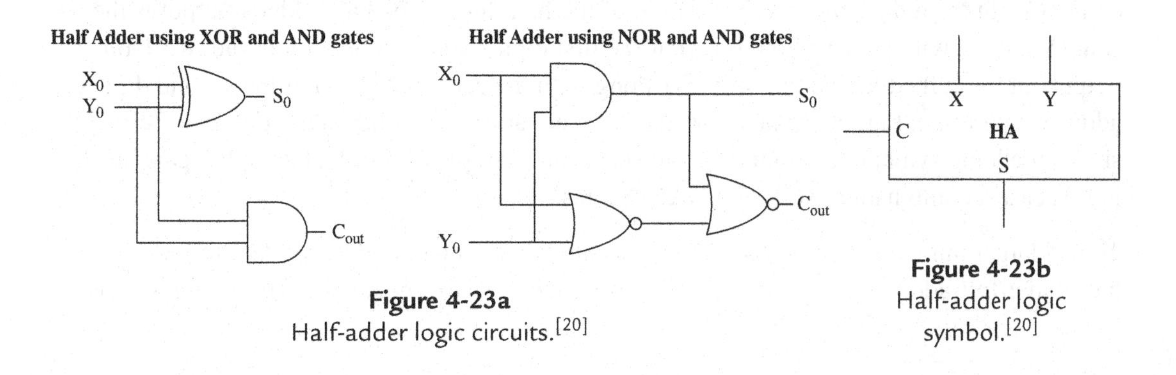


**Figure 4-23a**
Half-adder logic circuits.[20]

**Figure 4-23b**
Half-adder logic symbol.[20]

| X | Y | $C_{in}$ | S | $C_{out}$ |
|---|---|---|---|---|
| 0 | 0 | 0 | 0 | 0 |
| 0 | 0 | 1 | 1 | 0 |
| 0 | 1 | 0 | 1 | 0 |
| 0 | 1 | 1 | 0 | 1 |
| 1 | 0 | 0 | 1 | 0 |
| 1 | 0 | 1 | 0 | 1 |
| 1 | 1 | 0 | 0 | 1 |
| 1 | 1 | 1 | 1 | 1 |

Sum (S) = $XYC_{in} + XY'C_{in}' + X'YC_{in}' + X'Y'C_{in}'$
Carry Out (Cout) = $XY + X\,C_{in} = Y\,C_{in}$

**Figure 4-24a**
Full adder truth table and logic equations.[19]

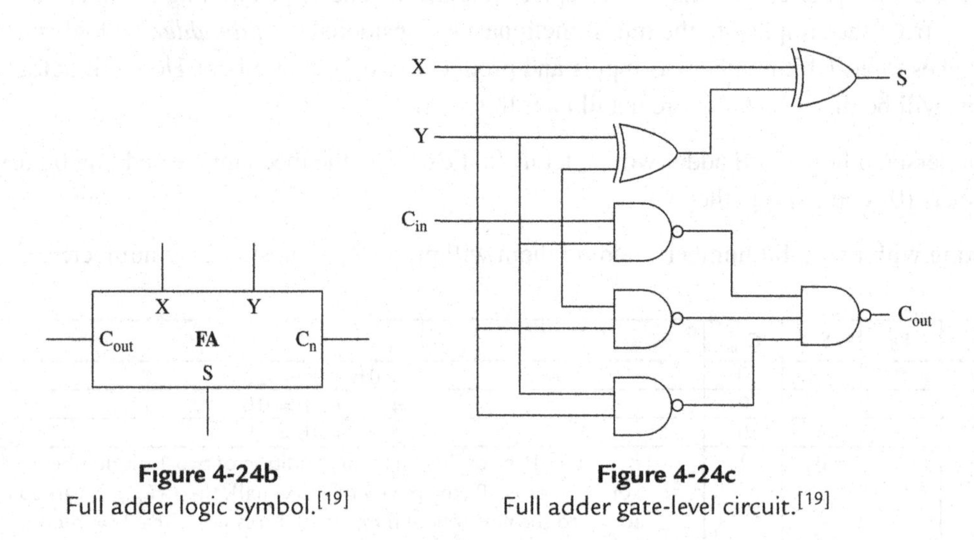


**Figure 4-24b**
Full adder logic symbol.[19]

**Figure 4-24c**
Full adder gate-level circuit.[19]

In order to add a larger number, the adder circuit must increase in complexity and this is where the full adder comes into play. When trying to add two-digit numbers, for example, a full adder must be used in conjunction with a half adder. The half adder takes care of adding the first digits of the two numbers to be added (i.e., $x_0$, $y_0$, ...)—the full adder's three 1-bit inputs are the second digits of the two numbers to be added (i.e., $x_1$, $y_1$, ...) along with the carry ($C_{in}$) in from the half adder's addition of the first digits. The half adder's output is the sum ($S_0$) along with the carry out ($C_{out}$) of the first digit's addition operation; the two 1-bit outputs of the full adder are the sum ($S_1$) along with the carry out ($C_{out}$) of the second digit's addition operation. Figure 4-24a shows the logic equations and truth table, Figure 4-24b shows the logic symbol, and Figure 4-24c shows an example of a full adder at the gate level, in this case a combination XOR and NAND gates.

To add larger numbers, additional full adders can then be integrated (cascaded) to the half-adder/full-adder hybrid circuit (see Figure 4-25). The example shown in Figure 4-25 is

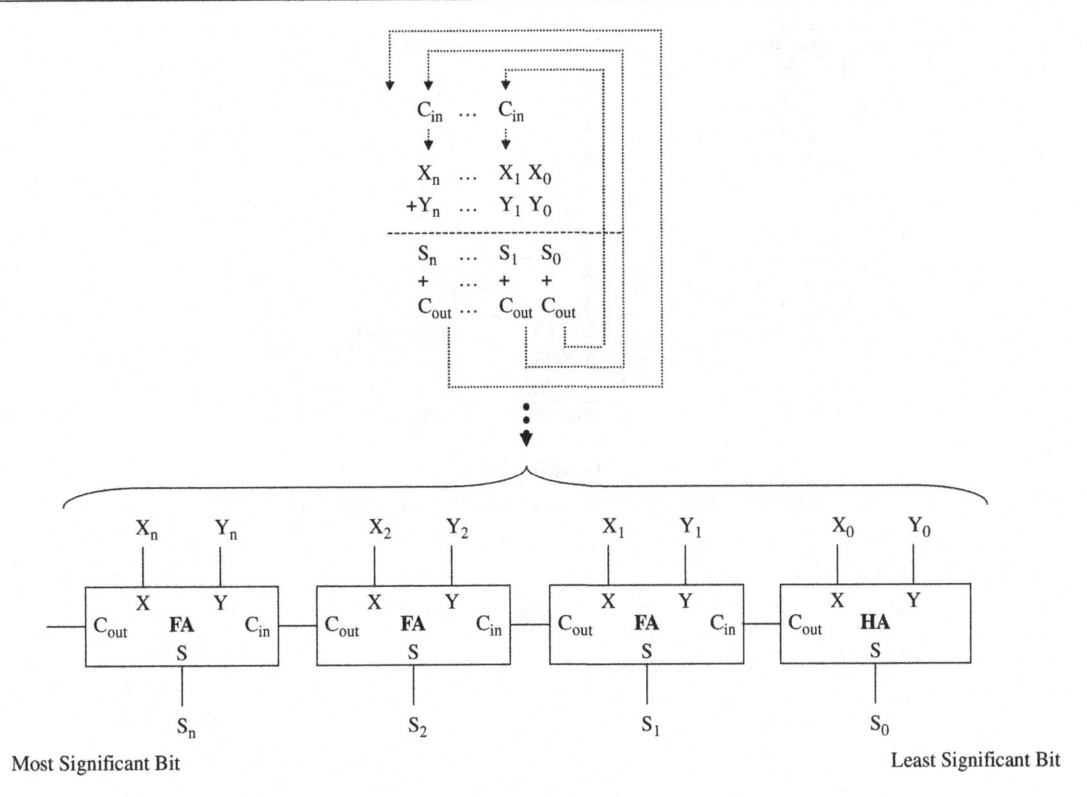


**Figure 4-25**
Cascaded adders.

the basis of the *ripple-carry adder* (one of many types of adders), in which "*n*" full adders are cascaded so the carry produced by the lower stages propagates (ripples) through to the higher stages in order for the addition operation to complete successfully.

Multifunction ALUs that provide addition operations, along with other mathematical and logical operations, are designed around the adder circuitry, with additional circuitry incorporated for performing subtraction, logical AND, logical OR, etc. (see Figure 4-26a). The logic diagram shown in Figure 4-26b is an example of two stages of an $n$-bit multifunction ALU. The circuit in Figure 4-26 is based on the *ripple-carry adder* that was just described. In the logic circuit in Figure 4-26b, control inputs $k_0$, $k_1$, $k_2$, and $c_{in}$ determine the function performed on the operand or operands. Operand inputs are $X = x_{n-1}, \ldots, x_1x_0$ and $Y = y_{n-1}, \ldots, y_1y_0$ and the output is sum $(S) = s_{n-1}, \ldots, s_1s_0$.

Where the ALU saves the generated results varies with different architectures. With the PowerPC shown in Figure 4-27, results are saved in a register called an *Accumulator.* Results

| Control Inputs | | | | Result | Function |
|---|---|---|---|---|---|
| $k_2$ | $k_1$ | $k_0$ | $c_{in}$ | | |
| 0 | 0 | 0 | 0 | S = X | Transfer X |
| 0 | 0 | 0 | 1 | S = X + 1 | Increment X |
| 0 | 0 | 1 | 0 | S = X + Y | Addition |
| 0 | 0 | 1 | 1 | S = X + Y + 1 | Add with Carry In |
| 0 | 1 | 0 | 0 | S = X − Y − 1 | Subtract with Borrow |
| 0 | 1 | 0 | 1 | S = X − Y | Subtraction |
| 0 | 1 | 1 | 0 | S = X − 1 | Decrement X |
| 0 | 1 | 1 | 1 | S = X | Transfer X |
| 1 | 0 | 0 | ... | S = X OR Y | Logical OR |
| 1 | 0 | 1 | ... | S = X XOR Y | Logical XOR |
| 1 | 1 | 0 | ... | S = X AND Y | Logical AND |
| 1 | 1 | 1 | ... | S = NOT X | Bit-wise Complement |

**Figure 4-26a**
Multifunction ALU truth table and logic equations.[19]

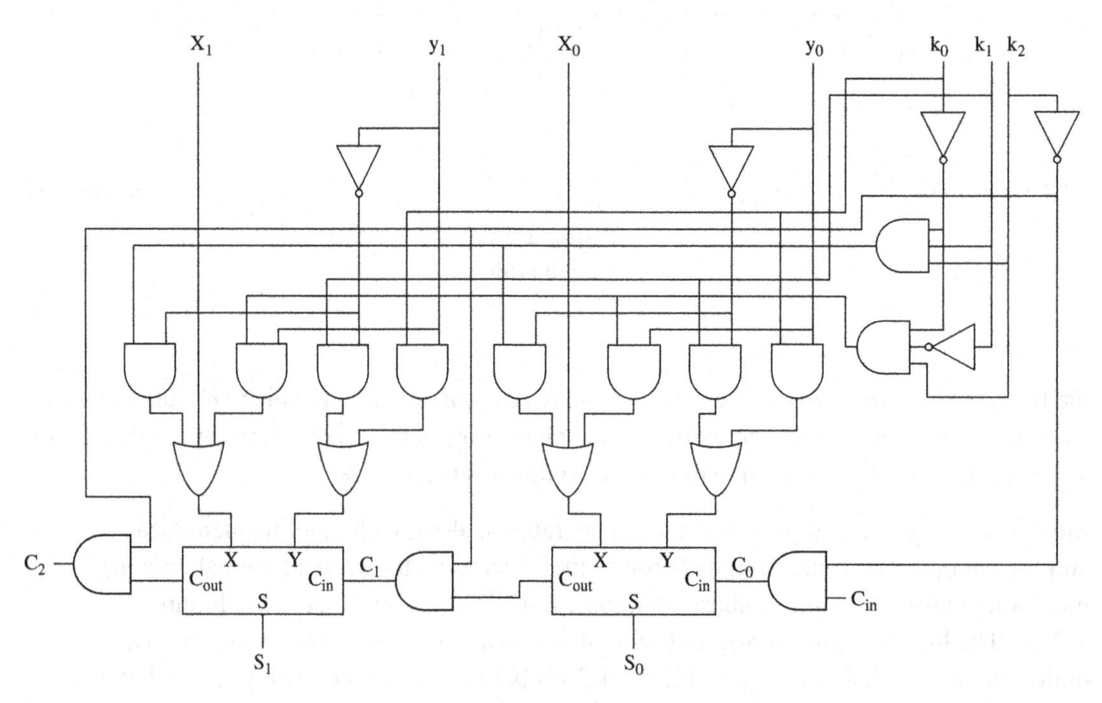


**Figure 4-26b**
Multifunction ALU gate-level circuitry.[19]

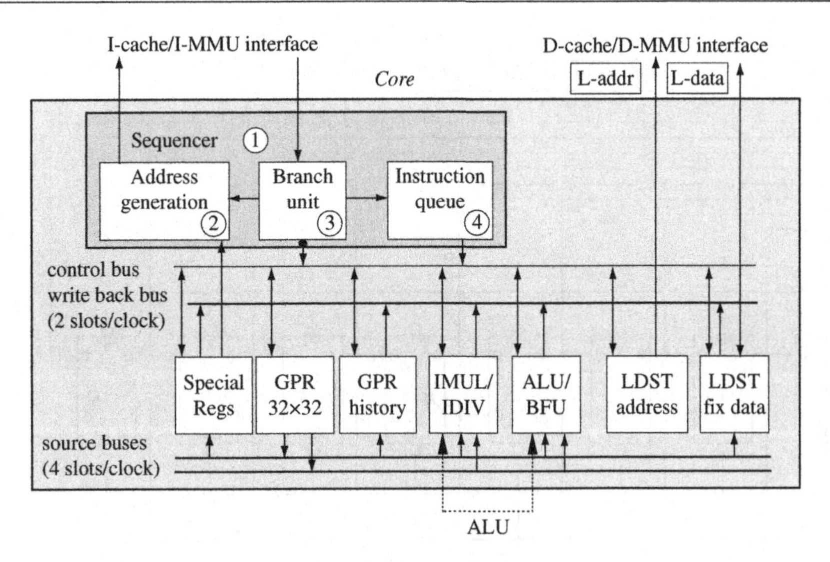


**Figure 4-27**

PowerPC core and the ALU.[15]

In the PowerPC core, the ALU is part of the "Fixed Point Unit" that implements all fixed-point instructions other than load/store instructions. The ALU is responsible for fixed-point logic, add, and subtract instruction implementation. In the case of the PowerPC, generated results of the ALU are stored in an Accumulator. Also, note that the PowerPC has an IMUL/IDIV unit (essentially another ALU) specifically for performing multiplication and division operations.

can also be saved in memory (on a stack or elsewhere) or in some hybrid combination of these locations.

*Registers*

Registers are simply a combination of various flip-flops that can be used to temporarily store data or to delay signals. A *storage register* is a form of fast programmable internal processor memory usually used to temporarily store, copy, and modify operands that are immediately or frequently used by the system. *Shift registers* delay signals by passing the signals between the various internal flip-flops with every clock pulse.

Registers are made up of a set of flip-flops that can be activated either individually or as a set. In fact, it is *the number of flip-flops in each register* that is actually used to describe a processor (e.g., a 32-bit processor has working registers that are 32 bits wide containing 32 flip-flops, a 16-bit processor has working registers that are 16 bits wide containing 16 flip-flops, etc.). The number of flip-flops within these registers also determines the width of the data buses used in the system. Figure 4-28 shows an example of how eight flip-flops could comprise an 8-bit register, and thus impact the size of the data bus.

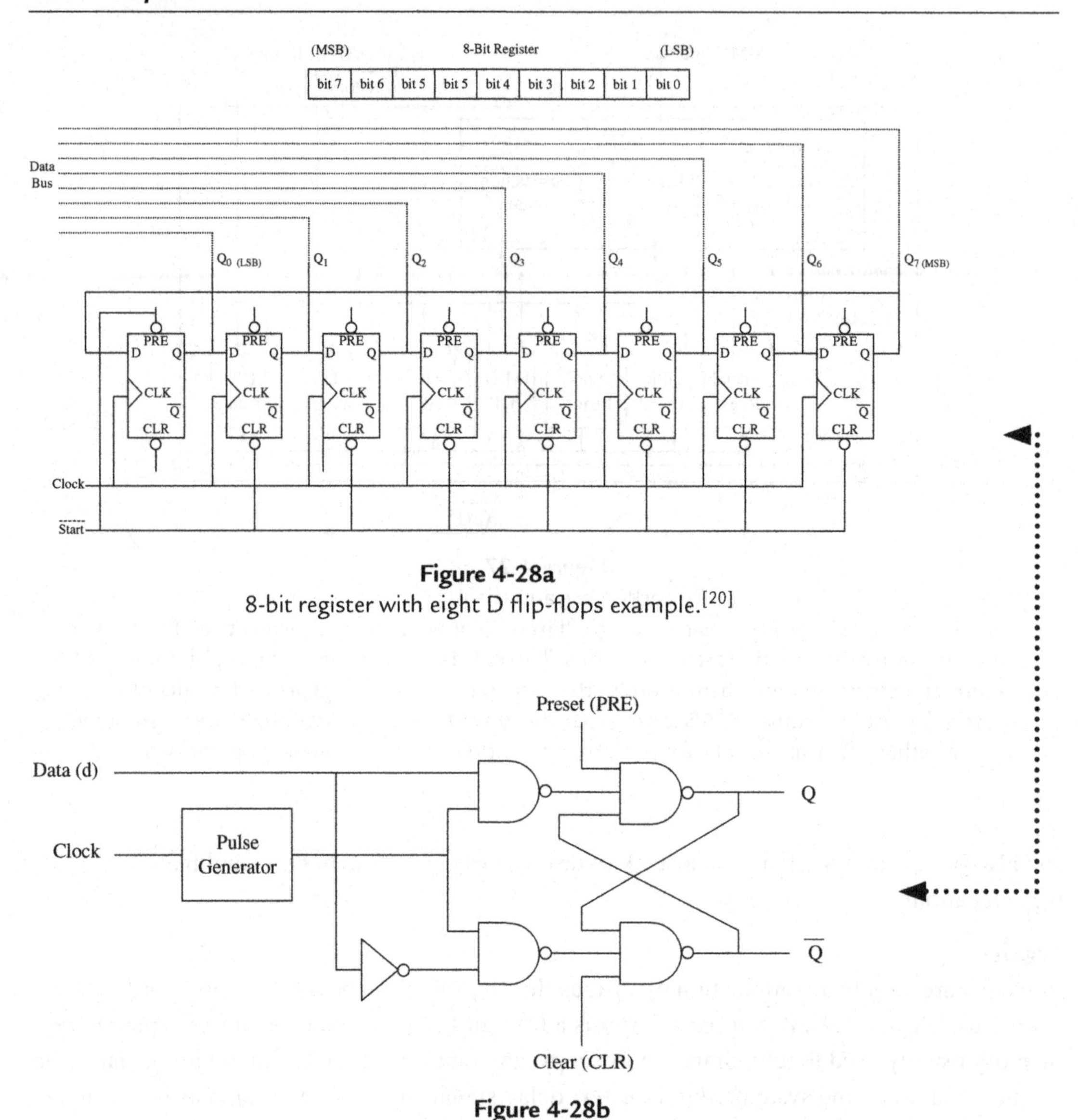


**Figure 4-28a**
8-bit register with eight D flip-flops example.[20]

**Figure 4-28b**
Example of gate level circuit of flip-flop.[20]

In short, registers are made up of one flip-flop for every bit being manipulated or stored by the register.

While ISA designs do not all use registers in the same way to process the data, storage typically falls under one of two categories: *general purpose* or *special purpose* (see Figure 4-29). *General-purpose registers* can be used to store and manipulate any type of data determined by the programmer, whereas *special-purpose registers* can *only* be used in a

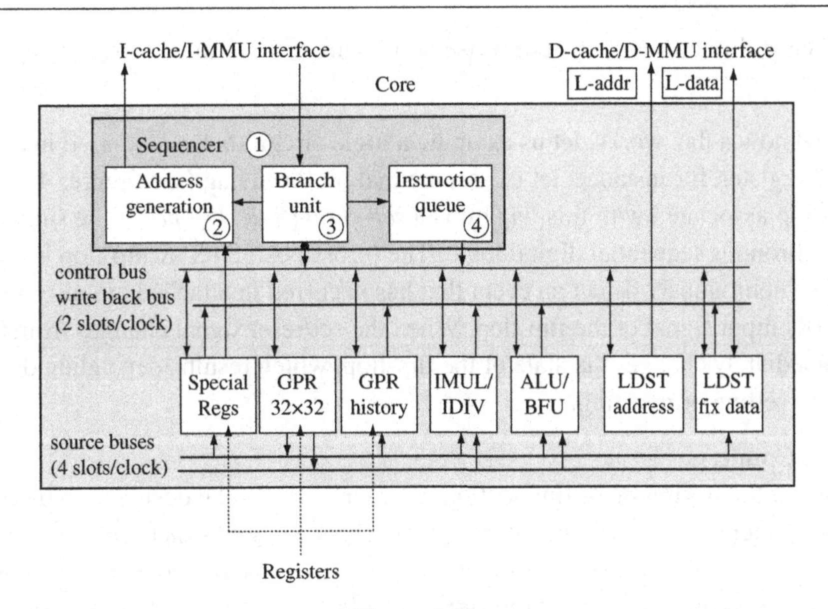


**Figure 4-29**

PowerPC core and register usage.[15]

The PowerPC Core has a "Register Unit" which contains all registers visible to a user. PowerPC processors generally have two types of registers: general-purpose and special-purpose (control) registers.

manner specified by the ISA, including holding results for specific types of computations, having predetermined *flags* (single bits within a register that can act and be controlled independently), acting as *counters* (registers that can be programmed to change states, i.e., increment, asynchronously or synchronously after a specified length of time), and controlling *I/O ports* (registers managing the external I/O pins connected to the body of the processor and to board I/O). Shift registers are inherently special purpose, because of their limited functionality.

The number of registers, the types of registers, and the size of the data that these registers can store (8-, 16-, 32-bit, etc.) vary depending on the CPU, according to the ISA definitions. In the cycle of fetching and executing instructions, the CPU's registers have to be fast, so as to quickly feed data to the ALU, for example, and to receive data from the CPU's internal data bus. Registers are also multiported so as to be able to both receive and transmit data to these CPU components. Some real-world examples of how some common registers in architectures can be designed, specifically flags and counters, are given below.

### Example 4.1: Flags

Flags are typically used to indicate to other circuitry that an event or a state change has occurred. In some architectures, flags can be grouped together into specific *flag registers*, while in other architectures, flags comprise some part of several different types of registers.

To understand how a flag works, let us examine a logic circuit that can be used in designing a flag. Given a register, for instance, let us assume that bit 0 is a flag (see Figures 4-30a and b) and the flip-flop associated with this flag bit is a *set–reset (S–R) flip-flop*—the simplest of data storage asynchronous sequential digital logic. The (cross NAND) S–R flip-flop is used in this example to asynchronously detect an event that has occurred in attached circuitry via the set (S) or reset (R) input signal of the flip-flop. When the set/reset signal changes from 0 to 1 or 1 to 0, it immediately changes the state of the flip-flop, which results, depending on the input, in the flip-flop setting or resetting.

### Example 4.2: Counters

As mentioned at the beginning of this section, registers can also be designed to be counters, programmed to increment or decrement either asynchronously or synchronously, such as with a processor's program counter (PC) or timers, which are essentially counters that count clock cycles. An asynchronous counter is a register whose flip-flops are not driven by the same

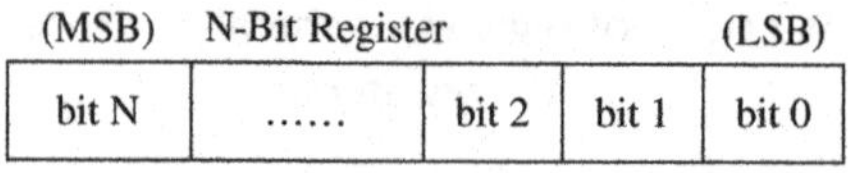

(MSB) N-Bit Register (LSB)

| bit N | ...... | bit 2 | bit 1 | bit 0 |
|---|---|---|---|---|

**Figure 4-30a**
N-bit register with flag and S-R flip-flop example.[20]

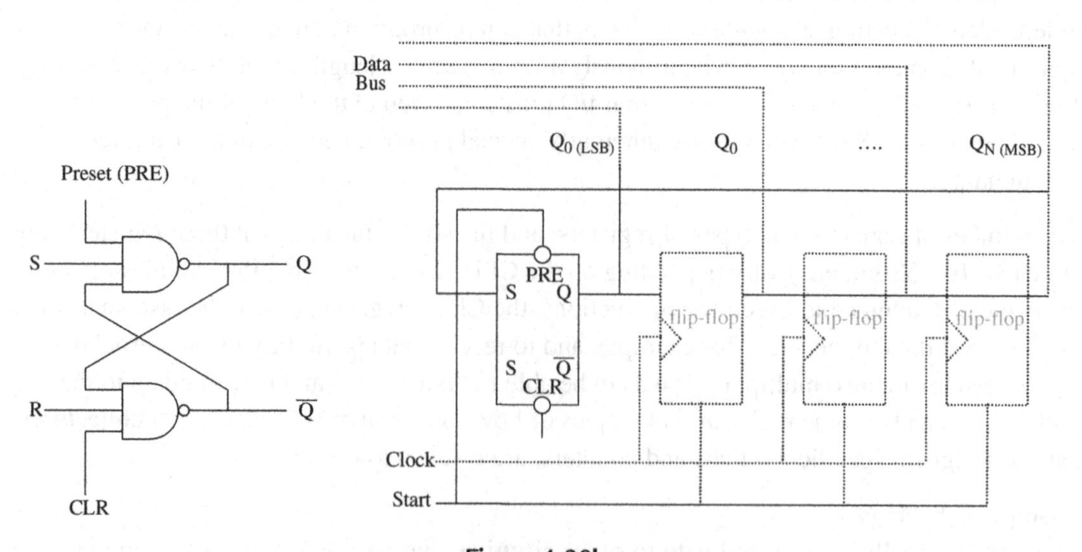


**Figure 4-30b**
SR flip-flop gate-level circuit example.[20]

central clock signal. Figure 4-31a shows an example of an 8-bit MOD-256 (modulus-256) asynchronous counter using JK flip-flops (which has 128 binary states—capable of counting between 0 and 255, 128 * 2 = 256). This counter is a binary counter, made up of 1s and 0s, with eight digits, one flip-flop per digit. It loops counting between 00000000 and 11111111, recycling back to 00000000 when 11111111 is reached, ready to start over with the count. Increasing the size of the counter—the maximum number of digits the counter can count to—is only a matter of adding an additional flip-flop for every additional digit.

All the flip flops of the counter are fixed in toggle mode; looking at the counter's truth table in Figure 4-31b under toggle mode, the flip-flop inputs (J and K) are both = 1 (HIGH). In toggle mode, the first flip-flop's output ($Q_0$) switches to the opposite of its current state at each active clock HIGH-to-LOW (falling) edge (see Figure 4-32).

As seen from Figure 4-32, the result of toggle mode is that $Q_0$, the output of the first flip-flop, has half the frequency of the CLK signal that was input into its flip-flop. $Q_0$ becomes the CLK signal for the next flip-flop in the counter. As seen from the timing diagram in Figure 4-33,

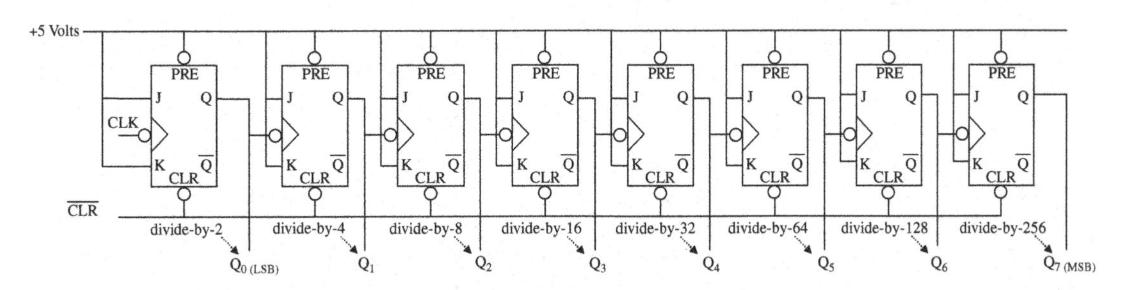


**Figure 4-31a**
8-bit MOD-256 asynchronous counter example.[20]

| $\overline{PRE}$ | $\overline{CLR}$ | CLK | J | K | Q | $\overline{Q}$ | Mode |
|---|---|---|---|---|---|---|---|
| 0 | 1 | x | x | x | 1 | 0 | Preset |
| 1 | 0 | x | x | x | 0 | 1 | Clear |
| 0 | 0 | x | x | x | 1 | 1 | Unused |
| 1 | 1 | – | 0 | 0 | $Q_0$ | $\overline{Q_0}$ | Hold |
| 1 | 1 | – | 0 | 1 | 0 | 0 | Reset |
| 1 | 1 | – | 1 | 0 | 0 | 0 | Set |
| 1 | 1 | – | 1 | 1 | $\overline{Q_0}$ | $Q_0$ | Toggle |
| 1 | 1 | –0.1 | 1 | 1 | $Q_0$ | $\overline{Q_0}$ | Hold |

**Figure 4-31b**
JK flip-flop truth table.[20]

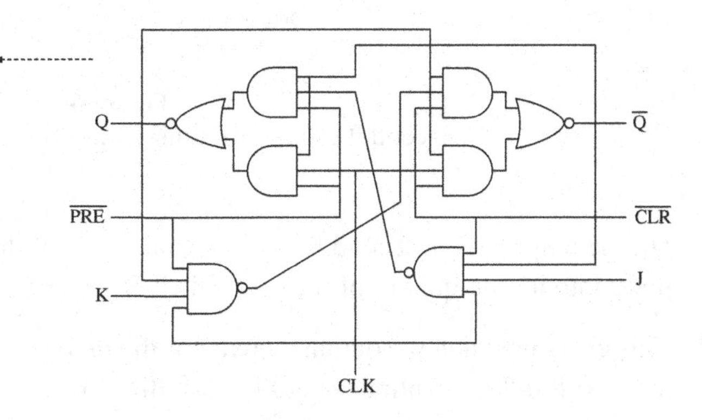


**Figure 4-31c**
JK flip-flop gate level diagram.[20]

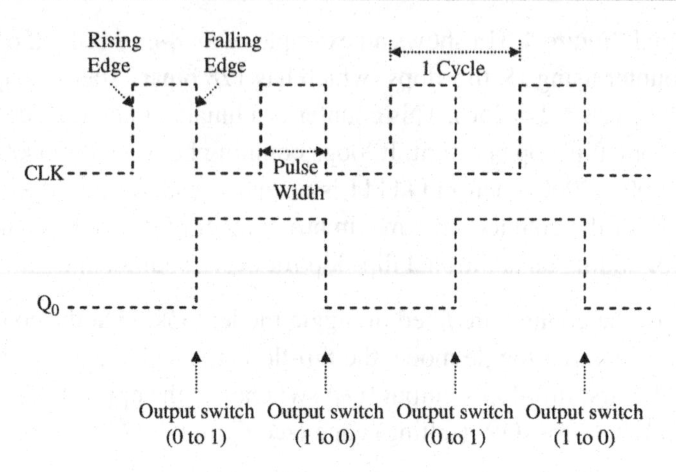


**Figure 4-32**
First flip-flop CLK timing waveform for MOD-256 counter.

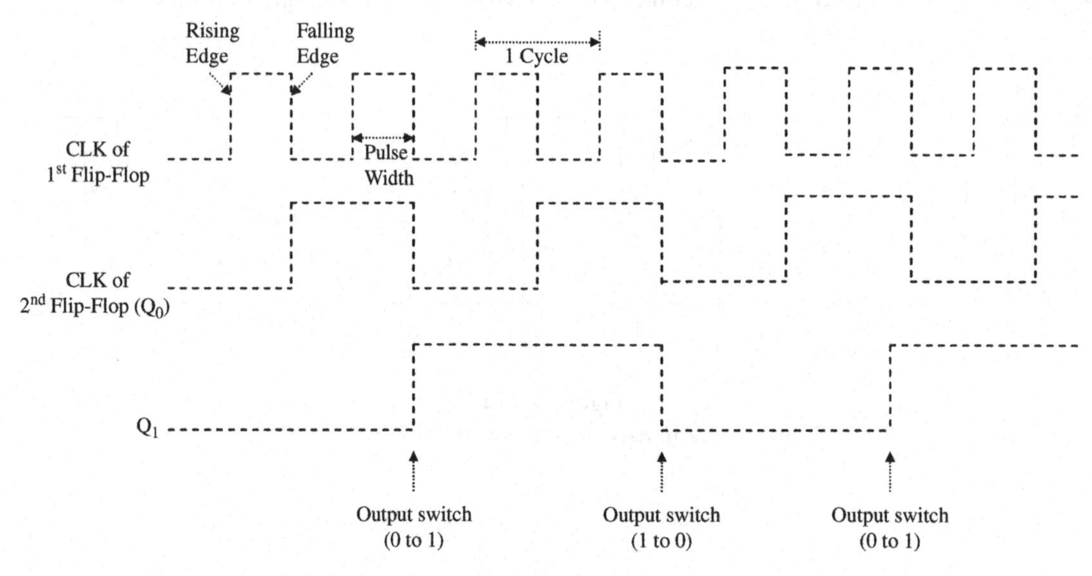


**Figure 4-33**
Second flip-flop CLK timing waveform for MOD-256 counter.

$Q_1$, the output of the second flip-flop signal, has half the frequency of the CLK signal that was input into it (one-quarter of the original CLK signal).

This cycle in which the output signals for the preceding flip-flops become the CLK signals for the next flip-flops continues until the last flip-flop is reached. The division of the CLK signal originally input into the first flip-flop can be seen in Figure 4-31a. The combination of output switching of all the flip-flops on the falling edges of the outputs of the previous

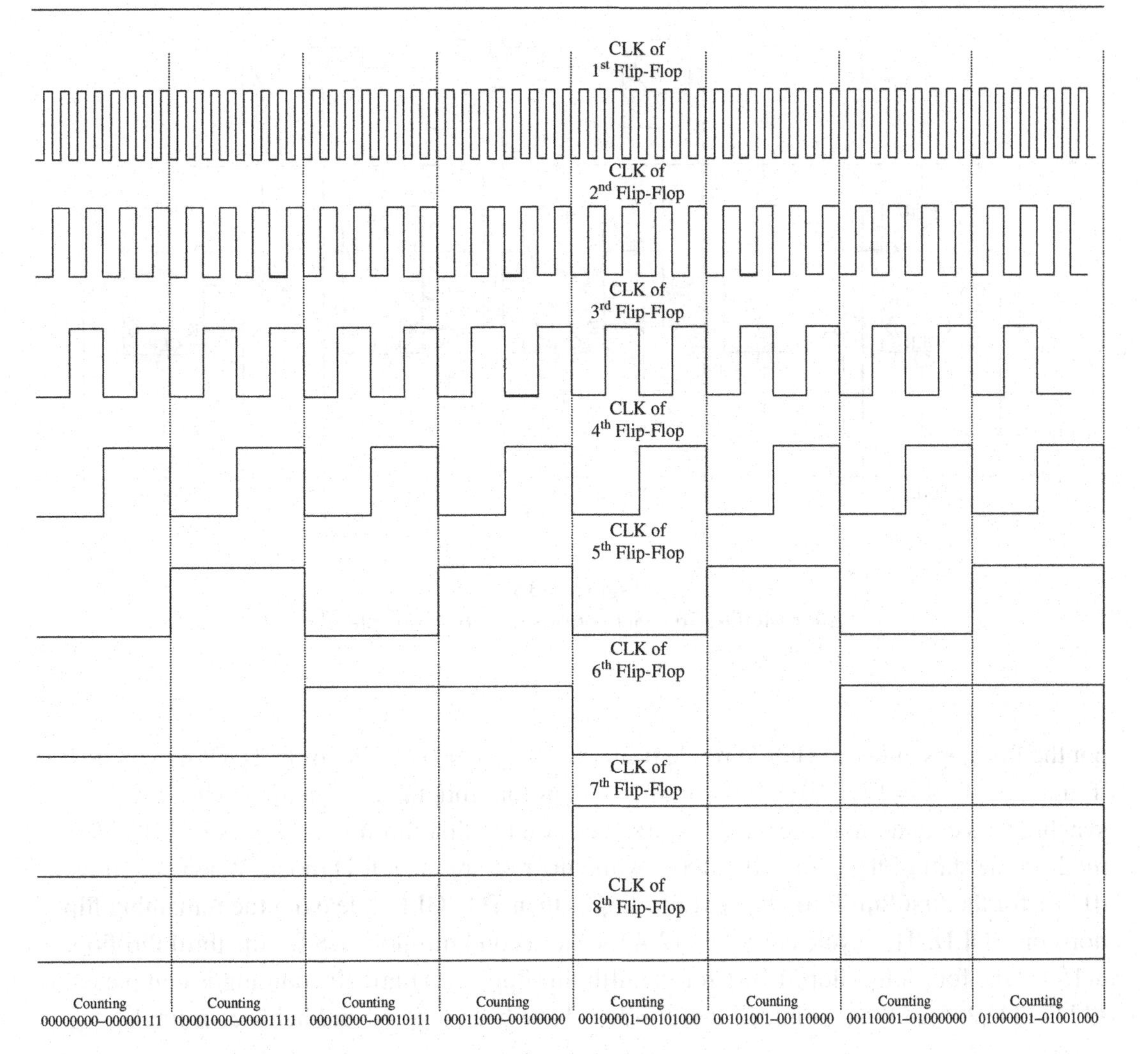


**Figure 4-34**
All flip-flop CLK timing waveforms for MOD-256 counter.

flip-flop, which acts as their CLK signals, is how the counter is able to count from 00000000 to 11111111 (see Figure 4-34).

With synchronous counters, all flip-flops within the counter are driven by a *common* clock input signal. Again using JK flip-flops, Figure 4-35 demonstrates how a MOD-256 synchronous counter circuitry differs from a MOD-256 asynchronous counter (the previous example).

The five additional AND gates (note that two of the five AND gates are not explicitly shown due to the scale of the diagram) in the synchronous counter example in Figure 4-35 serve to

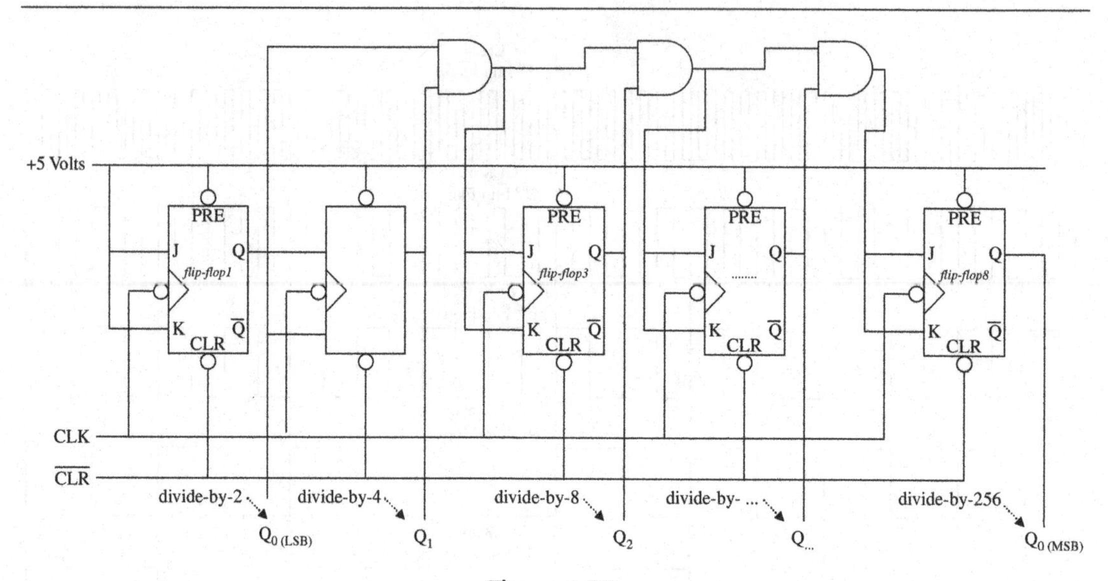


**Figure 4-35**
8-bit MOD-256 synchronous counter example.[20]

put the flip-flops either in HOLD mode if inputs J and K = 0 (LOW) or in TOGGLE mode if inputs J and K = 1 (HIGH). Refer to the JK flip-flop truth table in Figure 4-30b. The synchronous counter in this example works because the first flip-flop is always in TOGGLE mode at the start of the count 00000000, while the rest are in HOLD mode. When counting (0 to 1 for the first flip-flop), the next flip-flop is then TOGGLED, leaving the remaining flip-flops on HOLD. This cycle continues (2-4 for the second flip-flop, 4-8 for the third flip-flop, 8-15 for the fourth flip-flop, 15-31 for the fifth flip-flop, etc.) until all counting is completed to 11111111 (255). At that point, all the flip-flops have been toggled and held accordingly.

### *CU*

The CU is primarily responsible for generating timing signals, as well as controlling and coordinating the fetching, decoding, and execution of instructions in the CPU. After the instruction has been fetched from memory and decoded, the control unit then determines what operation will be performed by the ALU, and selects and writes signals appropriate to each functional unit within or outside of the CPU (memory, registers, ALU, etc.). To better understand how a processor's control unit functions, let's examine more closely the control unit of a PowerPC processor.

As shown in Figure 4-36, the PowerPC core's CU is called a "sequencer unit" and is the heart of the PowerPC core. The sequencer unit is responsible for managing the continuous cycle of fetching, decoding, and executing instructions while the PowerPC has power, including such tasks as:

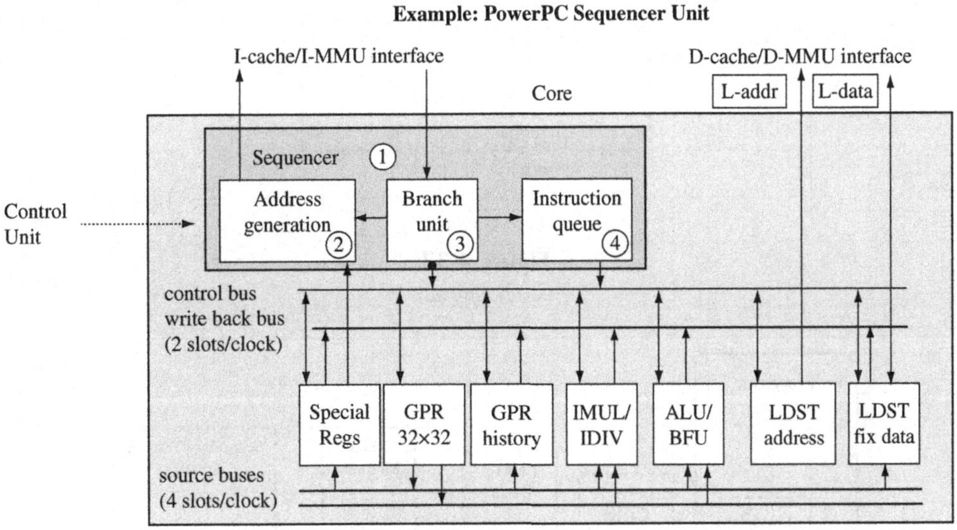

**Figure 4-36**
PowerPC core and the CU.[15]

- Providing the central control of the data and instruction flow among the other major units within the PowerPC core (CPU), such as registers, ALU, and buses.
- Implementing the basic instruction pipeline.
- Fetching instructions from memory to issue these instructions to available execution units.
- Maintaining a state history for handling exceptions.

Like many CUs, the PowerPC's sequencer unit isn't one physically separate, explicitly defined unit; rather, it is made up of several circuits distributed within the CPU that all work together to provide the managing capabilities. Within the sequencer unit these components are mainly an *address generation unit* (provides address of next instruction to be processed), a *branch prediction unit* (processes branch instructions), a *sequencer* (provides information and centralized control of instruction flow to the other control subunits), and an *instruction queue* (stores the next instructions to be processed and dispatches the next instructions in the queue to the appropriate execution unit).

### *CPU and System (Master) Clock*

A processor's execution is ultimately synchronized by an external *system* or *master clock*, located on the board. The master clock is an oscillator along with a few other components, such as a crystal. It produces a fixed frequency sequence of regular on/off pulse signals (square waves), as seen in Figure 4-37. The CU, along with several other components on an embedded board, depends on this master clock to function. Components are driven by either the actual level of the signal (a "0" or a "1"), the rising edge of a signal (the transition from "0" to "1"), and/or the falling edge of the signal (the transition from "1" to "0"). Different

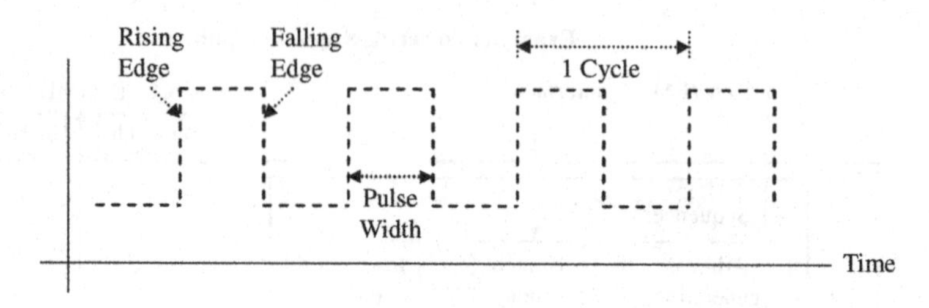


**Figure 4-37**
Clock signal.

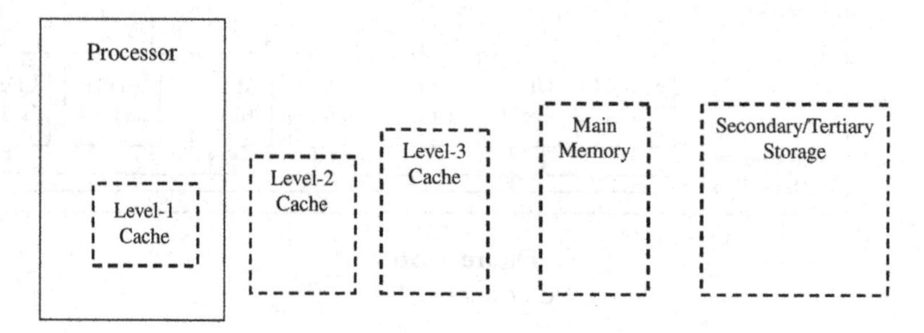


**Figure 4-38**
Memory hierarchy.

master clocks, depending on the circuitry, can run at a variety of frequencies, but typically must run so the slowest component on the board has its timing requirements met. In some cases, the master clock signal is divided by the components on the board to create other clock signals for their own use.

In the case of the CU, for instance, the signal produced by the master clock is usually divided or multiplied within the CPU's CU to generate at least one internal clock signal. The CU then uses internal clock signals to control and coordinate the fetching, decoding, and execution of instructions.

### 4.2.2 On-Chip Memory

*Note: The material covered in this section is very similar to that in Chapter 5, Board Memory, since, aside from certain types of memory and memory management components, the memory integrated within an IC is similar to memory located discretely on a board.*

The CPU goes to memory to get what it needs to process, because it is in memory that all of the data and instructions to be executed by the system are stored. Embedded platforms have a *memory hierarchy*, a collection of different types of memory, each with unique speeds, sizes, and usages (see Figure 4-38). Some of this memory can be physically integrated on the processor, such as registers, read-only memory (ROM), certain types of random access memory (RAM), and level 1 cache.

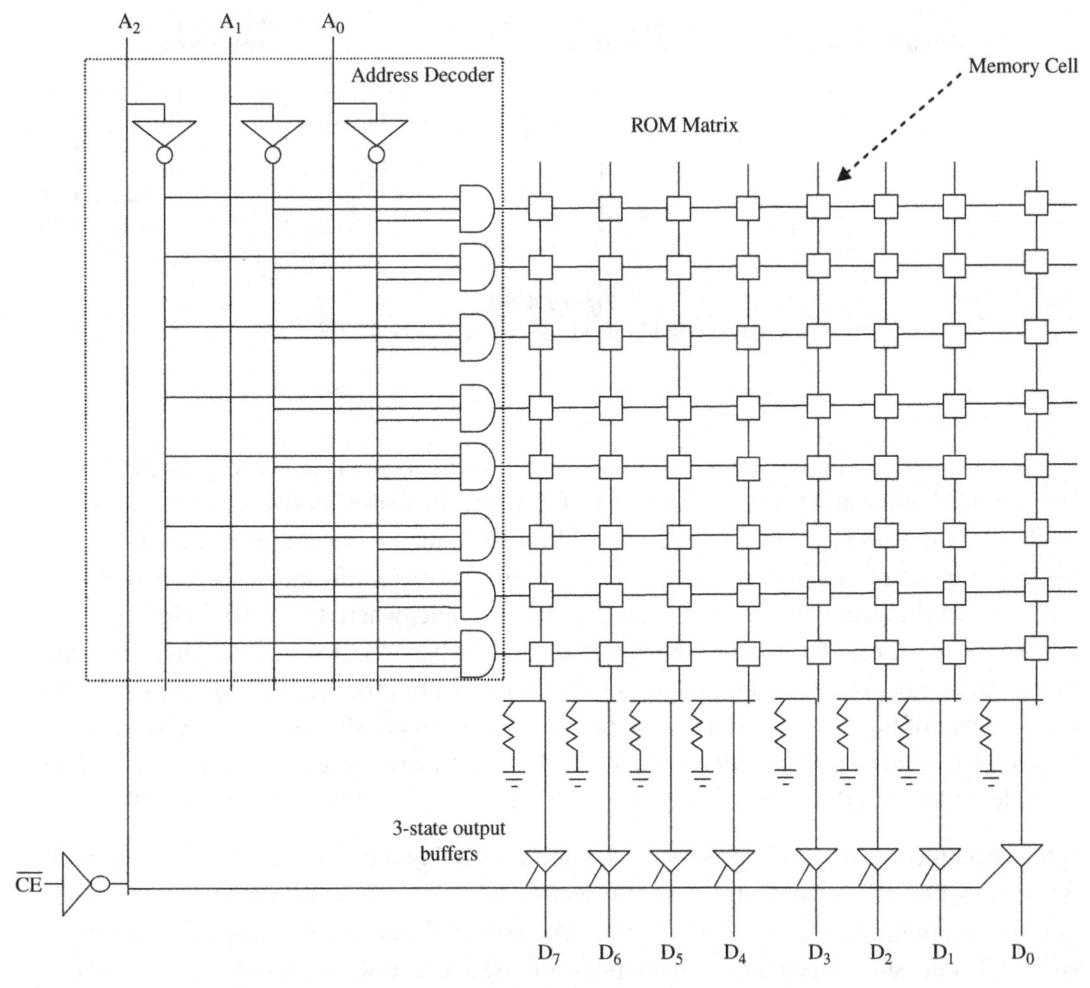


**Figure 4-39**
8 × 8 ROM logic circuit.[6]

### *ROM*

On-chip ROM is memory integrated into a processor that contains data or instructions that remain even when there is no power in the system, due to a small, longer-life battery, and therefore is considered to be *non-volatile memory*. The content of on-chip ROM usually can only be read by the system it is used in.

To get a clearer understanding of how ROM works, let's examine a sample logic circuit of 8 × 8 ROM, shown in Figure 4-39. This ROM includes three address lines ($\log_2 8$) for all eight words, meaning the 3-bit addresses ranging from 000 to 111 will each represent one of the eight bytes. *(Note: Different ROM designs can include a wide variety of addressing configurations for the exact same matrix size and this addressing scheme is just an example*

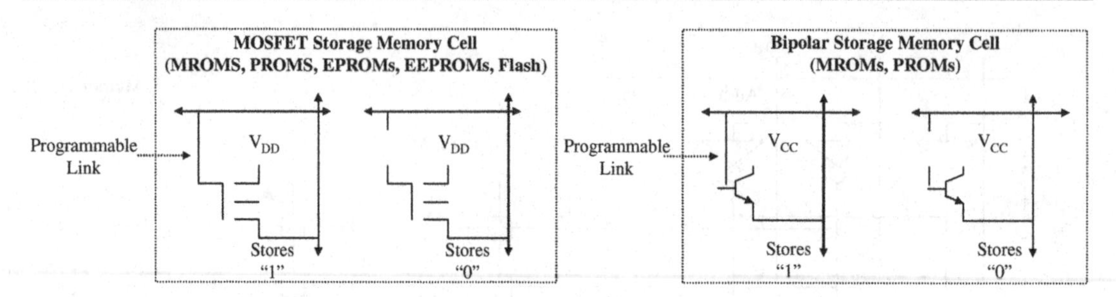


**Figure 4-40**
8 × 8 MOSFET and bipolar memory cells.[20]

*of one such scheme.)* $D_0$–$D_7$ are the output lines from which data is read—one output line for each bit. Adding additional rows to the ROM matrix increases its size in terms of the number of address spaces, whereas adding additional columns increases a ROM's data size (the number of bits per address it can store). ROM size specifications are represented in the real world identically to what is used in this example, where the matrix reference (8 × 8,16 K × 32, etc.) reflects the actual size of ROM where the first number, preceding the "×," is the number of addresses, and the second number, after the "×," reflects the size of the data (number of bits) at each address location (8 = byte, 16 = half word, 32 = word, etc.). Also, note that in some design documentation, the ROM matrix size may be summarized. For example, 16kB (kbytes) of ROM is 16K × 8 ROM, 32MB of ROM is 32M × 8 ROM.

In this example, the 8 × 8 ROM is an 8 × 8 matrix, meaning it can store eight different 8-bit bytes, or 64 bits of information. Every intersection of a row and column in this matrix is a memory location, called a *memory cell*. Each memory cell can contain either a bipolar or MOSFET transistor (depending on the type of ROM) and a fusible link (see Figure 4-40).

When a programmable link is in place, the transistor is biased ON, resulting in a "1" being stored. All ROM memory cells are typically manufactured in this configuration. When writing to ROM, a "0" is stored by breaking the programmable link. How links are broken depends on the type of ROM. How to read from a ROM depends on the ROM, but in this example, the chip enable (CE) line is toggled (HIGH to LOW) to allow the data stored to be output via $D_0$–$D_7$ after having received the 3-bit address requesting the row of data bits (see Figure 4-41).

Finally, the most common types of on-chip ROM include:

- *MROM* (mask ROM): ROM (with data content) that is permanently etched into the microchip during the manufacturing of the processor and cannot be modified later.
- *PROM* (programmable ROM), or OTPs (one-time programmables), which is a type of ROM that can be integrated on-chip, that is one-time programmable by a PROM programmer (in other words, it can be programmed outside the manufacturing factory).

| Gate | A2 | A1 | A0 | D7 | D6 | D5 | D4 | D3 | D2 | D1 | D0 |
|---|---|---|---|---|---|---|---|---|---|---|---|
| 1 | 0 | 0 | 0 | 1 | 1 | 1 | 1 | 0 | 1 | 1 | 1 |
| 2 | 0 | 0 | 1 | 1 | 1 | 0 | 1 | 1 | 1 | 0 | 1 |
| 3 | 0 | 1 | 0 | 0 | 1 | 1 | 1 | 1 | 0 | 1 | 1 |
| 4 | 0 | 1 | 1 | 0 | 0 | 1 | 0 | 1 | 1 | 1 | 1 |
| 5 | 1 | 0 | 0 | 1 | 1 | 1 | 1 | 1 | 1 | 1 | 1 |
| 6 | 1 | 0 | 1 | 1 | 1 | 1 | 0 | 0 | 0 | 0 | 1 |
| 7 | 1 | 1 | 0 | 0 | 1 | 1 | 1 | 1 | 1 | 1 | 1 |
| 8 | 1 | 1 | 1 | 1 | 0 | 1 | 1 | 1 | 1 | 1 | 0 |

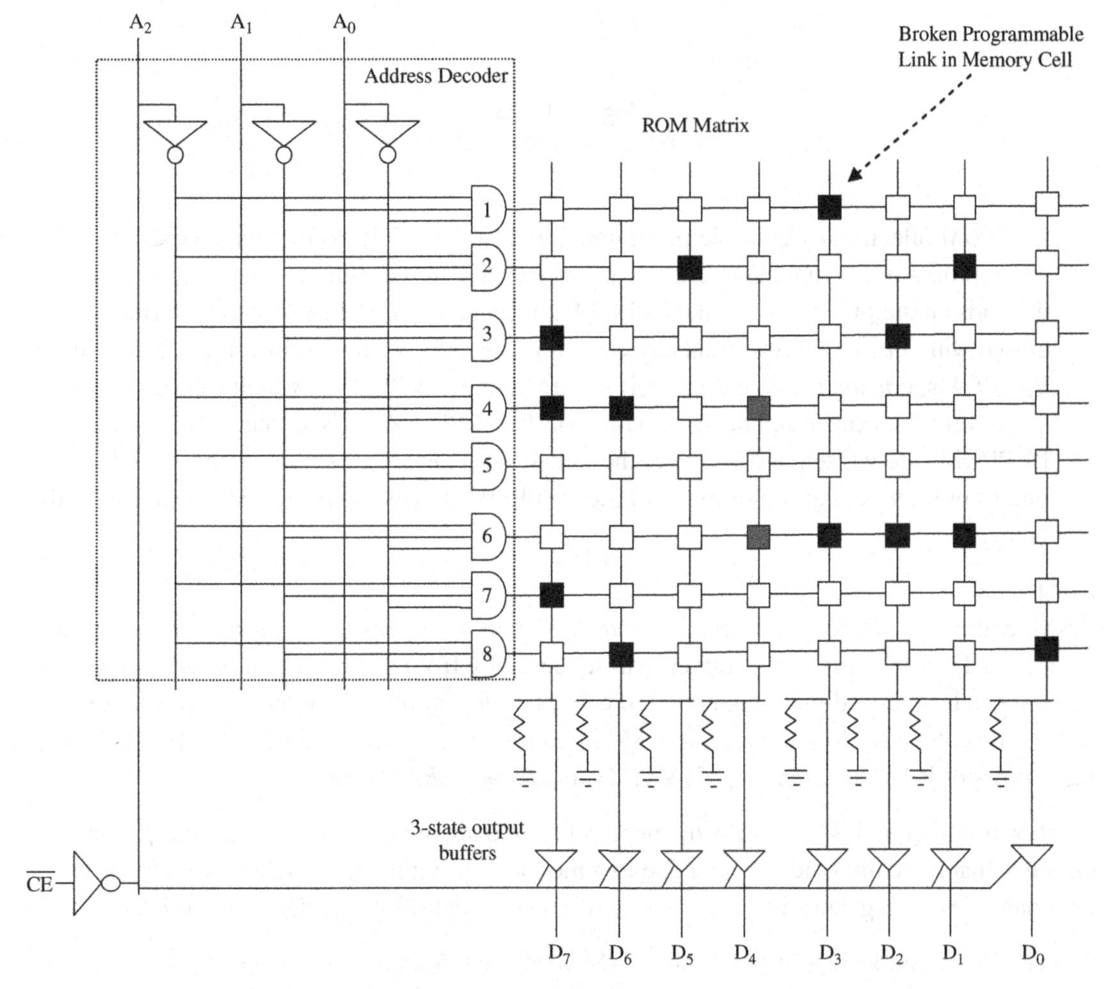


**Figure 4-41**
8 × 8 reading ROM circuit.[20]

- *EPROM* (erasable programmable ROM): ROM that can be integrated on a processor, in which content can be erased and reprogrammed more than once (the number of times erasure and reuse can occur depends on the processor). The content of EPROM is written to the device using special separate devices and erased, either selectively or in its entirety, using other devices that output intense ultraviolet light into the processor's built-in window.

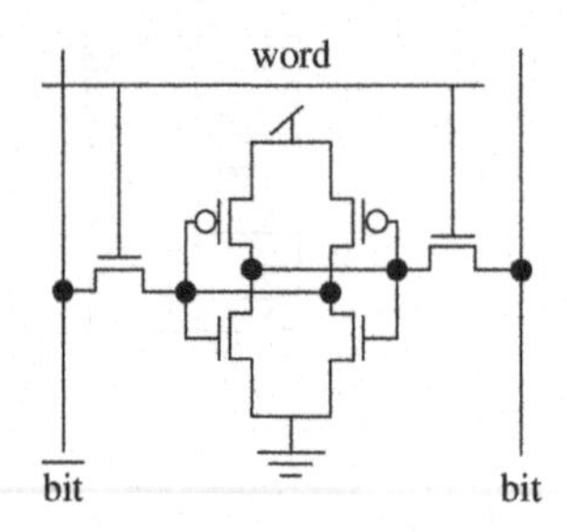


**Figure 4-42a**
Six-transistor SRAM cell.[25]

- *EEPROM* (electrically erasable programmable ROM): like EPROM, can be erased and reprogrammed more than once. The number of times erasure and reuse can occur depends on the processor. Unlike EPROMs, the content of EEPROM can be written and erased without using any special devices while the embedded system is functioning. With EEPROMs, erasing must be done in its entirety, unlike EPROMs, which can be erased selectively. A cheaper and faster variation of the EEPROM is *Flash* memory. Where EEPROMs are written and erased at the byte level, Flash can be written and erased in blocks or sectors (a group of bytes). Like EEPROM, Flash can be erased, while still in the embedded device.

*RAM*

RAM, commonly referred to as *main memory*, is memory in which any location within it can be accessed directly (randomly, rather than sequentially from some starting point) and whose content can be changed more than once (the number depending on the hardware). Unlike ROM, contents of RAM are erased if RAM loses power, meaning RAM is *volatile*. The two main types of RAM are *static RAM (SRAM)* and *dynamic RAM (DRAM)*.

As shown in Figure 4-42a, SRAM memory cells are made up of transistor-based flip-flop circuitry that typically holds its data due to a moving current being switched bidirectionally on a pair of inverting gates in the circuit, until power is cut off or the data is overwritten.

To get a clearer understanding of how SRAM works, let us examine a sample logic circuit of 4K × 8 SRAM shown in Figure 4-42b.

In this example, the 4K × 8 SRAM is a 4K × 8 matrix, meaning it can store 4096 (4 × 1024) different 8-bit bytes, or 32768 bits of information. As shown in the diagram below, 12 address lines ($A_0$–$A_{11}$) are needed to address all 4096 (000000000000b–111111111111b) possible addresses—one address line for every address digit of the address. In this example, the 4K × 8 SRAM is set up as a 64 × 64 array of rows and columns where addresses $A_0$–$A_5$ identify the row, and $A_6$–$A_{11}$ identify the column. As with ROM, every intersection of a row and column in the SRAM matrix is a memory cell, and in the case of SRAM memory

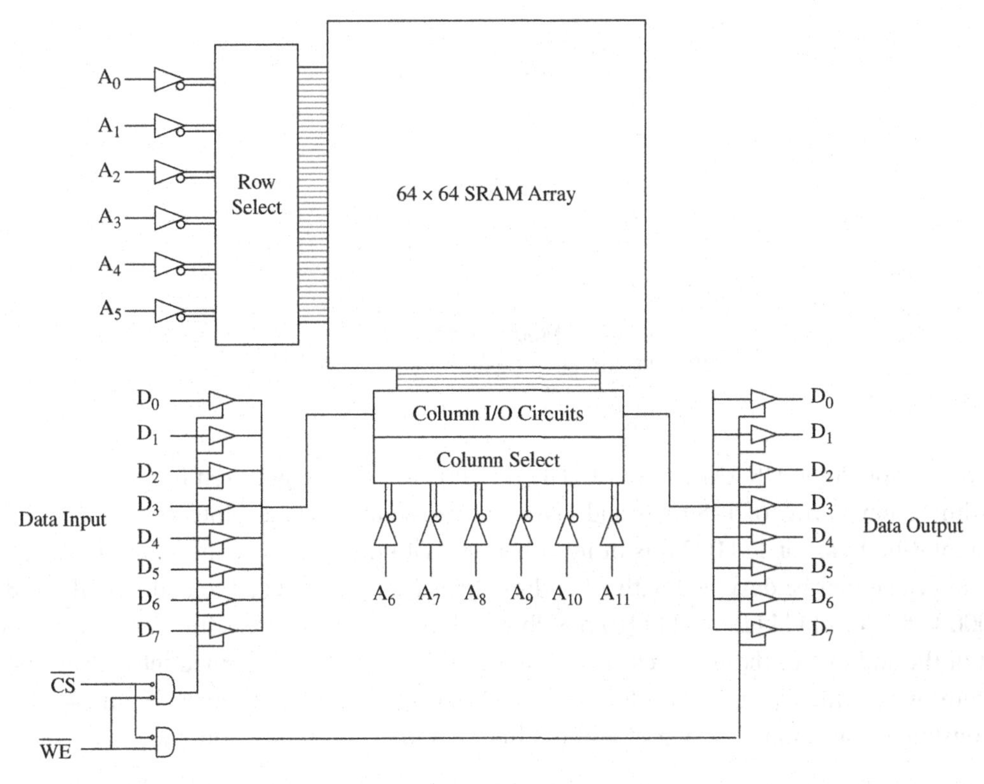


**Figure 4-42b**
4 K × 8 SRAM logic circuit.[20]

cells, they can contain flip-flop circuitry mainly based on semiconductor devices such as polysilicon load resistors, bipolar transistors, and/or CMOS transistors. There are eight output lines ($D_0$–$D_7$)—a byte for every byte stored at an address.

In this SRAM example, when the chip select (CS) is HIGH, then memory is in standby mode (no read or writes are occurring). When CS is toggled to LOW and write enable (WE) is LOW, then a byte of data is written through the data input lines ($D_0$–$D_7$) at the address indicated by the address lines. Given the same CS value (LOW) and WE is HIGH, then a byte of data is being read from the data output lines ($D_0$–$D_7$) at the address indicated by the address lines ($A_0$–$A_7$).

As shown in Figure 4-43, DRAM memory cells are circuits with *capacitors* that hold a charge in place—the charges or lack thereof reflecting data. DRAM capacitors need to be refreshed frequently with power in order to maintain their respective charges and to recharge capacitors after DRAM is read (reading DRAM discharges the capacitor). The cycle of discharging and recharging of memory cells is why this type of RAM is called dynamic.

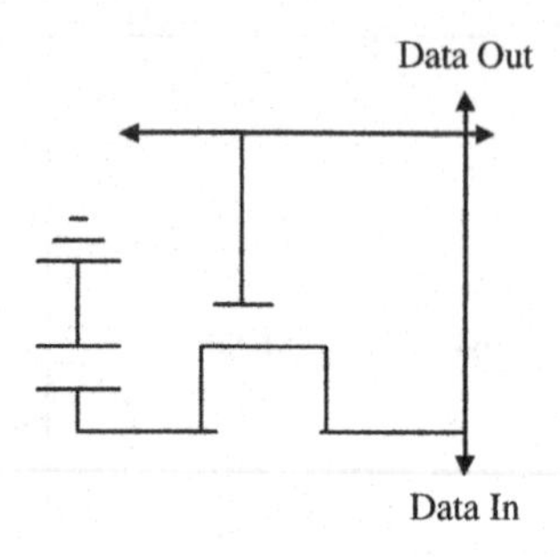


**Figure 4-43**
DRAM (capacitor-based) memory cell.[20]

Given a sample logic DRAM circuit of 16 K × 8, this RAM configuration is a two-dimensional array of 128 rows and 128 columns, meaning it can store 16 384 (16 × 1024) different 8-bit bytes or 131 072 bits of information. With this address configuration, larger DRAMs can either be designed with 14 address lines ($A_0$–$A_{13}$) needed to address all 16 384 (000000000000b–11111111111111b) possible addresses—one address line for every address digit of the address, or these address lines can be *multiplexed* (or combined into fewer lines to share) with some type of data selection circuit managing the shared lines. Figure 4-44 demonstrates how a multiplexing of address lines could occur in this example.

The 16 K × 8 DRAM is set up with addresses $A_0$–$A_6$ identifying the row and $A_7$–$A_{13}$ identifying the column. In this example, the Row Address Strobe (RAS) line is toggled (from HIGH to LOW) for $A_0$–$A_6$ to be transmitted and then the Column Address Strobe (CAS) line is toggled (from HIGH to LOW) for $A_7$–$A_7$ to be transmitted. After this point the memory cell is latched and ready to be written to or read from. There are eight output lines ($D_0$–$D_7$)—a byte for every byte stored at an address. When the WE input line is HIGH, data can be read from output lines $D_0$–$D_7$ and when WE is LOW, data can be written to input lines $D_0$–$D_7$.

One of the major differences between SRAM and DRAM lies in the makeup of the DRAM memory array itself. The capacitors in the memory array of DRAM are not able to hold a charge (data). The charge gradually dissipates over time, thus requiring some additional mechanism to *refresh* DRAM, in order to maintain the integrity of the data. This mechanism *reads* the data in DRAM before it is lost, via a sense amplification circuit that senses a charge stored within the memory cell, and *writes* it back onto the DRAM circuitry. Ironically, the process of reading the cell also discharges the capacitor, even though reading the cell in the first place is part of the process of correcting the problem of the capacitor gradually discharging in the first place. A *memory controller* (MEMC; see Section 5.4 on memory management for more information) in the embedded system typically manages a DRAM's recharging and discharging cycle by initiating refreshes and keeping track of the refresh sequence of events. It is this refresh cycling mechanism that discharges and recharges

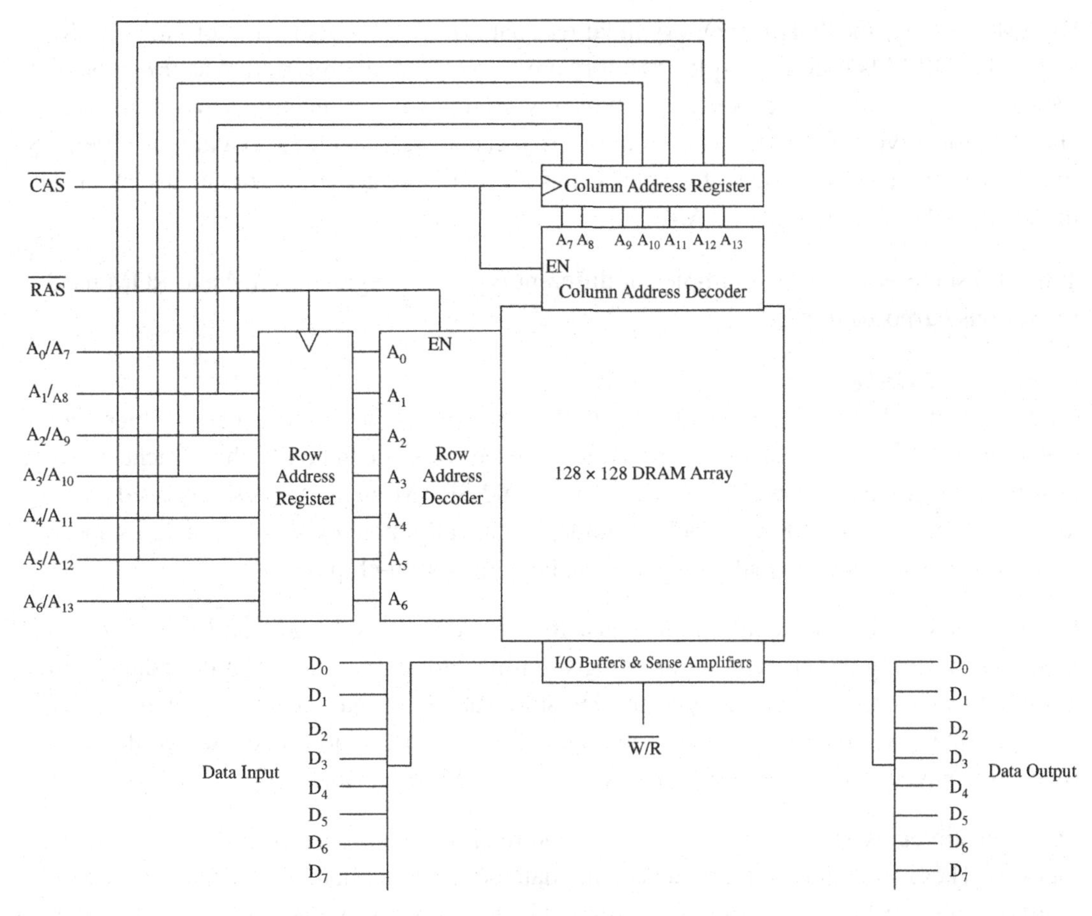


**Figure 4-44**
16 K × 8 SRAM logic circuit.[20]

memory cells that gives this type of RAM its name—"dynamic" RAM (DRAM)—and the fact that the charge in SRAM stays put is the basis for its name, "static" RAM (SRAM). It is this same additional recharge circuitry that makes DRAM slower in comparison to SRAM. *(Note: SRAM is usually slower than registers, because the transistors within the flip-flop are usually smaller and thus do not carry as much current as those typically used within registers.)*

SRAMs also usually consume less power than DRAMs, since no extra energy is needed for a refresh. On the flip side, DRAM is typically cheaper than SRAM, because of its capacitance-based design, in comparison to its SRAM flip-flop counterpart (more than one transistor). DRAM also can hold more data than SRAM, since DRAM circuitry is much smaller than SRAM circuitry and more DRAM circuitry can be integrated into an IC.

DRAM is usually the "main" memory in larger quantities, and is also used for video RAM and cache. DRAMs used for display memory are also commonly referred to as *frame buffers.* SRAM, because it is more expensive, is typically used in smaller quantities, but because it is also the fastest type of RAM, it is used in external cache (see Section 5.2) and video memory (when processing certain types of graphics, and given a more generous budget, a system can implement a better-performing RAM).

Table 4-3 summarizes some examples of different types of integrated RAM and ROM used for various purposes in ICs.

*Cache (Level 1 Cache)*

Cache is the level of memory between the CPU and main memory in the memory hierarchy (see Figure 4-45). Cache can be integrated into a processor or can be off-chip. Cache existing on-chip is commonly referred to as level 1 cache and SRAM memory is usually used as level 1 cache. Because (SRAM) cache memory is typically more expensive due to its speed, processors usually have a small amount of cache, whether on-chip or off-chip.

Using cache has become popular in response to systems that display a good *locality of reference*, meaning that these systems in a given time period access most of their data from a limited section of memory. Cache is used to store subsets of main memory that are used or accessed often. Some processors have one cache for both instructions and data, while other processors have separate on-chip caches for each. See Figure 4-46.

Different strategies are used when writing to and reading data from level 1 cache and main memory. These strategies include transferring data between memory and cache in either one-word or multiword *blocks*. These blocks are made up of data from main memory, as well as of tags, which represent the location of that data in main memory (called *tags*).

In the case of writing to memory, given some memory address from the CPU, this address is translated to determine its equivalent location in level 1 cache, since cache is a snapshot of a subset of memory. Writes must be done in both cache and main memory to ensure that cache and main memory are *consistent* (have the same value). The two most common write strategies to guarantee this are *write-through*, in which data is written to both cache and main memory every time, and *write-back*, in which data is initially only written into cache, and only when it is to be bumped and replaced from cache is it written into main memory.

When the CPU wants to read data from memory, level 1 cache is checked first. If the data is in cache, it is called a *cache hit*. The data is returned to the CPU and the memory access process is complete. If the data is not located in level 1 cache, it is called *cache miss*. Off-chip caches (if any) are then checked for the data desired. If this is a miss, then main memory is accessed to retrieve the data and return it to the CPU.

**Table 4-3: On-Chip Memory[26]**

| | Main Memory | Video Memory | Cache |
|---|---|---|---|
| SRAM | NA | RAMDAC (Random Access Memory Digital-to-Analog Converter) RAMDAC processors are used in video cards for display systems without true color, to convert digital image data into analog display data for analog displays, such as CRTs (cathode ray tubes). The built-in SRAM contains the color palette table that provides the RGB (Red/Green/Blue) on version values used by the DACs (digital-to-analog converters), also built into the RAMDAC, to change the digital image data into the analog signals for the display units. | SRAM has been used for both level 1 and level 2 caches. A type of SRAM, called BSRAM (Burst/SynchBurst Static Random Access Memory), that is synchronized with either the system clock or a cache bus clock, has been primarily used for level 2 cache memory (see Section 4.2). |
| DRAM | SDRAM (Synchronous Dynamic Random Access Memory) is DRAM that is synchronized with the microprocessor's clock speed (in MHz). Several types of SDRAMs are used in various systems, such as the JDEC SDRAM (JEDEC Synchronous Dynamic Random Access Memory), PC100 SDRAM (PC100 Synchronous Dynamic Random Access Memory), and DDR SDRAM (Double Data Rate Synchronous Dynamic Random Access Memory). ESDRAM (Enhanced Synchronous Dynamic Random Access Memory) is SDRAM that integrates SRAM within the SDRAM, allows for faster SDRAM (basically the faster SRAM portion of the ESDRAM is checked first for data, then if not found, the remaining SDRAM portion is searched). | RDRAM (On-Chip Rambus Dynamic Random Access Memory) and MDRAM (On-Chip Multibank Dynamic Random Access Memory) are DRAMs commonly used as display memory that store arrays of bit values (pixels of the image on the display). The resolution of the image is determined by the number of bits that have been defined per each pixel. | Enhanced Dynamic Random Access Memory (EDRAM) actually integrates SRAM within the DRAM, and is usually used as level 2 cache (see Section 4.2). The faster SRAM portion of EDRAM is searched first for the data, and if not found there, then the DRAM portion of EDRAM is searched. |

(*Continued*)

**Table 4-3: (Continued)**

| | Main Memory | Video Memory | Cache |
|---|---|---|---|
| | DRDRAM (Direct Rambus Dynamic Random Access Memory) and SLDRAM (SyncLink Dynamic Random Access Memory) are DRAMs whose bus signals can be integrated and accessed on one line, thus decreasing the access time (since synchronizing operations on multiple lines are not necessary). | FPM DRAM (Fast Page Mode Dynamic Random Access Memory), EDORAM/EDO DRAM (Data Output Random Access/Dynamic Random Access Memory), and BEDO DRAM (Data Burst Extended Data Output Dynamic Random Access Memory) ... | |
| | FRAM (Ferroelectric Random Access Memory) is non-volatile DRAM, meaning data isn't lost from DRAM when power is shut off. FRAM has a lower power requirement than other types of SRAM, DRAM, and some ROMs (Flash), and is targeted for smaller handheld devices (PDAs, phones, etc.). | | |
| | FPM DRAM (Fast Page Mode Dynamic Random Access Memory), EDORAM/EDO DRAM (Data Output Random Access/Dynamic Random Access Memory), and BEDO DRAM (Data Burst Extended Data Output Dynamic Random Access Memory) ... | | |

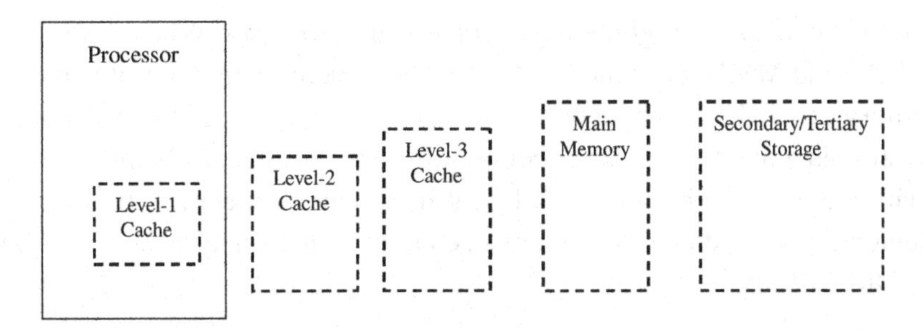


**Figure 4-45**
Level 1 cache in the memory hierarchy.

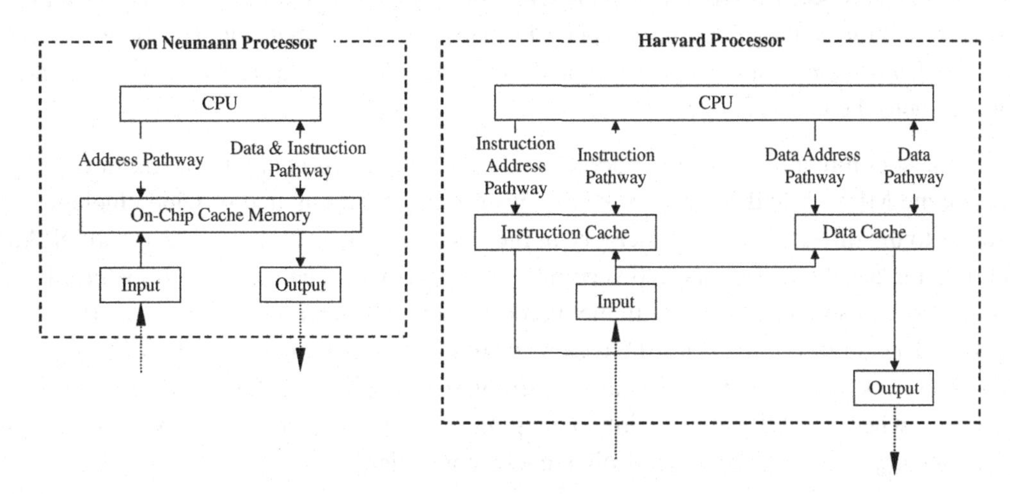


**Figure 4-46**
Level 1 cache in the von Neumann and Harvard models.

Data is usually stored in cache in one of three schemes:

- *Direct Mapped*: data in cache is located by its associated block address in memory (using the "tag" portion of the block).
- *Set Associative*: cache is divided into sets into which multiple blocks can be placed. Blocks are located according to an index field that maps into a cache's particular set.
- *Full Associative*: blocks are placed anywhere in cache and must be located by searching the entire cache every time.

In systems with *memory management units (MMUs)* to perform the translation of addresses (see next section "On-Chip Memory Management"), cache can be integrated between the CPU and the MMU, or the MMU and main memory. There are advantages and disadvantages to both methods of cache integration with an MMU, mostly surrounding the handling of *DMA (direct memory access)*, which is the direct access of off-chip main memory by slave processors on the board without going through the main processor. When cache is integrated between the CPU and MMU, only the CPU accesses to memory affect cache; therefore DMA writes to memory can make cache inconsistent with main memory unless CPU access to memory is restricted while DMA data is being transferred or cache is being kept updated by other units within the system besides the CPU. When cache is integrated between the MMU and main memory, more address translations need to be done, since cache is affected by both the CPU and DMA devices.

*On-Chip Memory Management*

Many different types of memory can be integrated into a system, and there are also differences in how software running on the CPU views memory addresses (*logical/virtual addresses*) and the actual *physical* memory addresses (the two-dimensional array or row and column). *Memory managers* are ICs designed to manage these issues and in some cases are integrated onto the master processor.

The two most common types of memory managers that are integrated into the master processor are MEMCs and MMUs. A *MEMC* is used to implement and provide glueless interfaces to the different types of memory in the system, such as cache, SRAM, and DRAM, synchronizing access to memory and verifying the integrity of the data being transferred. MEMCs access memory directly with the memory's own physical (two-dimensional) addresses. The controller manages the request from the master processor and accesses the appropriate banks, awaiting feedback and returning that feedback to the master processor. In some cases, where the MEMC is mainly managing one type of memory, it may be referred to by that memory's name (DRAM controller, cache controller, etc.).

*MMUs* are used to translate logical addresses into physical addresses (*memory mapping*), as well as handle memory security, control cache, handle bus arbitration between the CPU and memory, and generate appropriate exceptions. Figure 4-47 shows the MPC860, which has both an integrated MMU (in the core) and an integrated MEMC (in the system interface unit).

In the case of translated addresses, the MMU can use level 1 cache on the processor, or portions of cache allocated as buffers for caching address translations, commonly referred to as the translation lookaside buffer (TLB), to store the mappings of logical addresses to physical addresses. MMUs also must support the various schemes in translating addresses, mainly *segmentation*, *paging*, or some combination of *both* schemes. In general, segmentation is the division of logical memory into large variable-size sections, whereas paging is the dividing up of logical memory into smaller fixed-size units.

The memory protection schemes then provide shared, read/write, or read-only accessibility to the various pages and/or segments. If a memory access is not defined or allowed, an interrupt is typically triggered. An interrupt is also triggered if a page or segment isn't accessible during address translation (in the case of a paging scheme, a *page fault*, etc.). At that point the interrupt

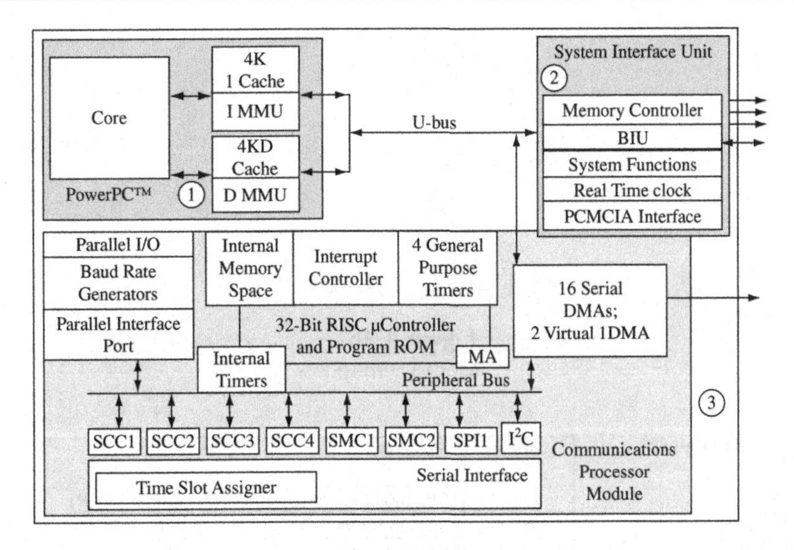


**Figure 4-47**

Memory management and the MPC860.[15] *© 2004 Freescale Semiconductor, Inc. Used by permission.*

would need to be handled (e.g., the page or segment would have to be retrieved from secondary memory). The scheme supporting segmentation and/or paging of the MMU typically depends on the software (i.e., the operating system (OS)). See Chapter 9 on OSs for more on virtual memory and how MMUs are used along with the system software to manage virtual memory.

*Memory Organization*

Memory organization includes not only the makeup of the memory hierarchy of the particular platform, but also the internal organization of memory—specifically what different portions of memory may or may not be used for, as well as how all the different types of memory are organized and accessed by the rest of the system. For example, some architectures may split memory so that a portion stores only instructions and another only stores data. The SHARC DSP contains integrated memory that is divided into separate *memory spaces* (sections of memory) for data and programs (instructions). In the case of the ARM architectures, some are based upon the von Neumann model (e.g., ARM7), which means it has one memory space for instructions and data, whereas other ARM architectures (i.e., ARM9) are based upon the Harvard model, meaning memory is divided into a section for data and a separate section for instructions.

The master processor, along with the software, treats memory as one large one-dimensional array, called a *memory map* (see Figure 4-48a). This map serves to clearly define what address or set of addresses are occupied by what components.

Within this memory map, an architecture may define multiple address spaces accessible to only certain types of information. For example, some processors may require at a specific

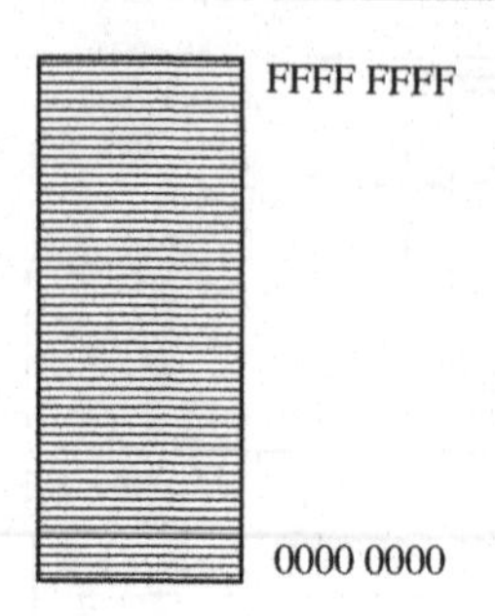


**Figure 4-48a**
Memory map.

FFFF FFFF

0000 0000

| Address Offset | Register | Size |
|---|---|---|
| 000 | SIU module configuration register (SIUMCR) | 32 bits |
| 004 | System Protection Control Register (SYPCR) | 32 bits |
| 008-00D | Reserved | 6 bytes |
| 00E | Software Service Register (SWSR) | 16 bits |
| 010 | SIU Interrupt Pending Register (SIPEND) | 32 bits |
| 014 | SIU Interrupt Mask Register (SIMASK) | 32 bits |
| 018 | SIU Interrupt Edge/Level Register (SIEL) | 32 bits |
| 01C | SIU Interrupt Vector Register (SIVEC) | 32 bits |
| 020 | Transfer Error Status Register (TESR) | 32 bits |
| .... | .... | .... |

**Figure 4-48b**
MPC860 registers within memory map.[15] *© 2004 Freescale Semiconductor, Inc. Used by permission.*

location—or given a random location—a set of offsets to be reserved as space for its own internal registers (see Figure 4-48b). The processor may also allow specific address spaces accessible to only internal I/O functionality, instructions (programs), and/or data.

### 4.2.3 Processor I/O

*Note: The material in this section is similar to material in Chapter 6, Board I/O, since aside from certain types of I/O or components of an I/O subsystem that are integrated on an IC versus discretely located on a board, the basics are essentially the same.*

I/O components of a processor are responsible for moving information to and from the processor's other components to any memory and I/O outside of the processor, on the board (see Figure 4-49). Processor I/O can either be input components that only bring information into the master processor, output components that bring information out of the master processor, or components that do both.

Virtually any electromechanical system, embedded and non-embedded, conventional (keyboard, mouse, etc.), as well as unconventional (power plants, human limbs, etc.), can be

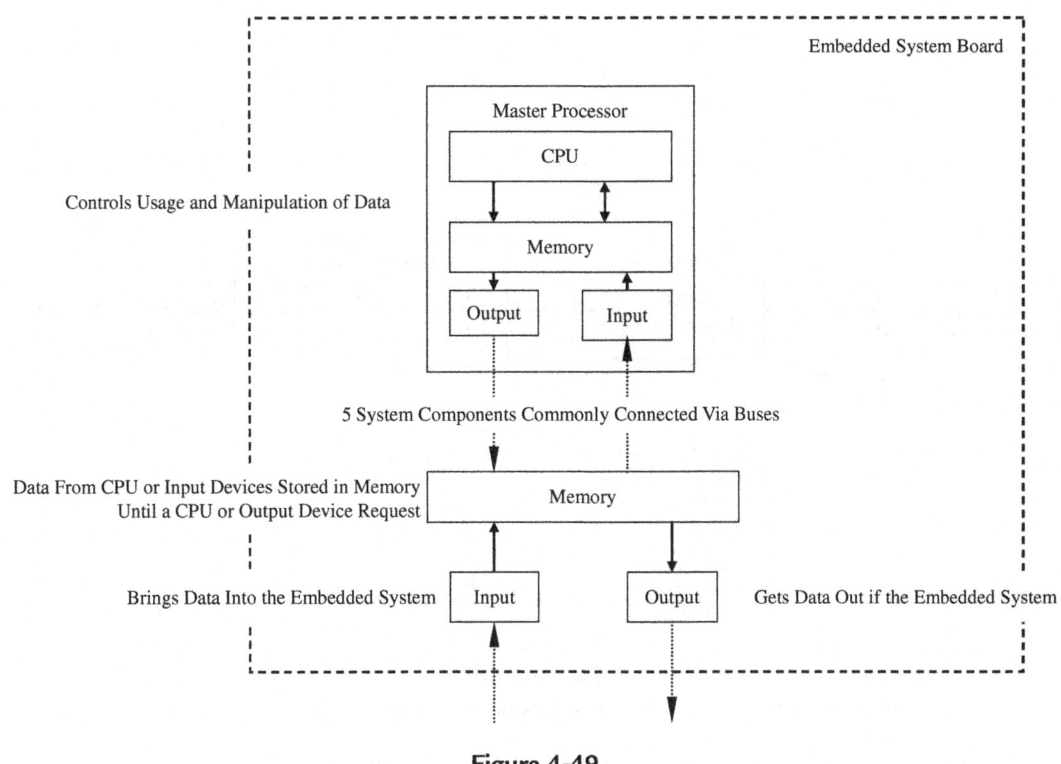


**Figure 4-49**
Processor I/O diagram.

connected to an embedded board and act as an I/O device. I/O is a high-level group that can be subdivided into smaller groups of either output devices, input devices, or devices that are both input and output devices. Output devices can receive data from board I/O components and display that data in some manner, such as printing it to paper, to a disk, or to a screen, or a blinking light-emitting diode (LED) for a person to see. An input device transmits data to board I/O components, such as a mouse, keyboard, or remote control. I/O devices can do both, such as a networking device that can transmit data to and from the Internet. An I/O device can be connected to an embedded board via a wired or wireless data transmission medium, such as a keyboard or remote control, or can be located on the embedded board itself, such as an LED.

Because I/O devices can be such a wide variety of electromechanical systems, ranging from simple circuits to another embedded system entirely, processor I/O components can be organized into categories based on the functions they support, the most common including:

- Networking and communications I/O (the physical layer of the OSI (Open Systems Interconnection) model; see Chapter 2).
- Input (keyboard, mouse, remote control, voice, etc.).

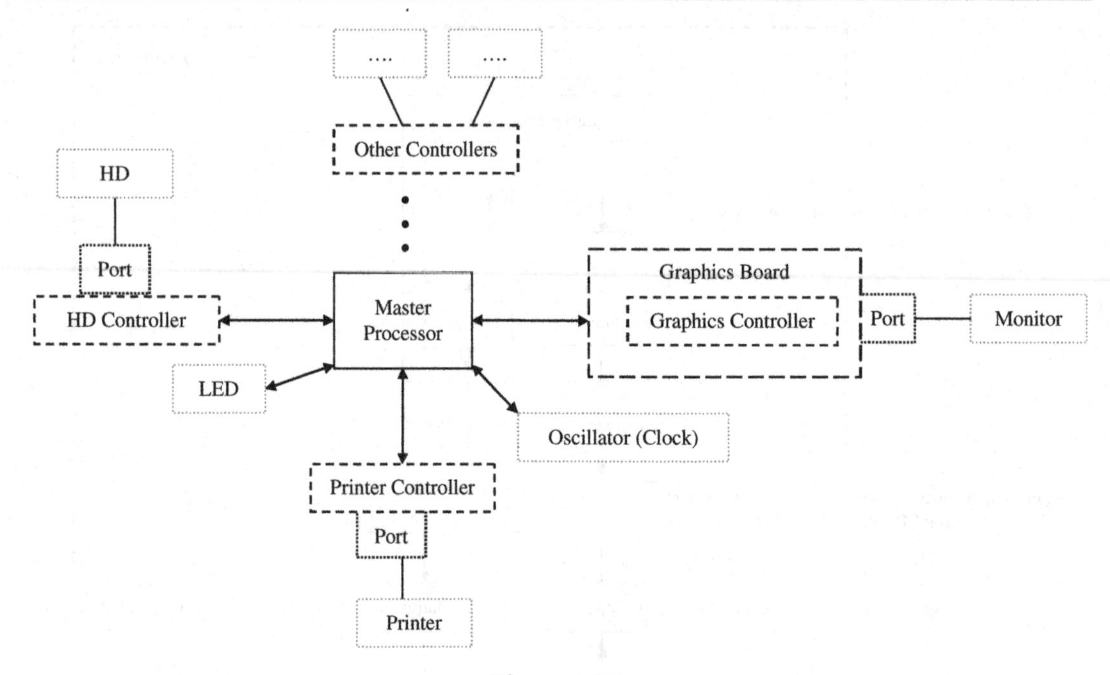


**Figure 4-50**
Ports and device controllers on an embedded board.

- Graphics and output I/O (touch screen, CRT, printers, LEDs, etc.).
- Storage I/O (optical disk controllers, magnetic disk controllers, magnetic tape controllers, etc.).
- Debugging I/O (Background Debug Mode (BDM), JTAG, serial port, parallel port, etc.).
- Real-time and miscellaneous I/O (timers/counters, analog-to-digital converters and digital-to-analog converters, key switches, etc.).

In short, an I/O subsystem can be as simple as a basic electronic circuit that connects the master processor directly to an I/O device (such as a master processor's I/O port to a clock or LED located on the board) to more complex I/O subsystem circuitry that includes several units, as shown in Figure 4-50. I/O hardware is typically made up of all or some combination of six main logical units:

- *Transmission medium*: wireless or wired medium connecting the I/O device to the embedded board for data communication and exchanges.
- *Communication port*: what the transmission medium connects to on the board, or, if a wireless system, what receives the wireless signal.
- *Communication interface*: manages data communication between master CPU and I/O device or I/O controller; also responsible for encoding data and decoding data to and from the logical level of an IC and the logical level of the I/O port. This interface can be integrated into the master processor, or can be a separate IC.

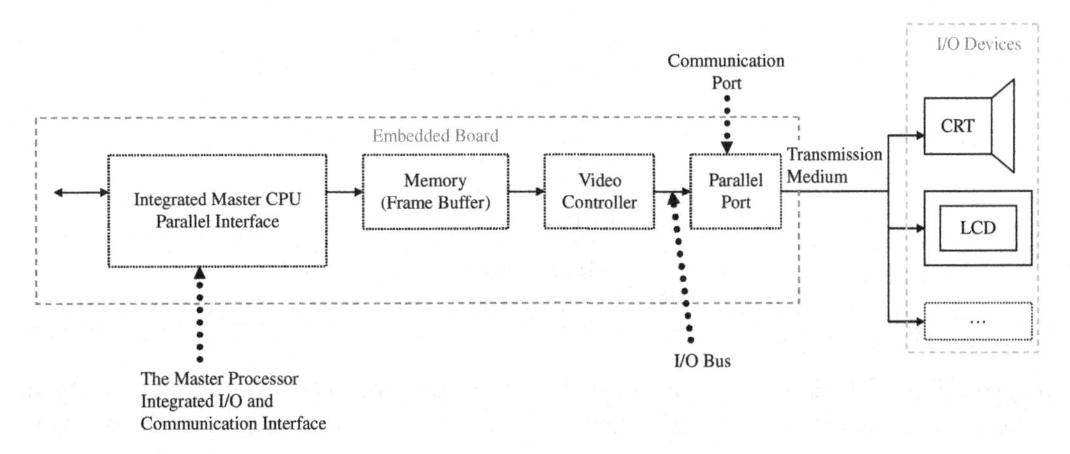


**Figure 4-51a**
Complex I/O subsystem.

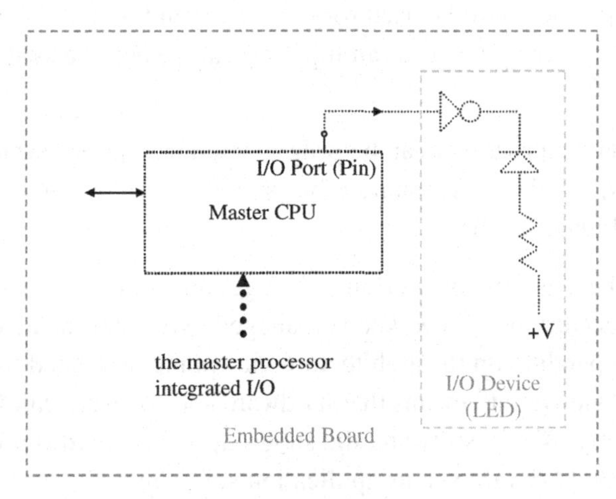


**Figure 4-51b**
Simple I/O subsystem.[29]

- *I/O controller*: slave processor that manages the I/O device.
- *I/O buses*: connection between the board I/O and master processor.
- Master processor integrated I/O.

This means that the I/O on the board can range from a complex combination of components, as shown in Figure 4-51a, to a few integrated I/O board components, as shown in Figure 4-51b.

Transmission mediums, buses, and board I/O are beyond the scope of this section, and are covered in Chapter 2 (transmission mediums), Chapter 7 (board buses), and Chapter 6 (board I/O), respectively. I/O controllers are a type of processor (see Section 4.1 *on* ISA Architecture Models). An I/O device can be connected directly to the master processor via I/O ports

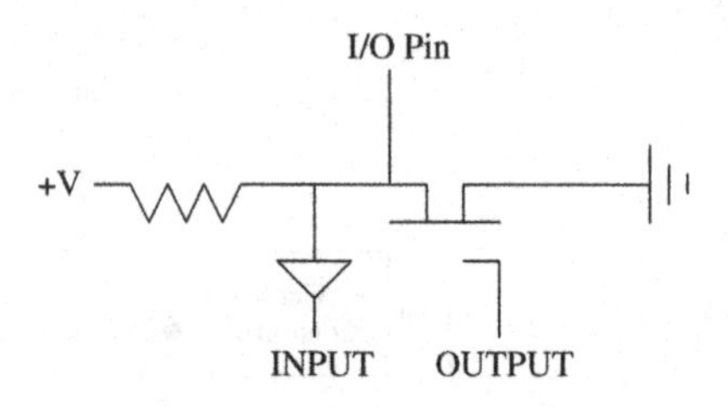


**Figure 4-52**
I/O port sample circuit.[23]

(processor pins) if the I/O devices are located on the board, or can be connected indirectly via a *communication interface* integrated into the master processor or a separate IC on the board.

As seen from the sample circuit in Figure 4-52, an I/O pin is typically connected to some type of current source and switching device. In this example it's a MOSFET transistor. This sample circuit allows for the pin to be used for both input and output. When the transistor is turned OFF (open switch), the pin acts as an input pin, and when the switch is ON it operates as an output port.

A pin or sets of pins on the processor can be programmed to support particular I/O functions (Ethernet port receiver, serial port transmitter, bus signals, etc.), through a master processor's control registers (see Figure 4-53).

Within the various I/O categories (networking, debugging, storage, etc.), processor I/O is typically subgrouped according to how data is managed. Note that the actual subgroups may be entirely different depending on the architecture viewpoint, as related to the embedded systems model. Here "viewpoint" means that hardware and software can view (and hence subgroup) I/O differently. Within software, the subgroups can even differ depending on the level of software (system software versus application software, OS versus device drivers, etc.). For example, in many OSs, I/O is considered to be either block or character I/O. Block I/O stores and transmits data in fixed block sizes, and is addressable only in blocks. Character I/O, on the other hand, manages data in streams of characters, the size of the character depending on the architecture, e.g., 1 byte).

From a hardware viewpoint, I/O manages (transmits and/or stores) data in *serial*, in *parallel*, or *both.*

#### *Managing I/O Data: Serial versus Parallel I/O*

##### Serial I/O

Processor I/O that can transmit and receive *serial* data is made up of components in which data is stored, transferred, and/or received *one bit at a time.* Serial I/O hardware is typically made up of some combination of the six main logical units outlined previously in Figure 4-50. Serial communication then includes within its I/O subsystem a *serial port* and a *serial interface.*

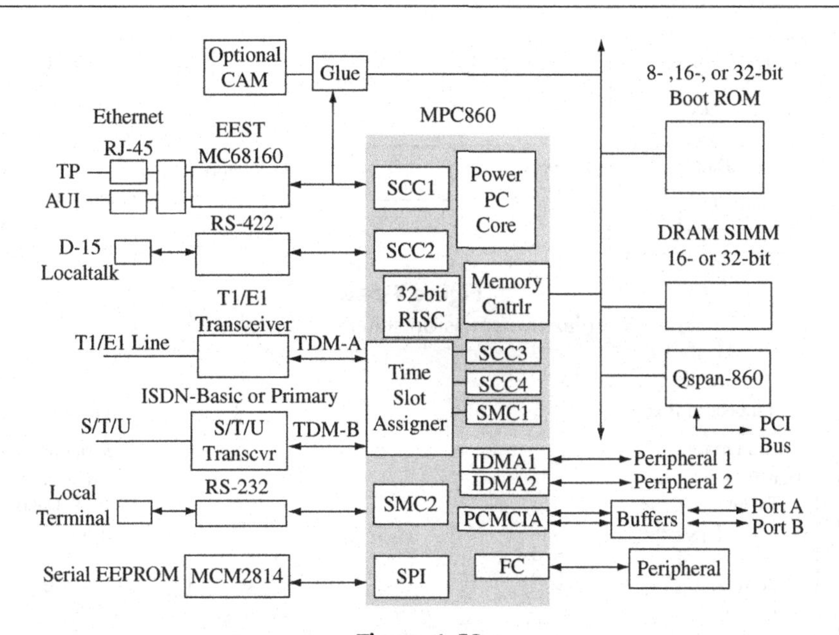


**Figure 4-53**
MPC860 reference platform and I/O.[24]
In the case of the MPC860, the I/O such as Ethernet and RS-232 are implemented by the SCC registers, RS-232 by SMC2, etc. The configuration of pins occurs in software and so will be discussed in Chapter 8. *© 2004 Freescale Semiconductor, Inc. Used by permission.*

*Serial interfaces* manage the serial data transmission and reception between the master CPU and either the I/O device or its controller. They include reception and transmission buffers to store and encode or decide the data they are responsible for transmitting either to the master CPU or an I/O device. In terms of serial data transmission and reception schemes, they generally differ as to what *direction* data can be transmitted and received, as well as in the actual *process* of how the data bits are transmitted (and thus received) within the data stream.

Data can be transmitted between two devices in one of three directions: in a one-way direction, in both directions but at separate times because they share the same transmission line, and in both directions simultaneously. A *simplex* scheme for serial I/O data communication is one in which a data stream can only be transmitted—and thus received—in one direction (see Figure 4-54a). A *half-duplex* scheme is one in which a data stream can be transmitted and received in either direction, but in only one direction at any one time (see Figure 4-54b). A *full-duplex* scheme is one in which a data stream can be transmitted and received in either direction, simultaneously (see Figure 4-54c).

Within the actual data stream, serial I/O transfers can occur either as a steady (continuous) stream at regular intervals regulated by the CPU's clock, referred to as a *synchronous* transfer, or intermittently at irregular (random) intervals, referred to as an *asynchronous* transfer.

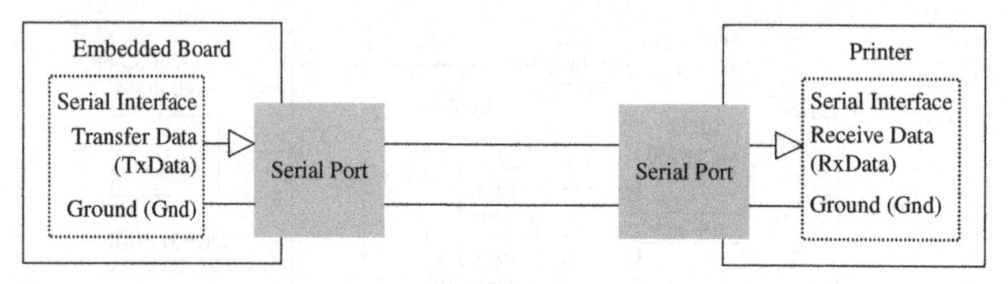


**Figure 4-54a**
Simplex transmission scheme example.[18]

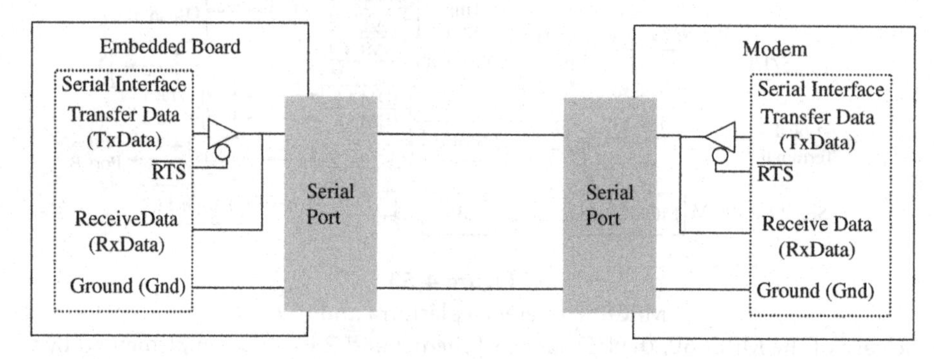


**Figure 4-54b**
Half-duplex transmission scheme example.[18]

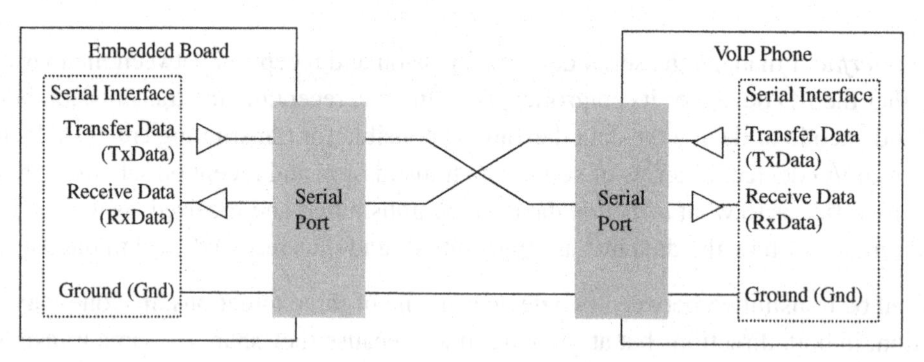


**Figure 4-54c**
Full-duplex transmission scheme example.[18]

In an asynchronous transfer (shown in Figure 4-55), the data being transmitted can be stored and modified within a serial interface's transmission buffer or registers. The serial interface at the transmitter divides the data stream into *packets* that typically range from either 4 to 8 or 5 to 9 bits, the number of bits per character. Each of these packets is then encapsulated in frames to be transmitted separately. The frames are packets that are modified before transmission by the serial interface to include a START bit at the start of the stream, and a

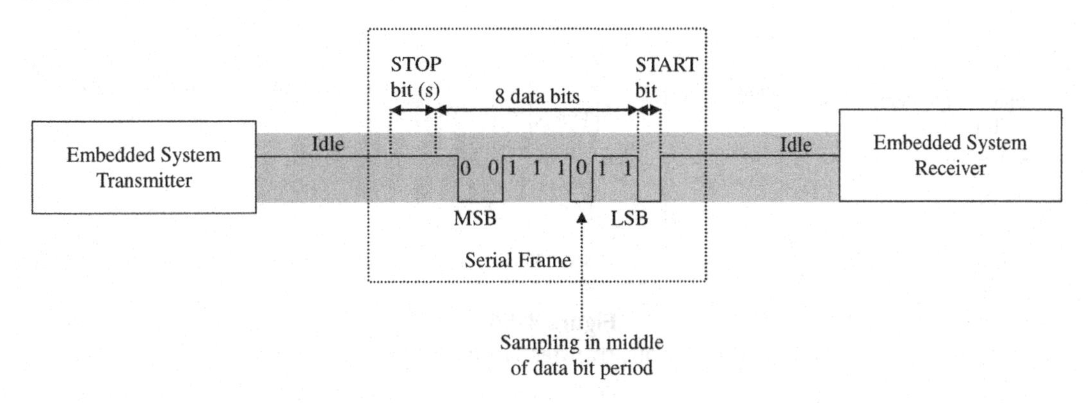


**Figure 4-55**
Asynchronous transfer sample diagram.

STOP bit or bits (i.e., can be 1, 1.5, or 2 bits in length to ensure a transition from "1" to "0" for the START bit of the next frame) at the end of the data stream being transmitted. Within the frame, after the data bits and before the STOP bit, a *parity* bit may also be appended. A START bit indicates the start of a frame, the STOP bit(s) indicates the end of a frame, and the parity is an optional bit used for very basic error checking. Basically, parity for a serial transmission can be NONE (for no parity bit and thus no error checking), EVEN (where the total number of bits set to "1" in the transmitted stream, excluding the START and STOP bits, needs to be an even number for the transmission to be a success), and ODD (where the total number of bits set to "1" in the transmitted stream, excluding the START and STOP bits, needs to be an odd number for the transmission to be a success).

Between the transmission of frames, the communication channel is kept in an idle state, meaning a logical level "1" or non-return to zero (NRZ) state is maintained.

The serial interface of the receiver then receives frames by synchronizing to the START bit of a frame, delays for a brief period, and then shifts in bits, one at a time, into its receive buffer until reaching the STOP bit(s). In order for asynchronous transmission to work, the *bit rate* (bandwidth) has to be synchronized in all serial interfaces involved in the communication. The *bit rate* is defined as:

Bit rate * number of actual data bits per frame/total number of bits per frame) * baud rate.

The baud rate is the total number of bits, regardless of type, per unit of time (kbits/s, Mbits/s, etc.) that can be transmitted.

Both the transmitter's serial interface and the receiver's serial interface synchronize with separate bit-rate clocks to sample data bits appropriately. At the transmitter, the clock starts when transmission of a new frame starts, and continues until the end of the frame so that the data stream is sent at intervals the receiver can process. At the receiving end, the clock starts

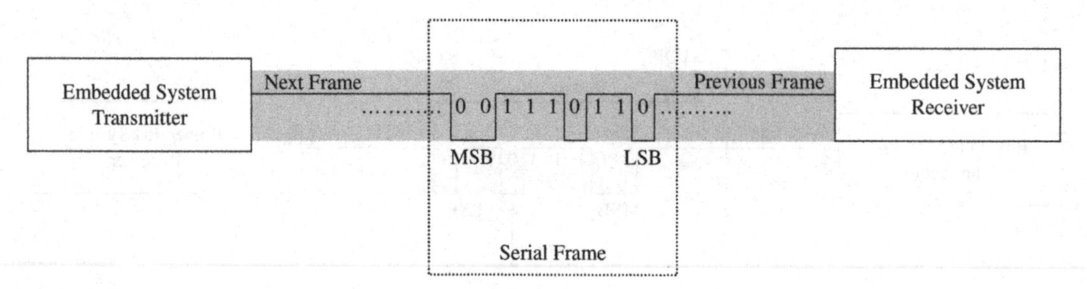


**Figure 4-56**
Synchronous transfer sample diagram.

with the reception of a new frame, delaying when appropriate (in accordance with the bit rate), sampling the middle of each data bit period of time, and then stopping when receiving the frame's STOP bit(s).

In a synchronous transmission (as shown in Figure 4-56), there are no START or STOP bits appended to the data stream and there is no idle period. As with asynchronous transmissions, the data rates on the receiving and transmitting have to be in sync. However, unlike the separate clocks used in an asynchronous transfer, the devices involved in a synchronous transmission are synchronizing off of one common clock which does not start and stop with each new frame (and on some boards there may be an entirely separate clock line for the serial interface to coordinate the transfer of bits). In some synchronous serial interfaces, if there is no separate clock line, the clock signal may even be transmitted along with the data bits.

The *UART (universal asynchronous receiver transmitter)* is an example of a serial interface that does asynchronous serial transmission, whereas *SPI (serial peripheral interface)* is an example of a synchronous serial interface. *(Note: Different architectures that integrate a UART or other types of serial interfaces can have varying names for the same type of interface, such as the MPC860, which has SMC (serial management controller) UARTs, for example. Review the relevant documentation to understand the specifics.)*

Serial interfaces can either be separate slave ICs on the board, or integrated onto the master processor. The serial interface transmits data to and from an I/O device via a *serial port* (see Chapter 6). Serial ports are serial communication (COM) interfaces that are typically used to interconnect off-board serial I/O devices to on-board serial board I/O. The serial interface is then responsible for converting data that is coming to and from the serial port at the logic level of the serial port into data that the logic circuitry of the master CPU can process.

### Processor Serial I/O Example 1: An Integrated UART

The UART is an example of a full-duplex serial interface that can be integrated into the master processor and that does asynchronous serial transmission. As mentioned earlier,

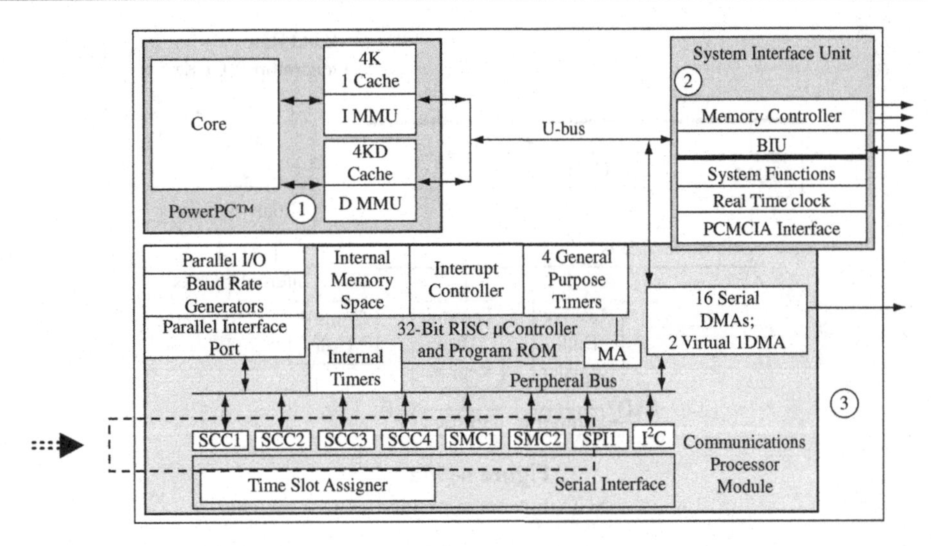


**Figure 4-57**

MPC860 UARTs.[24] *© 2004 Freescale Semiconductor, Inc. Used by permission.*

UARTs can exist in many variations and under many names; however, they are all based upon the same design—the original 8251 UART controller implemented in older PCs. A UART (or something like it) must exist on both sides of the communication channel, in the I/O device as well as on the embedded board, in order for this communication scheme to work.

In this example, we look at the MPC860 internal UART scheme since the MPC860 has more than one way to implement a UART. The MPC860 allows for two methods to configure a UART, either using an *SCC (serial communication controller)* or an *SMC (serial management controller)*. Both of these controllers reside in the PowerPC's Communication Processor Module (shown in Figure 4-57) and can be configured to support a variety of different communication schemes, such as Ethernet and HDLC, for the SCC, and transparent and GCI for SMCs. In this example, however, we are only examining both being configured and functioning as a UART.

### *MPC860 SCC in UART Mode*

As introduced at the start of this section, in an asynchronous transfer the data being transmitted can be stored and modified within a serial interface's transmission buffer. With the SCCs on the MPC860, there are two UART FIFO (First In First Out) buffers—one for receiving data for the processor and one for transmitting data to external I/O (see Figures 4-58a and b). Both buffers are typically allocated space in main memory.

As can be seen in Figures 4-58a and b, along with the reception and transmission buffers, there are control registers to define the baud rate, the number of bits per character, the parity, and the length of the stop bit, among other things. As shown in Figures 4-58a and b, as well

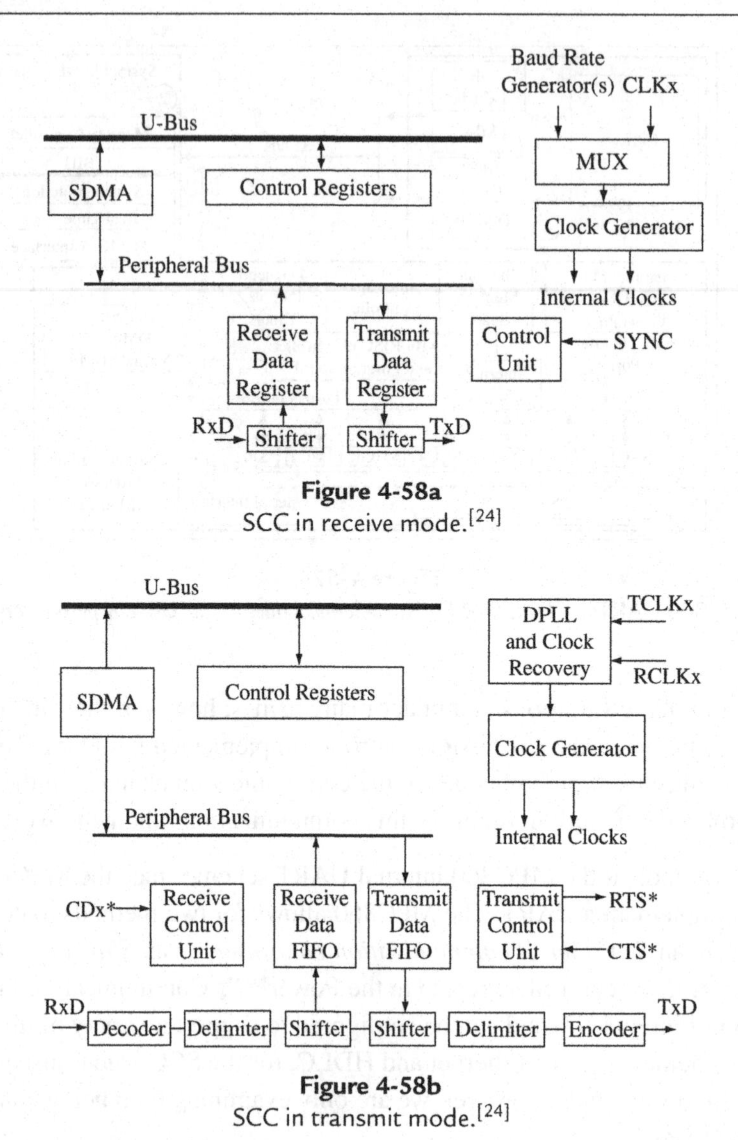


**Figure 4-58a**
SCC in receive mode.[24]



**Figure 4-58b**
SCC in transmit mode.[24]

as in Figure 4-59, there are five pins, extending out from the PowerPC chip, that the SCC is connected to for data transmission and reception: transmit (TxD), receive (RxD), carrier detect (CDx), collision on the transceiver (CTSx), and request-to-send (RTS). How these pins work together is described in the next few paragraphs.

In either receive or transmit modes, the internal SCC clock is activated. In asynchronous transfers, every UART has its own internal clock that, though unsynchronized with the clock in the UART of the external I/O device, is set at the same baud rate as that of the UART it is in communication with. The CDx is then asserted to allow the SCC to receive data or the collision on the CTSx is asserted to allow the SCC to transmit data.

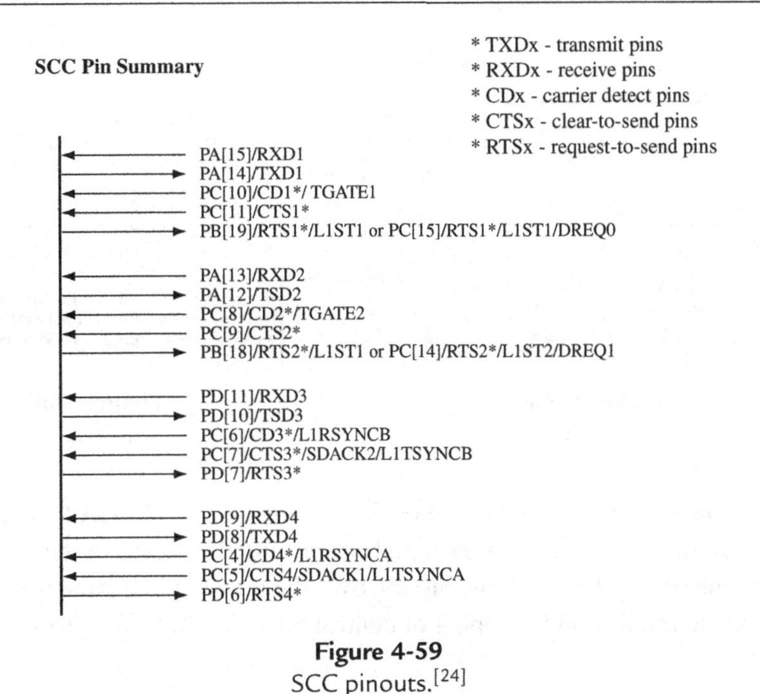


**Figure 4-59**
SCC pinouts.[24]

As mentioned, data is encapsulated into frames in asynchronous serial transmissions. When transmitting data, the SDMA transfers the data to the transmit FIFO and the RTS pin asserts (because it is a transmit control pin and asserts when data is loaded into the transmit FIFO). The data is then transferred (in parallel) to the shifter. The shifter shifts the data (in serial) into the delimiter, which appends the framing bits (start bits, stop bits, etc.). The frame is then sent to the encoder for encoding before transmission. In the case of an SCC receiving data, the framed data is then decoded by the decoder and sent to the delimiter to strip the received frame of non-data bits—start bit, stop bit(s), etc. The data is then shifted serially into the shifter, which transfers (in parallel) the received data into the receive data FIFO. Finally, the SDMA transfers the received data to another buffer for continued processing by the processor.

*MPC860 SMC in UART Mode*

As shown in Figure 4-60a, the internal design of the SMC differs greatly from the internal design of the SCC (shown in Figures 4-58a and b), and in fact has fewer capabilities than an SCC. An SMC has no encoder, decoder, delimiter, or receive/transmit FIFO buffers. It uses registers instead. As shown in Figure 4-60b, there are only three pins that an SMC is connected to: a transmit pin (SMTXDx), a receive pin (SMRXDx), and a sync signal pin (SMSYN). The sync pin is used in transparent transmissions to control receive and transmit operations.

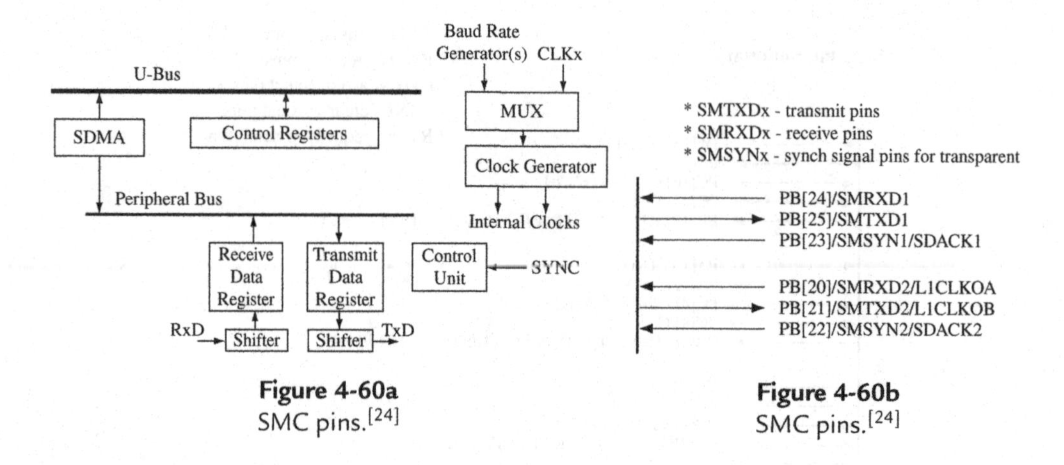


**Figure 4-60a**
SMC pins.[24]

**Figure 4-60b**
SMC pins.[24]

Data is received via the receive pin into the receive shifter and the SDMA then transfers the received data from the receive register. Data to be transmitted is stored in the transmit register, and then moved into the shifter for transmission over the transmit pin. Note that the SMC does not provide the framing and stripping of control bits (start bit, stop bit(s), etc.) that the SCC provides.

### Processor Serial I/O Example 2: An Integrated SPI

The SPI is an example of a full-duplex serial interface that can be integrated into the master processor and that does *synchronous* serial transmission. Like the UART, an SPI needs to exist on both sides of the communication channel (in the I/O device, as well as on the embedded board) in order for this communication scheme to work. In this example, we examine the MPC860 internal SPI, which resides in the PowerPC's Communication Processor Module (shown in Figure 4-61).

In a synchronous serial communication scheme, both devices are synchronized by the same clock signal generated by one of the communicating devices. In such a case, a master–slave relationship develops in which the master generates the clock signal which it and the slave device, adheres to. It is this relationship that is the basis of the four pins that the MPC860 SPI is connected to (as shown in Figure 4-62b): the master out/slave in or transmit (SPIMOSI), master in/slave out or receive (SPIMISO), clock (SPICLK), and slave select (SPISEL).

When the SPI operates in a master mode, it generates the clock signals, while in slave mode, it receives clock signals as input. SPIMOSI in master mode is an output pin, SPMISO in master mode is an input pin, and SPICLK supplies an output clock signal in master mode that synchronizes the shifting of received data over the SPIMISO pin or shifts out transmitted data over SPIMOSI. In slave mode, SPIMOSI is an input pin, SPIMISO is an output pin, and SPICLK receives a clock signal from the master synchronizing the shifting of data over the transmit and receive pins. The SPISEL is also relevant in slave mode, because it enables input into the slave.

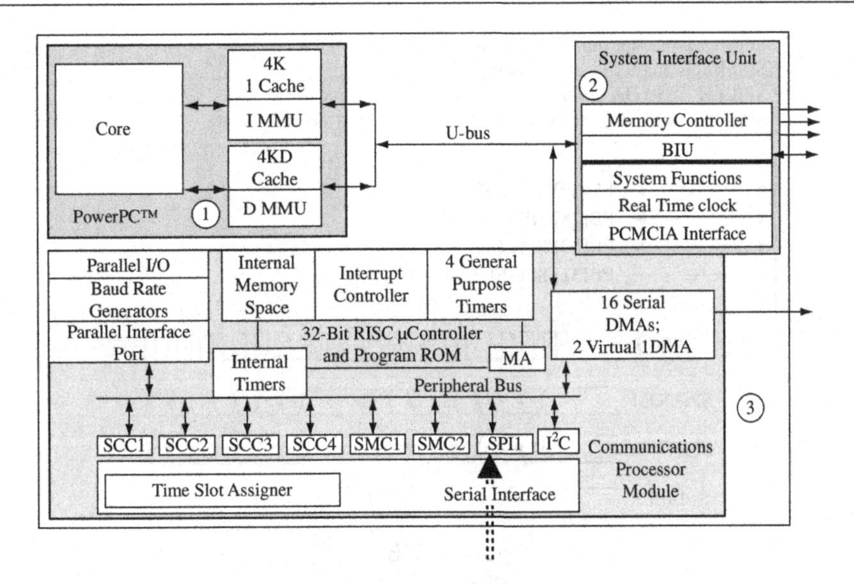


**Figure 4-61**
MPC860 SPI.[15] *© 2004 Freescale Semiconductor, Inc. Used by permission.*

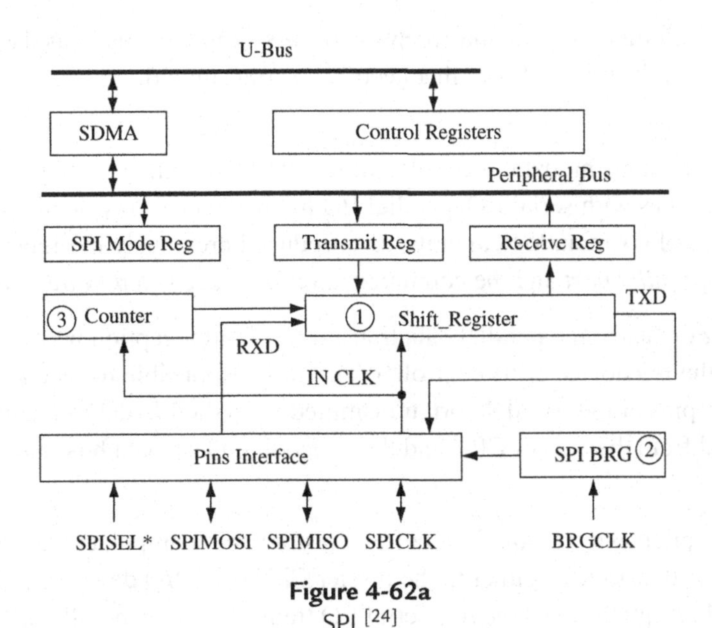


**Figure 4-62a**
SPI.[24]

How these pins work together, along with the internal components of the SPI, is shown in Figure 4-62a. Essentially, data is received or transmitted via one shift register. If data is received, it is then moved into a receive register. The SDMA then transfers the data into a receive buffer that usually resides in main memory. In the case of a data transmission, the SDMA moves the data to be transmitted from the transfer buffer in main memory to the

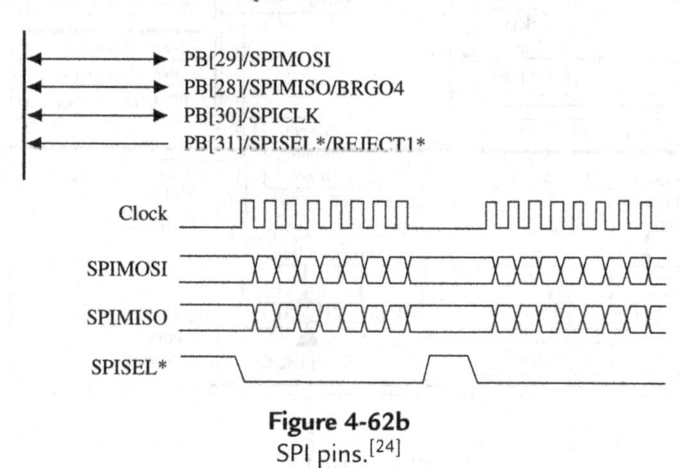


**Figure 4-62b**
SPI pins.[24]

transmit register. SPI transmission and reception occurs simultaneously; as data is received into the shift register, it shifts out data that needs to be transmitted.

*Parallel I/O*

I/O components that transmit data in parallel allow data to be transferred in multiple bits simultaneously. Just as with serial I/O, parallel I/O hardware is also typically made up of some combination of six main logical units, as introduced previously in Figure 4-50, except that the port is a *parallel port* and the communication interface is a *parallel interface.*

*Parallel interfaces* manage the parallel data transmission and reception between the master CPU and either the I/O device or its controller. They are responsible for decoding data bits received over the pins of the parallel port, transmitted from the I/O device, and receiving data being transmitted from the master CPU and then encoding these data bits onto the parallel port pins.

They include reception and transmission buffers to store and manipulate the data they are responsible for transmitting either to the master CPU or an I/O device. Parallel data transmission and reception schemes, like serial I/O transmission, generally differ in terms of what *direction* data can be transmitted and received, as well as the actual *process* of how the actual data bits are transmitted (and thus received) within the data stream. In the case of direction of transmission, as with serial I/O, parallel I/O uses simplex, half-duplex, or full-duplex modes. Again, like serial I/O, parallel I/O can be transmitted asynchronously or synchronously. Unlike serial I/O, parallel I/O does have a greater capacity to transmit data, because multiple bits can be transmitted or received simultaneously. Examples of I/O devices

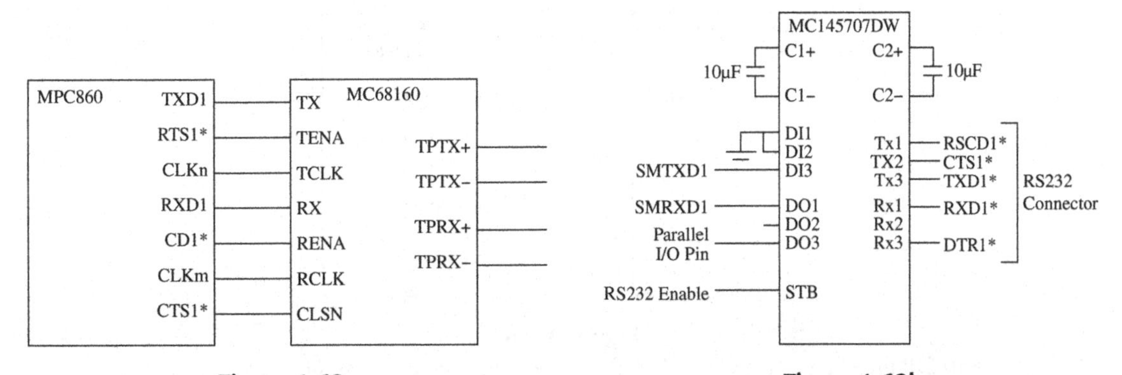


**Figure 4-63a**
MPC860 SCC UART interfaced to Ethernet controller.[24] *© 2004 Freescale Semiconductor, Inc. Used by permission.*

**Figure 4-63b**
MPC860 SMC interfaced to RS-232.[24] *© 2004 Freescale Semiconductor, Inc. Used by permission.*

that transfer and receive data in parallel include IEEE1284 controllers (for printer/display I/O devices), CRT ports, and SCSI (for storage I/O devices).

*Interfacing the Master Processor with an I/O Controller*

When the communication interface is integrated into the master processor, as is the case with the MPC860, it is a matter of connecting the identical pins for transmitting data and receiving data from the master processor to an I/O controller. The remaining control pins are then connected according to their function. In Figure 4-63a, for instance, the RTS (request to send) on the PowerPC is connected to transmit enable (TENA) on the Ethernet controller, since RTS is automatically asserted if data is loaded into the transmit FIFO, indicating to the controller that data is on its way. The CTS (collision on the transceiver) on the PowerPC is connected to the CLSN (clear to send) on the Ethernet controller and the CD (carrier detect) is connected to the RENA (receive enable) pin, since when either CD or CTS is asserted, a transmission or data reception can take place. If the controller does not clear to send or receive enable to indicate data is on its way to the PowerPC, no transmission or reception can take place. Figure 4-63b shows a MPC860 SMC interfaced to an RS-232 IC, which takes the SMC signals (transmit pin (SMTXDx) and receive pin (SMRXDx)) and maps them to an RS-232 port in this example.

Figure 4-63c shows an example of a PowerPC SPI in master mode interfaced with some slave IC, in which the SPIMISO (master in/slave out) is mapped to SPISO (SPI slave out). Since in master mode SPIMISO is an input port, SPIMOSI (master out/slave in) is mapped to SPISI (slave in). Since SPIMOSI in master mode is an output port, SPICLK is mapped to SPICK (clock) as both ICs are synchronized according to the same clock, and SPISEL is mapped to SPISS (slave select input) which is only relevant if the PowerPC is in slave mode. If it were the other way around (i.e., PowerPC in slave mode and slave IC in master mode), the interface would map identically.

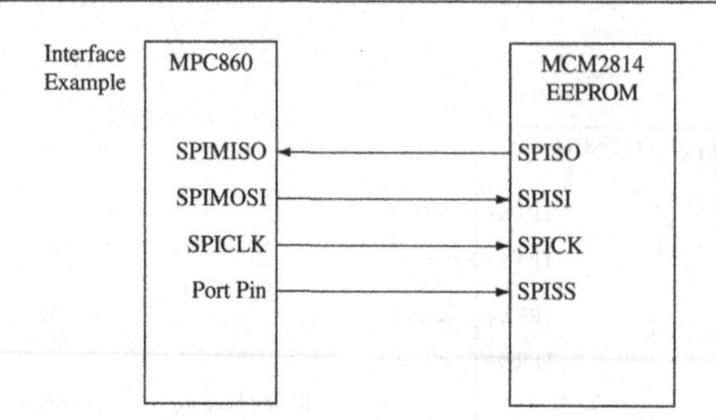


**Figure 4-63c**
MPC860 SPI interfaced to ROM.[24] *© 2004 Freescale Semiconductor, Inc. Used by permission.*

Finally, for a subsystem that contains an I/O controller to manage the I/O device, the interface between an I/O controller and master CPU (via a communications interface) is based on four requirements:

- *An ability for the master CPU to initialize and monitor the I/O controller.* I/O controllers can typically be configured via *control registers* and monitored via *status registers*. These registers are all located on the I/O controller itself. Control registers can be modified by the master processor to configure the I/O controller. Status registers are read-only registers in which the master processor can get information as to the state of the I/O controller. The master CPU uses these status and control registers to communicate and/or control attached I/O devices via the I/O controller.
- *A way for the master processor to request I/O.* The most common mechanisms used by the master processor to request I/O via the I/O controller are *special I/O instructions* (I/O mapped) in the ISA and *memory-mapped I/O*, in which the I/O controller registers have reserved spaces in main memory.
- *A way for the I/O device to contact the master CPU.* I/O controllers that have the ability to contact the master processor via an interrupt are referred to as interrupt driven I/O. Generally, an I/O device initiates an asynchronous interrupt requesting signaling to indicate (for example) that control and status registers can be read from or written to. The master CPU then uses its interrupt scheme to determine when an interrupt will be discovered.
- *Some mechanism for both to exchange data.* This refers to the process by which data is actually exchanged between the I/O controller and the master processor. In a *programmed transfer*, the master processor receives data from the I/O controller into its registers, and the CPU then transmits this data to memory. For memory-mapped I/O schemes, DMA (direct memory access) circuitry can be used to bypass the master CPU entirely. DMA has the ability to manage data transmissions or receptions directly to and from main memory and an I/O device. On some systems, DMA is integrated into the master processor, and on others there is a separate DMA controller.

*Note: More on I/O with examples of components will be covered in Chapter 6, Board I/O. Some (unintegrated) processors have I/O components that work in conjunction with board I/O, so there is a great deal of overlapping information.*

#### *Interrupts*

*Interrupts* are signals triggered by some event during the execution of an instruction stream by the master processor. This means they can be initiated asynchronously, for external hardware devices, resets, power failures, or synchronously for instruction-related activities such as system calls, or illegal instructions. These signals cause the master processor to stop executing the current instruction stream and start the process of *handling* (processing) the interrupt.

The three main types of interrupts are *software, internal hardware*, and *external hardware.* Software interrupts are explicitly triggered internally by some instruction within the current instruction stream being executed by the master processor. Internal hardware interrupts, on the other hand, are initiated by an event that is a result of a problem with the current instruction stream that is being executed by the master processor because of the features (or limitations) of the hardware, such as illegal math operations like overflow or divide-by-zero, debugging (single-stepping, breakpoints), invalid instructions (opcodes), etc. Interrupts that are raised (requested) by some internal event to the master processor (basically, software and internal hardware interrupts) are also commonly referred to as *exceptions* or *traps* (depending on the type of interrupt). Finally, external hardware interrupts are interrupts initiated by hardware other than the master CPU (board buses, I/O, etc.). What actually triggers an interrupt is typically determined by the software via register bits that activate or deactivate potential interrupt sources in the initialization device driver code.

For interrupts that are raised by external events, the master processor is either wired via an input pin(s), called an *IRQ (Interrupt Request Level)* pin or port, to outside intermediary hardware (i.e., interrupt controllers), or directly to other components on the board with dedicated interrupt ports that signal the master CPU when they want to raise the interrupt. These types of interrupts are triggered in one of two ways: *level-triggered* or *edge-triggered.* A level-triggered interrupt is initiated when the IRQ signal is at a certain level (i.e., HIGH or LOW; see Figure 4-64a). These interrupts are processed when the CPU finds a request for a level-triggered interrupt when sampling its IRQ line, such as at the end of processing each instruction.

An edge-triggered interrupt triggers when on its IRQ line (from LOW to HIGH/rising edge of signal or from HIGH to LOW/falling edge of signal; see Figure 4-64b). Once triggered, these interrupts latch into the CPU until processed.

Both types of interrupts have their strengths and drawbacks. With a level-triggered interrupt, as shown in the example in Figure 4-65a, if the request is being processed and has not been

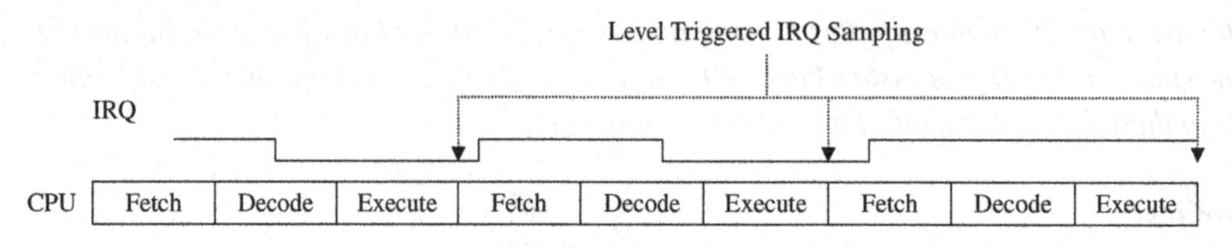


**Figure 4-64a**
Level-triggered interrupts.[23]

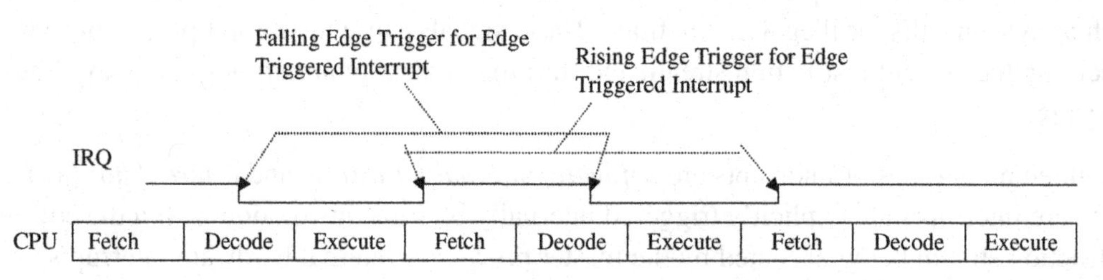


**Figure 4-64b**
Edge-triggered interrupts.[23]

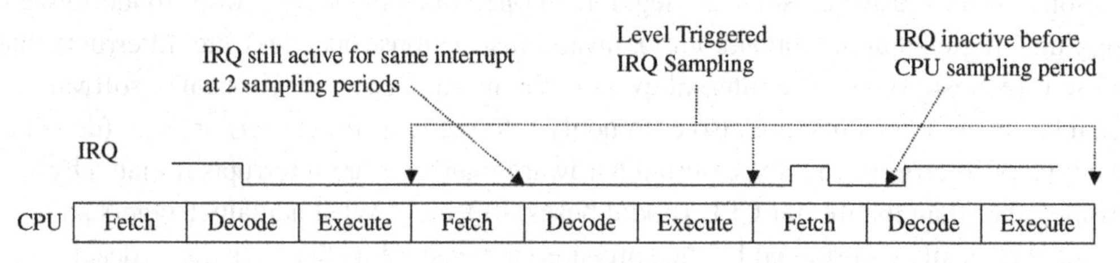


**Figure 4-65a**
Level-triggered interrupts drawbacks.[23]

disabled before the next sampling period, the CPU would try to service the same interrupt again. On the flip side, if the level-triggered interrupt were triggered and then disabled before the CPU's sample period, the CPU would never note its existence and would therefore never process it. Edge level interrupts can have problems if they share the same IRQ line, should they be triggered in the same manner at about the same time (say before the CPU could process the first interrupt), resulting in the CPU being able to detect only one of the interrupts (see Figure 4-65b).

Because of these drawbacks, level-triggered interrupts are generally recommended for interrupts that share IRQ lines, whereas edge-triggered interrupts are typically recommended for interrupt signals that are very short or very long.

At the point an IRQ of a master processor receives a signal that an interrupt has been raised, the interrupt is processed by the interrupt-handling mechanisms within the system. These

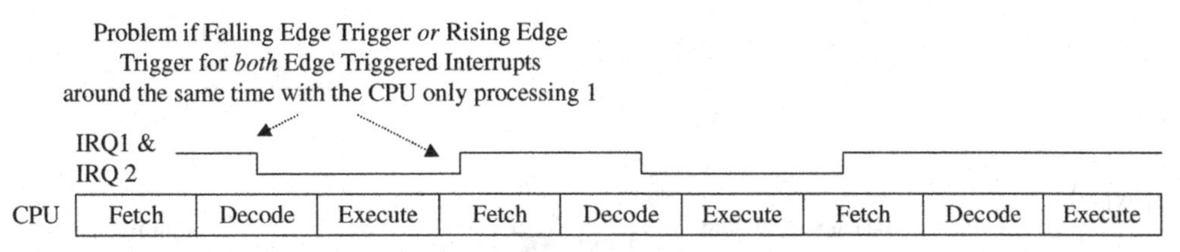


**Figure 4-65b**
Edge-triggered interrupts drawbacks.[23]

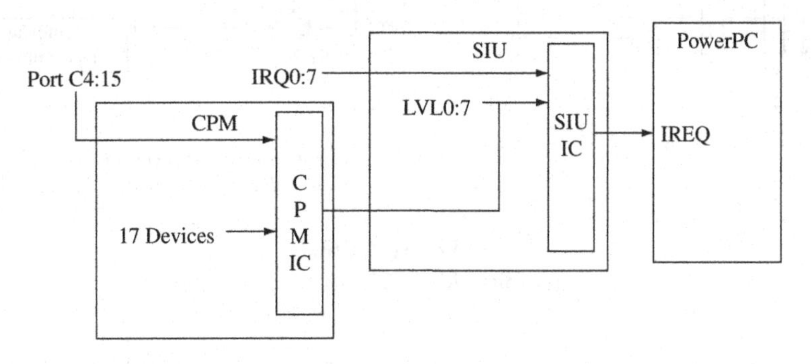


**Figure 4-66a**
Motorola/Freescale MPC860 interrupt controllers.[24] *© 2004 Freescale Semiconductor, Inc. Used by permission.*

mechanisms are made up of a combination of both hardware and software components. In terms of hardware, an *interrupt controller* can be integrated onto a board or within a processor to mediate interrupt transactions in conjunction with software. Architectures that include an interrupt controller within their interrupt-handling schemes include the 268/386 (x86) architectures that use two PICs (Intel's Programmable Interrupt Controller); MIPS32, which relies on an external interrupt controller; and the MPC860 (shown in Figure 4-66a), which integrates two interrupt controllers, one in the CPM and one in its SIU. For systems with no interrupt controller, such as the Mitsubishi M37267M8 TV microcontroller shown in Figure 4-66b), the interrupt request lines are connected directly to the master processor, and interrupt transactions are controlled via software and some internal circuitry, such as registers and/or counters.

*Interrupt acknowledgment (IACK)* is typically handled by the master processor when an external device triggers an interrupt. Because IACK cycles are a function of the local bus, the IACK function of the master CPU depends on interrupt policies of system buses, as well as the interrupt policies of components within the system that trigger the interrupts. With respect to the external device triggering an interrupt, the interrupt scheme depends on whether that device can provide an *interrupt vector* (a place in memory that holds the address of an interrupt's *ISR (Interrupt Service Routine)*, the software that the master CPU executes after

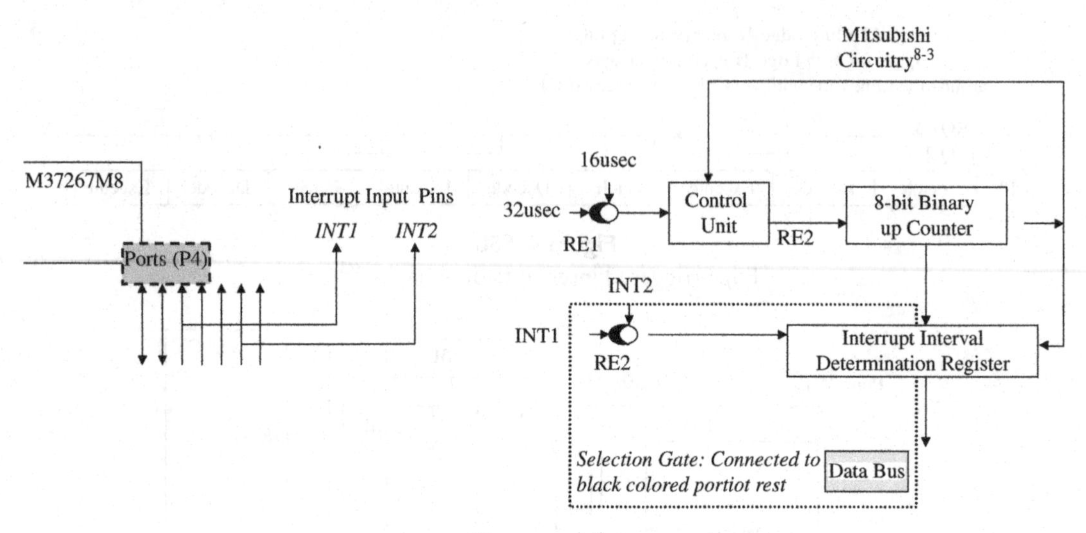


**Figure 4-66b**
Mitsubishi M37267M8 circuitry.[21]

the triggering of an interrupt). For devices that cannot provide the interrupt vector, referred to as *non-vectored* interrupts, master processors implement an *auto-vectored* interrupt scheme and acknowledgment is done via software.

An *interrupt vectored* scheme is implemented to support peripherals that can provide an interrupt vector over a bus and acknowledgment is automatic. Some IACK register on the master CPU informs the device, requesting the interrupt to stop requesting interrupt service, and provides the master processor with what it needs to process the correct interrupt (such as the interrupt number or vector number). Based upon the activation of an external interrupt pin, an interrupt controller's interrupt select register, a device's interrupt select register, or some combination of these, the master processor can determine which ISR to execute. After the ISR completes, the master processor resets the interrupt status by adjusting the bits in the processor's status register or an interrupt mask in the external interrupt controller. The interrupt request and acknowledge mechanisms are determined by the device requesting the interrupt (since it determines which interrupt service to trigger), the master processor, and the system bus protocols.

Keep in mind that this is a general introduction to interrupt handling, covering some of the key features found in a variety of schemes. The overall interrupt-handling scheme can vary widely from architecture to architecture. For example, PowerPC architectures implement an auto-vectored scheme, with no interrupt vector base register. The 68000 architecture supports both auto-vectored and interrupt vectored schemes, whereas MIPS32 architectures have no IACK cycle and so the interrupt handler handles the triggered interrupts.

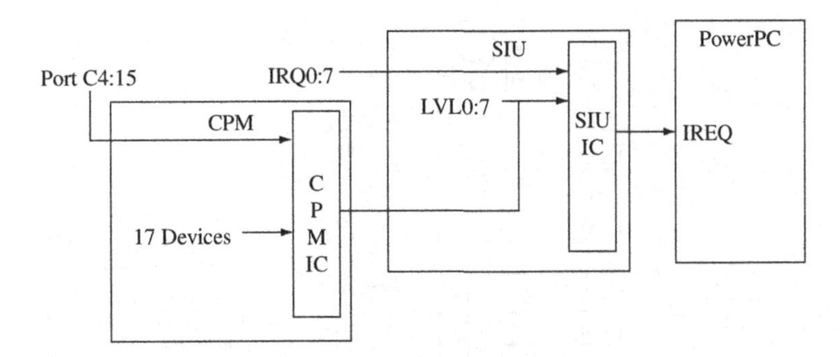


**Figure 4-67a**
Motorola/Freescale MPC860 interrupt pins and table.[24] *© 2004 Freescale Semiconductor, Inc. Used by permission.*

All available interrupts within a processor have an associated interrupt level, which is the priority of that interrupt within the system. Typically, interrupts starting at level "1" are the highest priority within the system and incrementally from there (2, 3, 4, etc.) the priorities of the associated interrupts decrease. Interrupts with higher levels (priorities) have precedence over any instruction stream being executed by the master processor. This means that not only do interrupts have precedence over the main program, but they also have precedence over interrupts with lower priorities as well.

The master processor's internal design determines the number and types of interrupts available, as well as the interrupt levels (priorities) supported within an embedded system. In Figure 4-67a, the MPC860 CPM, SIU, and PowerPC core all work together to implement interrupts on the MPC823 processor. The CPM allows for internal interrupts (two SCCs, two SMCs, SPI, I2C, PIP, general-purpose timers, two IDMAs, one SDMA, one RISC Timers) and 12 external pins of port C, and drives the interrupt levels on the SIU. The SIU receives interrupts from eight external pins (IRQ0-7), and eight internal sources, for a total of 16 sources of interrupts (one of which can be the CPM), and drives the IREQ input to the core. When the IREQ pin is asserted, external interrupt processing begins. The priority levels are shown in Figure 4-67b.

In another architecture, such as the 68000 (shown in Figures 4-68a and b), there are eight levels of interrupts (0–7) and interrupts at level 7 are the highest priority. The 68000 interrupt table (see Figure 4-68b) contains 256 32-bit vectors.

The M37267M8 architecture (shown in Figure 4-69a) allows for interrupts to be caused by 16 events (13 internal, two external, and one software) whose priorities and usages are summarized in Figure 4-69b.

Several different priority schemes are implemented in various architectures. These schemes commonly fall under one of three models: the *equal single level* (where the latest interrupt to

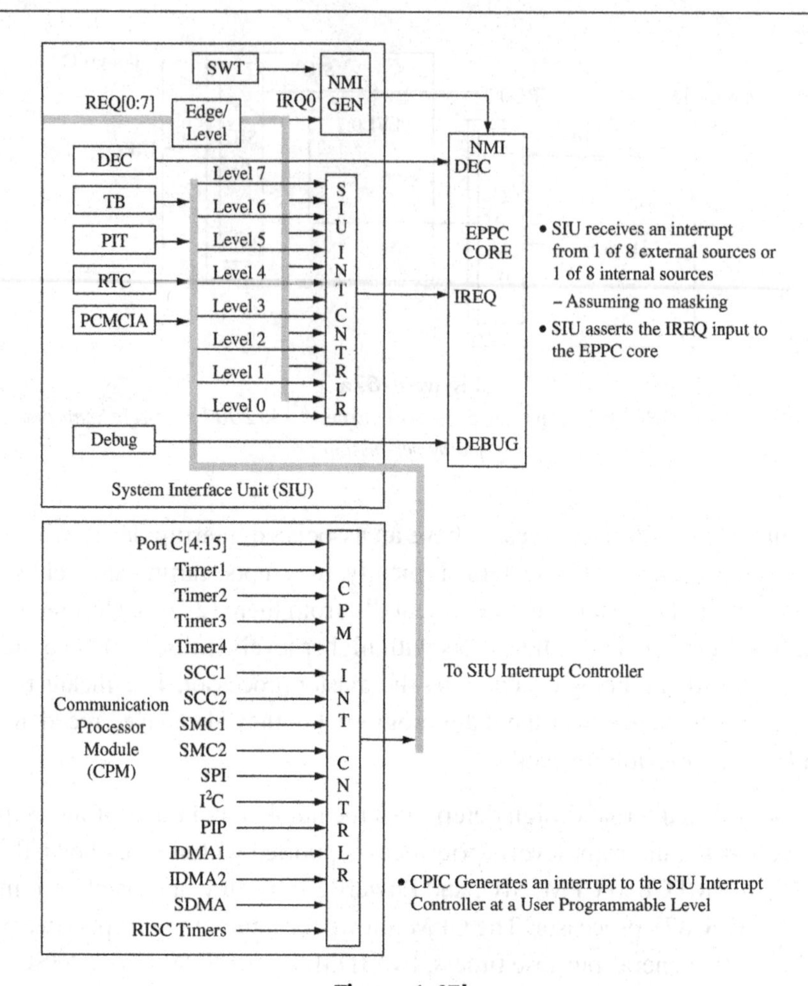


**Figure 4-67b**
Motorola/Freescale MPC860 interrupt levels.[24] *© 2004 Freescale Semiconductor, Inc. Used by permission.*

be triggered gets the CPU); the *static multilevel* (where priorities are assigned by a priority encoder, and the interrupt with the highest priority gets the CPU); and the *dynamic multilevel* (where a priority encoder assigns priorities and the priorities are reassigned when a new interrupt is triggered).

After the interrupt is acknowledged, the remaining interrupt-handling process as described above is typically handled via software and so the discussion of interrupts will be continued in Chapter 8, Device Drivers.

### 4.2.4 Processor Buses

Like the CPU buses, the processor's buses interconnect the processor's major internal components (in this case the CPU, memory, and I/O as shown in Figure 4-70), carrying signals between the different components.

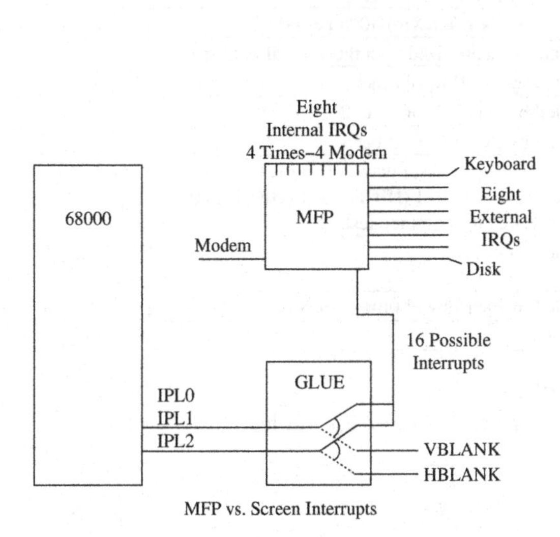


**Figure 4-68a**
Motorola/Freescale 68K IRQs.[22]

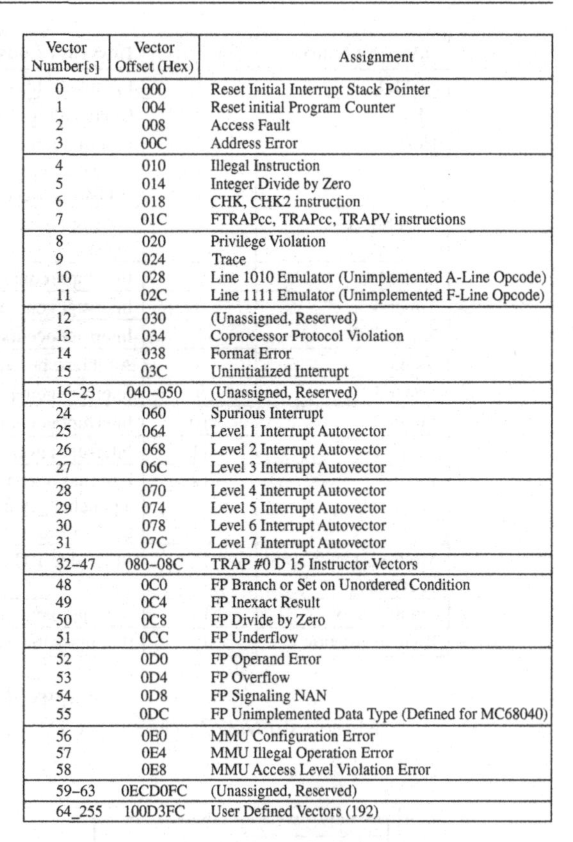

| Vector Number[s] | Vector Offset (Hex) | Assignment |
|---|---|---|
| 0 | 000 | Reset Initial Interrupt Stack Pointer |
| 1 | 004 | Reset initial Program Counter |
| 2 | 008 | Access Fault |
| 3 | 00C | Address Error |
| 4 | 010 | Illegal Instruction |
| 5 | 014 | Integer Divide by Zero |
| 6 | 018 | CHK, CHK2 instruction |
| 7 | 01C | FTRAPcc, TRAPcc, TRAPV instructions |
| 8 | 020 | Privilege Violation |
| 9 | 024 | Trace |
| 10 | 028 | Line 1010 Emulator (Unimplemented A-Line Opcode) |
| 11 | 02C | Line 1111 Emulator (Unimplemented F-Line Opcode) |
| 12 | 030 | (Unassigned, Reserved) |
| 13 | 034 | Coprocessor Protocol Violation |
| 14 | 038 | Format Error |
| 15 | 03C | Uninitialized Interrupt |
| 16–23 | 040–050 | (Unassigned, Reserved) |
| 24 | 060 | Spurious Interrupt |
| 25 | 064 | Level 1 Interrupt Autovector |
| 26 | 068 | Level 2 Interrupt Autovector |
| 27 | 06C | Level 3 Interrupt Autovector |
| 28 | 070 | Level 4 Interrupt Autovector |
| 29 | 074 | Level 5 Interrupt Autovector |
| 30 | 078 | Level 6 Interrupt Autovector |
| 31 | 07C | Level 7 Interrupt Autovector |
| 32–47 | 080–08C | TRAP #0 D 15 Instructor Vectors |
| 48 | 0C0 | FP Branch or Set on Unordered Condition |
| 49 | 0C4 | FP Inexact Result |
| 50 | 0C8 | FP Divide by Zero |
| 51 | 0CC | FP Underflow |
| 52 | 0D0 | FP Operand Error |
| 53 | 0D4 | FP Overflow |
| 54 | 0D8 | FP Signaling NAN |
| 55 | 0DC | FP Unimplemented Data Type (Defined for MC68040) |
| 56 | 0E0 | MMU Configuration Error |
| 57 | 0E4 | MMU Illegal Operation Error |
| 58 | 0E8 | MMU Access Level Violation Error |
| 59–63 | 0ECD0FC | (Unassigned, Reserved) |
| 64_255 | 100D3FC | User Defined Vectors (192) |

**Figure 4-68b**
Motorola/Freescale 68K IRQs table.[22]

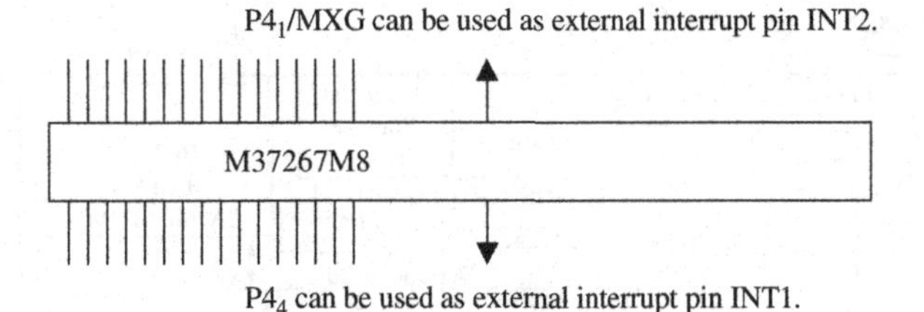


**Figure 4-69a**
Mitsubishi M37267M8 8-bit TV microcontroller interrupts.[21]

A key feature of processor buses is their *width*, which is the number of bits that can be transmitted at any one time. This can vary depending on both the buses implemented within the processor (x86 contains bus widths of 16/32/64, 68K has 8/16/32/64-bit buses, MIPS32 has 32 bit buses, etc.) as well as the ISA register size definitions. Each bus also has a bus *speed* (in MHz) that impacts the performance of the processor. Buses implemented in

| Interrupt Source | Priority | Interrupt Causes |
|---|---|---|
| RESET | 1 | (nonmaskable) |
| CRT | 2 | Occurs after character block display to CRT is completed |
| INT1 | 3 | External Interrupt ** the processor detects that the level of a pin changes from 0 (LOW) to 1 (HIGH), or 1 (HIGH) to 0 (LOW) and generates an interrupt request |
| Data Slicer | 4 | Interrupt occurs at end of line specified in caption position register |
| Serial I/O | 5 | Interrupt request from synchronous serial I/O function |
| Timer 4 | 6 | Interrupt generated by overflow of timer 4 |
| Xin & 4096 | 7 | Interrupt occurs regularly w/a f(Xin)/4096 period |
| Vsync | 8 | An interrupt request synchronized with the vertical sync signal |
| Timer 3 | 9 | Interrupt generated by overflow of timer 3 |
| Timer 2 | 10 | Interrupt generated by overflow of timer 2 |
| Timer 1 | 11 | Interrupt generated by overflow of timer 1 |
| INT2 | 12 | External Interrupt ** the processor detects that the level of a pin changes from 0 (LOW) to 1 (HIGH), or 1 (HIGH) to 0 (LOW) and generates an interrupt request |
| Multimaster $I^2C$ Bus interface | 13 | Related to $I^2C$ bus interface |
| Timer 5 & 6 | 14 | Interrupt generated by overflow of timer 5 or 6 |
| BRK instruction | 15 | (nonmaskable software) |

**Figure 4-69b**

Mitsubishi M37267M8 8-bit TV microcontroller interrupt table.[21]

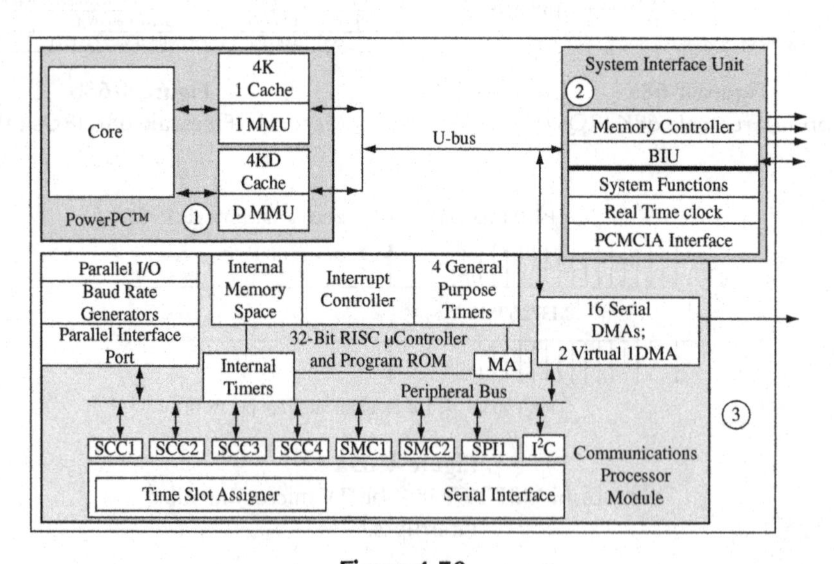


**Figure 4-70**

MPC860 processor buses.[15]

In the case of the MPC860, the processor buses include the U-bus interconnecting the system interface unit (SIU), the communications processor module (CPM), and the PowerPC core. Within the CPM, there is a peripheral bus, as well. Of course, this includes the buses within the CPU.

real-world processor designs include the U, peripheral, and CPM buses in the MPC8xx family of processors, and the C and X buses in the x86 Geode.

*Note: To avoid redundancy, buses are covered in more detail in Chapter 7, Board Buses, and more examples are provided there.*

## 4.3 Processor Performance

There are several measures of processor performance, but are all based upon the processor's behavior over a given length of time. One of the most common definitions of processor performance is a processor's *throughput*—the amount of work the CPU completes in a given period of time.

As discussed in Section 4.2.1, a processor's execution is ultimately synchronized by an external *system* or *master clock*, located on the board. The master clock is simply an oscillator producing a fixed frequency sequence of regular on/off pulse signals that is usually divided or multiplied within the CPU's CU to generate at least one internal clock signal running at a constant number of clock cycles per second, or *clock rate*, to control and coordinate the fetching, decoding, and execution of instructions. The CPU's clock rate is expressed in megahertz (MHz).

Using the clock rate, the CPU's *execution time*, which is the total time the processor takes to process some program in seconds per program (total number of bytes), can be calculated. From the clock rate, the length of time a CPU takes to complete a clock cycle is the inverse of the clock rate (1/clock rate), called the *clock period* or *cycle time* and expressed in seconds per cycle. The processor's clock rate or clock period is usually located in the processor's specification documentation.

Looking at the instruction set, the *CPI* (average number of clock cycles per instruction) can be determined in several ways. One way is to obtain the CPI for each instruction (from the processor's instruction set manual) and multiplying that by the frequency of that instruction, and then adding up the numbers for the total CPI.

$$\text{CPI} = \sum (\text{CPI per instruction} * \text{instruction frequency}).$$

At this point the total CPU's execution time can be determined by:

CPU execution time in seconds per program = (total number of instructions per program or instruction count) * (CPI in number of cycle cycles/instruction) * (clock period in seconds per cycle) = ((instruction count) * (CPI in number of cycle cycles/instruction))/(clock rate in MHz).

The processor's average execution rate, also referred to as *throughput* or *bandwidth*, reflects the amount of work the CPU does in a period of time and is the inverse of the CPU's execution time:

$$\text{CPU throughput (in bytes/s or MB/s)} = 1/\text{CPU execution time} = \text{CPU performance}.$$

Knowing the performance of two architectures (e.g., Geode and SA-1100), the *speedup* of one architecture over another can then be calculated as:

$$\text{Performance(Geode)/performance (SA} - 1100) = \text{execution time (SA} - 1100\text{)/execution time (Geode)} = \text{“}X\text{,”}$$

therefore Geode is "*X*" times faster than SA-1100.

Other definitions of performance besides throughput include:

- A processor's responsiveness, or *latency*: length of elapsed time a processor takes to respond to some event.
- A processor's *availability*: amount of time the processor runs normally without failure; *reliability*: average time between failures (MTBF); and *recoverability*: average time the CPU takes to recover from failure (MTTR).

On a final note, a processor's internal design determines a processor's clock rate and the CPI; thus a processor's performance depends on which ISA is implemented and how the ISA is implemented. For example, architectures that implement instruction-level parallelism ISA models have better performance over the application-specific and general-purpose-based processors because of the parallelism that occurs within these architectures. Performance can be improved because of the actual physical implementations of the ISA within the processor, such as implementing pipelining in the ALU. (*Note: There are variations on the full adder that provide additional performance improvements, such as the carry lookahead adder (CLA), carry completion adder, conditional sum adder, and carry select adder. In fact, some algorithms that can improve the performance of a processor do so by designing the ALU to be able to process logical and mathematical instructions at a higher throughput—a technique called pipelining.*) The increasing gap between the performance of processors and memory can be improved by cache algorithms that implement instruction and data prefetching (especially algorithms that make use of branch prediction to reduce stall time), and lockup-free caching. Basically, any design feature that allows for either an increase in the clock rate or decrease in the CPI will increase the overall performance of a processor.

### 4.3.1 Benchmarks

One of the most common performance measures used for processors in the embedded market is *millions of instructions per seconds (MIPS)*:

$$MIPS = instruction\ count/(CPU\ execution\ time * 10^6) = clock\ rate/(CPI * 10^6).$$

The MIPS performance measure gives the impression that faster processors have higher MIPS values, since part of the MIPS formula is inversely proportional to the CPU's execution time. However, MIPS can be misleading when making this assumption for a number of reasons, including:

- Instruction complexity and functionality aren't taken into consideration in the MIPS formula, so MIPS cannot compare the capabilities of processors with different ISAs.
- MIPS can vary on the same processor when running different programs (with varying instruction count and different types of instructions).

Software programs called *benchmarks* can be run on a processor to measure its performance; the performance discussion will continue in Section IV, Putting It All Together.

## 4.4 Reading a Processor's Datasheet

A processor's datasheet provides key areas of useful processor information.

**Author Note**

I don't assume that what I read from a vendor is 100% accurate, until I have seen the processor running and verified the features myself.

Datasheets exist for almost any component, both hardware and software, and the information they contain varies between vendors. Some datasheets are a couple of pages long and list only the main features of a system, while others contain over 100 pages of technical information. In this section, I have used the MPC860EC rev. 6.3 datasheet, which is 80 pages long, to summarize some of the key areas of useful information in a processor's datasheet. The reader can then use it as an example for reading other processor datasheets, which typically have similar overviews and technical information.

### 4.4.1 MPC860 Datasheet Example

*Section 2 of MPC860 Datasheet Example: Overview of the Processor's Features*

Figure 4-71a shows a block diagram of the MPC860, which is described in the datasheet's feature list shown in Figure 4-71b. As shown in the various shaded and unshaded sections of the overview, everything from the description of the physical IC packaging to the major features of the processor's internal memory scheme is summarized. The remaining sections of the datasheet also provide a wide variety of information, including providing recommendations as to how the MPC860 should be integrated onto a PCB such as: VDD

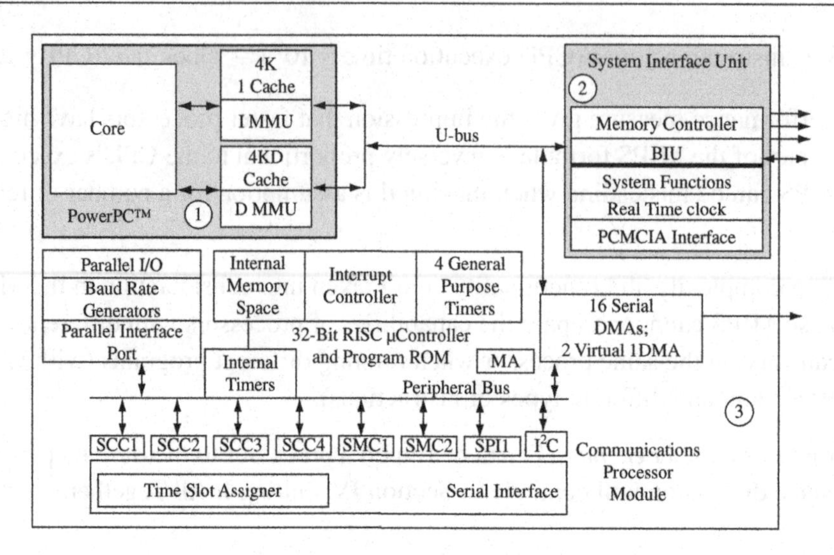


**Figure 4-71a**

MPC860 processor block diagram.[15] *© 2004 Freescale Semiconductor, Inc. Used by permission.*

pins should be provided with low-impedance paths to the board's supply, GND pins should be provided with low-impedance paths to ground, all unused inputs/signals that will be inputs during reset should be pulled up, to providing electrical specifications for the IEEE 1149.1 JTAG timings, the AC and DC electrical specifications for the CPM, the AC electrical specifications for the UTOPIA interface, the AC electrical specifications for the fast Ethernet controller (FEC), etc.

*Section 3 of MPC860 Datasheet Example: Maximum Tolerated Ratings*

This section of the MPC860 datasheet provides information on the maximum voltage and temperature ranges that this processor can be exposed to (for the MPC860, shown in Table 4-4a). The maximum tolerated temperature for processors is the maximum temperature a processor can withstand without damage, whereas the maximum tolerated voltage is the maximum voltage a processor can withstand without damage.

Different processors will have their own maximum tolerated voltage and power ratings (as seen from Table 4-4b the tables for the maximum temperatures and voltages of the NET + ARM processor).

*Section 4 of MPC860 Datasheet Example: Thermal Characteristics*

The thermal characteristics of a processor indicate what type of thermal design requirements need to be taken into consideration for using that processor on a particular board. Table 4-5 shows the thermal characteristics of the MPC860; more information on thermal management is contained in sections 5 and 7 of the Datasheet. A processor that exceeds the ranges of its absolute and functional temperature limits runs the risk of having logical errors, a degradation

**Datasheet Overview**

- Embedded single-issue, 32-bit PowerPC core (implementing the PowerPC architecture) with thirty-two 32-bit general-purpose registers (GPRs)
  - — The core performs branch prediction with conditional prefetch without conditional execution.

*On-chip Memory*

— 4- or 8-Kbyte data cache and 4- or 16-Kbyte instruction cache
- 16-Kbyte instruction caches are four-way, set-associative with 256 sets; 4-Kbyte instruction caches are two-way, set-associative with 128 sets.
- 8-Kbyte data caches are two-way, set-associative with 256 sets; 4-Kbyte data caches are two-way, set-associative with 128 sets.
- Cache coherency for both instruction and data caches is maintained on 128-bit (4-word) cache blocks.
- Caches are physically addressed, implement a least recently used (LRU) replacement algorithm, and are lockable on a cache block basis.

*Memory Management*

— MMUs with 32-entry TLB, fully-associative instruction, and data TLBs
— MMUs support multiple page sizes of 4-, 16-, and 512-Kbytes, and 8-Mbytes; 16 virtual address spaces and 16 protection groups

— Advanced on-chip-emulation debug mode

***External Data Bus Width and Support***

- Up to 32-bit data bus (dynamic bus sizing for 8, 16, and 32 bits)
- 32 address lines
- Operates at up to 80 MHz

*Memory Management*

- Memory controller (eight banks)
  - — Contains complete dynamic RAM (DRAM) controller
  - — Each bank can be a chip select or RAS to support a DRAM bank.
  - — Up to 15 wait states programmable per memory bank
  - — Glueless interface to DRAM, SIMMS, SRAM, EPROM, Flash EPROM, and other memory devices
  - — DRAM controller programmable to support most size and speed memory interfaces
  - — Four CAS lines, four WE lines, and one OE line
  - — Boot chip-select available at reset (options for 8-, 16-, or 32-bit memory)
  - — Variable block sizes (32 Kbyte to 256 Mbyte)
  - — Selectable write protection
  - — On-chip bus arbitration logic

***SIU features (timers, ports, etc.)***

- General-purpose timers
  - — Four 16-bit timers or two 32-bit timers
  - — Gate mode can enable/disable counting
  - — Interrupt can be masked on reference match and event capture.
- System integration unit (SIU)
  - — Bus monitor
  - — Software watchdog
  - — Periodic interrupt timer (PIT)
  - — Low-power stop mode
  - — Clock synthesizer
  - — Decrementer, time base, and real-time clock (RTC) from the PowerPC architecture
  - — Reset controller
  - — IEEE 1149.1 test access port (JTAG)

**Figure 4-71b1**

MPC860 overview from datasheet.[17] *© 2004 Freescale Semiconductor, Inc. Used by permission.*

**Datasheet Overview**

*Interrupt Scheme*

- Interrupts
  - — Seven external interrupt request (IRQ) lines
  - — 12 port pins with interrupt capability
  - — 23 internal interrupt sources
  - — Programmable priority between SCCs
  - — Programmable highest priority reques

***I/O NetworkingFeatures***

- 10/100 Mbps Ethernet support, fully compliant with the IEEE 802.3u Standard (not available when using ATM over UTOPIA interface)

- ATM support compliant with ATM forum UNI 4.0 specification
  - — Cell processing up to 50–70 Mbps at 50-MHz system clock
  - — Cell multiplexing/demultiplexing
  - — Support of AAL5 and AAL0 protocols on a per-VC basis. AAL0 support enables OAM and software implementation of other protocols.
  - — ATM pace control (APC) scheduler, providing direct support for constant bit rate (CBR) and unspecified bit rate (UBR) and providing control mechanisms enabling software support of available bit rate (ABR)
  - — Physical interface support for UTOPIA (10/100-Mbps is not supported with this interface) and byte-aligned serial (for example, T1/E1/ADSL)
  - — UTOPIA-mode ATM supports level-1 master with cell-level handshake, multi-PHY (up to four physical layer devices), connection to 25-, 51-, or 155-Mbps framers, and UTOPIA/system clock ratios of 1/2 or 1/3.
  - — Serial-mode ATM connection supports transmission convergence (TC) function for T1/E1/ADSL lines, cell delineation, cell payload scrambling/descrambling, automatic idle/unassigned cell insertion/stripping, header error control (HEC) generation, checking, and statistics.

***CPM Features***

- Communications processor module (CPM)
  - — RISC communications processor (CP)
  - — Communication-specific commands (for example, GRACEFUL –STOP-TRANSMIT, ENTER-HUNT-MODE, and RESTART-TRANSMIT)
  - — Supports continuous mode transmission and reception on all serial channels

*CPM Internal Memory and Memory Management*

- — Up to 8 Kbytes of dual-port RAM
- — 16 serial DMA (SDMA) channels

*CPM I/O*

- — Three parallel I/O registers with open-drain capability

- Four baud-rate generators (BRGs)
  - — Independent (can be tied to any SCC or SMC)
  - — Allows changes during operation
  - — Autobaud support option

*CPM I/O*

- Four serial communications controllers (SCCs)
  - — Ethernet/IEEE 802.3 optional on SCC1–4, supporting full 10-Mbps operation (available only on specially programmed devices)
  - — HDLC/SDLC (all channels supported at 2 Mbps)
  - — HDLC bus (implements an HDLC-based local area network (LAN))
  - — Asynchronous HDLC to support point-to-point protocol (PPP)
  - — AppleTalk
  - — Universal asynchronous receiver transmitter (UART)
  - — Synchronous UART
  - — Serial infrared (IrDA)
  - — Binary synchronous communication (BISYNC)
  - — Totally transparent (bit streams)
  - — Totally transparent (frame-based with optional cyclic redundancy check (CRC))

**Figure 4-71b2**
(Continued)

**Datasheet Overview**

***CPM Features***

*CPM I/O*

- Two SMCs (serial management channels)
  - — UART
  - — Transparent
  - — General circuit interface (GCI) controller
  - — Can be connected to the time-division multiplexed (TDM) channels
- One SPI (serial peripheral interface)
  - — Supports master and slave modes
  - — Supports multimaster operation on the same bus

*External Bus Support*

- One I2C (inter-integrated circuit) port
  - — Supports master and slave modes
  - — Multiple-master environment support

*CPM I/O*

- Time-slot assigner (TSA)
  - — Allows SCCs and SMCs to run in multiplexed and/or non-multiplexed operation
  - — Supports T1, CEPT, PCM highway, ISDN basic rate, ISDN primary rate, user defined
  - — 1- or 8-bit resolution
  - — Allows independent transmit and receive routing, frame synchronization, and clocking
  - — Allows dynamic changes
  - — Can be internally connected to six serial channels (four SCCs and two SMCs)
- Parallel interface port (PIP)
  - — Centronics interface support
  - — Supports fast connection between compatible ports on the MPC860 or the MC68360
- PCMCIA interface
  - — Master (socket) interface, release 2.1 compliant
  - — Supports two independent PCMCIA sockets
  - — Supports eight memory or I/O windows

- Lowpower support
  - — Full on—all units fully powered
  - — Doze—core functional units disabled except time base decrementer, PLL, memory controller, RTC, and CPM in lowpower standby
  - — Sleep—all units disabled except RTC and PIT, PLL active for fast wake up
  - — Deep sleep—all units disabled including PLL except RTC and PIT
  - — Power down mode—all units powered down except PLL, RTC, PIT, time base, and decrementer

*Debugging Support*

- Debug interface
  - — Eight comparators: four operate on instruction address, two operate on data address, and two operate on data
  - — Supports conditions: =.< >
  - — Each watchpoint can generate a break-point internally.

***Voltage Source/Power Information***

- 3.3-V operation with 5-V TTL compatibility except EXTAL and EXTCLK

*IC Packaging*

- 357-pin ball grid array (BGA) package

**Figure 4-71b3**
(Continued)

**Table 4-4a: MPC860 Processor Maximum Tolerated Voltage (GND = 0 V) and Temperature Ratings[17]**

| Rating | Symbol | Value | Unit |
|---|---|---|---|
| Supply voltage[1] | $V_{DDH}$ | -0.3 to 4.0 | V |
| | $V_{DDL}$ | -0.3 to 4.0 | V |
| | KAPWR | -0.3 to 4.0 | V |
| | VDDSYN | -0.3 to 4.0 | V |
| Input voltage[2] | $V_{in}$ | GND - 0.3 to VDDH | V |
| Temperature[3] (standard) | $T_{A(min)}$ | 0 | °C |
| | $T_{i(max)}$ | 95 | °C |
| Temperature[3] (extended) | $T_{A(min)}$ | -40 | °C |
| | $T_{i(max)}$ | 95 | °C |
| Storage temperature range | $T_{sig}$ | -55 to 150 | °C |

[1]The power supply of the device must start its ramp from 0.0 V.
[2]Functional operating conditions are provided with the DC electrical specifications in Table 4-4b. Absolute maximum ratings are stress ratings only; functional operation at the maxima is not guaranteed. Stress beyond those listed may affect device reliability or cause permanent damage to the device. *Caution: All inputs that tolerate 5 V cannot be more than 2.5 V greater than the supply voltage.* This restriction applies to power-up and normal operation (i.e., if the MPC860 is unpowered, voltage greater than 2.5 V must not be applied to its inputs).
[3]Minimum temperatures are guaranteed as ambient temperature, $T_A$. Maximum temperatures are guaranteed as junction temperature, $T_j$.


**Table 4-4b: NET + ARM Processor Maximum Tolerated Voltage And Temperature Ratings[17]**

| Characteristic | Symbol | Min | Max | Unit |
|---|---|---|---|---|
| Thermal resistance—junction-to-ambient | $\theta_{JA}$ | | 31 | °C/W |
| Operating junction temperature | $T_J$ | -40 | 100 | °C |
| Operating ambient temperature | $T_A$ | -40 | 85 | °C |
| Storage temperature | $T_{STG}$ | -60 | 150 | °C |
| Internal core power @ 3.3 V—cache enabled | $P_{INT}$ | | 15 | mW/MHz |
| Internal core power @ 3.3 V—cache enabled | $P_{INT}$ | | 9 | mW/MHz |

| Symbol | Parameter | Conditions | Min | Max | Unit |
|---|---|---|---|---|---|
| $V_{DD3}$ | DC supply voltage | Core and standard I/Os | -0.3 | 4.6 | V |
| $V_I$ | DC input voltage, 3.3V I/Os | | -0.3 | $V_{DD3}$+0.3, 4.6 max | V |
| $V_O$ | DC output voltage, 3.3V I/Os | | -0.3 | $V_{DD3}$+0.3, 4.6 max | V |
| TEMP | Operating free air temperature range | Industrial | -40 | +85 | °C |
| $T_{SIG}$ | Storage temperature | | -60 | 150 | °C |

**Table 4-5: MPC860 Processor Thermal Characteristics[17]**

| Rating | Environment | | Symbol | Revision A | Revisions B, C, and D | Unit |
|---|---|---|---|---|---|---|
| Junction-to-ambient[1] | Natural convection | Single-layer board (1s) | $R_{\theta JA}$[2] | 31 | 40 | °C/W |
| | | Four-layer board (2s2p) | $R_{\theta JMA}$[3] | 20 | 25 | |
| | Airflow (200 ft/min) | Single-layer board (1s) | $R_{\theta JMA}$[3] | 26 | 32 | |
| | | Four-layer board (2s2p) | $R_{\theta JMA}$[3] | 16 | 21 | |
| Junction-to-board[4] | | | $R_{\theta JB}$ | 8 | 15 | |
| Junction-to-case[5] | | | $R_{\theta JC}$ | 5 | 7 | |
| Junction-to-package top[6] | Natural convection | | $\Psi_{JT}$ | 1 | 2 | |
| | Airflow (200 ft/min) | | | 2 | 3 | |

[1]Junction temperature is a function of on-chip power dissipation, package thermal resistance, mounting site (board) temperature, ambient temperature, air flow, power dissipation of other components on the board, and board thermal resistance.
[2]Per SEMI G38-87 and JEDEC JESD51-2 with the single layer board horizontal.
[3]Per JEDEC JESD51-6 with the board horizontal.
[4]Thermal resistance between the die and the printed circuit board per JEDEC JESD51-8. Board temperature is measured on the top surface of the board near the package.
[5]Indicates the average thermal resistance between the die and the case top surface as measured by the cold plate method (MIL SPEC-883 Method 1012.1) with the cold plate temperature used for the case temperature. For exposed pad packages where the pad would be expected to be soldered, junction-to-case thermal resistance is a simulated value from the junction to the exposed pad without contact resistance.
[6]Thermal characterization parameter indicating the temperature difference between the package top and the junction temperature per JEDEC JESD51-2.


in performance, changes in operating characteristics, and/or even permanent physical destruction of the processor.

A processor's temperature is a result of the embedded board it resides on as well as its own thermal characteristics. A processor's thermal characteristics depend on the size and material used to package the IC, the type of interconnection to the embedded board, and the presence and type of mechanism used for cooling the processor (heat sink, heat pipes, thermoelectric cooling, liquid cooling, etc.), as well as the thermal constraints imposed on the processor by the embedded board, such as power density, thermal conductivity/airflow, local ambient temperature, and heat sink size.

*Section 5 of MPC860 Datasheet Example: Power Dissipation*

The thermal management of an embedded board includes the technologies, processes, and standards that must be implemented in order to remove heat that results from the power

**Table 4-6: MPC860 Processor Power Dissipation[17]**

| Die Revision | Frequency (MHz) | Typical[1] | Maximum[2] | Unit |
|---|---|---|---|---|
| A.3 and previous | 25 | 450 | 550 | mW |
| | 40 | 700 | 850 | mW |
| | 50 | 870 | 1050 | mW |
| B.1 and C.1 | 33 | 375 | TBD | mW |
| | 50 | 575 | TBD | mW |
| | 66 | 750 | TBD | mW |
| D.3 and D.4 (1:1 mode) | 50 | 656 | 735 | mW |
| | 66 | TBD | TBD | mW |
| D.3 and D.4 (2:1 mode) | 66 | 722 | 762 | mW |
| | 80 | 851 | 909 | mW |

[1]Typical power dissipation is measured at 3.3 V.
[2]Maximum power dissipation is measured at 3.5 V.


dissipation of a board component like a processor from individual components on an embedded board. The heat must be transferred in a controlled way to the board's cooling mechanism, which is a device that keeps the board components from overheating by insuring that the temperatures of board components stay within functional temperature limits.

A processor's power dissipation results in an increase in temperature relative to the temperature of a reference point, with the increase depending on the net *thermal resistance* (opposition to the flow of expended heat energy, specified as the degree of temperature rise per unit of power) between the junction the die within the processor package, and a reference point. In fact, one of the most important factors that determines how much power a processor can handle is thermal resistance (more on thermal resistance in section 7 of the Datasheet).

Table 4-6 provides MPC860's power dissipation for the processor running at a variety of frequencies, as well as in modes where the CPU and bus speeds are equal (1:1) or where CPU frequency is twice the bus speed (2:1).

Both the thermal characteristics for the MPC860, shown in Table 4-5, that indicate what maximum junction temperature this processor needs to stay below, as well as the power dissipation levels for the processor, shown in Table 4-6, will determine what reliable thermal mechanisms are needed on the board in order to maintain the PowerPC's junction temperature within acceptable limits. *(Note: Developing a reliable thermal solution for the entire board means the thermal requirements for* ***all*** *board components are taken into consideration, not just those of the processors.)*

*Section 6 of MPC860 Datasheet Example: DC Characteristics*

Table 4-7 outlines the electrical DC characteristics of the MPC860, which are the specific operating voltage ranges for this processor. Within this table, these characteristics generally are:

**Table 4-7: MPC860 Processor DC Characteristics[17]**

| Characteristic | Symbol | Min | Max | Unit |
|---|---|---|---|---|
| Operating voltage at 40 MHz or less | $V_{DDH}$, $V_{DDL}$, VDDSYN | 3.0 | 3.6 | V |
| | KAPWR (power-down mode) | 2.0 | 3.6 | V |
| | KAPWR (all other operating modes) | $V_{DDH} - 0.4$ | $V_{DDH}$ | V |
| Operating voltage greater than 40 MHz | $V_{DDH}$, $V_{DDL}$, KAPWR, VDDSYN | 3.135 | 3.465 | V |
| | KAPWR (power-down mode) | 2.0 | 3.6 | V |
| | KAPWR (all other operating modes) | $V_{DDH} - 0.4$ | $V_{DDH}$ | V |
| Input high voltage (all inputs except EXTAL and EXTCLK) | $V_{IH}$ | 2.0 | 5.5 | V |
| Input low voltage | $V_{IL}$ | GND | 0.8 | V |
| EXTAL, EXTCLK input high voltage | $V_{IHC}$ | $0.7 \times V_{DDH}$ | $V_{DDH} + 0.3$ | V |
| Input leakage current, $V_{in} = 5.5V$ (except TMS, TRST, DSCK, and DSDI pins) | $I_{in}$ | – | 100 | μA |
| Input leakage current, $V_{in} = 3.6V$ (except TMS, TRST, DSCK, and DSDI pins) | $I_{in}$ | – | 10 | μA |
| Input leakage current, $V_{in} = 0V$ (except TMS, TRST, DSCK, and DSDI pins) | $I_{in}$ | – | 10 | μA |
| Input capacitance[1] | $C_{in}$ | – | 20 | pF |
| Output high voltage, $I_{OH} = -2.0\,mA$, $V_{DDH} = 3.0V$ (except XTAL, XFC, and open-drain pins) | $V_{OH}$ | 2.4 | – | V |

(*Continued*)

**Table 4-7: (Continued)**

| Characteristic | Symbol | Min | Max | Unit |
|---|---|---|---|---|
| Output low voltage: IOL = 2.0 mA, CLKOUT<br>IOL = 3.2 mA[2]<br>IOL = 5.3 mA[3]<br>IOL = 7.0 mA,<br>TXD1/PA14, TXD2/PA12<br>IOL = 8.9 mA,<br>TS, TA, TEA, BI, BB,<br>FIRESET, SRESET | $V_{OL}$ | — | 0.5 | V |

[1]Input capacitance is periodically sampled.

[2]A(0:31), TSIZ0/ REG , TSIZ1, D(0:31), DP(0:3)/ IRQ (3:6), RD/ WR , BURST , RSV/IRQ2 , IP_B(0:1)/IWP(0:1)/VFLS(0:1), IP_B2/IOIS16_B/AT2, IP_B3/IWP2/VF2, IP_B4/LWP0/VF0, IP_B5/LWP1/VF1, IP_B6/DSDI/AT0, IP_B7/PTR/AT3, RXD1/PA15, RXD2/PA13, L1TXDB/PA11, L1RXDB/PA10, L1TXDA/PA9, L1RXDA/ PA8, TIN1/L1RCLKA/BRGO1/CLK1/PA7, BRGCLK1/ TOUTI /CLK2/PA6, TIN2/L1TCLKA/BRGO2/CLK3/PA5, TOUT2 /CLK4/PA4, TIN3/BRGO3/CLK5/PA3, BRGCLK2/L1RCLKB/ TOUT3 /CLK6/PA2, TIN4/BRGO4/CLK7/PA1, L1TCLKB/ TOUT4 /CLK8/PA0, REJCT1 / SPISEL /PB31, SPICLK/PB30, SPIMOSI/PB29, BRGO4/ SPIMJSO/PB28, BRGO1/I2CSDA/PB27, BRGO2/I2CSCL/PB26, SMTXD1/PB25, SMRXD1/PB24, SMSYN1 / SDACK1 /PB23, SMSYN2/ SDACK2 /PB22, SMTXD2/ L1CLKOB/PB21, SMRXD2/L1CLKOA/PB20, L1ST1/ RTS1 /PB19, L1ST2/ RTS2 /PB18, L1ST3/ LIRQB /PB17, L1ST4/ LIRQA /PB16, BRGO3/PB15, RSTRT1 /PB14, L1ST1/ RTS1 / DREQ0 /PC15, L1ST2/ RTS2 / DREQ1 /PC14, L1ST3/ LIRQB /PC13, L1ST4/ LIRQA /PC12, CTS1/PC11, TGATE1 / CD1 /PC10, CTS2 /PC9, TGATE2/ CD2/ /PC8, SDACK2 /L1TSYNCB/PC7, L1RSYNCB/PC6, SDACK1/L1TSYNCA/PC5, L1RSYNCA/PC4, PD15, PD14, PD13, PD12, PD11, PD10, PD9, PD8, PD5, PD6, PD7, PD4, PD3, MII_MDC, MII_TX_ER, MII_EN, MII_MDIO, MII_TXD[0:3].

[3]BDIP/GPL_B(5), BR, BG, FRZ/IRQ6, CS (0:5), CS (6)/ CE (1)_B, CS (7)/ CE (2)_B, WE0/BS_B0/IORD, WE1/BS_B1/IORD, WE2/BS_B2/PCOE, WE3/BS_B3/PCWE, BS_A(0:3), GPL_A0/GPL_B0, OE/GPL_A1/GPL_B1, GPL_A(2:3)/GPL_B(2:3)/CS(2:3), UPWAITA/GPL_A4, UPWAITB/ GPL_B4, GPL_A5, ALE_A, CE1_A, CE2_A, ALE_B/DSCK/AT1, OP(0:1), OP2/MODCK1/STS, OP3/MODCK2/DSDO, BADDR(28:30).

- The *operating voltage* (the first two entries in Table 4-7) of a processor is the voltage applied from the power supply to the power pin ($V_{DD}$, $V_{CC}$, etc.) on the processor.
- The *input high voltage* (the third entry in Table 4-7) is the voltage range for all input pins, except EXTAL and EXTLCK, at logic level high, where voltages exceeding the maximum value can damage the processor, whereas voltages less the minimum value are typically interpreted as a logic level low or undefined.
- The *input low voltage* (the fourth entry in Table 4-7) is the voltage range for all input pins at logic level low, where voltages below the minimum stated value can damage or cause the processor to behave unreliably, whereas voltages greater than the maximum value are typically interpreted as a logic level high or undefined.
- The *EXTAL and EXTLCK* input high voltage (the fifth entry in Table 4-7) are the maximum and minimum voltages for these two pins, and voltage values have to remain between these ranges to avoid damaging the processor.
- *The various input leakage currents* for different $V_{in}$ (entries 6–8 in Table 4-7) mean that when the input voltage is between the required range, a leakage current flows on various ports, except for pins TMS, TRST, DSCK, and DSDI.
- The *output high voltage* (the ninth entry in Table 4-7) states the minimum high-output voltage is not less than 2.4 V when the processor is sourcing a current of 2.0 mA, except on XTAL, XFC, and open-drain pins.
- The *output low voltage* (the last entry in Table 4-7) states the maximum low-output voltage is not higher than 5 V when the processor is sourcing various currents on various pins.

*Section 7 of MPC860 Datasheet Example: Thermal Calculation and Measurement*

As mentioned in section 5 of the Datasheet, thermal resistance is one of the most important factors that determine how much power a processor can handle. In this datasheet, the specified thermal parameters for the MPC860 (shown in Figure 4-72) are the thermal

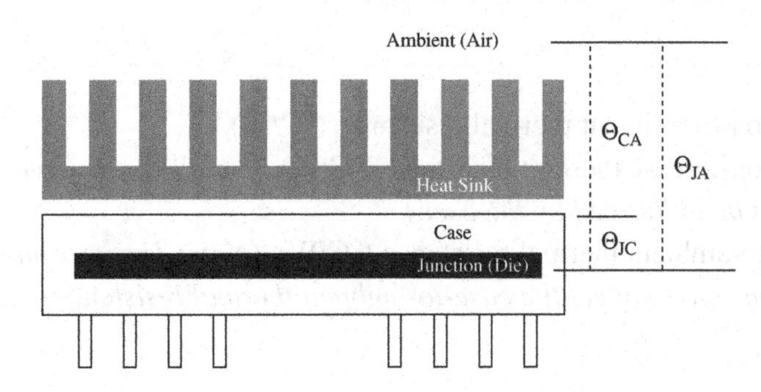


**Figure 4-72**
MPC860 processor thermal parameters.[17] *© 2004 Freescale Semiconductor, Inc. Used by permission.*

resistance estimates from junction-to-ambient ($R_{\theta JA}$), from junction-to-case ($R_{\theta JC}$), and from junction-to-board ($R_{\theta JB}$). For these equations, $P_D = (V_{DD} * I_{DD}) + P_{I/O}$ is assumed, where:

$P_D$ = Power dissipation in package.
$P_{I/O}$ = Power dissipation of the I/O drivers.
$V_{DD}$ = Supply voltage.
$I_{DD}$ = Supply current.

### Junction-to-Ambient Thermal Resistance

This is an industry standard value that provides an estimation of thermal performance. The processor's junction temperature, $T_J$ (which is the average temperature of the die within the package in degrees Celsius), can be obtained from the equation:

$$T_j = T_A + (R_{\theta JA} * P_D),$$

where:

$T_A$ = Ambient temperature (°C) is the temperature of the undistributed ambient air surrounding the package, usually measured at some fixed distance away from the processor package.
$R_{\theta JA}$ = Package junction-to-ambient thermal resistance (°C/W).
$P_D$ = Power dissipation in package.

### Junction-to-Case Thermal Resistance

The junction-to-case thermal resistance estimates thermal performance when a heat sink is used or where a substantial amount of heat is dissipated from the top of the processor's package. Typically, in these scenarios, the thermal resistance is expressed as:

$$R_{\theta JA} = R_{\theta JC} + R_{\theta CA} \text{(the sum of a junction} - \text{to} - \text{case thermal resistance and a case} - \text{to} - \text{ambient thermal resistance)},$$

where:

$R_{\theta JA}$ = Junction-to-ambient thermal resistance (°C/W).
$R_{\theta JC}$ = Junction-to-case thermal resistance (°C/W). *(Note: $R_{\theta JC}$ is device related and cannot be influenced by the user.)*
$R_{\theta CA}$ = Case-to-ambient thermal resistance (°C/W). *(Note: The user adjusts the thermal environment to affect the case-to-ambient thermal resistance, $R_{\theta CA}$.)*

### Junction-to-Board Thermal Resistance

The junction-to-board thermal resistance estimates thermal performance when most of the heat is conducted to the embedded board. Given a known board temperature and assuming heat lost to air can be negated, a junction temperature estimation can be made using the following equation:

$$T_J = T_B + (R_{\theta JB} * P_D),$$

where:

$T_B$ = Board temperature (°C).
$R_{\theta JA}$ = Junction-to-board thermal resistance (°C/W).
$P_D$ = Power dissipation in package.

When board temperature is not known, then a thermal simulation is recommended. When an actual prototype is available, the junction temperature can then be calculated via:

$$T_J = T_T + (\Psi_{JT} * P_D),$$

where:

$\Psi_{JT}$ = Thermal characterization parameter from a measurement of temperature from the top center of processor's packaging case.

## 4.5 Summary

This chapter discussed what an embedded processor is and what it is made up of. It began by introducing the concept of the ISA as the main differentiator between processors and went on to discuss what an ISA defines, as well as what common types of processors fall under a particular type of ISA model (application-specific, general-purpose, or instruction-level parallel). After the ISA discussion, the second main section of this chapter discussed how the features of an ISA are physically implemented in a processor. With this, the von Neumann model once again came into play, since the same major types of components that can be found on an embedded board can also be found at the IC (processor) level. Finally, this chapter wrapped up with a discussion on how processor performance is typically measured, as well as insight into how to read a processor's datasheet.

The next chapter, Chapter 5, *Board Memory*, introduces the hardware basics on board memory and also discusses the impact of board memory on an embedded system's performance.

## Chapter 4: Problems

1. [a] What is an ISA?
   [b] What features does an ISA define?
2. [a] Name and describe the three most common ISA models on which architectures are based.
   [b] Name and describe two types of ISAs that fall under each of the three ISA models.
   [c] Give four real-world processors that fall under the types of ISAs listed in [b].

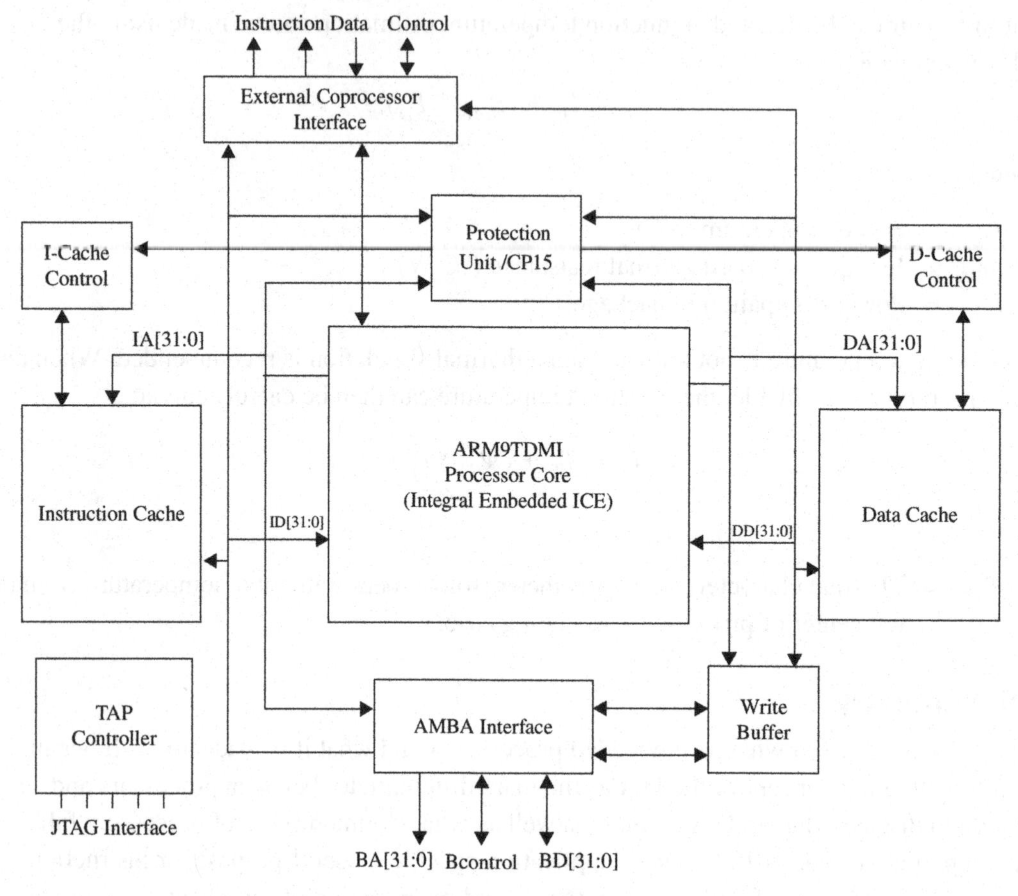


**Figure 4-73a**
ARM9 processor.[27]

3. What do the main components of a board and the internal design of a processor have in common in reference to the von Neumann model?
4. [T/F] The Harvard model is derived from the von Neumann model.
5. Indicate whether Figures 4-73a and b are von Neumann-based or Harvard-based processors. Explain your reasoning.
6. According to the von Neumann model, list and define the major components of the CPU.
7. [a] What is a register?
   [b] Name and describe the two most common types of registers.
8. What are the active electrical elements that registers are made of?
9. A processor's execution is ultimately synchronized by what board mechanism?
   A. System clock.
   B. Memory.
   C. I/O bus.

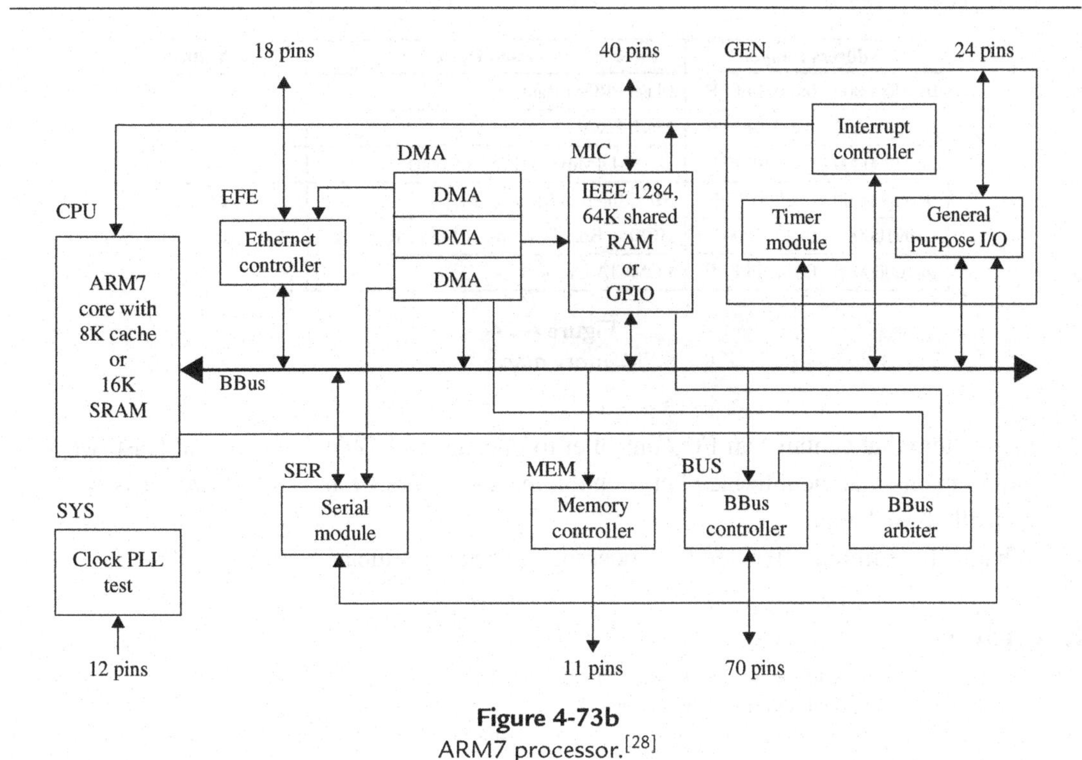


**Figure 4-73b**
ARM7 processor.[28]

D. Network slave controller.
E. None of the above.

10. Draw and describe the memory hierarchy of an embedded system.
11. What are the types of memory that can be integrated into a processor?
12. [a] What is the difference between ROM and RAM?
    [b] Give two examples of each.
13. [a] What are the three most common schemes used to store and retrieve data in cache?
    [b] What is the difference between a cache hit and cache miss?
14. Name and describe the two most common types of units that manage memory.
15. What is the difference between physical memory and logical memory?
16. [a] What is the memory map?
    [b] What is the memory makeup of a system with the memory map shown in Figure 4-74?
    [c] Which memory components shown in the memory map of Figure 4-74 are typically integrated into the master processor?
17. Name and describe the six logical units used to classify I/O hardware.
18. [a] What is the difference between serial and parallel I/O?
    [b] Give a real-world example of each.

| Address Range | Accessed Device | Port Width |
|---|---|---|
| 0×00000000 - 0×003FFFFF | Flash PROM Bank 1 | 32 |
| 0×00400000 - 0×007FFFFF | Flash PROM Bank 2 | 32 |
| 0×04000000 - 0×043FFFFF | DRAM 4Mbyte (1Meg ×32-bit) | 32 |
| 0×09000000 - 0×09003FFF | MPC Internal Memory Map | 32 |
| 0×09100000 - 0×09100003 | BCSR - Board Control & Status Register | 32 |
| 0×10000000 - 0×17FFFFFF | PCMCIA Channel | 16 |

**Figure 4-74**
Memory map.[24]

19. In a system that contains an I/O controller to manage an I/O device, name at least two requirements that the interface between the master processor and I/O controller is typically based upon.
20. What is the difference between a processor's execution time and throughput?

## Endnotes

[1] EnCore 400 Embedded Processor Reference Manual, Revision A, p. 9.
[2] Motorola, MPC8xx Instruction Set Manual, p. 28.
[3] MIPS Technologies, MIPS32™ Architecture for Programmers Volume II: The MIPS32™ Instruction Set, Revision 0.95, p. 91.
[4] MIPS Technologies, MIPS32™ Architecture for Programmers Volume II: The MIPS32™ Instruction Set, Revision 0.95, pp. 39 and 90.
[5] *ARM Architecture*, pp. 12 and 15, V. Pietikainen.
[6] *Practical Electronics for Inventors*, p. 538 P. Scherz. McGraw-Hill/TAB Electronic; 2000.
[7] Texas Instruments website: http://focus.ti.com/docs/apps/catalog/resources/blockdiagram.jhtml?appId = 178&bdId = 112.
[8] "A Highly Integrated MPEG-4 ASIC for SDCAM Application," p. 4, C.-Ta Lee, J. Zhu, Y. Liu and K.-Hu Tzou, Divio, Sunnyvale, CA.
[9] www.ajile.com
[10] National Semiconductor, Geode User's Manual, Revision 1.
[11] Net Silicon, Net + ARM40 Hardware Reference Guide.
[12] www.zoran.com
[13] www.infineon.com
[14] www.semiconductors.philips.com
[15] Freescale, MPC860 PowerQUICC User's Manual.
[16] National Semiconductor, Geode User's Manual, Revision 1.
[17] Freescale, MPC860EC Revision 6.3 Datasheet.
[18] *Embedded Microcomputer Systems*, J. W. Valvano, CL Engineering, 2nd edn, 2006.
[19] *The Electrical Engineering Handbook*, p. 1742, R. C. Dorf, IEEE Computer Society Press, 2nd edn, 1998.
[20] *Practical Electronics for Inventors*, P. Scherz. McGraw-Hill/TAB Electronic; 2000.
[21] Mitsubishi Electronics, M37267M8 Specification.
[22] Motorola, 68000 User's Manual.
[23] *Embedded Controller Hardware Design*, K. Arnold, Newnes.
[24] Freescale, MPC860 Training Manual.
[25] "Computer Organization and Programming," p. 14, D. Ramm.

[26] "This RAM, that RAM … Which is Which?," J. Robbins.
[27] "The ARM9 Family—High Performance Microprocessors for Embedded Applications," S. Segars, p. 5.
[28] Net Silicon, NET + 50/20M Hardware Reference, p. 3.
[29] *Computers As Components*, p. 206, W. Wolf, Morgan Kaufmann, 2nd edn, 2008.
[30] "Altium FPGA Design," www.altium.com, p. 6.

CHAPTER 5

# Board Memory

## In This Chapter

- Defining the various types of board memory
- Discussing memory management of on-board memory
- Discussing memory performance

### Why Should Defining and Understanding the Board's Memory Map Matter?

It is important for the reader to understand the differences in memory hardware in order to understand the implementation of overlying components, as well as how these components will behave on these various underlying technologies. For example, memory will impact the board's performance when that memory has a lower bandwidth than the master CPU. So, in this case, it is important to understand memory timing parameters (performance indicators) such as memory access times and refresh cycle times. This chapter focuses on the reader getting a better understanding of internal design, because determining the impact of memory on performance is based on the understanding of the internal design. This includes, for example:

- Utilizing independent instruction and data memory buffers and ports.
- Integrating bus signals into one line to decrease the time it takes to arbitrate the memory bus to access memory.
- Having more memory interface connections (pins), increasing transfer bandwidth.
- Having a higher signaling rate on memory interface connections (pins).
- Implementing a memory hierarchy, with multiple levels of cache.

Another is example is that an embedded system may utilize different hardware storage devices. The underlying technology within these different storage mediums available today is often quite different in terms of how they work, their performance, and how they physically store the data. Thus, by learning the features of the various hardware storage mediums available, for example, it will be much simpler for the reader to understand particular overlying components that are found on many embedded designs, and how to modify a particular design in support of a storage medium, as well as determine which overlying component is the best "fit" for the device. In other words, it is important for the reader to understand the relevant features of (a) storage medium(s)—and to use this understanding when analyzing the overall embedded implementation that needs to support the particular storage medium.

Memory hardware features, quirks, and/or limitations will dictate the type of overlying technology required and/or what modifications must be implemented in a particular embedded design to support this hardware. This includes the:

- Amount of memory (i.e., is there enough for run time needs?).
- Location of memory and how to reserve it.
- Performance (i.e., any gaps between processor and memory speeds).
- Internal design of memory.
- Type of available memory on the board (i.e., Flash versus RAM versus EEPROM).

*Note: Some of the material in this next section is similar to material in Chapter 4 covering on-chip memory, since the basics of memory operation are essentially the same whether the memory is integrated into an integrated circuit (IC) or located discretely on a board.*

As first introduced in Chapter 4, embedded platforms can have a *memory hierarchy*— a collection of different types of memory, each with unique speeds, sizes, and usages (see Figure 5-1). Some of this memory can be physically integrated on the processor, like registers and certain types of *primary* memory, which is memory connected directly to or integrated in the processor such as read-only memory (ROM), random access memory (RAM), and level 1 cache. Types of memory that can be integrated into a processor were also introduced in Chapter 4. In this chapter, it is memory which is typically located outside of the processor, or that can be either integrated into the processor or located outside the processor, that is discussed. This includes other types of primary memory, such as ROM, level 2+ cache, and main memory, and *secondary/tertiary* memory, which is memory that is connected to the board but not the master processor directly, such as CD-ROM, floppy drives, hard drives, and tape.

Primary memory is typically a part of a memory subsystem (shown in Figure 5-2) made up of three components:

- The memory IC.
- An address bus.
- A data bus.

In general, a memory IC is made up of three units: the *memory array*, the *address decoder*, and the *data interface*. The memory array is actually the physical memory that stores the data bits. While the master processor, and programmers, treat memory as a one-dimensional array, where each cell of the array is a row of bytes and the number of bits per row can vary, in reality physical memory is a two-dimensional array made up of *memory cells* addressed by a unique row and column, in which each cell can store 1 bit (as shown in Figure 5-3).

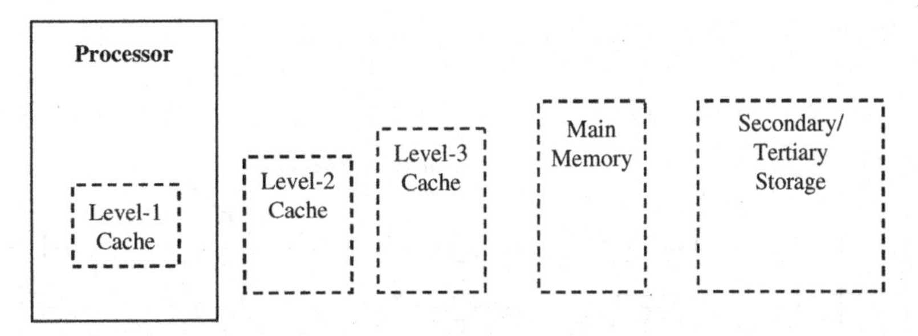


**Figure 5-1**
Memory hierarchy.

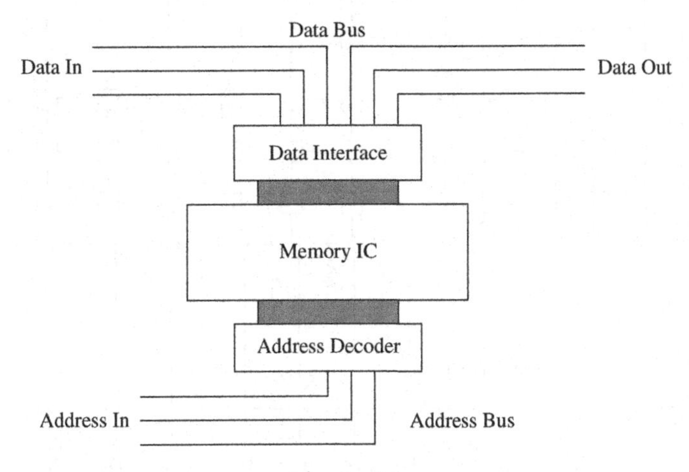


**Figure 5-2**
Hardware primary memory subsystem.

The locations of each of the cells within the two-dimensional memory array are commonly referred to as the *physical memory addresses*, made up of the column and row parameters. The main basic hardware building blocks of memory cells depend on the type of memory, to be discussed later in this chapter.

The remaining major component of a memory IC, the address decoder, locates the address of data within the memory array, based on information received over the address bus, and the data interface provides the data to the data bus for transmission. The address and data buses take address and data to and from the memory address decoder and data interface of the memory IC (buses are discussed in more detail in Chapter 7, Board Buses).

Memory ICs that can connect to a board come in a variety of packages, depending on the type of memory. Types of packages include dual in-line packages (DIPs), single in-line memory modules (SIMMs), and dual in-line memory modules (DIMMs). As shown in Figure 5-4a, DIPs

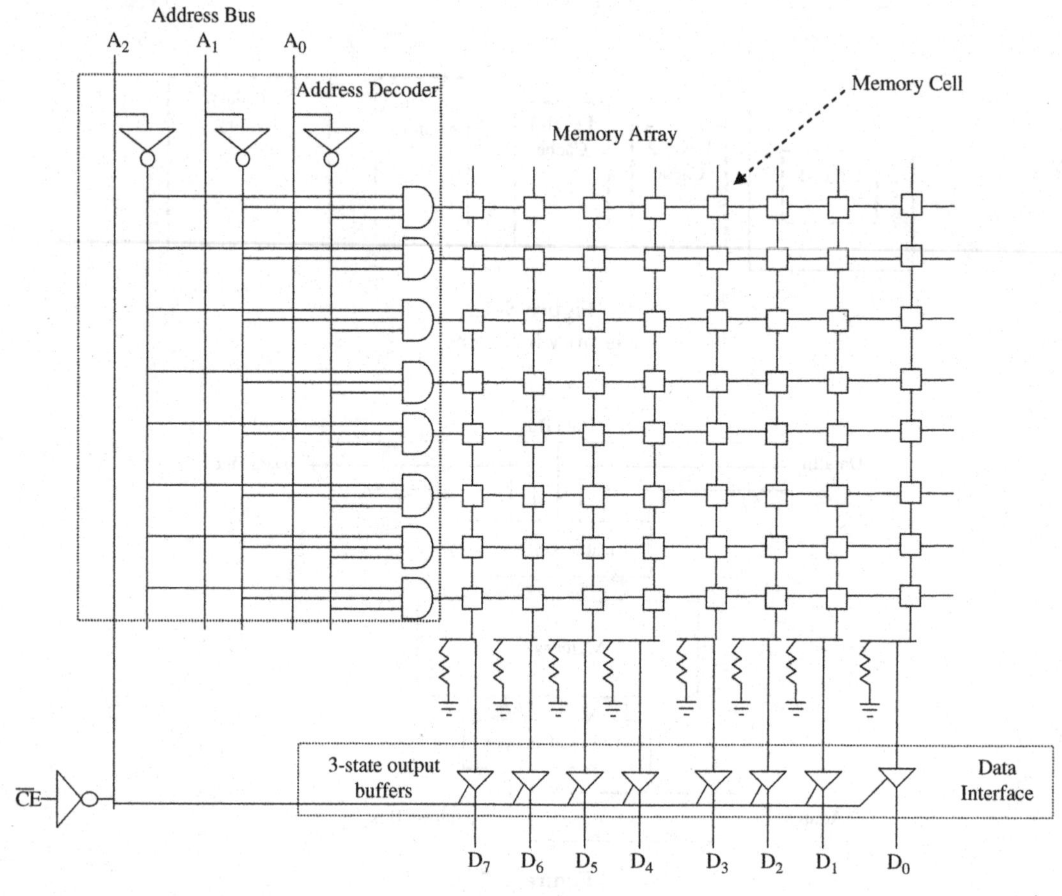


**Figure 5-3**
(ROM) memory array.[1]

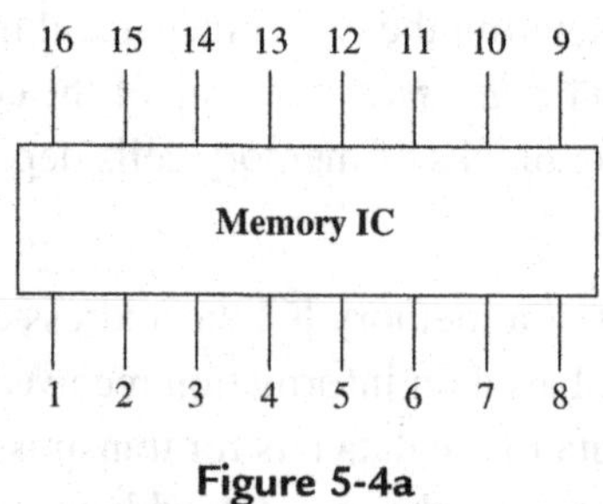


**Figure 5-4a**
DIP example.[1]

are packages enclosing the IC, made up of ceramic or plastic material, with pins protruding from two opposing sides of the package. The number of pins can vary between memory ICs, but actual pinouts of the various memory ICs have been standardized by JEDEC (Joint Electronic Device Engineering Committee) to simplify the process of interfacing external memory ICs to processors.

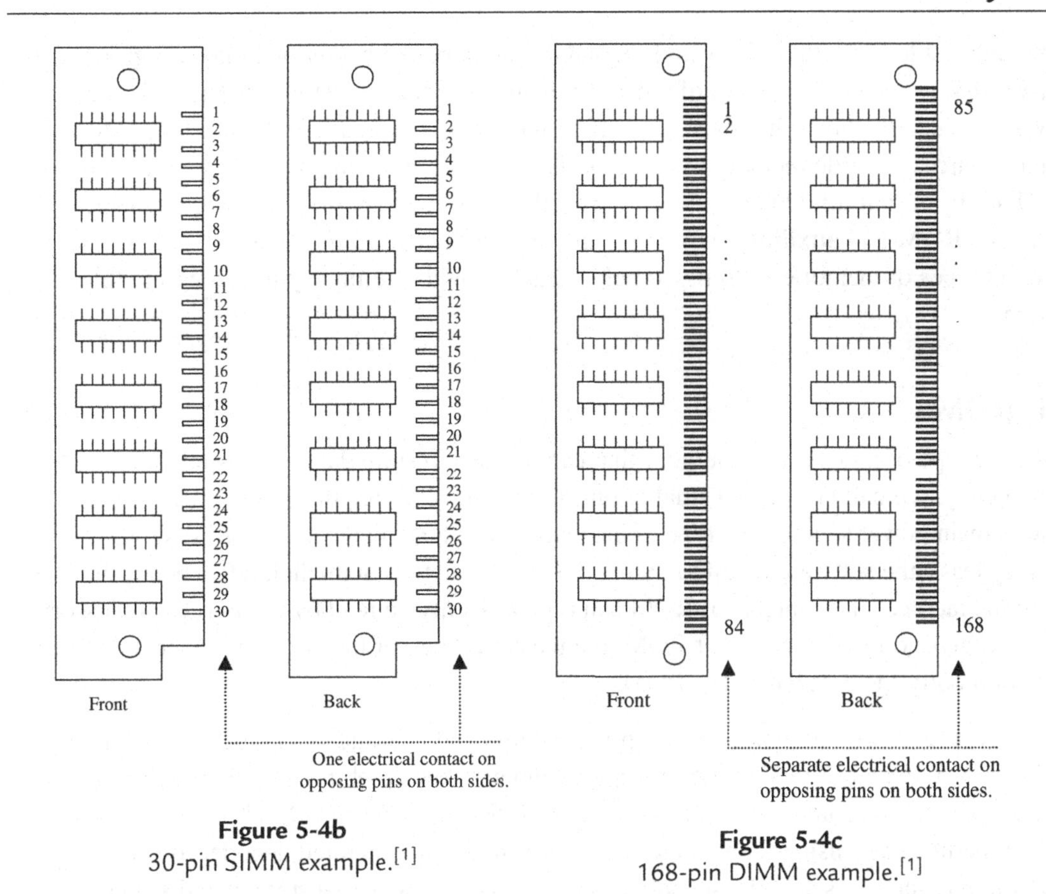


**Figure 5-4b**
30-pin SIMM example.[1]

**Figure 5-4c**
168-pin DIMM example.[1]

SIMMs and DIMMs (shown in Figures 5-4b and c) are minimodules (printed circuit boards (PCBs)) that hold several memory ICs. SIMMs and DIMMs have protruding pins from one side (both on the front and back) of the module that connect into a main embedded motherboard. The configurations of SIMMs and DIMMs can both vary in the size of the memory ICs on the module (256 kB, 1 MB, etc.). For example, a 256 K × 8 SIMM is a module providing 256 K (256 * 1025) addresses of 1 byte each. To support a 16-bit master processor, for example, two of these SIMMs would be needed; to support a 32-bit architecture, four SIMMs of this configuration would be needed, etc.

The number of pins protruding from SIMMs and DIMMs can vary as well (30, 72, 168 pins, etc.). The advantage of a SIMM or DIMM having more pins is that it allows for fewer modules needed to support larger architectures. So, for example, one 72-pin SIMM (256 K × 32) would replace the four 30-pin SIMMs (256 K × 8) for 32-bit architectures. Finally, the main difference between SIMMs and DIMMs is how the pins function on the module: on SIMMs the two pins on either side of the board are connected, creating one contact, whereas on DIMMs opposing pins are each independent contacts (see Figures 5-4b and c).

At the highest level, both primary and secondary memory can be divided into two groups, *non-volatile* or *volatile*. Non-volatile memory is memory that can store data after the main power source to the board has been shut off (usually due to a small, on-board, longer-life battery source). Volatile memory loses all of its "bits" when the main power source on the board has been shut off. On embedded boards, there are two types of *non-volatile memory* families—**ROM** and **auxiliary** memory—and one family of *volatile memory*, **RAM**. The different types of memories, discussed below, each provide a unique purpose within the system.

## 5.1 ROM

ROM is a type of non-volatile memory that can be used to store data on an embedded system permanently, typically through a smaller on-board battery source that is separate from the board's main power source. The type of data stored on ROM in an embedded system is (at the very least) the software required by the device to function in the field after being shipped out of the factory. The contents of ROM can typically only be read by the master processor; however, depending on the type of ROM, the master processor may or may not be able to erase or modify the data located in ROM.

Basically, a ROM circuit works by accepting column and row address inputs, as shown in Figure 5-5. Each cell (addressed by one column and row combination) stores a 1 or 0 depending on some voltage value. In fact, every ROM cell is designed to hold only either a 1 or 0 permanently using the voltage source attached. An integrated decoder uses the row/column inputs to select the specific ROM cell. While the actual storage and selection mechanisms are dependent on the type of components used in building the ROM (diodes, MOS, bipolar, etc.—the basic building blocks introduced in Chapter 3), all types of ROM can exist as external (chips) to the master CPU.

The circuit in Figure 5-5 includes three address lines ($\log_2 8$) for all eight words, meaning the 3-bit addresses ranging from 000 to 111 each represent one of the 8 bytes. *(Note: Different ROM designs can include a wide variety of addressing configurations for the exact same matrix size, and this addressing scheme is just an example of one such scheme.)* $D_0$–$D_7$ are the output lines from which data is read—one output line for each bit. Adding additional rows to the ROM matrix increases its size in terms of the number of address spaces, whereas adding additional columns increases a ROM's data size, or the number of bits per address it can store. ROM sizes are identified in the real world by a matrix reference (8 × 8, 16 K × 32, etc.), reflecting the actual size of ROM. The first number is the number of addresses, and the second number (after the "×") reflects the size of the data, or number of bits, at each address location (8 = 1 byte, 16 = half word, 32 = word, etc.). Also note that in some design documentation, the ROM matrix size may be summarized. For example, 16 kB (kbytes) of ROM is 16 K × 8 ROM, 32 MB of ROM is 32 M × 8 ROM.

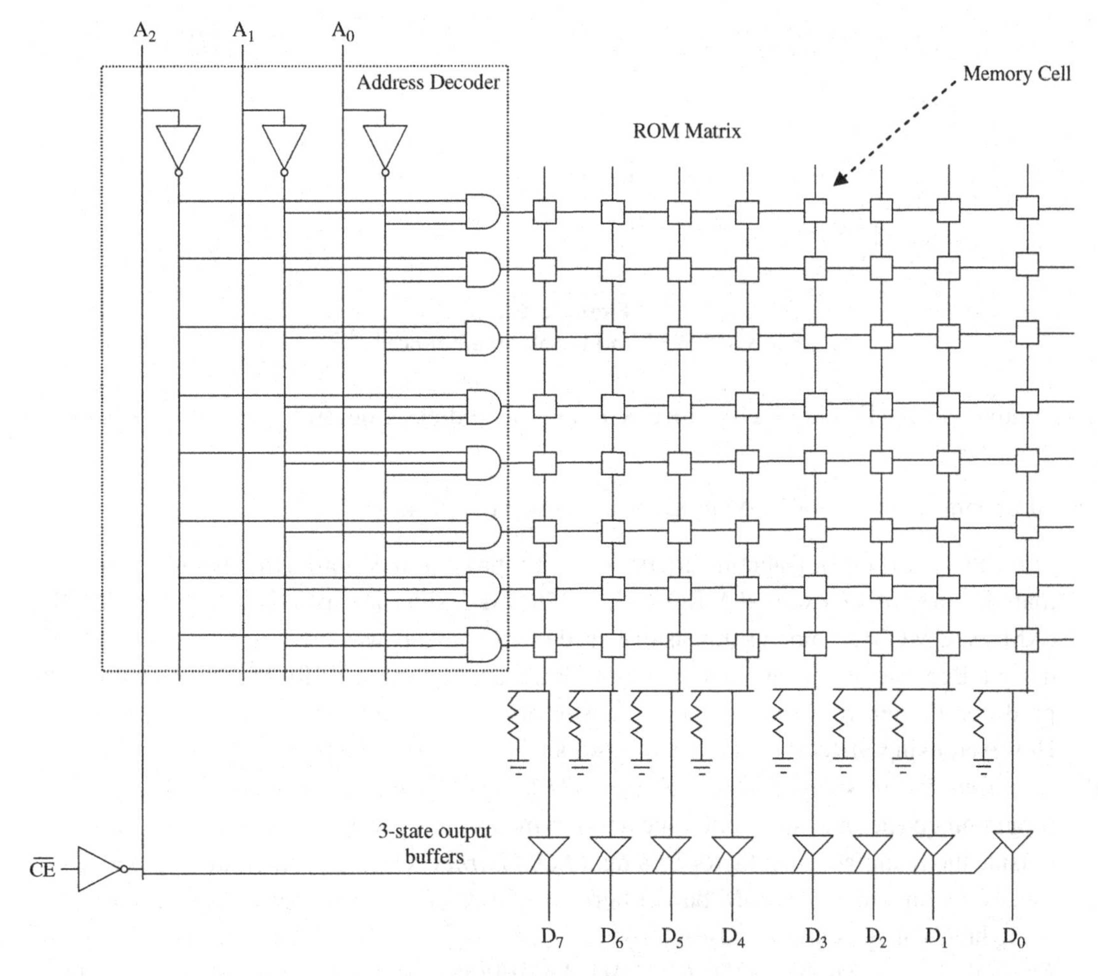


**Figure 5-5**
8 × 8 ROM logic circuit.[1]

In this example, the 8 × 8 ROM is an 8 × 8 matrix, meaning it can store eight different 8-bit words, or 64 bits of information. Every intersection of a row and column in this matrix is a memory location, called a *memory cell*. Each memory cell can contain either a bipolar or MOSFET transistor (depending on the type of ROM) and a fusible link (see Figure 5-6).

When a programmable link is in place, the transistor is biased ON, resulting in a "1" being stored. All ROM memory cells are typically manufactured in this configuration. When writing to ROM, a "0" is stored by breaking the programmable link. How links are broken depends on the type of ROM; this is discussed at the end of this section during the summary of different types of ROMs. How to read from a ROM depends on the ROM, but in this example, for instance, the chip enable (CE) line is toggled (i.e., HIGH to LOW) to allow the data stored

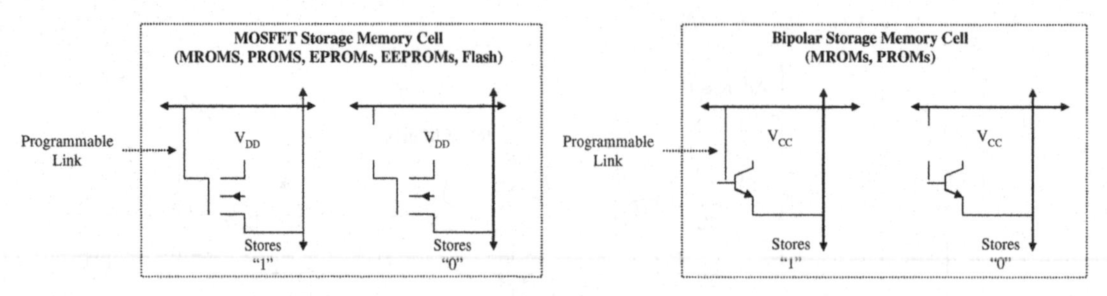


**Figure 5-6**
8 × 8 MOSFET and bipolar memory cells[1]

to be output via $D_0$–$D_7$ after having received the 3-bit address requesting the row of data bits (see Figure 5-7).

The most common types of ROM used on embedded boards are:

- *Mask ROM (MROM).* Data bits are permanently programmed into a microchip by the manufacturer of the external MROM chip. MROM designs are usually based upon MOS (NMOS, CMOS) or bipolar transistor-based circuitry. This was the original type of ROM design. Because of expensive setup costs for a manufacturer of MROMs, it is usually only produced in high volumes and there is a wait time of several weeks to several months. However, using MROMs in design of products is a cheaper solution.
- *One-Time Programmable ROM (OTP or OTPRom).* This type of ROM can only be programmed (permanently) one time as its name implies, but it can be programmed outside the manufacturing factory, using a *ROM burner*. OTPs are based upon bipolar transistors, in which the ROM burner burns out fuses of cells to program them to "1" using high voltage/current pulses.
- *Erasable Programmable ROM (EPROM).* An EPROM can be erased more than one time using a device that outputs intense short-wavelength, ultraviolet light into the EPROM package's built-in transparent window. (OTPs are one-time programmable EPROMs without the window to allow for erasure; the packaging without the window used in OTPs is cheaper.) EPROMs are made up of MOS (i.e., CMOS, NMOS) transistors whose extra "floating gate" (gate capacitance) is electrically charged, and the charge trapped, to store a "0" by the Romizer through "avalanche induced migration"—a method in which a high voltage is used to expose the floating gate. The floating gate is made up of a conductor floating within the insulator, which allows enough of a current flow to allow for electrons to be trapped within the gate, with the insulator of that gate preventing electron leakage. The floating gates are discharged via UV light, to store a "1" for instance. This is because the high-energy photons emitted by UV light provide enough energy for electrons to escape the insulating portion of the floating gate (remember from Chapter 3 that even the best of insulators, given the right circumstances, will conduct). The total number of erasures and rewrites is limited depending on the EPROM.

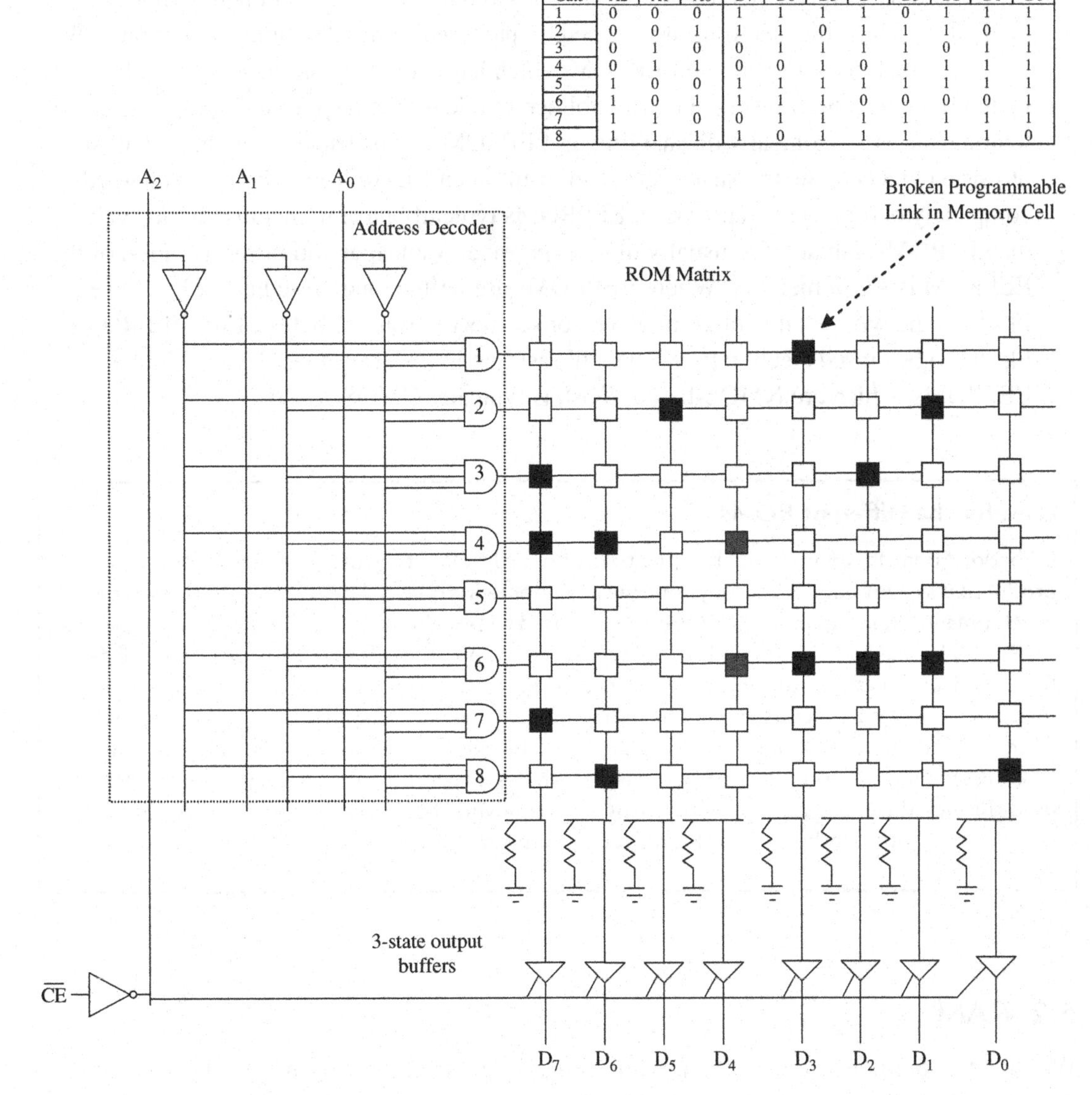

| Gate | A2 | A1 | A0 | D7 | D6 | D5 | D4 | D3 | D2 | D1 | D0 |
|---|---|---|---|---|---|---|---|---|---|---|---|
| 1 | 0 | 0 | 0 | 1 | 1 | 1 | 1 | 0 | 1 | 1 | 1 |
| 2 | 0 | 0 | 1 | 1 | 1 | 0 | 1 | 1 | 1 | 0 | 1 |
| 3 | 0 | 1 | 0 | 0 | 1 | 1 | 1 | 1 | 0 | 1 | 1 |
| 4 | 0 | 1 | 1 | 0 | 0 | 1 | 0 | 1 | 1 | 1 | 1 |
| 5 | 1 | 0 | 0 | 1 | 1 | 1 | 1 | 1 | 1 | 1 | 1 |
| 6 | 1 | 0 | 1 | 1 | 1 | 1 | 0 | 0 | 0 | 0 | 1 |
| 7 | 1 | 1 | 0 | 0 | 1 | 1 | 1 | 1 | 1 | 1 | 1 |
| 8 | 1 | 1 | 1 | 1 | 0 | 1 | 1 | 1 | 1 | 1 | 0 |



**Figure 5-7**
8 × 8 reading ROM circuit.[1]

- *Electrically Erasable Programmable ROM (EEPROM).* Like EPROM, EEPROMs can be erased and reprogrammed more than once. The number of times erasure and reuse occur depends on the EEPROMs. Unlike EPROMs, the content of EEPROM can be written and erased "in bytes" without using any special devices. In other words, the EEPROM can stay on its residing board, and the user can connect to the board interface to access and modify an EEPROM. EEPROMs are based upon NMOS transistor circuitry, except insulation of the floating gate in an EEPROM is thinner than that of the EPROM, and the method used to charge the floating gates is called the Fowler–Nordheim tunneling method (in which the electrons are trapped by passing through the thinnest section of the insulating material). Erasing an EEPROM which has been programmed electrically is a matter of using a high-reverse polarity voltage to release the trapped electrons within the floating gate. Electronically discharging an EEPROM can be tricky, though, in that any physical defects in the transistor gates can result in an EEPROM not being discharged completely before a new reprogram. EEPROMs typically have more erase/write cycles than EPROMs, but are also usually more expensive. A cheaper and faster variation of the EEPROM is *Flash* memory. Where EEPROMs are written and erased at the byte level, Flash can be written and erased in blocks or sectors (a group of bytes). Like EEPROM, Flash can be erased electrically, while still residing in the embedded device. Unlike EEPROMs, which are NMOS-based, Flash is typically CMOS-based.

**Uses for the Different ROMs**

Embedded boards can vary widely in the type of board ROMs they use, not only in the production system but even throughout the development process. For example, at the start of development, more expensive EPROMs may be used to test the software and hardware, whereas OTPs may be used at the end of development stages to provide different revisions of the code for a particular platform to various other groups (for testing/quality assurance, hardware, manufacturer of maskROMs, etc.). The ROMs actually used in mass production and deployed in an embedded system could be the maskROMs (the cheapest solution of the above family of ROM ICs). On more complex and expensive platforms, Flash memory may be the only ROM used through the entire device development and deployment process, or could be used in combination with another type of ROM, such as a boot maskROM.

## 5.2 RAM

With RAM, commonly referred to as *main memory*, any location within it can be accessed directly and randomly, rather than sequentially from some starting point, and its content can be changed more than once—the number of times depending on the hardware. Unlike ROM, contents of RAM are erased if the board loses power, meaning RAM is *volatile*. The two main types of RAM are *static RAM (SRAM)* and *dynamic RAM (DRAM)*.

As shown in Figure 5-8a, SRAM memory cells are made up of transistor-based flip-flop circuitry that typically holds its data, due to a moving current being switched bidirectionally on a pair of inverting gates in the circuit, until power is cut off or the data is overwritten. To get a clearer understanding of how SRAM works, let us examine a sample logic circuit of 4K × 8 SRAM shown in Figure 5-8b.

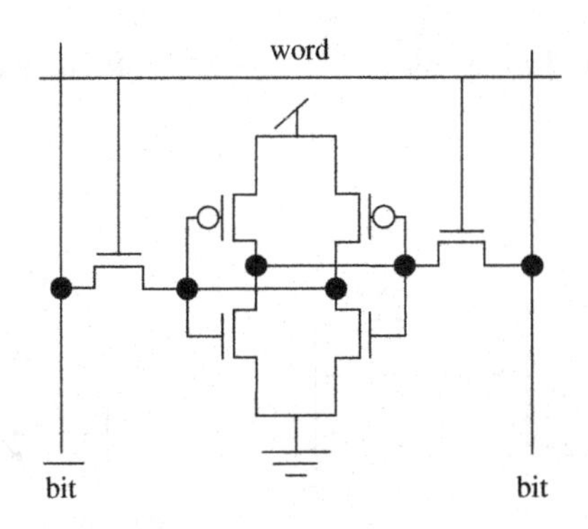


**Figure 5-8a**
Six-transistor SRAM cell.[2]

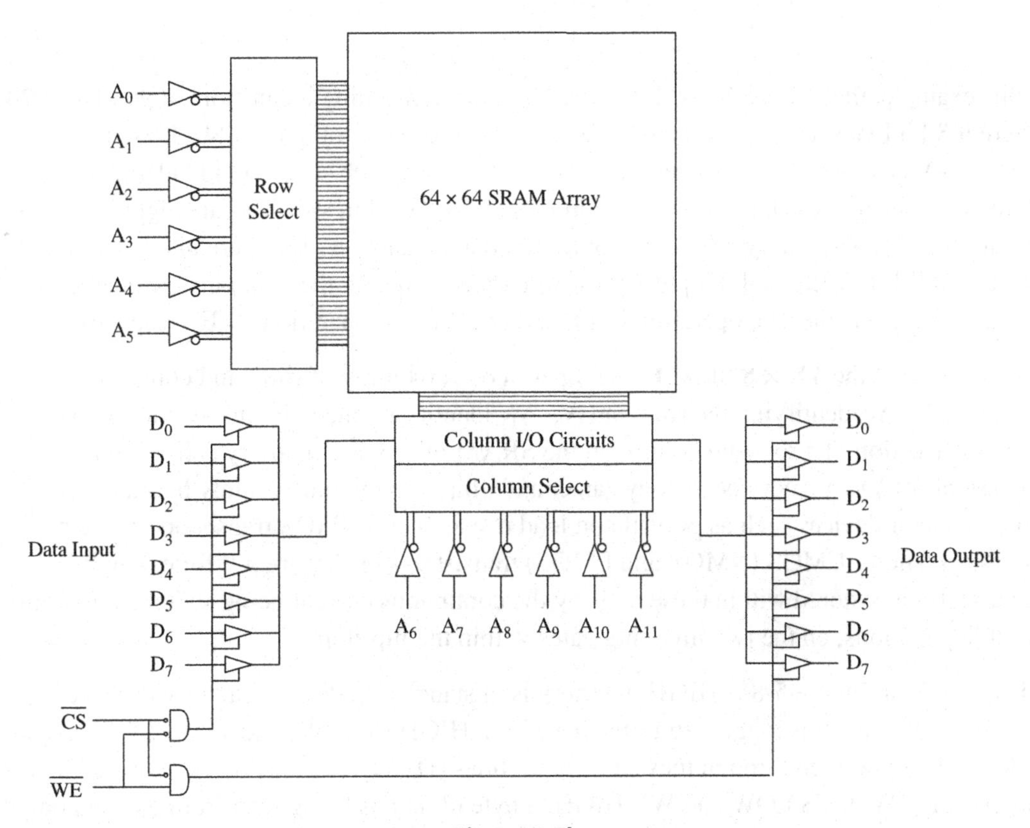


**Figure 5-8b**
4 K × 8 SRAM logic circuit.[1]

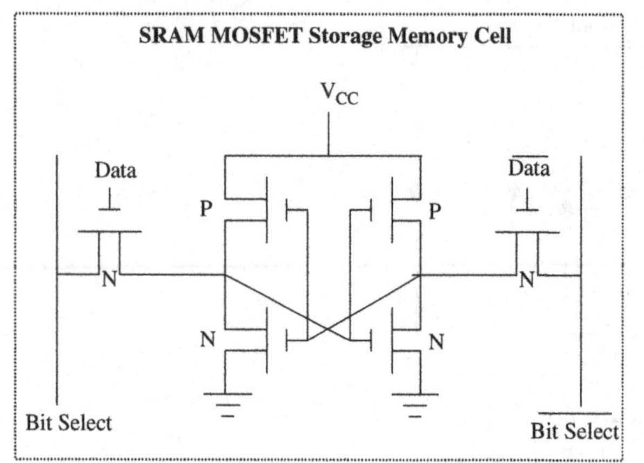


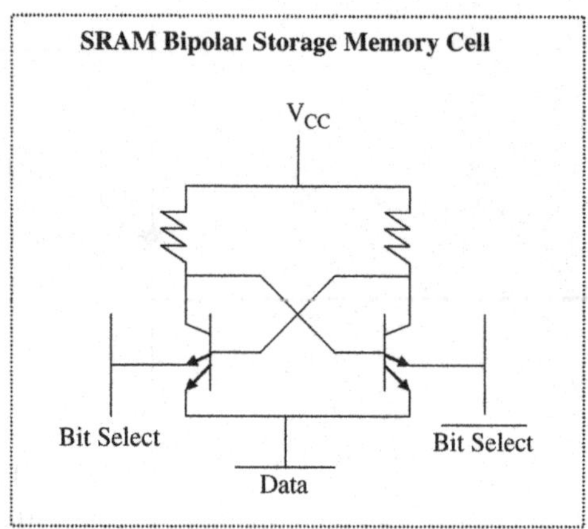


**Figure 5-9**
Flip-flop SRAM memory cell logic circuit example.[2]

In this example, the 4 K × 8 SRAM is a 4 K × 8 matrix, meaning it can store 4096 (4 × 1024) different 8-bit bytes, or 32768 bits of information. As shown in Figure 5-8b, 12 address lines ($A_0$–$A_{11}$) are needed to address all 4096 (000000000000b–111111111111b) possible addresses—one address line for every address digit of the address. There are eight input and output lines ($D_0$–$D_7$)—a byte for every byte stored at an address. There are also CS (chip select) and WE (write enable) input signals to indicate whether the data pins are enabled (CS) and to indicate whether the operation is a READ or WRITE operation (WE), respectively.

In this example, the 4 K × 8 SRAM is set up as a 64 × 64 array of rows and columns with addresses $A_0$–$A_5$ identifying the row, and $A_6$–$A_{11}$ identifying the column. As with ROM, every intersection of a row and column in the SRAM matrix is a memory cell—and in the case of SRAM memory cells, they can contain flip-flop circuitry mainly based on semiconductor devices such as polysilicon load resistors and NMOS transistors, bipolar transistors, and/or CMOS (NMOS and PMOS) transistors (see Figure 5-9 for example circuits). Data is stored within these cells by the continuous current being switched, in both possible directions, on the two inverting gates within the flip-flop.

When the CS in Figure 5-8 is HIGH, memory is in standby mode (no read or writes are occurring). When CS is toggled to LOW (i.e., from HIGH to LOW) and WE is LOW, a byte of data is being written through these data input lines ($D_0$–$D_7$) at the address indicated by the address lines. With CS LOW and WE HIGH, a byte of data is being read from the data output lines ($D_0$–$D_7$) at the address indicated by the address lines ($A_0$–$A_7$). The timing diagram in Figure 5-10 demonstrates how the different signals can function for a memory read and memory write in SRAM.

As shown in Figure 5-11a, DRAM memory cells are circuits with *capacitors* that hold a charge in place—the charges or lack thereof reflecting data. DRAM capacitors need to be

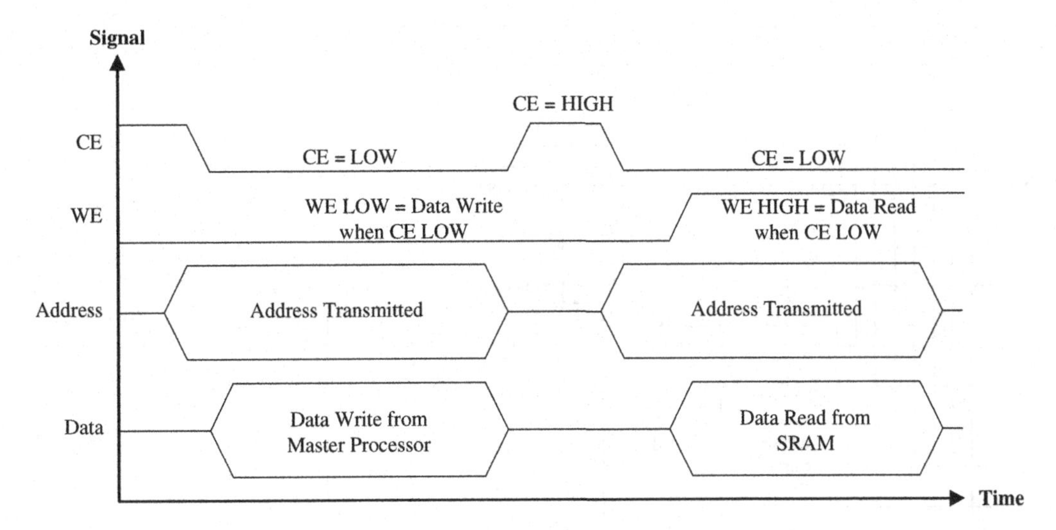


**Figure 5-10**
SRAM timing diagram.[1]

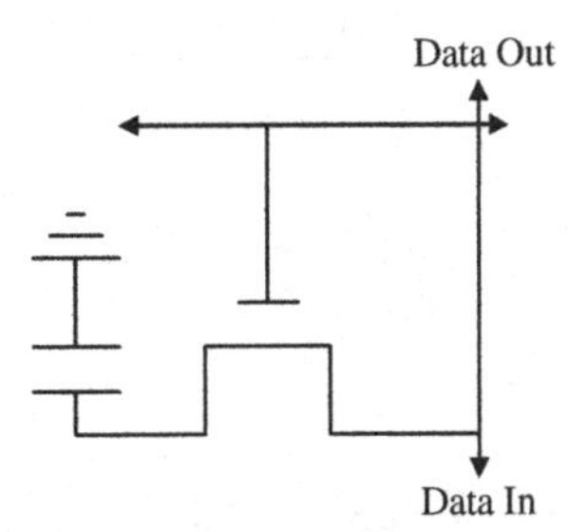


**Figure 5-11a**
DRAM (capacitor-based) memory cell.[1]

refreshed frequently with power in order to maintain their respective charges and to recharge capacitors after DRAM is read, since reading DRAM discharges the capacitor. The cycle of discharging and recharging of memory cells is why this type of RAM is called dynamic.

Let's look at a sample logic DRAM circuit of 16 K × 8. This RAM configuration is a two-dimensional array of 128 rows and 128 columns, meaning it can store 16384 (16 × 1024) different 8-bit bytes or 131072 bits of information. With this address configuration, larger DRAMs can be designed with 14 address lines ($A_0$–$A_{13}$) needed to address all 16384 (000000000000b–11111111111111b) possible addresses—one address line for every address digit of the address—or these address lines can be *multiplexed* (or combined into fewer lines to share) with some type of data selection circuit managing the shared lines. Figure 5-11b demonstrates how address lines could be multiplexed in this example.

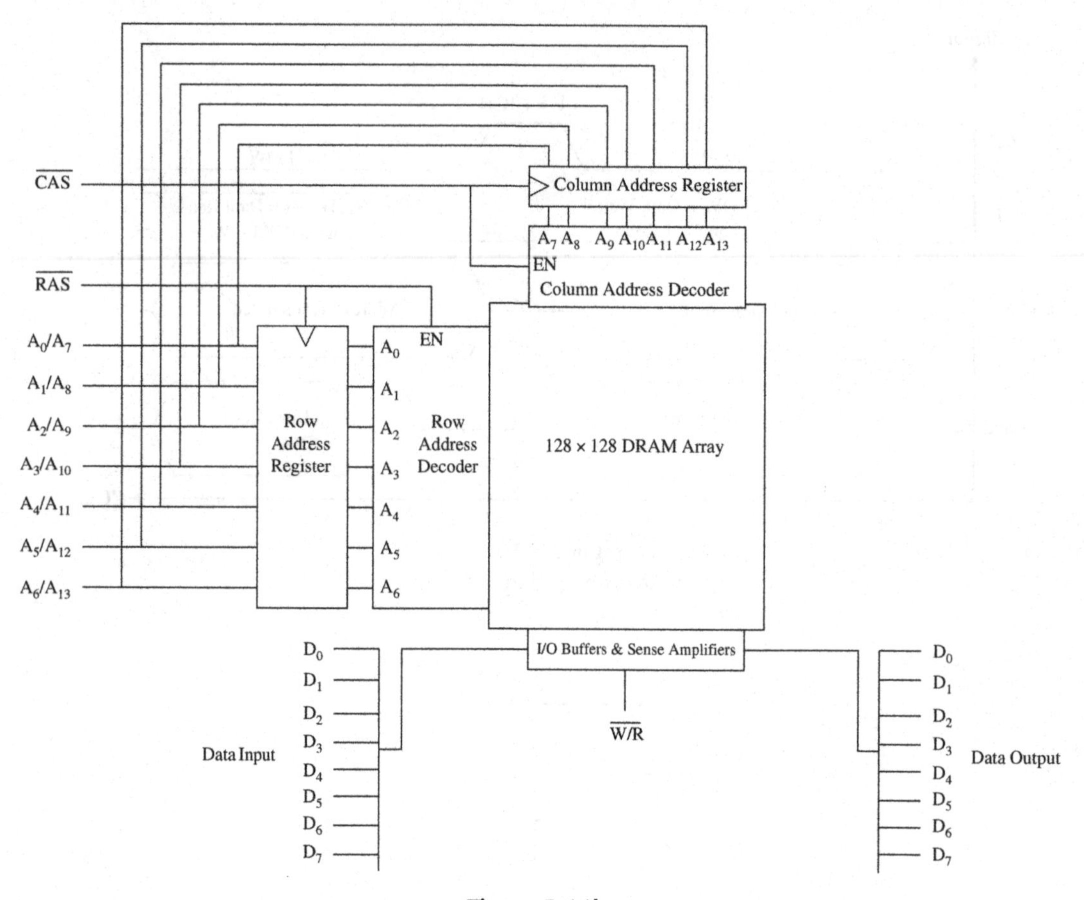


**Figure 5-11b**
16 K × 8 DRAM logic circuit.[1]

The 16 K × 8 DRAM is set up with addresses $A_0$–$A_6$ identifying the row and $A_7$–$A_{13}$ identifying the column. As shown in Figure 5-12a, the Row Address Strobe (RAS) line is toggled (i.e., from HIGH to LOW) for $A_0$–$A_6$ to be transmitted and then the Column Address Strobe (CAS) line is toggled (i.e., from HIGH to LOW) for $A_7$–$A_{13}$ to be transmitted. After this point the memory cell is latched and ready to be written to or read from. There are eight output lines ($D_0$–$D_7$)—a byte for every byte stored at an address. When the WE input line is HIGH, data can be read from output lines $D_0$–$D_7$; when WE is LOW, data can be written to input lines $D_0$–$D_7$. The timing diagrams in Figure 5-12 demonstrates how the different signals can function for a memory read and memory write in DRAM.

One of the major differences between SRAM and DRAM lies in the makeup of the DRAM memory array. The capacitors in the memory array of DRAM are not able to hold a charge (data). The charge gradually dissipates over time, thus requiring some additional mechanism to *refresh* DRAM, in order to maintain the integrity of the data. This mechanism *reads*

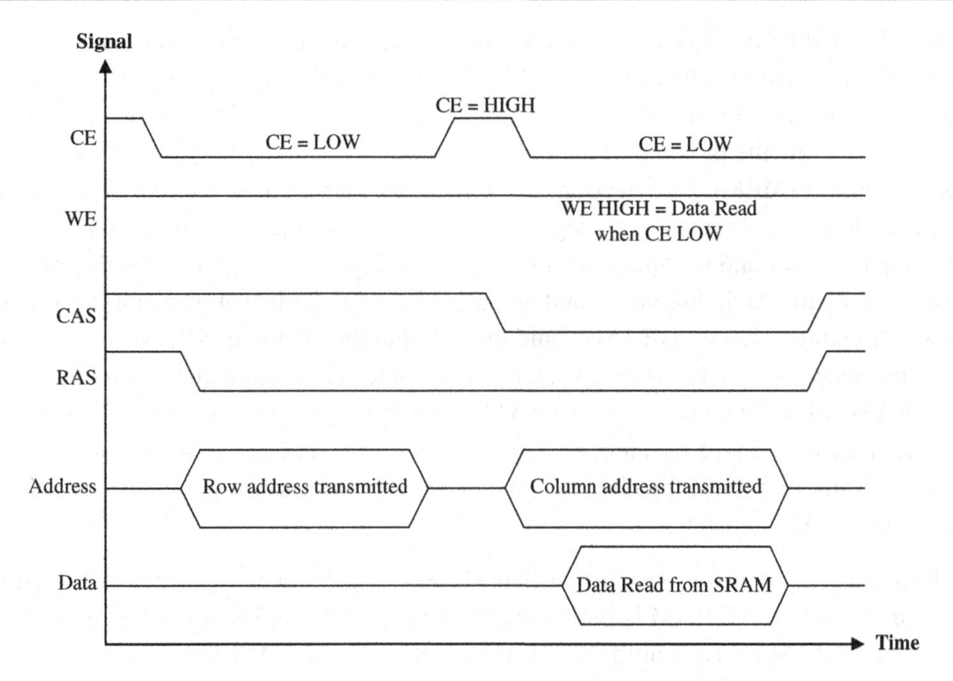


**Figure 5-12a**
DRAM read timing diagram.[1]

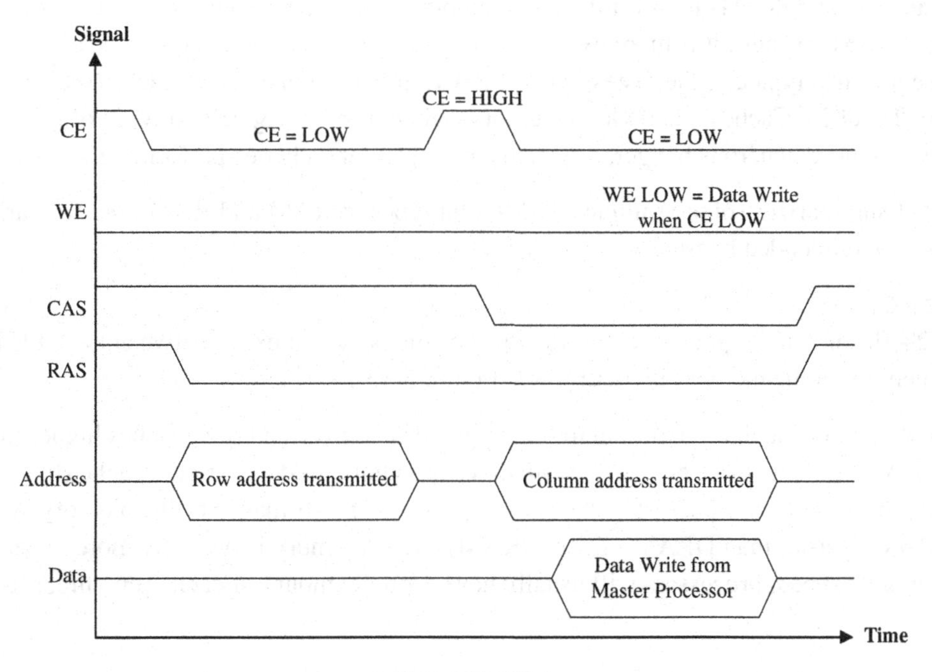


**Figure 5-12b**
DRAM write timing diagram.[1]

the data in DRAM before it is lost using a sense amplification circuit that senses a charge stored within the memory cell, and *writes* it back onto the DRAM circuitry. The process of reading the cell also discharges the capacitor (even though reading the cell is part of the process of correcting the problem of the capacitor gradually discharging in the first place). A *memory controller* (MEMC; see Section 5.4 on memory management for more information) in the embedded system typically manages a DRAM's recharging and discharging cycle by initiating refreshes and keeping track of the refresh sequence of events. It is this refresh cycling mechanism which discharges and recharges memory cells that gives this type of RAM its name—"dynamic" RAM (DRAM)—and the fact that the charge in SRAM stays put is the basis for its name, "static" RAM (SRAM). It is this same additional recharge circuitry which makes DRAM slower in comparison to SRAM. Note that one of the reasons SRAM is usually slower than registers (a type of integrated memory discussed in Chapter 4) is that when the transistors within the SRAM flip-flop are smaller, they do not carry as much current as those typically used within registers.

SRAMs also usually consume less power than DRAMs, since no extra energy is needed for a refresh. On the flip side, DRAM is typically cheaper than SRAM, because of its capacitance-based design. DRAM also can hold more data than SRAM, since DRAM circuitry is much smaller then SRAM circuitry and more DRAM circuitry can be integrated into an IC.

DRAM is usually the "main" memory in larger quantities, as well as being used for video RAM and cache. DRAMs used for display memory are also commonly referred to as *frame buffers*. SRAM, because it is more expensive, is typically used in small quantities, but because it is also typically the fastest type of RAM, it is used in external cache (see next section "Level 2+ Caches") and video memory (where processing certain types of graphics, and given a more generous budget, a system can implement a better-performing RAM).

Table 5-1 summarizes some examples of different types of RAM and ROM used for various purposes on embedded boards.

*Level 2+ Caches*

Level 2+ (level 2 and higher) cache is the level of memory that exists between the CPU and main memory in the memory hierarchy (see Figure 5-13).

In this section, cache that is external to the processor is introduced (i.e., caches higher than level 1). As shown in Table 5-1, SRAM memory is usually used as external cache (like level 1 cache), because the purpose of cache is to improve the performance of the memory system, and SRAM is faster than DRAM. Since (SRAM) cache memory is typically more expensive because of its speed, processors will usually have a small amount of cache (on-chip, off-chip, or both).

Using cache became popular in response to systems that displayed a good locality of reference, meaning that these systems, in a given time period, accessed most of their data

**Table 5-1: Board memory[4]**

| | Main Memory | Video Memory | Cache |
|---|---|---|---|
| SRAM | … | … | BSRAM (Burst/SynchBurst Static Random Access Memory) is a type of SRAM that is synchronized with either the system clock or a cache bus clock. |
| DRAM | SDRAM (Synchronous Dynamic Random Access Memory) is DRAM that is synchronized with the microprocessor's clock speed (in MHz). Several types of SDRAMs are used in various systems, such as the JDEC SDRAM (JEDEC Synchronous Dynamic Random Access Memory), PC100 SDRAM (PC100 Synchronous Dynamic Random Access Memory), and DDR SDRAM (Double Data Rate Synchronous Dynamic Random Access Memory). ESDRAM (Enhanced Synchronous Dynamic Random Access Memory) is SDRAM that integrates SRAM within the SDRAM, allowing for faster SDRAM (basically the faster SRAM portion of the ESDRAM is checked first for data, and then if not found, the remaining SDRAM portion is searched). | RDRAM (On-Chip Rambus Dynamic Random Access Memory) and MDRAM (On-Chip Multibank Dynamic Random Access Memory) are DRAMs commonly used as display memory that store arrays of bit values (pixels of the image on the display). The resolution of the image is determined by the number of bits that have been defined per each pixel. | Enhanced Dynamic Random Access Memory (EDRAM) actually integrates SRAM within the DRAM and is usually used as level 2 cache (see Section 2.1). The faster SRAM portion of EDRAM is searched first for the data, and if not found there, then the DRAM portion of EDRAM is searched. |

(*Continued*)

**Table 5-1: (Continued)**

| | Main Memory | Video Memory | Cache |
|---|---|---|---|
| | DRDRAM (Direct Rambus Dynamic Random Access Memory) and SLDRAM (SyncLink Dynamic Random Access Memory) are DRAMs whose bus signals (see Chapter 7, *Memory Buses*, for more information) can be integrated and accessed on one line, thus decreasing the access time (since synchronizing operations on multiple lines are not necessary). | Video RAM (VRAM) is DRAM in which the refresh buffer is duplicated and connected to the outside world as a second, serial I/O port. A line of data can be fetched from the memory in parallel, just as is done in a refresh, and then be read out serially. If the RAM contains pixel values, then this sequence nicely corresponds to a scan line on a monitor and facilitates display generation. At the same time, the master processor can access the RAM normally with almost no interference. | … |
| | FRAM (Ferroelectric Random Access Memory) is non-volatile DRAM, meaning data isn't lost from DRAM when power is shut off. FRAM has a lower power requirement then other types of SRAM, DRAM, and some ROMs (Flash), and is targeted for smaller handheld devices (PDAs, phones, etc.). | FPM DRAM (Fast Page Mode Dynamic Random Access Memory), EDORAM/ EDO DRAM (Data Output Random Access/Dynamic Random Access Memory),and BEDO DRAM (Data Burst Extended Data Output Dynamic Random Access Memory) … | … |
| | FPM DRAM (Fast Page Mode Dynamic Random Access Memory), EDORAM/ EDO DRAM (Data Output Random Access/Dynamic Random Access Memory), and BEDO DRAM (Data Burst Extended Data Output Dynamic Random Access Memory) … | … | … |

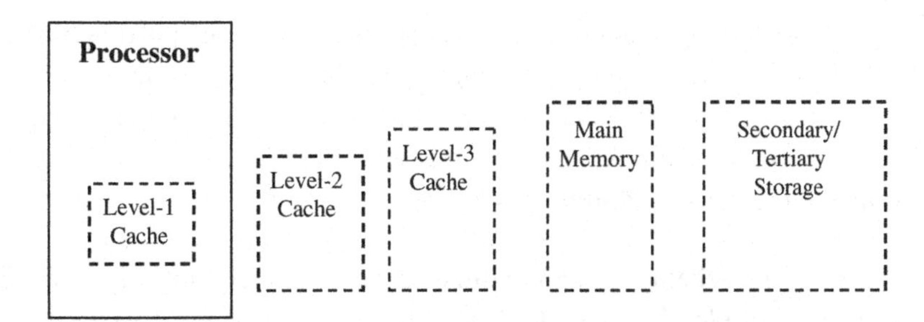


**Figure 5-13**
Level 2+ cache in the memory hierarchy.

from a limited section of memory. Basically, cache is used to store subsets of main memory that are used or accessed often, capitalizing on the locality of reference and making main memory seem to execute faster. Because cache holds copies of what is in main memory, it gives the illusion to the master processor that it is operating from main memory even if actually operating from cache.

There are different strategies used when writing to and reading data from a set of memory addresses, called the *working set*, to and from cache. One-word or multiword *blocks* are used to transfer data between memory and cache. These blocks are made up of data from main memory, as well as of tags, which represent the location of that data in main memory (called *tags*).

When writing to memory, the memory address from the CPU is translated to determine its equivalent location in level 1 cache, given that cache is a snapshot of a subset of memory. Writes must be done in both cache and main memory to ensure that cache and main memory are *consistent* (have the same value). The two most common write strategies to guarantee this are *write-through*, in which data is written to both cache and main memory every time, and *write-back*, in which data is initially only written into cache, and only when it is to be bumped and replaced from cache will it be written into main memory.

When the CPU wants to read data from memory, level 1 cache is checked first. If the data is in cache, it is called a *cache hit*, the data is returned to the CPU and the memory access process is complete. If the data is not located in level 1 cache, it is called *cache miss*. External off-chip caches are then checked and if there is a miss there also, then on to main memory to retrieve and return the data to the CPU.

Data is usually stored in cache in one of three schemes: *direct mapped*, *set associative*, or *full associative*. In the direct mapped cache scheme, addresses in cache are divided into sections called *blocks*. Every block is made up of the data, a *valid tag* (flag indicating if block is valid), and a *tag* indicating the memory address(es) represented by the block. In this scheme, data is located by its associated block address in memory, using the "tag" portion of the block. The tag is derived from the actual memory address, and is made up of three sections: a tag, an index, and an offset. The index value indicates the block, the offset value is the offset of the desired address within the block, and the tag is used to make a comparison with the actual address tag to insure the correct address was located.

The set associative cache scheme is one in which cache is divided into sections called *sets* and within each set, multiple blocks are located at the set level. The set associative scheme is implemented at the set-level. At the block level, the direct-mapped scheme is used. Essentially, all sets are checked for the desired address via a universal broadcast request. The desired block is then located according to a tag that maps into a cache's particular set. The full associative cache scheme, like the set associative cache scheme, is also composed of blocks. In the full associative scheme, however, blocks are placed anywhere in cache and must be located by searching the entire cache every time.

As with any scheme, each of the cache schemes has its strengths and drawbacks. Whereas the set associative and full associative schemes are slower than the direct mapped, the direct mapped cache scheme runs into performance problems when the block sizes get too big. On the flip side, the cache and full associative schemes are less predictable than the direct mapped cache scheme, since their algorithms are more complex.

Finally, the actual cache swapping scheme is determined by the architecture. The most common cache selection and replacement schemes include:

- *Optimal*, using future reference time, swapping out pages that won't be used in the near future.
- *Least Recently Used (LRU)*, which swaps out pages that were used the least recently.
- *First In First Out (FIFO)*, another scheme that, as its name implies, swaps out the pages that are the oldest, regardless of how often they are accessed in the system. While a simpler algorithm then LRU, FIFO is much less efficient.
- *Not Recently Used (NRU)*, which swaps out pages that were not used within a certain time period.
- *Second Chance*, a FIFO scheme with a reference bit, if "0" will be swapped out (a reference bit is set to "1" when access occurs, and reset to "0" after the check).
- *Clock Paging*, pages being replaced according to clock (how long they have been in memory), in clock order, if they haven't been accessed (a reference bit is set to "1" when access occurs, and reset to "0" after the check).

On a final note, these selection and replacement algorithms are not only limited to swapping data in and out of cache, but can be implemented via software for other types of memory swapping (e.g., operating system (OS) memory management covered in Chapter 9).

**Managing Cache**

In systems with *memory management units (MMUs)* to perform the translation of addresses (see Section 5.4), cache can be integrated between the master processor and the MMU, or the MMU and main memory. There are advantages and disadvantages to both methods of cache integration with an MMU, mostly surrounding the handling of *DMA (direct memory access)* devices that allow data to access off-chip main memory directly without going through the main processor. (DMA is discussed in Chapter 6, *Board I/O.*) When cache is integrated between the master processor and MMU, only the master processor access to memory affects cache; therefore, DMA writes to memory can make cache inconsistent with main memory unless master processor access to memory is restricted while DMA data is being transferred or cache is being kept updated by other units within the system besides the master processor. When cache is integrated between the MMU and main memory, more address translations must be done, since cache is affected by both the master processor and DMA devices.

In some systems, a MEMC may be used to manage a system with external cache (data requests and writes, for instance). More details on MEMCs can be found in Section 5.4.

## 5.3 Auxiliary Memory

As mentioned at the start of this chapter, certain types of memory can be connected directly to the master processor, such as RAM, ROM, and cache, while other types of memory, called *secondary* memory, are connected to the master processor indirectly via another device. This type of memory, as shown in Figure 5-14, is the external secondary memory and tertiary memory and is commonly referred to as *auxiliary* or *storage* memory. Auxiliary memory is typically nonvolatile memory used to store larger amounts of regular, archival, and/or backups of data, for longer periods of time to indefinitely.

Auxiliary memory can only be accessed by a device that is plugged into an embedded board, such as the disks in a hard drive, the *CD* via a CD-ROM, a *floppy disk* via a floppy drive, or

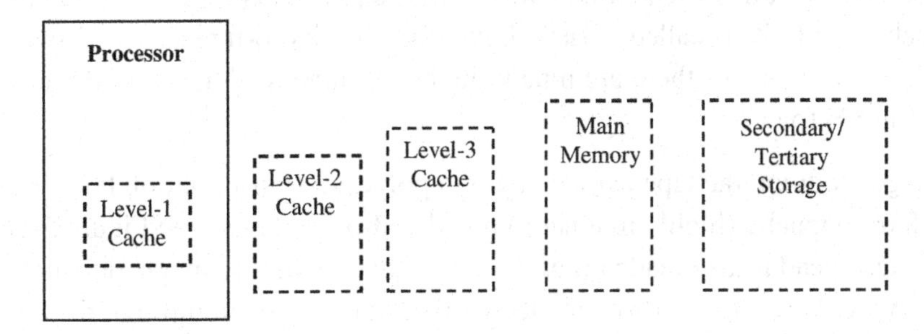


**Figure 5-14**
Auxiliary memory in the memory hierarchy.

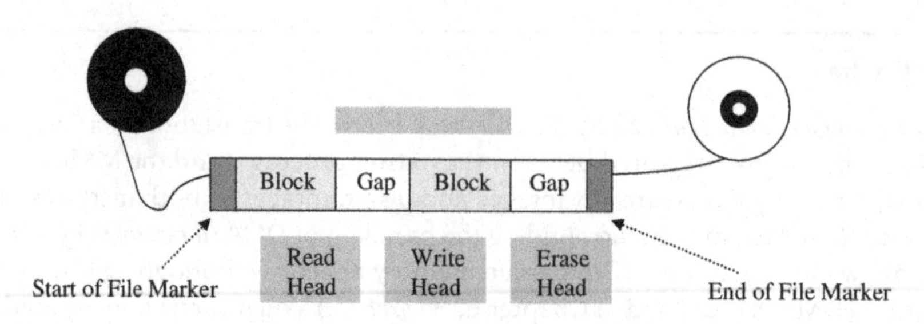


**Figure 5-15a**
Sequential access tape drive.[3]

*magnetic tape* via a magnetic tape drive. The auxiliary devices used to access auxiliary memory are typically classified as I/O devices, and are discussed in Chapter 6 in more detail. It is the auxiliary memories which plug into or are inserted within these I/O devices, that the master CPU can access that is discussed in this section. Auxiliary memory is typically classified by how its data is accessed (read and written) by its associated auxiliary device: sequential access in which data can only be accessed in sequential order; random access in which any data can be accessed directly; or direct access, which is both sequential and random access schemes combined.

Magnetic tape is a *sequential* type of memory, meaning that data can only be accessed in sequential order, and the information is stored on the tape in a sequence of rows, where sets of rows form blocks. The only data that can be accessed at any moment in time is the data in contact with the read/write/erase head(s) of the tape drive. When the read/write/erase head(s) is positioned at the beginning of the tape, the access time for retrieving data is dependent upon the location of that data on the tape, because all the data before the requested data must be accessed before retrieving the desired data. Figures 5-15a and b show an example of how magnetic tape works. Markers on the tape indicate the start and end of the tape. Within the tape, markers also indicate the start and end of files. The data with each file is divided into blocks, separated by gaps (of no data) to allow the hardware to accelerate (e.g., to begin operating) and slow down when needed. Within each block, data is separated into rows, where each row is a "bit" of the entire data width (i.e., of 9 bits for byte-sized data + 1 parity bit) and each row of bits is called a track. Each track has its own read/write/erase head(s), meaning for the nine tracks there are nine write heads, nine read heads, and nine erase heads. Refer to Figure 5-15a.

In this storage medium, the tape is made up of a polyester transportation layer with an overlying ferromagnetic (highly magnetic) powdered-oxide layer (see Figure 5-16). The read/write/erase head is also made up of materials that are highly magnetic (such as iron and cobalt). To write data onto the tape, an electrical current is passed through the magnetic coils of the write head, creating a magnetic leakage field within the air gap of the head. This field is what magnetizes the tape and reversing the current flowing through the write head's air

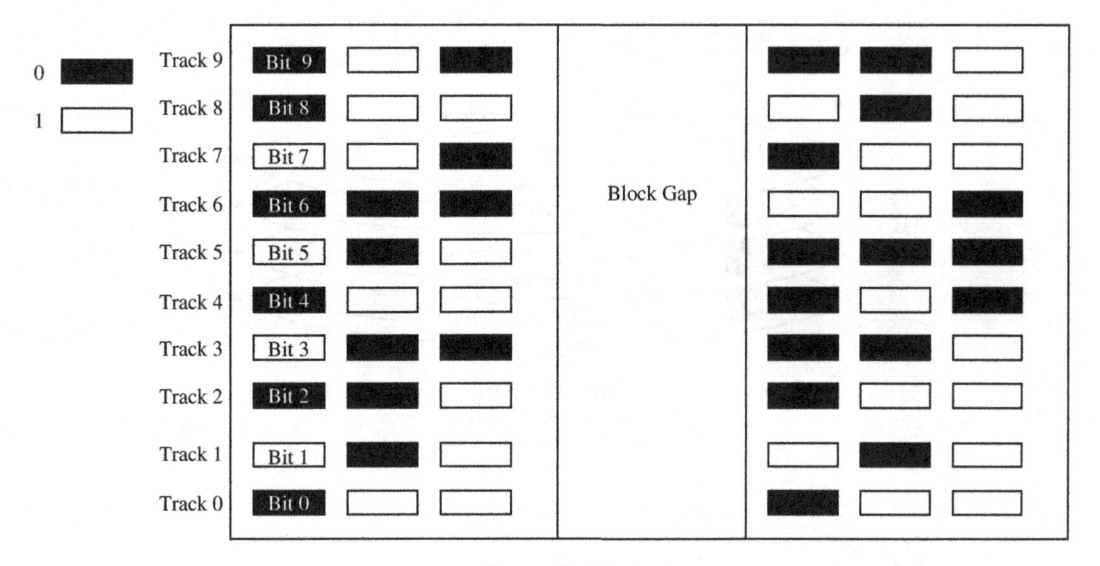


**Figure 5-15b**
Tape drive block.[3]

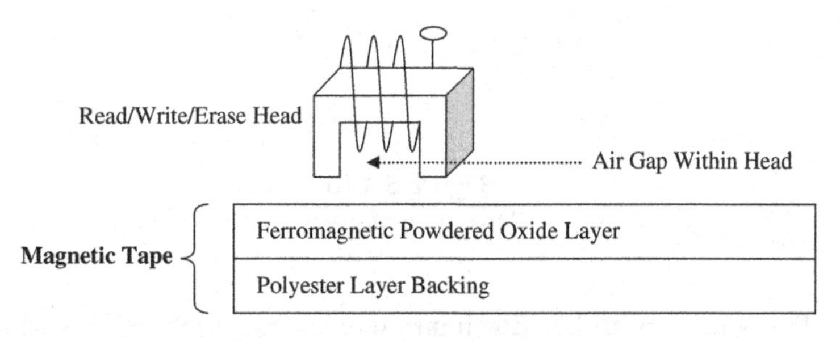


**Figure 5-16**
Magnetic tape.[3]

gap reverses the polarity of the magnetic field on the tape. The data is written in magnetized islands of rows when the tape passes the head, where a "0" is no change in magnetic polarity of the oxide layer on the tape, and a "1" is a polarity change of that layer. To read the data from a magnetic tape, a voltage is induced into the magnetic coil of the read head by the tape passing over it, which then is translated to a "0" or "1" depending on the polarity of the magnetic field on the tape. An erase head, for example, would then demagnetize the tape.

As shown in Figure 5-17a, a hard drive has multiple *platters*, which are metal disks covered with a magnetic material (film) to record data. Every platter contains multiple *tracks*, shown in Figure 5-17b. These are separate concentric rings representing separate sections for

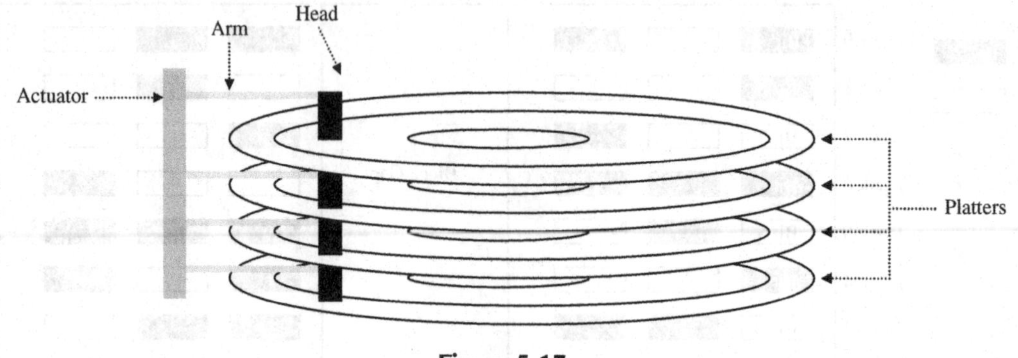


**Figure 5-17a**
Internals of hard drive.[3]

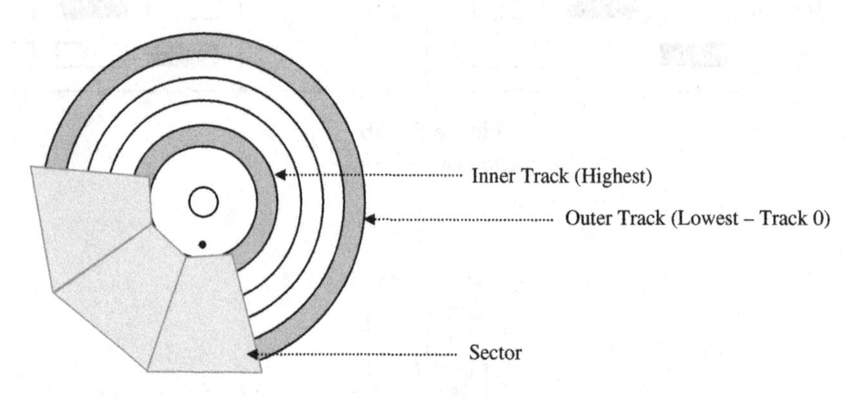


**Figure 5-17b**
Hard drive platter.[3]

recording data. Every track is broken down into *sectors*, basic subsections which can be read or written to simultaneously.

Depending on the size of the hard drive, it can have multiple *heads*, electromagnets used to record data to and read data from the platters via switchable magnetic fields. The head is supported by the disk *arm* that moves around via an *actuator*, which positions the head at the appropriate location to store, retrieve, and/or delete data.

A hard disk is an example of a memory that uses the direct access memory scheme, where a combination of random access and sequential access schemes is used to retrieve and store data. On each track, data is then stored sequentially. The read/write head(s) can be moved randomly to access the right track and the sectors of each track are then accessed sequentially to locate the appropriate data.

Like the disks in a hard drive, a compact disk is broken down into tracks and sectors (see Figure 5-18).

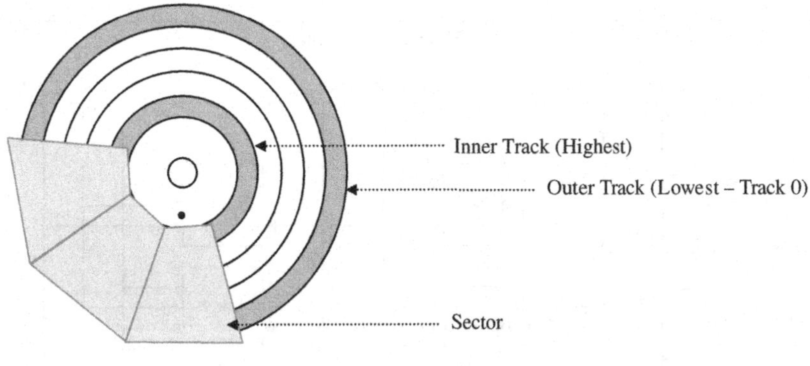


**Figure 5-18**
CD.[3]

The key difference between the platters in a hard drive and a CD is that the film on the purely optical CD isn't magnetic, but an ultrathin optical metal material. Also, where in a hard drive electromagnets are used to read and write data to a platter, lasers are used to read and write data to a CD. Another key difference between the hard disk device and a CD is that data can be read and written to the platters of the disk multiple times, whereas the CD can only be written to one time (with a high intensity laser) and read from (via a low intensity laser) multiple times. There are optical disks that can be erased, whose film is made up of magnetic and optical metal material. These disks are read, written, and erased via a combination of manipulating lasers and magnetic fields.

The main difference between primary and secondary memory lies in how they interact with the master processor. The master processor is directly connected to primary memory, and can only access the data directly that is in primary memory. Any other data that the master processor wants to access (such as that in secondary memory) must be transmitted to primary memory first before it is accessible to the master processor. Secondary memory is typically controlled by some intermediary device and is not directly accessible by the master processor.

The various access schemes, such as random access, sequential access, or direct access, can be used in either primary or secondary memory designs. However, since primary memories typically need to be faster, they usually employ a random access scheme, which is normally the faster of the access schemes. However, the circuitry required for this type of access method makes primary memory larger and more expensive, and consume more power than secondary memory.

## 5.4 Memory Management of External Memory

There are several different types of memory that can be integrated into a system, and there are also differences in how software running on the CPU views logical/virtual memory addresses and the actual *physical* memory addresses—the two-dimensional array or row and

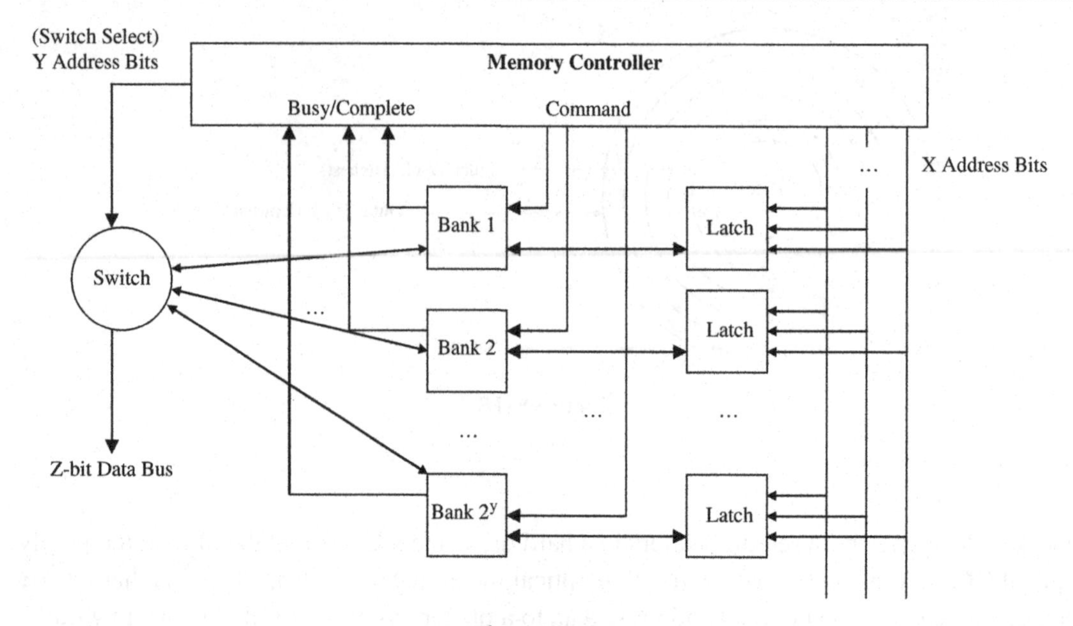


**Figure 5-19**
MEMC sample circuit.[3]

column. *Memory managers* are ICs designed to manage these issues. In some cases, they are integrated onto the master processor.

The two most common types of memory managers found on an embedded board are MEMCs and MMUs. A *MEMC*, shown in Figure 5-19, is used to implement and provide glueless interfaces to the different types of memory in the system, such as SRAM and DRAM, synchronizing access to memory and verifying the integrity of the data being transferred. MEMCs access memory directly with the memory's own physical two-dimensional addresses. The controller manages the request from the master processor and accesses the appropriate banks, awaiting feedback and returning that feedback to the master processor. In some cases, where the MEMC is mainly managing one type of memory, it may be referred to by that memory's name (DRAM controller, cache controller, etc.).

*MMUs* mainly allow for the flexibility in a system of having a larger *virtual memory* (abstract) space within an actual smaller physical memory. An MMU, shown in Figure 5-20, can exist outside the master processor and is used to translate logical (virtual) addresses into physical addresses (*memory mapping*), as well as handle memory security (memory protection), controlling cache, handling bus arbitration between the CPU and memory, and generating appropriate exceptions.

In the case of translated addresses, the MMU can use level 1 cache or portions of cache allocated as buffers for caching address translations, commonly referred to as the translation

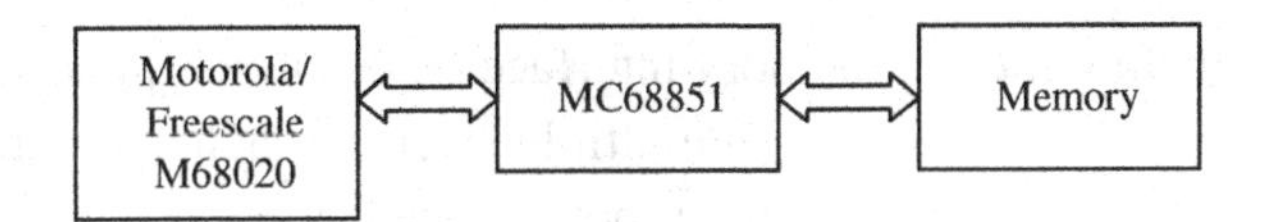


**Figure 5-20**
Motorola/Freescale M68020 external memory management.

lookaside buffer (TLB), on the processor to store the mappings of logical addresses to physical addresses. MMUs also must support the various schemes in translating addresses, mainly *segmentation*, *paging*, or some combination of *both* schemes. In general, segmentation is the division of logical memory into large variable size sections, whereas paging is the dividing up of logical memory into smaller fixed size units (more on segmentation and paging in Chapter 9). When both schemes are implemented, logical memory is first divided into segments, and segments are then divided into pages.

The memory protection schemes then provide shared, read/write or read-only accessibility to the various pages and/or segments. If a memory access is not defined or allowed, an interrupt is typically triggered. An interrupt is also triggered if a page or segment isn't accessible during address translation—for example, in the case of a paging scheme, or a *page fault*. At that point the interrupt would need to be handled (e.g., the page or segment would have to be retrieved from secondary memory).

The scheme supporting segmentation or paging of the MMU typically depends on the software (i.e., the OS). See Chapter 9 on embedded OSs for more on virtual memory and how MMUs can be used along with the system software to manage virtual memory.

## 5.5 Board Memory and Performance

As discussed in Chapter 4, one of the most common measures of a processor's performance is its *throughput* (bandwidth), or the CPU's average execution rate. The performance throughput can be negatively impacted by main memory especially, since the DRAM used for main memory can have a much lower bandwidth than that of the processors. There are specific timing parameters associated with memory (memory access times, refresh cycle times for DRAM, etc.) that act as indicators of memory performance.

Solutions for improving the bandwidth of main memory include:

- Integrating a Harvard-based architecture, with separate instruction and data memory buffers and ports, for systems that expect to perform a high number of memory accesses and computations on a large amount of data.
- Using DRAMs, such as DRDRAM and SLDRAM, that integrate bus signals into one line, to decrease the time it takes to arbitrate the memory bus to access memory.
- Using more memory interface connections (pins), increasing transfer bandwidth.

- Using a higher signaling rate on memory interface connections (pins).
- Implementing a *memory hierarchy* with multiple levels of cache, which has faster memory access times than those of other types of memory.

Memory hierarchies (shown in Figure 5-1) were designed in part to improve performance. This is because memory access during execution of programs tends not to be random, and exhibits good localities of reference. This means that systems, in a given time period, access most of their data from a limited section of memory (locality in space) or access the same data again within that given period of time (locality in time). Thus, faster memory (usually SRAM), called cache, was integrated into a memory system for this type of data to be stored and accessed by the CPU. This integration of different types of memories is referred to as the *memory hierarchy.* It is important that the memory hierarchy be effective, since the master processor spends most of its time accessing memory in order to process the applicable data. The memory hierarchy can be evaluated by calculating how many cycles are spent (wasted) due to memory latency or throughput problems, where:

Memory stall cycles = instruction count * memory references/instruction
* cache miss rate * cache miss penalty.

In short, memory performance can be improved by:

- Introducing cache, which means fewer slower DRAM accesses with a decrease in the average main memory access time; non-blocking cache will especially decrease any cache miss penalties. *(Note: With the introduction of cache, the average total memory access time = (cache hit time + (cache miss rate * cache miss penalty)) + (% cache misses * average main memory access time), where (cache hit time + (cache miss rate * cache miss penalty)) = average cache access time.)*
- Reducing the cache miss rate, by increasing cache block sizes or implementing prefetching (hardware or software)—a technique by which data and/or instructions theoretically needed in the future are transferred from main memory and stored in cache.
- Implementing pipelining, which is the process of breaking down the various functions associated with accessing memory into steps, and overlapping some of these steps. While pipelining doesn't help latency (the time it takes to execute one instruction), it does help to increase throughput, by decreasing the time it takes for cache writes, for example, and thus reducing cache write "hit" times. The pipeline rate is limited only by its slowest pipeline stage.
- Increasing the number of smaller multilevel caches rather than having one big cache, since smaller caches reduce the cache's miss penalty and average access time (hit time), whereas a larger cache has a longer cycle time and for pipe stages in an implemented pipeline.
- Integrating main memory onto the master processor, which is cheaper as well, since on-chip bandwidth is usually cheaper than pin bandwidth.

## 5.6 Summary

This chapter introduced some of the basic hardware concepts involving memory that are typically found on an embedded board, specifically the different types of board memory and the basic electrical elements that are used in building them. While there are few fundamental differences between memory located on a board and memory integrated into a processor, there are certain types of memory that can be, or are only, located outside the master processor on the embedded board itself—certain types of ROM and RAM (summarized in Table 5-1), as well as auxiliary memory. This chapter ended with an introduction of some of the key performance issues that revolve around board memory.

The next chapter, *Chapter 6, Board I/O*, discusses a variety of hardware I/O that can be found on an embedded board.

## Chapter 5: Problems

1. Draw and describe the memory hierarchy of an embedded system.
2. Which memory component(s) in a memory hierarchy is typically located on the board, outside the master processor?
   A. Level 2 cache.
   B. Main memory.
   C. Secondary memory.
   D. All of the above.
   E. None of the above.
3. [a] What is ROM?
   [b] Name and describe three types of ROM.
4. [a] What is RAM?
   [b] Name and describe three types of RAM.
5. [a] Draw examples of ROM, SRAM, and DRAM memory cells.
   [b] Describe the main differences between these memory cells.
6. [T/F] SRAM is usually used in external cache, because SRAM is slower than DRAM.
7. What type of memory is typically used as main memory?
8. [a] What is the difference between level 1, level 2, and level 3 cache?
   [b] How do they all work together in a system?
9. [a] What are the three most common schemes used to store and retrieve data in cache?
   [b] What is the difference between a cache hit and a cache miss?
10. Name and describe at least four cache swapping schemes.
11. [a] What is auxiliary memory?
    [b] List four examples of auxiliary memory.
12. [T/F] Auxiliary memory is typically classified according to how data is accessed.

| Address Range | Accessed Device | Port Width |
|---|---|---|
| 0x00000000 - 0x003FFFFF | Flash PROM Bank 1 | 32 |
| 0x00400000 - 0x007FFFFF | Flash PROM Bank 2 | 32 |
| 0x04000000 - 0x043FFFFF | DRAM 4Mbyte (1Meg×32-bit) | 32 |
| 0x09000000 - 0x09003FFF | MPC Internal Memory Map | 32 |
| 0x09100000 - 0x09100003 | BCSR - Board Control & Status Register | 32 |
| 0x10000000 - 0x17FFFFFF | PCMCIA Channel | 16 |

**Figure 5-21**
Memory map.[5]

13. [a] Name and define three data accessibility schemes commonly implemented in auxiliary memory.
    [b] Provide a real-world example that falls under each scheme.
14. Finish the sentence: MMUs and MEMCs not integrated into the master processor are typically implemented in:
    A. Separate slave ICs.
    B. Software.
    C. Buses.
    D. All of the above.
    E. None of the above.
15. [a] What is the difference between an MMU and an MEMC?
    [b] Can one embedded system incorporate both? Why or why not?
16. What is the difference between physical memory and logical memory?
17. [a] What is a memory map?
    [b] What is the memory makeup of a system with the memory map shown in Figure 5-21?
    [c] Which memory components shown in the memory map of Figure 5-21 are typically located on the board outside the master processor?
18. How can memory impact the performance of a system?
19. Define five ways in which the bandwidth of main memory and/or the overall performance of a memory subsystem can be improved.

## Endnotes


[1] *Practical Electronics for Inventors*, P. Scherz, McGraw-Hill/TAB Electronic; 2000.
[2] *Computer Organization and Programming*, p. 14, D. Ramm, MIT Press, 2002.
[3] *The Electrical Engineering Handbook*, R. C. Dorf, IEEE Computer Society Press, 2nd edn, 1998.
[4] "This RAM, that RAM … Which is Which?," J. Robbins.
[5] Freescale, MPC860 Training Manual.

CHAPTER 6

# Board I/O

## In This Chapter

- Introducing board I/O
- Discussing differences between serial and parallel I/O
- Introducing interfacing to I/O
- Discussing I/O performance

It is critical to care about the input/output (I/O) on the target hardware, specifically the I/O subsystems consisting of some combination of transmission medium, ports and interfaces, I/O controllers, buses, and the master processor integrated I/O. This is because the I/O subsystem will greatly impact the entire embedded system's performance in terms of throughput, execution time, and response time. Thus, it is important for the reader to gain a strong foundation in understanding board I/O hardware that includes:

- The data rates of the I/O devices.
- How to synchronize the speed of the master processor to the speeds of I/O.
- How I/O and the master processor communicate.

*Note*: The material in this section is similar to the material in Section 4.2.3, Processor I/O, since, aside from certain types of I/O or components of an I/O subsystem that are integrated on an integrated circuit (IC) versus discretely located on a board, the basics are essentially the same.

I/O components on a board are responsible for moving information into and out of the board to I/O devices connected to an embedded system. Board I/O can consist of input components, which only bring information from an input device to the master processor; output components, which take information out of the master processor to an output device; or components that do both (see Figure 6-1).

Virtually any electromechanical system, both embedded and non-embedded, whether conventional or unconventional, can be connected to an embedded board and act as an I/O device. I/O is a high-level group that can be subdivided into smaller groups of output devices, input devices, and devices that are both input and output devices. Output devices receive data from board I/O components and display that data in some manner, such as printing it to paper, to a disk, or to a screen or a blinking light-emitting diode (LED) for a person to see. An input device such as a mouse, keyboard, or remote control transmits data to board I/O components. Some I/O devices can do both, such as a networking device that can transmit data to and from

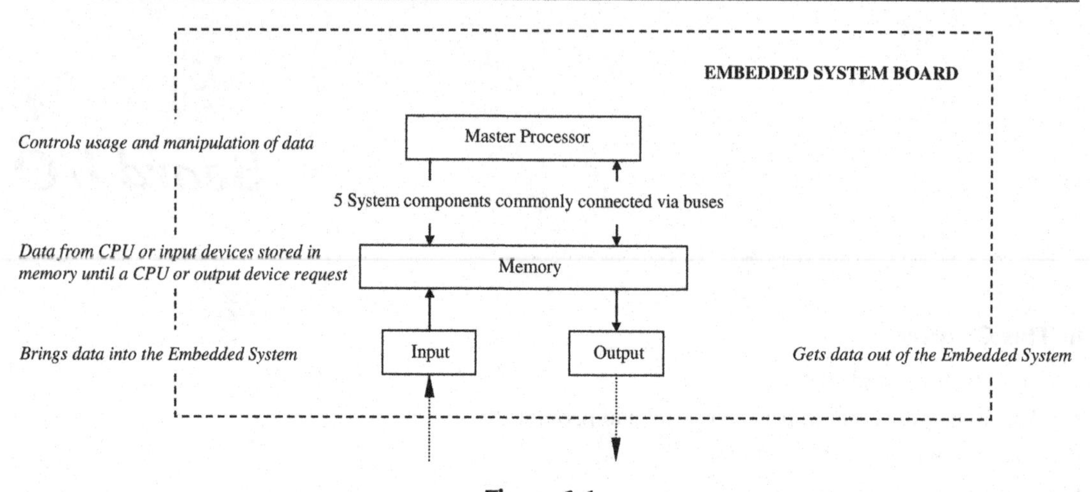


**Figure 6-1**
Von Neumann-based I/O block diagram.[1]

the Internet. An I/O device can be connected to an embedded board via a wired or wireless data transmission medium such as a keyboard or remote control, or can be located on the embedded board itself, such as an LED.

Because I/O devices are so varied, ranging from simple circuits to other complete embedded systems, board I/O components can fall under one or more of several different categories, the most common including:

- Networking and communications I/O (the physical layer of the OSI (Open Systems Interconnection) model; see Chapter 2).
- Input (keyboard, mouse, remote control, vocal, etc.).
- Graphics and output I/O (touch screen, CRT, printers, LEDs, etc.).
- Storage I/O (optical disk controllers, magnetic disk controllers, magnetic tape controllers, etc.).
- Debugging I/O (Background Debug Mode (BDM), JTAG, serial port, parallel port, etc.).
- Real time and miscellaneous I/O (timers/counters, analog-to-digital converters and digital-to-analog converters, key switches, etc.).

In short, board I/O can be as simple as a basic electronic circuit that connects the master processor directly to an I/O device, such as a master processor's I/O port to a clock or LED located on the board, to more complex I/O subsystem circuitry that includes several units, as shown in Figure 6-2. I/O hardware is typically made up of all or some combination of six main logical units:

- *Transmission medium*: wireless or wired medium connecting the I/O device to the embedded board for data communication and exchanges.
- *Communication port*: what the transmission medium connects to on the board or, if a wireless system, what receives the wireless signal.

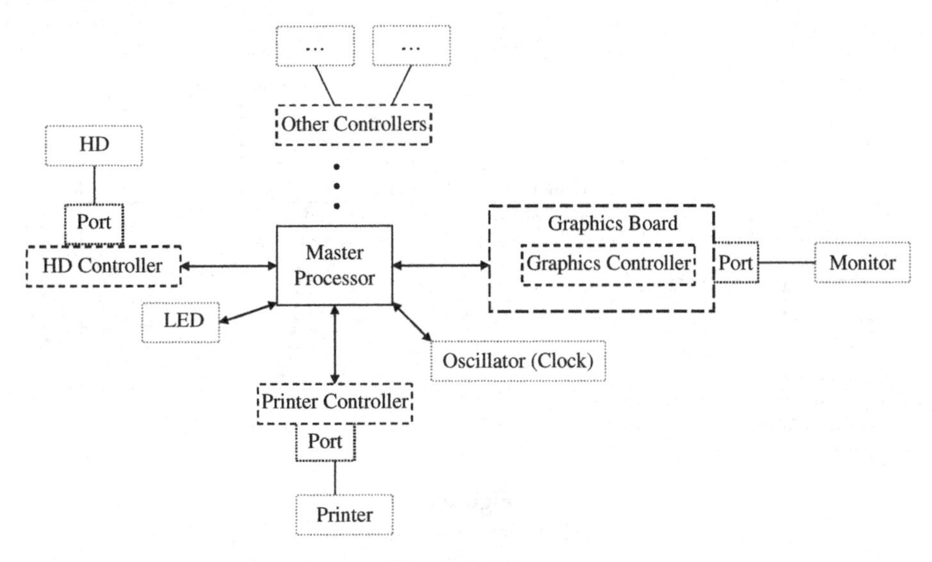


**Figure 6-2**
Ports and device controllers on an embedded board.

- *Communication interface*: manages data communication between master CPU and I/O device or I/O controller; also responsible for encoding data and decoding data to and from the logical level of an IC and the logical level of the I/O port. This interface can be integrated into the master processor, or can be a separate IC.
- *I/O controller*: slave processor that manages the I/O device.
- *I/O buses*: connection between the board I/O and master processor.
- Master processor integrated I/O.

The I/O on a board can thus range from a complex combination of components, as shown in Figure 6-3a, to a few integrated I/O board components, as shown in Figure 6-3b.

The actual makeup of an I/O system implemented on an embedded board, whether using connectors and ports or using an I/O device controller, is dependent on the type of I/O device connected to, or located on, the embedded board. This means that, while other factors such as reliability and expandability are important in designing an I/O subsystem, what mainly dictates the details behind an I/O design are the *features* of the I/O device—its purpose within the system—and the *performance* of the I/O subsystem, discussed in Section 6.3. Transmission mediums, buses, and master processor I/O are beyond the scope of this section, and are covered in Chapter 2 (transmission mediums), Chapter 7 (board buses), and Chapter 4 (processors). An I/O controller is essentially a type of processor, so again see Chapter 4 for more details.

Within the various I/O categories (networking, debugging, storage, etc.) board I/O is typically subgrouped according to how data is managed (transmitted). Note that the actual subgroups

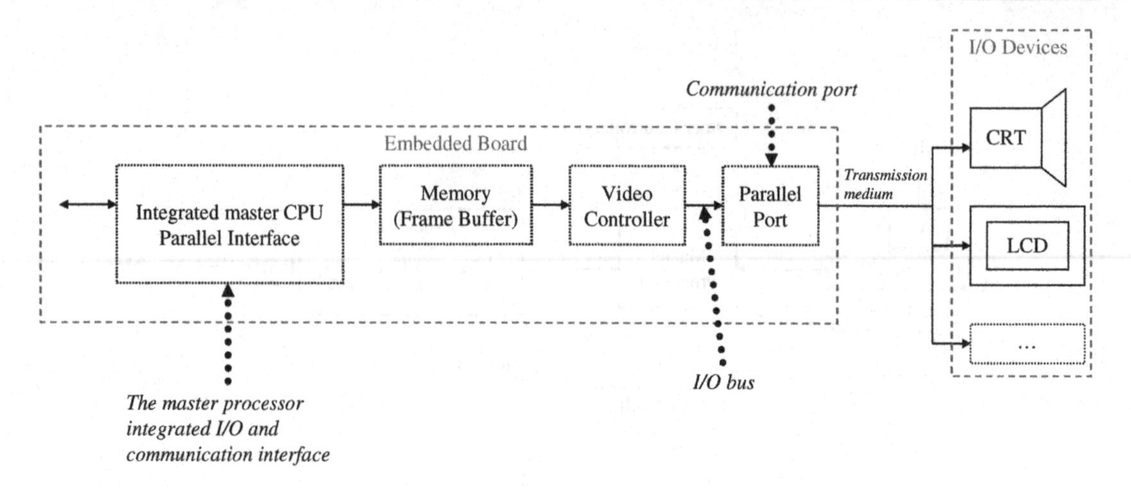


**Figure 6-3a**
Complex I/O subsystem.

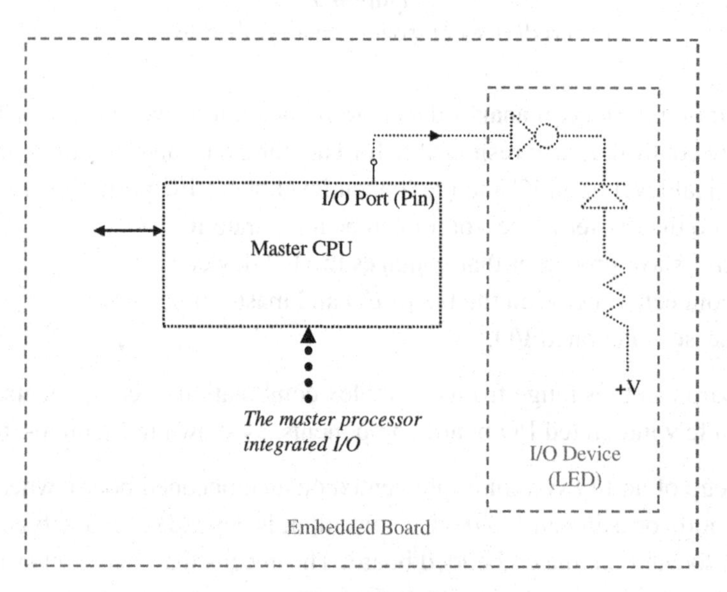


**Figure 6-3b**
Simple I/O subsystem.[2]

may be entirely different depending on the architecture viewpoint, as related to the embedded systems model. "Viewpoint" means that hardware and software can view, and hence subgroup, board I/O differently. Within software, the subgroups can even differ depending on the level of software (system software versus application software, operating system (OS) versus device drivers, etc.). For example, in many OSs board I/O is considered either as block or character I/O. In short, block I/O manages data in fixed block sizes and is addressable only in blocks. Character I/O, on the other hand, manages data in streams of characters, the size of the character depending on the architecture (e.g., 1 byte).

From a hardware viewpoint, I/O manages (transmits and/or stores) data in *serial*, in *parallel*, or *both.*

## 6.1 Managing Data: Serial versus Parallel I/O

### 6.1.1 Serial I/O

Board I/O that can transmit and receive data in *serial* is made up of components in which data (characters) are stored, transferred, and received *one bit at a time.* Serial I/O hardware is typically made up of some combination of the six main logical units outlined at the start of this chapter. Serial communication includes within its I/O subsystem a *serial port* and a *serial interface.*

*Serial interfaces* manage the serial data transmission and reception between the master CPU and either the I/O device or its controller. They include reception and transmission buffers to store and encode or decode the data they are responsible for transmitting either to the master CPU or an I/O device. Serial data transmission and reception schemes generally differ in terms of the direction data can be transmitted and received, as well as the actual transmission/reception process—in other words, how the data bits are transmitted and received within the data stream.

Data can be transmitted between two devices in one of three directions: in a one-way direction, in both directions but at separate times because they share the same transmission line, and in both directions simultaneously. Serial I/O data communication that uses a *simplex* scheme is one in which a data stream can only be transmitted—and thus received—in one direction (see Figure 6-4a). A *half-duplex* scheme is one in which a data stream can be transmitted and received in either direction, but in only one direction at any one time (see Figure 6-4b). A *full-duplex* scheme is one in which a data stream can be transmitted and received in either direction simultaneously (see Figure 6-4c).

Within the actual data stream, serial I/O transfers can occur either as a steady (continuous) stream at regular intervals regulated by the CPU's clock, referred to as a *synchronous* transfer, or intermittently at irregular (random) intervals, referred to as an *asynchronous* transfer.

In an asynchronous transfer (shown in Figure 6-5), the data being transmitted is typically stored and modified within a serial interface's transmission buffer. The serial interface at the transmitter divides the data stream into groups, called *packets*, that typically range from either 4 to 8 bits per character or 5 to 9 bits per character. Each of these packets is then encapsulated in frames to be transmitted separately. The frames are packets modified (before transmission) by the serial interface to include a START bit at the start of the stream, and a STOP bit or bits (this can be 1, 1.5, or 2 bits in length to ensure a transition from "1" to "0" for the START bit of the next frame) at the end of the data stream being transmitted. Within the frame, after the

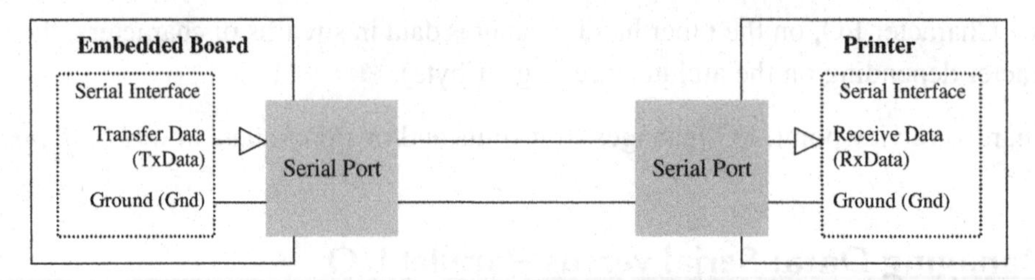


**Figure 6-4a**
Simplex transmission scheme example.[3]

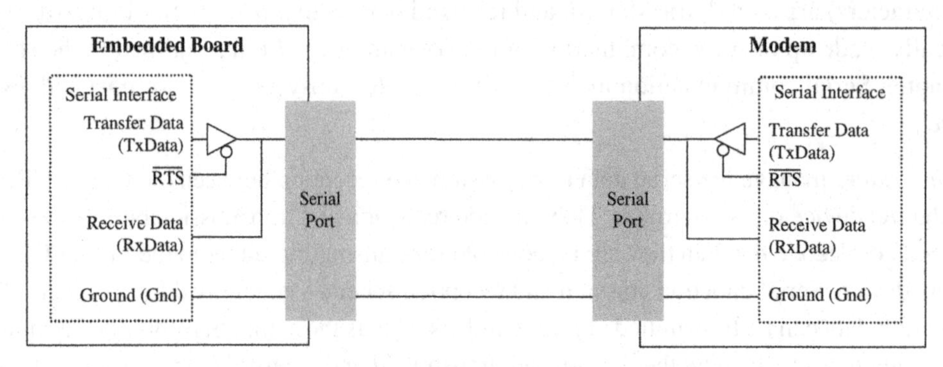


**Figure 6-4b**
Half-duplex transmission scheme example.[3]

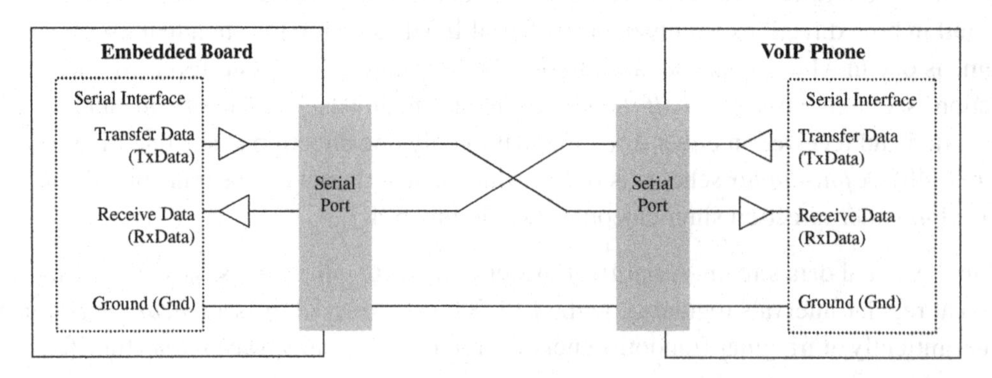


**Figure 6-4c**
Full-duplex transmission scheme example.[3]

data bits and before the STOP bit, a *parity* bit may also be appended. A START bit indicates the start of a frame, the STOP bit(s) indicates the end of a frame, and the parity is an optional bit used for very basic error checking. Basically, parity for a serial transmission can be NONE, for no parity bit and thus no error checking; EVEN, where the total number of bits set to "1" in the transmitted stream, excluding the START and STOP bits, must be an even

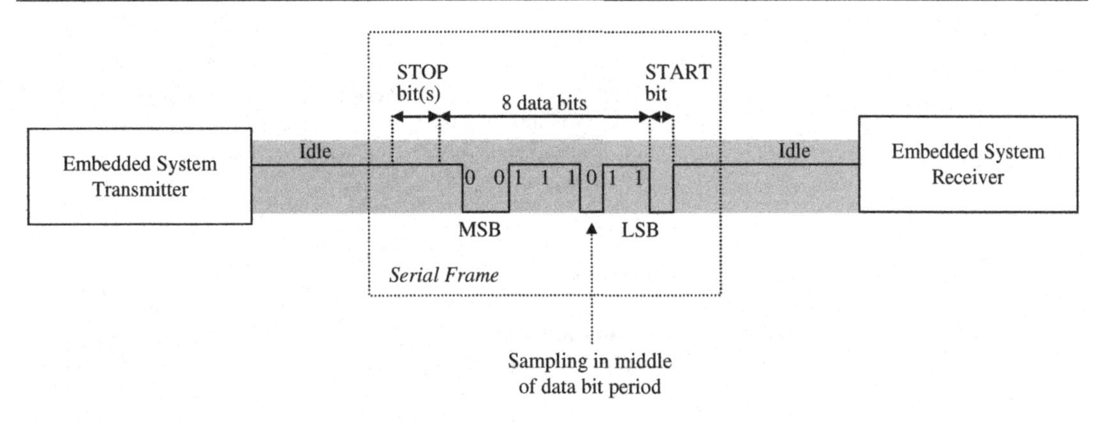


**Figure 6-5**
Asynchronous transfer sample diagram.

number in order for the transmission to be a success; and ODD, where the total number of bits set to "1" in the transmitted stream, excluding the START and STOP bits, must be an odd number in order for the transmission to be a success.

Between the transmission of frames, the communication channel is kept in an idle state, meaning a logical level "1" or non-return to zero (NRZ) state is maintained.

The serial interface of the receiver then receives frames by synchronizing to the START bit of a frame, delays for a brief period, and then shifts in bits, one at a time, into its receive buffer until reaching the STOP bit(s). In order for asynchronous transmission to work, the *bit rate* (bandwidth) has to be synchronized in all serial interfaces involved in the communication. The *bit rate* is defined as:

Bit rate * number of actual data bits per frame/total number of bits per frame) * baud rate.

The baud rate is the total number of bits, regardless of type, per unit of time (kbits/s, Mbits/s, etc.) that can be transmitted.

Both the transmitter's serial interface and the receiver's serial interface synchronize with separate bit-rate clocks to sample data bits appropriately. At the transmitter, the clock starts when transmission of a new frame starts, and continues until the end of the frame so that the data stream is sent at intervals the receiver can process. At the receiving end, the clock starts with the reception of a new frame, delaying when appropriate (in accordance with the bit rate), sampling the middle of each data bit period of time, and then stopping when receiving the frame's STOP bit(s).

In a synchronous transmission (as shown in Figure 6-6), there are no START or STOP bits appended to the data stream and there is no idle period. As with asynchronous transmissions, the data rates for receiving and transmitting must be in sync. However, unlike the separate

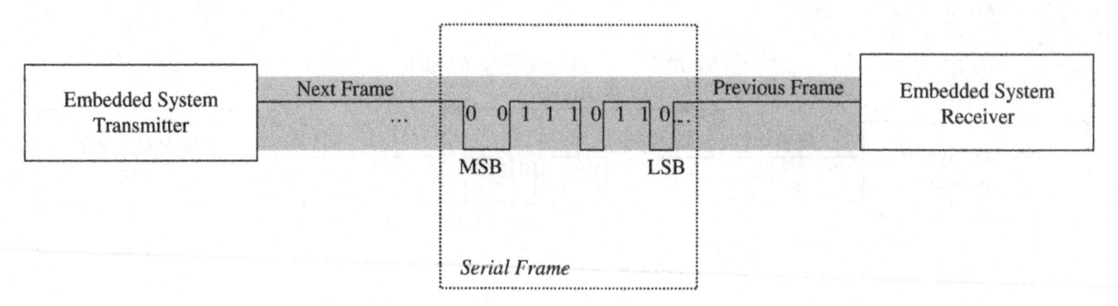


**Figure 6-6**
Synchronous transfer sample diagram.

clocks used in an asynchronous transfer, the devices involved in a synchronous transmission are synchronizing off of one common clock, which does not start and stop with each new frame. On some boards, there may be an entirely separate clock line for the serial interface to coordinate the transfer of bits. In some synchronous serial interfaces, if there is no separate clock line, the clock signal may even be transmitted along with the data bits.

The *UART (universal asynchronous receiver transmitter)* is an example of a serial interface that does asynchronous serial transmission, whereas *SPI (serial peripheral interface)* is an example of a synchronous serial interface. *(Note: Different architectures that integrate a UART or other types of serial interfaces may have different names and types for the same type of interface, such as the MPC860 which has SMC (serial management controller) UARTs, for example. Review the relevant documentation to understand the specifics.)*

Serial interfaces can either be separate slave ICs on the board, or integrated onto the master processor. The serial interface transmits data to and from an I/O device via a *serial port* (shown in Figures 6-4a–c). Serial ports are serial communication (COM) interfaces that are typically used to interconnect off-board serial I/O devices to on-board serial board I/O. The serial interface is then responsible for converting data that is coming to and from the serial port at the logic level of the serial port into data that the logic circuitry of the master CPU can process.

One of the most common serial communication protocols defining how the serial port is designed and what signals are associated with the different bus lines is *RS-232*.

### Serial I/O Example 1: Networking and Communications: RS-232

One of the most widely implemented serial I/O protocols for either synchronous or asynchronous transmission is the *RS-232* or EIA-232 (Electronic Industries Association-232), which is primarily based upon the Electronic Industries Association family of standards. These standards define the major components of any RS-232-based system, which is implemented almost entirely in hardware.

The hardware components can all be mapped to the physical layer of the OSI model (see Figure 6-7). The firmware (software) required to enable RS-232 functionality maps to the lower portion of the data-link, but will not be discussed in this section (see Chapter 8).

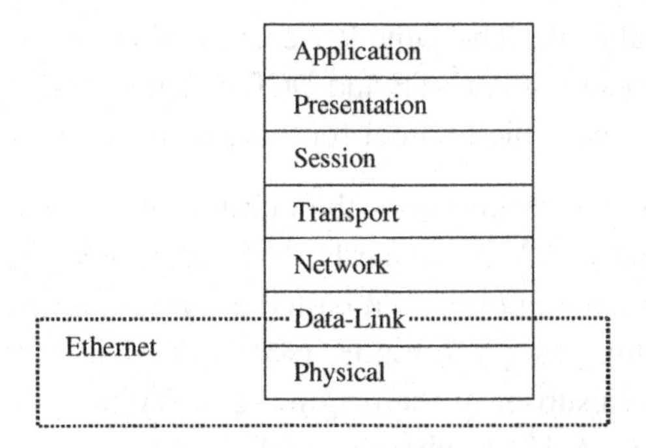


**Figure 6-7**
OSI model.

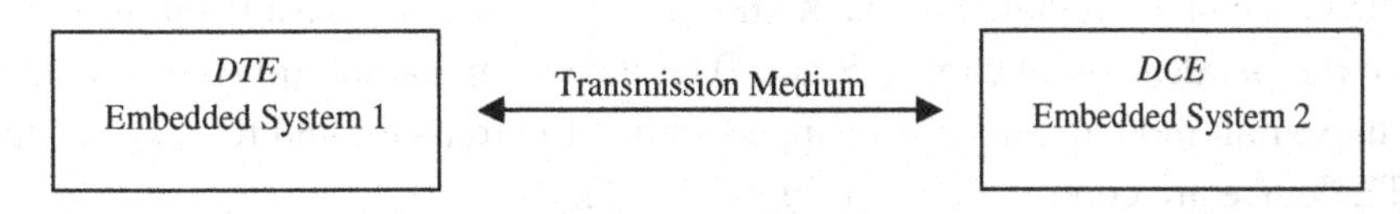


**Figure 6-8**
Serial network diagram.

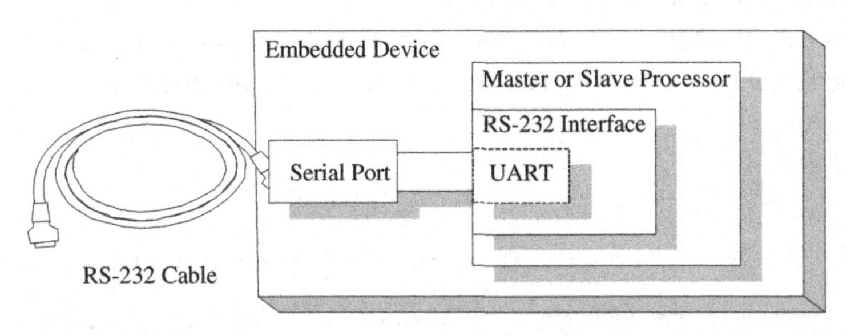


**Figure 6-9**
Serial components block diagram.

According to the EIA-232 standards, RS-232-compatible devices (shown in Figure 6-8) are called either DTE (Data Terminal Equipment) or DCE (Data Circuit-terminating Equipment). DTE devices, such as a PC or embedded board, are the initiators of a serial communication. DCE is the device that the DTE wants to communicate with, such as an I/O device connected to the embedded board.

The core of the RS-232 specification is called the *RS-232 interface* (see Figure 6-9). The RS-232 interface defines the details of the serial port and the signals along with some additional circuitry that maps signals from a synchronous serial interface (such as SPI) or an asynchronous serial interface (such as UART) to the serial port, and by extension to the I/O device itself. By defining the details of the serial port, RS-232 also defines the transmission medium, which is the serial cable. The same RS-232 interface must exist on both sides of a serial communication transmission (DTE and DCE or embedded board and I/O device), connected by an RS-232 serial cable in order for this scheme to work.

The actual physics behind the serial port—the number of signals and their definitions—differs among the different EIA232 standards. The parent RS-232 standard defines a total of 25 signals, along with a connector, called a DB25 connector, on either end of a wired transmission medium, shown in Figure 6-10a. The EIA RS-232 Standard EIA574 defines only nine signals (a subset of the original 25) that are compatible with a DB9 connector (shown in Figure 6-10b), whereas the EIA561 standard defines eight signals (again a subset of the original RS-232 25 signals) compatible with an RJ-45 connector (see Figure 6-10c).

Two DTE devices can interconnect to each other using an internal wiring variation on serial cables called *null modem* serial cables. Since DTE devices transmit and receive data on the same pins, these null modem pins are swapped so that the transmit and receive connections on each DTE device are coordinated (see Figure 6-11).

*Example: Motorola/Freescale MPC823 FADS Board RS-232 System Model*

The serial interface on the Motorola/Freescale FADS board (a platform for hardware and software development around the MPC8xx family of processors) is integrated in the master processor, in this case the MPC823. To understand the other major serial component located on the board, the serial port, one only has to read the board's hardware manual.

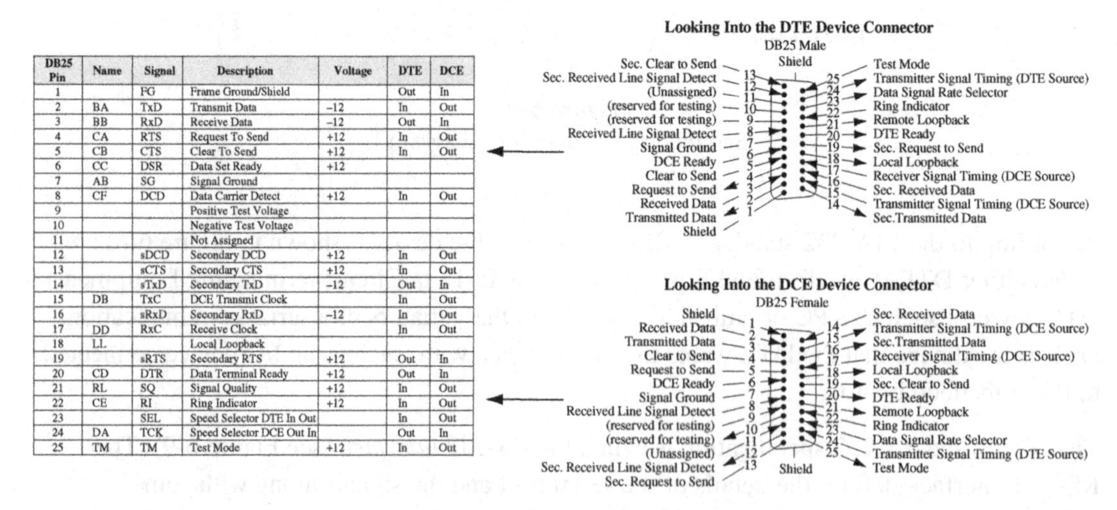

| DB25 Pin | Name | Signal | Description | Voltage | DTE | DCE |
|---|---|---|---|---|---|---|
| 1 | | FG | Frame Ground/Shield | | Out | In |
| 2 | BA | TxD | Transmit Data | –12 | In | Out |
| 3 | BB | RxD | Receive Data | –12 | Out | In |
| 4 | CA | RTS | Request To Send | +12 | In | Out |
| 5 | CB | CTS | Clear To Send | +12 | In | Out |
| 6 | CC | DSR | Data Set Ready | +12 | | |
| 7 | AB | SG | Signal Ground | | | |
| 8 | CF | DCD | Data Carrier Detect | +12 | In | Out |
| 9 | | | Positive Test Voltage | | | |
| 10 | | | Negative Test Voltage | | | |
| 11 | | | Not Assigned | | | |
| 12 | | sDCD | Secondary DCD | +12 | In | Out |
| 13 | | sCTS | Secondary CTS | +12 | In | Out |
| 14 | | sTxD | Secondary TxD | –12 | Out | In |
| 15 | DB | TxC | DCE Transmit Clock | | In | Out |
| 16 | | sRxD | Secondary RxD | –12 | In | Out |
| 17 | DD | RxC | Receive Clock | | In | Out |
| 18 | LL | | Local Loopback | | | |
| 19 | | sRTS | Secondary RTS | +12 | Out | In |
| 20 | CD | DTR | Data Terminal Ready | +12 | Out | In |
| 21 | RL | SQ | Signal Quality | +12 | In | Out |
| 22 | CE | RI | Ring Indicator | +12 | In | Out |
| 23 | | SEL | Speed Selector DTE In Out | | In | Out |
| 24 | DA | TCK | Speed Selector DCE Out In | | Out | In |
| 25 | TM | TM | Test Mode | +12 | In | Out |



**Figure 6-10a**
RS-232 signals and DB25 connector.

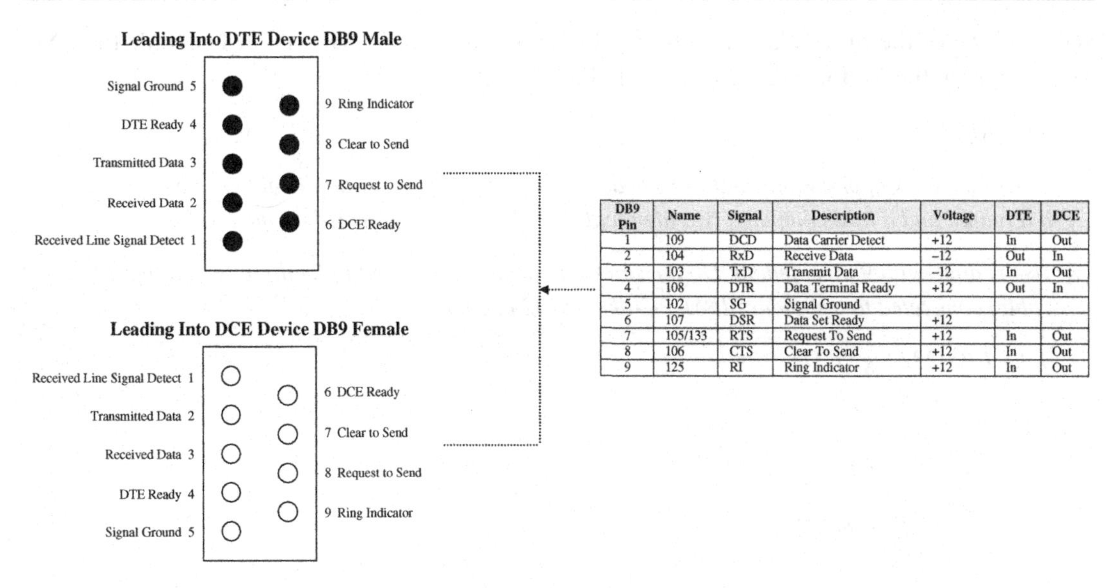


| DB9 Pin | Name | Signal | Description | Voltage | DTE | DCE |
|---|---|---|---|---|---|---|
| 1 | 109 | DCD | Data Carrier Detect | +12 | In | Out |
| 2 | 104 | RxD | Receive Data | −12 | Out | In |
| 3 | 103 | TxD | Transmit Data | −12 | In | Out |
| 4 | 108 | DTR | Data Terminal Ready | +12 | Out | In |
| 5 | 102 | SG | Signal Ground | | | |
| 6 | 107 | DSR | Data Set Ready | +12 | | |
| 7 | 105/133 | RTS | Request To Send | +12 | In | Out |
| 8 | 106 | CTS | Clear To Send | +12 | In | Out |
| 9 | 125 | RI | Ring Indicator | +12 | In | Out |

**Figure 6-10b**
RS-232 signals and DB9 connector.[4]

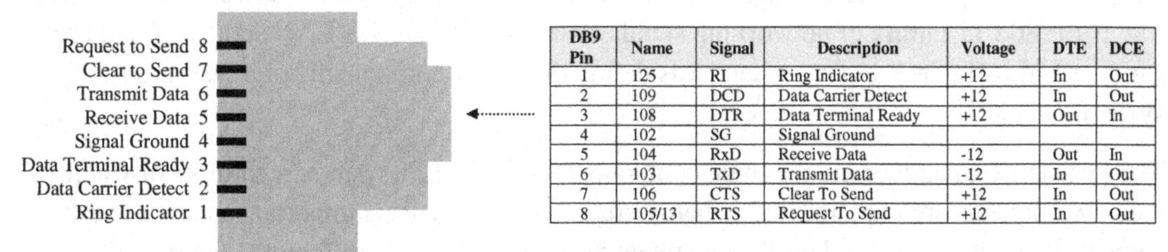


| DB9 Pin | Name | Signal | Description | Voltage | DTE | DCE |
|---|---|---|---|---|---|---|
| 1 | 125 | RI | Ring Indicator | +12 | In | Out |
| 2 | 109 | DCD | Data Carrier Detect | +12 | In | Out |
| 3 | 108 | DTR | Data Terminal Ready | +12 | Out | In |
| 4 | 102 | SG | Signal Ground | | | |
| 5 | 104 | RxD | Receive Data | -12 | Out | In |
| 6 | 103 | TxD | Transmit Data | -12 | In | Out |
| 7 | 106 | CTS | Clear To Send | +12 | In | Out |
| 8 | 105/13 | RTS | Request To Send | +12 | In | Out |

**Figure 6-10c**
RS-232 signals and RJ-45 connector.[4]

DCD 1
TX 2
RX 3
DTR 4
GND 5
6 DSR
7 RTS
8 CTS
9 NC

**Figure 6-11**
RS-232 serial port connector.

Section 4.9.3 of the Motorola/Freescale 8xxFADS User's Manual (Revision 1) details the RS-232 system on the Motorola/Freescale FADS board:

> *4.9.3 RS232 Ports*
>
> *To assist user's applications and to provide convenient communication channels with both a terminal and a host computer, two identical RS232 ports are provided on the FADS. …*
>
> *Use is done with 9 pins, female D-type stack connector, configured to be directly (via a flat cable) connected to a standard IBM-PC-like RS232 connector.*
>
> *4.9.3.1 RS-232 Signal Description*
>
> *In the following list:*
>
> - *DCD (O)—Data Carrier Detect*
> - *TX (O)—Transmit Data*
> - …

From this manual, we can see that the FADS RS-232 port definition is based upon the EIA-574 DB9 DCE female device connector definition.

#### Serial I/O Example 2: Networking and Communications: IEEE 802.11 Wireless LAN

The IEEE 802.11 family of networking standards are serial wireless LAN standards and are summarized in Table 6-1. These standards define the major components of a wireless LAN system.

The first step is to understand the main components of an 802.11 system, regardless of whether these components are implemented in hardware or software. This is important since different embedded architectures and boards implement 802.11 components differently. On most platforms today, 802.11 standards are made up of root components that are implemented almost entirely in hardware. The hardware components can all be mapped to the physical layer of the OSI model, as shown in Figure 6-12. Any software required to enable 802.11 functionality maps to the lower section of the OSI data-link layer, but will not be discussed in this section.

Off-the-shelf wireless hardware modules supporting one or some combination or the 802.11 standards (802.11a, 802.11b, 802.11g, etc.), have, in many ways, complicated the efforts to commit to one wireless LAN standard. These modules also come in a wide variety of forms, including embedded processor sets, PCMCIA, Compact Flash, and PCI formats. In general, as shown in Figures 6-13a and b, embedded boards need to either integrate 802.11 functionality as a slave controller or into the master chip, or the board needs to support one of the standard connectors for the other forms (PCI, PCMCIA, Compact Flash, etc.). This means that 802.11 chipset vendors can either (1) produce or port their PC Card firmware for an 802.11 embedded solution, which can be used for lower volume/more expensive devices

**Table 6-1: 802.11 Standards**

| IEEE 802.11 Standard | Description |
|---|---|
| **802.11-1999** Root Standard for Information Technology—Telecommunications and information exchange between systems—Local and Metropolitan Area networks—Specific requirements—Part 11: Wireless LAN Medium Access Control (MAC) and Physical Layer (PHY) specifications | The 802.11 standard was the first attempt to define how wireless data from a network should be sent. The standard defines operations and interfaces at the MAC and PHY levels in a TCP/IP network. There are three PHY layer interfaces defined (one IR and two radio: Frequency-Hopping Spread Spectrum (FHSS) and Direct Sequence Spread Spectrum (DSSS)) and the three do not interoperate. Use CSMA/CA (carrier sense multiple access with collision avoidance) as the basic medium access scheme for link sharing, phase-shift keying (PSK) for modulation. |
| **802.11a-1999** "WiFi5" Amendment 1: High Speed Physical Layer in the 5 GHz Band | Operates at radiofrequencies between 5 and 6 GHz to prevent interference with many consumer devices. Uses CSMA/CA (carrier sense multiple access with collision avoidance) as the basic medium access scheme for link sharing. As opposed to PSK, it uses a modulation scheme known as orthogonal frequency-division multiplexing (OFDM) that provides data rates as high as 54 Mbps maximum. |
| **802.11b-1999** "WiFi" Supplement to 802.11-1999, Wireless LAN MAC and PHY specifications: Higher Speed Physical Layer (PHY) Extension in the 2.4 GHz Band | Backward compatible with 802.11. 11 Mbps speed, one single PHY layer (DSSS), uses CSMA/CA (carrier sense multiple access with collision avoidance) as the basic medium access scheme for link sharing and complementary code keying (CCK), which allows higher data rates and is less susceptible to multipath-propagation interference. |
| **802.11b-1999/Cor1-2001** Amendment 2: Higher-Speed Physical Layer (PHY) Extension in the 2.4 GHz Band—Corrigendum 1 | To correct deficiencies in the MIB definition of 802.11b. |
| **802.11c** IEEE Standard for Information Technology—Telecommunications and information exchange between systems—Local Area networks—Media access control (MAC) bridges—Supplement for support by IEEE 802.11 | Designated in 1998 to add a subclass under 2.5 Support of the Internal Sublayer Service by specific MAC Procedures to cover bridge operation with IEEE 802.11 MACs. Allows the use of 802.11 access points to bridge across networks within relatively short distances from each other (i.e., where there was a solid wall dividing a wired network). |
| **802.11d-2001** Amendment to IEEE 802.11-1999 (ISO/IEC 8802-11) Specification for Operation in Additional Regulatory Domains | Internationalization—defines the physical layer requirements (channelization, hopping patterns, new values for current MIB attributes, and other requirements) to extend the operation of 802.11 WLANs to new regulatory domains (countries). |
| **802.11e** Amendment to STANDARD [for] Information Technology—Telecommunications and information exchange between systems—Local and Metropolitan Area networks—Specific requirements—Part 11: Wireless LAN Medium Access Control (MAC) and Physical Layer (PHY) Specifications: Medium Access Method (MAC) Quality of Service Enhancements | Enhance the 802.11 MAC to improve and manage Quality of Service, provide classes of service, and efficiency enhancements in the areas of the Distributed Coordination Function (DCF) and Point Coordination Function (PCF). Defining a series of extensions to 802.11 networking to allow for QoS operation. (i.e., to allow for adaptation for streaming audio or video via a preallocated dependable portion of the bandwidth.) |

(*Continued*)

**Table 6-1: (Continued)**

| IEEE 802.11 Standard | Description |
|---|---|
| **802.11f-2003** IEEE Recommended Practice for Multi-Vendor Access Point Interoperability via an Inter-Access Point Protocol Across Distribution Systems Supporting IEEE 802.11 Operation | Standard to enable handoffs (constant operation while the mobile terminal is actually moving) to be done in such a way as to work across access points from a number of vendors. Includes recommended practices for an Inter-Access Point Protocol (IAPP) which provides the necessary capabilities to achieve multi-vendor Access Point interoperability across a Distribution System supporting IEEE P802.11 Wireless LAN Links. This IAPP will be developed for the following environment(s): (1) A Distribution System consisting of IEEE 802 LAN components supporting an IETF IP environment. (2) Others as deemed appropriate. |
| **802.11g-2003** Amendment 4: Further Higher-Speed Physical Layer Extension in the 2.4 GHz Band | A higher speed(s) PHY extension to 802.11b—offering wireless transmission over relatively short distances at up to 54 Mbps compared to the maximum 11 Mbps of the 802.11b standard and operates in the 2.4 GHz range Uses CSMA/CA (carrier sense multiple access with collision avoidance) as the basic medium access scheme for link sharing. |
| **802.11h-2001** Spectrum and Transmit Power Management Extensions in the 5 GHz Band in Europe | Enhancing the 802.11 MAC standard and 802.11a high-speed PHY in the 5 GHz band supplement to the standard; to add indoor and outdoor channel selection for 5 GHz license exempt bands in Europe; and to enhance channel energy measurement and reporting mechanisms to improve spectrum and transmit power management (per CEPT and subsequent EU committee or body ruling incorporating CEPT Recommendation ERC 99/23). Looking into the tradeoffs involved in creating reduced-power transmission modes for networking in the 5 GHz space—essentially allowing 802.11a to be used by handheld computers and other devices with limited battery power available to them. Also, examining the possibility of allowing access points to reduce power to shape the geometry of a wireless network and reduce interference outside of the desired influence of such a network. |
| **802.11i** Amendment to STANDARD [for] Information Technology—Telecommunications and information exchange between systems—Local and Metropolitan Area networks—Specific requirements—Part 11: Wireless LAN Medium Access Control (MAC) and Physical Layer (PHY) Specifications: Medium Access Method (MAC) Security Enhancements | Enhances the 802.11 MAC to enhance security and authentication mechanisms, and improving the PHY-level security that is used on these networks. |

(*Continued*)

**Table 6-1: (Continued)**

| IEEE 802.11 Standard | Description |
|---|---|
| **802.11j** Amendment to STANDARD [for] Information Technology—Telecommunications and information exchange between systems—Local and Metropolitan Area networks—Specific requirements—Part 11: Wireless LAN Medium Access Control (MAC) and Physical Layer (PHY) Specifications: 4.9 GHz-5 GHz Operation in Japan | The scope of the project is to enhance the 802.11 standard and amendments, to add channel selection for 4.9 and 5 GHz in Japan to additionally conform to the Japanese rules for radio operation, to obtain Japanese regulatory approval by enhancing the current 802.11 MAC and 802.11a PHY to additionally operate in newly available Japanese 4.9 and 5 GHz bands. |
| **802.11k** Amendment to STANDARD [for] Information Technology—Telecommunications and information exchange between systems—Local and Metropolitan Area networks—Specific requirements—Part 11: Wireless LAN Medium Access Control (MAC) and Physical Layer (PHY) Specifications: Radio Resource Measurement of Wireless LANs | This project will define Radio Resource Measurement enhancements to provide interfaces to higher layers for radio and network measurements. |
| **802.11ma** Standard for Information Technology—Telecommunications and information exchange between systems—Local and Metropolitan Area networks—Specific requirements—Part 11: Wireless LAN Medium Access Control (MAC) and Physical Layer (PHY) Specifications—Amendment x: Technical Corrections and Clarifications | Incorporates accumulated maintenance changes (editorial and technical corrections) into 802.11-1999, 2003 edition (incorporating 802.11a-1999, 802.11b-1999, 802.11b-1999 corrigendum 1-2001, and 802.11d-2001). |
| **802.11n** Amendment to STANDARD [for] Information Technology—Telecommunications and information exchange between systems—Local and Metropolitan Area networks—Specific requirements—Part 11: Wireless LAN Medium Access Control (MAC) and Physical Layer (PHY) Specifications: Enhancements for Higher Throughput | The scope of this project is to define an amendment that shall define standardized modifications to both the 802.11 PHY and the 802.11 MAC so that modes of operation can be enabled that are capable of much higher throughputs, with a maximum throughput of at least 100 Mbps, as measured at the MAC data service access point (SAP). |

or during product development, or (2) the same vendor's chipset on a standard PC card could be placed on the embedded board, which can be used for devices that will be manufactured in larger volumes.

On top of the 802.11 chipset integration, an embedded board design needs to take into consideration wireless LAN antenna placement and signal transmission requirements. The designer must ensure that there are no obstructions to prevent receiving and transmitting data. When 802.11 is not integrated into the master CPU, such as with the System-on-Chip (SoC)

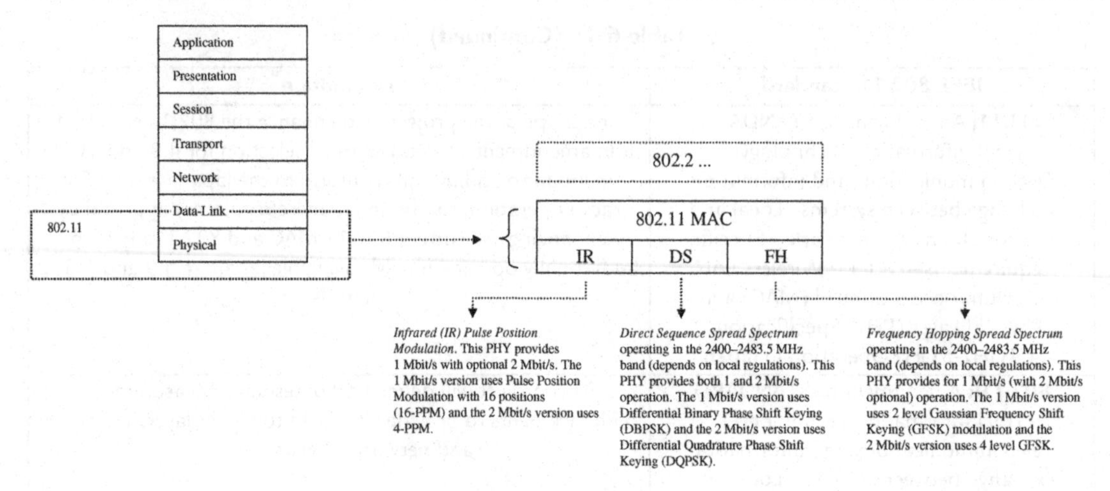


**Figure 6-12**
OSI model.[5]

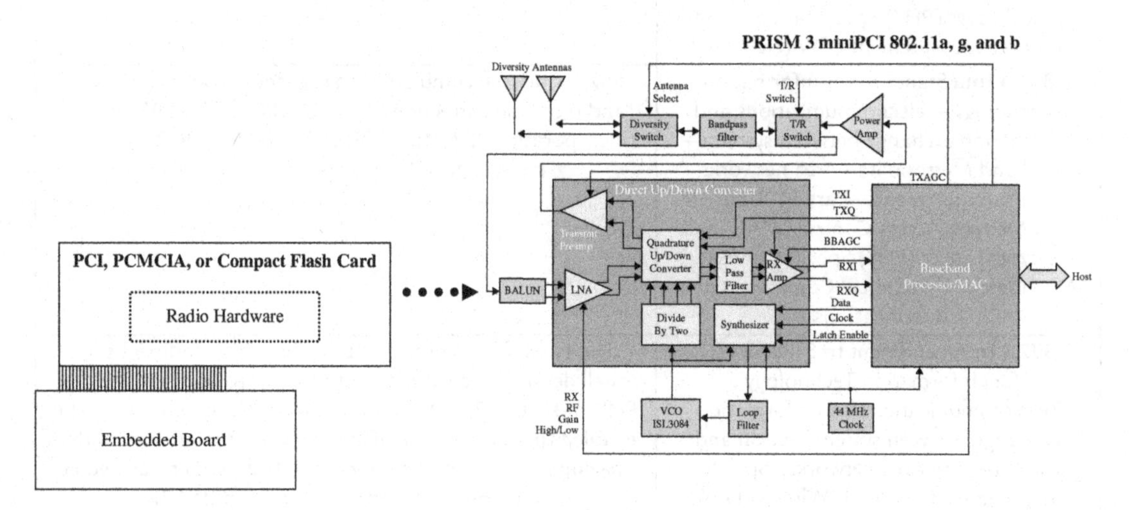


**Figure 6-13a**
802.11 sample hardware configurations with PCI card.[6]

shown in Figure 6-13b, the interface between the master CPU and the 802.11 board hardware also needs to be designed.

## 6.1.2 Parallel I/O

Components that transmit data in parallel are devices which can transfer data in multiple bits simultaneously. Just as with serial I/O, parallel I/O hardware is also typically made up of some combination of six main logical units, as introduced at the start of this chapter, except that the port is a *parallel port* and the communication interface is a *parallel interface.*

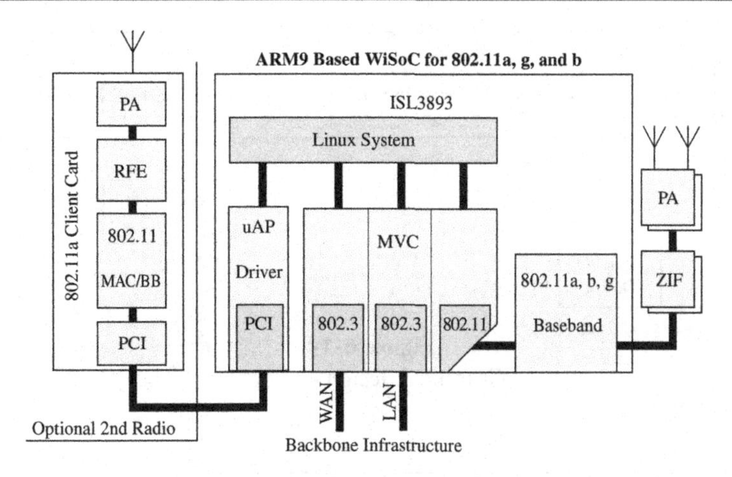


**Figure 6-13b**
802.11 sample hardware configurations with SoC.[7]

*Parallel interfaces* manage the parallel data transmission and reception between the master CPU and either the I/O device or its controller. They are responsible for decoding data bits received over the pins of the parallel port (transmitted from the I/O device), and receiving data being transmitted from the master CPU and then encoding these data bits onto the parallel port pins.

They include reception and transmission buffers to store and manipulate the data being transferred. In terms of parallel data transmission and reception schemes, like serial I/O transmission, they generally differ in terms of what *direction* data can be transmitted and received in, as well as the actual *process* of transmitting/receiving data bits within the data stream. In the case of direction of transmission, as with serial I/O, parallel I/O uses simplex, half-duplex, or full-duplex modes. Also, as with serial I/O, parallel I/O devices can transmit data asynchronously or synchronously. However, parallel I/O does have a greater capacity to transmit data than serial I/O, because multiple bits can be transmitted or received simultaneously. Examples of board I/O that transfer and receive data in parallel include: IEEE 1284 controllers (for printer/display I/O devices, see Example 3), CRT ports, and SCSI (for storage I/O devices). A protocol that can potentially support both parallel and serial I/O is Ethernet, presented in Example 4.

#### *Parallel I/O Example 3: "Parallel" Output and Graphics I/O*

Technically, the models and images that are created, stored, and manipulated in an embedded system are the graphics. There are typically three logical components (engines) of I/O graphics on an embedded board, as shown in Figure 6-14:

- *Geometric engine*: responsible for defining what an object is. This includes implementing color models, an object's physical geometry, material and lighting properties, etc.

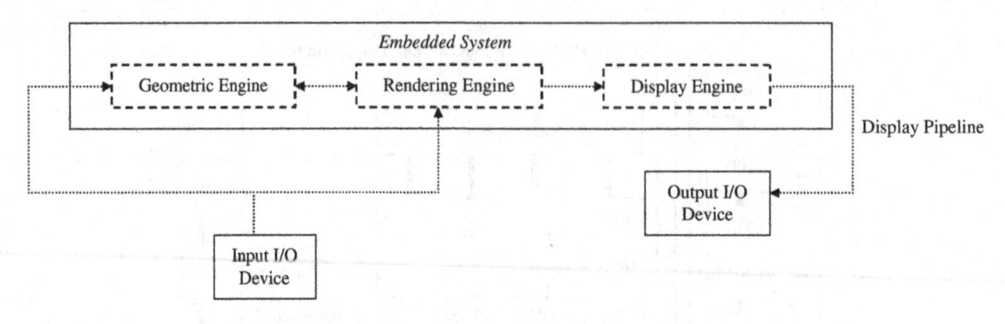


**Figure 6-14**
Graphical design engines.

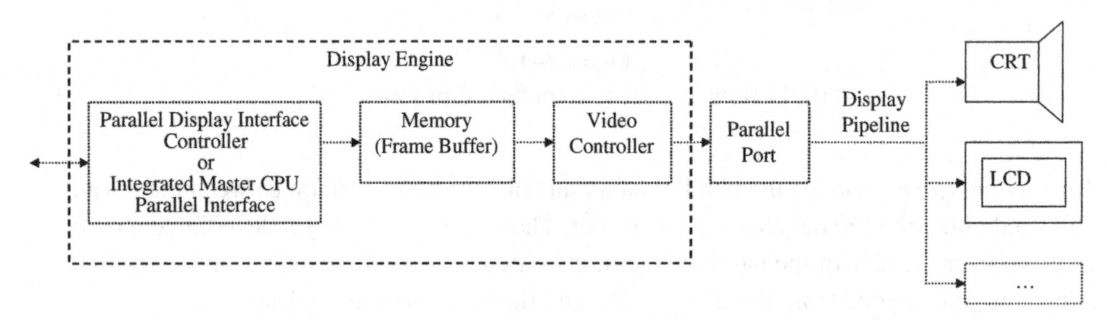


**Figure 6-15a**
Display engine of softcopy (video) graphics example.

- *Rendering engine*: responsible for capturing the description of objects. This includes providing functionality in support of geometric transformations, projections, drawing, mapping, shading, illumination, etc.
- *Raster and display engine*: responsible for physically displaying the object. It is in this engine that the output I/O hardware comes into play.

An embedded system can output graphics via softcopy (video) or hardcopy (on paper) means. The contents of the display pipeline differ according to whether the output I/O device outputs hard or soft graphics, so the display engine differs accordingly, as shown in Figures 6-15a and b.

The actual parallel port configuration differs from standard to standard in terms of the number of signals and the required cable. For example, on the Net Silicon's NET + ARM50 embedded board (see Figure 6-16), the master processor (an ARM7-based architecture) has an integrated IEEE 1284 interface, a configurable MIC controller integrated in the master processor, to transmit parallel I/O over four on-board parallel ports.

The IEEE 1284 specification defines a 40-signal port, but on the NET + ARM50 board, data and control signals are multiplexed to minimize the master processor's pin count. Aside from eight data signals DATA[8:1] ($D_0$–$D_7$), IEEE 1284 control signals include:

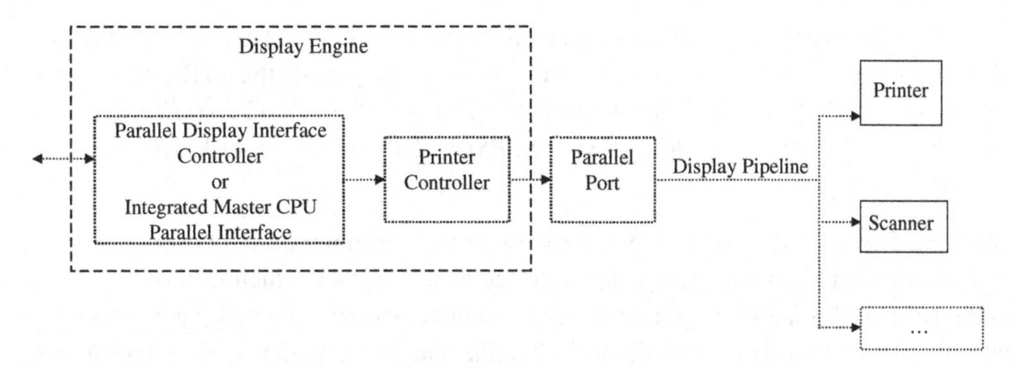


**Figure 6-15b**
Display engine of hardcopy graphics example.

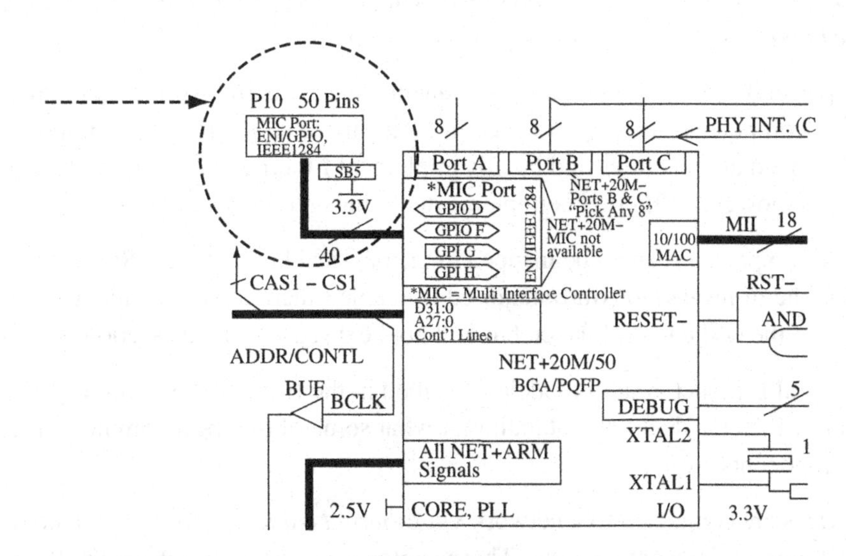


**Figure 6-16**
NET + ARM50 embedded board parallel I/O.[8]

- "*PDIR*, which is used for bidirectional modes and defines the direction of the external data transceiver. Its state is directly controlled by the BIDIR bit in the IEEE 1284 Control register (0 state, data is driven from the external transceiver towards 1285, the cable, and in the 1 state, data is received from the cable).
- *PIO*, which is controlled by firmware. Its state is directly controlled by the PIO bit in the IEEE 1284 Control register.
- *LOOPBACK*, which configures the port in external loopback mode and can be used to control the mux line in the external FCT646 devices (set to 1, the FCT646 transceivers drive inbound data from the input latch and not the real-time cable interface). Its state is directly controlled by the LOOP bit in the IEEE 1284 Control register. The LOOP strobe

signal is responsible for writing outbound data into the inbound latch (completing the loop back path). The LOOP strobe signal is an inverted copy of the STROBE* signal.

- *STROBE** (nSTROBE), *AUTOFD** (nAUTOFEED), *INIT** (nINIT), *HSELECT** (nSELECTIN), **ACK* (nACK), *BUSY, PE, PSELECT* (SELECT), **FAULT* (nERROR), …"[2]

### Parallel and Serial I/O Example 4: Networking and Communications—Ethernet

One of the most widely implemented LAN protocols is Ethernet, which is primarily based upon the IEEE 802.3 family of standards. These standards define the major components of any Ethernet system. Thus, in order to fully understand an Ethernet system design, you first need to understand the IEEE specifications. *(Remember: This is not a book about Ethernet and there is a lot more involved than what is covered here. This example is about understanding a networking protocol and then being able to understand the design of a system based upon a networking protocol, such as Ethernet.)*

The first step is understanding the main components of an Ethernet system, regardless of whether these components are implemented in hardware or software. This is important since different embedded architectures and boards implement Ethernet components differently. On most platforms, however, Ethernet is implemented almost entirely in hardware.

The hardware components can all be mapped to the physical layer of the OSI model (see Figure 6-17). The firmware (software) required to enable Ethernet functionality maps to the lower section of the OSI data-link layer, but will not be discussed in this section (see Chapter 8).

There are several Ethernet system models described in the IEEE 802.3 specification, so let us look at a few to get a clear understanding of what some of the most common Ethernet hardware components are.

Ethernet devices are connected to a network via *Ethernet cables*: thick coax (coaxial), thin coax, twisted-pair, or fiber-optic cables. These cables are commonly referred to by their IEEE

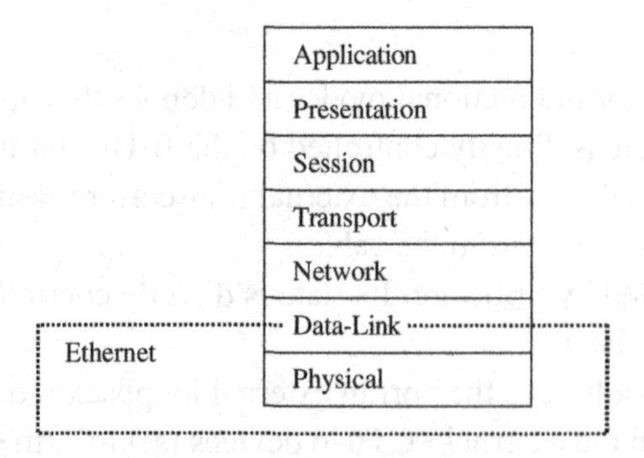


**Figure 6-17**
OSI model.

names. These names are made up of three components: the data transmission rate, the type of signaling used, and either the cable type or cable length.

For example, a 10Base-T cable is an Ethernet cable that handles a data transmission rate of 10 Mbps (million bits per second), will only carry Ethernet signals (baseband signaling), and is a twisted-pair cable. A 100Base-F cable is an Ethernet cable that handles a data transmission rate of 100 Mbps, supports baseband signaling, and is a fiber-optic cable. Thick or thin coax cables transmit at speeds of 10 Mbps, support baseband signaling, but differ in the length of maximum segments cut for these cables (500 m for thick coax, 200 m for thin coax). Thus, these thick coax cables are called 10Base-5 (short for 500) and thin coax cables are called 10Base-2 (short for 200).

The Ethernet cable must then be connected to the embedded device. The type of cable, along with the board I/O (communication interface, communication port, etc.), determines whether the Ethernet I/O transmission is *serial* or *parallel*. The *Medium-Dependent Interface (MDI)* is the network port on the board into which the Ethernet cable plugs. Different MDIs exist for the different types of Ethernet cables. For example, a 10Base-T cable has an RJ-45 jack as the MDI. In the system model of Figure 6-18, the MDI is an integrated part of the transceiver.

A transceiver is the physical device that receives and transmits the data bits; in this case it is the *Medium Attachment Unit (MAU)*. The MAU contains not only the MDI, but the *Physical Medium Attachment (PMA)* component as well. It is the PMA which "contains the functions for transmission, reception, and" depending on the transceiver, "collision detection, clock recovery and skew alignment" (p. 25, IEEE 802.3). Basically the PMA serializes (breaks down into a bit

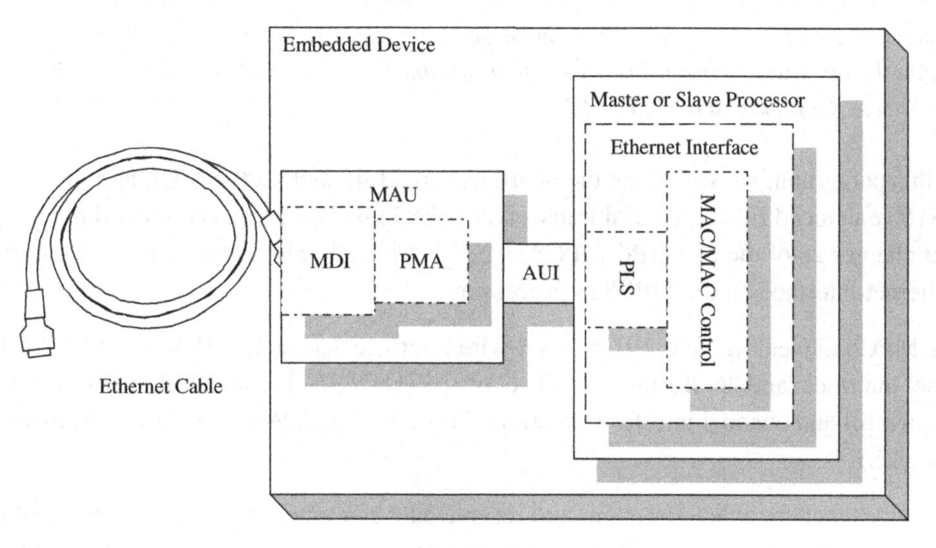


**Figure 6-18**
Ethernet components diagram.

stream) code groups received for transmission over the transmission medium or deserializes bits received from the transmission medium and converts these bits into code groups.

The transceiver is then connected to an *Attachment Unit Interface (AUI)*, which carries the encoded signals between an MAU and the Ethernet interface in a processor. Specifically, the AUI is defined for up to 10 Mbps Ethernet devices, and specifies the connection between the MAU and the *Physical Layer Signaling (PLS)* sublayer (signal characteristics, connectors, cable length, etc.).

The *Ethernet interface* can exist on a master or slave processor, and contains the remaining Ethernet hardware and software components. The PLS component monitors the transmission medium and provides a carrier sense signal to the *Media Access Control (MAC)* component. It is the MAC that initiates the transmission of data, so it checks the carrier signal before initiating a transmission, to avoid contention with other data over the transmission medium.

Let us start by looking at an embedded board for an example of this type of Ethernet system.

**Ethernet Example 1: Motorola/Freescale MPC823 FADS Board Ethernet System Model**
Section 4.9.1 of the Motorola/Freescale 8xxFADS User's Manual (Revision 1) details the Ethernet system on the Motorola/Freescale FADS board:

> *4.9.1 Ethernet Port*
>
> *The MPC8xxFADS has an Ethernet port with a 10Base-T interface. The communication port on which this resides is determined according to the MPC8xx type whose routing is on the daughter board. The Ethernet port uses an MC68160 EEST 10Base-T transceiver.*
>
> *You can also use the Ethernet SCC pins, which are on the expansion connectors of the daughter board and on the communication port expansion connector (P8) of the motherboard. The Ethernet transceiver can be disabled or enabled at any time by writing a 1 or a 0 to the EthEn bit in the BCSR1.*

From this paragraph, we know that the board has an RJ-45 jack as the MDI, and the MC68160 enhanced Ethernet serial transceiver (EEST) is the MAU. The second paragraph, as well as chapter 28 of the PowerPC MPC823 User's Manual, tells us more about the AUI and the Ethernet interface on the MPC823 processor.

On the MPC823 (see Figure 6-19), a seven-wire interface acts as the AUI. The SCC2 is the Ethernet interface, and "performs the full set of IEEE 802.3/Ethernet CSMA/CD media access control and channel interface functions." (See MPC823 PowerPC User's Manual, pp. 16–312.)

LAN devices that are able to transmit and receive data at a much higher rate than 10 Mbps implement a different combination of Ethernet components. The IEEE 802.3u Fast Ethernet (100 Mbps data rate) and the IEEE 802.3z Gigabit Ethernet (1000 Mbps data rate) systems

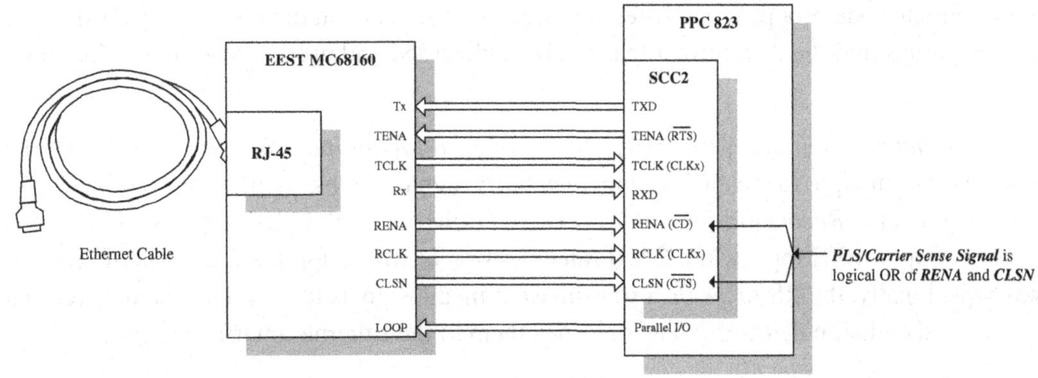

**Figure 6-19**
MPC823 Ethernet diagram. *© 2004 Freescale Semiconductor, Inc. Used by permission.*

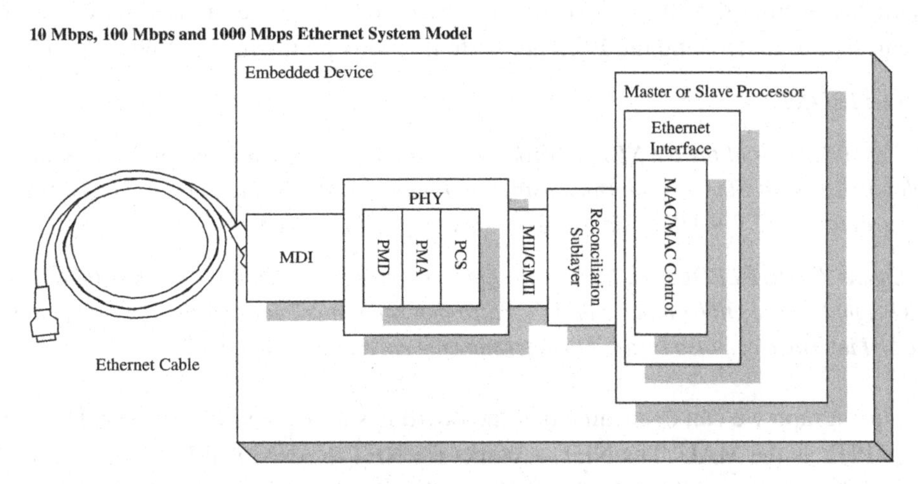


**Figure 6-20**
Ethernet diagram.

evolved from the original Ethernet system model (described in the previous section), and are based on the system model in Figure 6-20.

The MDI in this system is connected to the transceiver, not a part of the transceiver (as in the previous system model). The *Physical Layer Device (PHY)* transceiver in this system contains three components: the PMA (same as on the MAU transceiver in the 1/10 Mbps system model), the *Physical Coding Sublayer (PCS)*, and the *Physical Medium Dependent (PMD)*.

The PMD is the interface between the PMA and the transmission medium (through the MDI). The PMD is responsible for receiving serialized bits from the PMD and converting it to the appropriate signals for the transmission medium (optical signals for a fiber optic, etc.). When transmitting to the PMA, the PCS is responsible for encoding the data to be transmitted into the appropriate code group. When receiving the code groups from the PMA, the PCS decodes the code groups into the data format that can be understood and processed by upper Ethernet layers.

The *Media Independent Interface (MII)* and *the Gigabit Media Independent Interface (GMII)* are similar in principle to the AUI, except they carry signals (transparently) between the transceiver and the *Reconciliation Sublayer (RS)*. Furthermore, the MII supports a LAN data rate of up to 100 Mbps, while GMII (an extension of MII) supports data rates of up to 1000 Mps. Finally, the RS maps PLS transmission media signals to two status signals (carrier presence and collision detection), and provides them to the Ethernet interface.

#### Ethernet Example 2: Net Silicon ARM7 (6127001) Development Board Ethernet System Model

The NET + Works 6127001 Development Board Jumper and Component Guide from NetSilicon has an Ethernet interface section on their ARM based reference board, and from this we can start to understand the Ethernet system on this platform (see Figure 6-21).

> *Ethernet Interface*
>
> *The 10/100 version of the 3V NET + Works Hardware Development Board provides a full-duplex 10/100Mbit Ethernet Interface using the Enable 3V PHY chip. The Enable 3V PHY interfaces to the NET + ARM chip using the standard MII interface.*
>
> *The Enable 3V PHY LEDL (link indicator) signal is connected to the NET + ARM PORTC6 GPIO signal. The PORT6 input can be used to determine the current Ethernet link status (The MII interface can also be used to determine the current Ethernet link status) …*

From this paragraph we can determine that the board has an RJ-45 jack as the MDI, and the Enable 3V PHY is the MAU. The NET + Works for NET + ARM Hardware Reference Guide (section 5: *Ethernet Controller Interface*) tells us that the ARM7 based ASIC integrates an Ethernet controller, and that the Ethernet Interface is actually composed of two parts: the

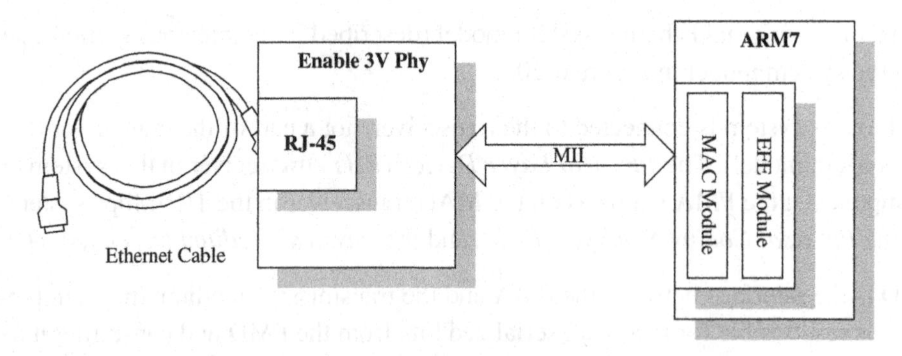


**Figure 6-21**
NET + ARM Ethernet block diagram.

Ethernet Front End (EFE) and the MAC modules. Finally, section 1.3 of this manual tells us the RS is integrated into the MII.

#### Ethernet Example 3: Adastra Neptune x86 Board Ethernet System Model

While both the ARM and PowerPC platforms integrate the Ethernet interface into the main processor, this x86 platform has a separate slave processor for this functionality. According to the Neptune User's Manual Revision A.2, the Ethernet controller the ("MAC Am79C791 10/100 Controller") connects to two different transceivers, with each connected to either an AUI or MII for supporting various transmission media (see Figure 6-22).

## 6.2 Interfacing the I/O Components

As discussed at the start of this chapter, I/O hardware is made up of all or some combination of integrated master processor I/O, I/O controllers, a communications interface, a communication port, I/O buses, and a transmission medium (see Figure 6-23).

All of these components are interfaced (connected) and communication mechanisms implemented via hardware, software, or both to allow for successful integration and function.

### *6.2.1 Interfacing the I/O Device with the Embedded Board*

For off-board I/O devices, such as keyboards, mice, LCDs, or printers, a transmission medium is used to interconnect the I/O device to an embedded board via a *communication port*. Aside from the I/O schemes implemented on the board (serial versus parallel), whether the medium

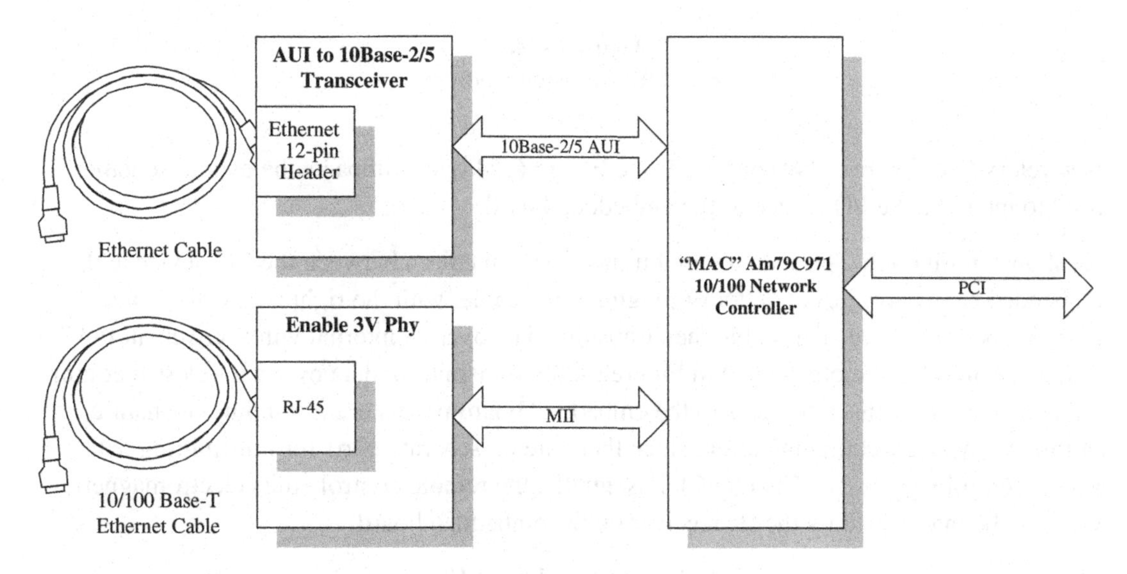


**Figure 6-22**
x86 Ethernet diagram.

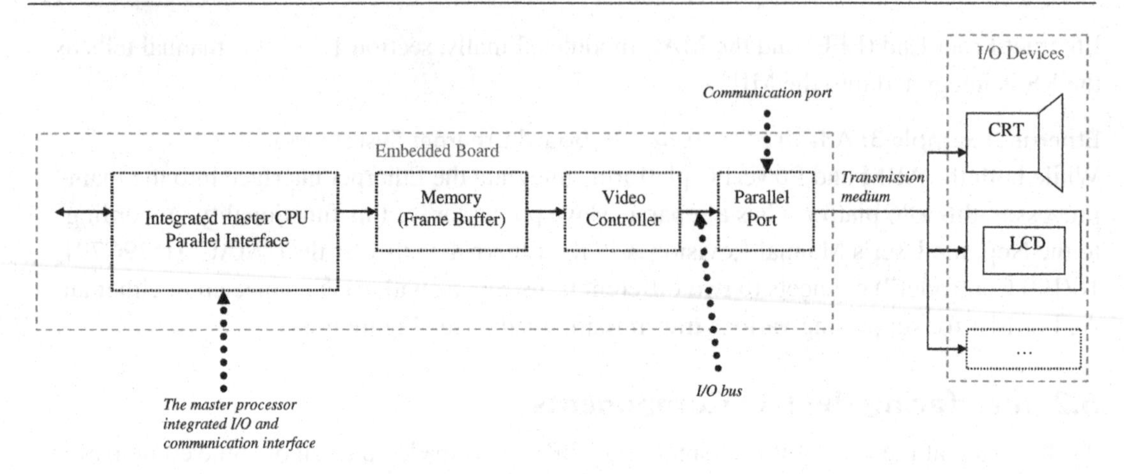


**Figure 6-23**
Sample I/O subsystem.

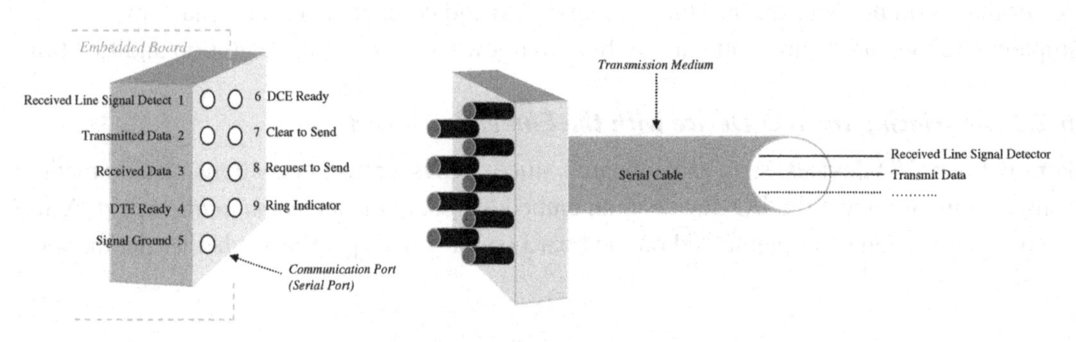


**Figure 6-24a**
Wired transmission medium.

is wireless (see Figure 6-24b) or wired (see Figure 6-24a) also impacts the overall scheme used to interface the I/O device to the embedded board.

As shown in Figure 6-24a, with a wired transmission medium between the I/O device and embedded board, it is just a matter of plugging in a cable, with the right connector head, to the embedded board. This cable then transmits data over its internal wires. Given an I/O device, such as the remote control in Figure 6-24b, transmitting data over a wireless medium, understanding how this interfaces to the embedded board means understanding the nature of infrared wireless communication, since there are no separate ports for transmitting data versus control signals (see Chapter 2). Essentially, the remote control emits electromagnetic waves to be intercepted by the IR receiver on the embedded board.

The communication port would then be interfaced to an I/O controller, a communication interface controller, or the master processor (with an integrated communication interface) via

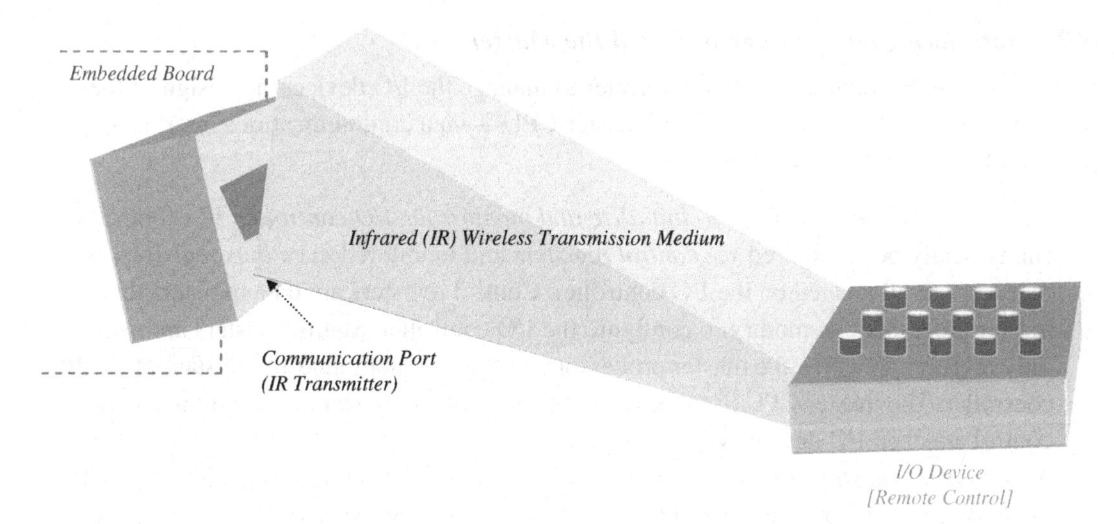


**Figure 6-24b**
Wireless transmission medium.

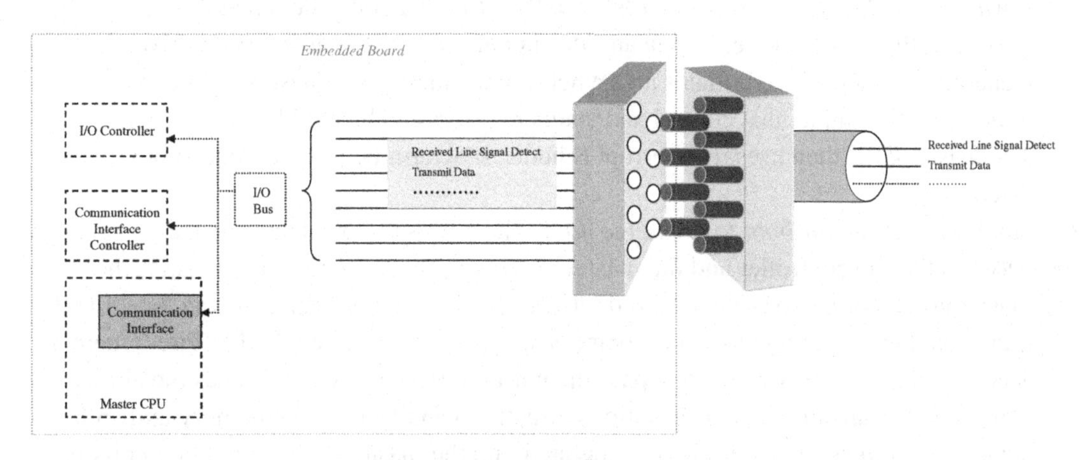


**Figure 6-25**
Interfacing communication port to other board I/O.

an *I/O bus* on the embedded board (see Figure 6-25). An I/O bus is essentially a collection of wires transmitting the data.

In short, an I/O device can be connected directly to the master processor via *I/O ports* (processor pins) if the I/O devices are located on the board, or can be connected indirectly using a *communication interface* integrated into the master processor or a separate IC on the board and the *communication port*. The communication interface itself is what is either connected directly to the I/O device, or the device's I/O controller. For off-board I/O devices, the relative board I/O components are interconnected via I/O buses.

### 6.2.2 Interfacing an I/O Controller and the Master CPU

In a subsystem that contains an I/O controller to manage the I/O device, the design of the interface between the I/O controller and master CPU—via a communications interface—is based on four requirements:

- *An ability of the master CPU to initialize and monitor the I/O controller.* I/O controllers can typically be configured via *control registers* and monitored via *status registers*. These registers are all located on the I/O controller. Control registers are data registers that the master processor can modify to configure the I/O controller. Status registers are read-only registers in which the master processor can get information as to the state of the I/O controller. The master CPU uses these status and control registers to communicate and/or control attached I/O devices via the I/O controller.
- *A way for the master processor to request I/O.* The most common mechanisms used by the master processor to request I/O via the I/O controller are *special I/O instructions* (I/O mapped) in the Instruction Set Architecture (ISA) and *memory-mapped I/O*, in which the I/O controller registers have reserved spaces in main memory.
- *A way for the I/O device to contact the master CPU.* I/O controllers that have the ability to contact the master processor via an interrupt are referred to as interrupt driven I/O. Generally, an I/O device initiates an asynchronous interrupt requesting signaling to indicate, for example, that control and status registers can be read from or written to. The master CPU then uses its interrupt scheme to determine when an interrupt will be discovered.
- *Some mechanism for both to exchange data.* This refers to how data is actually exchanged between the I/O controller and the master processor. In a *programmed transfer*, the master processor receives data from the I/O controller into its registers, and the CPU then transmits this data to memory. For memory-mapped I/O schemes, DMA (direct memory access) circuitry can be used to bypass the master CPU entirely. DMA has the ability to manage data transmissions or receptions directly to and from main memory and an I/O device. On some systems, DMA is integrated into the master processor, and on others there is a separate DMA controller. Essentially, DMA requests control of the bus from the master processor.

## 6.3 I/O and Performance

I/O performance is one of the most important issues of an embedded design. I/O can negatively impact performance by *bottlenecking the entire system*. In order to understand the type of performance hurdles I/O must overcome, it is important to understand that, with the wide variety of I/O devices, each device will have its own unique qualities. Thus, in a proper design, the engineer will have taken these unique qualities on a case by case basis into consideration. Some of the most important shared features of I/O that can negatively impact board performance include:

- *The data rates of the I/O devices.* I/O devices on one board can vary in data rates from a handful of characters per second with a keyboard or a mouse to transmission in Mbytes per second (networking, tape, disk).
- *The speed of the master processor.* Master processors can have clocks rates anywhere from tens of MHz to hundreds of MHz. Given an I/O device with an extremely slow data rate, a master CPU could have executed thousands of times more data in the time period that the I/O needs to process a handful of *bits* of data. With extremely fast I/O, a master processor would not even be able to process anything before the I/O device is ready to move forward.
- *How to synchronize the speed of the master processor to the speeds of I/O.* Given the extreme ranges of performance, a realistic scheme must be implemented that allows for either the I/O or master processor to process data successfully regardless of how different their speeds. Otherwise, with an I/O device processing data much slower than the master processor transmits, for instance, data would be lost by the I/O device. If the device is not ready, it could hang the entire system if there is no mechanism to handle this situation.
- *How I/O and the master processor communicate.* This includes whether there is an intermediate dedicated I/O controller between the master CPU and I/O device that manages I/O for the master processor, thus freeing up the CPU to process data more efficiently. Relative to an I/O controller, it becomes a question of whether the communication scheme is interrupt driven, polled, or memory-mapped (with dedicated DMA to, again, free up the master CPU). If interrupt-driven, for example, can I/O devices interrupt other I/O, or would devices on the queue have to wait until previous devices finished their turn, no matter how slow?

To improve I/O performance and prevent bottlenecks, board designers need to examine the various I/O and master processor communication schemes to ensure that every device can be managed successfully via one of the available schemes. For example, to synchronize slower I/O devices and the master CPU, status flags or interrupts can be made available for all ICs so that they can communicate their status to each other when processing data. Another example occurs when I/O devices are faster than the master CPU. In this case, some type of interface (i.e., DMA) that allows these devices to bypass the master processor altogether could be an alternative.

The most common units measuring performance relative to I/O include:

- *Throughput* of the various I/O components: the maximum amount of data per unit time that can be processed, in bytes per second. This value can vary for different components. The components with the *lowest* throughput are what drives the performance of the whole system.
- *Execution time* of an I/O component: the amount of time it takes to process all of the data it is provided with.
- The *response time* or *delay* time of an I/O component: the amount of time between a request to process data and the time the actual component begins processing.

In order to accurately determine the type of performance to measure, the benchmark has to match how the I/O functions within the system. If the board will be accessing and processing several larger stored data files, benchmarks will be needed to measure the *throughput* between memory and secondary/tertiary storage medium. If the access is to files that are very small, then *response time* is the critical performance measure, since execution times would be very fast for small files, and the I/O rate would depend on the number of storage accesses per second, including delays. In the end, the performance measured would need to reflect how the system would actually be used, in order for any benchmark to be of use.

## 6.4 Summary

In this chapter, an I/O subsystem was introduced as being some combination of a transmission medium, communication port, a communication interface, an I/O controller, I/O buses, and the master processor's integrated I/O. Of this subsystem, a communication port, communication interface if not integrated into the master processor, I/O controller, and I/O buses are the board I/O of the system. This chapter also discussed the integration of various I/O components within the subsystem in relation to each other. Networking schemes (RS-232, Ethernet, and IEEE 802.11) were provided as serial and parallel transmission I/O examples, as well as a graphics example being given for parallel transmission. Finally, this chapter discussed the impact of board I/O on an embedded system's performance.

Next, Chapter 7, *Board Buses*, discusses the types of buses that can be found on an embedded board, and provides real-world examples of board bus hardware implementations.

## Chapter 6: Problems

1. [a] What is the purpose of I/O on a board?
   [b] List five categories of board I/O, with two real-world examples under each category.
2. Name and describe the six logical units into which I/O hardware can be classified.
3. In Figures 6-26a and b, indicate what I/O components fall under what I/O logical unit.
4. [a] What is the difference between serial and parallel I/O?
   [b] Give a real-world I/O example of each.
5. [a] What is the difference between simplex, half-duplex, and full-duplex transmission?
   [b] Indicate which transmission scheme is shown in Figures 6-27a–c.
6. [a] What is asynchronous transfer of serial data?
   [b] Draw a diagram that describes how asynchronous transfer of serial data works.
7. The baud rate is:
   A. The bandwidth of the serial interface.
   B. The total number of bits that can be transmitted.
   C. The total number of bits per unit time that can be transmitted.
   D. None of the above.
8. [a] What is the bit rate of a serial interface?
   [b] Write the equation.

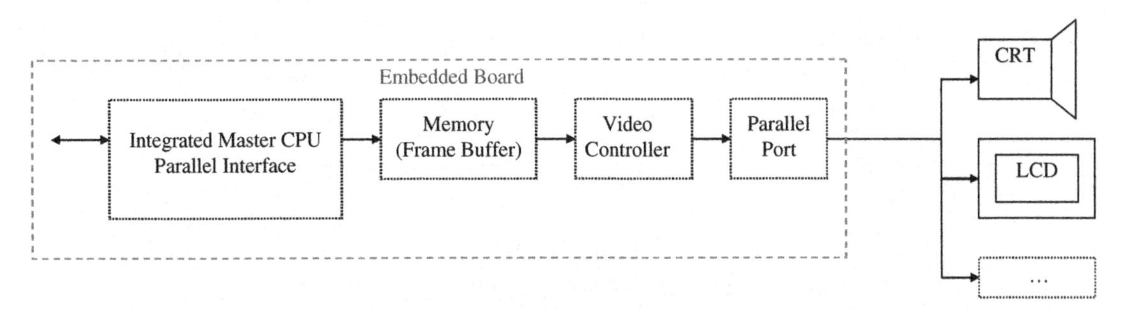


**Figure 6-26a**
Complex I/O subsystem.

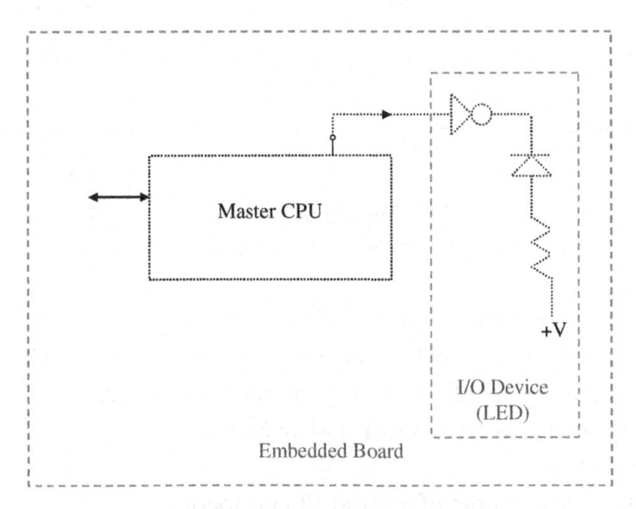


**Figure 6-26b**
Simple I/O subsystem.[2]

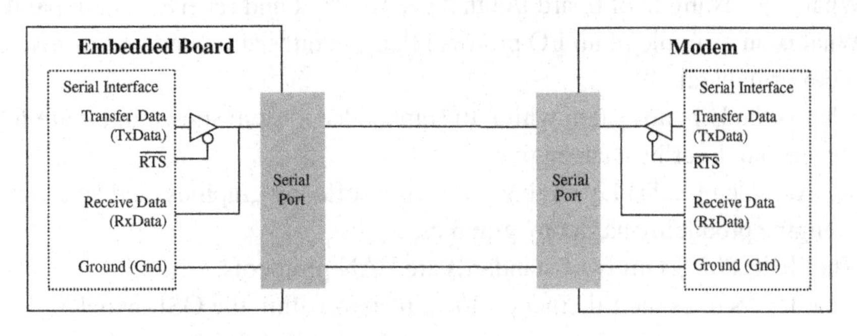


**Figure 6-27a**
Transmission scheme example.[3]

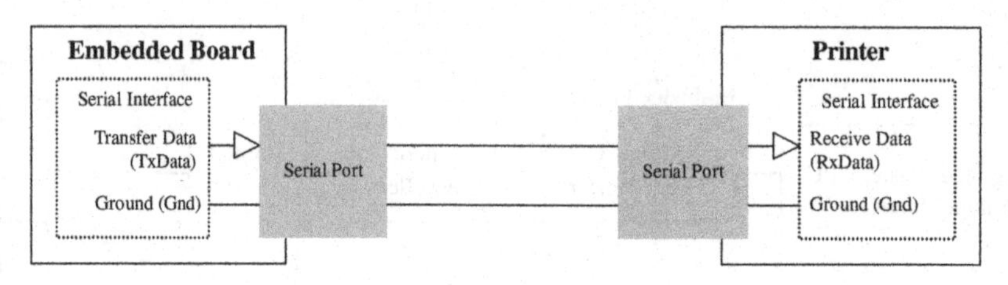


**Figure 6-27b**
Transmission scheme example.[3]

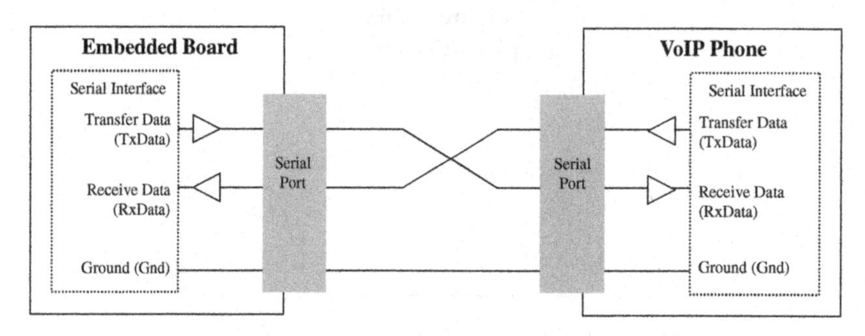


**Figure 6-27c**
Transmission scheme example.[3]

9. [a] What is synchronous transfer of serial data?
   [b] Draw and describe how synchronous transfer of serial data works.
10. [T/F] A UART is an example of a synchronous serial interface.
11. What is the difference between a UART and an SPI?
12. [a] What is a serial port?
    [b] Give a real-world example of a serial I/O protocol.
    [c] Draw a block diagram of the major components defined by this protocol, and define these components.
13. Where in the OSI model would the hardware components of an I/O interface map?
14. [a] What is an example of board I/O that can transmit and receive data in parallel?
    [b] What is an example of an I/O protocol that can either transmit and receive data in serial or in parallel?
15. [a] What is the I/O subsystem within the embedded system shown in Figure 6-28?
    [b] Define and describe each engine.
16. Draw an example of a display engine producing softcopy graphics, and an example of a display engine producing hardcopy graphics.
17. [T/F] The IEEE 802.3 family of standards are LAN protocols.
18. [a] What layers does the Ethernet protocol map to within the OSI model?
    [b] Draw and define the major components of a 10 Mbps Ethernet subsystem.

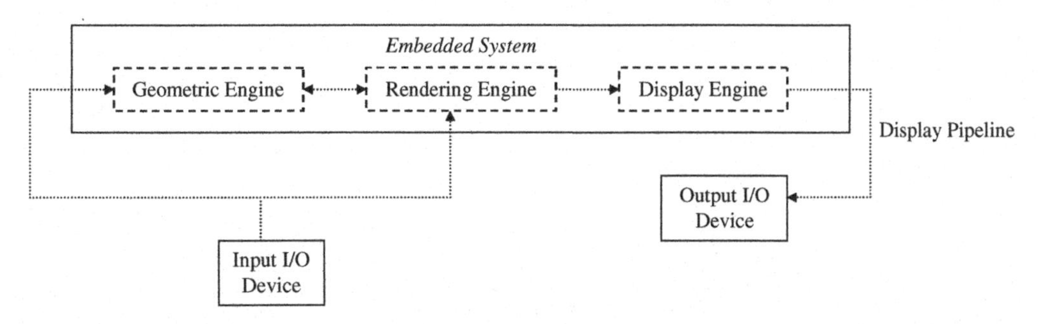


**Figure 6-28**
Graphical design engines.

19. For a system that contains an I/O controller to manage an I/O device, name at least two requirements that the interface between the master processor and I/O controller is typically based upon.
20. How can board I/O negatively impact a system's performance?
21. If there is no mechanism to synchronize the speed differences between I/O devices and the master CPU, then:
    A. Data can be lost.
    B. Nothing can go wrong.
    C. The entire system could crash.
    D. A and C only.
    E. None of the above.

## Endnotes

[1] *Foundations of Computer Architecture*, H. Malcolm. Additional references include: *Computer Organization and Architecture*, W. Stallings, Prentice Hall, 4th edn, 1995. *Structured Computer Organization*, A. S. Tanenbaum, Prentice Hall, 3rd edn, 1990; *Computer Architecture*, R. J. Baron and L. Higbie, Addison-Wesley, 1992; *MIPS RISC Architecture*, G. Kane and J. Heinrich, Prentice Hall, 1992; *Computer Organization and Design: The Hardware/Software Interface*, D. A. Patterson and J. L. Hennessy, Morgan Kaufmann, 3rd edn, 2005.
[2] *Computers As Components*, p. 206, W. Wolf, Morgan Kaufmann, 2 edn, 2008.
[3] *Embedded Microcomputer Systems*, J. W. Valvano, CL Engineering, 2nd edn, 2006.
[4] http://www.camiresearch.com/Data_Com_Basics/RS232_standard.html#anchor1155222
[5] http://grouper.ieee.org/groups/802/11/
[6] Conexant, PRISM 3 Product Brief.
[7] Conexant, PRISM APDK Product Brief.
[8] Net Silicon, NET + ARM50 Block Diagram.

CHAPTER 7

# Board Buses

**In This Chapter**

- Defining the different types of buses
- Discussing bus arbitration and handshaking schemes
- Introducing $I^2C$ and PCI bus examples

All of the other major components that make up an embedded board—the master processor, I/O (input/output) components, and memory—are interconnected via *buses* on the embedded board. As defined earlier, a bus is simply a collection of wires carrying various data signals, addresses, and control signals (clock signals, requests, acknowledgements, data type, etc.) between all of the other major components on the embedded board, which include the I/O subsystems, memory subsystem, and the master processor. On embedded boards, at least one bus interconnects the other major components in the system (see Figure 7-1).

But why does the reader need to pay so much attention to board buses? Why is it important to gain an understanding of material in this chapter—such as bus arbitration, handshaking, signal lines, and timing?

This is because bus *performance* is typically *measured* via bandwidth, where both *physical design* and associated *protocols* matter. For example:

- The simpler the bus handshaking scheme, the higher the bandwidth.
- The shorter the bus, the fewer connected devices, and the more data lines typically means the faster the bus and the higher its bandwidth.
- More bus lines means the more data that can be physically transmitted at any one time, in parallel.
- The bigger the bus width, the fewer the delays and the greater the bandwidth.

On more complex boards, multiple buses can be integrated on one board (see Figure 7-2). For embedded boards with several buses connecting components that need to intercommunicate, *bridges* on the board connect the various buses and carry information from one bus to another. In Figure 7-2, the PowerManna PCI bridge is one such example. A bridge can automatically provide a transparent mapping of address information when data is transferred from one bus to another, and implement different control signal requirements for various buses (e.g., acknowledgment cycles) as well as modify the data being transmitted if any transfer protocols differ bus to bus. For instance, if the byte ordering differs, the bridge can handle the byte swapping.

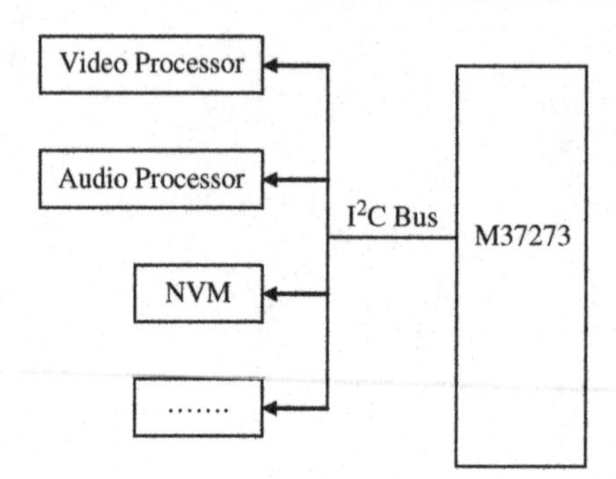


**Figure 7-1**
General bus structure.

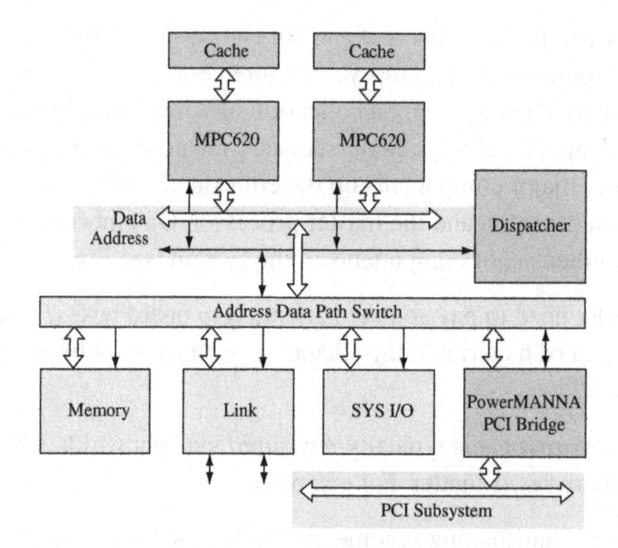


**Figure 7-2**
MPC620 board with bridge.[1] *© 2004 Freescale Semiconductor, Inc. Used by permission.*

Board buses typically fall under one of three main categories: *system buses*, *backplane buses*, or *I/O buses*. *System buses* (also referred to as "main," "local," or "processor-memory" buses) interconnect external main memory and cache to the master CPU and/or any bridges to the other buses. System buses are typically shorter, higher speed, custom buses. *Backplane buses* are also typically faster buses that interconnect memory, the master processor, and I/O, all on one bus. *I/O buses*, also referred to as "expansion," "external," or "host" buses, in effect act as *extensions* of the system bus to connect the remaining components to the master CPU, to each other, to the system bus via a bridge, and/or to the embedded system itself, via an I/O communication port. I/O buses are typically standardized buses that can be either shorter, higher speed buses such as PCI and USB, or longer, slower buses such as SCSI.

The major difference between system buses and I/O buses is the possible presence of *IRQ (interrupt request)* control signals on an I/O bus. There are a variety of ways I/O and the master processor can communicate, and interrupts are one of the most common methods. An IRQ line allows for I/O devices on a bus to indicate to the master processor that an event has taken place or an operation has been completed by a signal on that IRQ bus line. Different I/O buses can have different impacts on interrupt schemes. An ISA bus, for example, requires that each card that generates interrupts must be assigned its own unique IRQ value (via setting switches or jumpers on the card). The PCI bus, on the other hand, allows two or more I/O cards to share the same IRQ value.

Within each bus category, buses can be further divided into whether the bus is *expandable* or *non-expandable. An expandable* bus (PCMCIA, PCI, IDE, SCSI, USB, etc.) is one in which additional components can be plugged into the board on-the-fly, whereas a *non-expandable* bus (DIB, VME, $I^2C$, etc.) is one in which additional components cannot be simply plugged into the board and then communicate over that bus to the other components.

While systems implementing expandable buses are more flexible because components can be added ad-hoc to the bus and work "out of the box," expandable buses tend to be more expensive to implement. If the board is not initially designed with all of the possible types of components that could be added in the future in mind, performance can be negatively impacted by the addition of too many "draining" or poorly designed components onto the expandable bus.

## 7.1 Bus Arbitration and Timing

Associated with every bus is some type of *protocol* that defines how devices gain access to the bus (arbitration), the rules attached devices must follow to communicate over the bus (handshaking), and the signals associated with the various bus lines.

Board devices obtain access to a bus using a *bus arbitration* scheme. Bus arbitration is based upon devices being classified as either *master* devices (devices that can initiate a bus transaction) or *slave* devices (devices which can only gain access to a bus in response to a master device's request). The simplest arbitration scheme is for only one device on the board—the master processor—to be allowed to be master, while all other components are slave devices. In this case, no arbitration is necessary when there can only be one master.

For buses that allow for multiple masters, some have an *arbitrator* (separate hardware circuitry) that determines under what circumstances a master gets control of the bus. There are several bus arbitration schemes used for embedded buses, the most common being *dynamic central parallel, centralized serial* (daisy-chain), and *distributed self-selection.*

Dynamic central parallel arbitration (shown in Figure 7-3a) is a scheme in which the arbitrator is centrally located. All bus masters connect to the central arbitrator. In this scheme, masters are then granted access to the bus via a *FIFO* (First In First Out—see Figure 7-3b) or

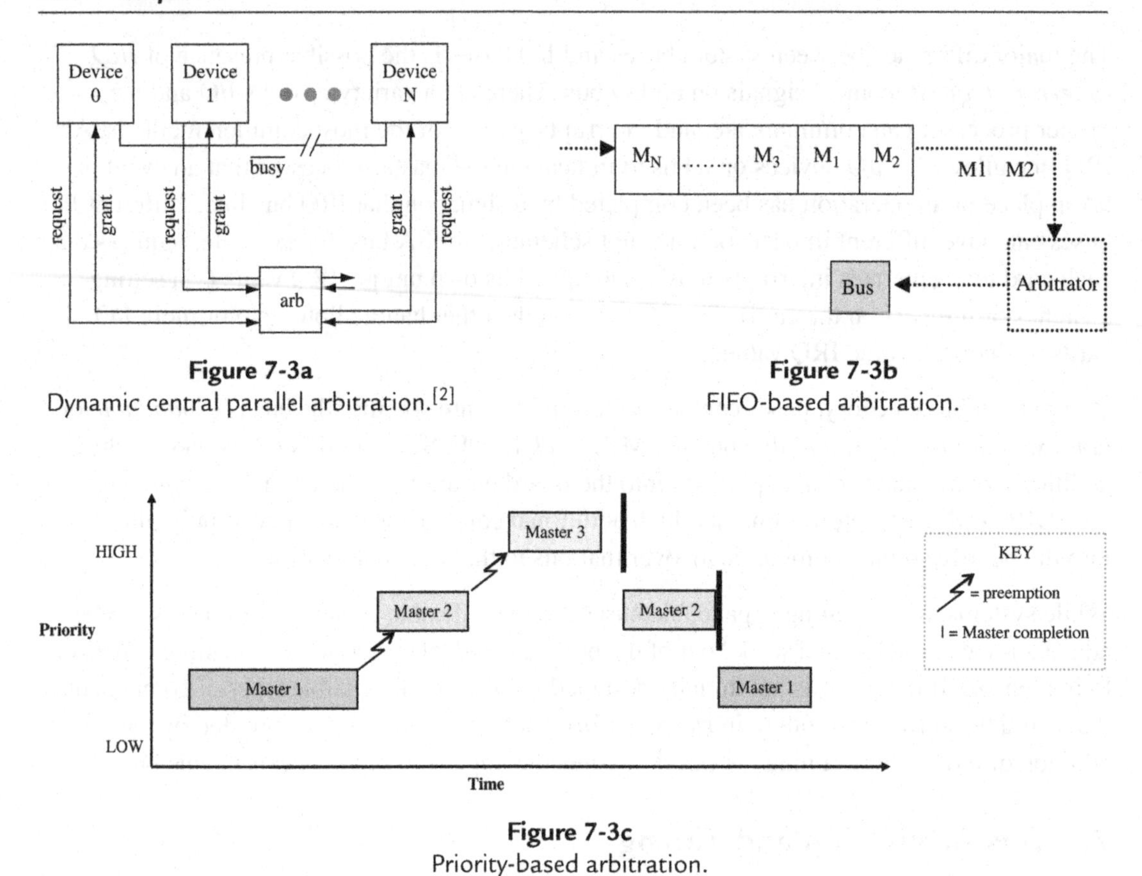


**Figure 7-3a**
Dynamic central parallel arbitration.[2]

**Figure 7-3b**
FIFO-based arbitration.

**Figure 7-3c**
Priority-based arbitration.

*priority-based* system (see Figure 7-3c). The FIFO algorithm implements some type of FIFO queue that stores a list of master devices ready to use the bus in the order of bus requests. Master devices are added at the end of the queue, and are allowed access to the bus from the start of the queue. One main drawback is the possibility of the arbitrator not intervening if a single master at the front of the queue maintains control of the bus, never completing and not allowing other masters to access the bus.

The priority arbitration scheme differentiates between masters based upon their relative importance to each other and the system. Basically, every master device is assigned a priority, which acts as an indicator of order of precedence within the system. If the arbitrator implements a *pre-emption* priority-based scheme, the master with the highest priority always can pre-empt lower priority master devices when they want access to the bus, meaning a master currently accessing the bus can be forced to relinquish it by the arbitrator if a higher priority master wants the bus. Figure 7-3c shows three master devices (1, 2, and 3, where master 1 is the lowest priority device and master 3 is the highest); master 3 pre-empts master 2, and master 2 pre-empts master 1 for the bus.

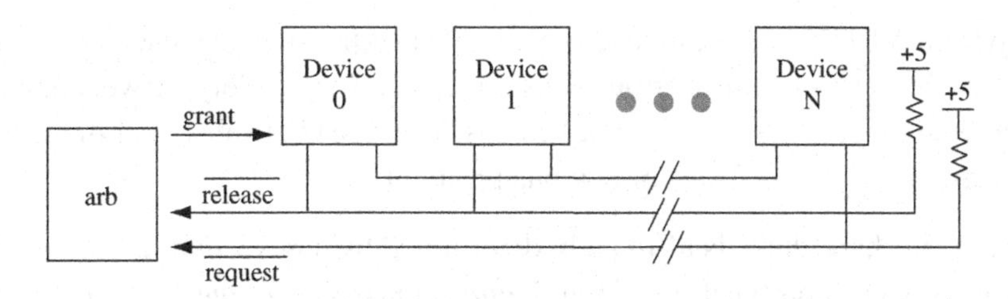


**Figure 7-4**
Centralized serial/daisy-chain arbitration.[2]

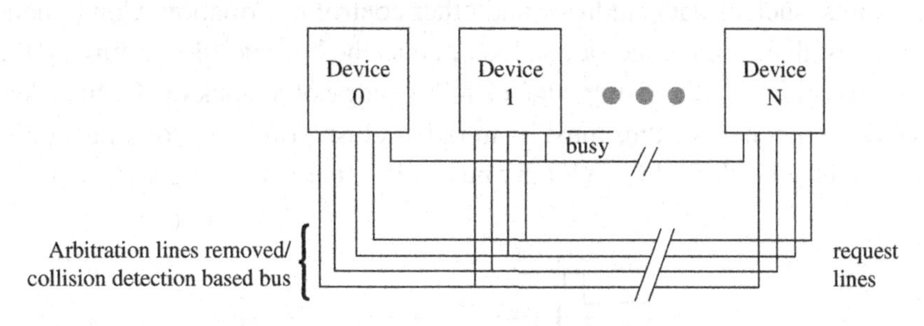


**Figure 7-5**
Distributed arbitration via self-selection.[2]

Central-serialized arbitration, also referred to as daisy-chain arbitration, is a scheme in which the arbitrator is connected to all masters, and the masters are connected in serial. Regardless of which master makes the request for the bus, the first master in the chain is granted the bus and passes the "bus grant" on to the next master in the chain if/when the bus is no longer needed (see Figure 7-4).

There are also distributed arbitration schemes, which means there is no central arbitrator and no additional circuitry, as shown in Figure 7-5. In these schemes, masters arbitrate themselves by trading priority information to determine if a higher priority master is making a request for the bus or even by removing all arbitration lines and waiting to see if there is a collision on the bus, which means that the bus is busy with more than one master trying to use it.

Again, depending on the bus, bus arbitrators can grant a bus to a master *atomically* (until that master is finished with its transmission) or allow for *split* transmissions, where the arbitrator can pre-empt devices in the middle of transactions, switching between masters to allow other masters to have bus access.

Once a master device is granted the bus, only two devices—a master and another device in slave mode—communicate over that bus at any given time. There are only two types of transactions that a bus device can do—READ (receive) and/or WRITE (transmit). These transactions can take place either between two processors (e.g., a master and I/O controller) or processor and memory (e.g., a master and memory). Within each type of transaction, whether READ or WRITE, there can also be several specific rules that each device needs to follow in order to complete a transaction. These rules can vary widely between the types of devices communicating, as well as from bus to bus. These sets of rules, commonly referred to as the bus handshake, form the basis of any bus protocol.

The basis of any bus handshake is ultimately determined by a bus's *timing scheme.* Buses are based upon one or some combination of *synchronous* or *asynchronous* bus timing schemes, which allow for components attached to the bus to synchronize their transmissions. A synchronous bus (such as that shown in Figure 7-6) includes a *clock signal* among the other signals it transmits, such as data, address, and other control information. Components using a synchronous bus all are run at the same clock rate as the bus and (depending on the bus) data is transmitted either on the rising edge or falling edge of a clock cycle. In order for this scheme to work, components either must be in rather close proximity for a faster clock rate or the clock rate must be slowed for a longer bus. A bus that is too long with a clock rate

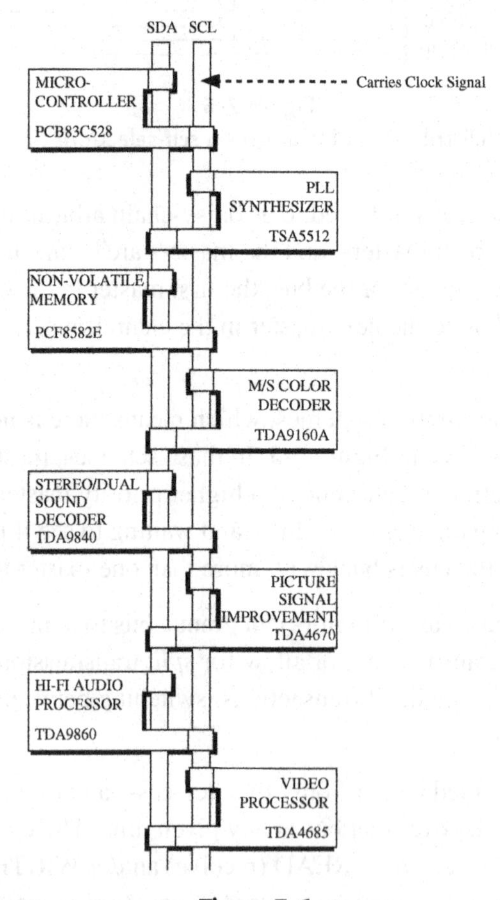


**Figure 7-6**
$I^2C$ bus with SCL clock.[3]

that is too fast (or even too many components attached to the bus) will cause a skew in the synchronization of transmissions, because transmissions in such systems won't be in sync with the clock. In short, this means that faster buses typically use a synchronous bus timing scheme.

An asynchronous bus, such as the one shown in Figure 7-7, transmits no clock signal, but transmits other (non-clock based) "handshaking" signals instead, such as request and acknowledgment signals. While the asynchronous scheme is more complex for devices having to coordinate request commands, reply commands, etc., an asynchronous bus has no problem with the length of the bus or a larger number of components communicating over the bus, because a clock is not the basis for synchronizing communication. An asynchronous bus, however, does need some other "synchronizer" to manage the exchange of information, and to interlock the communication.

The two most basic protocols that start any bus handshaking are the master indicating or requesting a transaction (a READ or WRITE) and the slave responding to the transaction indication or request (e.g., an acknowledgment/ACK or enquiry/ENQ). The basis of these two protocols are *control* signals transmitted either via a dedicated control bus line or over a data line. Whether it's a request for data at a memory location, or the value of an I/O controller's control or status registers, if the slave responds in the affirmative to the master device's transaction request, then either an *address* of the data involved in the transaction is exchanged via a dedicated address bus line or data line, or this address is transmitted as

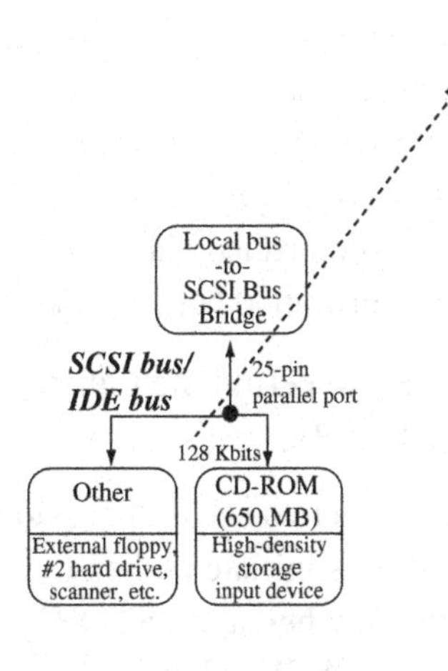


*The SCSI specification defines 50 bus signals, half of which are tied to ground. The 18 SCSI bus signals that are relevant to understanding SCSI transactions are shown below. Nine of these signals are used to initiate and control transactions, and nine are used for data transfer (8 data bits plus a parity bit).*

| Signal | Name | Description |
|---|---|---|
| /BSY | Busy | This signal indicates that the bus is in use. |
| /SEL | Select | The initiator uses this signal to select a target. |
| /C/D | Control/Data | The target uses this signal to indicate whether the information being transferred is control information (signal asserted) or data (signal negated). |
| /I/O | Input/Output | The target uses this signal to specify the direction of the data movement with respect to the initiator. When the signal is asserted, data flows to the initiator; when negated, data flows to the target. |
| /MSG | Message | This signal is used by the target during the message phase. |
| /REQ | Request | The target uses this signal to start a request/acknowledge handshake. |
| /ACK | Acknowledge | This signal is used by the initiator to end a request/acknowledge handshake. |
| /ATN | Attention | The initiator uses this signal to inform the target that the initiator has a message ready. The target retrieves the message, at its convenience, by transitioning to a message-out bus phase. |
| /RST | Reset | This signal is used to clear all devices and operations from the bus, and force the bus into the bus free phase. The Macintosh computer asserts this signal at startup. SCSI peripheral devices should never assert this signal. |
| /DB0-/DB7, /DBP | Data | There are eight data signals, numbered 0 to 7, and the parity signal. Macintosh computers generate proper SCSI parity, but the original SCSI Manager does not detect parity errors in SCSI transactions. |

**Figure 7-7**
SCSI bus.[4]

part of the same transmission as the initial transaction request. If the address is valid, then a data exchange takes place over a data line (plus or minus a variety of acknowledgments over other lines or multiplexed into the same stream). Again, note that handshaking protocols vary with different buses. For example, where one bus requires the transmission of enquiries and/or acknowledgments with every transmission, other buses may simply allow the broadcast of master transmissions to all bus (slave) devices, and only the slave device related to the transaction transmits data back to the sender. Another example of differences between handshaking protocols might be that, instead of a complex exchange of control signal information being required, a clock could be the basis of all handshaking.

Buses can also incorporate a variety of transferring mode schemes, which dictate how the bus transfers the data. The most common schemes are *single*, where an address transmission precedes every word transmission of data, and *blocked*, where the address is transmitted only once for multiple words of data. A blocked transferring scheme can increase the bandwidth of a bus (without the added space and time for retransmitting the same address) and is sometimes referred to as burst transfer scheme. It is commonly used in certain types of memory transactions, such as cache transactions. A blocked scheme, however, can negatively impact bus performance in that other devices may have to wait longer to access the bus. Some of the strengths of the single transmission scheme include not requiring slave devices to have buffers to store addresses and the multiple words of data associated with the address, as well as not having to handle any problems that could arise with multiple words of data either arriving out of order or not directly associated with an address.

### Non-Expandable Bus: $I^2C$ Bus Example

The $I^2C$ (inter-IC) bus interconnects processors that have incorporated an $I^2C$ on-chip interface, allowing direct communication between these processors over the bus. A master–slave relationship between these processors exists at all times, with the master acting as a master transmitter or master receiver. As shown in Figure 7-8, the $I^2C$ bus is a two-wire bus with one serial data line (SDA) and one serial clock line (SCL). The processors connected via $I^2C$ are each addressable by a unique address that is part of the data stream transmitted between devices.

The $I^2C$ master initiates data transfer and generates the clock signals to permit the transfer. Basically, the SCL just cycles between HIGH and LOW (see Figure 7-9).

The master then uses the SDA line (as SCL is cycling) to transmit data to a slave. A session is started and terminated as shown in Figure 7-10, where a "START" is initiated when the master pulls the SDA port (pin) LOW while the SCL signal is HIGH, whereas a "STOP" condition is initiated when the master pulls the SDA port HIGH when SCL is HIGH.

With regard to the transmission of data, the $I^2C$ bus is a serial, 8-bit bus. This means that, while there is no limit on the number of bytes that can be transmitted in a session, only one byte (8 bits) of data will be moved at any one time, 1 bit at a time (serially). How this

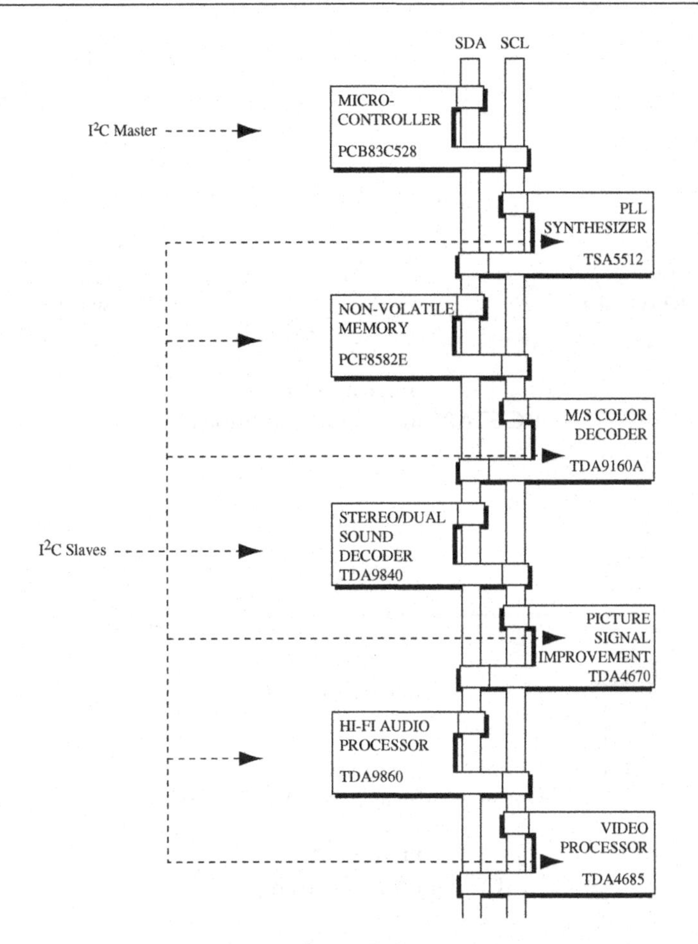


**Figure 7-8**
Sample analog TV board.[3]

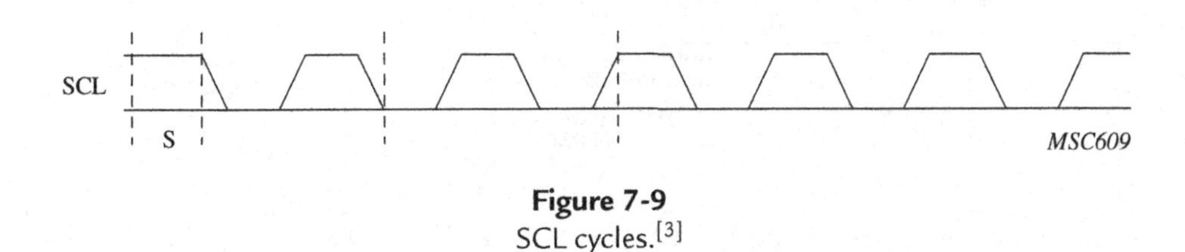


**Figure 7-9**
SCL cycles.[3]

translates into using the SDA and SCL signals is that a data bit is "read" whenever the SCL signal moves from HIGH to LOW, edge to edge. If the SDA signal is HIGH at the point of an edge, then the data bit is read as a "1." If the SDA signal is LOW, the data bit is read as a "0." An example of byte "00000001" transfer is shown in Figure 7-11a, while Figure 7-11b shows an example of a complete transfer session.

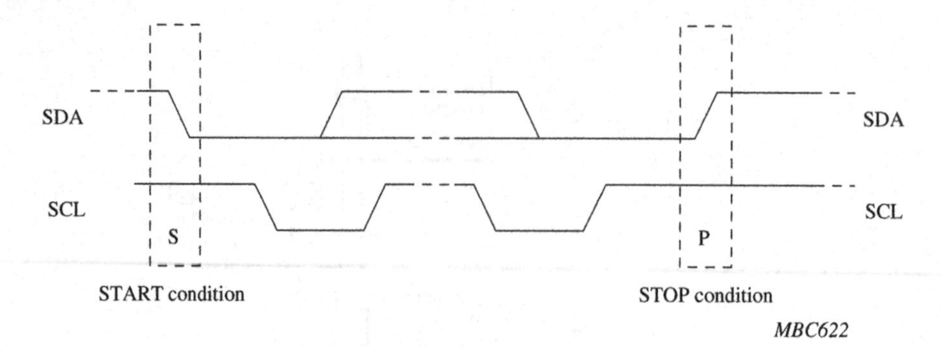


**Figure 7-10**
I²C START and STOP conditions.[3]

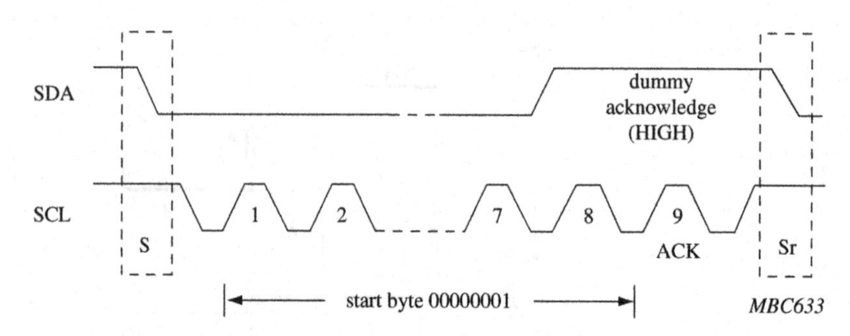


**Figure 7-11a**
I²C data transfer example.[3]

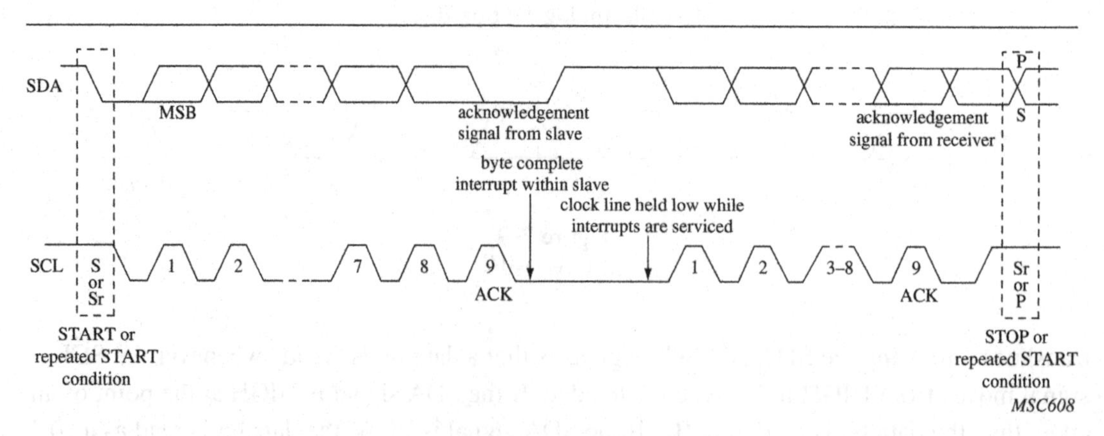


**Figure 7-11b**
I²C complete transfer diagram.

### PCI (Peripheral Component Interconnect) Bus Example: Expandable

The latest PCI specification at the time of writing, PCI Local Bus Specification Revision 2.1, defines requirements (mechanical, electrical, timing, protocols, etc.) of a PCI bus implementation. PCI is a synchronous bus, meaning that it synchronizes communication using a clock. The latest standard defines a PCI bus design with at least a 33-MHz clock (up to 66 MHz) and a bus width of at least 32 bits (up to 64 bits), giving a possible minimum throughput of approximately 132 Mbytes/s ((33 MHz*32 bits)/8) and up to 528 Mbytes/s maximum with 64-bit transfers given a 66-MHz clock. PCI runs at either of these clock speeds, regardless of the clock speeds at which the components attached to it are running.

As shown in Figure 7-12, the PCI bus has two connection interfaces: an internal PCI interface that connects it to the main board (to bridges, processors, etc.) via EIDE channels, and the expansion PCI interface, which consists of the slots into which PCI adaptor cards (audio, video, etc.) plug. The expansion interface is what makes PCI an expandable bus; it allows for hardware to be plugged into the bus, and for the entire system to automatically adjust and operate correctly.

Under the 32-bit implementation, the PCI bus is made up of 49 lines carrying multiplexed data and address signals (32 pins), as well as other control signals implemented via the remaining 17 pins (see table in Figure 7-12).

Because the PCI bus allows for multiple bus masters (*initiators* of a bus transaction), it implements a *dynamic centralized, parallel* arbitration scheme (see Figure 7-13). The PCI's arbitration scheme basically uses the REQ# and GNT# signals to facilitate communication between initiators and bus arbitrators. Every master has its own REQ# and GNT# pin,

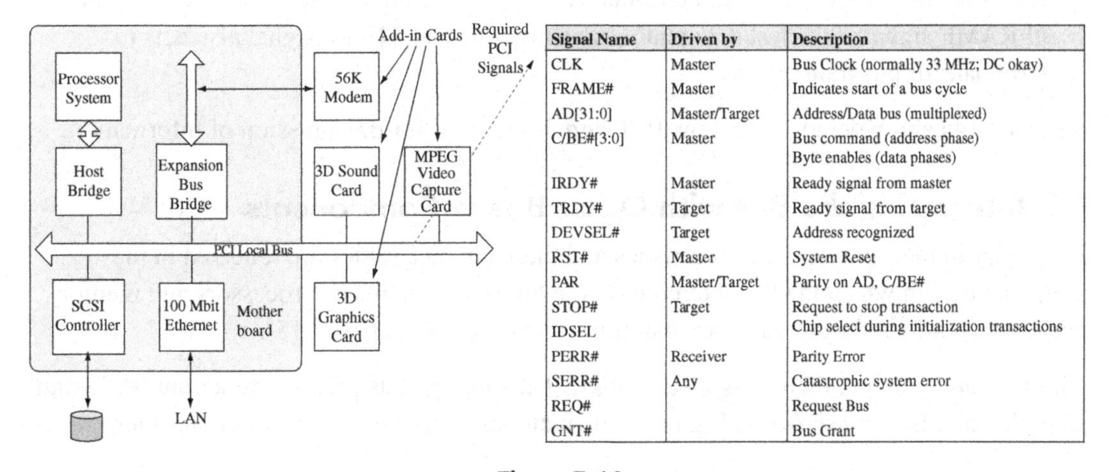


| Signal Name | Driven by | Description |
|---|---|---|
| CLK | Master | Bus Clock (normally 33 MHz; DC okay) |
| FRAME# | Master | Indicates start of a bus cycle |
| AD[31:0] | Master/Target | Address/Data bus (multiplexed) |
| C/BE#[3:0] | Master | Bus command (address phase)<br>Byte enables (data phases) |
| IRDY# | Master | Ready signal from master |
| TRDY# | Target | Ready signal from target |
| DEVSEL# | Target | Address recognized |
| RST# | Master | System Reset |
| PAR | Master/Target | Parity on AD, C/BE# |
| STOP# | Target | Request to stop transaction |
| IDSEL | | Chip select during initialization transactions |
| PERR# | Receiver | Parity Error |
| SERR# | Any | Catastrophic system error |
| REQ# | | Request Bus |
| GNT# | | Bus Grant |

**Figure 7-12**
PCI bus.[5]

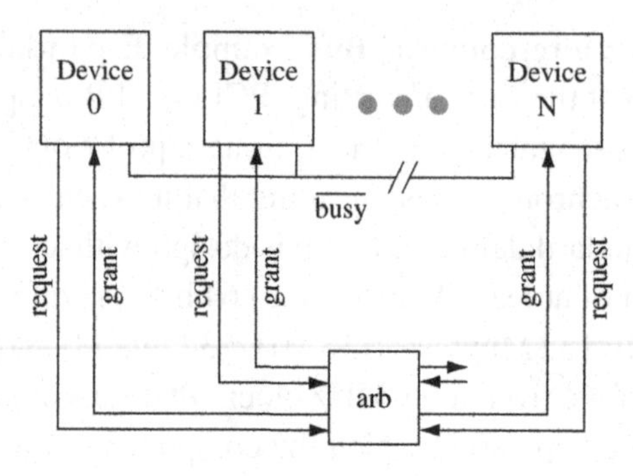


**Figure 7-13**
PCI arbitration scheme.[2]

allowing the arbitrator to implement a fair arbitration scheme, as well as determining the next target to be granted the bus while the current initiator is transmitting data.

In general, a PCI transaction is made up of five steps:

1. An initiator makes a bus request by asserting a REQ# signal to the central arbitrator.
2. The central arbitrator does a bus grant to the initiator by asserting GNT# signal.
3. The address phase begins when the initiator activates the FRAME# signal, and then sets the C/BE[3:0]# signals to define the type of data transfer (memory or I/O read or write). The initiator then transmits the address via the AD[31:0] signals at the next clock edge.
4. After the transmission of the address, the next clock edge starts the one or more data phases (the transmission of data). Data is also transferred via the AD[31:0] signals. The C/BE[3:0], along with IRDY# and #TRDY signals, indicate if transmitted data is valid.
5. Either the initiator or target can terminate a bus transfer through the deassertion of the #FRAME signal at the last data phase transmission. The STOP# signal also acts to terminate all bus transactions

Figures 7-14a and b demonstrate how PCI signals are used for transmission of information.

## 7.2 Integrating the Bus with Other Board Components

Buses vary in their physical characteristics and these characteristics are reflected in the components with which the bus interconnects, mainly the pinouts of processors and memory chips, which reflect the signals a bus can transmit (shown in Figure 7-15).

Within an architecture, there may also be logic that supports bus protocol functionality. As an example, the MPC860 shown in Figure 7-16a includes an integrated I²C bus controller.

As discussed earlier this chapter, the I²C bus is a bus with two signals: SDA and SCL, both of which are shown in the internal block diagram of the PowerPC I²C controller in Figure 7-16b. Because I²C is a synchronous bus, a baud rate generator within the controller supplies a clock

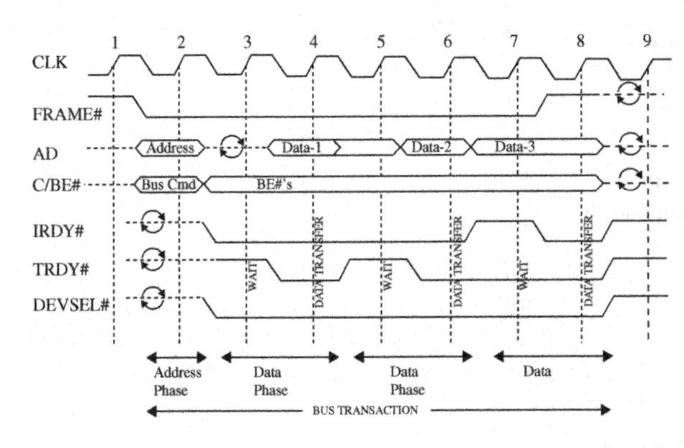


CLK Cycle 1 – The bus is idle.

CLK Cycle 2 – The initiator asserts a valid address and places a read command on the C/BE# signals.

** Start of address phase.**

CLK Cycle 3 – The initiator tri-states the address in preparation for the target driving read data. The initiator now drives valid byte enable information on the C/BE# signals. The initiator asserts IRDY# low indicating it is ready to capture read data. The target asserts DEVSEL# low (in this cycle or the next) as an acknowledgment it has positively decoded the address. The target drives TRDY# high indicating it is not yet providing valid read data.

CLK Cycle 4 – The target provides valid data and asserts TRDY# low indicating to the initiator that data is valid. IRDY# and TRDY# are both low during this cycle causing a data transfer to take place.

**Start of first data phase occurs, and the initiator captures the data.**

CLK Cycle 5 – The target deasserts TRDY# high indicating it needs more time to prepare the next data transfer.

CLK Cycle 6 – Both IRDY# and TRDY# are low.

**Start of next data phase occurs, and the initiator captures the data provided by the target.**

CLK Cycle 7 – The target provides valid data for the third data phase, but the initiator indicates it is not ready by deasserting IRDY# high.

CLK Cycle 8 – The initiator re-asserts IRDY# low to complete the third data phase. The initiator drives FRAME# high indicating this is the final data phase (master termination).

**Final data phase occurs; the initiator captures the data provided by the target, and terminates.**

CLK Cycle 9 – FRAME#, AD, and C/BE# are tri-stated, as IRDY#, TRDY#, and DEVSEL# are driven inactive high for one cycle prior to being tri-stated.

**Figure 7-14a**
PCI read example.[5]

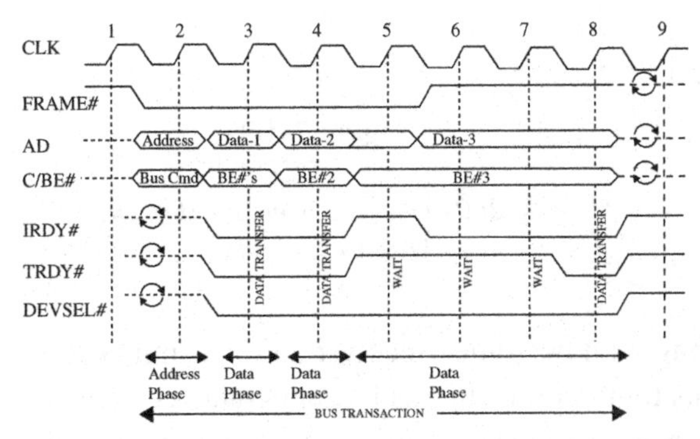


CLK Cycle 1 – The bus is idle.

CLK Cycle 2 – The initiator asserts a valid address and places a write command on the C/BE# signals.

**Start of address phase.**

CLK Cycle 3 – The initiator drives valid write data and byte enable signals. The initiator asserts IRDY# low indicating valid write data is available. The target asserts DEVSEL# low as an acknowledgment it has positively decoded the address (the target may not assert TRDY# before DEVSEL#). The target drives TRDY# low indicating it is ready to capture data. Both IRDY# and TRDY# are low.

**First data phase occurs with target capturing write data.**

CLK Cycle 4 – The initiator provides new data and byte enables. Both IRDY# and TRDY# are low.

**Next data phase occurs with target capturing write data.**

CLK Cycle 5 – The initiator deasserts IRDY# indicating it is not ready to provide the next data. The target deasserts TRDY# indicating it is not ready to capture the next data.

CLK Cycle 6 – The initiator provides the next valid data and asserts IRDY# low. The initiator drives FRAME# high indicating this is the final data phase (master termination). The target is still not ready and keeps TRDY# high.

CLK Cycle 7 – The target is still not ready and keeps TRDY# high.

CLK Cycle 8 – The target becomes ready and asserts TRDY# low. Both IRDY# and TRDY# are low.

**Final data phase occurs with target capturing write data.**

CLK Cycle 9 – FRAME#, AD, and C/BE# are tri-stated, as IRDY#, TRDY#, and DEVSEL# are driven inactive high for one cycle prior to being tri-stated.

**Figure 7-14b**
PCI write example.[5]

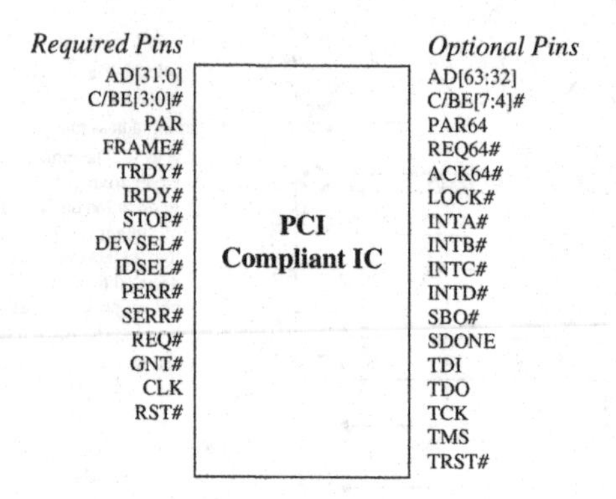


**Figure 7-15**
PCI-compliant IC.[5]

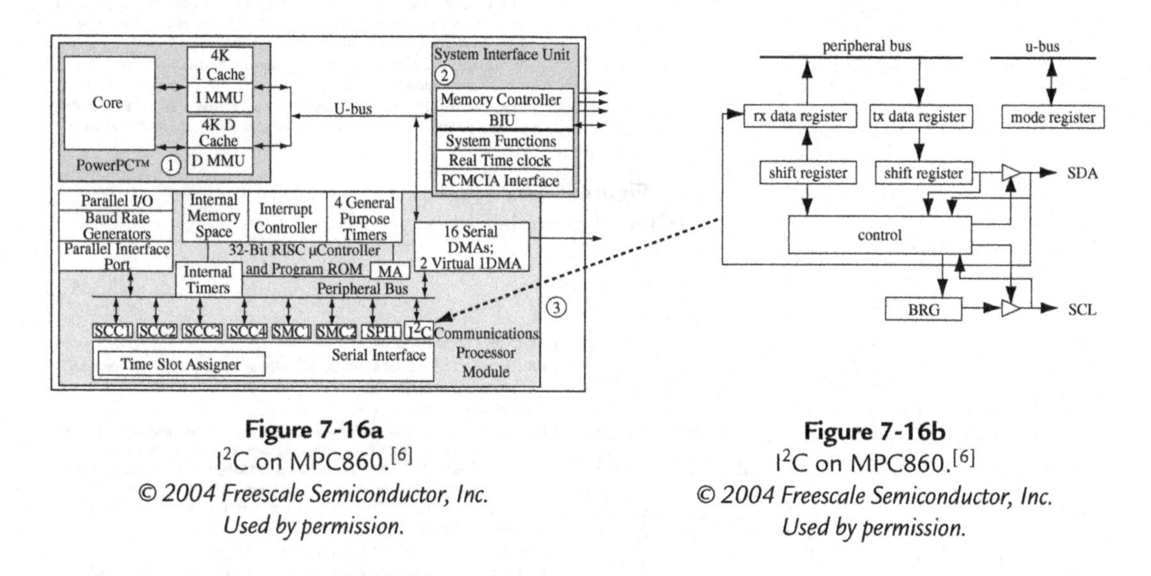


**Figure 7-16a**
I²C on MPC860.[6]


**Figure 7-16b**
I²C on MPC860.[6]


signal if the PowerPC is acting as a master, along with two units (receiver and transmitter) covering the processing and management of bus transactions. In this I²C integrated controller, address and data information is transmitted over the bus via the transmit data register and out of the shift register. When the MPC860 receives data, data is transmitted into the receive data register via a shift register.

## 7.3 Bus Performance

A bus's performance is typically measured by its *bandwidth*—the amount of data a bus can transfer for a given length of time. A bus's design—both physical design and its associated protocols—will impact its performance. In terms of protocols, for example, the simpler the handshaking scheme the higher the bandwidth (fewer "send enquiry," "wait for acknowledgment," etc., steps). The actual physical design of the bus (its length, the number of lines, the number of supported devices, etc.) limits or enhances its performance. The shorter the bus, the fewer connected devices, and the more data lines, typically the faster the bus and the higher its bandwidth.

The number of bus lines and how the bus lines are used (e.g., whether there are separate lines for each signal or whether multiple signals multiplex over fewer shared lines ) are additional factors that impact bus bandwidth. The more bus lines (wires), the more data that can be physically transmitted at any one time, in parallel. Fewer lines mean more data has to share access to these lines for transmission, resulting in less data being transmitted at any one time. Relative to cost, note that an increase in conducting material on the board, in this case the wires of the bus, increases the cost of the board. Note, however, that multiplexing lines will introduce delays on either end of the transmission, because of the logic required on either end of the bus to multiplex and demultiplex signals that are made up of different kinds of information.

Another contributing factor to a bus's bandwidth is the number of data bits a bus can transmit in a given bus *cycle* (transaction); this is the *bus width.* Buses typically have a bandwidth of some binary power of 2, such as 1 ($2^0$) for buses with a serial bus width, 8 ($2^3$) bit, 16 ($2^4$) bit, 32 ($2^5$) bit, etc. As an example, given 32 bits of data that need to be transmitted, if a particular bus has a width of 8 bits, then the data is divided and sent in four separate transmissions; if the bus width is 16 bits, then there are two separate packets to transmit; a 32-bit data bus transmits one packet, and serial means that only 1 bit at any one time can be transmitted. The bus width limits the bandwidth of a bus because it limits the number of data bits that can be transmitted in any one transaction. Delays can occur in each transmission session, because of handshaking (acknowledgment sequences), bus traffic, and different clock frequencies of the communicating components that put components in the system in delaying situations, such as a *wait state* (a time-out period). These delays increase as the number of data packets that need to be transmitted increases. Thus, the bigger the bus width, the fewer the delays, and the greater the bandwidth (throughput).

For buses with more complex handshaking protocols, the transferring scheme implemented can greatly impact performance. A block transfer scheme allows for greater bandwidth over the single transfer scheme, because of the fewer handshaking exchanges per blocks versus single words, bytes (or whatever) of data. On the flip side, block transfers can add to the latency due to devices waiting longer for bus access, since a block transfer-based transaction lasts longer than a single transfer-based transaction. A common solution for this type of latency is a bus that allows for *split transactions*, where the bus is released during the handshaking, such as while waiting for a reply to acknowledgement. This allows for other transactions to take place, and allows the bus not to have to remain idle waiting for devices of one transaction. However, it does add to the latency of the original transaction by requiring that the bus be acquired more than once for a single transaction.

## 7.4 Summary

This chapter introduced some of the fundamental concepts behind how board buses function, specifically the different types of buses and the protocols associated with transmitting over a bus. Two real-world examples were provided—the $I^2C$ bus (a non-expandable bus) and the PCI bus (an expandable bus)—to demonstrate some of the bus fundamentals, such as bus handshaking, arbitration, and timing. This chapter concluded with a discussion on the impact of buses on an embedded system's performance.

Next, Chapter 8, *Device Drivers*, introduces the lowest level software found on an embedded board. This chapter is the first of Section III, which discusses the major software components of an embedded design.

## Chapter 7: Problems

1. [a] What is a bus?
   [b] What is the purpose of a bus?
2. What component on a board interconnects different buses, carrying information from one bus to another?
   A. CDROM drive.
   B. MMU.
   C. Bridge.
   D. All of the above.
   E. None of the above.
3. [a] Define and describe the three categories under which board buses typically fall.
   [b] Provide real-world examples of each type of bus.
4. [a] What is the difference between an expandable bus and a non-expandable bus?
   [b] What are some strengths and drawbacks of each?
   [c] Provide real-world examples of expandable and non-expandable buses.
5. A bus protocol defines:
   A. The bus arbitration scheme.
   B. The bus handshaking scheme.
   C. The signals associated with the bus lines.
   D. A and B only.
   E. All of the above.
6. What is the difference between a bus master and a bus slave?
7. [a] Name and describe three common bus arbitration schemes.
   [b] What is a bus arbitrator?
8. What is the difference between a FIFO-based bus granting and a priority-based bus granting scheme?
9. [a] What is bus handshaking?
   [b] What is the basis of any bus handshake?

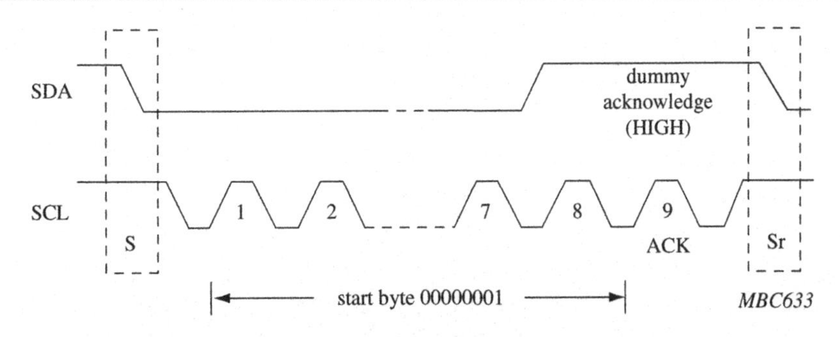


**Figure 7-17**
$I^2C$ data transfer example.[3]

10. [Select] Buses are based on timing schemes that are:
    A. Synchronous.
    B. Asynchronous.
    C. A and B.
    D. All of the above.
11. [T/F] An asynchronous bus transmits a clock signal along with the other types of signals being transmitted.
12. [a] What is a transferring mode scheme?
    [b] Name and describe two of the most common transferring mode schemes.
13. What is the $I^2C$ bus?
14. Draw a timing diagram of an $I^2C$ bus START and STOP condition.
15. Given the timing diagram in Figure 7-17, explain how the start byte "00000001" is being transmitted relative to the SDA and SCL signals.
16. What are the five general steps of a PCI bus transaction?
17. [a] What is the difference between bus bandwidth and bus width?
    [b] What is bus bandwidth a measurement of?
18. Name three physical and/or associated protocol features of a bus that can impact the performance of a bus.

## Endnotes

[1] http://www.first.gmd.de/PowerMANNA/PciBus.html
[2] *Computers As Components*, W. Wolf, Morgan Kaufmann, 2nd edn, 2008.
[3] "The $I^2C$ Bus Specification," Version 2.1, Philips Semiconductor.
[4] www.scsita.org
[5] *PCI Bus Demystified*, D. Abbott, Newnes, 2nd edn, 2004.
[6] Freescale, MPC860 Training Manual.

SECTION III

# Embedded Software Introduction

## Embedded Software Introduction

Section III takes a look at embedded software using the Embedded Systems Model as a reference. It discusses the possible permutations of software sublayers that can exist within an embedded system. Basically, embedded software can be divided into two general classes: systems software and application software. Systems software is any software that supports the applications, such as device drivers, operating systems, and middleware. Application software is the upper-level software that defines the function and purpose of the embedded device, and which handles most of the interaction with users and administrators. In the next three chapters, real-world examples of components within the software sublayers will be presented from an architectural level down to the pseudocode level. Although this is not a programming book, including pseudocode along with the architectural discussion is important, because it allows the reader to understand how requirements and standards evolve from theory into a software flow. The pseudocode is provided to give the reader a visual aid for understanding the software behind different software layer components.

The structure of this section is based upon the software sublayers that can be implemented in an embedded system. Chapter 8 discusses device drivers, Chapter 9 discusses operating systems and board support packages, and Chapter 10 introduces middleware and application software.

On a final important note, for readers who do "not" have a strong technical hardware foundation and have skipped Section II: it is important to remember the implicit wisdom shared in this book, which is that one of the most powerful approaches to understanding the key fundamentals of any embedded systems architecture and design is to take the systems approach. This means having a solid technical foundation via defining and understanding all required hardware components that underlie embedded software.

So, *why not* go back to Section II and start with understanding the hardware?

Because some of the most common mistakes programmers designing complex embedded systems make that lead to costly delays and problems, include:

- Being intimidated by the embedded hardware and tools.
- Treating all embedded hardware like it is a PC-Windows Desktop.
- Waiting for the hardware.
- Using PCs in place of "available" embedded hardware to do development and testing.
- NOT using embedded hardware similar to production hardware, mainly similar I/O (input/output), processing power, and memory.

Developing software for embedded hardware is *not* the same as developing software for a PC or a larger computer system, especially when it comes to adding the additional layer of complexity when introducing overlying software components discussed in this next Section III. The embedded systems boards used as real-world examples in this book demonstrate this point of how drastically embedded boards can vary in design—meaning that each of these board examples will vary widely in terms of the embedded software that can be supported. This is because the major hardware components are different, from the type of master processor to the available memory to the I/O devices. At the start of a project, target system hardware requirements will ultimately depend on the software, especially complex systems that contain an operating system, middleware components, in addition to the overlying application software.

Thus, the reader must, for example, learn to read the hardware schematics and datasheets (as discussed in Section II) to understand and verify all the major components found on an embedded board. This is to ensure that that the processor design is powerful enough to support the requirements of the software stack, the embedded hardware contains the required I/O, and the hardware has enough of the *right* type of memory.

CHAPTER 8

# Device Drivers

## In This Chapter

- Defining device drivers
- Discussing the difference between architecture-specific and board-specific drivers
- Providing several examples of different types of device drivers

Most embedded hardware requires some type of software initialization and management. The software that directly interfaces with and controls this hardware is called a *device driver*. All embedded systems that require software have, at the very least, device driver software in their system software layer. Device drivers are the software libraries that initialize the hardware and manage access to the hardware by higher layers of software. Device drivers are the liaison between the hardware and the operating system, middleware, and application layers. (See Figure 8-1.)

The reader must always check the details about the particular hardware if the hardware component is not 100% identical to what is currently supported by the embedded system. Never assume existing device drivers in the embedded system will be compatible for a particular hardware part—even if the hardware is the same type of hardware that the embedded device currently supports! So, it is very important when trying to understand device driver libraries that:

- Different types of hardware will have different device driver requirements that need to be met.
- Even the same type of hardware, such as Flash memory, that are created by different manufacturers can require substantially different device driver software libraries to support within the embedded device.

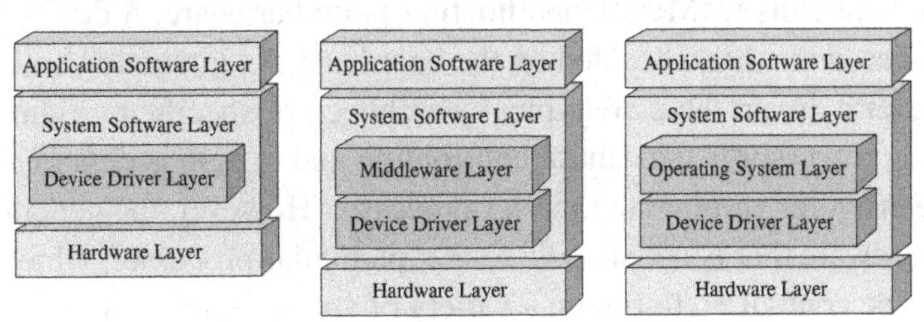


**Figure 8-1**
Embedded Systems Model and Device Drivers.

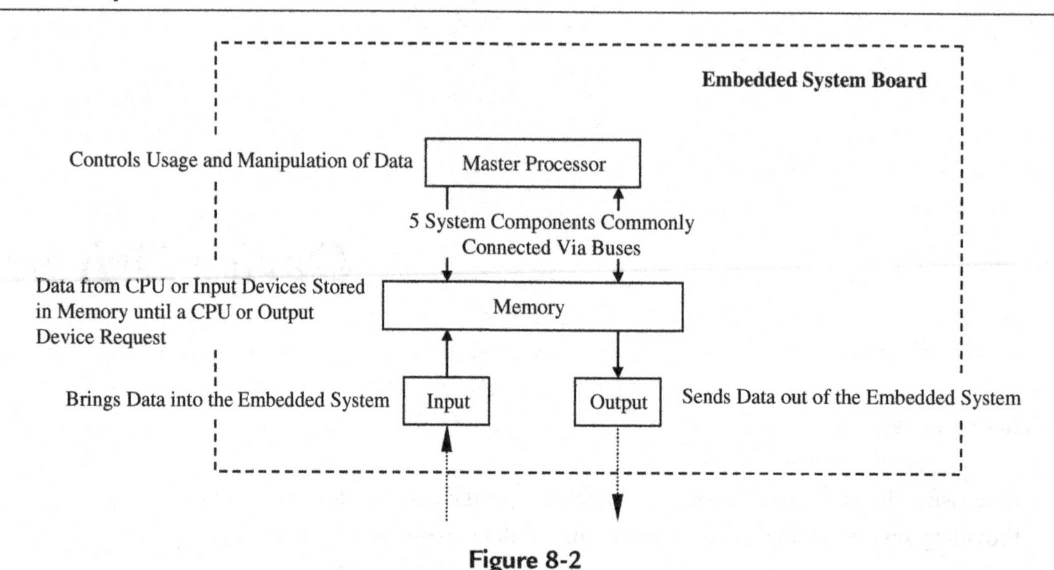


**Figure 8-2**
Embedded System Board Organization.[1]
Based upon the von Neumann architecture model (also referred to as the Princeton architecture).

The types of hardware components needing the support of device drivers vary from board to board, but they can be categorized according to the von Neumann model approach introduced in Chapter 3 (see Figure 8-2). The von Neumann model can be used as a software model as well as a hardware model in determining what device drivers are required within a particular platform. Specifically, this can include drivers for the *master processor* architecture-specific functionality, *memory* and memory management drivers, *bus* initialization and transaction drivers, and I/O (input/output) initialization and control drivers (such as for networking, graphics, input devices, storage devices, or debugging I/O) both at the board and master CPU level.

Device drivers are typically considered either *architecture-specific* or *generic*. A device driver that is *architecture-specific* manages the hardware that is integrated into the master processor (the architecture). Examples of architecture-specific drivers that initialize and enable components within a master processor include on-chip memory, integrated memory managers (memory management units (MMUs)), and floating-point hardware. A device driver that is *generic* manages hardware that is located on the board and not integrated onto the master processor. In a generic driver, there are typically architecture-specific portions of source code, because the master processor is the central control unit and to gain access to anything on the board usually means going through the master processor. However, the generic driver also manages board hardware that is not specific to that particular processor, which means that a generic driver can be configured to run on a variety of architectures that contain the related board hardware for which the driver is written. Generic drivers include code that initializes and manages access to the remaining major components of the board, including board buses

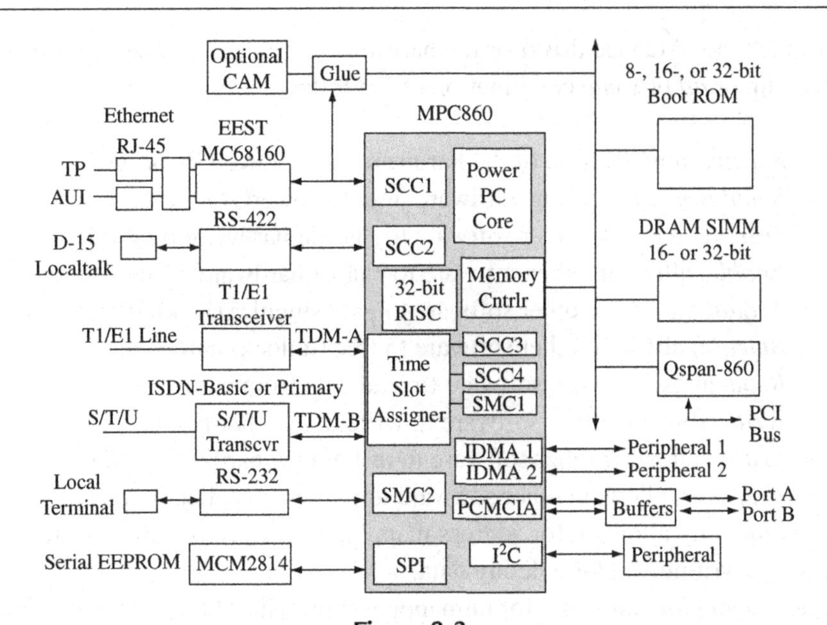


**Figure 8-3a**

MPC860 Hardware Block Diagram.[2]

*© Freescale Semiconductor, Inc. Used by permission.*

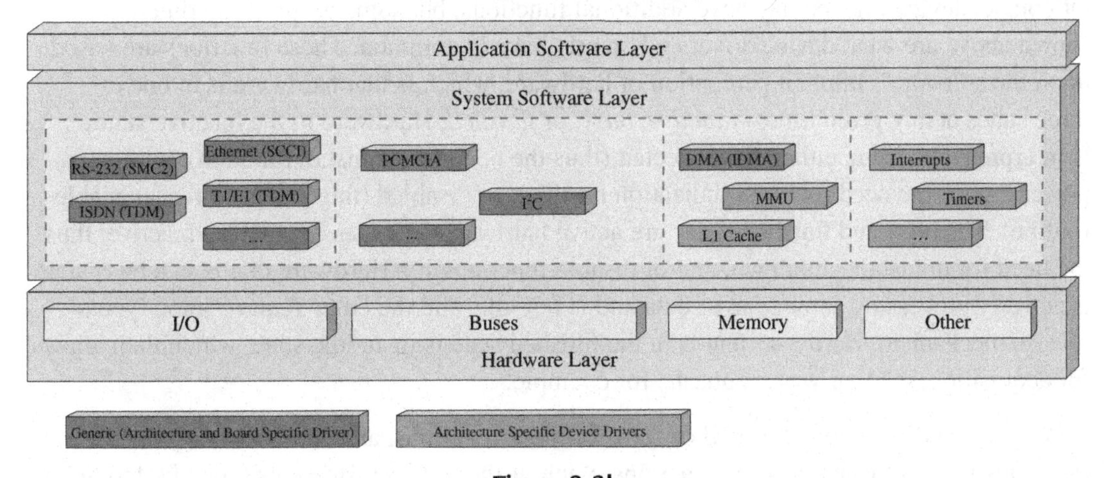


**Figure 8-3b**

MPC860 Architecture-Specific Device Driver System Stack.

*© Freescale Semiconductor, Inc. Used by permission.*

($I^2C$, PCI, PCMCIA, etc.), off-chip memory (controllers, level 2+ cache, Flash, etc.), and off-chip I/O (Ethernet, RS-232, display, mouse, etc.).

Figure 8-3a shows a hardware block diagram of an MPC860-based board and Figure 8-3b shows a systems diagram that includes examples of MPC860 processor-specific device drivers, as well as generic device drivers.

Regardless of the type of device driver or the hardware it manages, all device drivers are generally made up of *all* or *some* combination of the following functions:

- *Hardware Startup*: initialization of the hardware upon PowerON or reset.
- *Hardware Shutdown*: configuring hardware into its PowerOFF state.
- *Hardware Disable*: allowing other software to disable hardware on-the-fly.
- *Hardware Enable*: allowing other software to enable hardware on-the-fly.
- *Hardware Acquire*: allowing other software to gain singular (locking) access to hardware.
- *Hardware Release*: allowing other software to free (unlock) hardware.
- *Hardware Read*: allowing other software to read data from hardware.
- *Hardware Write*: allowing other software to write data to hardware.
- *Hardware Install*: allowing other software to install new hardware on-the-fly.
- *Hardware Uninstall*: allowing other software to remove installed hardware on-the-fly.
- *Hardware Mapping*: allowing for address mapping to and from hardware storage devices when reading, writing, and/or deleting data.
- *Hardware Unmapping*: allowing for unmapping (removing) blocks of data from hardware storage devices.

Of course, device drivers may have additional functions, but some or all of the functions shown above are what device drivers inherently have in common. These functions are based upon the software's implicit perception of hardware, which is that hardware is in one of three states at any given time—*inactive*, *busy*, or *finished*. Hardware in the inactive state is interpreted as being either disconnected (thus the need for an install function), without power (hence the need for an initialization routine), or disabled (thus the need for an enable routine). The busy and finished states are active hardware states, as opposed to inactive; thus the need for uninstall, shutdown, and/or disable functionality. Hardware that is in a busy state is actively processing some type of data and is not idle, and thus may require some type of release mechanism. Hardware that is in the finished state is in an idle state, which then allows for acquisition, read, or write requests, for example.

Again, device drivers may have all or some of these functions, and can integrate some of these functions into single larger functions. Each of these driver functions typically has code that interfaces directly to the hardware and code that interfaces to higher layers of software. In some cases, the distinction between these layers is clear, while in other drivers, the code is tightly integrated (see Figure 8-4).

On a final note, depending on the master processor, different types of software can execute in different modes, the most common being *supervisory* and *user* modes. These modes essentially differ in terms of what system components the software is allowed access to, with software running in supervisory mode having more access (privileges) than software running in user mode. Device driver code typically runs in supervisory mode.

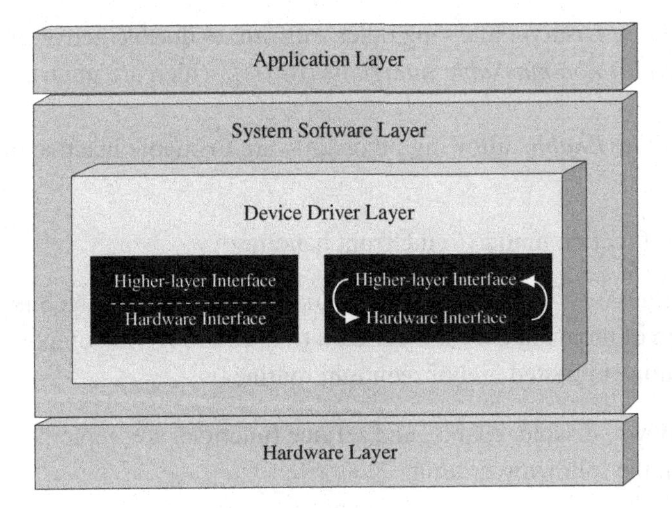


**Figure 8-4**
Driver Code Layers.

The next several sections provide real-world examples of device drivers that demonstrate how device driver functions can be written and how they can work. By studying these examples, the reader should be able to look at any board and figure out relatively quickly what possible device drivers need to be included in that system, by examining the hardware and going through a checklist, using the von Neumann model as a tool for keeping track of the types of hardware that might require device drivers. While not discussed in this chapter, later chapters will describe how device drivers are integrated into more complex software systems.

## 8.1 Example 1: Device Drivers for Interrupt Handling

As discussed previously, *interrupts* are signals triggered by some event during the execution of an instruction stream by the master processor. What this means is that interrupts can be initiated *asynchronously*, for external hardware devices, resets, power failures, etc., or *synchronously*, for instruction-related activities such as system calls or illegal instructions. These signals cause the master processor to stop executing the current instruction stream and start the process of *handling* (processing) the interrupt.

The software that handles interrupts on the master processor and manages interrupt hardware mechanisms (i.e., the interrupt controller) consists of the *device drivers* for interrupt handling. At least four of the 10 functions from the list of device driver functionality introduced at the start of this chapter are supported by interrupt-handling device drivers, including:

- *Interrupt Handling Startup*: initialization of the interrupt hardware (interrupt controller, activating interrupts, etc.) upon PowerON or reset.
- *Interrupt Handling Shutdown*: configuring interrupt hardware (interrupt controller, deactivating interrupts, etc.) into its PowerOFF state.

- *Interrupt Handling Disable*: allowing other software to disable active interrupts on-the-fly (not allowed for *non-maskable interrupts (NMIs)*, which are interrupts that cannot be disabled).
- *Interrupt Handling Enable*: allowing other software to enable inactive interrupts on-the-fly.

Plus one additional function unique to interrupt handling:

- *Interrupt Handler Servicing*: the interrupt handling code itself, which is executed after the interruption of the main execution stream (this can range in complexity from a simple non-nested routine to nested and/or reentrant routines).

How startup, shutdown, disable, enable, and service functions are implemented in software usually depends on the following criteria:

- The types, number, and priority levels of interrupts available (determined by the interrupt hardware mechanisms on-chip and on-board).
- How interrupts are triggered.
- The interrupt policies of components within the system that trigger interrupts, and the services provided by the master CPU processing the interrupts.

*Note: The material in the following paragraphs is similar to material found in* Section 4.2.3 *on interrupts.*

The three main types of interrupts are *software*, *internal hardware*, and *external hardware*. Software interrupts are explicitly triggered internally by some instruction within the current instruction stream being executed by the master processor. Internal hardware interrupts, on the other hand, are initiated by an event that is a result of a problem with the current instruction stream that is being executed by the master processor because of the features (or limitations) of the hardware, such as illegal math operations (overflow, divide-by-zero), debugging (single-stepping, breakpoints), and invalid instructions (opcodes). Interrupts that are raised (requested) by some internal event to the master processor (basically, software and internal hardware interrupts) are also commonly referred to as *exceptions* or *traps*. Exceptions are internally generated hardware interrupts triggered by errors that are detected by the master processor during software execution, such as invalid data or a divide by zero. How exceptions are prioritized and processed is determined by the architecture. Traps are software interrupts specifically generated by the software, via an exception instruction. Finally, external hardware interrupts are interrupts initiated by hardware other than the master CPU (board buses, I/O, etc.).

For interrupts that are raised by external events, the master processor is either wired via an input pin(s) called an *IRQ (Interrupt Request Level)* pin or port, to outside intermediary hardware (e.g., interrupt controllers), or directly to other components on the board with

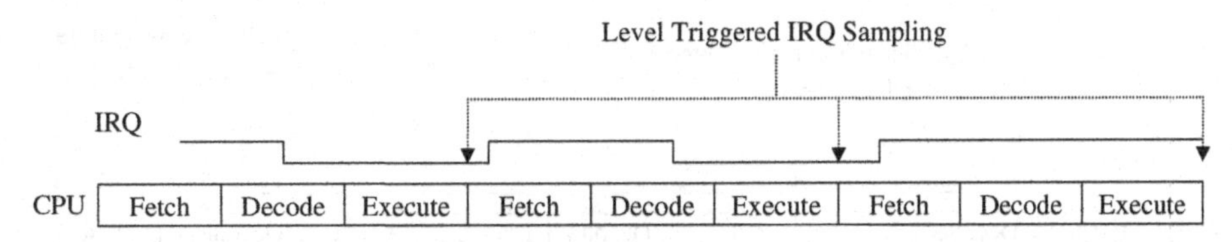


**Figure 8-5a**
Level-Triggered Interrupts.[3]

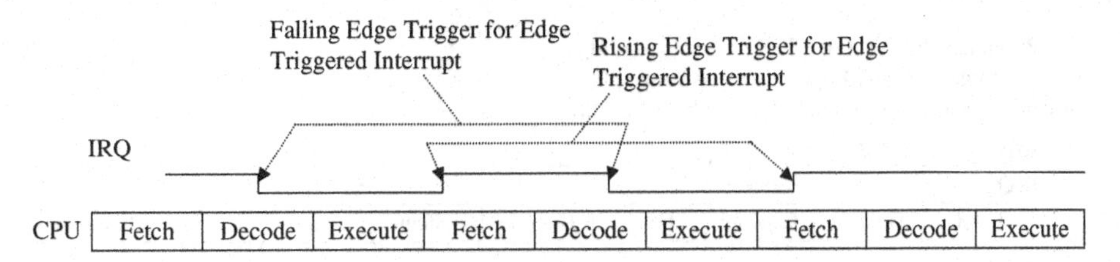


**Figure 8-5b**
Edge-Triggered Interrupts.[3]

dedicated interrupt ports, that signal the master CPU when they want to raise the interrupt. These types of interrupts are triggered in one of two ways: *level-triggered* or *edge-triggered*. A level-triggered interrupt is initiated when its IRQ signal is at a certain level (i.e., HIGH or LOW; see Figure 8-5a). These interrupts are processed when the CPU finds a request for a level-triggered interrupt when sampling its IRQ line, such as at the end of processing each instruction.

Edge-triggered interrupts are triggered when a change occurs on the IRQ line (from LOW to HIGH/rising edge of signal or from HIGH to LOW/falling edge of signal; see Figure 8-5b). Once triggered, these interrupts latch into the CPU until processed.

Both types of interrupts have their strengths and drawbacks. With a level-triggered interrupt, as shown in the example in Figure 8-6a, if the request is being processed and has not been disabled before the next sampling period, the CPU will try to service the same interrupt again. On the flip side, if the level-triggered interrupt were triggered and then disabled before the CPU's sample period, the CPU would never note its existence and would therefore never process it. Edge-triggered interrupts could have problems if they share the same IRQ line, if they were triggered in the same manner at about the same time (say before the CPU could process the first interrupt), resulting in the CPU being able to detect only one of the interrupts (see Figure 8-6b).

Because of these drawbacks, level-triggered interrupts are generally recommended for interrupts that share IRQ lines, whereas edge-triggered interrupts are typically recommended for interrupt signals that are very short or very long.

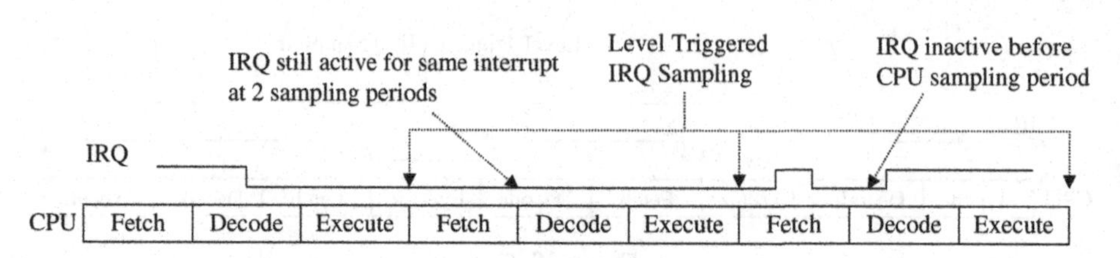


**Figure 8-6a**
Level-Triggered Interrupts Drawbacks.[3]

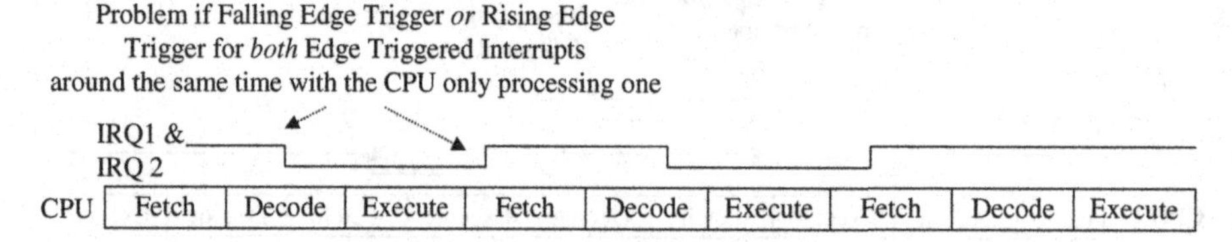


**Figure 8-6b**
Edge-Triggered Interrupts Drawbacks.[3]

At the point an IRQ of a master processor receives a signal that an interrupt has been raised, the interrupt is processed by the interrupt-handling mechanisms within the system. These mechanisms are made up of a combination of both hardware and software components. In terms of hardware, an *interrupt controller* can be integrated onto a board, or within a processor, to mediate interrupt transactions in conjunction with software. Architectures that include an interrupt controller within their interrupt-handling schemes include the 268/386 (x86) architectures, which use two PICs (Intel's Programmable Interrupt Controller); MIPS32, which relies on an external interrupt controller; and the MPC860 (shown in Figure 8-7a), which integrates two interrupt controllers, one in the CPM and one in its SIU. For systems with no interrupt controller, such as the Mitsubishi M37267M8 TV microcontroller shown in Figure 8-7b, the interrupt request lines are connected directly to the master processor, and interrupt transactions are controlled via software and some internal circuitry, such as registers and/or counters.

*Interrupt acknowledgment (IACK)* is typically handled by the master processor when an external device triggers an interrupt. Because IACK cycles are a function of the local bus, the IACK function of the master CPU depends on interrupt policies of system buses, as well as the interrupt policies of components within the system that trigger the interrupts. With respect to the external device triggering an interrupt, the interrupt scheme depends on whether that device can provide an *interrupt vector* (a place in memory that holds the address of an interrupt's *ISR (Interrupt Service Routine)*, the software that the master CPU executes after the triggering of an interrupt). For devices that cannot provide an interrupt vector, referred to as *non-vectored* interrupts, master processors implement an *auto-vectored* interrupt scheme in

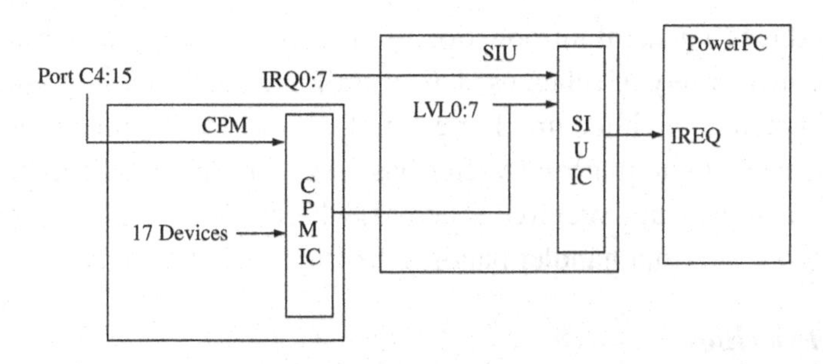


**Figure 8-7a**
Motorola/Freescale MPC860 Interrupt Controllers.[4]
*© Freescale Semiconductor, Inc. Used by permission.*

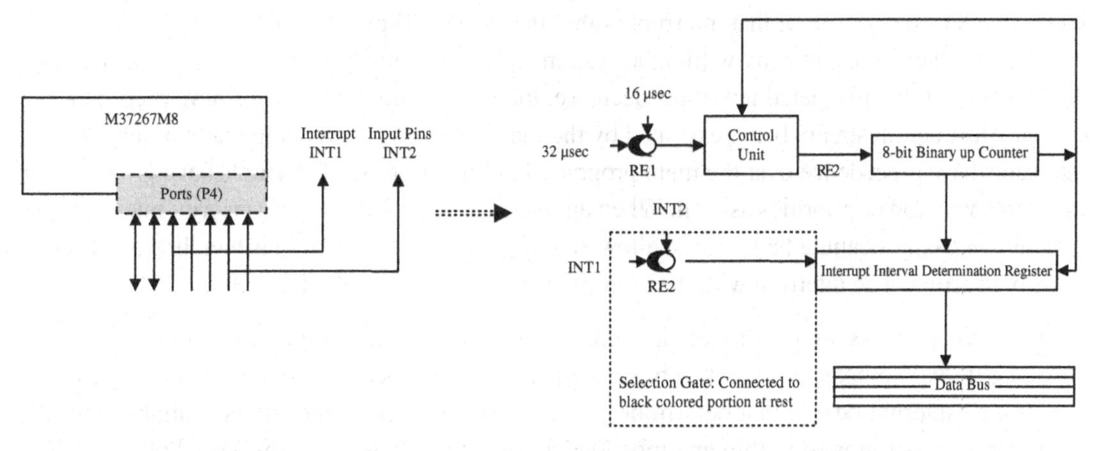


**Figure 8-7b**
Mitsubishi M37267M8 Circuitry.[5]

which one ISR is shared by the non-vectored interrupts; determining which specific interrupt to handle, interrupt acknowledgment, etc., are all handled by the ISR software.

An *interrupt-vectored* scheme is implemented to support peripherals that can provide an interrupt vector over a bus and where acknowledgment is automatic. An IACK-related register on the master CPU informs the device requesting the interrupt to stop requesting interrupt service, and provides what the master processor needs to process the correct interrupt (such as the interrupt number and vector number). Based upon the activation of an external interrupt pin, an interrupt controller's interrupt select register, a device's interrupt select register, or some combination of the above, the master processor can determine which ISR to execute. After the ISR completes, the master processor resets the interrupt status by adjusting the bits in the processor's status register or an interrupt mask in the external interrupt controller. The interrupt request and acknowledgment mechanisms are determined by the device requesting the interrupt (since it determines which interrupt service to trigger), the master processor, and the system bus protocols.

Keep in mind that this is a general introduction to interrupt handling, covering some of the key features found in a variety of schemes. The overall interrupt-handling scheme can vary widely from architecture to architecture. For example, PowerPC architectures implement an auto-vectored scheme, with no interrupt vector base register. The 68000 architecture supports both auto-vectored and interrupt-vectored schemes, whereas MIPS32 architectures have no IACK cycle and so the interrupt handler handles the triggered interrupts.

### 8.1.1 Interrupt Priorities

Because there are potentially multiple components on an embedded board that may need to request interrupts, the scheme that manages all of the different types of interrupts is *priority-based.* This means that all available interrupts within a processor have an associated interrupt level, which is the priority of that interrupt within the system. Typically, interrupts starting at level "1" are the highest priority within the system and incrementally from there (2, 3, 4, etc.) the priorities of the associated interrupts decrease. Interrupts with higher levels have precedence over any instruction stream being executed by the master processor, meaning that not only do interrupts have precedence over the main program, but higher priority interrupts have priority over interrupts with lower priorities as well. When an interrupt is triggered, lower priority interrupts are typically *masked,* meaning they are not allowed to trigger when the system is handling a higher-priority interrupt. The interrupt with the highest priority is usually called an *NMI.*

How the components are prioritized depends on the IRQ line they are connected to, in the case of external devices, or what has been assigned by the processor design. It is the master processor's internal design that determines the number of external interrupts available and the interrupt levels supported within an embedded system. In Figure 8-8a, the MPC860 CPM, SIU, and PowerPC Core all work together to implement interrupts on the MPC823 processor. The CPM allows for internal interrupts (two SCCs, two SMCs, SPI, $I^2C$, PIP, general-purpose timers, two IDMAs, SDMA, RISC Timer) and 12 external pins of port C, and it drives the interrupt levels on the SIU. The SIU receives interrupts from eight external pins (IRQ0–7) and eight internal sources, for a total of 16 sources of interrupts, one of which can be the CPM, and drives the IREQ input to the Core. When the IREQ pin is asserted, external interrupt processing begins. The priority levels are shown in Figure 8-8b.

In another processor, such as the 68000 (shown in Figures 8-9a and b), there are eight levels of interrupts (0–7), where interrupts at level 7 have the highest priority. The 68000 interrupt table (see Figure 8-9b) contains 256 32-bit vectors.

The M37267M8 architecture (shown in Figure 8-10a) allows for interrupts to be caused by 16 events (13 internal, two external, and one software), whose priorities and usages are summarized in Figure 8-10b.

Several different priority schemes are implemented in the various architectures. These schemes commonly fall under one of three models: the *equal single level,* where the latest

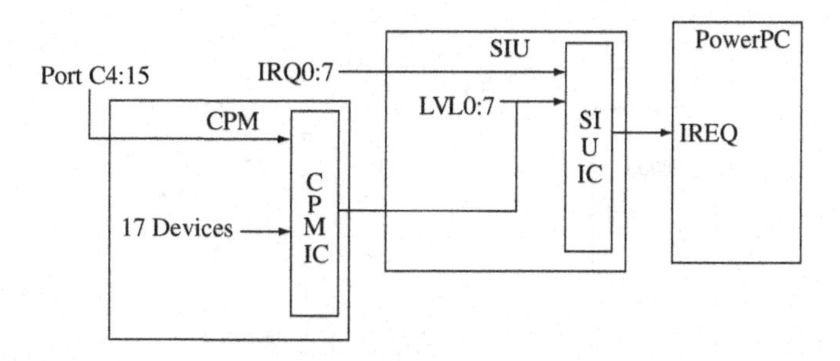


**Figure 8-8a**
Motorola/Freescale MPC860 Interrupt pins and table.[4]
*© Freescale Semiconductor, Inc. Used by permission.*

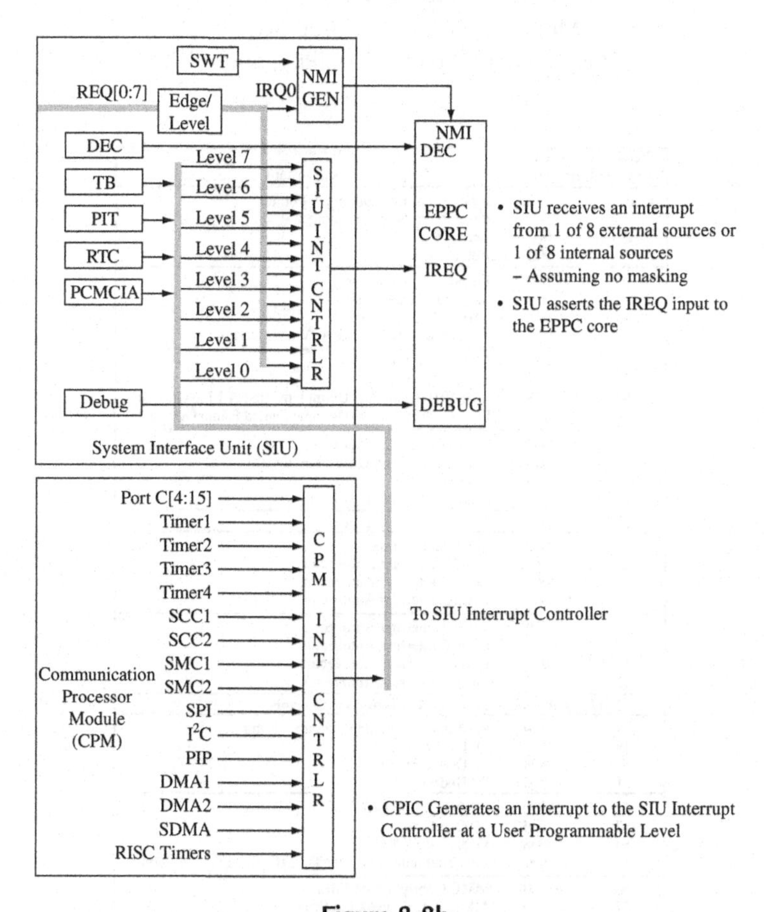


**Figure 8-8b**
Motorola/Freescale MPC860 Interrupt Levels.[4]
*© Freescale Semiconductor, Inc. Used by permission.*

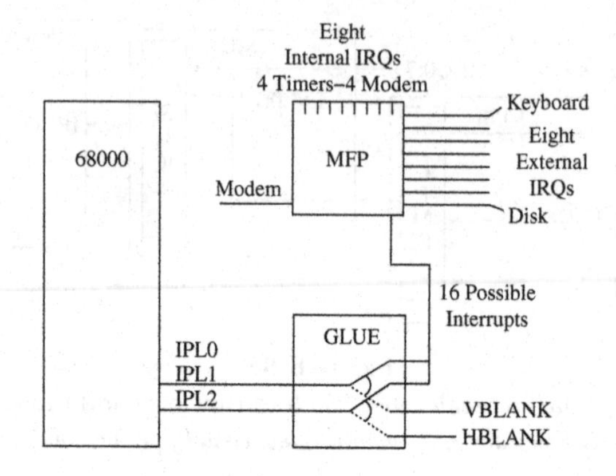


**Figure 8-9a**
Motorola/Freescale 68000 IRQs.[6]
There are 3 IRQ pins: IPL0, IPL1, and IPL2.

| Vector Number[s] | Vector Offset (Hex) | Assignment |
|---|---|---|
| 0 | 000 | Reset Initial Interrupt Stack Pointer |
| 1 | 004 | Reset initial Program Counter |
| 2 | 008 | Access Fault |
| 3 | 00C | Address Error |
| 4 | 010 | Illegal Instruction |
| 5 | 014 | Integer Divide by Zero |
| 6 | 018 | CHK, CHK2 instruction |
| 7 | 01C | FTRAPcc, TRAPcc, TRAPV instructions |
| 8 | 020 | Privilege Violation |
| 9 | 024 | Trace |
| 10 | 028 | Line 1010 Emulator (Unimplemented A-Line Opcode) |
| 11 | 02C | Line 1111 Emulator (Unimplemented F-line Opcode) |
| 12 | 030 | (Unassigned, Reserved) |
| 13 | 034 | Coprocessor Protocol Violation |
| 14 | 038 | Format Error |
| 15 | 03C | Uninitialized Interrupt |
| 16–23 | 040–050 | (Unassigned, Reserved) |
| 24 | 060 | Spurious Interrupt |
| 25 | 064 | Level 1 Interrupt Autovector |
| 26 | 068 | Level 2 Interrupt Autovector |
| 27 | 06C | Level 3 Interrupt Autovector |
| 28 | 070 | Level 4 Interrupt Autovector |
| 29 | 074 | Level 5 Interrupt Autovector |
| 30 | 078 | Level 6 Interrupt Autovector |
| 31 | 07C | Level 7 Interrupt Autovector |
| 32–47 | 080–08C | TRAP #0 D 15 Instructor Vectors |
| 48 | 0C0 | FP Branch or Set on Unordered Condition |
| 49 | 0C4 | FP Inexact Result |
| 50 | 0C8 | FP Divide by Zero |
| 51 | 0CC | FP Underflow |
| 52 | 0D0 | FP Operand Error |
| 53 | 0D4 | FP Overflow |
| 54 | 0D8 | FP Signaling NAN |
| 55 | 0DC | FP Unimplemented Data Type (Defined for MC68040) |
| 56 | 0E0 | MMU Configuration Error |
| 57 | 0E4 | MMU Illegal Operation Error |
| 58 | 0E8 | MMU Access Level Violation Error |
| 59–63 | 0ECD0FC | (Unassigned, Reserved) |
| 64–255 | 100D3FC | User DefinedVectors (192) |

**Figure 8-9b**
Motorola/Freescale 68000 IRQs Interrupt Table.[6]

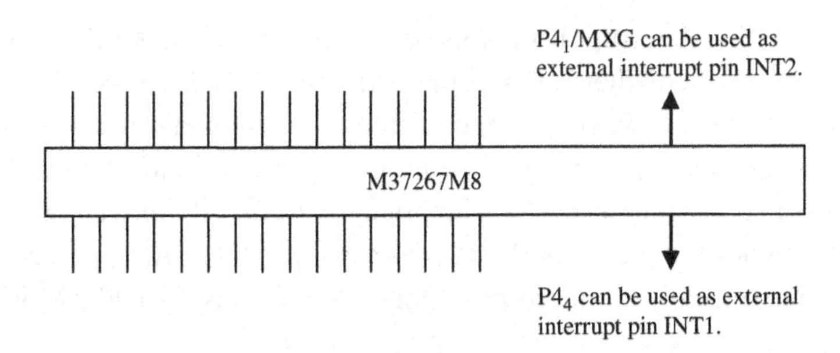
P4$_1$/MXG can be used as external interrupt pin INT2.
M37267M8
P4$_4$ can be used as external interrupt pin INT1.

**Figure 8-10a**
Mitsubishi M37267M8 8-bit TV Microcontroller Interrupts.[5]

| Interrupt Source | Priority | Interrupt Causes |
|---|---|---|
| RESET | 1 | (nonmaskable) |
| CRT | 2 | Occurs after character block display to CRT is completed |
| INT1 | 3 | External Interrupt ** the processor detects that the level of a pin changes from 0 (LOW) to 1 (HIGH), or 1(HIGH) to 0 (LOW) and generates an interrupt request |
| Data Slicer | 4 | Interrupt occurs at end of line specified in caption position register |
| Serial I/O | 5 | Interrupt request from synchronous serial I/O function |
| Timer 4 | 6 | Interrupt generated by overflow of timer 4 |
| Xin & 4096 | 7 | Interrupt occurs regularly with a f(Xin)/4096 period |
| Vsync | 8 | Interrupt request synchronized with the vertical sync signal |
| Timer 3 | 9 | Interrupt generated by overflow of timer 3 |
| Timer 2 | 10 | Interrupt generated by overflow of timer 2 |
| Timer 1 | 11 | Interrupt generated by overflow of timer 1 |
| INT2 | 12 | External Interrupt ** the processor detects that the level of a pin changes from 0 (LOW) to 1 (HIGH), or 1 (HIGH) to 0 (LOW) and generates an interrupt request |
| Multimaster I$^2$C Bus interface | 13 | Related to I$^2$C bus interface |
| Timer 5 & 6 | 14 | Interrupt generated by overflow of timer 5 or 6 |
| BRK instruction | 15 | (nonmaskable software) |

**Figure 8-10b**
Mitsubishi M37267M8 8-bit TV Microcontroller Interrupt table.[5]

interrupt to be triggered gets the CPU; the *static multilevel*, where priorities are assigned by a priority encoder, and the interrupt with the highest priority gets the CPU; and the *dynamic multilevel*, where a priority encoder assigns priorities and the priorities are reassigned when a new interrupt is triggered.

### *8.1.2 Context Switching*

After the hardware mechanisms have determined which interrupt to handle and have acknowledged the interrupt, the current instruction stream is halted and a *context switch* is performed, a process in which the master processor switches from executing the current instruction stream to another set of instructions. This alternate set of instructions being executed as the result of an interrupt is the *ISR* or *interrupt handler*. An ISR is simply a fast, short program that is executed when an interrupt is triggered. The specific ISR executed for a particular interrupt depends on whether a non-vectored or vectored scheme is in place. In the case of a non-vectored interrupt, a memory location contains the start of an ISR that the *PC (program counter)* or some similar mechanism branches to for all non-vectored interrupts. The ISR code then determines the source of the interrupt and provides the appropriate processing. In a vectored scheme, typically an interrupt vector table contains the address of the ISR.

The steps involved in an interrupt context switch include stopping the current program's execution of instructions, saving the context information (registers, the PC, or similar mechanism that indicates where the processor should jump back to after executing the ISR) onto a stack, either dedicated or shared with other system software, and perhaps the disabling of other interrupts. After the master processor finishes executing the ISR, it context switches back to the original instruction stream that had been interrupted, using the context information as a guide.

The *interrupt services* provided by device driver code, based upon the mechanisms discussed above, include *enabling/disabling* interrupts through an interrupt control register on the master CPU or the disabling of the interrupt controller, *connecting* the ISRs to the interrupt table, providing interrupt levels and vector numbers to peripherals, providing address and control data to corresponding registers, etc. Additional services implemented in interrupt access drivers include the *locking/unlocking* of interrupts, and the implementation of the actual ISRs. The pseudocode in the following example shows interrupt handling initialization and access drivers that act as the basis of interrupt services (in the CPM and SIU) on the MPC860.

### 8.1.3 Interrupt Device Driver Pseudocode Examples

The following pseudocode examples demonstrate the implementation of various interrupt-handling routines on the MPC860, specifically startup, shutdown, disable, enable, and interrupt servicing functions in reference to this architecture. These examples show how interrupt handling can be implemented on a more complex architecture like the MPC860, and this in turn can be used as a guide to understand how to write interrupt-handling drivers on other processors that are as complex or less complex than this one.

*Interrupt Handling Startup (Initialization) MPC860*
*Overview of initializing interrupts on MPC860 (in both CPM and SIU)*

1. Initializing CPM Interrupts in MPC860 Example
   - 1.1. Setting Interrupt Priorities via CICR.
   - 1.2. Setting individual enable bit for interrupts via CIMR.
   - 1.3. Initializing SIU Interrupts via SIU Mask Register including setting the SIU bit associated with the level that the CPM uses to assert an interrupt.
   - 1.4. Set Master Enable bit for all CPM interrupts.

2. Initializing SIU Interrupts on MPC860 Example
   2.1. Initializing the SIEL Register to select the edge-triggered or level-triggered interrupt handling for external interrupts and whether processor can exit/wakeup from low power mode.
   2.2. If not done, initializing SIU Interrupts via SIU Mask Register including setting the SIU bit associated with the level that the CPM uses to assert an interrupt.

** Enabling all interrupts via MPC860 "mtspr" instruction next step—see Interrupt Handling Enable **

```
// Initializing CPM for interrupts - four-step process
// ***** step 1 *****
// initializing the 24-bit CICR (see Figure 8-11), setting priorities and the
interrupt
// levels. Interrupt Request Level, or IRL[0:2] allows a user to program the
priority
// request level of the CPM interrupt with any number from level 0 (highest
priority)
// through level 7 (lowest priority).
```

```
...
int RESERVED94 = 0xFF000000;    // bits 0-7 reserved, all set to 1

// the PowerPC SCCs are prioritized relative to each other. Each SCxP field is
representative
// of a priority for each SCC where SCdP is the lowest and ScaP is the highest
priority.
// Each SCxP field is made up of 2 bits (0-3), one for each SCC, where 0d (00b) = SCC1,
// 1d (01b) = SCC2, 2d (10b) = SCC3, and 3d (11b) = SCC4. See Figure 8-11b.

int CICR.SCdP = 0x00C00000;    // bits 8-9 both = 1, SCC4 = lowest priority
int CICR.SCcP = 0x00000000;    // bits 10-11, both = 0, SCC1 = 2nd to lowest
                                  priority
int CICR.SCbP = 0x00040000;    // bits 12-13,=01b, SCC2 2nd highest priority
int CICR.SCaP = 0x00020000;    // bits 14-15,=10b, SCC3 highest priority

// IRL0_IRL2 is a 3-bit configuration parameter called the Interrupt Request
Level - it
// allows a user to program the priority request level of the CPM interrupt with
bits
// 16-18 with a value of 0-7 in terms of its priority mapping within the SIU.
In this
// example, it is a priority 7 since all 3 bits set to 1.
int CICR.IRL0 = 0x00008000;    // interrupt request level 0 (bit 16)=1
int CICR.IRL1 = 0x00004000;    // interrupt request level 1 (bit 17)=1
int CICR.IRL2 = 0x00002000;    // interrupt request level 2 (bit 18)=1
```

```
// HP0-HP 4 are five bits (19-23) used to represent one of the CPM Interrupt
Controller
// interrupt sources (shown in Figure 8-8b) as being the highest priority source
relative to
// their bit location in the CIPR register - see Figure 8-11c. In this example,
HP0-HP4
//=11111b (31d) so highest external priority source to the PowerPC core is PC15
int CICR.HP0 = 0x00001000;     /* Highest priority */
int CICR.HP1 = 0x00000800;     /* Highest priority */
int CICR.HP2 = 0x00000400;     /* Highest priority */
int CICR.HP3 = 0x00000200;     /* Highest priority */
int CICR.HP4 = 0x00000100;     /* Highest priority */

// IEN bit 24 - Master enable for CPM interrupts - not enabled here - see step 4

int RESERVED95 = 0x0000007E;  // bits 25-30 reserved, all set to 1

int CICR.SPS = 0x00000001;     // Spread priority scheme in which SCCs are spread
                               // out by priority in interrupt table, rather than
                               grouped
                               // by priority at the top of the table
```

CICR – CPM Interrupt Configuration Register

| 0 | 1 | 2 | 3 | 4 | 5 | 6 | 7 | 8 | 9 | 10 | 11 | 12 | 13 | 14 | 15 |
|---|---|---|---|---|---|---|---|---|---|---|---|---|---|---|---|
| | | | | | | | | SCdP | | SCcP | | SCbP | | SCaP | |

| 16 | 17 | 18 | 19 | 20 | 21 | 22 | 23 | 24 | 25 | 26 | 27 | 28 | 29 | 30 | 31 |
|---|---|---|---|---|---|---|---|---|---|---|---|---|---|---|---|
| IRL0_IRL2 | | | HP0_HP4 | | | | | IEN | - | | | | | | SPS |

**Figure 8-11a**
CICR Register.[2]

| | | Highest | | Lowest | |
|---|---|---|---|---|---|
| SCC | Code | SCaP | SCbP | SCcP | SCdP |
| SCC1 | 00 | | | 00 | |
| SCC2 | 01 | | 01 | | |
| SCC3 | 10 | 10 | | | |
| SCC4 | 11 | | | | 11 |

**Figure 8-11b**
SCC Priorities.[2]

CIPR - CPM Interrupt Pending Register

| 0 | 1 | 2 | 3 | 4 | 5 | 6 | 7 | 8 | 9 | 10 | 11 | 12 | 13 | 14 | 15 |
|---|---|---|---|---|---|---|---|---|---|---|---|---|---|---|---|
| PC15 | SCC1 | SCC2 | SCC3 | SCC4 | PC14 | Timer 1 | PC13 | PC12 | SDMA | IDMA 1 | IDMA 2 | - | Timer 2 | R_TT | I2C |

| 16 | 17 | 18 | 19 | 20 | 21 | 22 | 23 | 24 | 25 | 26 | 27 | 28 | 29 | 30 | 31 |
|---|---|---|---|---|---|---|---|---|---|---|---|---|---|---|---|
| PC11 | PC10 | - | Timer 3 | PC9 | PC8 | PC7 | - | Timer 4 | PC6 | SPI | SMC1 | SMC2 /PIP | PC5 | PC4 | - |

**Figure 8-11c**
CIPR Register.[2]

CIPR - CPM Interrupt Mask Register

| 0 | 1 | 2 | 3 | 4 | 5 | 6 | 7 | 8 | 9 | 10 | 11 | 12 | 13 | 14 | 15 |
|---|---|---|---|---|---|---|---|---|---|---|---|---|---|---|---|
| PC15 | SCC1 | SCC2 | SCC3 | SCC4 | PC14 | Timer 1 | PC13 | PC12 | SDMA | IDMA 1 | IDMA 2 | - | Timer 2 | R_TT | I2C |

| 16 | 17 | 18 | 19 | 20 | 21 | 22 | 23 | 24 | 25 | 26 | 27 | 28 | 29 | 30 | 31 |
|---|---|---|---|---|---|---|---|---|---|---|---|---|---|---|---|
| PC11 | PC10 | - | Timer 3 | PC9 | PC8 | PC7 | - | Timer 4 | PC6 | SPI | SMC1 | SMC2 /PIP | PC5 | PC4 | - |

**Figure 8-12**
CIMR Register.[2]

```
// ***** step 2 *****
// initializing the 32-bit CIMR (see Figure 8-12), CIMR bits correspond to CMP
// Interrupt Sources indicated in CIPR (see Figure 8-11c), by setting the bits
// associated with the desired interrupt sources in the CIMR register (each bit
// corresponds to a CPM interrupt source).
```

```
int CIMR.PC15 = 0x80000000;          // PC15 (Bit 0) set to 1, interrupt source
                                     enabled
int CIMR.SCC1 = 0x40000000;          // SCC1 (Bit 1) set to 1, interrupt source
                                     enabled
int CIMR.SCC2 = 0x20000000;          // SCC2 (Bit 2) set to 1, interrupt source
                                     enabled
int CIMR.SCC4 = 0x08000000;          // SCC4 (Bit 4) set to 1, interrupt source
                                     enabled
int CIMR.PC14 = 0x04000000;          // PC14 (Bit 5) set to 1, interrupt source
                                     enabled
int CIMR.TIMER1 = 0x02000000;        // Timer1 (Bit 6) set to 1, interrupt source
                                     enabled
int CIMR.PC13 = 0x01000000;          // PC13 (Bit 7) set to 1, interrupt source
                                     enabled
int CIMR.PC12 = 0x00800000;          // PC12 (Bit 8) set to 1, interrupt source
                                     enabled
int CIMR.SDMA = 0x00400000;          // SDMA (Bit 9) set to 1, interrupt source
                                     enabled
int CIMR.IDMA1 = 0x00200000;          // IDMA1 (Bit 10) set to 1, interrupt source
                                      enabled
int CIMR.IDMA2 = 0x00100000;         // IDMA2 (Bit 11) set to 1, interrupt source
                                     enabled
int RESERVED100 = 0x00080000;        // unused bit 12
int CIMR.TIMER2 = 0x00040000;        // Timer2 (Bit 13) set to 1, interrupt source
                                     enabled
int CIMR.R.TT = 0x00020000;          // R-TT (Bit 14) set to 1, interrupt source
                                     enabled
int CIMR.I2C = 0x00010000;           // I2C (Bit 15) set to 1, interrupt source
                                     enabled
int CIMR.PC11 = 0x00008000;          // PC11 (Bit 16) set to 1, interrupt source
                                     enabled
```

```
int CIMR.PC10 = 0x00004000;         // PC10 (Bit 17) set to 1, interrupt source
                                    enabled
int RESERVED101 = 0x00002000;       // unused bit 18
int CIMR.TIMER3 = 0x00001000;       // Timer3 (Bit 19) set to 1, interrupt source
                                    enabled
int CIMR.PC9 = 0x00000800;          // PC9 (Bit 20) set to 1, interrupt source
                                    enabled
int CIMR.PC8 = 0x00000400;          // PC8 (Bit 21) set to 1, interrupt source
                                    enabled
int CIMR.PC7 = 0x00000200;          // PC7 (Bit 22) set to 1, interrupt source
                                    enabled
int RESERVED102 = 0x00000100;       // unused bit 23
int CIMR.TIMER4 = 0x00000080;       // Timer4 (Bit 24) set to 1, interrupt source
                                    enabled
int CIMR.PC6 = 0x00000040;          // PC6 (Bit 25) set to 1, interrupt source
                                    enabled
int CIMR.SPI = 0x00000020;          // SPI (Bit 26) set to 1, interrupt source
                                    enabled
int CIMR.SMC1 = 0x00000010;         // SMC1 (Bit 27) set to 1, interrupt source
                                    enabled
int CIMR.SMC2-PIP = 0x00000008;     // SMC2/PIP (Bit 28) set to 1, interrupt source
                                    enabled

int CIMR.PC5 = 0x00000004;          // PC5 (Bit 29) set to 1, interrupt source
                                    enabled
int CIMR.PC4 = 0x00000002;          // PC4 (Bit 30) set to 1, interrupt source
                                    enabled
int RESERVED103 = 0x00000001;       // unused bit 31
```

```
// ***** step 3 *****
// Initializing the SIU Interrupt Mask Register (see Figure 8-13) including setting
the SIU
// bit associated with the level that the CPM uses to assert an interrupt.
```

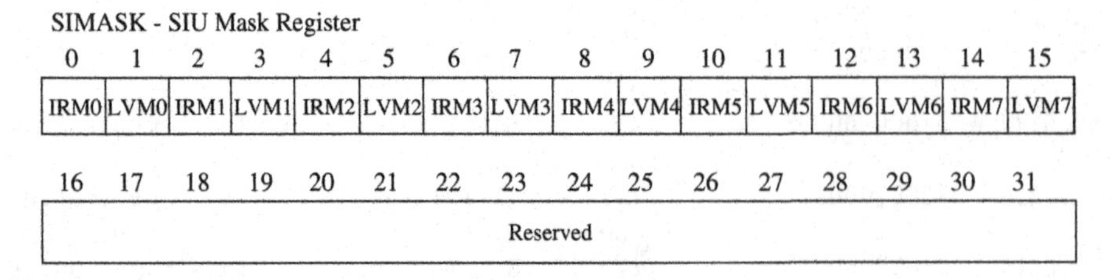


**Figure 8-13**
SIMASK Register.[2]

```
int SIMASK.IRM0 = 0x80000000;     // enable external interrupt input level 0
int SIMASK.LVM0 = 0x40000000;     // enable internal interrupt input level 0
int SIMASK.IRM1 = 0x20000000;     // enable external interrupt input level 1
int SIMASK.LVM1 = 0x10000000;     // enable internal interrupt input level 1
int SIMASK.IRM2 = 0x08000000;     // enable external interrupt input level 2
int SIMASK.LVM2 = 0x04000000;     // enable internal interrupt input level 2
int SIMASK.IRM3 = 0x02000000;     // enable external interrupt input level 3
int SIMASK.LVM3 = 0x01000000;     // enable internal interrupt input level 3
int SIMASK.IRM4 = 0x00800000;     // enable external interrupt input level 4
int SIMASK.LVM4 = 0x00400000;     // enable internal interrupt input level 4
int SIMASK.IRM5 = 0x00200000;     // enable external interrupt input level 5
int SIMASK.LVM5 = 0x00100000;     // enable internal interrupt input level 5
int SIMASK.IRM6 = 0x00080000;     // enable external interrupt input level 6
int SIMASK.LVM6 = 0x00040000;     // enable internal interrupt input level 6
int SIMASK.IRM7 = 0x00020000;     // enable external interrupt input level 7
int SIMASK.LVM7 = 0x00010000;     // enable internal interrupt input level 7
int RESERVED6 = 0x0000FFFF;       // unused bits 16-31
```

```
// ***** step 4 *****
```

```
// IEN bit 24 of CICR register - Master enable for CPM interrupts
int CICR.IEN = 0x00000080;     // interrupts enabled IEN = 1
```

```
// Initializing SIU for interrupts - two-step process

// ***** step 1 *****
// Initializing the SIEL Register (see Figure 8-14) to select the edge-triggered (set
to 1
// for falling edge indicating interrupt request) or level-triggered (set to 0 for a 0
logic
// level indicating interrupt request) interrupt handling for external interrupts
(bits
// 0, 2, 4, 6, 8, 10, 12, 14) and whether processor can exit/wakeup from low power
mode
// (bits 1, 3, 5, 7, 9, 11, 13, 15). Set to 0 is NO, set to 1 is Yes
```

SIEL - SIU Interrupt Edge Level Mask Register

| 0 | 1 | 2 | 3 | 4 | 5 | 6 | 7 | 8 | 9 | 10 | 11 | 12 | 13 | 14 | 15 |
|---|---|---|---|---|---|---|---|---|---|---|---|---|---|---|---|
| ED0 | WM0 | ED1 | WM1 | ED2 | WM2 | ED3 | WM3 | ED4 | WM4 | ED5 | WM5 | ED6 | WM6 | ED7 | WM7 |

| 16 | 17 | 18 | 19 | 20 | 21 | 22 | 23 | 24 | 25 | 26 | 27 | 28 | 29 | 30 | 31 |
|---|---|---|---|---|---|---|---|---|---|---|---|---|---|---|---|
| Reserved | | | | | | | | | | | | | | | |

**Figure 8-14**
SIEL Register.[2]

```
int SIEL.ED0 = 0x80000000;      // interrupt level 0 (falling) edge-triggered
int SIEL.WM0 = 0x40000000;      // IRQ at interrupt level 0 allows CPU to exit from
                                low
                                // power mode
int SIEL.ED1 = 0x20000000;      // interrupt level 1 (falling) edge-triggered
int SIEL.WM1 = 0x10000000;      // IRQ at interrupt level 1 allows CPU to exit from
                                low
                                // power mode
int SIEL.ED2 = 0x08000000;      // interrupt level 2 (falling) edge-triggered
int SIEL.WM2 = 0x04000000;      // IRQ at interrupt level 2 allows CPU to exit from
                                low
                                // power mode
int SIEL.ED3 = 0x02000000;      // interrupt level 3 (falling) edge-triggered
int SIEL.WM3 = 0x01000000;      // IRQ at interrupt level 3 allows CPU to exit from
                                low
                                // power mode
int SIEL.ED4 = 0x00800000;      // interrupt level 4 (falling) edge-triggered
int SIEL.WM4 = 0x00400000;      // IRQ at interrupt level 4 allows CPU to exit from
                                low
                                // power mode
int SIEL.ED5 = 0x00200000;      // interrupt level 5 (falling) edge-triggered
int SIEL.WM5 = 0x00100000;      // IRQ at interrupt level 5 allows CPU to exit from
                                low
                                // power mode
int SIEL.ED6 = 0x00080000;      // interrupt level 6 (falling) edge-triggered
int SIEL.WM6 = 0x00040000;      // IRQ at interrupt level 6 allows CPU to exit from
                                low
                                // power mode
int SIEL.ED7 = 0x00020000;      // interrupt level 7 (falling) edge-triggered
int SIEL.WM7 = 0x00010000;      // IRQ at interrupt level 7 allows CPU to exit from
                                low
                                // power mode
int RESERVED7 = 0x0000FFFF;     // bits 16-31 unused
```

```
// ***** step 2 *****
// Initializing SIMASK register - done in step 3 of initializing CPM.
```

### *Interrupt Handling Shutdown on MPC860*

There essentially is no shutdown process for interrupt handling on the MPC860, other than perhaps disabling interrupts during the process.

```
// Essentially disabling all interrupts via IEN bit 24 of CICR - Master disable for
CPM
// interrupts
CICR.IEN="CICR.IEN" AND "0";    // interrupts disabled IEN = 0
```

*Interrupt Handling Disable on MPC860*

```
// To disable specific interrupt means modifying the SIMASK, so disabling the
external
// interrupt at level 7 (IRQ7) for example is done by clearing bit 14
SIMASK.IRM7="SIMASK.IRM7" AND "0";    // disable external interrupt input level 7

// disabling of all interrupts takes effect with the mtspr instruction.
mtspr 82,0;                           // disable interrupts via mtspr (move to
                                      special purpose register)
   // instruction
```

*Interrupt Handling Enable on MPC860*

```
// specific enabling of particular interrupts done in initialization section of
this example -
// so the interrupt enable of all interrupts takes effect with the mtspr
instruction.

mtspr 80,0;                          // enable interrupts via mtspr (move to
special purpose
                                     // register) instruction

// in review, to enable specific interrupt means modifying the SIMASK, so enabling
the
// external interrupt at level 7 (IRQ7) for example is done by setting bit 14
SIMASK.IRM7="SIMASK.IRM7" OR "1";    // enable external interrupt input level 7
```

*Interrupt Handling Servicing on MPC860*

In general, this ISR (and most ISRs) essentially disables interrupts first, saves the context information, processes the interrupt, restores the context information, and then enables interrupts.

```
InterruptServiceRoutineExample ()
{
   …
   // disable interrupts
   disableInterrupts();          // mtspr 82,0;
   // save registers
   saveState();
   // read which interrupt from SI Vector Register (SIVEC)
   interruptCode = SIVEC.IC;

   // if IRQ 7 then execute
   if (interruptCode = IRQ7) {
   …
```

```
    // If an IRQx is edge-triggered, then clear the service bit in the SI Pending
    Register
    // by putting a "1".
    SIPEND.IRQ7 = SIPEND.IRQ7 OR "1";
    // main process
    ...
    }                                      // endif IRQ7

    // restore registers
    restoreState();
    // re-enable interrupts
    enableInterrupts();     // mtspr 80,0;
}
```

### 8.1.4 Interrupt Handling and Performance

The performance of an embedded design is affected by the *latencies* (delays) involved with the interrupt-handling scheme. The interrupt *latency* is essentially the time from when an interrupt is triggered until its ISR starts executing. The master CPU, under normal circumstances, accounts for a lot of overhead for the time it takes to process the interrupt request and acknowledge the interrupt, obtaining an interrupt vector (in a vectored scheme), and context switching to the ISR. In the case when a lower-priority interrupt is triggered during the processing of a higher priority interrupt, or a higher priority interrupt is triggered during the processing of a lower priority interrupt, the interrupt latency for the original lower priority interrupt increases to include the time in which the higher priority interrupt is handled (essentially how long the lower priority interrupt is disabled). Figure 8-15 summarizes the variables that impact interrupt latency.

Within the ISR itself, additional overhead is caused by the context information being stored at the start of the ISR and retrieved at the end of the ISR. The time to context switch back to the original instruction stream that the CPU was executing before the interrupt was triggered also adds to the overall interrupt execution time. While the hardware aspects of interrupt handling (the context switching, processing interrupt requests, etc.) are beyond the software's control, the overhead related to when the context information is saved, as well as how the ISR is written both in terms of the programming language used and the size, are under the software's control. Smaller ISRs, or ISRs written in a lower-level language like assembly, as opposed to larger ISRs or ISRs written in higher-level languages like Java, or saving/retrieving less context information at the start and end of an ISR, can all decrease the interrupt handling execution time and increase performance.

## 8.2 Example 2: Memory Device Drivers

While in reality all types of physical memory are two-dimensional arrays (matrices) made up of cells addressed by a unique row and column, the master processor and programmers view

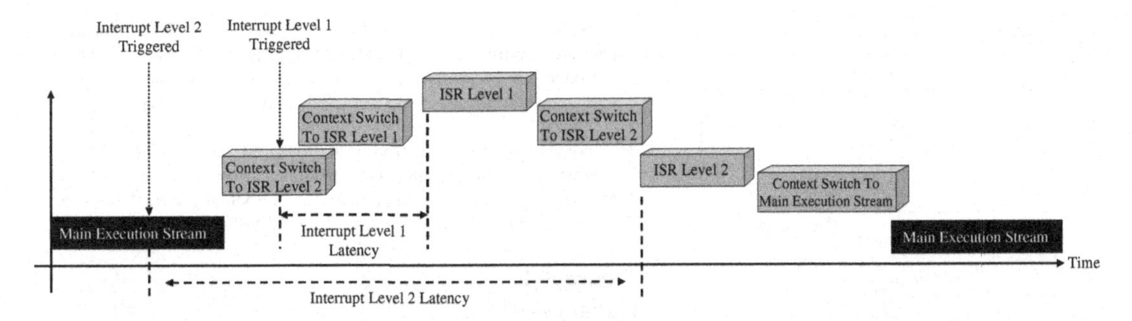


**Figure 8-15**
Interrupt Latency.

| Address Range | Accessed Device | Port Width |
|---|---|---|
| 0x00000000 - 0x003FFFFF | Flash PROM Bank 1 | 32 |
| 0x00400000 - 0x007FFFFF | Flash PROM Bank 2 | 32 |
| 0x04000000 - 0x043FFFFF | DRAM 4 Mbyte (1Meg × 32-bit) | 32 |
| 0x09000000 - 0x09003FFF | MPC Internal Memory Map | 32 |
| 0x09100000 - 0x09100003 | BCSR - Board Control & Status Register | 32 |
| 0x10000000 - 0x17FFFFFF | PCMCIA Channel | 16 |

**Figure 8-16**
Sample Memory Map.[4]

memory as a large one-dimensional array, commonly referred to as the *memory map* (see Figure 8-16). In the memory map, each cell of the array is a row of bytes (8 bits) and the number of bytes per row depends on the width of the data bus (8-, 16-, 32-, 64-bit, etc.). This, in turn, depends on the width of the registers of the master architecture. When physical memory is referenced from the software's point-of-view it is commonly referred to as *logical* memory and its most basic unit is the byte. Logical memory is made up of all the physical memory (registers, ROM, and RAM) in the entire embedded system.

The software must provide the processors in the system with the ability to access various portions of the memory map. The software involved in managing the memory on the master processor and on the board, as well as managing memory hardware mechanisms, consists of the *device drivers* for the management of the overall memory subsystem. The memory subsystem includes all types of memory management components, such as memory controllers and MMU, as well as the types of memory in the memory map, such as registers, cache, ROM, and DRAM. All or some combination of six of the 10 device driver functions from the list of device driver functionality introduced at the start of this chapter are commonly implemented, including:

- *Memory Subsystem Startup*: initialization of the hardware upon PowerON or reset (initialize translation lookaside buffers (TLBs) for MMU, initialize/configure MMU).
- *Memory Subsystem Shutdown*: configuring hardware into its PowerOFF state. *(Note: Under the MPC860, there is no necessary shutdown sequence for the memory subsystem, so pseudocode examples are not shown.)*

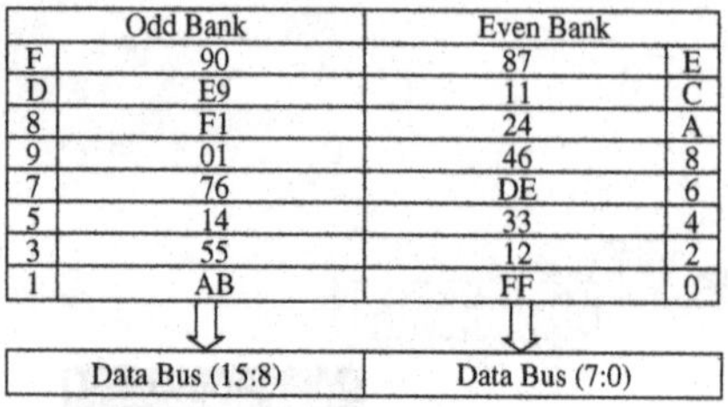

| | Odd Bank | Even Bank | |
|---|---|---|---|
| F | 90 | 87 | E |
| D | E9 | 11 | C |
| 8 | F1 | 24 | A |
| 9 | 01 | 46 | 8 |
| 7 | 76 | DE | 6 |
| 5 | 14 | 33 | 4 |
| 3 | 55 | 12 | 2 |
| 1 | AB | FF | 0 |
| | ⇩ | ⇩ | |
| | Data Bus (15:8) | Data Bus (7:0) | |

In little-endian mode if a byte is read from address "0", an "FF" is returned; if 2 bytes are read from address 0, then (reading from the lowest byte which is furthest to the LEFT in little-endian mode) an "ABFF" is returned. If 4 bytes (32-bits) are read from address 0, then a "5512ABFF" is returned.

In big-endian mode if a byte is read from address "0", an "FF" is returned; if 2 bytes are read from address 0, then (reading from the lowest byte which is furthest to the RIGHT in big-endian mode) an "FFAB" is returned. If 4 bytes (32-bits) are read from address 0, then a "1255FFAB" is returned.

**Figure 8-17**
Endianess.[4]

- *Memory Subsystem Disable*: allowing other software to disable hardware on-the-fly (disabling cache).
- *Memory Subsystem Enable*: allowing other software to enable hardware on-the-fly (enable cache).
- *Memory Subsystem Write*: storing in memory a byte or set of bytes (i.e., in cache, ROM, and main memory).
- *Memory Subsystem Read*: retrieving from memory a "copy" of the data in the form of a byte or set of bytes (i.e., in cache, ROM, and main memory).

Regardless of what type of data is being read or written, all data within memory is managed as a sequence of bytes. While one memory access is limited to the size of the data bus, certain architectures manage access to larger *blocks* (a contiguous set of bytes) of data, called *segments*, and thus implement a more complex address translation scheme in which the logical address provided via software is made up of a *segment number* (address of start of segment) and *offset* (within a segment) which is used to determine the physical address of the memory location.

The order in which bytes are retrieved or stored in memory depends on the *byte ordering* scheme of an architecture. The two possible byte ordering schemes are *little-endian* and *big-endian*. In little-endian mode, bytes (or "bits" with 1-byte (8-bit) schemes) are retrieved and stored in the order of the lowest byte first, meaning the lowest byte is furthest to the left. In big-endian mode bytes are accessed in the order of the highest byte first, meaning that the lowest byte is furthest to the right (see Figure 8-17).

What is important regarding memory and byte ordering is that performance can be greatly impacted if data requested isn't aligned in memory according to the byte ordering scheme defined by the architecture. As shown in Figure 8-17, memory is either soldered into or plugged into an area on the embedded board, called memory *banks*. While the configuration and number of banks can vary from platform to platform, memory addresses are aligned in an odd or even bank format. If data is aligned in little-endian mode, data taken from address "0" in an even bank is "ABFF," and as such is an aligned memory access. So, given a 16-bit data bus, only one memory access is needed. But if data were to be taken from address "1" (an odd

| Address Range | Accessed Device | Port Width |
|---|---|---|
| 0x00000000 - 0x003FFFFF | Flash PROM Bank 1 | 32 |
| 0x00400000 - 0x007FFFFF | Flash PROM Bank 2 | 32 |
| 0x04000000 - 0x043FFFFF | DRAM 4 Mbyte (1Meg x 32-bit) | 32 |
| 0x09000000 - 0x09003FFF | MPC Internal Memory Map | 32 |
| 0x09100000 - 0x09100003 | BCSR - Board Control & Status Register | 32 |
| 0x10000000 - 0x17FFFFFF | PCMCIA Channel | 16 |

**Figure 8-18**
Sample Memory Map.[4]

bank) in a memory aligned as shown in Figure 8-17, the little-endian ordering scheme should retrieve "12AB" data. This would require two memory accesses, one to read the AB, the odd byte, and one to read "12," the even byte, as well as some mechanism within the processor or in driver code to perform additional work to align them as "12AB." Accessing data in memory that is aligned according to the byte ordering scheme can result in access times at least twice as fast.

Finally, how memory is actually accessed by the software will, in the end, depend on the programming language used to write the software. For example, assembly language has various architecture-specific addressing modes that are unique to an architecture and Java allows modifications of memory through objects.

### *8.2.1 Memory Management Device Driver Pseudocode Examples*

The following pseudocode demonstrates implementation of various memory management routines on the MPC860, specifically startup, disable, enable, and writing/erasing functions in reference to the architecture. These examples demonstrate how memory management can be implemented on a more complex architecture, and this in turn can serve as a guide to understanding how to write memory management drivers on other processors that are as complex or less complex than the MPC860 architecture.

#### *Memory Subsystem Startup (Initialization) on MPC860*

In the sample memory map in Figure 8-18, the first two banks are 8 MB of Flash, then 4 MB of DRAM, followed by 1 MB for the internal memory map and control/status registers. The remainder of the map represents 4 MB of an additional PCMCIA card. The main memory subsystem components that are initialized in this example are the physical memory chips themselves (i.e., Flash, DRAM) which in the case of the MPC860 are initialized via a memory controller, configuring the internal memory map (registers and dual-port RAM), as well as configuring the MMU.

##### Initializing the Memory Controller and Connected ROM/RAM

The MPC860 memory controller (shown in Figure 8-19) is responsible for the control of up to eight memory banks, interfacing to SRAM, EPROM, Flash EPROM, various DRAM devices,

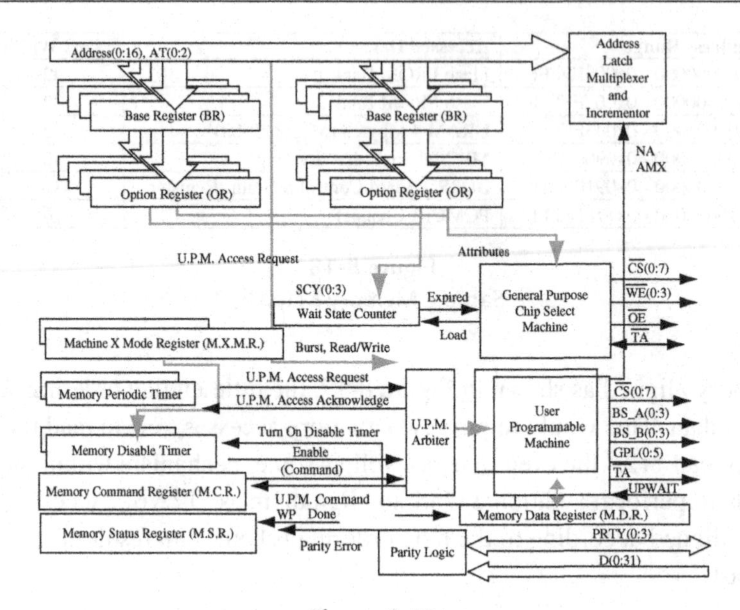


**Figure 8-19**
MPC860 Integrated Memory Controller.[4]
*© Freescale Semiconductor, Inc. Used by permission.*

and other peripherals (i.e., PCMCIA). Thus, in this example of the MPC860, on-board memory (Flash, SRAM, DRAM, etc.) is initialized by initializing the memory controller.

The memory controller has two different types of subunits, the general-purpose chip-select machine (GPCM) and the user-programmable machines (UPMs); these subunits exist to connect to certain types of memory. The GPCM is designed to interface to SRAM, EPROM, Flash EPROM, and other peripherals (such as PCMCIA), whereas the UPMs are designed to interface to a wide variety of memory, including DRAMs. The pinouts of the MPC860's memory controller reflect the different signals that connect these subunits to the various types of memory (see Figures 8-20a–c). For every chip select (CS), there is an associated memory bank.

With every new access request to external memory, the memory controller determines whether the associated address falls into one of the eight address ranges (one for each bank) defined by the eight base registers (which specify the start address of each bank) and option registers (which specify the bank length) pairs (see Figure 8-21). If it does, the memory access is processed by either the GPCM or one of the UPMs, depending on the type of memory located in the memory bank that contains the desired address.

Because each memory bank has a pair of base and option registers (BR0/OR0–BR7/OR7), they need to be configured in the memory controller initialization drivers. The base register (BR) fields are made up of a 16-bit start address BA (bits 0-16); AT (bits 17–19)

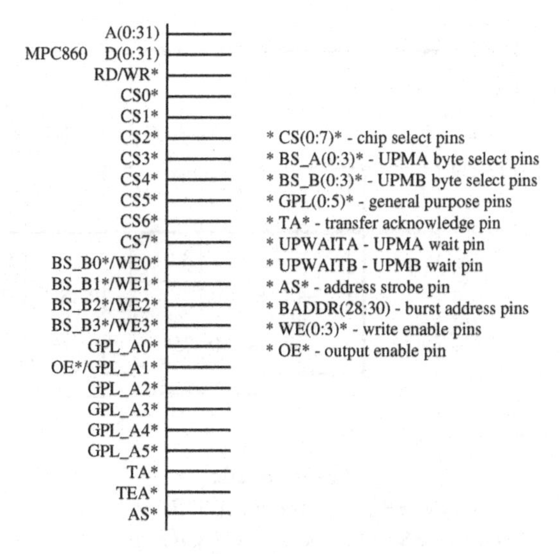


**Figure 8-20a**
Memory Controller Pins.[4]
*© Freescale Semiconductor, Inc. Used by permission.*

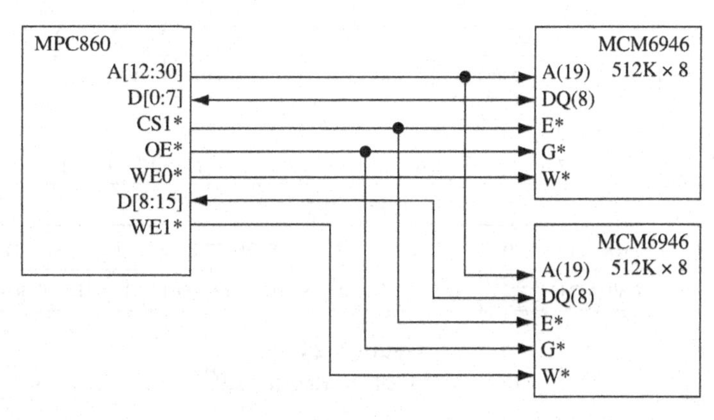


**Figure 8-20b**
PowerPC Connected to SRAM.[4]
*© Freescale Semiconductor, Inc. Used by permission.*

specifies the *address type* (allows sections of memory space to be limited to only one particular type of data), a port size (8-, 16-, 32-bit); a parity checking bit; a bit to write protect the bank (allowing for read-only or read/write access to data); a memory controller machine selection set of bits (for GPCM or one of the UPMs); and a bit indicating if the bank is valid. The option register (OR) fields are made up of bits of control information for configuring the GPCM and UPMs accessing and addressing scheme (burst accesses, masking, multiplexing, etc.).

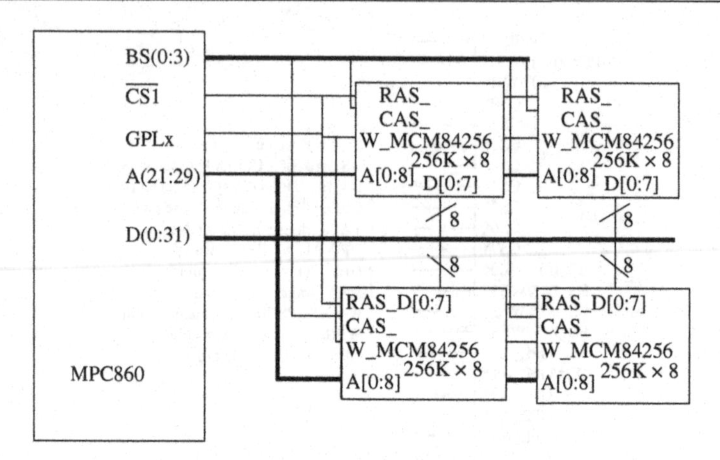


**Figure 8-20c**
PowerPC Connected to DRAM.[4]
*© Freescale Semiconductor, Inc. Used by permission.*

BRx - Base Register

| 0 | 1 | 2 | 3 | 4 | 5 | 6 | 7 | 8 | 9 | 10 | 11 | 12 | 13 | 14 | 15 |
|---|---|---|---|---|---|---|---|---|---|---|---|---|---|---|---|
| BA0 – BA15 | | | | | | | | | | | | | | | |

| 16 | 17 | 18 | 19 | 20 | 21 | 22 | 23 | 24 | 25 | 26 | 27 | 28 | 29 | 30 | 31 |
|---|---|---|---|---|---|---|---|---|---|---|---|---|---|---|---|

| BA16 | AT0_AT2 | PS0_PS1 | PARE | WP | MS0_MS1 | Reserved | V |
|---|---|---|---|---|---|---|---|

ORx - Option Register

| 0 | 1 | 2 | 3 | 4 | 5 | 6 | 7 | 8 | 9 | 10 | 11 | 12 | 13 | 14 | 15 |
|---|---|---|---|---|---|---|---|---|---|---|---|---|---|---|---|
| AM0 – AM15 | | | | | | | | | | | | | | | |

| 16 | 17 | 18 | 19 | 20 | 21 | 22 | 23 | 24 | 25 | 26 | 27 | 28 | 29 | 30 | 31 |
|---|---|---|---|---|---|---|---|---|---|---|---|---|---|---|---|

| AM16 | ATM0_ATM2 | CSNT/ SAM | ACS0_ACS1 | BI | SCY0_SCY3 | SETA | TRLX | EHTR | Res |
|---|---|---|---|---|---|---|---|---|---|

**Figure 8-21**
Base and Option Registers.[2]

The type of memory located in the various banks, and connected to the appropriate CS, can then be initialized for access via these registers. So, given the memory map example in Figure 8-18, the pseudocode for configuring the first two banks (of 4 MB of Flash each), and the third bank (4 MB of DRAM) would be as follows:

*Note: Length initialized by looking up the length in the table below, and entering 1 s from bit 0 to bit position indicating that length, and entering 0 s into the remaining bits.*

| 0 | 1 | 2 | 3 | 4 | 5 | 6 | 7 | 8 | 9 | 10 | 11 | 12 | 13 | 14 | 15 | 16 |
|---|---|---|---|---|---|---|---|---|---|---|---|---|---|---|---|---|
| 2 G | 1 G | 512 M | 256 M | 128 M | 64 M | 32 M | 16 M | 8 M | 4 M | 2 M | 1 M | 512 K | 256 K | 128 K | 64 K | 32 K |

```
...
// OR for Bank 0-4 MB of Flash, 0x1FF8 for bits AM (bits 0-16) OR0 = 0x1FF80954;
// Bank 0 - Flash starting at address 0x00000000 for bits BA (bits 0-16),
configured for
// GPCM, 32-bit
BR0 = 0x00000001;

// OR for Bank 1-4 MB of Flash, 0x1FF8 for bits AM (bits 0-16) OR1 = 0x1FF80954;
// Bank 1-4 MB of Flash on CS1 starting at address 0x00400000, configured for GPCM,
// 32-bit
BR1 = 0x00400001;

// OR for Bank 2-4 MB of DRAM, 0x1FF8 for bits AM (bits 0-16) OR2 =
// 0x1FF80800; Bank 2-4 MB of DRAM on CS2 starting at address 0x04000000,
// configured for UPMA, 32-bit
BR2 = 0x04000081;

// OR for Bank 3 for BCSR OR3 = 0xFFFF8110; Bank 3 - Board Control and Status
// Registers from address 0x09100000
BR3 = 0x09100001;
...
```

So, to initialize the memory controller, the base and option registers are initialized to reflect the types of memory in its banks. While no additional GPCM registers need initialization, for memory managed by the UPMA or UPMB, at the very least, the memory periodic timer prescaler register (MPTPR) is initialized for the required refresh timeout (i.e., for DRAM), and the related memory mode register (MAMR or MBMR) for configuring the UPMs needs initialization. The core of every UPM is a (64×32 bit) RAM array that specifies the specific type of accesses (logical values) to be transmitted to the UPM managed memory chips for a given clock cycle. The RAM array is initialized via the memory command register (MCR), which is specifically used during initialization to read from and write to the RAM array, and the memory data register (MDR), which stores the data the MCR uses to write to or read from the RAM array (see sample pseudocode below).[3]

```
...
// set periodic timer prescaler to divide by 8
MPTPR = 0x0800;                                    // 16-bit register

// periodic timer prescaler value for DRAM refresh period (see the PowerPC manual
for calculation), timer enable, ...
MAMR = 0xC0A21114;

// 64-Word UPM RAM Array content example - the values in this table were generated
using the
```

```
// UPM860 software available on the Motorola/Freescale Netcomm Web site.
UpmRamARRY:
// 6 WORDS - DRAM 70ns - single read. (offset 0 in upm RAM)
.long 0x0fffcc24, 0x0fffcc04, 0x0cffcc04, 0x00ffcc04, 0x00ffcc00, 0x37ffcc47
// 2 WORDs - offsets 6-7 not used
.long 0xffffffff, 0xffffffff
// 14 WORDs - DRAM 70ns - burst read. (offset 8 in upm RAM)
.long 0x0fffcc24, 0x0fffcc04, 0x08ffcc04, 0x00ffcc04, 0x00ffcc08, 0x0cffcc44,
.long 0x00ffec0c, 0x03ffec00, 0x00ffec44, 0x00ffcc08, 0x0cffcc44,
.long 0x00ffec04, 0x00ffec00, 0x3fffec47
// 2 WORDs - offsets 16-17 not used
.long 0xffffffff, 0xffffffff
// 5 WORDs - DRAM 70ns - single write. (offset 18 in upm RAM)
.long 0x0fafcc24, 0x0fafcc04, 0x08afcc04, 0x00afcc00, 0x37ffcc47
// 3 WORDs - offsets 1d-1f not used
.long 0xffffffff, 0xffffffff, 0xffffffff
// 10 WORDs - DRAM 70ns - burst write. (offset 20 in upm RAM)
.long 0x0fafcc24, 0x0fafcc04, 0x08afcc00, 0x07afcc4c, 0x08afcc00, 0x07afcc4c,
.long 0x08afcc00, 0x07afcc4c, 0x08afcc00, 0x37afcc47
// 6 WORDs - offsets 2a-2f not used
.long 0xffffffff, 0xffffffff, 0xffffffff, 0xffffffff, 0xffffffff, 0xffffffff
// 7 WORDs - refresh 70ns. (offset 30 in upm RAM)
.long 0xe0ffcc84, 0x00ffcc04, 0x00ffcc04, 0x0fffcc04, 0x7fffcc04, 0xffffcc86,
.long 0xffffcc05
// 5 WORDs - offsets 37-3b not used
.long 0xffffffff, 0xffffffff, 0xffffffff, 0xffffffff, 0xffffffff
// 1 WORD - exception. (offset 3c in upm RAM)
.long 0x33ffcc07
// 3 WORDs - offset 3d-3f not used
.long 0xffffffff, 0xffffffff, 0x40004650
UpmRAMArrayEnd:

// Write To UPM Ram Array
Index = 0
Loop While Index<64
{
MDR = UPMRamArray[Index];                // store data to MDR
MCR = 0x0000;                            // issue "Write" command to MCR register
                                         to store what is in MDR in RAM Array
Index = Index + 1;
}                                        // end loop
…
```

Initializing the Internal Memory Map on the MPC860

The MPC860's internal memory map contains the architecture's special purpose registers (SPRs), as well as dual-port RAM, also referred to as parameter RAM, that contain the buffers of the various integrated components, such as Ethernet or $I^2C$. On the MPC860, it is simply a matter of configuring one of these SPRs, the Internal Memory Map Register (IMMR) shown in

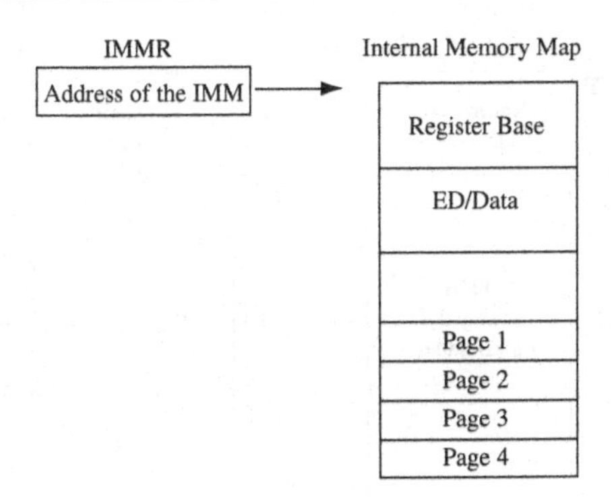


**Figure 8-22**
IMMR.[4]

Figure 8-22, to contain the base address of the internal memory map, as well as some factory-related information on the specific MPC860 processor (part number and mask number).

In the case of the sample memory map used in this section, the internal memory map starts at 0×09000000, so in pseudocode form, the IMMR would be set to this value via the "mfspr" or "mtspr" commands:

```
mtspr 0x090000FF    // the top 16 bits are the address, bits 16-23 are the part
                     number
                    // (0x00 in this example) and bits 24-31 is the mask number
                    // (0xFF in this example).
```

### Initializing the MMU on the MPC860

The MPC860 uses the MMUs to manage the board's virtual memory management scheme, providing logical/effective to physical/real address translations, cache control (instruction MMU and instruction cache, data MMU and data cache), and memory access protections. The MPC860 MMU (shown in Figure 8-23a) allows support for a 4GB uniform (user) address space that can be divided into pages of a variety of sizes, specifically 4kB, 16kB, 512kB, or 8MB, that can be individually protected and mapped to physical memory.

Using the smallest page size a virtual address space can be divided into on the MPC860 (4kB), a translation table—also commonly referred to as the *memory map* or *page table*—would contain a million address translation entries, one for each 4kB page in the 4GB address space. The MPC860 MMU does not manage the entire translation table at one time (in fact, most MMUs do not). This is because embedded boards do not typically have 4GB of physical memory that needs to be managed at one time. It would be very time consuming for an MMU to update a million entries with every update to virtual memory by the software, and an MMU would need to use a lot of faster (and more expensive) on-chip memory in order to store a memory map of such a size. So, as a result, the MPC860 MMU contains small caches within it to store a subset of this memory map. These caches are referred to as TLBs (shown in

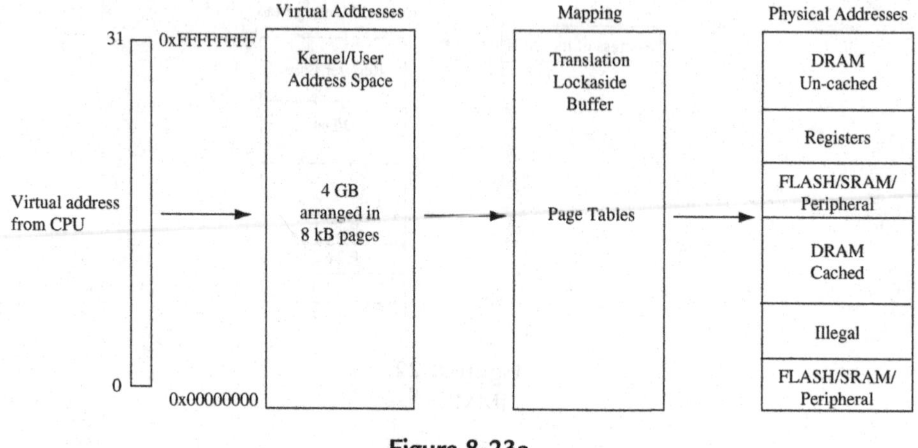


**Figure 8-23a**
TLB within VM Scheme.[4]

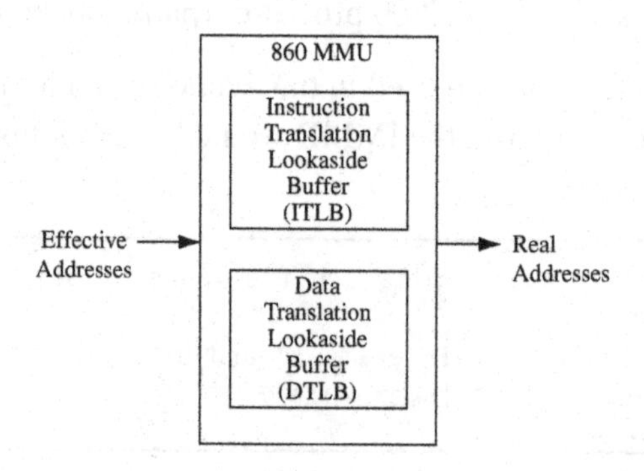


**Figure 8-23b**
TLB.[4]

Figure 8-23b—one instruction and one data) and are part of the MMU's initialization sequence. In the case of the MPC860, the TLBs are 32-entry and fully associative caches. The entire memory map is stored in cheaper off-chip main memory as a two-level tree of data structures that define the physical memory layout of the board and their corresponding effective memory address.

The TLB is how the MMU translates (maps) logical/virtual addresses to physical addresses. When the software attempts to access a part of the memory map not within the TLB, a *TLB miss* occurs, which is essentially a trap requiring the system software (through an exception handler) to load the required translation entry into the TLB. The system software that loads the new entry into the TLB does so through a process called a *tablewalk*. This is basically the process of traversing the MPC860's two-level memory map tree in main memory to locate the desired entry to be loaded in the TLB. The first level of the PowerPC's multilevel translation table scheme (its translation table structure uses one level 1 table and one or more level 2 tables) refers to a page table entry in the page table of the second level. There are 1024

Table 8-1: Level 1 and 2 Entries[4]

| Page Size | No. of Pages per Segment | Number of Entries in Level 2 Table | Level 2 Table Size (Bytes) |
|---|---|---|---|
| 8 MB | 0.5 | 1 | 4 |
| 512 kB | 8 | 8 | 32 |
| 16 kB | 256 | 1024 | 4096 |
| 4 kB | 1024 | 1024 | 4096 |

entries, where each entry is 4 bytes (24 bits) and represents a segment of virtual memory that is 4 MB in size. The format of an entry in the level 1 table is made up of a valid bit field (indicating that the 4 MB respective segment is valid), a level 2 base address field (if valid bit is set, pointer to base address of the level 2 table which represents the associated 4 MB segment of virtual memory), and several attribute fields describing the various attributes of the associated memory segment.

Within each level 2 table, every entry represents the pages of the respective virtual memory segment. The number of entries of a level 2 table depends on the defined virtual memory page size (4 kB, 16 kB, 512 kB, or 8 MB); see Table 8-1. The larger the virtual memory page size, the less memory used for level 2 translation tables, since there are fewer entries in the translation tables (e.g., a 16 MB physical memory space can be mapped using 2 × 8 MB pages (2048 bytes in the level 1 table and a 2 × 4 in the level 2 table for a total of 2056 bytes) or 4096 × 4 kB pages (2048 bytes in the level 1 table and a 4 × 4096 in the level 2 table for a total of 18 432 bytes)).

In the MPC860's TLB scheme, the desired entry location is derived from the incoming effective memory address. The location of the entry within the TLB sets is specifically determined by the index field(s) derived from the incoming logical memory address. The format of the 32-bit logical (effective) address generated by the PowerPC Core differs depending on the page size. For a 4 kB page, the effective address is made up of a 10-bit level 1 index, a 10-bit level 2 index, and a 12-bit page offset (see Figure 8-24a). For a 16 kB page, the page offset becomes 14 bits and the level 2 index is 8-bits (see Figure 8-24b). For a 512 kB page, the page offset is 19 bits and the level 2 index is then 3 bits long (see Figure 8-24c); for an 8 MB page, the page offset is 23 bits long, there is no level 2 index, and the level 1 index is 9 bits long (see Figure 8-24d).

The page offset of the 4 kB effective address format is 12 bits wide to accommodate the offset within the 4 kB (0 × 0000 to 0 × 0FFF) pages. The page offset of the 16 kB effective address format is 14 bits wide to accommodate the offset within the 16 kB (0 × 0000 to 0 × 3FFF) pages. The page offset of the 512 kB effective address format is 19 bits wide to accommodate the offset within the 512 kB (0 × 0000 to 0 × 7FFFF) pages and the page offset of the 8 MB effective address format is 23 bits wide to accommodate the offset within the 8 MB (0×0000 to 0×7FFFF8) pages.

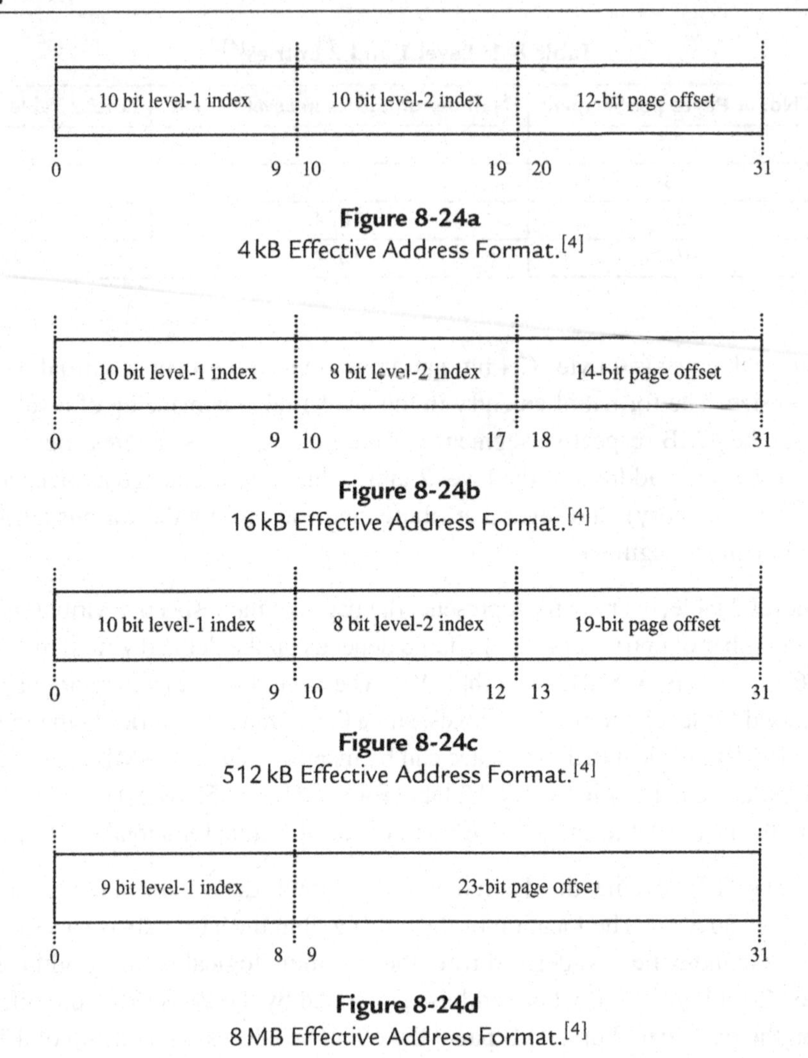


**Figure 8-24a**
4 kB Effective Address Format.[4]



**Figure 8-24b**
16 kB Effective Address Format.[4]



**Figure 8-24c**
512 kB Effective Address Format.[4]



**Figure 8-24d**
8 MB Effective Address Format.[4]

In short, the MMU uses these effective address fields (level 1 index, level 2 index, and offset) in conjunction with other registers, TLB, translation tables, and the tablewalk process to determine the associated physical address (see Figure 8-25).

The MMU initialization sequence involves initializing the MMU registers and translation table entries. The initial steps include initializing the MMU Instruction Control Register (MI_CTR) and the Data Control Registers (MD_CTR) shown in Figures 8-26a and b. The fields in both registers are generally the same, most of which are related to memory protection.

Initializing translation table entries is a matter of configuring two memory locations (level 1 and level 2 descriptors), and three register pairs, one for data and one for instructions, in each pair, for a total of six registers. This equals one each of an Effective Page Number (EPN) register, Tablewalk Control (TWC) register, and Real Page Number (RPN) register.

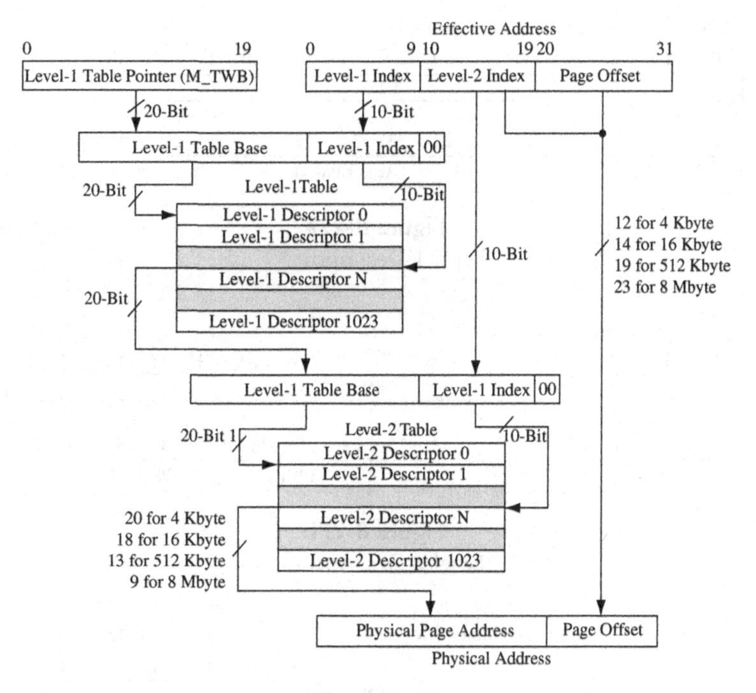


**Figure 8-25**
Level 2 Translation Table for 4 kB page Scheme.[4]

MI_CTR - MMU Instruction Control Register

| 0 | 1 | 2 | 3 | 4 | 5 | 6 | 7–15 |
|---|---|---|---|---|---|---|---|
| GPM | PPM | CI DEF | Res | RS V4I | Res | PPCS | Reserved |

| 16–18 | 19–23 | 24–31 |
|---|---|---|
| Res | ITLB_INDX | Reserved |

**Figure 8-26a**
MI_CTR.[2]

MD_CTR - MMU Data Control Register

| 0 | 1 | 2 | 3 | 4 | 5 | 6 | 7–15 |
|---|---|---|---|---|---|---|---|
| GPM | PPM | CI DEF | WT DEF | RS V4D | TW AM | PPCS | Reserved |

| 16–18 | 19–23 | 24–31 |
|---|---|---|
| Res | DTLB_INDX | Reserved |

**Figure 8-26b**
MD_CR.[2]

The level 1 descriptor (see Figure 8-27a) defines the fields of the level 1 translation table entries, such as the Level 2 Base Address (L2BA), the access protection group, and page size. The level 2 page descriptor (see Figure 8-27b) defines the fields of the level 2 translation table entries, such as: the physical page number, page valid bit, and page protection.

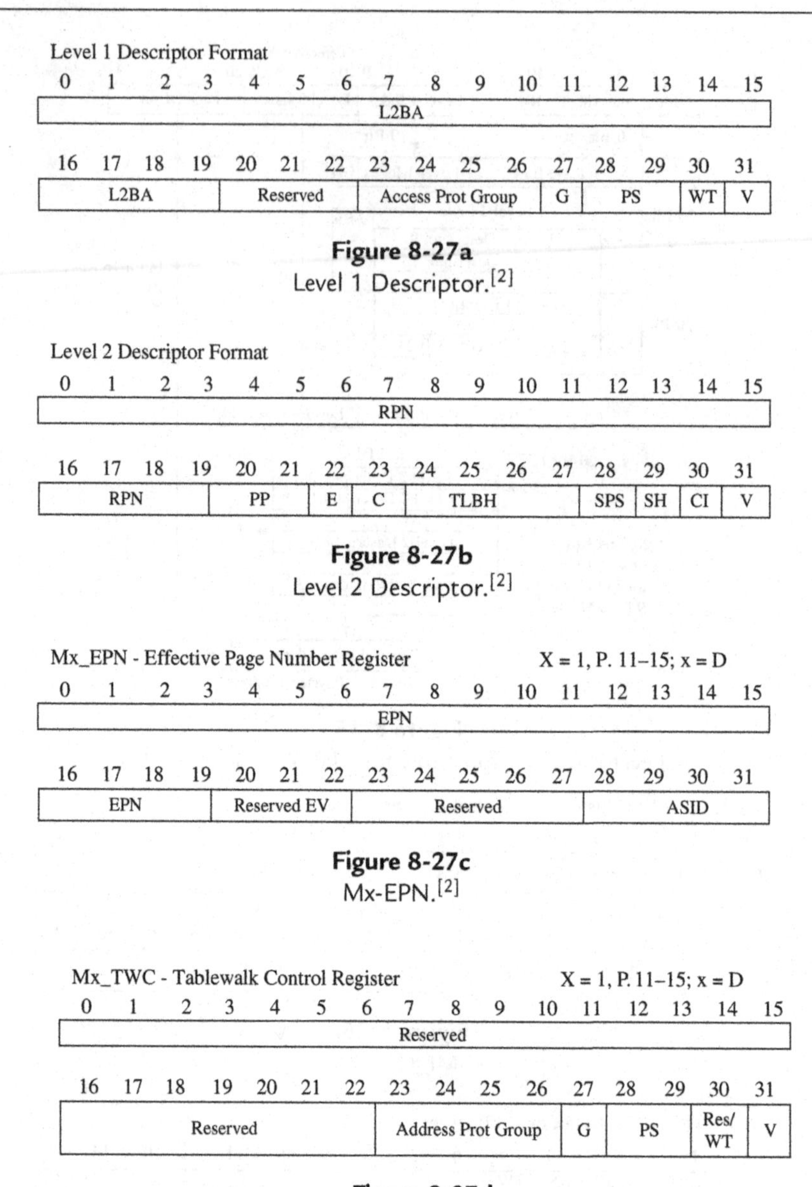


**Figure 8-27a**
Level 1 Descriptor.[2]

**Figure 8-27b**
Level 2 Descriptor.[2]

**Figure 8-27c**
Mx-EPN.[2]

**Figure 8-27d**
Mx-TWC.[2]

The registers shown in Figures 8-27c–e are essentially TLB source registers used to load entries into the TLBs. The Effective Page Number (EPN) registers contain the effective address to be loaded into a TLB entry. The Tablewalk Control (TWC) registers contain the attributes of the effective address entry to be loaded into the TLB (page size, access protection, etc.), and the Real Page Number (RPN) registers contain the physical address and attributes of the page to be loaded into the TLB.

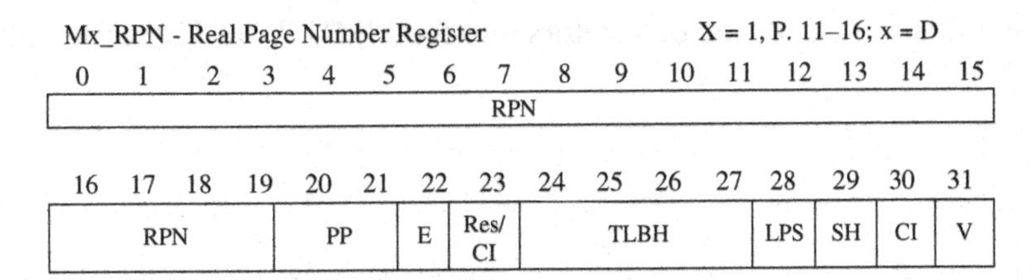


**Figure 8-27e**
Mx-RPN.[2]

An example of an MMU initialization sequence on the MPC860 is pseudocoded below.

```
// Invalidating TLB entries
tlbia;    // the MPC860's instruction to invalidate entries within the TLBs, also
the
                                    // "tlbie" can be used

// Initializing the MMU Instruction Control Register
...
MI_CTR.fld.all = 0;                 // clear all fields of register so group
                                    protection mode =
   // PowerPC mode, page protection mode is page resolution, etc.
MI_CTR.fld.CIDEF = 1;               // instruction cache inhibit default when MMU
                                    disabled
...
// Initializing the MMU Data Control Register
...
MD_CTR.fld.all = 0;                 // clear all fields of register so group
                                    protection mode =
                                    // PowerPC mode, page protection mode is page
resolution, etc.
MD_CTR.fld.TWAM = 1;                // tablewalk assist mode = 4 kbyte page hardware
                                    assist
MD_CTR.fld.CIDEF = 1;               // data cache inhibit default when MMU disabled
...
```

Move to Exception Vector Table the Data and Instruction TLB Miss and Error ISRs *(MMU interrupt vector locations shown in table below).*[4]

| Offset (hex) | Interrupt Type |
|---|---|
| 01100 | Implementation Dependent Instruction TLB Miss |
| 01200 | Implementation Dependent Data TLB Miss |
| 01300 | Implementation Dependent Instruction TLB Error |
| 01400 | Implementation Dependent Data TLB Error |

With a TLB miss, an ISR loads the descriptors into the MMU. Data TLB Reload ISR example:

```
...
// put next code into address, incrementing vector by 4 after each line, i.e.,
"mtspr
// M_TW, r0"="07CH, 011H, 013H, 0A6H", so put integer 0x7C1113A6H at vector
// 0x1200 and increment vector by 4;
install start of ISR at vector address offset = 0x1200;

// save general purpose register into MMU tablewalk special register
mtspr M_TW, GPR;

mfspr GPR, M_TWB;           // load GPR with address of level one descriptor
lwz GPR, (GPR);             // load level one page entry

// save level 2 base pointer and level 1 # attributes into DMMU tablewalk control
// register
mtspr MD_TWC, GPR;

// load GPR with level 2 pointer while taking into account the page size
mfspr GPR, MD_TWC;

lwz GPR, (GPR);             // load level 2 page entry
mtspr MD_RPN, GPR;          // write TLB entry into real page number register

// restore GPR from tablewalk special register return to main execution stream;
mfspr GPR, M_TW;
...
```

Instruction TLB Reload ISR example:

```
// put next code into address, incrementing vector by 4 after each line,
i.e., "mtspr
// M_TW, r0"="07CH, 011H, 013H, 0A6H", so put integer 0x7C1113A6H at vector
// 0x1100 and increment vector by 4;
install start of ISR at vector address offset = 0x1100;
...
// save general purpose register into MMU tablewalk special register
mtspr M_TW, GPR;

mfspr GPR, SRR0      // load GPR with instruction miss effective address
mtspr MD_EPN, GPR    // save instruction miss effective address in MD_EPN
mfspr GPR, M_TWO     // load GPR with address of level one descriptor
lwz GPR, (GPR)       // load level one page entry
mtspr MI_TWC, GPR    // save level one attributes
mtspr MD_TWC, GPR    // save level two base pointer
```

```
// load R1 with level two pointer while taking into account the page size
mfspr GPR, MD_TWC

lwz GPR, (GPR)        // load level two page entry
mtspr MI_RPN, GPR     // write TLB entry
mfspr GPR, M_TW       // restore R1

return to main execution stream;

// Initialize L1 table pointer and clear L1 table, i.e., MMU tables/TLBs 043F0000-
// 043FFFFF
Level1_Table_Base_Pointer = 0 × 043F0000;

index:= 0;
WHILE ((index MOD 1024) is NOT = 0) DO
Level1 Table Entry at Level1_Table_Base_Pointer + index=0;
index = index + 1;
end WHILE;
...
```

Initialize translation table entries and map in desired segments in level 1 table and pages in level 2 tables. For example, given the physical memory map in Figure 8-28, the L1 and L2 descriptors would need to be configured for Flash, DRAM, etc.

| Address Range | Accessed Device | Port Width |
|---|---|---|
| 0x00000000 - 0x003FFFFF | Flash PROM Bank 1 | 32 |
| 0x00400000 - 0x007FFFFF | Flash PROM Bank 2 | 32 |
| 0x04000000 - 0x043FFFFF | DRAM 4 Mbyte (1Meg × 32-bit)it) | 32 |
| 0x09000000 - 0x09003FFF | MPC Internal Memory Map | 32 |
| 0x09100000 - 0x09100003 | BCSR - Board Control & Status Register | 32 |
| 0x10000000 - 0x17FFFFFF | PCMCIA Channel | 16 |

**Figure 8-28a**
Physical Memory Map.[4]

| PS | # | Used for... | Address Range | CI | WT | S/U | R/W | SH |
|---|---|---|---|---|---|---|---|---|
| 8M | 1 | Monitor & trans. tbls | 0x0 - 0x7FFFFF | N | Y | S | R/O | Y |
| 512K | 2 | Stack & scratchpad | 0x40000000 - 0x40FFFFF | N | N | S | R/W | Y |
| 512K | 1 | CPM data buffers | 0x4100000 - 0x417FFFF | Y | - | S | R/W | Y |
| 512K | 5 | Prob. prog. & data | 0x4180000 - 0x43FFFFF | N | N | S/U | R/W | Y |
| 16K | 1 | MPC int mem. map | 0x9000000 - Y | - | S | R/W | Y | |
| 16K | 1 | Board config. regs | 0x9100000 - 0x9103FFF | Y | - | S | R/W | Y |
| 8M | 16 | PCMCIA | 0x10000000 - 0x17FFFFFF | Y | - | S | R/W | Y |

**Figure 8-28b**
L1/L2 Configuration.[4]

```
// i.e., Initialize entry for and Map in 8 MB of Flash at 0x00000000, adding entry
into L1 table, and
// adding a level 2 table for every L1 segment - as shown in Figure 8-28b, page
size is 8 MB, cache is
// not inhibited, marked as write-through, used in supervisor mode, read only, and
shared.

// 8 MB Flash

…
Level2_Table_Base_Pointer = Level1_Table_Base_Pointer +
size of L1 Table (i.e., 1024);
L1desc(Level1_Table_Base_Pointer + L1Index).fld.BA = Level2_Table_Base_Pointer;
L1desc(Level1_Table_Base_Pointer + L1Index).fld.PS = 11b;    // page size = 8MB

// Writethrough attribute = 1 writethrough cache policy region
L1desc.fld(Level1_Table_Base_Pointer + L1Index).WT = 1;
L1desc(Level1_Table_Base_Pointer + L1Index).fld.PS = 1;      // page size = 512KB

// level-one segment valid bit = 1 segment valid
L1desc(Level1_Table_Base_Pointer + L1Index).fld.V = 1;

// for every segment in L1 table, there is an entire level2 table
L2index:=0;
WHILE (L2index<# Pages in L1Table Segment) DO
L2desc[Level2_Table_Base_Pointer + L2index * 4].fld.RPN = physical page number;
L2desc[Level2_Table_Base_Pointer + L2index * 4].fld.CI = 0;  // Cache Inhibit
                                                                Bit = 0

…
L2index = L2index + 1;
end WHILE;

// i.e., Map in 4 MB of DRAM at 0x04000000, as shown in Figure 8-29b, divided into
eight,
// 512KB pages. Cache is enabled, and is in copy-back mode, supervisor mode,
supports
// reading and writing, and it is shared.

…
Level2_Table_Base_Pointer = Level2_Table_Base_Pointer +
Size of L2Table for 8MB Flash;
L1desc(Level1_Table_Base_Pointer + L1Index).fld.BA = Level2_Table_Base_Pointer;
L1desc(Level1_Table_Base_Pointer + L1Index).fld.PS = 01b;    // page size = 512KB

// Writethrough Attribute = 0 copyback cache policy region
L1desc.fld(Level1_Table_Base_Pointer + L1Index).WT = 0;
L1desc(Level1_Table_Base_Pointer + L1Index).fld.PS = 1;      // page size = 512KB
```

```
// Level 1 segment valid bit = 1 segment valid
L1desc(Level1_Table_Base_Pointer + L1Index).fld.V = 1;
...

// Initializing Effective Page Number Register
loadMx_EPN(mx_epn.all);

// Initializing the Tablewalk Control Register Descriptor
load Mx_TWC(L1desc.all);

// Initializing the Mx_RPN Descriptor
load Mx_RPN (L2desc.all);
...
```

At this point the MMU and caches can be enabled (see Memory Subsystem Enable section).

### *Memory Subsystem Disable on MPC860*

```
// Disable MMU - The MPC860 powers up with the MMUs in disabled mode, but to
// disable translation IR and DR bits need to be cleared.
...
rms msr ir 0; rms msr dr 0;    // disable translation
...

// Disable caches
...

// Disable caches (0100b in bits 4-7, IC_CST[CMD] and DC_CST[CMD] registers)
addis r31,r0,0x0400
mtspr DC_CST,r31
mtspr IC_CST,r31
...
```

### *Memory Subsystem Enable on MPC860*

```
// Enable MMU via setting IR and DR bits and "mtmsr" command on MPC860
...
ori r3,r3,0x0030;        // set the IR and DR bits
mtmsr r3;                // enable translation
isync;
...

// Enable caches
...
```

```
addis r31,r0,0x0a00      // unlock all in both caches
mtspr DC_CST,r31
mtspr IC_CST,r31
addis r31,r0,0x0c00      // invalidate all in both caches
mtspr DC_CST,r31
mtspr IC_CST,r31

// Enable caches (0010b in bits 4-7, IC_CST[CMD] and DC_CST[CMD] registers)
addis r31,r0,0x0200
mtspr DC_CST,r31
mtspr IC_CST,r31
...
```

*Memory Subsystem Writing/Erasing Flash*

While reading from Flash is the same as reading from RAM, accessing Flash for writing or erasing is typically much more complicated. Flash memory is divided into blocks, called sectors, where each sector is the smallest unit that can be erased. While Flash chips differ in the process required to perform a write or erase, the general handshaking is similar to the pseudocode examples below for the Am29F160D Flash chip. The Flash erase function notifies the Flash chip of the impending operation, sends the command to erase the sector, and then loops, polling the Flash chip to determine when it completes. At the end of the erase function, the Flash is then set to standard read mode. The write routine is similar to that of the erase function, except the command is transmitted to perform a write to a sector, rather than an erase.

```
...

// The address at which the Flash devices are mapped
int FlashStartAddress = 0x00000000;

int FlashSize = 0x00800000;      // The size of the Flash devices in bytes,
                                    i.e., 8MB.

// Flash memory block offset table from the Flash base of the various sectors, as
well as
// the corresponding sizes.
BlockOffsetTable={{ 0x00000000, 0x00008000 }, { 0x00008000, 0x00004000 },
  { 0x0000C000, 0x00004000 }, { 0x00010000, 0x00010000 },
  { 0x00020000, 0x00020000 }, { 0x00040000, 0x00020000 },
  { 0x00060000, 0x00020000 }, { 0x00080000, 0x00020000 }, ...};

// Flash write pseudocode example
FlashErase (int startAddress, int offset) {
```

```
...
// Erase sector commands
Flash [startAddress + (0x0555 << 2)] = 0x00AA00AA;      // unlock 1 Flash command
Flash [startAddress + (0x02AA << 2)] = 0x00550055;      // unlock 2 Flash command
Flash [startAddress + (0x0555 << 2)] = 0x00800080);     // erase setup Flash
                                                          command
Flash [startAddress + (0x0555 << 2)] = 0x00AA00AA;      // unlock 1 Flash command
Flash [startAddress + (0x02AA << 2)] = 0x00550055;      // unlock 2 Flash command
Flash [startAddress + offset] = 0x00300030;             // set Flash sector erase
                                                          command

// Poll for completion: avg. block erase time is 700ms, worst-case block erase
  time
// is 15s
int poll;
int loopIndex = 0;
while (loopIndex < 500) {
for (int i = 0; i<500 * 3000; i++);
poll = Flash(startAddr + offset);
if ((poll AND 0x00800080) = 0x00800080 OR
(poll AND 0x00200020) = 0x00200020) {
exit loop;
}
loopIndex++;
}

// exit
Flash (startAddr) = 0x00F000F0;                         // read reset command
Flash(startAddr + offset) == 0xFFFFFFFF;
}
```

## 8.3 Example 3: On-Board Bus Device Drivers

As discussed in Chapter 7, associated with every bus is/are (1) some type of *protocol* that defines how devices gain access to the bus (arbitration), (2) the rules attached devices must follow to communicate over the bus (handshaking), and (3) the signals associated with the various bus lines. Bus protocol is supported by the bus device drivers, which commonly include all or some combination of all of the 10 functions from the list of device driver functionality introduced at the start of this chapter, including:

- *Bus Startup*: initialization of the bus upon PowerON or reset.
- *Bus Shutdown*: configuring bus into its PowerOFF state.
- *Bus Disable*: allowing other software to disable bus on-the-fly.
- *Bus Enable*: allowing other software to enable bus on-the-fly.
- *Bus Acquire*: allowing other software to gain singular (locking) access to bus.

- *Bus Release*: allowing other software to free (unlock) bus.
- *Bus Read*: allowing other software to read data from bus.
- *Bus Write*: allowing other software to write data to bus.
- *Bus Install*: allowing other software to install new bus device on-the-fly for expandable buses.
- *Bus Uninstall*: allowing other software to remove installed bus device on-the-fly for expandable buses.

Which of the routines are implemented and how they are implemented depends on the actual bus. The pseudocode below is an example of an $I^2C$ bus initialization routine provided as an example of a bus startup (initialization) device driver on the MPC860.

### 8.3.1 On-Board Bus Device Driver Pseudocode Examples

The following pseudocode gives an example of implementing a bus initialization routine on the MPC860, specifically the startup function in reference to the architecture. These examples demonstrate how bus management can be implemented on a more complex architecture, and this can be used as a guide to understand how to write bus management drivers on other processors of equal or lesser complexity than the MPC860 architecture. Other driver routines have not been pseudocoded, because the same concepts apply here as in Sections 8.1 and 8.2—essentially, looking in the architecture and bus documentation for the mechanisms that enable a bus, disable a bus, acquire a bus, etc.

#### *$I^2C$ Bus Startup (Initialization) on the MPC860*

The $I^2C$ (inter-IC) protocol is a serial bus with one serial data line (SDA) and one serial clock line (SCL). With the $I^2C$ protocol, all devices attached to the bus have a unique address (identifier), and this identifier is part of the data stream transmitted over the SDL line.

The components on the master processor that support the $I^2C$ protocol are what need initialization. In the case of the MPC860, there is an integrated $I^2C$ controller on the master processor (see Figure 8-29). The $I^2C$ controller is made up transmitter registers, receiver registers, a baud rate generator, and a control unit. The baud rate generator generates the clock signals when the $I^2C$ controller acts as the $I^2C$ bus master—if in slave mode, the controller uses the clock signal received from the master. In reception mode, data is transmitted from the SDA line into the control unit, through the shift register, which in turn transmits the data to the receive data register. The data that will be transmitted over the $I^2C$ bus from the PPC is initially stored in the transmit data register and transferred out through the shift register to the control unit and over the SDA line. Initializing the $I^2C$ bus on the MPC860 means initializing the $I^2C$ SDA and SCL pins, many of the $I^2C$ registers, some of the parameter RAM, and the associated buffer descriptors.

The MPC860 $I^2C$ SDA and SCL pins are configured via the Port B general purpose I/O port (see Figures 8-30a and b). Because the I/O pins can support multiple functions, the specific function

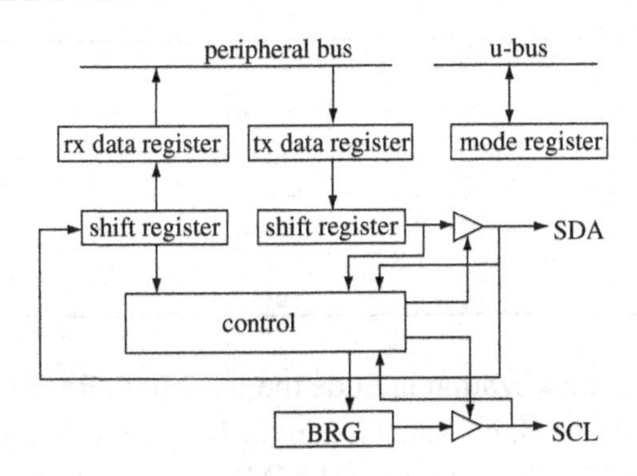


**Figure 8-29**
$I^2C$ Controller on MPC860.[4]


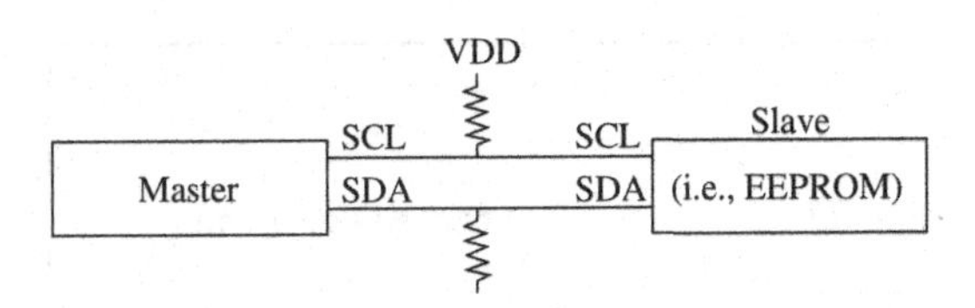


**Figure 8-30a**
SDA and SCL pins on MPC860.[4]


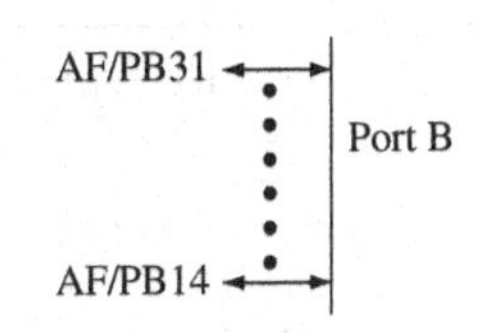


**Figure 8-30b**
MPC860 Port B Pins.[4]


| | 14 | 15 | 16 | 17 | 18 | 19 | 20 | 21 | 22 | 23 | 24 | 25 | 26 | 27 | 28 | 29 | 30 | 31 |
|---|---|---|---|---|---|---|---|---|---|---|---|---|---|---|---|---|---|---|
| PBDAT | D | D | D | D | D | D | D | D | D | D | D | D | D | D | D | D | D | D |
| PBODR | OD | OD | OD | OD | OD | OD | OD | OD | OD | OD | OD | OD | OD | OD | OD | OD | OD | OD |
| PBDIR | DR | DR | DR | DR | DR | DR | DR | DR | DR | DR | DR | DR | DR | DR | DR | DR | DR | DR |
| PBPAR | DD | DD | DD | DD | DD | DD | DD | DD | DD | DD | DD | DD | DD | DD | DD | DD | DD | DD |

**Figure 8-30c**
MPC860 Port B Register.[4]


a pin will support needs to be configured via port B's registers (shown in Figure 8-30c). Port B has four read/write (16-bit) control registers: the Port B Data Register (PBDAT), the Port B Open Drain Register (PBODR), the Port B Direction Register (PBDIR), and the Port B Pin Assignment Register (PBPAR). In general, the PBDAT register contains the data on the pin, the PBODR configures the pin for open drain or active output, the PBDIR configures the pin as either an input or output pin, and the PBPAR assigns the pin its function ($I^2C$, general purpose I/O, etc.).

An example of initializing the SDA and SCL pins on the MPC860 is given in the pseudocode below.

```
...
immr = immr & 0xFFFF0000;                 // MPC8xx internal register map
// Configure Port B pins to enable SDA and SCL
immr->pbpar=(pbpar) OR (0x00000030);      // set to dedicated I2C
immr->pbdir=(pbdir) OR (0x00000030);      // enable I2CSDA and I2CSCL as outputs
...
```

The I²C registers that need initialization include the I²C Mode Register (I2MOD), I²C Address Register (I2ADD), the Baud Rate Generator Register (I2BRG), the I²C Event Register (I2CER), and the I²C Mask Register (I2CMR) shown in Figures 8-31a–e).

An example of I²C register initialization pseudocode is as follows:

```
/* I2C Registers Initialization Sequence */
...
// Disable I2C before initializing it, LSB character order for transmission and
reception,
// I2C clock not filtered, clock division factor of 32, etc.
```

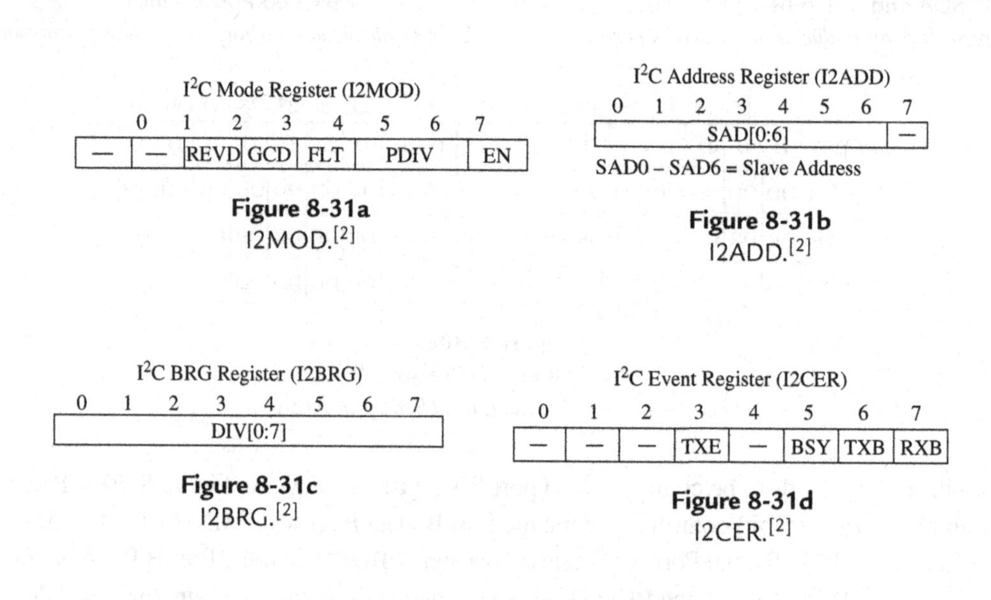


**Figure 8-31a**
I2MOD.[2]

**Figure 8-31b**
I2ADD.[2]

**Figure 8-31c**
I2BRG.[2]

**Figure 8-31d**
I2CER.[2]

**Figure 8-31e**
I2CMR.[2]

```
immr->i2mod = 0x00;
immr->i2add = 0x80;    // I2C MPC860 address = 0x80
immr->i2brg = 0x20;    // divide ratio of BRG divider
immr->i2cer = 0x17;    // clear out I2C events by setting relevant bits to "1"
immr->i2cmr = 0x17;    // enable interrupts from I2C in corresponding I2CER
immr->i2mod = 0x01;    // enable I2C bus
...
```

Five of the 15 field I²C parameter RAM need to be configured in the initialization of I²C on the MPC860. They include the receive function code register (RFCR), the transmit function code register (TFCR), and the maximum receive buffer length register (MRBLR), the base value of the receive buffer descriptor array (Rbase), and the base value of the transmit buffer descriptor array (Tbase) shown in Figure 8-32.

See the following pseudocode for an example of I²C parameter RAM initialization:

```
// I2C Parameter RAM Initialization
...

// specifies for reception big endian or true little endian byte ordering and
channel # 0
immr->I2Cpram.rfcr = 0x10;

// specifies for reception big endian or true little endian byte ordering and
channel # 0
immr->I2Cpram.tfcr = 0x10;
immr->I2Cpram.mrblr = 0x0100;    // the maximum length of I2C receive buffer
immr->I2Cpram.rbase = 0x0400;    // point RBASE to first RX BD
immr->I2Cpram.tbase = 0x04F8;    // point TBASE to TX BD

...
```

Data to be transmitted or received via the I²C controller (within the CPM of the PowerPC) is input into buffers which the transmit and receive buffer descriptors refer to. The first half word (16 bits) of the transmit and receive buffer contain status and control bits (as shown in Figures 8-33a and b). The next 16 bits contain the length of the buffer.

In both buffers the Wrap (W) bit indicates whether this buffer descriptor is the final descriptor in the buffer descriptor table (when set to 1, the I²C controller returns to the first buffer in the buffer descriptor ring). The Interrupt (I) bit indicates whether the I²C controller issues an interrupt when this buffer is closed. The Last bit (L) indicates whether this buffer contains the last character of the message. The CM bit indicates whether the I²C controller clears the Empty (E) bit of the reception buffer or Ready (R) bit of the transmission buffer when it is finished with this buffer. The Continuous Mode (CM) bit refers to continuous mode in which, if a single buffer descriptor is used, continuous reception from a slave I²C device is allowed.

| Offset[1] | Name | Width | Description |
|---|---|---|---|
| 0x00 | RBASE | Hword | Rx/TxBD table base address. Indicate where the BD tables begin in the dual-port RAM. Setting Rx/TxBD[W] in the last BD in each BD table determines how many BDs are allocated for the Tx and Rx sections of the $I^2C$. Initialize RBASE/TBASE before enabling the $I^2C$. Furthermore, do not configure BD tables of the $I^2C$ to overlap any other active controller's parameter RAM. RBASE and TBASE should be divisible by eight. |
| 0x02 | TBASE | Hword | |
| 0x04 | RFCR | Byte | Rx/Tx function code. Contains the value to appear on AT[1–3] when the associated SDMA channel accesses memory. Also controls the byte-ordering convention for transfers. |
| 0x05 | TFCR | Byte | |
| 0x06 | MRBLR | Hword | Maximum receive buffer length. Defines the maximum number of bytes the $I^2C$ receiver writes to a receive buffer before moving to the next buffer. The receiver writes fewer bytes to the buffer than the MRBLR value if an error or end-of-frame occurs. Receive buffers should not be smaller than MRBLR. Transmit buffers are unaffected by MRBLR and can vary in length; the number of bytes to be sent is specified in TxBD[Data Length].<br><br>MRBLR is not intended to be changed while the $I^2C$ is operating. However, it can be changed in a single bus cycle with one 16-bit move (not two 8-bit bus cycles back-to-back). The change takes effect when the CP moves control to the next RxBD. To guarantee the exact RxBD on which the change occurs, change MRBLR only while the $I^2C$ receiver is disabled. MRBLR should be greater than zero. |
| 0x08 | RSTATE | Word | Rx internal state. Reserved for CPM use. |
| 0x0C | RPTR | Word | Rx internal data pointer[2] is updated by the SDMA channels to show the next address in the buffer to be accessed. |
| 0x10 | RBPTR | Hword | RxBD pointer. Points to the next descriptor the receiver transfers data to when it is in an idle state or to the current descriptor during frame processing for each $I^2C$ channel. After a reset or when the end of the descriptor table is reached, the CP initializes RBPTR to the value in RBASE. Most applications should not write RBPTR, but it can be modified when the receiver is disabled or when no receive buffer is used. |
| 0x12 | RCOUNT | Hword | Rx internal byte count[2] is a down-count value that is initialized with the MRBLR value and decremented with every byte the SDMA channels write. |
| 0x14 | RTEMP | Word | Rx temp. Reserved for CPM use. |
| 0x18 | TSTATE | Word | Tx internal state. Reserved for CPM use. |
| 0x1C | TPTR | Word | Tx internal data pointer[2] is updated by the SDMA channels to show the next address in the buffer to be accessed. |
| 0x20 | TBPTR | Hword | TxBD pointer. Points to the next descriptor that the transmitter transfers data from when it is in an idle state or to the current descriptor during frame transmission. After a reset or when the end of the descriptor table is reached, the CPM initializes TBPTR to the value in TBASE. Most applications should not write TBPTR, but it can be modified when the transmitter is disabled or when no transmit buffer is used. |
| 0x22 | TCOUNT | Hword | Tx internal byte count[2] is a down-count value initialized with TxBD [Data Length] and decremented with every byte read by the SDMA channels. |
| 0x24 | TTEMP | Word | Tx temp. Reserved for CP use. |
| 0x28-0x 2F | — | | Used for $I^2C$/SPI relocation. |

[1] As programmed in $I_2C$_BASE, the default value is IMMR + 0x3C80.

[2] Normally, these parameters need not be accessed.

**Figure 8-32**
$I^2C$ Parameter RAM.[4]

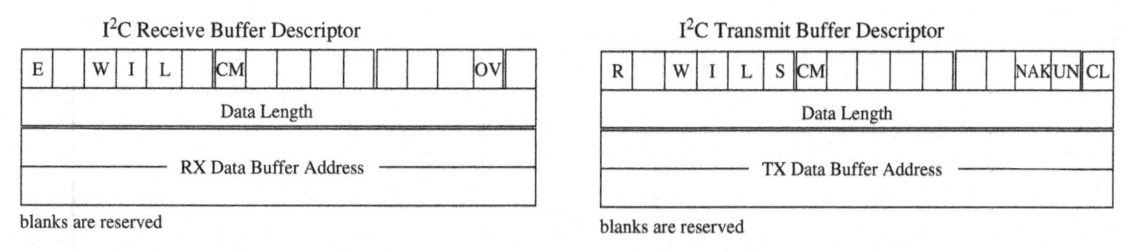


**Figure 8-33a**
Receive Buffer Descriptor.[2]

**Figure 8-33b**
Transmit Buffer Descriptor.[2]

In the case of the transmission buffer, the Ready (R) bit indicates whether the buffer associated with this descriptor is ready for transmission. The Transmit Start Condition (S) bit indicates whether a start condition is transmitted before transmitting the first byte of this buffer. The NAK bit indicates that the I²C aborted the transmission because the last transmitted byte did not receive an acknowledgement. The Under-run Condition (UN) bit indicates that the controller encountered an under-run condition while transmitting the associated data buffer. The Collision (CL) bit indicates that the I²C controller aborted transmission because the transmitter lost while arbitrating for the bus. In the case of the reception buffer, the Empty (E) bit indicates if the data buffer associated with this buffer descriptor is empty and the Over-run (OV) bit indicates whether an overrun occurred during data reception.

An example of I²C buffer descriptor initialization pseudocode would look as follows:

```
// I2C Buffer Descriptor Initialization
...
// 10 reception buffers initialized
index = 0;
While (index < 9) do
{
// E = 1, W = 0, I = 1, L = 0, OV = 0
immr->udata_bd ->rxbd[index].cstatus = 0x9000;
immr->bd ->rxbd[index].length = 0;    // buffer empty
immr->bd ->rxbd[index].addr=...
index = index+1;
}

// last receive buffer initialized
immr->bd->rxbd[9].cstatus = 0xb000;    // E = 1, W = 1, I = 1, L = 0, OV = 0
immr->bd ->rxbd[9].length = 0;         // buffer empty
immr->udata_bd ->rxbd[9].addr=...;

// transmission buffer
immr->bd ->txbd.length = 0x0010;    // transmission buffer 2 bytes long
```

```
// R = 1, W = 1, I = 0, L = 1, S = 1, NAK = 0, UN = 0, CL = 0
immr->bd->txbd.cstatus = 0xAC00;

immr->udata_bd ->txbd.bd_addr=...;

/* Put address and message in TX buffer */
...

// Issue Init RX & TX Parameters Command for I2C via CPM command register CPCR.
while(immr->cpcr & (0x0001));        // loop until ready to issue command
immr->cpcr=(0x0011);                 // issue command
while(immr->cpcr & (0x0001));        // loop until command processed
...
```

## 8.4 Board I/O Driver Examples

The board I/O subsystem components that require some form of software management include the components integrated on the master processor, as well as an I/O slave controller, if one exists. The I/O controllers have a set of status and control registers used to control the processor and check on its status. Depending on the I/O subsystem, commonly all or some combination of all of the 10 functions from the list of device driver functionality introduced at the start of this chapter are typically implemented in I/O drivers, including:

- *I/O Startup*: initialization of the I/O upon PowerON or reset.
- *I/O Shutdown*: configuring I/O into its PowerOFF state.
- *I/O Disable*: allowing other software to disable I/O on-the-fly.
- *I/O Enable*: allowing other software to enable I/O on-the-fly.
- *I/O Acquire*: allowing other software gain singular (locking) access to I/O.
- *I/O Release*: allowing other software to free (unlock) I/O.
- *I/O Read*: allowing other software to read data from I/O.
- *I/O Write*: allowing other software to write data to I/O.
- *I/O Install*: allowing other software to install new I/O on-the-fly.
- *I/O Uninstall*: allowing other software to remove installed I/O on-the-fly.

The Ethernet and RS232 I/O initialization routines for the PowerPC and ARM architectures are provided as examples of I/O startup (initialization) device drivers. These examples are to demonstrate how I/O can be implemented on more complex architectures, such as PowerPC and ARM, and this in turn can be used as a guide to understand how to write I/O drivers on other processors that are as complex or less complex than the PowerPC and ARM architectures. Other I/O driver routines were not pseudocoded in this chapter, because the same concepts apply here as in Sections 8.1 and 8.2. In short, it is up to the responsible developer to study the architecture and I/O device documentation for the mechanisms used to read from an I/O device, write to an I/O device, enable an I/O device, etc.

### 8.4.1 Example 4: Initializing an Ethernet Driver

Continuing the networking example from Chapter 6, the example used here will be the widely implemented LAN protocol Ethernet, which is primarily based upon the IEEE 802.3 family of standards.

As shown in Figure 8-34, the software required to enable Ethernet functionality maps to the lower section of the OSI (Open Systems Interconnection) data-link layer. The hardware components can all be mapped to the physical layer of the OSI model, but will not be discussed in this section (see Section II).

As mentioned in Section II, the Ethernet component that can be integrated onto the master processor is called the *Ethernet Interface*. The only firmware (software) that is implemented is in the Ethernet interface. The software is dependent on how the hardware supports two main components of the IEEE802.3 Ethernet protocol: the *media access management* and *data encapsulation*.

#### *Data Encapsulation (Ethernet Frame)*

In an Ethernet LAN, all devices connected via Ethernet cables can be set up as a bus or star topology (see Figure 8-35).

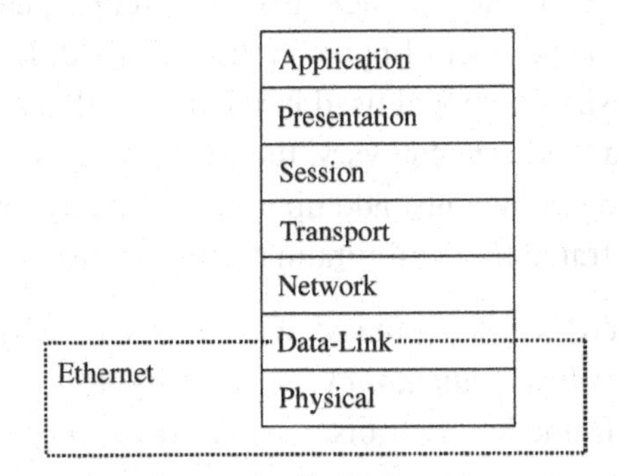


**Figure 8-34**
OSI Model.

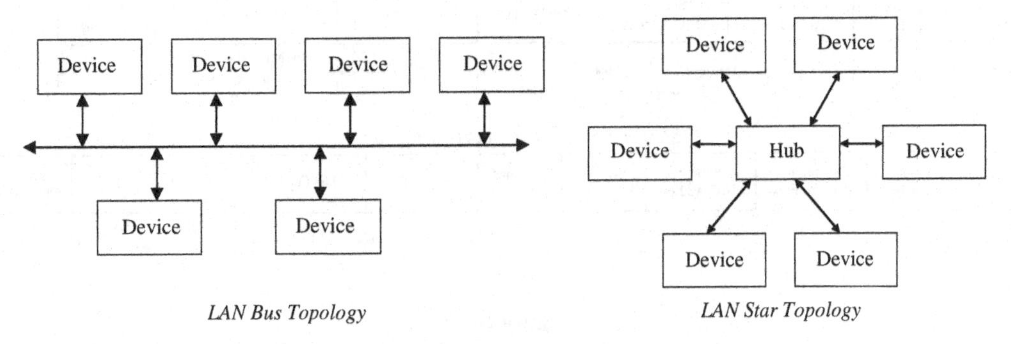


**Figure 8-35**
Ethernet Topologies.

In these topologies, all devices share the same signaling system. After a device checks for LAN activity and determines after a certain period there is none, the device then transmits its Ethernet signals serially. The signals are then received by all other devices attached to the LAN—thus the need for an "Ethernet frame," which contains the data as well as the information needed to communicate to each device which device the data is actually intended for.

Ethernet devices encapsulate data they want to transmit or receive into what are called "Ethernet frames." The Ethernet frame (as defined by IEEE 802.3) is made of up a series of bits, each grouped into fields. Multiple Ethernet frame formats are available, depending on the features of the LAN. Two such frames (see the IEEE 802.3 specification for a description of all defined frames) are shown in Figure 8-36.

The *preamble* bytes tell devices on the LAN that a signal is being sent. They are followed by "10101011" to indicate the *start* of a *frame*. The *media access control (MAC) addresses* in the Ethernet frame are physical addresses unique to each Ethernet interface in a device, so every device has one. When the frame is received by a device, its data-link layer looks at the destination address of the frame. If the address doesn't match its own MAC address, the device disregards the rest of the frame.

The *Data* field can vary in size. If the data field is less than or equal to 1500 then the *Length/Type* field indicates the number of bytes in the data field. If the data field is greater than 1500, then the type of MAC protocol used in the device that sent the frame is defined in Length/Type. While the data field size can vary, the MAC Addresses, the Length/Type, the Data, Pad, and Error checking fields must add up to be at least 64 bytes long. If not, the *Pad* field is used to bring up the frame to its minimum required length.

The *Error checking* field is created using the MAC Addresses, Length/Type, Data Field, and Pad fields. A 4-byte *CRC (cyclical redundancy check)* value is calculated from these fields and stored at the end of the frame before transmission. At the receiving device, the value is recalculated, and, if it doesn't match, the frame is discarded.

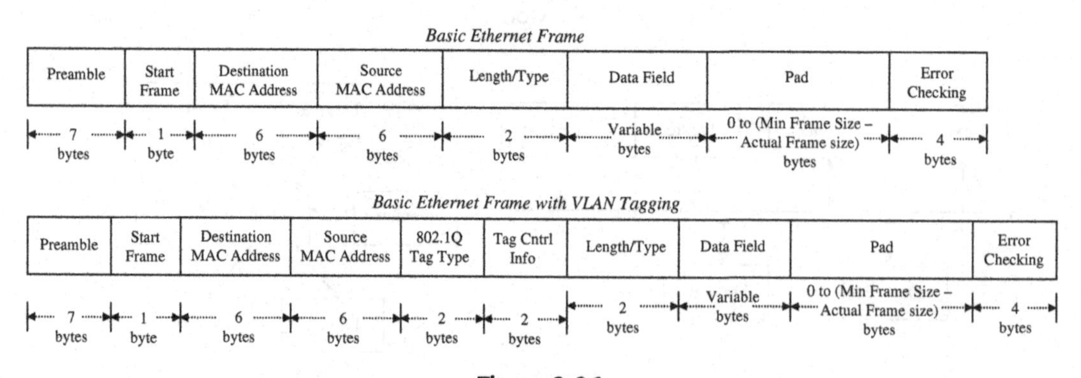


**Figure 8-36**
Ethernet Frames.[7]

Finally, remaining frame formats in the Ethernet specification are extensions of the basic frame. The VLAN (virtual local-area network) tagging frame shown above is an example of one of these extended frames, and contains two additional fields: *802.1Q tag type* and *Tag Control Information*. The *802.1Q tag type* is always set to 0×8100 and serves as an indicator that there is a VLAN tag following this field, and not the Length/Type field, which in this format is shifted 4-bytes over within the frame. The *Tag Control Information* is actually made up of three fields: the *user priority field* (UPF), the *canonical format indicator* (CFI), and the *VLAN identifier* (VID). The UPF is a 3-bit field that assigns a priority level to the frame. The CFI is a 1-bit field to indicate whether there is a Routing Information Field (RIF) in the frame, while the remaining 12 bits is the VID, which identifies which VLAN this frame belongs to. Note that while the VLAN protocol is actually defined in the IEEE 802.1Q specification, it's the IEEE 802.3ac specification that defines the Ethernet-specific implementation details of the VLAN protocol.

#### *Media Access Management*

Every device on the LAN has an equal right to transmit signals over the medium, so there have to be rules that ensure every device gets a fair chance to transmit data. Should more than one device transmit data at the same time, these rules must also allow the device a way to recover from the data colliding. This is where the two MAC protocols come in: the IEEE 802.3 *Half-Duplex* Carrier Sense Multiple Access/Collision Detect (CDMA/CD) and the IEEE 802.3x *Full-Duplex Ethernet* protocols. These protocols, implemented in the Ethernet interface, dictate how these devices behave when sharing a common transmission medium.

Half-Duplex CDMA/CD capability in an Ethernet device means that a device can either receive or transmit signals over the same communication line, but not do both (transmit and receive) at the same time. Basically, a Half-Duplex CDMA/CD (also known as the MAC sublayer) in the device can both transmit and receive data, from a higher layer or from the physical layer in the device. In other words, the MAC sublayer functions in two modes: transmission (data received from higher layer, processed, then passed to physical layer) or reception (data received from physical layer, processed, then passed to higher layer). The transmit data encapsulation (TDE) component and the transmit media access management (TMAM) components provide the transmission mode functionality, while the receive media access management (RMAM) and the receive data decapsulation (RDD) components provide the reception mode functionality.

#### *CDMA/CD (MAC Sublayer) Transmission Mode*

When the MAC sublayer receives data from a higher layer to transmit to the physical layer, the TDE component first creates the Ethernet frame, which is then passed to the TMAM component. Then, the TMAM component waits for a certain period of time to ensure the transmission line is quiet, and that no other devices are currently transmitting. When the TMAM component has determined that the transmission line is quiet, it transmits (via the physical layer) the data frame over the transmission medium, in the form of bits, one bit at a

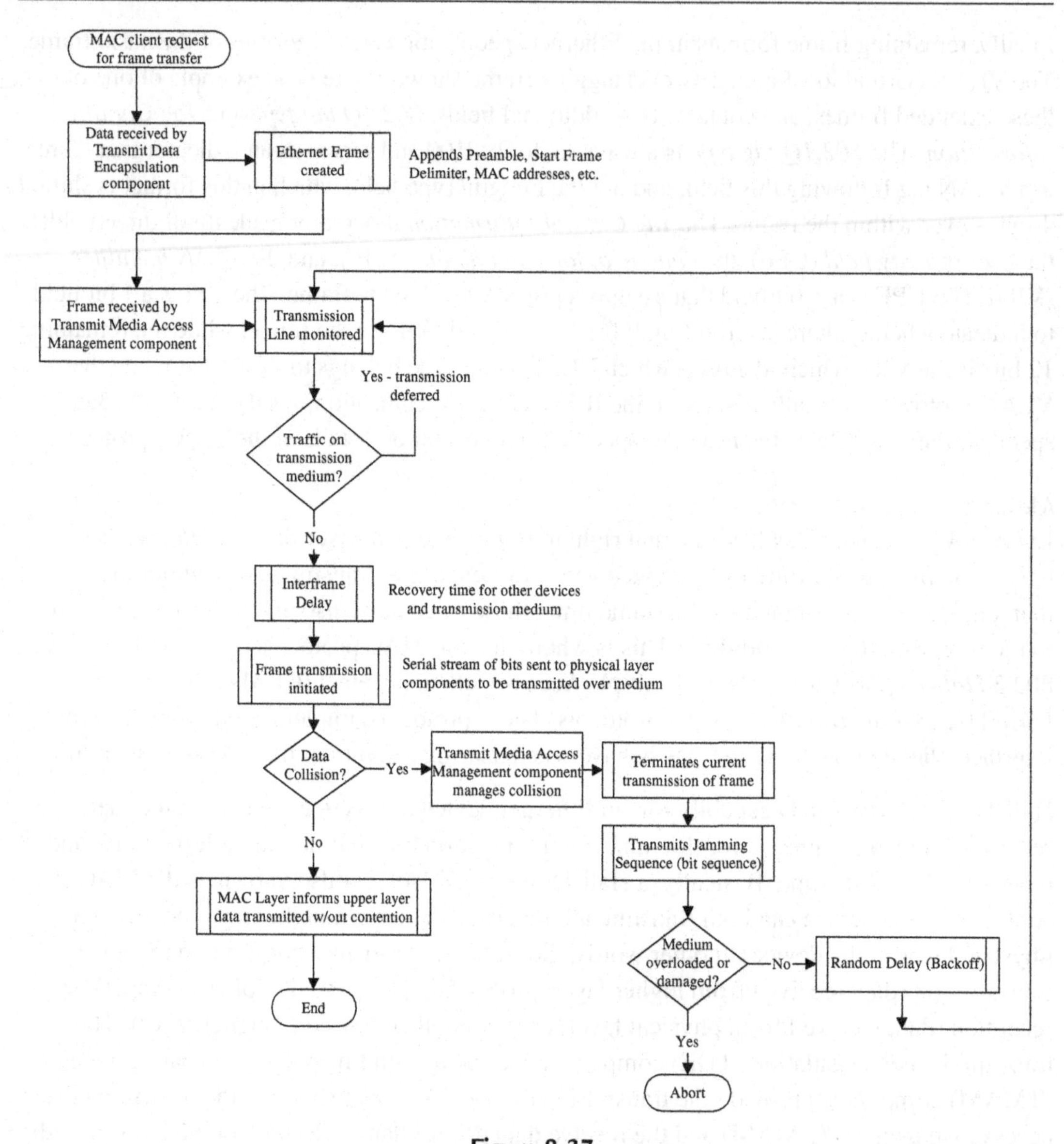


**Figure 8-37**
High-Level Flow Chart of MAC Layer Processing a MAC Client's request to Transmit a Frame.[7]

time (serially). If the TMAM component of this device learns that its data has collided with other data on the transmission line, it transmits a series of bits for a predefined period to let all devices on the system know that a collision has occurred. The TMAM component then stops all transmission for another period of time, before attempting to retransmit the frame again.

Figure 8-37 shows a high-level flow chart of the MAC layer processing a MAC client's (an upper layer) request to transmit a frame.

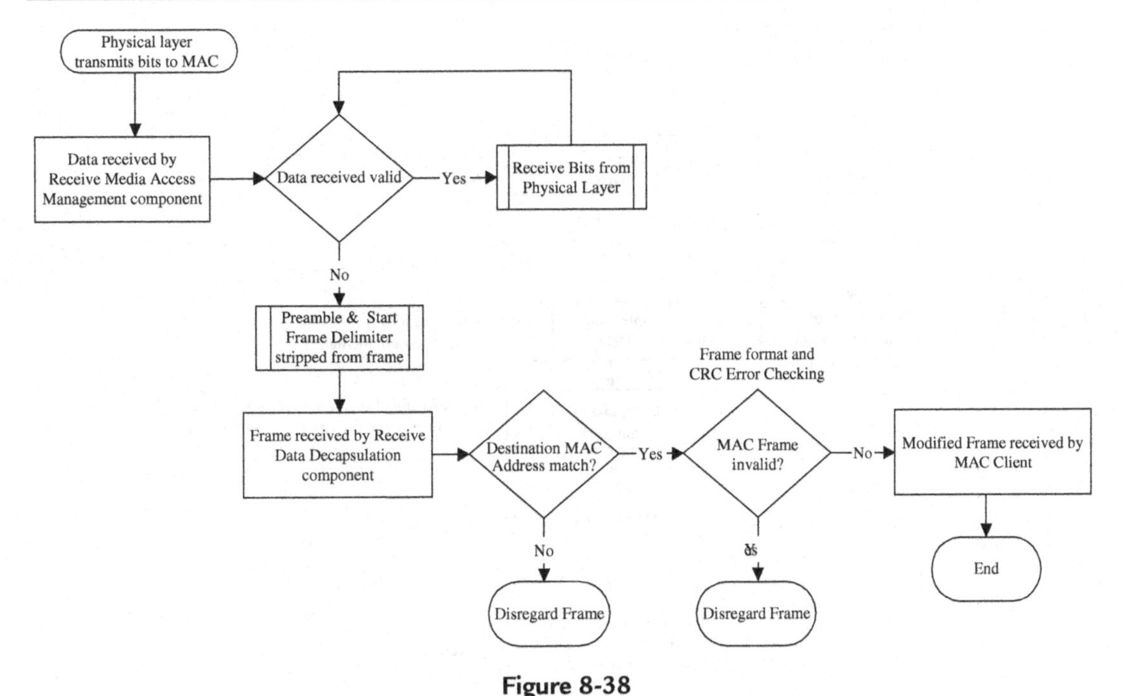


**Figure 8-38**

High-Level Flow Chart of MAC Layer Processing incoming bits from the Physical Layer.[7]

*CDMA/CD (MAC Sublayer) Reception Mode*

When the MAC sublayer receives the stream of bits from the physical layer, to be later transmitted to a MAC client, the MAC sublayer RMAM component receives these bits from the physical layer as a "frame." Note that, as the bits are being received by the RMAM component, the first two fields (preamble and start frame delimiter) are disregarded. When the physical layer ceases transmission, the frame is then passed to the RDD component for processing. It is this component that compares the MAC Destination Address field in this frame to the MAC Address of the device. The RDD component also checks to ensure the fields of the frame are properly aligned, and executes the CRC Error Checking to ensure the frame wasn't damaged en route to the device (the Error Checking field is stripped from the frame). If everything checks out, the RDD component then transmits the remainder of the frame, with an additional status field appended, to the MAC Client.

Figure 8-38 shows a high-level flow chart of the MAC layer processing incoming bits from the physical layer:

It is not uncommon to find that half-duplex capable devices are also full-duplex capable. This is because only a subset of the MAC sublayer protocols implemented in half-duplex are needed for full-duplex operation. Basically, a full-duplex capable device can receive and transmit signals over the same communication media line at the same time. Thus, the throughput in a full-duplex LAN is double that of a half-duplex system.

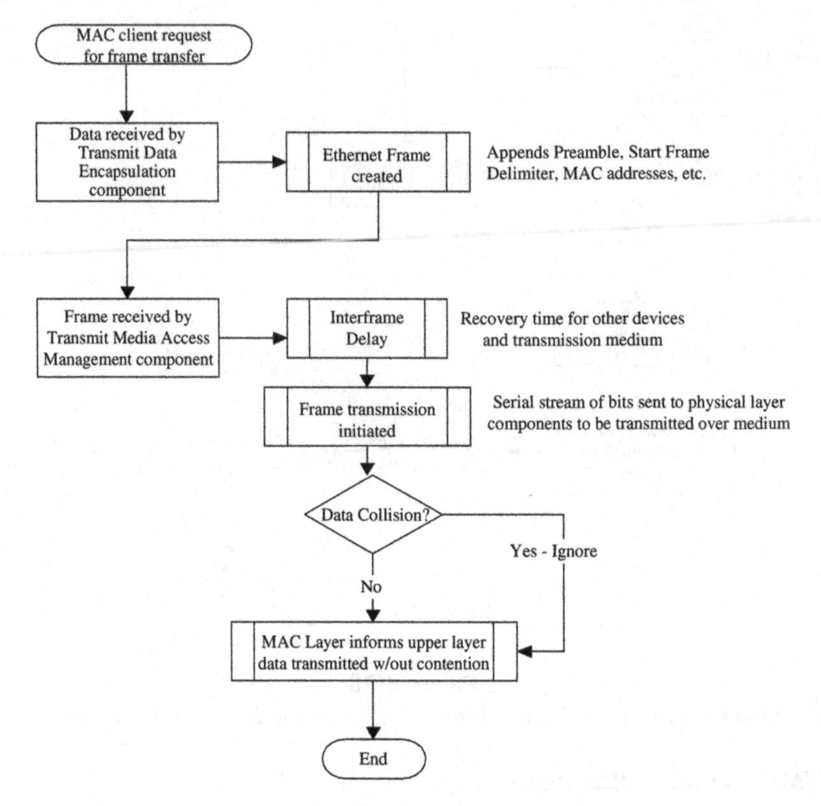


**Figure 8-39**
Flow Chart of High-Level Functions of Full-Duplex in Transmission Mode.[7]

The transmission medium in a full-duplex system must also be capable of supporting simultaneous reception and transmission without interference. For example, 10Base-5, 10Base-2, 10Base-FX, etc., are cables that *do not* support full-duplex, while 10/100/1000Base-T, 100Base-FX, etc., meet full-duplex media specification requirements.

Full-duplex operation in a LAN is restricted to connecting only two devices, and both devices must be capable and configured for full duplex operation. While it is restricting to only allow point to point links, the efficiency of the link in a full-duplex system is actually improved. Having only two devices eliminates the potential for collisions, and eliminates any need for the CDMA/CD algorithms implemented in a half-duplex capable device. Thus, while the reception algorithm is the same for both full and half duplex, Figure 8-39 flowcharts the high-level functions of full-duplex in transmission mode.

Now that you have a definition of all components (hardware and software) that make up an Ethernet system, let's take a look at how architecture-specific Ethernet components are implemented via software on various reference platforms.

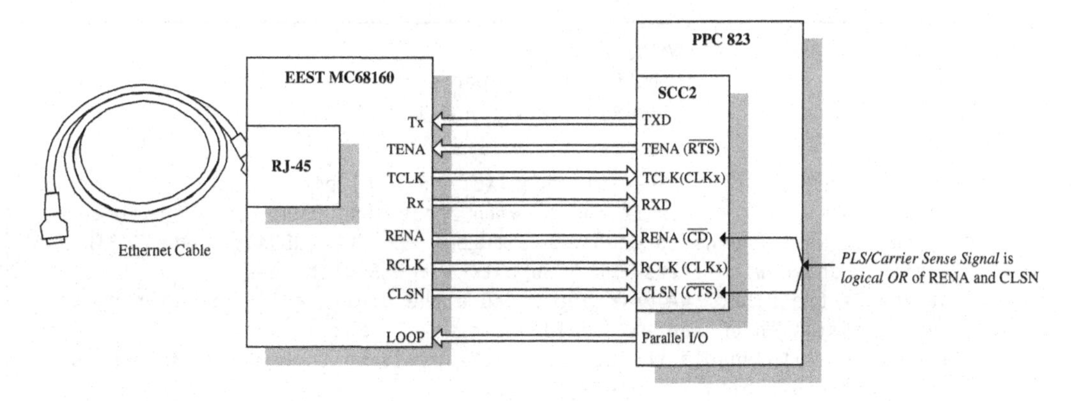


**Figure 8-40**
MPC823 Ethernet Block Diagram.[2]
*© Freescale Semiconductor, Inc. Used by permission.*

*Motorola/Freescale MPC823 Ethernet Example*

Figure 8-40 is a diagram of a MPC823 connected to Ethernet hardware components on the board (see Section II for more information on Ethernet hardware components).

A good starting point for understanding how Ethernet runs on a MPC823 is section 16 in the 2000 MPC823 User's Manual on the MPC823 component that handles networking and communications, called the CPM (Communication Processor Module). It is here that we learn that configuring the MPC823 to implement Ethernet is done through serial communication controllers (SCCs).

*16.9 THE SERIAL COMMUNICATION CONTROLLERS*

The MPC823 has two serial communication controllers (SCC2 and SCC3) that can be configured independently to implement different protocols. They can be used to implement bridging functions, routers, and gateways, and interface with a wide variety of standard WANs, LANs, and proprietary networks ...

The serial communication controllers do not include the physical interface, but they form the logic that formats and manipulates the data obtained from the physical interface. Many functions of the serial communication controllers are common to (among other protocols) the Ethernet controller. The serial communication controllers' main features include support for full 10Mbps Ethernet/IEEE 802.3.

Section 16.9.22 in the MPC823 User's Manual discusses in detail the features of the Serial Communication Controller in Ethernet mode, including full-duplex operation support. In fact, what actually needs to be implemented in software to initialize and configure Ethernet on the PPC823 can be based on the Ethernet programming example in section 16.9.23.7.

*16.9.23.7 SCC2 ETHERNET PROGRAMMING EXAMPLE*

The following is an example initialization sequence for the SCC2 in Ethernet mode. The CLK1 pin is used for the Ethernet receiver and the CLK2 pin is used for the transmitter.

1. Configure the port A pins to enable the TXD1 and RXD1 pins. Write PAPAR bits 12 and 13 with ones, PADIR bits 12 and 13 with zeros, and PAODR bit 13 with zero.
2. Configure the Port C pins to enable CTS2(CLSN) and CD2 (RENA). Write PCPAR and PCDIR bits 9 and 8 with zeros and PCSO bits 9 and 8 with ones.
3. Do not enable the RTS2(TENA) pin yet because the pin is still functioning as RTS and transmission on the LAN could accidentally begin.
4. Configure port A to enable the CLK1 and CLK2 pins. Write PAPAR bits 7 and 6 with ones and PADIR bits 7 and 6 with zeros.
5. Connect the CLK1 and CLK2 pins to SCC2 using the serial interface. Write the R2CS field in the SICR to 101 and the T2CS field to 100.
6. Connect SCC2 to the NMSI and clear the SC2 bit in the SICR.
7. Initialize the SDMA configuration register (SDCR) to 0x0001.
8. Write RBASE and TBASE in the SCC2 parameter RAM to point to the RX buffer descriptor and TX buffer descriptor in the dual-port RAM. Assuming one RX buffer descriptor at the beginning of dual-port RAM and one TX buffer descriptor following that RX buffer descriptor, write RBASE with 0x2000 and TBASE with 0x2008.
9. Program the CPCR to execute the INIT RX BD PARAMETER command for this channel.
10. Write RFCR and TFCR with 0x18 for normal operation.
11. Write MRBLR with the maximum number of bytes per receive buffer. For this case assume 1,520 bytes, so MRBLR = 0x05F0. In this example, the user wants to receive an entire frame into one buffer, so the MRBLR value is chosen to be the first value larger than 1,518 that is evenly divisible by four.
12. Write C_PRES with 0xFFFFFFFF to comply with 32-bit CCITT-CRC.
13. Write C_MASK with 0xDEBB20E3 to comply with 32-bit CDITT-CRC.
14. Clear CRCEC, ALEC, and DISFC for clarity.
15. Write PAD with 0x8888 for the pad value.
16. Write RET_LIM with 0x000F.
17. Write MFLR with 0x05EE to make the maximum frame size 1,518 bytes.
18. Write MINFLR with 0x0040 to make the minimum frame size 64 bytes.
19. Write MAXD1 and MAXD2 with 0x005EE to make the maximum DMA count 1,518 bytes.
20. Clear GADDR1-GADDR4. The group hash table is not used.
21. Write PADDR1_H with 0x0380, PADDR1_M with 0x12E0, and PADDR1_L with 0x5634 to configure the physical address 8003E0123456.
22. Write P_Per with 0x000. It is not used.
23. Clear IADDR1-IADDR4. The individual hash table is not used.
24. Clear TADDR_H, TADDR_M, and TADDR_L for clarity.
25. Initialize the RX buffer descriptor and assume the RX data buffer is at 0x00001000 main memory. Write 0xB000 to Rx_BD_Status, 0x0000 to Rx_BD_Length (optional) and 0x00001000 to Rx_BD_Pointer.
26. Initialize the TX buffer descriptor and assume the TX data frame is at 0x00002000 main memory and contains fourteen 8-bit characters (destination and source addresses plus the type field). Write 0xFC00 to Tx_BD_Status, add PAD to the frame and generate a CRC. Then write 0x000D to Tx_BD_Length and 0x00002000to Tx_BD_Pointer.

```
27. Write 0xFFFF to the SCCE-Ethernet to clear any previous events.
28. Write 0x001A to the SCCM-Ethernet to enable the TXE, RXF, and TXB interrupts.
29. Write 0x20000000 to the CIMR so that SCC2 can generate a system interrupt. The
    CICR must also be initialized.
30. Write 0x00000000 to the GSMR_H to enable normal operation of all modes.
31. Write 0x1088000C to the GSMR_L to configure the CTS2 (CLSN) and CD2 (RENA)
    pins to automatically control transmission and reception (DIAG field) and the
    Ethernet mode. TCI is set to allow more setup time for the EEST to receive
    the MPC82 transmit data. TPL and TPP are set for Ethernet requirements. The
    DPLL is not used with Ethernet. Notice that the transmitter (ENT) and receiver
    (ENR) have not been enabled yet.
32. Write 0xD555 to the DSR.
33. Set the PSMR-SCC Ethernet to 0x0A0A to configure 32-bit CRC, promiscuous mode
    and begin searching for the start frame delimiter 22 bits after RENA.
34. Enable the TENA pin (RTS2). Since the MODE field of the GMSR_L is written to
    Ethernet, the TENA signal is low. Write PCPAR bit 14 with a one and PCDIR bit
    14 with a zero.
35. Write 0x1088003C to the GSMR_L register to enable the SCC2 transmitter and
    receiver. This additional write ensures that the ENT and ENR bits are enabled
    last.

NOTE: After 14 bytes and the 46 bytes of automatic pad (plus the 4 bytes of CRC)
are transmitted, the TX buffer descriptor is closed. Additionally, the receive
buffer is closed after a frame is received. Any data received after 1,520 bytes or
a single frame causes a busy (out-of-buffer) condition since only one RX buffer
descriptor is prepared.
```

It is from section 16.9.23.7 that the Ethernet initialization device driver source code can be written. It is also from this section that it can be determined how Ethernet on the MPC823 is configured to be *interrupt driven*. The actual initialization sequence can be divided into seven major functions: disabling SCC2, configuring ports for Ethernet transmission and reception, initializing buffers, initializing parameter RAM, initializing interrupts, initializing registers, and starting Ethernet (see pseudocode below).

```
MPC823 Ethernet Driver Pseudocode


// disabling SCC2
   // Clear GSMR_L[ENR] to disable the receiver
   GSMR_L = GSMR_L & 0x00000020
   // Issue Init Stop TX Command for the SCC
   Execute Command (GRACEFUL_STOP_TX)
   // clear GSLM_L[ENT] to indicate that transmission has stopped
   GSMR_L = GSMR_L & 0x00000010
```

```
-=-=-=-=
// Configure port A to enable TXD1 and RXD1 - step 1 from user's manual
PADIR = PADIR & 0xFFF3     // Set PAPAR[12,13]
PAPAR = PAPAR | 0x000C     // clear PADIR[12,13]
PAODR = PAODR & 0xFFF7     // clear PAODR[12]

// Configure port C to enable CLSN and RENA - step 2 from user's manual
PCDIR = PCDIR & 0xFF3F     // clear PCDIR[8,9]
PCPAR = PCPAR & 0xFF3F     // Clear PCPAR[8,9]
PCSO = PCSO | 0x00C0       // set PCSO[8,9]

// step 3 - do nothing now

// configure port A to enable the CLK2 and CLK4 pins - step 4 from user's manual
PAPAR = PAPAR | 0x0A00     // set PAPAR[6] (CLK2) and PAPAR[4] (CLK4).
PADIR = PADIR & 0xF5FF     // clear PADIR[4] and PADIR[6]. (All 16-bit)

// Initializing the SI Clock Route Register (SICR) for SCC2.
// Set SICR[R2CS] to 111 and Set SICR[T2CS] to 101, Connect SCC2 to NMSI and Clear
SICR[SC2] - steps 5 & 6 from user's manual
SICR = SICR & 0xFFFFBFFF
SICR = SICR | 0x00003800
SICR=(SICR & 0xFFFFF8FF) | 0x00000500

// Initializing the SDMA configuration register - step 7
SDCR = 0x01                // Set SDCR to 0x1 (SDCR is 32-bit) - step 7 from user's
                           manual

// Write RBASE in the SCC1 parameter RAM to point to the RxBD table and the TxBD
table in the
// dual-port RAM and specify the
// size of the respective buffer descriptor pools - step 8 user's manual
RBase = 0x00 (for example)
RxSize = 1500 bytes (for example)
TBase = 0x02 (for example)
TxSize = 1500 bytes (for example)
Index = 0
While (index < RxSize) do
{
// Set up one receive buffer descriptor that tells the communication processor that
the next packet is
// ready to be received - similar to step 25
// Set up one transmit buffer descriptor that tells the communication processor
that the next packet is
// ready to be transmitted - similar to step 26
index = index + 1}

// Program the CPCR to execute the INIT_RX_AND_TX_PARAMS - deviation from step 9 in
user's
// guide
```

```
execute Command(INIT_RX_AND_TX_PARAMS)
// Write RFCR and TFCR with 0x10 for normal operation (all 8-bits) or 0x18 for
normal operation
// and Motorola/Freescale byte ordering - step 10 from user's manual
RFCR = 0x10
TFCR = 0x10

// Write MRBLR with the maximum number of bytes per receive buffer and assume 16
bytes - step
// 11 user's manual
MRBLR = 1520

// Write C_PRES with 0xFFFFFFFF to comply with the 32 bit CRC-CCITT - step 12
user's manual
C_PRES = 0xFFFFFFFF

// Write C_MASK with 0xDEBB20E3 to comply with the 16 bit CRC-CCITT - step 13 user's
// manual
C_MASK = 0xDEBB20E3

// Clear CRCEC, ALEC, and DISFC for clarity - step 14 user's manual
CRCEC = 0x0
ALEC = 0x0
DISFC = 0x0

// Write PAD with 0x8888 for the PAD value - step 15 user's manual
PAD = 0x8888

// Write RET_LIM to specify how many retries (with 0x000F for example) - step 16
RET_LIM = 0x000F

// Write MFLR with 0x05EE to make the maximum frame size 1518 bytes - step 17
MFLR = 0x05EE

// Write MINFLR with 0x0040 to make the minimum frame size 64 bytes - step 18
MINFLR = 0x0040

// Write MAXD1 and MAXD2 with 0x05F0 to make the maximum DMA count 1520 bytes -
step 19
MAXD1 = 0x05F0
MAXD2 = 0x05F0

// Clear GADDR1-GADDR4. The group hash table is not used - step 20
GADDR1 = 0x0
GADDR2 = 0x0
GADDR3 = 0x0
GADDR4 = 0x0

// Write PADDR1_H, PADDR1_M and PADDR1_L with the 48-bit station address - step 21
stationAddr="embedded device's Ethernet address" = (for example) 8003E0123456
```

```
PADDR1_H = 0x0380 ["80 03" of the station address]
PADDR1_M = 0x12E0 ["E0 12" of the station address]
PADDR1_L = 0x5634 ["34 56" of the station address]

// Clear P_PER. It is not used - step 22
P_PER = 0x0

// Clear IADDR1-IADDR4. The individual hash table is not used - step 23
IADDR1 = 0x0
IADDR2 = 0x0
IADDR3 = 0x0
IADDR4 = 0x0

// Clear TADDR_H, TADDR_M and TADDR_L for clarity - step 24
groupAddress = "embedded device's group address" = no group address for example
TADDR_H = 0 [similar as step 21 high byte reversed]
TADDR_M = 0 [middle byte reversed]
TADDR_L = 0 [low byte reversed]

// Initialize the RxBD and assume that Rx data buffer is at 0x00001000. Write
0xB000 to
// RxBD[Status and Control] Write 0x0000 to RxBD[Data Length]
// Write 0x00001000 to RxDB[BufferPointer] - step 25
RxBD[Status and Control] is the status of the buffer = 0xB000
Rx data buffer is the byte array the communication processor can use to store the
incoming packet in.
= 0x00001000
Save Buffer and Buffer Length in Memory, Then Save Status

// Initialize the TxBD and assume that Tx data buffer is at 0x00002000 Write 0xFC00 to
// TxBD[Status and Control] Write 0x0000 to TxBD[Data Length]

// Write 0x00002000 to TxDB[BufferPointer] - step 26

TxBD[Status and Control] is the status of the buffer = 0xFC00
Tx data buffer is the byte array the communication processor can use to store the
outgoing packet in.
= 0x00002000
Save Buffer and Buffer Length in Memory, Then Save Status

// Write 0xFFFF to the SCCE-Transparent to clear any previous events - step 27
user's manual
SCCE = 0xFFFF

// Initialize the SCCM-Transparent (SCC mask register) depending on the interrupts
required of the
// SCCE[TXB, TXE, RXB, RXF] interrupts possible. - step 28 user's manual
 // Write 0x001B to the SCCM for generating TXB, TXE, RXB, RXF interrupts (all events).
 // Write 0x0018 to the SCCM for generating TXE and RXF Interrupts (errors).
 // Write 0x0000 to the SCCM in order to mask all interrupts.
 SCCM = 0x0000
```

```
// Initialize CICR, and Write to the CIMR so that SCC2 can generate a system
interrupt. - step 29
CIMR = 0x200000000
CICR = 0x001B9F80

// Write 0x00000000 to the GSMR_H to enable normal operation of all modes - step 30
user's manual
GSMR_H = 0x0

// GSMR_L: 0x1088000C: TCI = 1, TPL = 0b100, TPP = 0b01, MODE = 1100 to configure the
// CTS2 and CD2 pins to automatically control transmission and reception (DIAG field).
Normal
// operation of the transmit clock is used. Notice that the transmitter (ENT) and
receiver (ENR) are
// not enabled yet. - step 31 user's manual
GSMR_L = 0x1088000C

// Write 0xD555 to the DSR - step 32
DSR = 0xD555

// Set PSMR-SCC Ethernet to configure 32-bit CRC - step 33
   // 0x080A: IAM = 0, CRC = 10 (32-bit), PRO = 0, NIB = 101
   // 0x0A0A: IAM = 0, CRC = 10 (32-bit), PRO = 1, NIB = 101
   // 0x088A: IAM = 0, CRC = 10 (32-bit), PRO = 0, SBT = 1, NIB = 101
   // 0x180A: HBC = 1, IAM = 0, CRC = 10 (32-bit), PRO = 0, NIB = 101
PSMR = 0x080A

// Enable the TENA pin (RTS2) Since the MODE field of the GSMR_L is written to
Ethernet, the
// TENA signal is low. Write PCPAR bit 14 with a one and PCDIR bit 14 with a
// zero - step 34
PCPAR = PCPAR | 0x0001
PCDIR = PCDIR & 0xFFFE

// Write 0x1088003C to the GSMR_L register to enable the SCC2 transmitter and
receiver. - step 35
GSMR_L = 0x1088003C

-=-=-=-

// Start the transmitter and the receiver
// After initializing the buffer descriptors, program the CPCR to execute an INIT
RX AND TX
// PARAMS command for this channel.
      Execute Command(Cp.INIT_RX_AND_TX_PARAMS)
/
/ Set GSMR_L[ENR] and GSMR_L[ENT] to enable the receiver and the transmitter
      GSMR_L = GSMR_L | 0x00000020 | 0x00000010

// END OF MPC823 ETHERNET INITIALIZATION SEQUENCE - now when appropriate inter-
// rupt triggered, data is moved to or from transmit/receive buffers
```

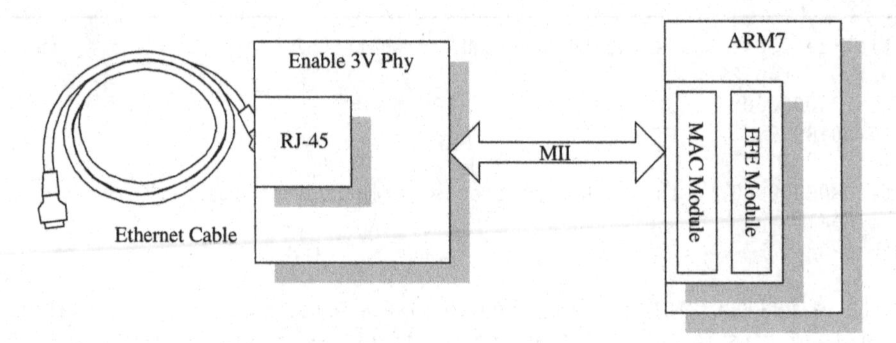


**Figure 8-41**
NET + ARM Ethernet Block Diagram.[8]

*NetSilicon NET+ARM40 Ethernet Example*

Figure 8-41 shows a diagram of a NET+ARM connected to Ethernet hardware components on the board (see Section II for more information on Ethernet hardware components).

Like the MPC823, the NET+ARM40 Ethernet protocol is configured to have full-duplex support, as well as be *interrupt driven*. However, unlike the MPC823, the NET+ARM's initialization sequence is simpler and can be divided into three major functions: performing reset of Ethernet processor, initializing buffers, and enabling DMA channels (see NET+ARM Hardware User's Guide for NET+ARM 15/40 and pseudocode below).

```
NET + ARM40 Pseudocode

...
// Perform a low level reset of the NCC Ethernet chip
// determine MII type
MIIAR = MIIAR & 0xFFFF0000 | 0x0402
MIICR = MIICR | 0x1
// wait until current PHY operation completes

if using MII
{
// set PCSCR according to poll count - 0x00000007 (>= 6), 0x00000003 (< 6)
// enable autonegotiation
}
else {                              // ENDEC MODE
EGCR = 0x0000C004
// set PCSCR according to poll count - 0x00000207 (>= 6), 0x00000203 (< 6)
// set EGCR to correct mode if automan jumper removed from board
}
```

```
// clear transfer and receive registers by reading values
get LCC
get EDC
get MCC
get SHRTFC
get LNGFC
get AEC
get CRCEC
get CEC

// Inter-packet Gap Delay = 0.96us for MII and 9.6us for 10BaseT
if using MII then {
B2BIPGGTR = 0x15
NB2BIPGGTR = 0x0C12
} else {
B2BIPGGTR = 0x5D
NB2BIPGGTR = 0x365A);
}

MACCR = 0x0000000D

// Perform a low level reset of the NCC Ethernet chip continued

// Set SAFR = 3: PRO Enable Promiscuous Mode (receive ALL packets), 2: PRM Accept
ALL
// multicast packets, 1: PRA Accept multicast packets using Hash
// Table, 0 : BROAD Accept ALL broadcast packets
SAFR = 0x00000001

// load Ethernet address into addresses 0xFF8005C0-0xFF8005C8
// load MCA hash table into addresses 0xFF8005D0-0xFF8005DC

STLCR = 0x00000006
If using MII{
   // Set EGCR according to what rev - 0xC0F10000 (rev<4), 0xC0F10000 (PNA support
   disabled)
else {
   // ENDEC mode
EGCR = 0xC0C08014}

// Initialize buffer descriptors
   // setup Rx and Tx buffer descriptors
   DMABDP1A = "receive buffer descriptors"
       DMABDP2 = "transmit buffer descriptors"

// enable Ethernet DMA channels
// setup the interrupts for receive channels
DMASR1A = DMASR1A & 0xFF0FFFFF | (NCIE | ECIE | NRIE | CAIE)
```

```
// setup the interrupts for transmit channels
DMASR2 = DMASR2 & 0xFF0FFFFF | (ECIE | CAIE)

   // Turn each channel on

   If MII is 100Mbps then {

                              DMACR1A = DMACR1A & 0xFCFFFFFF | 0x02000000
                              }

DMACR1A = DMACR1A & 0xC3FFFFFF | 0x80000000
   If MII is 100Mbps then {
       DMACR2 = DMACR2 & 0xFCFFFFFF | 0x02000000
       }
else if MII is 10Mbps{
       DMACR2 = DMACR2 & 0xFCFFFFFF
       }
DMACR2 = DMACR2 & 0xC3FFFFFF | 0x84000000

// Enable the interrupts for each channel
DMASR1A = DMASR1A | NCIP | ECIP | NRIP | CAIP
DMASR2 = DMASR2 | NCIP | ECIP | NRIP | CAIP

// END OF NET+ARM ETHERNET INITIALIZATION SEQUENCE - now when appropriate
// interrupt triggered, data is moved to or from transmit/receive buffers
```

### 8.4.2 Example 5: Initializing an RS-232 Driver

One of the most widely implemented asynchronous serial I/O protocols is the *RS-232* or EIA-232 (Electronic Industries Association-232), which is primarily based upon the Electronic Industries Association family of standards. These standards define the major components of any RS-232 based system, which is implemented almost entirely in hardware.

The firmware (software) required to enable RS-232 functionality maps to the lower section of the OSI data-link layer. The hardware components can all be mapped to the physical layer of the OSI model (see Figure 8-42), but will not be discussed in this section (see Section II).

As mentioned in Chapter 6, the RS-232 component that can be integrated on the master processor is called the *RS-232 Interface*, which can be configured for synchronous or asynchronous transmission. For example, in the case of asynchronous transmission, the only firmware (software) that is implemented for RS-232 is in a component called the *UART (universal asynchronous transmitter receiver)*, which implements the serial data transmission (see Figure 8-43).

Data is transmitted asynchronously over RS-232 in a stream of bits that are traveling at a constant rate. The frame processed by the UART is in the format shown in Figure 8-44.

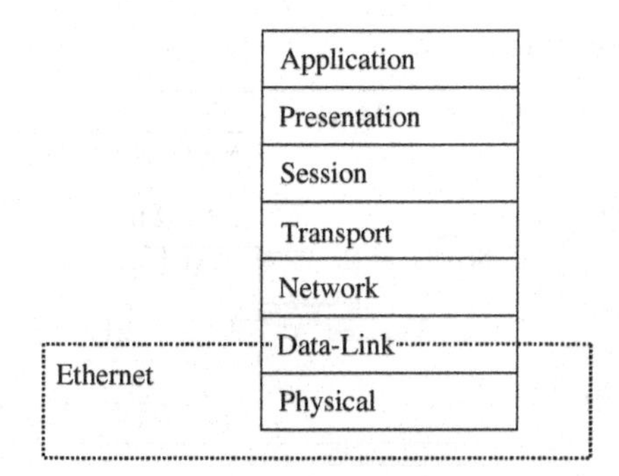


**Figure 8-42**
OSI model.

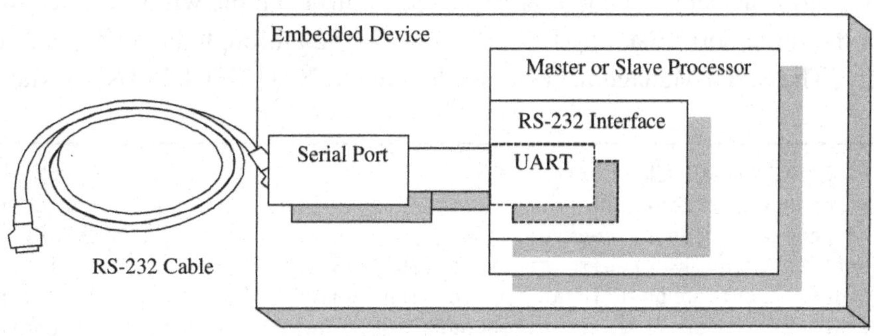


**Figure 8-43**
RS-232 Hardware Diagram.[7]

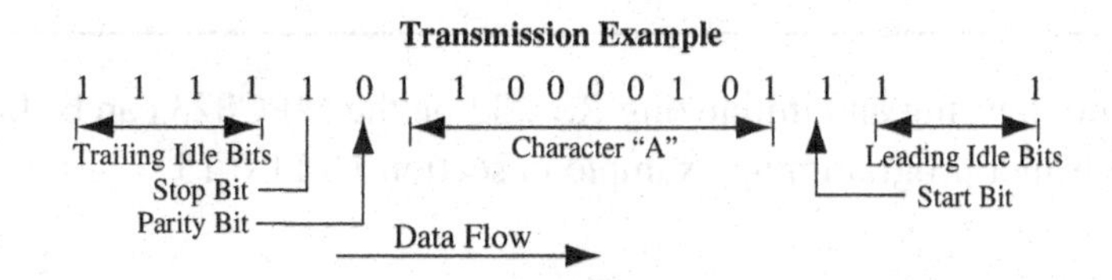


**Figure 8-44**
RS-232 Frame Diagram.[7]
The RS232 protocol defines frames as having: one start bit, seven or eight data bits, one parity bit, and one or two stop bits.

### *Motorola/Freescale MPC823 RS-232 Example*

Figure 8-45 shows a MPC823 connected to RS-232 hardware components on the board (see Section II for more information on the other hardware components).

There are different integrated components on a MPC823 that can be configured into UART mode, such as SCC2 and the SMCs (serial management controllers). SCC2 was discussed in

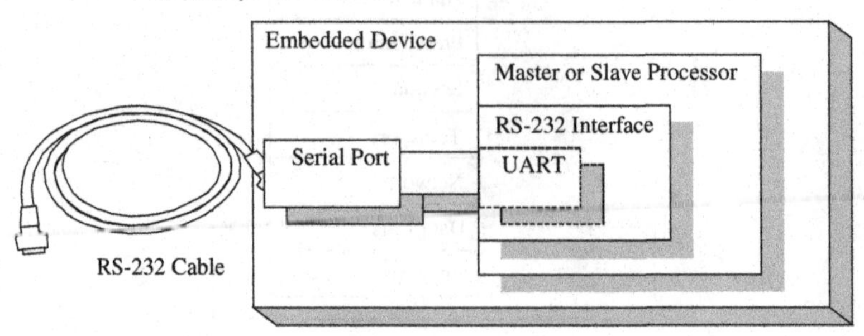


**Figure 8-45**
MPC823 RS-232 Block Diagram.[9]
*© Freescale Semiconductor, Inc. Used by permission.*

the previous section as being enabled for Ethernet, so this example will look at configuring an SMC for the serial port. Enabling RS-232 on a MPC823 through the SMCs is discussed in section 16.11, The Serial Management Controllers in the 2000 MPC823 User's Manual.

```
16.11 THE SERIAL MANAGEMENT CONTROLLERS
The serial management controllers (SMCs) consist of two full-duplex ports that
can be independently configured to support any one of three protocols—UART,
Transparent, or general-circuit interface (GCI). Simple UART operation is used to
provide a debug/monitor port in an application, which allows a serial communication
controller (SCCx) to be free for other purposes. The serial management controller
clock can be derived from one of four internal baud rate generators or from a 16x
external clock pin.

...
```

The software for configuring and initializing RS-232 on the MPC823 can be based upon the SMC1 UART controller programming example in section 16.11.6.15.

```
16.11.6.15 SMC1 UART CONTROLLER PROGRAMMING EXAMPLE
The following is an initialization sequence for 9,600baud, 8 data bits, no parity,
and 1 stop bit operation of an SMC1 UART controller assuming a 25MHz system
frequency. BRG1 and SMC1 are used.

1. Configure the port B pins to enable SMTXD1 and SMRXD1. Write PBPAR bits 25 and
   24 with ones and then PBDIR and PBODR bits 25 and 24 with zeros.
2. Configure the BRG1. Write 0x010144 to BRGC1. The DIV16 bit is not used and
   divider is 162 (decimal). The resulting BRG1 clock is 16x the preferred bit
   rate of SMC1 UART controller.
```

3. Connect the BRG1 clock to SMC1 using the serial interface. Write the SMC1 bit SIMODE with a D and the SMC1CS field in SIMODE register with 0x000.
4. Write RBASE and TBASE in the SMC1 parameter RAM to point to the RX buffer descriptor and TX buffer descriptor in the dual-port RAM. Assuming one RX buffer descriptor at the beginning of dual-port RAM and one TX buffer descriptor following that RX buffer descriptor, write RBASE with 0x2000 and TBASE with 0x2008.
5. Program the CPCR to execute the INIT RX AND TX PARAMS command. Write 0x0091 to the CPCR.
6. Write 0x0001 to the SDCR to initialize the SDMA configuration register.
7. Write 0x18 to the RFCR and TFCR for normal operation.
8. Write MRBLR with the maximum number of bytes per receive buffer. Assume 16 bytes, so MRBLR = 0x0010.
9. Write MAX_IDL with 0x0000 in the SMC1 UART parameter RAM for clarity.
10. Clear BRKLN and BRKEC in the SMC1 UART parameter RAM for clarity.
11. Set BRKCR to 0x0001, so that if a STOP TRANSMIT command is issued, one bit character is sent.
12. Initialize the RX buffer descriptor. Assume the RX data buffer is at 0x00001000 in main memory. Write 0xB000 to RX_BD_Status. 0x0000 to RX_BD_Length (not required), and 0x00001000 to RX_BD_Pointer.
13. Initialize the TX buffer descriptor. Assume the TX data buffer is at 0x00002000 in main memory and contains five 8-bit characters. Then write 0xB000 to TX_BD_Status, 0x0005 to TX_BD_Length, and 0x00002000 to TX_BD_Pointer.
14. Write 0xFF to the SMCE-UART register to clear any previous events.
15. Write 0x17 to the SMCM-UART register to enable all possible serial management controller interrupts.
16. Write 0x00000010 to the CIMR to SMC1 can generate a system interrupt. The CICR must also be initialized.
17. Write 0x4820 to SMCMR to configure normal operation (not loopback), 8-bit characters, no parity, 1 stop bit. Notice that the transmitter and receiver are not enabled yet.
18. Write 0x4823 to SMCMR to enable the SMC1 transmitter and receiver. This additional write ensures that the TEN and REN bits are enabled last.

NOTE: After 5 bytes are transmitted, the TX buffer descriptor is closed. The receive buffer is closed after 16 bytes are received. Any data received after 16 bytes causes a busy (out-of-buffers) condition since only one RX buffer descriptor is prepared.

Similarly to the Ethernet implementation, MPC823 serial driver is configured to be *interrupt driven,* and its initialization sequence can also be divided into seven major functions: disabling SMC1, setting up ports and the baud rate generator, initializing buffers, setting up parameter RAM, initializing interrupts, setting registers, and enabling SMC1 to transmit/receive (see the following pseudocode).

```
MPC823 Serial Driver Pseudocode
…

// disabling SMC1

// Clear SMCMR[REN] to disable the receiver
        SMCMR = SMCMR & 0×0002
   // Issue Init Stop TX Command for the SCC
              execute command(STOP_TX)
   // clear SMCMR[TEN] to indicate that transmission has stopped
        SMCMR = SMCMR & 0×0002

-=-=-

// Configure port B pins to enable SMTXD1 and SMRXD1. Write PBPAR bits 25 and 24
with ones
// and then PBDIR bits 25 and 24 with zeros - step 1 user's manual
PBPAR = PBPAR | 0×000000C0
PBDIR= PBDIR & 0xFFFFFF3F
PBODR = PBODR & 0xFFFFFF3F

// Configure BRG1 - BRGC: 0×10000 - EN = 1-25 MHZ : BRGC: 0×010144 - EN = 1, CD = 162
// (b10100010), DIV16 = 0 (9600)
// BRGC: 0×010288 - EN = 1, CD = 324 (b101000100), DIV16 = 0 (4800)
// 40 Mhz : BRGC: 0×010207 - EN = 1, CD = 259 (b1 0000 0011), DIV16 = 0
// (9600) - step 2 user's manual

BRGC= BRGC | 0×010000

// Connect the BRG1 (Baud rate generator) to the SMC. Set the SIMODE[SMCx] and the
// SIMODE[SMC1CS] depending on baude rate generator where SIMODE[SMC1] =
// SIMODE[16], and SIMODE[SMC1CS]=SIMODE[17-19] - step 3 user's manual

SIMODE = SIMODE & 0xFFFF0FFF | 0×1000

// Write RBASE and TBASE in the SCM parameter RAM to point to the RxBD table and
the TxBD
// table in the dual-port RAM - step 4

RBase = 0×00 (for example)
RxSize = 128 bytes (for example)
TBase = 0×02 (for example)
TxSize = 128 bytes (for example)
Index = 0
While (index<RxSize) do
{

// Set up one receive buffer descriptor that tells the communication processor that
the next packet is
```

```
// ready to be received - similar to step 12
// Set up one transmit buffer descriptor that tells the communication processor
that the next packet is
// ready to be transmitted - similar to step 13
index = index + 1}
// Program the CPCR to execute the INIT RX AND TX PARAMS command - step 5
execute Command(INIT_RX_AND_TX_PARAMS)

// Initialize the SDMA configuration register, Set SDCR to 0x1 (SDCR is 32-bit) -
step 6 user's
// manual
SDCR =0x01

// Set RFCR,TFCR - Rx,Tx Function Code, Initialize to 0x10 for normal operation
(All 8-bits),
// Initialize to 0x18 for normal operation and Motorola/Freescale byte ordering -
step 7
RFCR = 0x10
TFCR = 0x10

// Set MRBLR - Max. Receive Buffer Length, assuming 16 bytes (multiple of 4) - step 8
MRBLR = 0x0010

// Write MAX_IDL (Maximum idle character) with 0x0000 in the SMC1 UART parameter
RAM to
// disable the MAX_IDL functionality - step 9
MAX_IDL = 0

// Clear BRKLN and BRKEC in the SMC1 UART parameter RAM for clarity - step 10
BRKLN = 0
BRKEC = 0

// Set BRKCR to 0x01 - so that if a STOP TRANSMIT command is issued, one break
character is
// sent - step 11

BRKCR = 0x01
```

## 8.5 Summary

This chapter discussed device drivers, the type of software needed to manage the hardware in an embedded system. The chapter also introduced a general set of device driver routines, which make up most device drivers. Interrupt handling (on the PowerPC platform), memory management (on the PowerPC platform), $I^2C$ bus (on a PowerPC-based platform), and I/O (Ethernet and RS-232 on PowerPC and ARM-based platforms) were real-world examples provided, along with pseudocode to demonstrate how device driver functionality can be implemented.

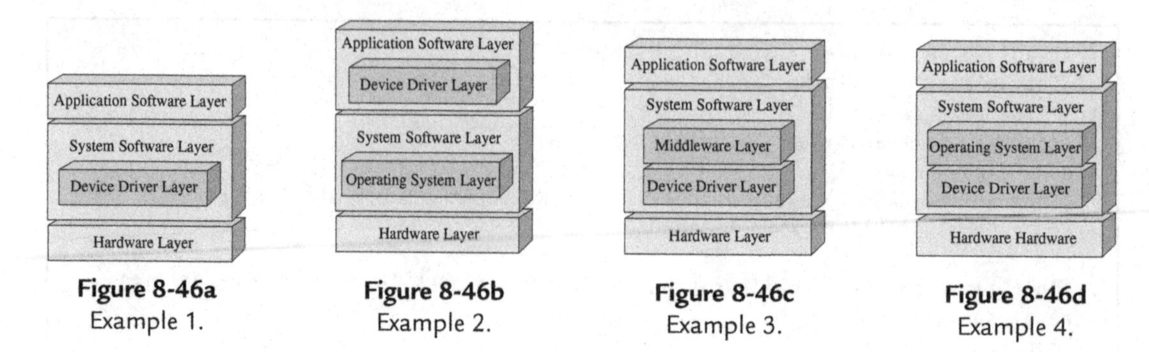


**Figure 8-46a**
Example 1.

**Figure 8-46b**
Example 2.

**Figure 8-46c**
Example 3.

**Figure 8-46d**
Example 4.

The next chapter, Chapter 9, *Embedded Operating Systems*, is an introduction to the technical fundamentals of embedded operating systems and their function within a design.

## Chapter 8: Problems

1. What is a device driver?
2. Which of Figures 8-46a–d is incorrect in terms of mapping device driver software into the Embedded Systems Model?
3. [a] What is the difference between an architecture-specific device driver and a generic device driver?
   [b] Give two examples of each.
4. Define at least 10 types of device drivers that would be needed based on the block diagram shown in Figure 8-47. Data sheet information is on the CD under Chapter 3 file "sbcARM7."
5. List and describe five types of device driver functions.
6. Finish the sentence: The software's implicit perception of hardware is that it exists in one of three states at any given time:
   A. Inactive, finished, or busy.
   B. Inactive, finished, or broken.
   C. Fixed, finished, or busy.
   D. Fixed, inactive, or broken.
   E. None of the above.
7. [T/F] On master processors that offer different modes in which different types of software can execute, device drivers usually do not run in supervisory mode.
8. [a] What is an interrupt?
   [b] How can interrupts be initiated?
9. Name and describe four examples of device driver functions that can be implemented for interrupt handling.
10. [a] What are the three main types of interrupts?
    [b] List examples in which each type is triggered.

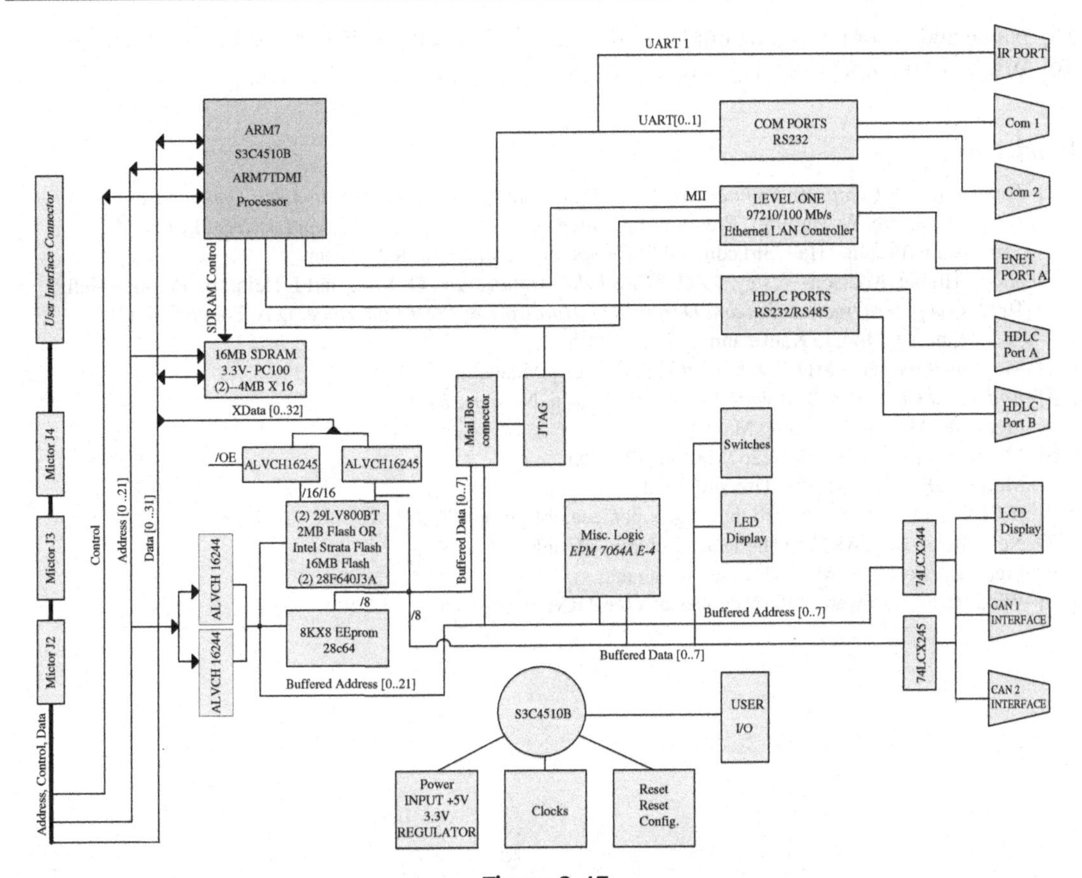


**Figure 8-47**
ARM Board Block Diagram.[10]

11. [a] What is the difference between a level-triggered interrupt and an edge-triggered interrupt?
    [b] What are some strengths and drawbacks of each?
12. An IACK is:
    A. An interrupt controller.
    B. An IRQ port.
    C. An interrupt acknowledgement.
    D. None of the above.
13. [T/F] An ISR is executed before an interrupt is triggered.
14. What is the difference between an auto-vectored and an interrupt-vectored scheme?
15. Name and describe four examples of device driver functions that can be implemented for managing memory.
16. [a] What is byte ordering?
    [b] Name and describe the possible byte ordering schemes.
17. Name and describe four examples of device driver functions that can be implemented for bus protocols.

18. Name and describe four examples of device driver functions that can be implemented for I/O.
19. Where in the OSI model are the Ethernet and serial device drivers mapped to?

## Endnotes


[1] *Foundations of Computer Architecture*, H. Malcolm. Additional references include: *Computer Organization and Architecture*, W. Stallings, Prentice Hall, 4th edn, 1995; *Structured Computer Organization*, A.S. Tanenbaum, Prentice Hall, 3rd edn, 1990; *Computer Architecture*, R.J. Baron and L. Higbie, Addison-Wesley, 1992; *MIPS RISC Architecture*, G. Kane and J. Heinrich, Prentice Hall, 1992; *Computer Organization and Design: The Hardware/Software Interface*, D.A. Patterson and J.L. Hennessy, Morgan Kaufmann, 3rd edn, 2005.

[2] Motorola/Freescale, MPC860 PowerQUICC User's Manual.

[3] *Embedded Controller Hardware Design,* K. Arnold, Newnes, 2001.

[4] Freescale, MPC860 Training Manual.

[5] Mitsubishi Electronics, M37267M8 Specification.

[6] Motorola/Freescale, 68000 Users Manual.

[7] IEEE802.3 Ethernet Standard, http://grouper.ieee.org/groups/802/3/.

[8] Net Silicon, Net+ARM40 Hardware Reference Guide.

[9] Freescale, PowerPC MPC823 User's Manual.

[10] Wind River, Hardware Reference Designs for ARM7 Datasheet.

CHAPTER 9

# Embedded Operating Systems

## In This Chapter

- Defining OS
- Discussing process management, scheduling, and intertask communication
- Introducing memory management at the OS level
- Discussing I/O management in OSs

An operating system (OS) is an optional part of an embedded device's system software stack, meaning that not all embedded systems have one. OSs can be used on any processor (*Instruction Set Architecture (ISA)*) to which the OS has been ported. As shown in Figure 9-1, an OS either sits over the hardware, over the device driver layer, or over a BSP (Board Support Package, which will be discussed in Section 9.7).

The OS is a set of software libraries that serves two main purposes in an embedded system: providing an abstraction layer for software on top of the OS to be less dependent on hardware, making the development of middleware and applications that sit on top of the OS easier, and managing the various system hardware and software resources to ensure the entire system operates efficiently and reliably. While embedded OSs vary in what components they possess, all OSs have a *kernel* at the very least. The kernel is a component that contains the main functionality of the OS, specifically all

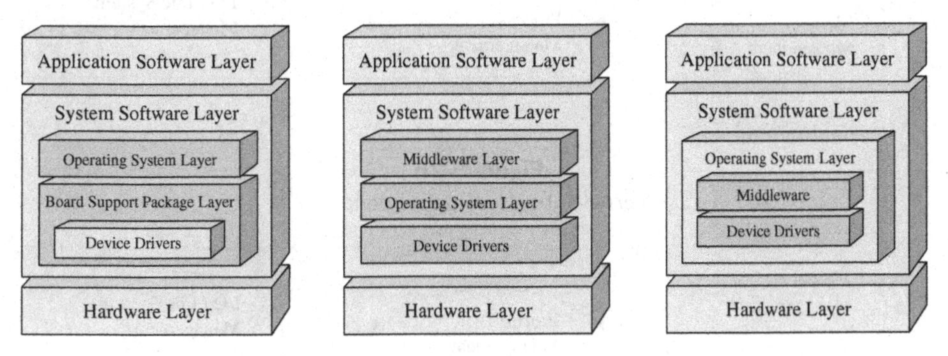


**Figure 9-1**
OSs and the Embedded Systems Model.

or some combination of features and their interdependencies, shown in Figures 9-2a–e, including:

- *Process Management*: how the OS manages and views other software in the embedded system (via *processes*—more in Section 9.2, Multitasking and Process Management). A subfunction typically found within process management is *interrupt and error detection management*. The multiple interrupts and/or traps generated by the various processes need to be managed efficiently, so that they are handled correctly and the processes that triggered them are properly tracked.
- *Memory Management*: the embedded system's memory space is shared by all the different processes, so that access and allocation of portions of the memory space need to be managed (more in Section 9.3, Memory Management). Within memory management,

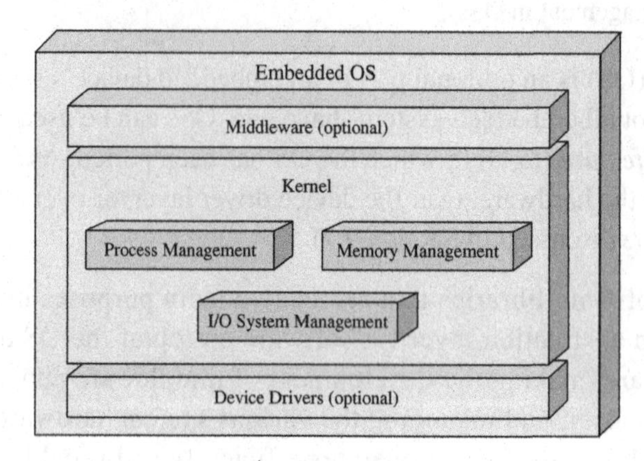


**Figure 9-2a**
General OS model.

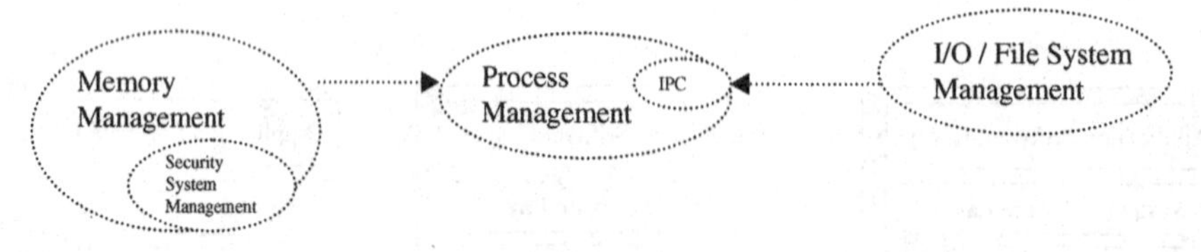


**Figure 9-2b**
Kernel subsystem dependencies.

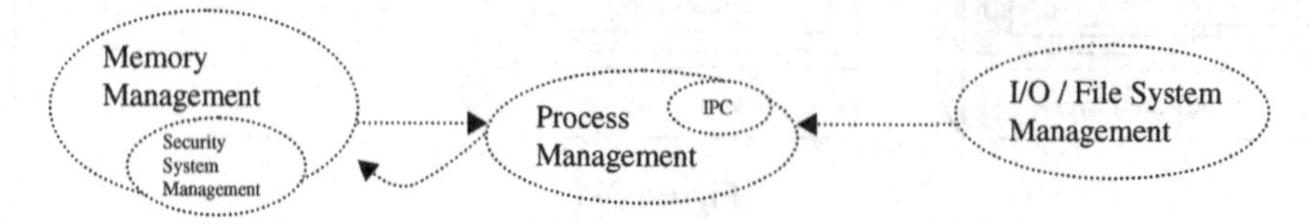


**Figure 9-2c**
Kernel subsystem dependencies.

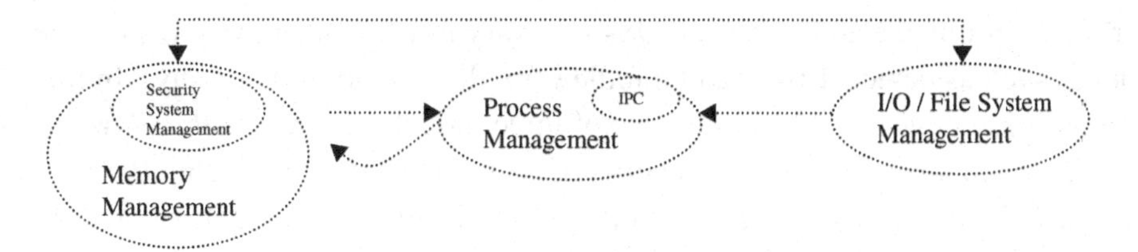


**Figure 9-2d**
Kernel subsystem dependencies.

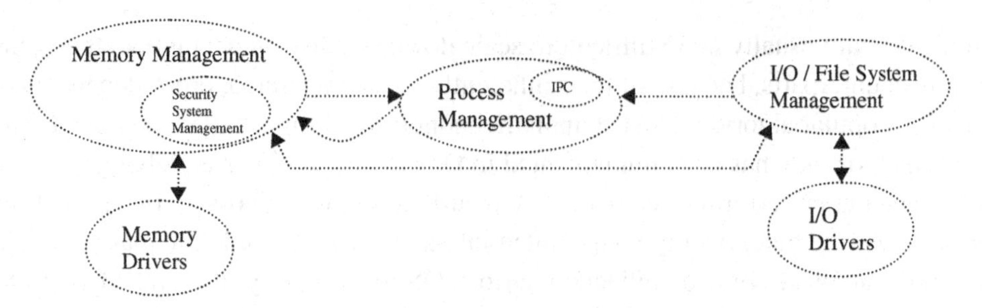


**Figure 9-2e**
Kernel subsystem dependencies.

other subfunctions such as *security system management* allow for portions of the embedded system sensitive to disruptions that can result in the disabling of the system, to remain secure from unfriendly, or badly written, higher-layer software.

- *I/O System Management*: I/O devices also need to be shared among the various processes and so, just as with memory, access and allocation of an I/O device need to be managed (more in Section 9.4, I/O and File System Management). Through I/O system management, *file system management* can also be provided as a method of storing and managing data in the forms of files.

Because of the way in which an OS manages the software in a system, using processes, the process management component is the most central subsystem in an OS. All other OS subsystems depend on the process management unit.

Since all code must be loaded into main memory (random access memory (RAM) or cache) for the master CPU to execute, with boot code and data located in non-volatile memory (read-only memory (ROM), Flash, etc.), the process management subsystem is equally dependent on the memory management subsystem.

I/O management, for example, could include networking I/O to interface with the memory manager in the case of a network file system (NFS).

Outside the kernel, the Memory Management and I/O Management subsystems then rely on the device drivers, and vice-versa, to access the hardware.

Whether inside or outside an OS kernel, OSs also vary in what other system software components, such as device drivers and middleware, they incorporate (if any). In fact, most embedded OSs are typically based upon one of three models, the *monolithic*, *layered*, or *microkernel* (*client/server*) design. In general, these models differ according to the internal design of the OS's kernel, as well as what other system software has been incorporated into the OS. In a *monolithic* OS, middleware and device driver functionality is typically integrated into the OS along with the kernel. This type of OS is a single executable file containing all of these components (see Figure 9-3).

Monolithic OSs are usually more difficult to scale down, modify, or debug than their other OS architecture counterparts, because of their inherently large, integrated, cross-dependent nature. Thus, a more popular algorithm, based upon the monolithic design, called the *monolithic-modularized* algorithm, has been implemented in OSs to allow for easier debugging, scalability, and better performance over the standard monolithic approach. In a monolithic-modularized OS, the functionality is integrated into a single executable file that is made up of *modules*, separate pieces of code reflecting various OS functionality. The embedded Linux OS is an example of a monolithic-based OS, whose main modules are shown in Figure 9-4. The Jbed RTOS, MicroC/OS-II, and PDOS are all examples of embedded monolithic OSs.

In the *layered* design, the OS is divided into hierarchical layers (0,…, *N*), where upper layers are dependent on the functionality provided by the lower layers. Like the monolithic design, layered OSs are a single large file that includes device drivers and middleware (see Figure 9-5). While the layered OS can be simpler to develop and maintain than a monolithic design, the APIs (application program interfaces) provided at each layer create additional overhead that can impact size and performance. DOS-C (FreeDOS), DOS/eRTOS, and VRTX are all examples of a layered OS.

An OS that is stripped down to minimal functionality, commonly only process and memory management subunits as shown in Figure 9-6, is called a *client/server* OS or a *microkernel*.

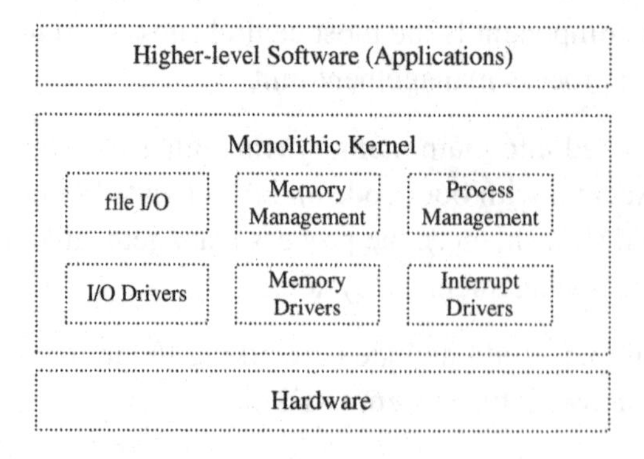


**Figure 9-3**
Monolithic OS block diagram.

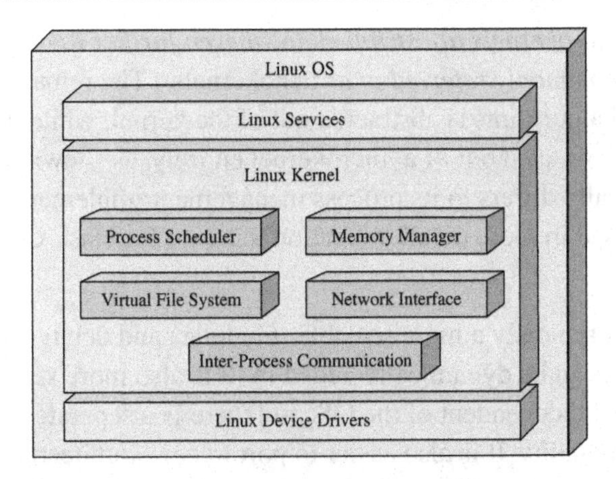


**Figure 9-4**
Linux OS block diagram.

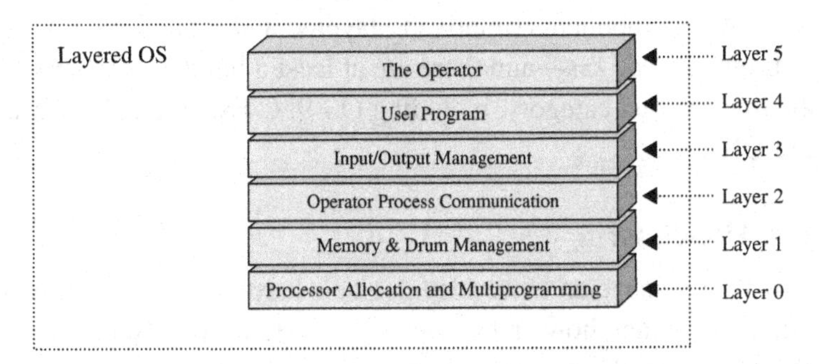


**Figure 9-5**
Layered OS block diagram.

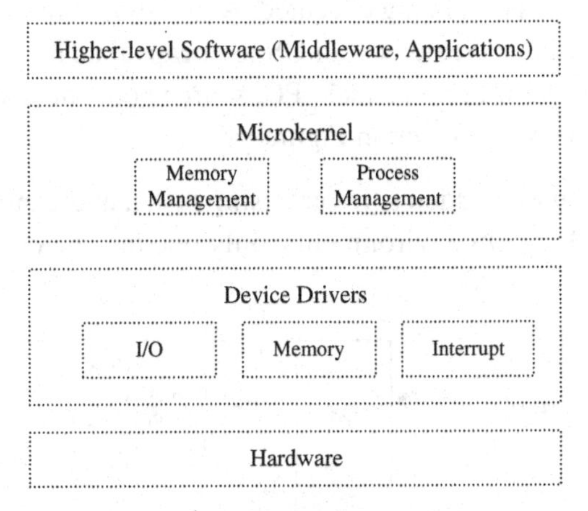


**Figure 9-6**
Microkernel-based OS block diagram.

*(Note: A subclass of microkernels are stripped down even further to only process management functionality and are commonly referred to as* nanokernels.*)* The remaining functionality typical of other kernel algorithms is abstracted out of the kernel, while device drivers, for instance, are usually abstracted out of a microkernel entirely, as shown in Figure 9-6. A microkernel also typically differs in its process management implementation over other types of OSs. This is discussed in more detail in Section 9.2.3 on Intertask Communication and Synchronization.

The microkernel OS is typically a more scalable (modular) and debuggable design, since additional components can be dynamically added in. It is also more secure since much of the functionality is now independent of the OS, and there is a separate memory space for client and server functionality. It is also easier to port to new architectures. However, this model may be slower than other OS architectures, such as the monolithic, because of the communication paradigm between the microkernel components and other "kernel-like" components. Overhead is also added when switching between the kernel and the other OS components and non-OS components (relative to layered and monolithic OS designs). Most of the off-the-shelf embedded OSs—and there are at least a hundred of them—have kernels that fall under the microkernel category, including OS-9, C Executive, VxWorks, CMX-RTX, Nucleus Plus, and QNX.

## 9.1 What Is a Process?

To understand how OSs manage an embedded device's hardware and software resources, the reader must first understand how an OS views the system. An OS differentiates between a program and the executing of a program. A program is simply a *passive, static* sequence of instructions that could represent a system's hardware and software resources. The actual execution of a program is an *active, dynamic* event in which various properties change relative to time and the instruction being executed. A *process* (commonly referred to as a *task* in many embedded OSs) is created by an OS to encapsulate all the information that is involved in the executing of a program (stack, PC, source code, data, etc.). This means that a program is only part of a task, as shown in Figure 9-7.

Embedded OSs manage all embedded software using tasks, and can be either *unitasking* or *multitasking*. In unitasking OS environments, only one task can exist at any given time,

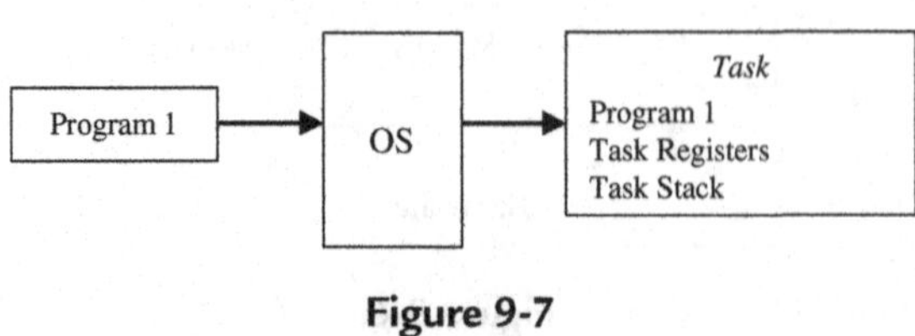


**Figure 9-7**
OS task.

whereas in a multitasking OS, multiple tasks are allowed to exist simultaneously. Unitasking OSs typically don't require as complex a task management facility as a multitasking OS. In a multitasking environment, the added complexity of allowing multiple existing tasks requires that each process remain independent of the others and not affect any other without the specific programming to do so. This multitasking model provides each process with more security, which is not needed in a unitasking environment. Multitasking can actually provide a more organized way for a complex embedded system to function. In a multitasking environment, system activities are divided up into simpler, separate components, or the same activities can be running in multiple processes simultaneously, as shown in Figure 9-8.

Some multitasking OSs also provide *threads* (*lightweight processes*) as an additional, alternative means for encapsulating an instance of a program. Threads are created within the context of a task (meaning a thread is bound to a task) and, depending on the OS, the task can own one or more threads. A thread is a sequential execution stream within its task. Unlike tasks, which have their own independent memory spaces that are inaccessible to other tasks, threads of a task share the same resources (working directories, files, I/O devices, global data, address space, program code, etc.), but have their own PCs, stack, and scheduling information (PC, SP, stack, registers, etc.) to allow for the instructions they are executing to be scheduled independently. Since threads are created within the context of the same task and can share the same memory space, they can allow for simpler communication and coordination relative to tasks. This is because a task can contain at least one thread executing one program in one address space or can contain many threads executing different portions of one program in one address space (see Figure 9-9), needing no intertask communication mechanisms. This is discussed in more detail in Section 9.2.3 Intertask Communication and Synchronization. Also, in the case of shared resources, multiple threads are typically less expensive than creating multiple tasks to do the same work.

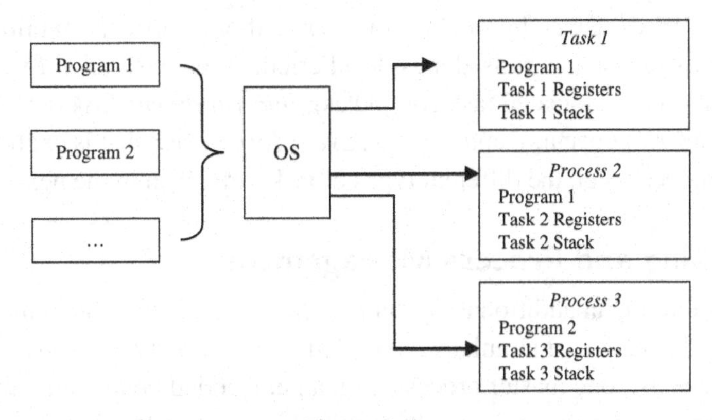


**Figure 9-8**
Multitasking OS.

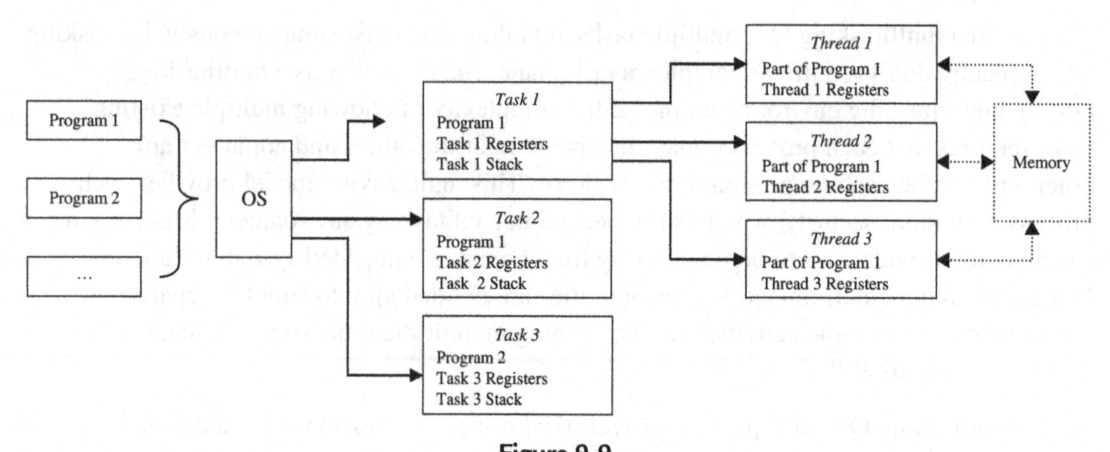


**Figure 9-9**
Tasks and threads.

Usually, programmers define a separate task (or thread) for each of the system's distinct activities to simplify all the actions of that activity into a single stream of events, rather than a complex set of overlapping events. However, it is generally left up to the programmer as to how many tasks are used to represent a system's activity and, if threads are available, if and how they are used within the context of tasks.

DOS-C is an example of a unitasking embedded OS, whereas VxWorks (Wind River), embedded Linux (Timesys), and Jbed (Esmertec) are examples of multitasking OSs. Even within multitasking OSs, the designs can vary widely. Traditional versions of VxWorks (i.e., 5.x and 6.x) have one type of task, each of which implements one "thread of execution," whereas another VxWorks version called VxWorks653 is made up of a more complex multitasking scheme that integrates some combination of a module OS and instantiations of partition OSs. Timesys Linux has two types of tasks, the Linux fork and the Periodic task, whereas Jbed provides six different types of tasks that run alongside threads: *OneshotTimer Task* (a task that is run only once), *PeriodicTimer Task* (a task that is run after a particular set time interval), *HarmonicEvent Task* (a task that runs alongside a Periodic timer task), *JoinEvent Task* (a task that is set to run when an associated task completes), *InterruptEvent Task* (a task that is run when a hardware interrupt occurs), and the *UserEvent Task* (a task that is explicitly triggered by another task). More details on the different types of tasks are given in the next section.

## 9.2 Multitasking and Process Management

Multitasking OSs require an additional mechanism over unitasking OSs to manage and synchronize tasks that can exist simultaneously. This is because even when an OS allows multiple tasks to coexist, one master processor on an embedded board can only execute one task or thread at any given time. As a result, multitasking embedded OSs must find some way of allocating each task a certain amount of time to use the master CPU and switching the master processor between the various tasks. It is by accomplishing this through task *implementation*, *scheduling*, *synchronization*, and *intertask communication* mechanisms that an OS successfully gives the illusion of a single processor simultaneously running multiple tasks (see Figure 9-10).

### 9.2.1 Process Implementation

In multitasking embedded OSs, tasks are structured as a hierarchy of parent and child tasks, and when an embedded kernel starts up only one task exists (as shown in Figure 9-11). It is from this first task that all others are created. *(Note: The first task is also created by the programmer in the system's initialization code, which will be discussed in more detail in Chapter 12.)*

Task creation in embedded OSs is primarily based upon two models, *fork/exec* (which derived from the IEEE/ISO POSIX 1003.1 standard) and *spawn* (which is derived from fork/exec). Since the spawn model is based upon the fork/exec model, the methods of creating tasks under both models are similar. All tasks create their child tasks through fork/exec or spawn system calls. After the system call, the OS gains control and creates the *Task Control Block (TCB)*, also referred to as a *Process Control Block (PCB)* in some OSs, that contains OS

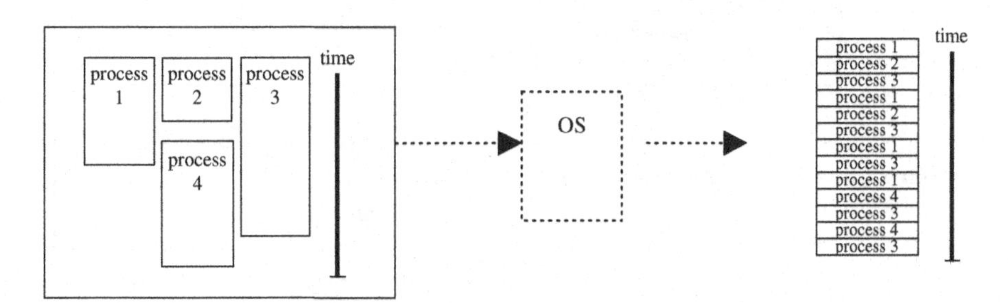


**Figure 9-10**
Interleaving tasks.

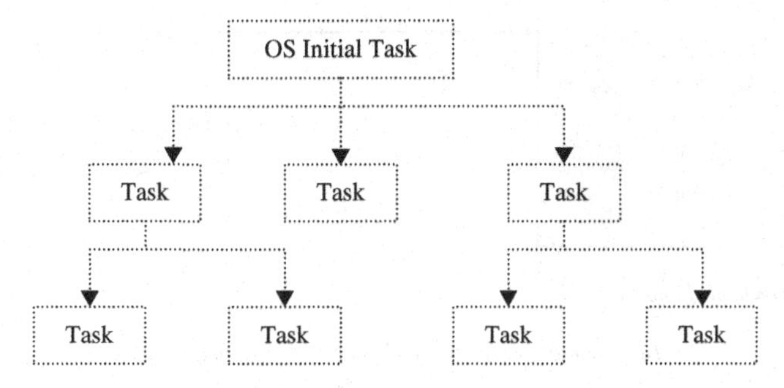


**Figure 9-11**
Task hierarchy.

control information, such as task ID, task state, task priority, and error status, and CPU context information, such as registers, for that particular task. At this point, memory is allocated for the new child task, including for its TCB, any parameters passed with the system call and the code to be executed by the child task. After the task is set up to run, the system call returns and the OS releases control back to the main program.

The main difference between the fork/exec and spawn models is how memory is allocated for the new child task. Under the fork/exec model, as shown in Figure 9-12, the "fork" call creates a copy of the parent task's memory space in what is allocated for the child task, thus allowing the child task to *inherit* various properties, such as program code and variables, from the parent task. Because the parent task's entire memory space is duplicated for the child task, two copies of the parent task's program code are in memory—one for the parent and one belonging to the child. The "exec" call is used to explicitly remove from the child task's memory space any references to the parent's program and sets the new program code belonging to the child task to run.

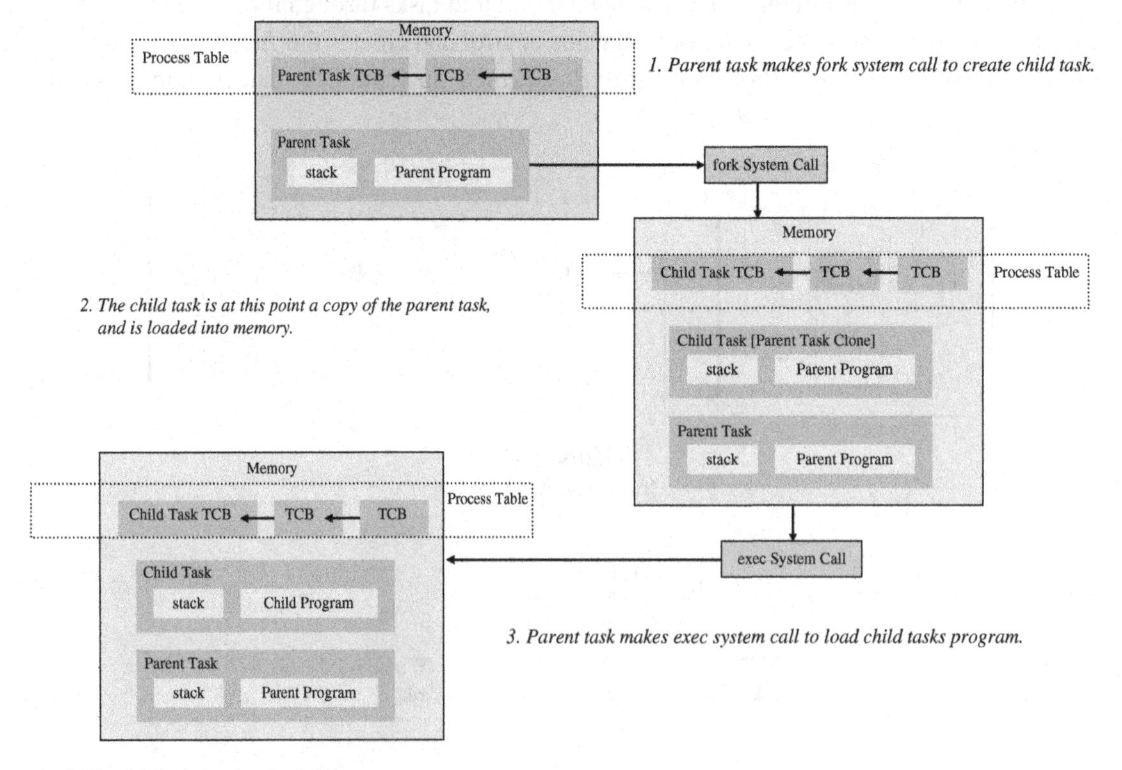


**Figure 9-12**
Fork/exec process creation.

The spawn model, on the other hand, creates an entirely new address space for the child task. The spawn system call allows for the new program and arguments to be defined for the child task. This allows for the child task's program to be loaded and executed immediately at the time of its creation (see Figure 9-13).

Both process creation models have their strengths and drawbacks. Under the spawn approach, there are no duplicate memory spaces to be created and destroyed, and then new space allocated, as is the case with the fork/exec model. The advantages of the fork/exec model, however, include the efficiency gained by the child task inheriting properties from the parent task, and then having the flexibility to change the child task's environment afterwards. In Examples 1, 2, and 3, real-world embedded OSs are shown along with their process creation techniques.

### Example 1: Creating a Task in VxWorks[1]

The two major steps of spawn task creation form the basis of creating tasks in VxWorks. The VxWorks system called "taskSpawn" is based upon the POSIX spawn model, and it is what creates, initializes, and activates a new (child) task.

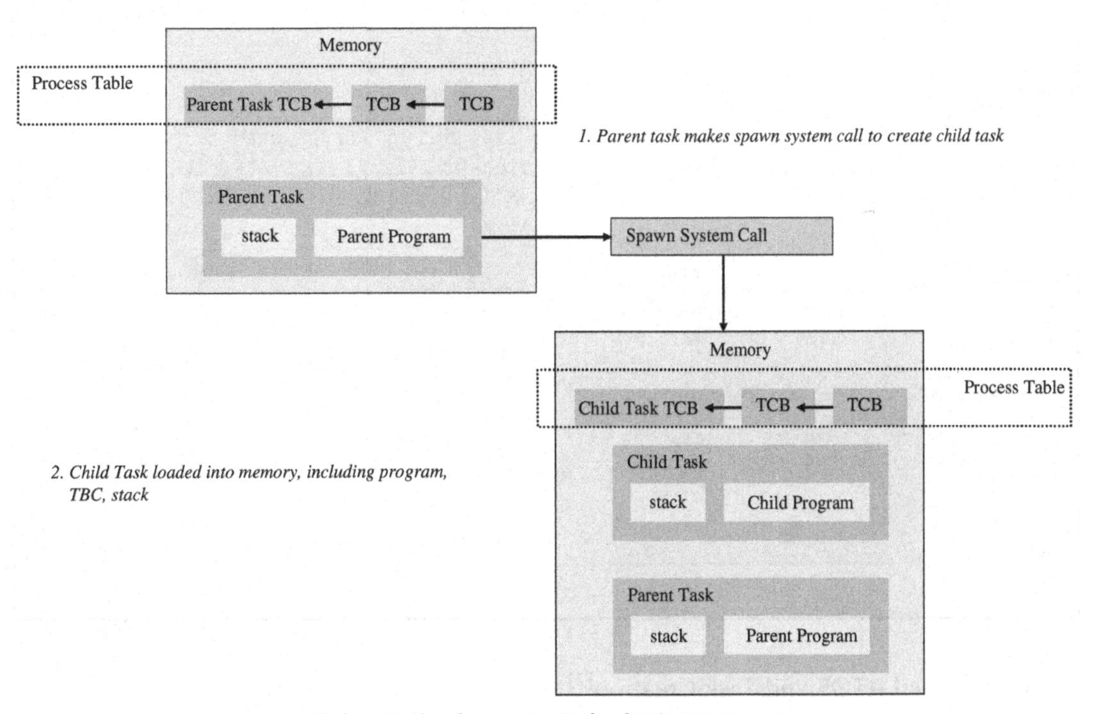


**Figure 9-13**
Spawn process creation.

```
int taskSpawn(
{Task Name},
{Task Priority 0-255, related to scheduling; this will be discussed in the next
section},
{Task Options - VX_FP_TASK, execute with floating point coprocessor
VX_PRIVATE_ENV, execute task with private environment
VX_UNBREAKABLE, disable breakpoints for task
VX_NO_STACK_FILL, do not fill task stack with 0xEE}
{Stack Size}
{Task address of entry point of program in memory - initial PC value}
{Up to 10 arguments for task program entry routine})
```

After the spawn system call, an image of the child task (including TCB, stack, and program) is allocated into memory. Below is a pseudocode example of task creation in the VxWorks RTOS where a parent task "spawns" a child task software timer.

**Task Creation VxWorks Pseudocode**

```
// parent task that enables software timer
void parentTask(void)
{
...
if sampleSoftware Clock NOT running {
    /"newSWClkId" is a unique integer value assigned by kernel when task is created
    newSWClkId = taskSpawn ("sampleSoftwareClock", 255, VX_NO_STACK_FILL, 3000,
                     (FUNCPTR) minuteClock, 0, 0, 0, 0, 0, 0, 0, 0, 0, 0);
    ....
}

// child task program Software Clock
void minuteClock (void) {
   integer seconds;
   while (softwareClock is RUNNING) {
       seconds = 0;
       while (seconds < 60) {
           seconds = seconds + 1;
       }
...
}
```

### Example 2: Jbed RTOS and Task Creation[2]

In Jbed, there is more than one way to create a task, because in Java there is more than one way to create a Java thread—and in Jbed, tasks are extensions of Java threads. One of the most common methods of creating a task in Jbed is through the "task" routines, one of which is:

```
public Task(long duration,
         long allowance,
         long deadline,
         RealtimeEvent event)
Throws AdmissionFailure
```

Task creation in Jbed is based upon a variation of the spawn model, called *spawn threading*. Spawn threading is spawning, but typically with less overhead and with tasks sharing the same memory space. Below is a pseudocode example of task creation of a Oneshot task, one of Jbed's six different types of tasks, in the Jbed RTOS where a parent task "spawns" a child task software timer that runs only one time.

**Task Creation Jbed Pseudocode**

```
// Define a class that implements the Runnable interface for the software clock
public class ChildTask implements Runnable{
       // child task program Software Clock
       public void run () {
              integer seconds;
              while (softwareClock is RUNNING) {
                      seconds = 0;
                      while (seconds < 60) {
                             seconds = seconds + 1;
              }
              ...
       }
}

// parent task that enables software timer
void parentTask(void)
{
...
if sampleSoftware Clock NOT running {
       try{
              DURATION,
              ALLOWANCE,
              DEADLINE,
              OneshotTimer);
       }catch(AdmissionFailure error){
              Print Error Message ("Task creation failed");
   }
}
....
}
```

The creation and initialization of the Task object is the Jbed (Java) equivalent of a TCB. The task object, along with all objects in Jbed, is located in Jbed's heap (in Java, there is only one heap for all objects). Each task in Jbed is also allocated its own stack to store primitive data types and object references.

Example 3: Embedded Linux and Fork/Exec[3]
In embedded Linux, all process creation is based upon the fork/exec model:

```
int fork (void)                        void exec (...)
```

In Linux, a new "child" process can be created with the fork system call (shown above), which creates an almost identical copy of the parent process. What differentiates the parent task from the child is the process ID—the process ID of the child process is returned to the parent, whereas a value of "0" is what the child process believes its process ID to be.

```
#include <sys/types.h>
#include <unistd.h>

void program(void)
{
    processId child_processId;
        /* create a duplicate: child process */
        child_processId = fork();

        if (child_processId == -1) {
            ERROR;
        }
        else if (child_processId == 0) {
            run_childProcess();
        }
        else {
            run_parentParent();
        }
```

The exec function call can then be used to switch to the child's program code.

```
int program (char* program, char** arg_list)
{
    processed child_processId;
```

```
    /* Duplicate this process */
    child_processId = fork ();

if (child_pId ! = 0)

    /* This is the parent process */
    return child_processId;
    else
    {
    /* Execute PROGRAM, searching for it in the path */
    execvp (program, arg_list);

    /* execvp returns only if an error occurs */
    fprintf (stderr, "Error in execvp\n");
    abort (); }
    }
}
```

| Call | Description |
|---|---|
| exit() | Terminates the calling task and frees memory (task stacks and task control blocks only). |
| taskDelete() | Terminates a specified task and frees memory (task stacks and task control blocks only).* |
| taskSafe() | Protects the calling task from deletion. |
| taskUnsafe() | Undoes a taskSafe() (makes the calling task available for deletion). |

* Memory that is allocated by the task during its execution is *not* freed when the task is terminated.

```
void vxWorksTaskDelete (int taskId)
  {
    int localTaskId = taskIdFigure (taskId);

   /* no such task ID */
   if (localTaskId == ERROR)
       printf ("Error: ask not found.\n");
   else if (localTaskId == 0)
       printf ("Error: The shell can't delete itself.\n");
   else if (taskDelete (localTaskId) != OK)
       printf ("Error");
  }
```

**Figure 9-14a**
VxWorks and Spawn task deleted.[4]

Tasks can terminate for a number of different reasons, such as normal completion, hardware problems such as lack of memory, and software problems such as invalid instructions. After a task has been terminated, it must be removed from the system so that it doesn't waste resources, or even keep the system in limbo. In *deleting* tasks, an OS deallocates any memory allocated for the task (TCBs, variables, executed code, etc.). In the case of a parent task being deleted, all related child tasks are also deleted or moved under another parent, and any shared system resources are released (see Figure 9-14a).

When a task is deleted in VxWorks, other tasks are not notified and any resources such as memory allocated to the task are not freed—it is the responsibility of the programmer to manage the deletion of tasks using the subroutines below.

```
#include <stdio.h>
#include <stdlib.h>

main ()
{...
if (fork == 0)
  exit (10);
....
}
```

**Figure 9-14b**
Embedded Linux and fork/exec task deleted.[3]

In Linux, processes are deleted with the *void exit(int status)* system call, which deletes the process and removes any kernel references to process (updates flags, removes processes from queues, releases data structures, updates parent-child relationships, etc.). Under Linux, child processes of a deleted process become children of the main *init* parent process (see Figure 9-14b).

Because Jbed is based upon the Java model, a garbage collector (GC) is responsible for deleting a task and removing any unused code from memory once the task has stopped running. Jbed uses a non-blocking mark-and-sweep garbage collection algorithm, which marks all objects still being used by the system and deletes (sweeps) all unmarked objects in memory.

In addition to creating and deleting tasks, an OS typically provides the ability to *suspend* a task (meaning temporarily blocking a task from executing) and *resume* a task (meaning any blocking of the task's ability to execute is removed). These two additional functions are provided by the OS to support task *states*. A task's state is the activity (if any) that is going on with that task once it has been created, but has not been deleted. OSs usually define a task as being in one of three states:

- *READY*: the process is ready to be executed at any time, but is waiting for permission to use the CPU.
- *RUNNING*: the process has been given permission to use the CPU, and can execute.
- *BLOCKED or WAITING*: the process is waiting for some external event to occur before it can be "ready" to "run."

OSs usually implement separate READY and BLOCKED/WAITING "queues" containing tasks (their TCBs) that are in the relative state (see Figure 9-15). Only one task at any one time can be in the RUNNING state, so no queue is needed for tasks in the RUNNING state.

Based upon these three states (READY, BLOCKED, and RUNNING), most OSs have some process state transition model similar to the state diagram in Figure 9-16. In this

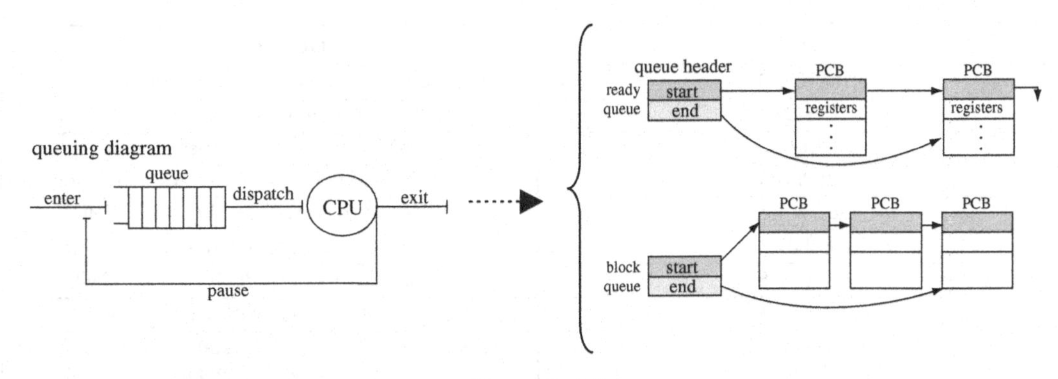


**Figure 9-15**
Task states and queues.[4]

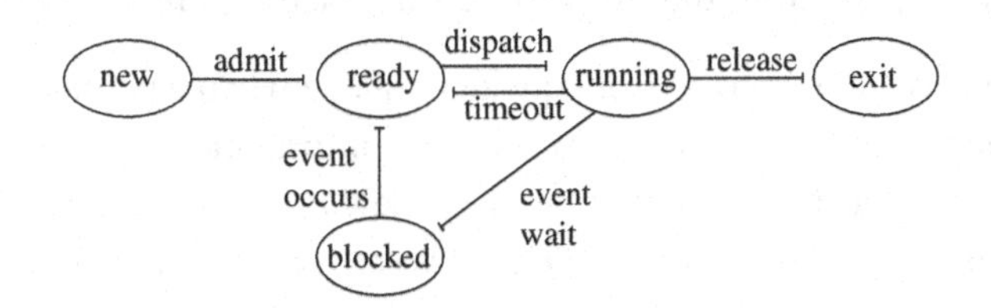


**Figure 9-16**
Task state diagram.[2]

diagram, the "New" state indicates a task that has been created, and the "Exit" state is a task that has terminated (suspended or stopped running). The other three states are defined above (READY, RUNNING, and BLOCKED). The state transitions (according to Figure 9-16) are New → READY (where a task has entered the ready queue and can be scheduled for running), READY → RUNNING (based on the kernel's scheduling algorithm, the task has been selected to run), RUNNING → READY (the task has finished its turn with the CPU and is returned to the ready queue for the next time around), RUNNING → BLOCKED (some event has occurred to move the task into the blocked queue, not to run until the event has occurred or been resolved), and BLOCKED → READY (whatever blocked task was waiting for has occurred and task is moved back to the ready queue).

When a task is moved from one of the queues (READY or BLOCKED/WAITING) into the RUNNING state, it is called a *context switch*. Examples 4, 5, and 6 give real-world examples of OSs and their state management schemes.

### Example 4: VxWorks Wind Kernel and States[5]

Other than the RUNNING state, VxWorks implements nine variations of the READY and BLOCKED/WAITING states, as shown in the following table and state diagram.

| State | Description |
|---|---|
| STATE + 1 | The state of the task with an inherited priority |
| READY | Task in READY state |
| DELAY | Task in BLOCKED state for a specific time period |
| SUSPEND | Task is BLOCKED usually used for debugging |
| DELAY + S | Task is in two states: DELAY & SUSPEND |
| PEND | Task in BLOCKED state due to a busy resource |
| PEND + S | Task is in two states: PEND & SUSPEND |
| PEND + T | Task is in PEND state with a timeout value |
| PEND + S + T | Task is in two states: PEND state with a timeout value and SUSPEND |

Under VxWorks, separate ready, pending, and delay state queues exist to store the TCB information of a task that is within that respective state (see Figure 9-17a2).

A task's TCB is modified and is moved from queue to queue when a context switch occurs. When the Wind kernel *context switches* between two tasks, the information of the task currently running is saved in its TCB, while the TCB information of the new task to be executed is loaded for the CPU to begin executing. The Wind kernel contains two types of context switches: *synchronous*, which occurs when the running task blocks itself (through pending, delaying, or suspending), and *asynchronous*, which occurs when the running task is blocked due to an external interrupt.

| State | Description |
|---|---|
| STATE + 1 | The state of the task with an inherited priority |
| READY | Task in READY state |
| DELAY | Task in BLOCKED state for a specific time period |
| SUSPEND | Task is BLOCKED, usually used for debugging |
| DELAY + S | Task is in 2 states: DELAY & SUSPEND |
| PEND | Task in BLOCKED state due to a busy resource |
| PEND + S | Task is in 2 states: PEND & SUSPEND |
| PEND + T | Task is in PEND state with a timeout value |
| PEND + S + T | Task is in 2 states: PEND state with a timeout value and SUSPEND |

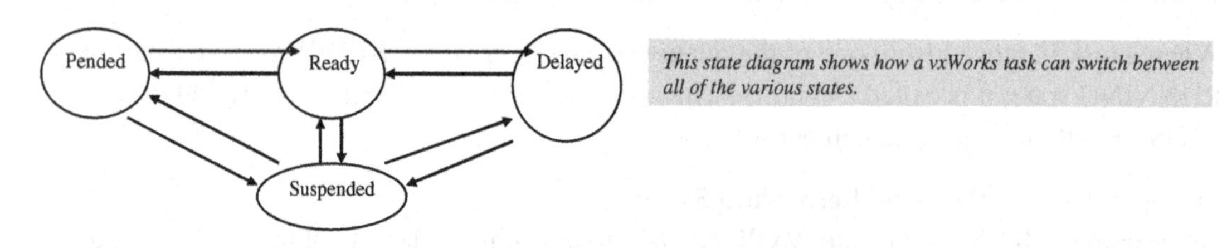
Pended
Ready
Delayed
Suspended
*This state diagram shows how a vxWorks task can switch between all of the various states.*

**Figure 9-17a1**
State diagram for VxWorks tasks.[5]

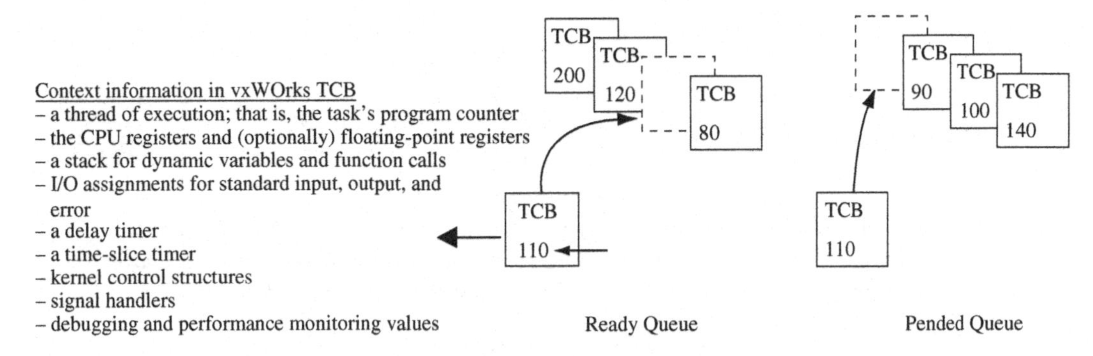


**Figure 9-17a2**
VxWorks tasks and queues.[4]

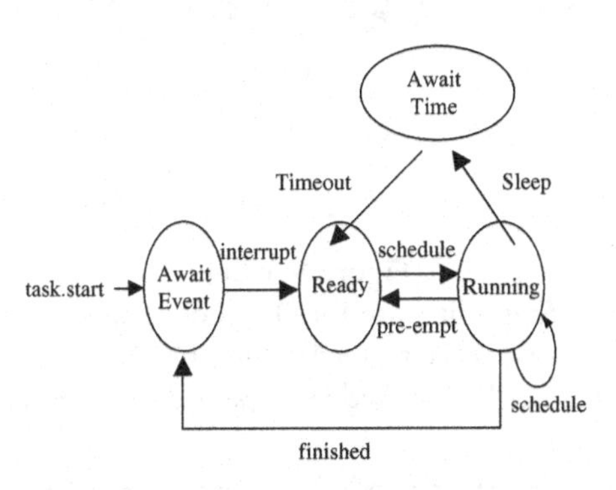


**Figure 9-17b1**
State diagram for Jbed Interrupt tasks.[6]
This state diagram shows some possible states for Interrupt tasks. Basically, an interrupt task is in an Await Event state until a hardware interrupt occurs—at which point the Jbed scheduler moves an Interrupt task into the READY state to await its turn to run.

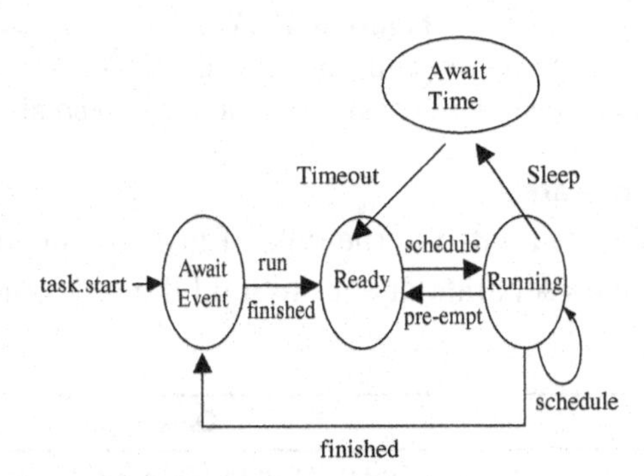


**Figure 9-17b2**
State diagram for Jbed Joined tasks.[6]
This state diagram shows some possible states for Joined tasks. Like the Interrupt task, the Joined task is in an Await Event state until an associated task has finished running—at which point the Jbed scheduler moves a Joined task into the READY state to await its turn to run. At any time, the Joined task can enter a timed waiting period.

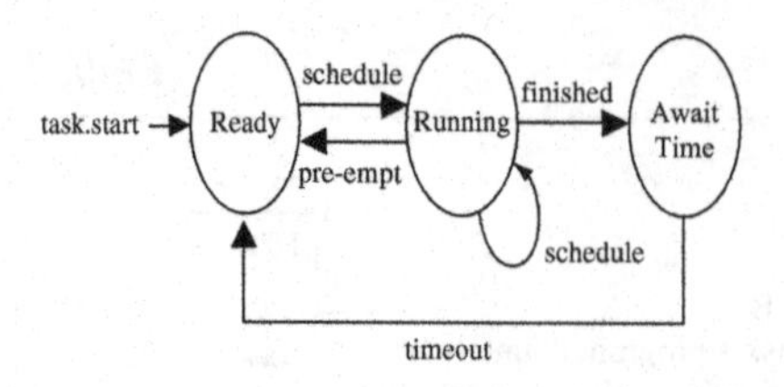


**Figure 9-17b3**
State diagram for Periodic tasks.[6]
This state diagram shows some possible states for Periodic tasks. A Periodic task runs continuously at certain intervals and gets moved into the Await Time state after every run to await that interval before being put into the ready state.

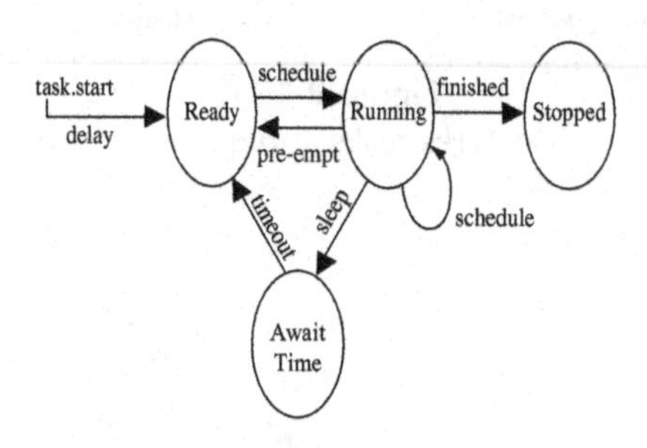


**Figure 9-17b4**
State diagram for Oneshot tasks.[6]
This state diagram shows some possible states for Oneshot tasks. A Oneshot task can either run once and then end (stop) or be blocked for a period of time before actually running.

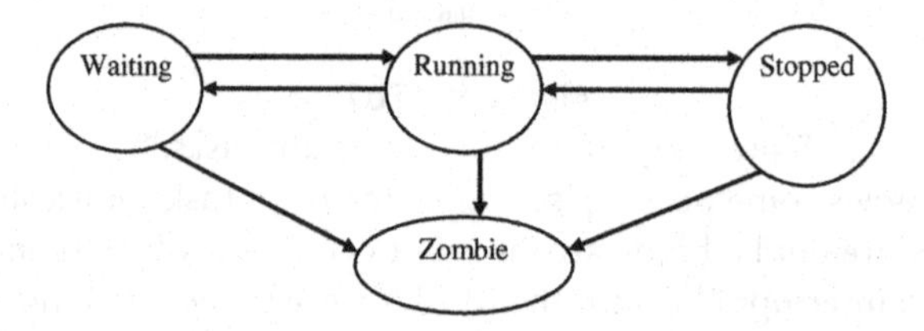


**Figure 9-17c1**
State diagram for Linux tasks.[3]
This state diagram shows how a Linux task can switch between all of the various states.

### Example 5: Jbed Kernel and States[6]

In Jbed, some states of tasks are related to the type of task, as shown in the table and state diagrams below. Jbed also uses separate queues to hold the task objects that are in the various states.

| State | Description |
|---|---|
| RUNNING | For all types of tasks, task is currently executing |
| READY | For all types of tasks, task in READY state |
| STOP | In Oneshot tasks, task has completed execution |
| AWAIT TIME | For all types of tasks, task in BLOCKED state for a specific time period |
| AWAIT EVENT | In Interrupt and Joined tasks, BLOCKED while waiting for some event to occur |

### Example 6: Embedded Linux and States

In Linux, RUNNING combines the traditional READY and RUNNING states, while there are three variations of the BLOCKED state.

| State | Description |
|---|---|
| RUNNING | Task is in either the RUNNING or READY state |
| WAITING | Task in BLOCKED state waiting for a specific resource or event |
| STOPPED | Task is BLOCKED, usually used for debugging |
| ZOMBIE | Task is BLOCKED and no longer needed |

Under Linux, a process's context information is saved in a PCB called the task_struct shown in Figure 9-17c2. Shown boldface in the figure is an entry in the task_struct containing a Linux process's state. In Linux there are separate queues that contain the task_struct (PCB) information for the process with that respective state.

```
struct task_struct
{
                    ....
// -1 unrunnable, 0 runnable, >0 stopped
 volatile long    state;
//number of clock ticks left to run in this scheduling slice, decremented by a timer.
 long              counter;
 // the process' static priority, only changed through well-known system calls like nice, POSIX.1b
 // sched_setparam, or 4.4BSD/SVR4 setpriority.
 long              priority;
 unsigned            long signal;
 //  bitmap of masked signals
 unsigned            long blocked;
// per process flags, defined below
 unsigned            long flags;
 int errno;
 // hardware debugging registers
 long              debugreg[8];
 struct exec_domain   *exec_domain;
 struct linux_binfmt  *binfmt;
 struct task_struct   *next_task, *prev_task;
 struct task_struct   *next_run, *prev_run;
 unsigned long        saved_kernel_stack;
 unsigned long        kernel_stack_page;
 int             exit_code, exit_signal;
  unsigned long        personality;
 int             dumpable:1;
 int             did_exec:1;
 int             pid;
 int             pgrp;
 int             tty_old_pgrp;
 int             session;
 // boolean value for session group leader
 int             leader;
 int             groups[NGROUPS];
 // pointers to (original) parent process, youngest child, younger sibling, older sibling, respectively.  (p->father
 // can be replaced with  p->p_pptr->pid)
 struct task_struct  *p_opptr, *p_pptr, *p_cptr,
                 *p_ysptr, *p_osptr;
struct wait_queue    *wait_chldexit;
 unsigned short      uid,euid,suid,fsuid;
 unsigned short      gid,egid,sgid,fsgid;
 unsigned long       timeout;
// the scheduling policy, specifies which scheduling class the task belongs to, such as : SCHED_OTHER
// (traditional UNIX process), SCHED_FIFO (POSIX.1b FIFO realtime process - A FIFO realtime process will
// run until either a) it blocks on I/O, b) it explicitly yields the CPU or c) it is pre-empted by another realtime
// process with a higher p->rt_priority value.) and SCHED_RR (POSIX round-robin realtime process –
// SCHED_RR is the same as SCHED_FIFO, except that when its timeslice expires it goes back to the end of the
// run queue).
 unsigned long       policy;
//realtime priority
 unsigned long       rt_priority;
 unsigned long       it_real_value, it_prof_value, it_virt_value;
 unsigned long       it_real_incr, it_prof_incr, it_virt_incr;
 structtimer_list    real_timer;
 long             utime, stime, cutime, cstime, start_time;
// mm fault and swap info: this can arguably be seen as either mm-specific or thread-specific */
 unsigned long       min_flt, maj_flt, nswap, cmin_flt, cmaj_flt, cnswap;
 int swappable:1;
 unsigned long       swap_address;
 // old value of maj_flt
 unsigned long       old_maj_flt;
 // page fault count of the last time
 unsigned long       dec_flt;
// number of pages to swap on next pass
 unsigned long       swap_cnt
// limits
 struct rlimit       rlim[RLIM_NLIMITS];
 unsigned short      used_math;
 char              comm[16];
// file system info
 int              link_count;
// NULL if no tty
 struct tty_struct    *tty;
// ipc stuff
 struct sem_undo      *semundo;
 struct sem_queue     *semsleeping;
// ldt for this task - used by Wine.  IfNULL, default_ldt is used
 struct desc_struct *ldt;
// tss for this task
 struct thread_struct tss;
// file system information
 structfs_struct     *fs;
// open file information
 struct files_struct  *files;
// memory management info
 struct mm_struct     *mm;
// signal handlers
 struct signal_struct *sig;
#ifdef __SMP__
 int              processor;
 int              last_processor;
 int              lock_depth;    /* Lock depth.
                          We can context switch in and out
                          of holding a syscall kernel lock... */
#endif
}
```

**Figure 9-17c2**

Task structure.[15]

### 9.2.2 Process Scheduling

In a multitasking system, a mechanism within an OS, called a *scheduler* (shown in Figure 9-18), is responsible for determining the order and the duration of tasks to run on the CPU. The scheduler selects which tasks will be in what states (READY, RUNNING, or BLOCKED), as well as loading and saving the TCB information for each task. On some OSs the same scheduler allocates the CPU to a process that is loaded into memory and ready to run, while in other OSs a *dispatcher* (a separate scheduler) is responsible for the actual allocation of the CPU to the process.

There are many scheduling algorithms implemented in embedded OSs, and every design has its strengths and tradeoffs. The key factors that impact the effectiveness and performance of a scheduling algorithm include its *response time* (time for scheduler to make the context switch to a ready task and includes waiting time of task in ready queue), *turnaround time* (the time it takes for a process to complete running), *overhead* (the time and data needed to determine which tasks will run next), and *fairness* (what are the determining factors as to which processes get to run). A scheduler needs to balance utilizing the system's resources, keeping the CPU, I/O, as busy as possible, with task *throughput*, processing as many tasks as possible in a given amount of time. Especially in the case of fairness, the scheduler has to ensure that task *starvation*, where a task never gets to run, doesn't occur when trying to achieve a maximum task throughput.

In the embedded OS market, scheduling algorithms implemented in embedded OSs typically fall under two approaches: *non-pre-emptive* and *pre-emptive* scheduling. Under non-pre-emptive scheduling, tasks are given control of the master CPU until they have finished execution, regardless of the length of time or the importance of the other tasks that are waiting. Scheduling algorithms based upon the non-pre-emptive approach include:

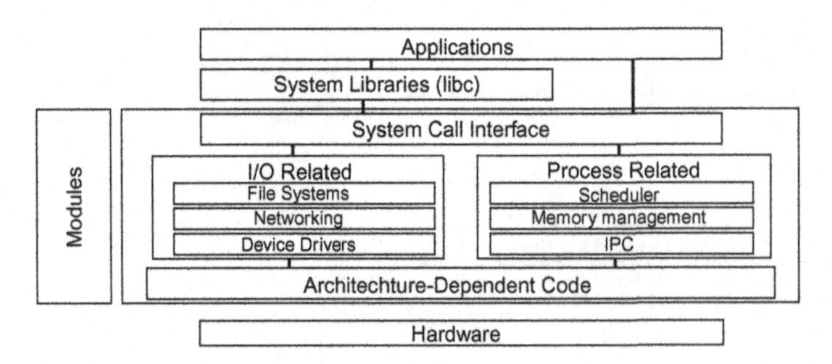


**Figure 9-18**
OS Block diagram and the scheduler.[3]

- *First Come First Served (FCFS)/Run-To-Completion*, where tasks in the READY queue are executed in the order they entered the queue and where these tasks are run until completion when they are READY to be run (see Figure 9-19). Here, non-pre-emptive means there is no BLOCKED queue in an FCFS scheduling design. The response time of a FCFS algorithm is typically slower than other algorithms (i.e., especially if longer processes are in front of the queue requiring that other processes wait their turn), which then becomes a fairness issue since short processes at the end of the queue get penalized for the longer ones in front. With this design, however, starvation is not possible.
- *Shortest Process Next (SPN)/Run-To-Completion*, where tasks in the READY queue are executed in the order in which the tasks with the shortest execution time are executed first (see Figure 9-20). The SPN algorithm has faster response times for shorter processes. However, then the longer processes are penalized by having to wait until all the shorter processes in the queue have run. In this scenario, starvation can occur to longer processes if the ready queue is continually filled with shorter processes. The overhead is higher than that of FCFS, since the calculation and storing of run times for the processes in the ready queue must occur.
- *Co-operative*, where the tasks themselves run until they tell the OS when they can be context switched (for I/O, etc.). This algorithm can be implemented with the FCFS or SPN algorithms, rather than the run-to-completion scenario, but starvation could still occur with SPN if shorter processes were designed not to "cooperate," for example (see Figure 9-21).

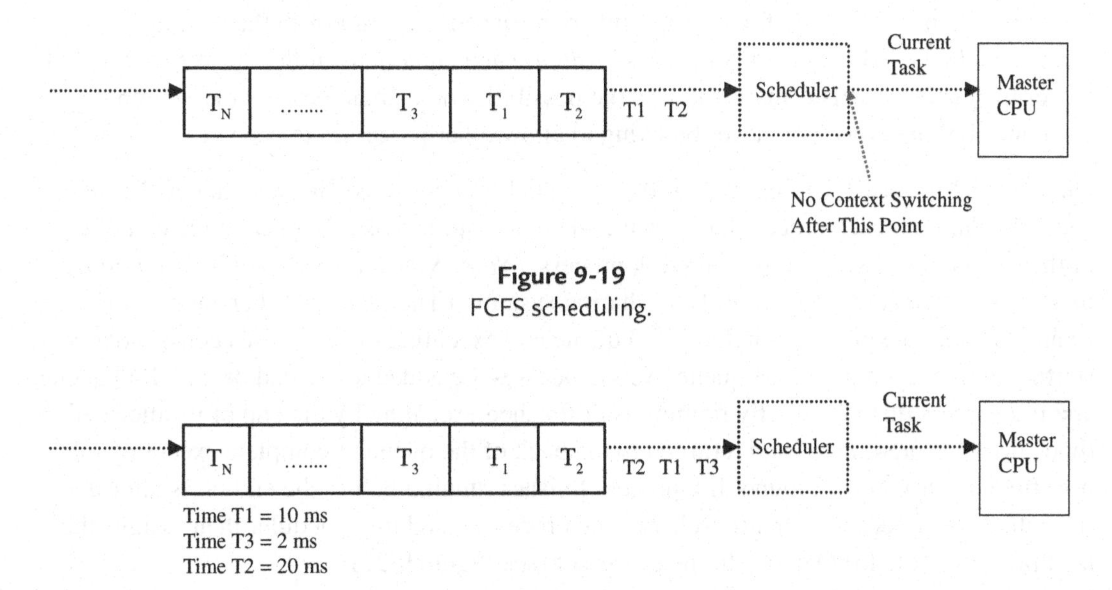


**Figure 9-19**
FCFS scheduling.

**Figure 9-20**
Shortest process next scheduling.

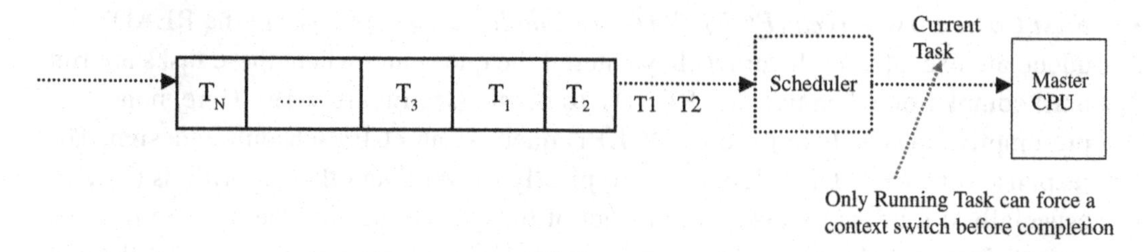


**Figure 9-21**
Cooperative scheduling.

Non-pre-emptive algorithms can be riskier to support since an assumption must be made that no one task will execute in an infinite loop, shutting out all other tasks from the master CPU. However, OSs that support non-pre-emptive algorithms don't force a context-switch before a task is ready, and the overhead of saving and restoration of accurate task information when switching between tasks that have not finished execution is only an issue if the non-pre-emptive scheduler implements a cooperative scheduling mechanism. In *pre-emptive scheduling*, on the other hand, the OS forces a context-switch on a task, whether or not a running task has completed executing or is cooperating with the context switch. Common scheduling algorithms based upon the pre-emptive approach include: *Round Robin/FIFO (First In First Out) scheduling*, *priority (pre-emptive) scheduling*, and *EDF (Earliest Deadline First)/Clock Driven scheduling*.

- *Round Robin/FIFO Scheduling*. The Round Robin/FIFO algorithm implements a FIFO queue that stores *ready* processes (processes ready to be executed). Processes are added to the queue at the end of the queue and are retrieved to be *run* from the start of the queue. In the FIFO system, all processes are treated equally regardless of their workload or interactivity. This is mainly due to the possibility of a single process maintaining control of the processor, never blocking to allow other processes to execute.

Under round-robin scheduling, each process in the FIFO queue is allocated an equal *time slice* (the duration each process has to run), where an interrupt is generated at the end of each of these intervals to start the pre-emption process. *(Note: Scheduling algorithms that allocate time slices, are also referred to as* time-sharing systems.*)* The scheduler then takes turns rotating among the processes in the FIFO queue and executing the processes consecutively, starting at the beginning of the queue. New processes are added to the end of the FIFO queue, and if a process that is currently running isn't finished executing by the end of its allocated time slice, it is pre-empted and returned to the back of the queue to complete executing the next time its turn comes around. If a process finishes running before the end of its allocated time slice, the process voluntarily releases the processor, and the scheduler then assigns the next process of the FIFO queue to the processor (see Figure 9-22).

While Round Robin/FIFO scheduling ensures the equal treatment of processes, drawbacks surface when various processes have heavier workloads and are constantly pre-empted, thus

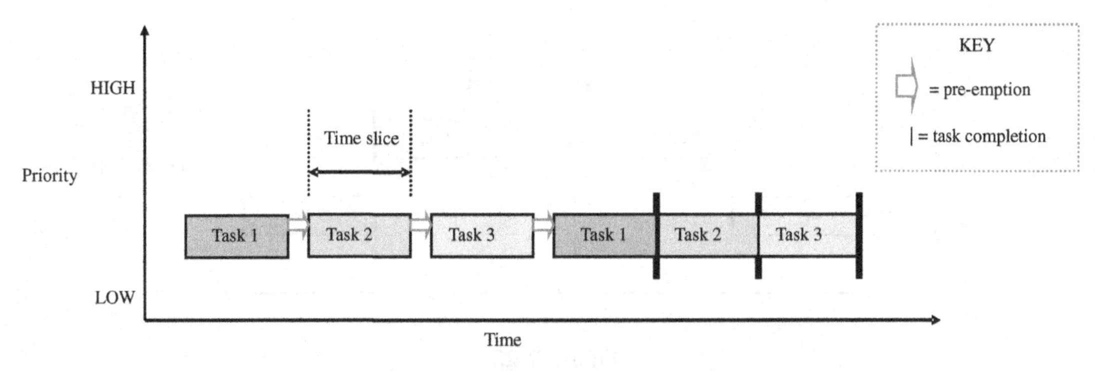


**Figure 9-22**
Round Robin/FIFO scheduling.[7]

creating more context switching overhead. Another issue occurs when processes in the queue are interacting with other processes (such as when waiting for the completion of another process for data) and are continuously pre-empted from completing any work until the other process of the queue has finished its run. The throughput depends on the time slice. If the time slice is too small, then there are many context switches, while too large a time slice isn't much different from a non-pre-emptive approach, like FCFS. Starvation is not possible with the round-robin implementation.

- *Priority (Pre-Emptive) Scheduling*. The priority pre-emptive scheduling algorithm differentiates between processes based upon their relative importance to each other and the system. Every process is assigned a priority, which acts as an indicator of orders of precedence within the system. The processes with the highest priority always pre-empt lower priority processes when they want to run, meaning a running task can be forced to block by the scheduler if a higher priority task becomes ready to run. Figure 9-23 shows three tasks (1, 2, and 3, where task 1 is the lowest priority task and task 3 is the highest), and task 3 pre-empts task 2 and task 2 pre-empts task 1.

While this scheduling method resolves some of the problems associated with round-robin/FIFO scheduling in dealing with processes that interact or have varying workloads, new problems can arise in priority scheduling including:

- *Process starvation*: a continuous stream of high priority processes keep lower priority processes from ever running. Typically resolved by aging lower priority processes (as these processes spend more time on queue, increase their priority levels).
- *Priority inversion*: higher priority processes may be blocked waiting for lower priority processes to execute, and processes with priorities in between have a higher priority in running, thus the lower priority as well as higher priority processes don't run (see Figure 9-24).
- How to *determine the priorities* of various processes. Typically, the more important the task, the higher the priority it should be assigned. For tasks that are equally

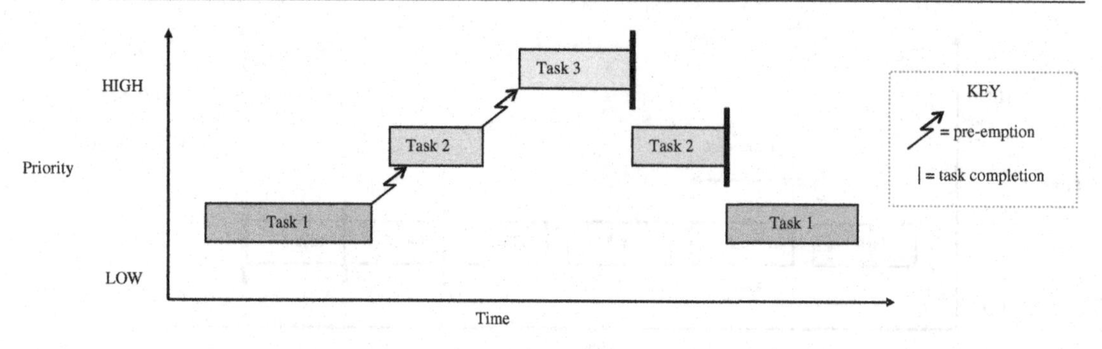


**Figure 9-23**
Pre-emptive priority scheduling.[8]

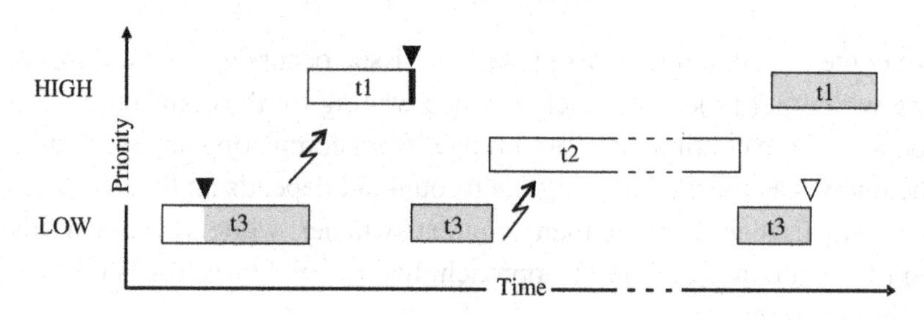


**Figure 9-24**
Priority inversion.[8]

important, one technique that can be used to assign task priorities is the *Rate Monotonic Scheduling (RMS)* scheme, in which tasks are assigned a priority based upon how often they execute within the system. The premise behind this model is that, given a pre-emptive scheduler and a set of tasks that are completely independent (no shared data or resources) and are run periodically (meaning run at regular time intervals), the more often a task is executed within this set, the higher its priority should be. The RMS Theorem says that if the above assumptions are met for a scheduler and a set of "*n*" tasks, all timing deadlines will be met if the inequality $\Sigma E_i/T_i \leq n(2^{1/n} - 1)$ is verified, where *i* is the Periodic task, *n* is the number of Periodic tasks, $T_i$ is the execution period of task *i*, $E_i$ is the worst-case execution time of task *i*, and $E_i/T_i$ is the fraction of CPU time required to execute task *i*. So, given two tasks that have been prioritized according to their periods, where the shortest period task has been assigned the highest priority, the "$n(2^{1/n} - 1)$" portion of the inequality would equal approximately 0.828, meaning the CPU utilization of these tasks should not exceed about 82.8% in order to meet all hard deadlines. For 100 tasks that have been prioritized according to their periods, where the shorter period tasks have been assigned the higher priorities, CPU utilization of these tasks should not exceed approximately 69.6% (100 ∗ (21/100 – 1)) in order to meet all deadlines.

**Real-World Advice**

*To Benefit Most from a Fixed-Priority Pre-Emptive OS*

Algorithms for assigning priorities to OS tasks are typically classified as fixed-priority where tasks are assigned priorities at design time and do not change through the lifecycle of the task, dynamic-priority where priorities are assigned to tasks at runtime, or some combination of both algorithms. Many commercial OSs typically support only the fixed-priority algorithms, since it is the least complex scheme to implement. The key to utilizing the fixed-priority scheme is:

- To assign the priorities of tasks according to their periods, so that the shorter the periods, the higher the priorities.
- To assign priorities using a fixed-priority algorithm (like the Rate Monotonic Algorithm, the basis of RMS) to assign fixed priorities to tasks and as a tool to quickly to determine if a set of tasks is schedulable.
- To understand that in the case when the inequality of a fixed-priority algorithm, like RMS, is not met, an analysis of the specific task set is required. RMS is a tool that allows for assuming that deadlines would be met in most cases if the total CPU utilization is below the limit ("most" cases meaning there are tasks that are not schedulable via any fixed-priority scheme). It is possible for a set of tasks to still be schedulable in spite of having a total CPU utilization above the limit given by the inequality. Thus, an analysis of each task's period and execution time needs to be done in order to determine if the set can meet required deadlines.
- To realize that a major constraint of fixed-priority scheduling is that it is not always possible to completely utilize the master CPU 100%. If the goal is 100% utilization of the CPU when using fixed priorities, then tasks should be assigned harmonic periods, meaning a task's period should be an exact multiple of all other tasks with shorter periods.

*Based on the article "Introduction to Rate Monotonic Scheduling," M. Barr,* Embedded Systems Programming, *February 2002.*

- *EDF/Clock Driven Scheduling*. As shown in Figure 9-25, the EDF/Clock Driven algorithm schedules priorities to processes according to three parameters: *frequency* (number of times a process is run), *deadline* (when processes execution needs to be completed), and *duration* (time it takes to execute the process). While the EDF algorithm allows for timing constraints to be verified and enforced (basically guaranteed deadlines for all tasks), the difficulty is defining an exact duration for various processes. Usually, an average estimate is the best that can be done for each process.

*Pre-Emptive Scheduling and the Real-Time Operating System (RTOS)*

One of the biggest differentiators between the scheduling algorithms implemented within embedded OSs is whether the algorithm guarantees its tasks will meet execution time deadlines. If tasks always meet their deadlines (as shown in the first two graphs in Figure 9-26) and related execution times are predictable (deterministic), the OS is referred to as an *RTOS*.

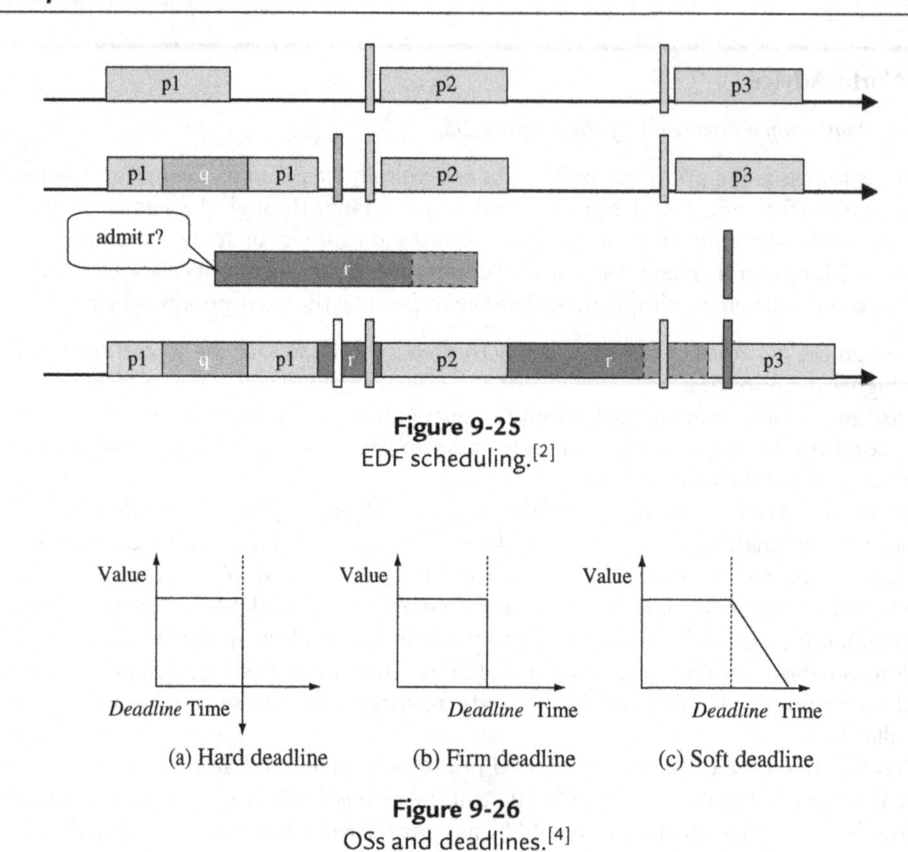


**Figure 9-25**
EDF scheduling.[2]

**Figure 9-26**
OSs and deadlines.[4]

Pre-emptive scheduling must be one of the algorithms implemented within RTOS schedulers, since tasks with real-time requirements have to be allowed to pre-empt other tasks. RTOS schedulers also make use of their own array of *timers*, ultimately based upon the system clock, to manage and meet their hard deadlines.

Whether an RTOS or a non-RTOS in terms of scheduling, all will vary in their implemented scheduling schemes. For example, VxWorks (Wind River) is a priority-based and round-robin scheme, Jbed (Esmertec) is an EDF scheme, and Linux (Timesys) is a priority-based scheme. Examples 7, 8, and 9 examine further the scheduling algorithms incorporated into these embedded off-the-shelf OSs.

#### Example 7: VxWorks Scheduling

The Wind scheduler is based upon both pre-emptive priority and round-robin real-time scheduling algorithms. As shown in Figure 9-27a1, round-robin scheduling can be teamed with pre-emptive priority scheduling to allow for tasks of the *same priority* to share the master processor, as well as allow higher priority tasks to pre-empt for the CPU.

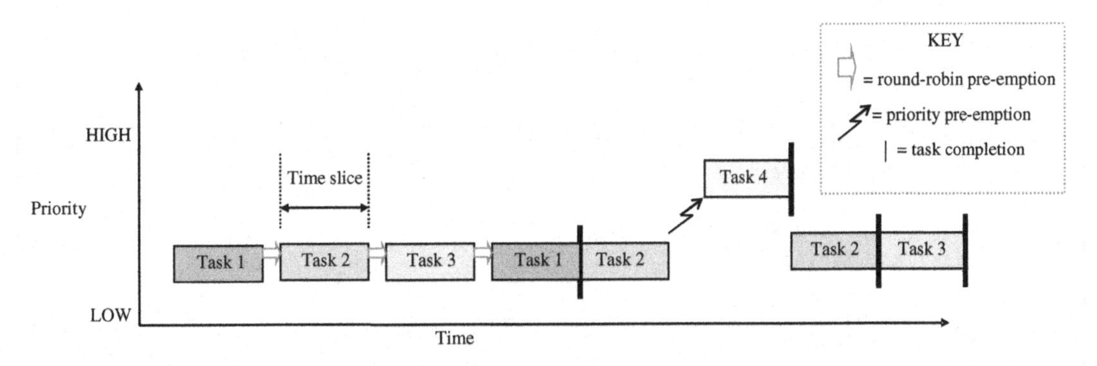


**Figure 9-27a1**
Pre-emptive priority scheduling augmented with round-robin scheduling.[7]

Without round-robin scheduling, tasks of equal priority in VxWorks would never pre-empt each other, which can be a problem if a programmer designs one of these tasks to run in an infinite loop. However, the pre-emptive priority scheduling allows VxWorks its real-time capabilities, since tasks can be programmed never to miss a deadline by giving them the higher priorities to pre-empt all other tasks. Tasks are assigned priorities via the "taskSpawn" command at the time of task creation:

```
int taskSpawn(
{Task Name},
{Task Priority 0-255, related to scheduling and will be discussed in the next
section},
{Task Options - VX_FP_TASK, execute with floating point coprocessor
       VX_PRIVATE_ENV, execute task with private environment
       VX_UNBREAKABLE, disable breakpoints for task
       VX_NO_STACK_FILL, do not fill task stack with 0xEE}
{Task address of entry point of program in memory - initial PC value}
{Up to 10 arguments for task program entry routine})
```

### Example 8: Jbed and EDF Scheduling

Under the Jbed RTOS, all six types of tasks have the three variables—"duration," "allowance," and "deadline"—when the task is created for the EDF scheduler to schedule all tasks, as shown in the method (Java subroutine) calls below.

```
public Task(
       long duration,
       long allowance,
       long deadline,
       RealtimeEvent event)
    Throws AdmissionFailure
```

```
Public Task (java.lang.String name,
       long duration,
       long allowance,
       long deadline,
       RealtimeEvent event)
   Throws AdmissionFailure
Public Task (java.lang.Runnable target,
       java.lang.String name,
       long duration,
       long allowance,
       long deadline,
       RealtimeEvent event)
   Throws AdmissionFailure
```

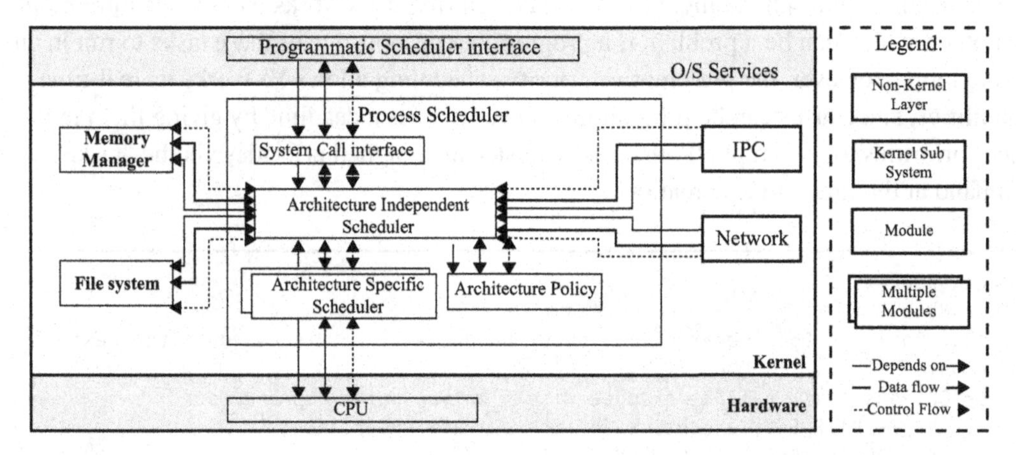


**Figure 9-27b1**
Embedded Linux block diagram.[9]

### Example 9: TimeSys Embedded Linux Priority-Based Scheduling

As shown in Figure 9-27b1, the embedded Linux kernel has a scheduler that is made up of four modules:[9]

- *System call interface module*: acts as the interface between user processes and any functionality explicitly exported by the kernel.
- *Scheduling policy module*: determines which processes have access to the CPU.
- *Architecture specific scheduler module*: an abstraction layer that interfaces with the hardware (i.e., communicating with CPU and the memory manager to suspend or resume processes).
- *Architecture independent scheduler module*: an abstraction layer that interfaces between the scheduling policy module and the architecture specific module.

The scheduling policy module implements a "priority-based" scheduling algorithm. While most Linux kernels and their derivatives are non-pre-emptable, have no rescheduling, and are not real-time, Timesys' Linux scheduler is priority-based, but has been modified to allow for real-time capabilities. Timesys has modified the traditional Linux's standard software timers, which are too coarsely grained to be suitable for use in most real-time applications because they rely on the kernel's jiffy timer, and implements high-resolution clocks and timers based on a hardware timer. The scheduler maintains a table listing all of the tasks within the entire system and any state information associated with the tasks. Under Linux, the total number of tasks allowed is only limited to the size of physical memory available. A dynamically allocated linked list of a task structure, whose fields that are relevant to scheduling are highlighted in Figure 9-27b2, represents all tasks in this table.

After a process has been created in Linux, through the fork or fork/exec commands, for instance, its priority is set via the setpriority command.

```
int setpriority(int which, int who, int prio);
   which = PRIO_PROCESS, PRIO_PGRP, or PRIO_USER_
   who = interpreted relative to which
   prio = priority value in the range -20 to 20
```

### 9.2.3 Intertask Communication and Synchronization

Different tasks in an embedded system typically must share the same hardware and software resources or may rely on each other in order to function correctly. For these reasons, embedded OSs provide different mechanisms that allow for tasks in a multitasking system to intercommunicate and synchronize their behavior so as to coordinate their functions, avoid problems, and allow tasks to run simultaneously in harmony.

Embedded OSs with multiple intercommunicating processes commonly implement interprocess communication (IPC) and synchronization algorithms based upon one or some combination of *memory sharing*, *message passing*, and *signaling* mechanisms.

With the *shared data* model shown in Figure 9-28, processes communicate via access to shared areas of memory in which variables modified by one process are accessible to all processes.

While accessing shared data as a means to communicate is a simple approach, the major issue of *race conditions* can arise. A race condition occurs when a process that is accessing shared variables is pre-empted before completing a modification access, thus affecting the integrity of shared variables. To counter this issue, portions of processes that access shared data, called *critical sections*, can be earmarked for *mutual exclusion* (or *Mutex* for short). Mutex mechanisms allow shared memory to be locked up by the process accessing it, giving that process exclusive access to shared data. Various mutual exclusion mechanisms can be

```
struct task_struct
{    ....
              // -1 unrunnable, 0 runnable, >0 stopped
 volatile long      state;

// number of clock ticks left to run in this scheduling slice, decremented
by a timer.
 long           counter;

 // the process' static priority, only changed through well-known system
calls like nice, POSIX.1b
 // sched_setparam, or 4.4BSD/SVR4 setpriority.
 long           priority;

 unsigned       long signal;

 //  bitmap of masked signals
 unsigned       long blocked;

 // per process flags, defined below
 unsigned       long flags;
 int errno;

 // hardware debugging registers
 long           debugreg[8];
 struct exec_domain  *exec_domain;
 struct linux_binfmt *binfmt;
 struct task_struct  *next_task, *prev_task;
 struct task_struct  *next_run, *prev_run;
 unsigned long     saved_kernel_stack;
 unsigned long     kernel_stack_page;
 int            exit_code, exit_signal;
  unsigned long     personality;
 int            dumpable:1;
 int            did_exec:1;
 int            pid;
 int            pgrp;
 int            tty_old_pgrp;
 int            session;
 // boolean value for session group leader
 int            leader;
 int            groups[NGROUPS];

 // pointers to (original) parent process, youngest child, younger sibling,
// older sibling, respectively.  (p->father  can be replaced with  p->p_pptr->pid)
 struct task_struct  *p_opptr, *p_pptr, *p_cptr,
                *p_ysptr, *p_osptr;
 struct wait_queue   *wait_chldexit;
 unsigned short    uid,euid,suid,fsuid;
 unsigned short    gid,egid,sgid,fsgid;
 unsigned long     timeout;

// the scheduling policy, specifies which scheduling class the task belongs to,
// such as : SCHED_OTHER (traditional UNIX process), SCHED_FIFO
// (POSIX.1b FIFO realtime process - A FIFO realtime process will
//run until either a) it blocks on I/O, b) it explicitly yields the CPU or c) it is
// pre-empted by another realtime process with a higher p->rt_priority value.)
// and SCHED_RR (POSIX round-robin realtime process –
//SCHED_RR is the same as SCHED_FIFO, except that when its timeslice
// expires it goes back to the end of the run queue).
 unsigned long     policy;

//realtime priority
 unsigned long     rt_priority;

 unsigned long     it_real_value, it_prof_value, it_virt_value;
 unsigned long     it_real_incr, it_prof_incr, it_virt_incr;
 struct timer_list  real_timer;
 long           utime, stime, cutime, cstime, start_time;

// mm fault and swap info: this can arguably be seen as either  mm-
specific or thread-specific */
 unsigned long     min_flt, maj_flt, nswap, cmin_flt, cmaj_flt,
cnswap;
 int swappable:1;
 unsigned long     swap_address;

 // old value of maj_flt
 unsigned long     old_maj_flt;

 // page fault count of the last time
 unsigned long     dec_flt;

 // number of pages to swap on next pass
 unsigned long     swap_cnt;

//limits
 struct rlimit      rlim[RLIM_NLIMITS];
 unsigned short     used_math;
 char            comm[16];

// file system info
 int            link_count;

// NULL if no tty
 struct tty_struct    *tty;

// ipc stuff
 struct sem_undo     *semundo;
 struct sem_queue    *semsleeping;

// ldt for this task - used by Wine.  If NULL, default_ldt is used
 struct desc_struct *ldt;

// tss for this task
 struct thread_struct tss;

// filesystem information
 struct fs_struct    *fs;

// open file information
 struct files_struct  *files;

// memory management info
 struct mm_struct    *mm;

// signal handlers
 struct signal_struct *sig;
#ifdef __SMP__
 int            processor;
 int            last_processor;
 int            lock_depth;   /* Lock depth.
                     We can context switch in and out
                     of holding a syscall kernel lock... */
#endif
                 .....
}
```

**Figure 9-27b2**

Task structure.[15]

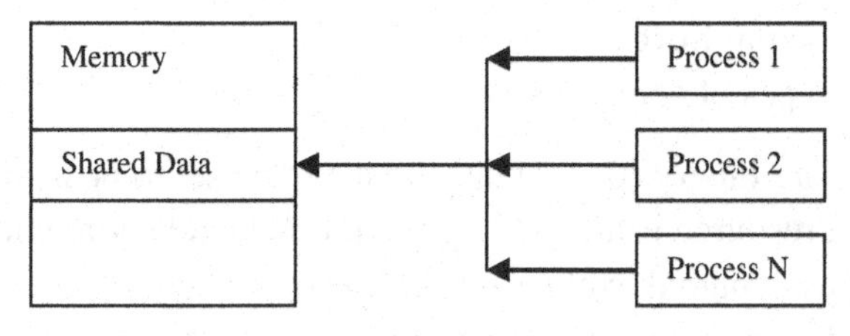


**Figure 9-28**
Memory sharing.

```
FuncA ()
    {
        int lock = intLock ();
        .
        . critical region that cannot be interrupted
        .
        intUnlock (lock);
    }
```

**Figure 9-29**
VxWorks processor-assisted locks.[10]

implemented not only for coordinating access to shared memory, but for coordinating access to other shared system resources as well. Mutual exclusion techniques for *synchronizing* tasks that wish to concurrently access shared data can include:

- *Processor-assisted locks* for tasks accessing shared data that are scheduled such that no other tasks can pre-empt them; the only other mechanisms that could force a context switch are interrupts. Disabling interrupts while executing code in the critical section would avoid a race condition scenario if the interrupt handlers access the same data. Figure 9-29 demonstrates this processor-assisted lock of disabling interrupts as implemented in VxWorks. VxWorks provides an interrupt locking and unlocking function for users to implement in tasks. Another possible processor-assisted lock is the "test-and-set-instruction" mechanism (also referred to as the *condition variable* scheme). Under this mechanism, the setting and testing of a register flag (condition) is an atomic function, a process that cannot be interrupted, and this flag is tested by any process that wants to access a critical section. In short, both the interrupt disabling and the condition variable type of locking schemes guarantee a process exclusive access to memory, where nothing can pre-empt the access to shared data and the system cannot respond to any other event for the duration of the access.
- *Semaphores*, which can be used to lock access to shared memory (mutual exclusion) and also can be used to coordinate running processes with outside events (synchronization). The semaphore functions are *atomic* functions, and are usually invoked through system calls by the process. Example 10 demonstrates semaphores provided by VxWorks.

### Example 10: VxWorks Semaphores

VxWorks defines three types of semaphores:

- *Binary* semaphores are binary (0 or 1) flags that can be set to be available or unavailable. Only the associated resource is affected by the mutual exclusion when a binary semaphore is used as a mutual exclusion mechanism (whereas processor assisted locks, for instance, can affect other unrelated resources within the system). A binary semaphore is initially set = 1 (full) to show the resource is available. Tasks check the binary semaphore of a resource when wanting access and, if available, then take the associated semaphore when accessing a resource (setting the binary semaphore = 0), and then give it back when finishing with a resource (setting the binary semaphore = 1). When a binary semaphore is used for task synchronization, it is initially set equal to 0 (empty), because it acts as an event other tasks are waiting for. Other tasks that need to run in a particular sequence then wait (block) for the binary semaphore to be equal to 1 (until the event occurs) to take the semaphore from the original task and set it back to 0. The VxWorks pseudocode example below demonstrates how binary semaphores can be used in VxWorks for task synchronization.

```
#include "VxWorks.h"
#include "semLib.h"
#include "arch/arch/ivarch.h" /* replace arch with architecture type */

SEM_ID syncSem; /* ID of sync semaphore */

init (int someIntNum)
{
   /* connect interrupt service routine */
      intConnect (INUM_TO_IVEC (someIntNum), eventInterruptSvcRout, 0);

   /* create semaphore */
   syncSem = semBCreate (SEM_Q_FIFO, SEM_EMPTY);

   /* spawn task used for synchronization. */
   taskSpawn ("sample", 100, 0, 20000, task1, 0,0,0,0,0,0,0,0,0,0);
}
task1 (void)
{
   …
   semTake (syncSem, WAIT_FOREVER); /* wait for event to occur */
   printf ("task 1 got the semaphore\n");
   … /* process event */
}

eventInterruptSvcRout (void)
```

```
{
…
   semGive (syncSem); /* let task 1 process event */
   …
}
[4]
```

- *Mutual exclusion* semaphores are binary semaphores that can only be used for mutual exclusion issues that can arise within the VxWorks scheduling model, such as priority inversion, deletion safety (ensuring that tasks that are accessing a critical section and blocking other tasks aren't unexpectedly deleted), and recursive access to resources. Below is a pseudocode example of a mutual exclusion semaphore used recursively by a task's subroutines.

```
/* Function A requires access to a resource which it acquires by taking
* mySem;
* Function A may also need to call function B, which also requires mySem:
*/
/* includes */
#include "VxWorks.h"
#include "semLib.h"
SEM_ID mySem;

/* Create a mutual-exclusion semaphore. */
init ()
{
   mySem = semMCreate (SEM_Q_PRIORITY);
}

funcA ()
{
   semTake (mySem, WAIT_FOREVER);
   printf ("funcA: Got mutual-exclusion semaphore\n");
   …
   funcB ();
   semGive (mySem);
   printf ("funcA: Released mutual-exclusion semaphore\n");
}

   funcB ()
{
   semTake (mySem, WAIT_FOREVER);
```

```
    printf ("funcB: Got mutual-exclusion semaphore\n");
    ...
    semGive (mySem);
    printf ("funcB: Releases mutual-exclusion semaphore\n");
}
[4]
```

- *Counting* semaphores are positive integer counters with two related functions: incrementing and decrementing. Counting semaphores are typically used to manage multiple copies of resources. Tasks that need access to resources decrement the value of the semaphore; when tasks relinquish a resource, the value of the semaphore is incremented. When the semaphore reaches a value of "0," any task waiting for the related access is blocked until another task gives back the semaphore.

```
/* includes */
#include "VxWorks.h"
#include "semLib.h"
SEM_ID mySem;

/* Create a counting semaphore. */ init ()
{
    mySem = semCCreate (SEM_Q_FIFO,0);
}

...
[4]
```

On a final note, with mutual exclusion algorithms, only one process can have access to shared memory at any one time, basically having a *lock* on the memory accesses. If more than one process blocks waiting for their turn to access shared memory, and relying on data from each other, a *deadlock* can occur (such as priority inversion in priority based scheduling). Thus, embedded OSs have to be able to provide deadlock-avoidance mechanisms as well as deadlock-recovery mechanisms. As shown in the examples above, in VxWorks, semaphores are used to avoid and prevent deadlocks.

Intertask communication via *message passing* is an algorithm in which *messages* (made up of data bits) are sent via *message queues* between processes. The OS defines the protocols for process addressing and authentication to ensure that messages are delivered to processes reliably, as well as the number of messages that can go into a queue and the message sizes. As shown in Figure 9-30, under this scheme, OS tasks send messages to a message queue, or receive messages from a queue to communicate.

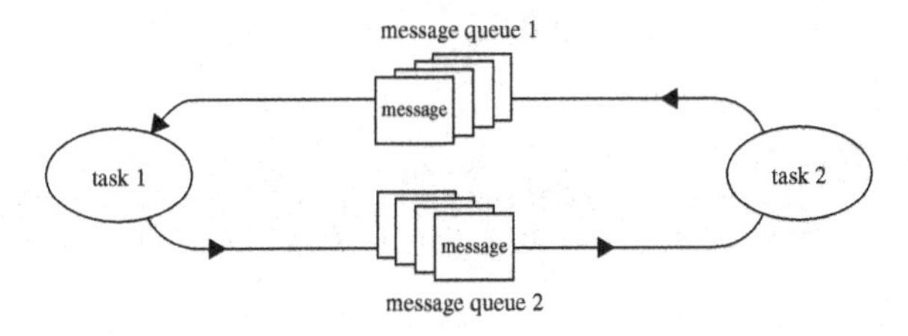


**Figure 9-30**
Message queues.[4]
The wind kernel supports two types of signal interface: UNIX BSD-style and POSIX-compatible signals.

Microkernel-based OSs typically use the message passing scheme as their main synchronization mechanism. Example 11 demonstrates message passing in more detail, as implemented in VxWorks.

### Example 11: Message Passing in VxWorks[4]

VxWorks allows for intertask communication via message passing queues to store data transmitted between different tasks or an *interrupt service routine (ISR)*. VxWorks provides the programmer four system calls to allow for the development of this scheme:

| Call | Description |
|---|---|
| msgQCreate() | Allocates and initializes a message queue |
| msgQDelete() | Terminates and frees a message queue |
| msgQSend() | Sends a message to a message queue |
| msgQReceive() | Receives a message from a message queue |

These routines can then be used in an embedded application, as shown in the source code example below, to allow for tasks to intercommunicate:

```
/* In this example, task t1 creates the message queue and sends a message
* to task t2. Task t2 receives the message from the queue and simply
* displays the message.
*/
/* includes */
#include "VxWorks.h"
#include "msgQLib.h"
/* defines */
#define MAX_MSGS (10)
#define MAX_MSG_LEN (100)
MSG_Q_ID myMsgQId;
```

```
task2 (void)
{
    char msgBuf[MAX_MSG_LEN];
    /* get message from queue; if necessary wait until msg is available */
    if (msgQReceive(myMsgQId, msgBuf, MAX_MSG_LEN, WAIT_FOREVER) == ERROR)
    return (ERROR);
    /* display message */
    printf ("Message from task 1:\n%s\n", msgBuf);
}

#define MESSAGE "Greetings from Task 1" task1 (void)
{
    /* create message queue */
    if ((myMsgQId = msgQCreate (MAX_MSGS, MAX_MSG_LEN, MSG_Q_PRIORITY)) == NULL)
    return (ERROR);
    /* send a normal priority message, blocking if queue is full */
    if (msgQSend (myMsgQId, MESSAGE, sizeof (MESSAGE), WAIT_FOREVER, MSG_PRI_
NORMAL) == ERROR)
    return (ERROR);
    }
[4]
```

*Signals and Interrupt Handling (Management) at the Kernel Level*

Signals are indicators to a task that an asynchronous event has been generated by some external event (other processes, hardware on the board, timers, etc.) or some internal event (problems with the instructions being executed, etc.). When a task receives a signal, it suspends executing the current instruction stream and context switches to a signal handler (another set of instructions). The signal handler is typically executed within the task's context (stack) and runs in the place of the signaled task when it is the signaled task's turn to be scheduled to execute.

The *wind* kernal supports two types of signal interface: UNIX BSD-style and POSIX-compatible signals (see Figure 9-31).

Signals are typically used for interrupt handling in an OS, because of their asynchronous nature. When a signal is raised, a resource's availability is unpredictable. However, signals can be used for general intertask communication, but are implemented so that the possibility of a signal handler blocking or a deadlock occurring is avoided. The other intertask communication mechanisms (shared memory, message queues, etc.), along with signals, can be used for ISR-to-Task level communication, as well.

When signals are used as the OS abstraction for interrupts and the signal handling routine becomes analogous to an ISR, the OS manages the interrupt table, which contains the interrupt and information about its corresponding ISR, as well as provides a system call

| BSD 4.3 | POSIX 1003.1 |
|---|---|
| sigmask( ) | sigemptyset( ), sigfillset( ), sigaddset( ), sigdelset( ), sigismember( ) |
| sigblock( ) | sigprocmask( ) |
| sigsetmask( ) | sigprocmask( ) |
| pause( ) | sigsuspend( ) |
| sigvec( ) | sigaction( ) |
| (none) | sigpending( ) |
| signal( ) | signal( ) |
| kill( ) | kill( ) |

**Figure 9-31**
VxWorks signaling mechanism.[4]

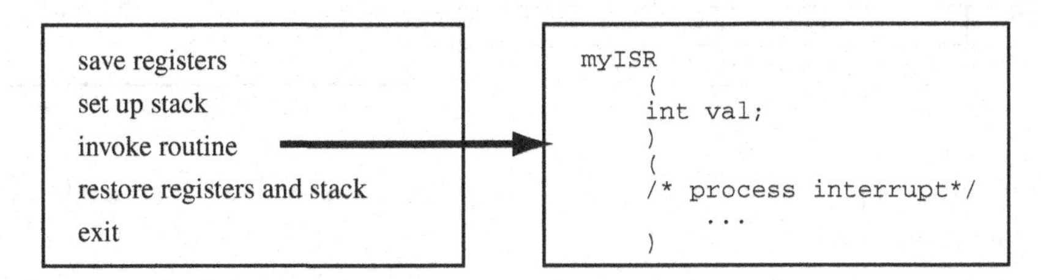


**Figure 9-32**
OS interrupt subroutine.[4]

(subroutine) with parameters that that can be used by the programmer. At the same time, the OS protects the integrity of the interrupt table and ISRs, because this code is executed in kernel/supervisor mode. The general process that occurs when a process receives a signal generated by an interrupt and an interrupt handler is called is shown in Figure 9-32.

As mentioned in previous chapters, the architecture determines the interrupt model of an embedded system (that is, the number of interrupts and interrupt types). The interrupt device drivers initialize and provide access to interrupts for higher layer of software. The OS then provides the *signal* IPC mechanism to allow for its processes to work with interrupts, as well as being able to provide various interrupt subroutines that abstracts out the device driver.

While all OSs have some sort of interrupt scheme, this will vary depending on the architecture they are running on, since architectures differ in their own interrupt schemes. Other variables include *interrupt latency/response*, the time between the actual initiation of an interrupt and the execution of the ISR code, and *interrupt recovery*, the time it takes to switch back to the interrupted task. Example 12 shows an interrupt scheme of a real-world embedded RTOS.

### Example 12: Interrupt Handling in VxWorks

Except for architectures that do not allow for a separate interrupt stack (and thus the stack of the interrupted task is used), ISRs use the same interrupt stack, which is initialized and configured at system start-up, outside the context of the interrupting task. Table 9-1

**Table 9-1: Interrupt Routines in VxWorks[4]**

| Call | Description |
|---|---|
| intConnect() | Connects a C routine to an interrupt vector |
| intContext() | Returns TRUE if called from interrupt level |
| intCount() | Gets the current interrupt nesting depth |
| intLevelSet() | Sets the processor interrupt mask level |
| intLock() | Disables interrupts |
| intUnlock() | Re-enables interrupts |
| intVecBaseSet() | Sets the vector base address |
| intVecBaseGet() | Gets the vector base address |
| intVecSet() | Sets an exception vector |
| intVecGet() | Gets an exception vector |

```
/* This routine initializes the
serial driver, sets up interrupt
vectors,
* and performs hardware
initialization of the serial ports.
*/
void InitSerialPort (void)
{
initSerialPort():
(void) intConnect (INUM_TO_IVEC
(INT_NUM_SCC), serialInt, 0);
...
}
```

summarizes the interrupt routines provided in VxWorks, along with a pseudocode example of using one of these routines.

## 9.3 Memory Management

As mentioned earlier in this chapter, a kernel manages program code within an embedded system via tasks. The kernel must also have some system of loading and executing tasks within the system, since the CPU only executes task code that is in cache or RAM. With multiple tasks sharing the same memory space, an OS needs a security system mechanism to protect task code from other independent tasks. Also, since an OS must reside in the same memory space as the tasks it is managing, the protection mechanism needs to include managing its own code in memory and protecting it from the task code it is managing. It is these functions, and more, that are the responsibility of the memory management components of an OS. In general, a kernel's memory management responsibilities include:

- Managing the mapping between logical (physical) memory and task memory references.
- Determining which processes to load into the available memory space.
- Allocating and deallocating of memory for processes that make up the system.
- Supporting memory allocation and deallocation of code requests (within a process), such as the C language "alloc" and "dealloc" functions, or specific buffer allocation and deallocation routines.
- Tracking the memory usage of system components.
- Ensuring cache coherency (for systems with cache).
- Ensuring process memory protection.

As introduced in Chapters 5 and 8, physical memory is composed of two-dimensional arrays made up of cells addressed by a unique row and column, in which each cell can store 1 bit.

Again, the OS treats memory as one large one-dimensional array, called a *memory map*. Either a hardware component integrated in the master CPU or on the board does the conversion between logical and physical addresses (such as a *memory management unit (MMU)*), or it must be handled via the OS.

How OSs manage the logical memory space differs from OS to OS, but kernels generally run kernel code in a separate memory space from processes running higher level code (i.e., middleware and application layer code). Each of these memory spaces (*kernel* containing kernel code and *user* containing the higher-level processes) is managed differently. In fact, most OS processes typically run in one of two modes: *kernel mode* and *user mode*, depending on the routines being executed. Kernel routines run in *kernel mode* (also referred to as *supervisor mode*), in a different memory space and level than higher layers of software such as middleware or applications. Typically, these higher layers of software run in *user mode*, and can only access anything running in kernel mode via *system calls*, the higher-level interfaces to the kernel's subroutines. The kernel manages memory for both itself and user processes.

### 9.3.1 User Memory Space

Because multiple processes are sharing the same physical memory when being loaded into RAM for processing, there also must be some protection mechanism so processes cannot inadvertently affect each other when being swapped in and out of a single physical memory space. These issues are typically resolved by the OS through memory "swapping," where partitions of memory are *swapped* in and out of memory at runtime. The most common partitions of memory used in swapping are *segments* (fragmentation of processes from within) and *pages* (fragmentation of logical memory as a whole). Segmentation and paging not only simplify the swapping—memory allocation and deallocation—of tasks in memory, but allow for code reuse and memory protection, as well as providing the foundation for *virtual memory*. Virtual memory is a mechanism managed by the OS to allow a device's limited memory space to be shared by multiple competing "user" tasks, in essence enlarging the device's actual physical memory space into a larger "virtual" memory space.

#### Segmentation

As mentioned earlier in this chapter, a process encapsulates all the information that is involved in executing a program, including source code, stack, and data. All of the different types of information within a process are divided into "logical" memory units of variable sizes, called *segments*. A segment is a set of logical addresses containing the same type of information. Segment addresses are logical addresses that start at 0, and are made up of a *segment number*, which indicates the base address of the segment, and a *segment offset*, which defines the actual physical memory address. Segments are independently protected, meaning they have assigned accessibility characteristics, such as *shared* (where other processes can access that segment), *read-only*, or *read/write*.

Most OSs typically allow processes to have all or some combination of five types of information within segments: text (or code) segment, data segment, BSS (block started by symbol) segment, stack segment, and the heap segment. A *text* segment is a memory space containing the source code. A *data* segment is a memory space containing the source code's initialized variables (data). A *BSS* segment is a statically allocated memory space containing the source code's un-initialized variable (data). The data, text, and BSS segments are all fixed in size at compile time, and are as such *static* segments; it is these three segments that typically are part of the *executable file.* Executable files can differ in what segments they are composed of, but in general they contain a header, and different sections that represent the types of segments, including name, permissions, etc., where a segment can be made up of one or more sections. The OS creates a task's image by *memory mapping* the contents of the executable file, meaning loading and interpreting the segments (sections) reflected in the executable into memory. There are several executable file formats supported by embedded OSs, the most common including:

- *ELF* (Executable and Linking Format): UNIX-based, includes some combination of an ELF header, the program header table, the section header table, the ELF sections, and the ELF segments. Linux (Timesys) and VxWorks (WRS) are examples of OSs that support ELF. (See Figure 9-33.)
- *Class* (Java byte code): a class file describes one Java class in detail in the form of a stream of 8-bit bytes (hence the name "byte code"). Instead of segments, elements of the class file are called *items.* The Java class file format contains the class description, as well as how that class is connected to other classes. The main components of a class file are a symbol table (with constants), declaration of fields, method implementations (code), and symbolic references (where other classes references are located). The Jbed RTOS is an example that supports the Java byte code format. (See Figure 9-34.)

| Linking View | Execution View |
|---|---|
| ELF header | ELF header |
| Program header table Optional | Program header table |
| section 1 | Segment 1 |
| ... | |
| section n | Segment 2 |
| ... | ... |
| ... | ... |
| Section header table | Section header table (optional) |

**Figure 9-33**
ELF executable file format.[11]

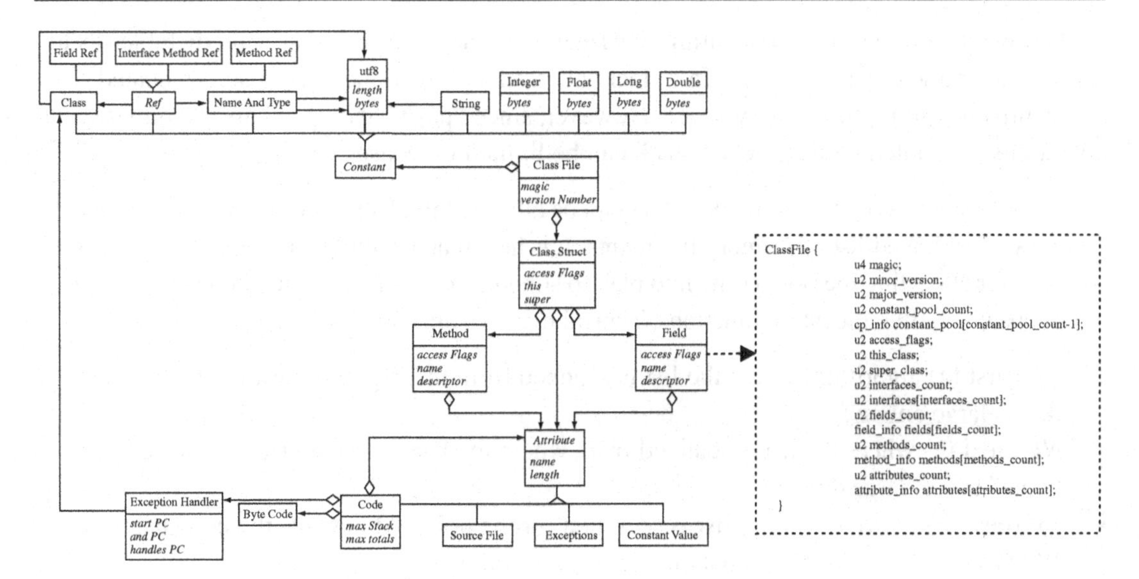


**Figure 9-34**
Class executable file format.[12]

| Offset | Size | Field | Description |
|---|---|---|---|
| 0 | 2 | Machine | Number identifying type of target machine. |
| 2 | 2 | Number of Sections | Number of sections; indicates size of the Section Table, which immediately follows the headers. |
| 4 | 4 | Time/Date Stamp | Time and date the file was created. |
| 8 | 4 | Pointer to Symbol | Offset, within the COFF file, of the symbol table. |
| 12 | 4 | Number of Symbols | Number of entries in the symbol table. This data can be used in locating the string table, which immediately follows the symbol table. |
| 16 | 2 | Optional Header | Size of the optional header, which is Size included for executable files but not object files. An object file should have a value of 0 here. |
| 18 | 2 | Characteristics | Flags indicating attributes of the file. |

**Figure 9-35**
Class executable file format.[13]

- *COFF* (Common Object File Format): a class file format which (among other things) defines an image file that contains file headers that include a file signature, COFF Header, an Optional Header, and also object files that contain only the COFF Header. Figure 9-35 shows an example of the information stored in a COFF header. WinCE[MS] is an example of an embedded OS that supports the COFF executable file format.

The *stack* and *heap* segments, on the other hand, are not fixed at compile time, and can change in size at runtime and so are *dynamic* allocation components. A *stack* segment is a section of memory that is structured as a LIFO (Last In First Out) queue, where data is "pushed" onto the stack or "popped" off of the stack (push and pop are the only two operations associated with a stack). Stacks are typically used as a simple and efficient method within a program for allocating and freeing memory for data that is predictable (local variables, parameter passing, etc.). In a stack, all used and freed memory space is located consecutively within the memory space. However, since "push" and "pop" are the only two operations associated with a stack, a stack can be limited in its uses.

A *heap* segment is a section of memory that can be allocated in blocks at runtime, and is typically set up as a free linked-list of memory fragments. It is here that a kernel's memory management facilities for allocating memory come into play to support the "malloc" C function (for example) or OS-specific buffer allocation functions. Typical memory allocation schemes include:

- *FF* (first fit) algorithm, where the list is scanned from the beginning for the first "hole" that is large enough.
- *NF* (next fit) where the list is scanned from where the last search ended for the next "hole" that is large enough.
- *BF* (best fit) where the entire list is searched for the hole that best fits the new data.
- *WF* (worst fit), which places data in the largest available "hole."
- *QF* (quick fit) where a list is kept of memory sizes and allocation is done from this information.
- *The buddy system* where blocks are allocated in sizes of powers of 2. When a block is deallocated, it is then merged with contiguous blocks.

The method by which memory that is no longer needed within a heap is freed depends on the OS. Some OSs provide a GC that automatically reclaims unused memory (garbage collection algorithms include generational, copying, and mark-and-sweep; see Figures 9-36a–c). Other OSs require that the programmer explicitly free memory through a system call (i.e., in support of the "free" C function). With the latter technique, the programmer has to be aware

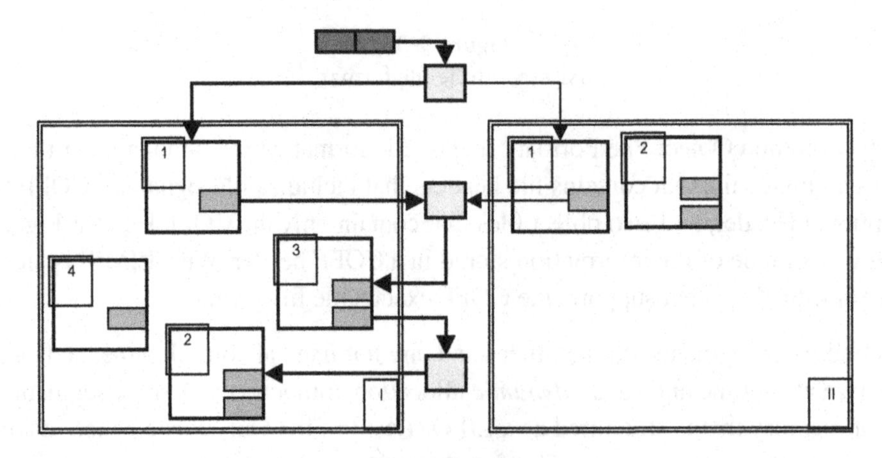


**Figure 9-36a**
Copying GC diagram.[2]

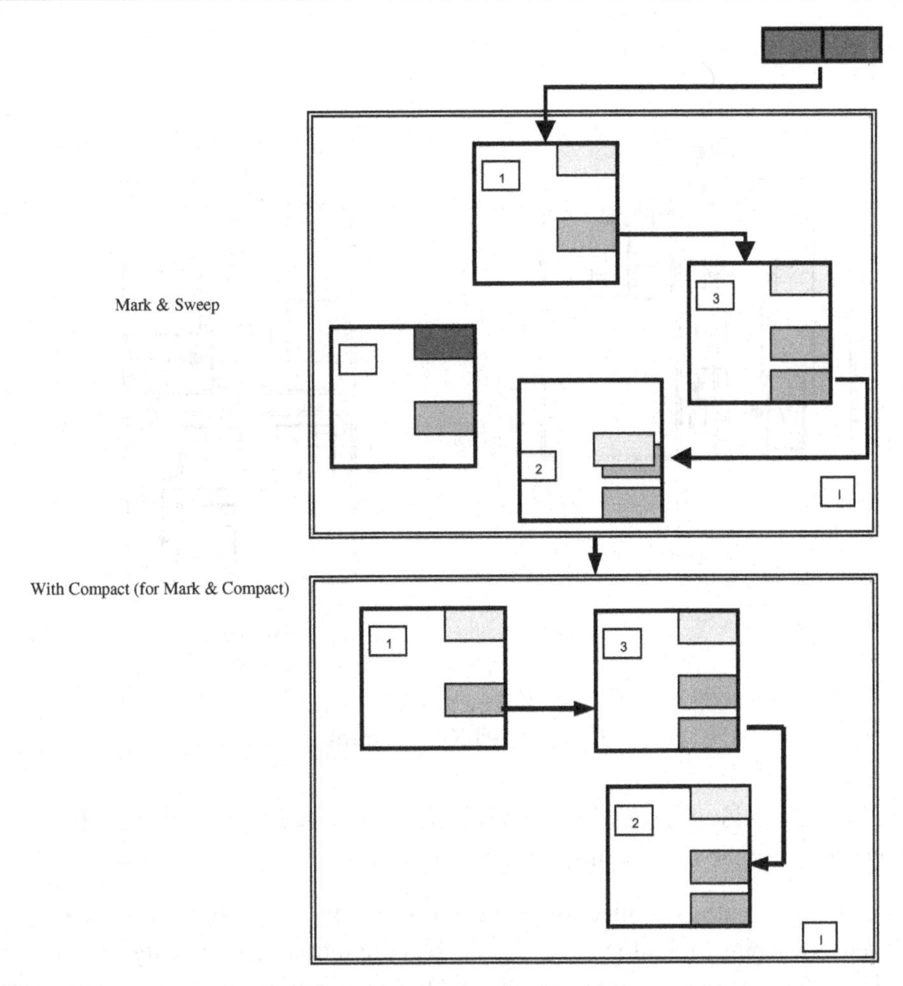


**Figure 9-36b**
Mark-and-sweep and mark-and-compact GC diagram.[2]

of the potential problem of memory leaks, where memory is lost because it has been allocated but is no longer in use and has been forgotten, which is less likely to happen with a GC.

Another problem occurs when allocated and freed memory cause memory fragmentation, where available memory in the heap is spread out in a number of holes, making it more difficult to allocate memory of the required size. In this case, a memory compaction algorithm must be implemented if the allocation/deallocation algorithms cause a lot of fragmentation. This problem can be demonstrated by examining garbage collection algorithms.

The copying garbage collection algorithm works by copying referenced objects to a different part of memory and then freeing up the original memory space. This algorithm uses a larger memory area to work and usually cannot be interrupted during the copy (it blocks

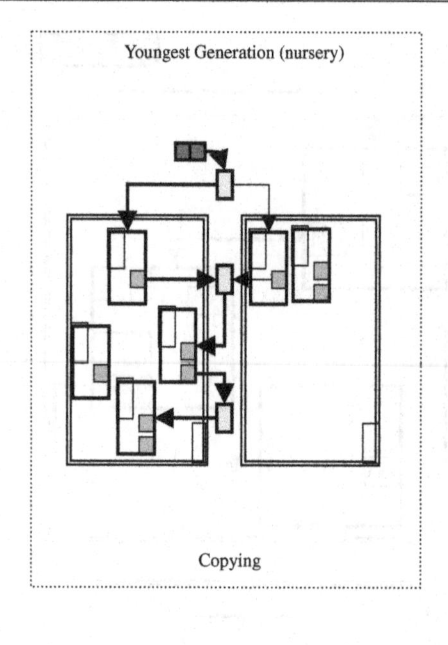


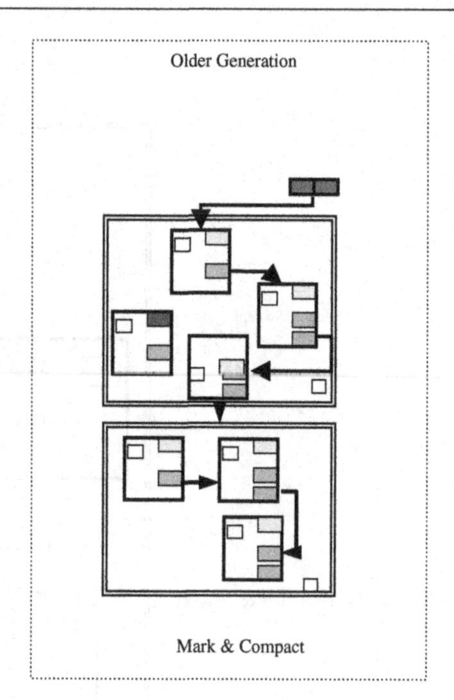


**Figure 9-36c**
Generational GC diagram.

the systems). However, it does ensure that what memory is used, is used efficiently by compacting objects in the new memory space.

The mark-and-sweep garbage collection algorithm works by "marking" all objects that are used and then "sweeping" (deallocating) objects that are unmarked. This algorithm is usually non-blocking, so the system can interrupt the GC to execute other functions when necessary. However, it doesn't compact memory the way a copying GC would, leading to memory fragmentation with small, unusable holes possibly existing where deallocated objects used to exist. With a mark-and-sweep GC, an additional memory compacting algorithm could be implemented making it a mark (sweep)-and-compact algorithm.

Finally, the generational garbage collection algorithm separates objects into groups, called *generations*, according to when they were allocated in memory. This algorithm assumes that most objects that are allocated are short-lived; thus copying or compacting the remaining objects with longer lifetimes is a waste of time. So, it is objects in the younger generation group that are cleaned up more frequently than objects in the older generation groups. Objects can also be moved from a younger generation to an older generation group. Each generational GC also may employ different algorithms to deallocate objects within each generational group, such as the copying algorithm or mark-and-sweep algorithms described above. Compaction algorithms would be needed in both generations to avoid fragmentation problems.

Finally, heaps are typically used by a program when allocation and deletion of variables are unpredictable (linked lists, complex structures, etc.). However, heaps aren't as simple or as efficient as stacks. As mentioned, how memory in a heap is allocated and deallocated is typically affected by the programming language the OS is based upon, such as a C-based OS using "malloc" to allocate memory in a heap and "free" to deallocate memory or a Java-based OS having a GC. Pseudocode Examples 13, 14, and 15 demonstrate how heap space can be allocated and deallocated under various embedded OSs.

### Example 13: VxWorks Memory Management and Segmentation

VxWorks tasks are made up of text, data, and BSS static segments, as well as each task having its own stack.

The VxWorks system call "taskSpawn" is based upon the POSIX spawn model, and is what creates, initializes, and activates a new (child) task. After the spawn system call, an image of the child task (including TCB, stack, and program) is allocated into memory. In the pseudocode below, the code itself is the text segment, data segments are any initialized variables, and the BSS segments are the uninitialized variables (seconds, etc.). In the taskSpawn system call, the task stack size is 3000 bytes and is not filled with 0xEE because of the VX_NO_ STACK_FILL parameter in the system call.

**Task Creation VxWorks Pseudocode**

```
// parent task that enables software timer
void parentTask(void)
{
...
if sampleSoftware Clock NOT running {
    /"newSWClkId" is a unique integer value assigned by kernel when task is created
    newSWClkId = taskSpawn ("sampleSoftwareClock", 255, VX_NO_STACK_FILL, 3000,
    (FUNCPTR) minuteClock, 0, 0, 0, 0, 0, 0, 0, 0, 0, 0);
    ...
}

// child task program Software Clock
void minuteClock(void) {
    integer seconds;
    while (softwareClock is RUNNING) {
              seconds = 0;
              while (seconds < 60) {
                        seconds = seconds + 1;
    }
    ...
}
[4]
```

Heap space for VxWorks tasks is allocated by using the C-language malloc/new system calls to dynamically allocate memory. There is no GC in VxWorks, so the programmer must deallocate memory manually via the free() system call.

```
/* The following code is an example of a driver that performs address
* translations. It attempts to allocate a cache-safe buffer, fill it, and
* then write it out to the device. It uses CACHE_DMA_FLUSH to make sure
* the data is current. The driver then reads in new data and uses
* CACHE_DMA_INVALIDATE to guarantee cache coherency. */
#include "VxWorks.h"
#include "cacheLib.h"
#include "myExample.h"

STATUS myDmaExample (void)
{
void * pMyBuf;
void * pPhysAddr;
/* allocate cache safe buffers if possible */
if ((pMyBuf = cacheDmaMalloc (MY_BUF_SIZE)) == NULL)
return (ERROR);
… fill buffer with useful information …
/* flush cache entry before data is written to device */
CACHE_DMA_FLUSH (pMyBuf, MY_BUF_SIZE);
/* convert virtual address to physical */
pPhysAddr = CACHE_DMA_VIRT_TO_PHYS (pMyBuf);
/* program device to read data from RAM */
myBufToDev (pPhysAddr);
… wait for DMA to complete …
… ready to read new data …
/* program device to write data to RAM */
myDevToBuf (pPhysAddr);
… wait for transfer to complete …
/* convert physical to virtual address */
pMyBuf = CACHE_DMA_PHYS_TO_VIRT (pPhysAddr);
/* invalidate buffer */
CACHE_DMA_INVALIDATE (pMyBuf, MY_BUF_SIZE);
… use data …
/* when done free memory */
if (cacheDmaFree (pMyBuf) == ERROR)
return (ERROR);
return (OK);
}
[4]
```

### Example 14: Jbed Memory Management and Segmentation

In Java, memory is allocated in the Java heap via the "new" keyword (e.g., unlike the "malloc" in C). However, there are a set of interfaces defined in some Java standards, called *JNIs (Java Native Interfaces)*, that allow for C and/or assembly code to be integrated within Java code, so in essence, the "malloc" is available if JNI is supported. For memory deallocation, as specified by the Java standard, is done via a GC.

Jbed is a Java-based OS, and as such supports "new" for heap allocation.

```
public void CreateOneshotTask(){
    // Task execution time values
    final long DURATION = 100L; // run method takes < 100µs
    final long ALLOWANCE = 0L; // no DurationOverflow handling
    final long DEADLINE = 1000L;// complete within 1000µs
    Runnable target; // Task's executable code
    OneshotTimer taskType;
    Task task;

    // Create a Runnable object
    target = new MyTask();

    // Create Oneshot tasktype with no delay
    taskType = new OneshotTimer(0L);

                                                     Memory allocation in Java
    // Create the task
    try{
    task = new Task(target,
    DURATION, ALLOWANCE, DEADLINE,
    taskType);
    }catch(AdmissionFailure e){
       System.out.println("Task creation failed");
    return;
    }
    [2]
```

Memory deallocation is handled automatically in the heap via a Jbed GC based upon the mark-and-sweep algorithm (which is non-blocking and is what allows Jbed to be an RTOS). The GC can be run as a reoccurring task, or can be run by calling a "runGarbageCollector" method.

### Example 15: Linux Memory Management and Segmentation

Linux processes are made up of text, data, and BSS static segments; in addition, each process has its own stack (which is created with the fork system call). Heap space for Linux tasks are allocated via the C-language malloc/new system calls to dynamically allocate memory. There is no GC in Linux, so the programmer must deallocate memory manually via the free() system call.

```
void *mem_allocator (void *arg)
{
   int i;
   int thread_id = *(int *)arg;
   int start = POOL_SIZE * thread_id;
   int end = POOL_SIZE * (thread_id + 1);
   if(verbose_flag) {
      printf("Releaser %i works on memory pool %i to %i\n",
      thread_id, start, end);
   printf("Releaser %i started…\n", thread_id);
   }
   while(!done_flag) {
      /* find first NULL slot */
      for (i = start; i < end; ++i) {
         if (NULL == mem_pool[i]) {
            mem_pool[i] = malloc(1024);
            if (debug_flag)
               printf("Allocate %i: slot %i\n", thread_id, i);
               break;
            }
            }
   }
   pthread_exit(0);
}
void *mem_releaser(void *arg)
{
   int i;
   int loops = 0;
   int check_interval = 100;
   int thread_id = *(int *)arg;
   int start = POOL_SIZE * thread_id;
   int end = POOL_SIZE * (thread_id + 1);
   if(verbose_flag) {
      printf("Allocator %i works on memory pool %i to %i\n", thread_id, start, end);
      printf("Allocator %i started…\n", thread_id);
   }

   while(!done_flag) {
      /* find non-NULL slot */
      for (i = start; i < end; ++i) {
         if (NULL!= mem_pool[i]) {
            void *ptr = mem_pool[i];
            mem_pool[i] = NULL;
            free(ptr);
            ++counters[thread_id];
            if (debug_flag)
               printf("Releaser %i: slot %i\n", thread_id, i);
            break;
```

```
            }
        }
        ++loops;
        if ((0 == loops % check_interval) &&
            (elapsed_time(&begin) > run_time)) {
                done_flag = 1;
                break;
        }
    }
    pthread_exit(0);
}
[3]
```

*Paging and Virtual Memory*

Either with or without segmentation, some OSs divide logical memory into some number of fixed-size partitions, called *blocks*, *frames*, *pages*, or *some combination of a few or all of these*. For example, with OSs that divide memory into frames, the logical address is a compromise of a frame number and offset. The user memory space can then, also, be divided into pages, where page sizes are typically equal to *frame* sizes.

When a process is loaded in its entirety into memory (in the form of pages), its pages may not be located within a contiguous set of frames. Every process has an associated process table that tracks its pages, and each page's corresponding frames in memory. The logical address spaces generated are unique for each process, even though multiple processes share the same physical memory space. Logical address spaces are typically made up of a page-frame number, which indicates the start of that page, and an offset of an actual memory location within that page. In essence, the logical address is the sum of the page number and the offset. (See Figure 9-37.)

An OS may start by *prepaging*, or loading the pages needed to get started, and then implementing the scheme of *demand paging* where processes have no pages in memory and pages are only loaded into RAM when a *page fault* (an error occurring when attempting to access a page not in RAM) occurs. When a page fault occurs, the OS takes over and loads the needed page into memory, updates page tables, and then the instruction that triggered the page fault in the first place is re-executed. This scheme is based upon Knuth's Locality of Reference theory, which estimates that 90% of a system's time is spent on processing just 10% of code.

Dividing up logical memory into pages aids the OS in more easily managing tasks being relocated in and out of various types of memory in the memory hierarchy, a process called *swapping*. Common page selection and replacement schemes to determine which pages are swapped include:

- *Optimal*: using future reference time, swapping out pages that won't be used in the near future.

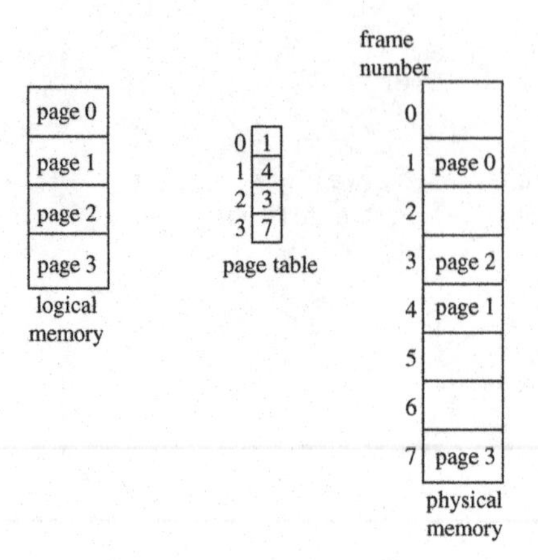


**Figure 9-37**
Paging.[3]

- *Least Recently Used (LRU)*: which swaps out pages that have been used the least recently.
- *FIFO*: which as its name implies, swaps out the pages that are the oldest (regardless of how often it is accessed) in the system. While a simpler algorithm then LRU, FIFO is much less efficient.
- *Not Recently Used (NRU)*: swaps out pages that were not used within a certain time period.
- *Second Chance*: FIFO scheme with a reference bit, if "0" will be swapped out (a reference bit is set to "1" when access occurs, and reset to "0" after the check).
- *Clock Paging*: pages replaced according to clock (how long they have been in memory), in clock order, if they haven't been accessed (a reference bit is set to "1" when access occurs, and reset to "0" after the check).

While every OS has its own swap algorithm, all are trying to reduce the possibility of *thrashing*, a situation in which a system's resources are drained by the OS constantly swapping in and out data from memory. To avoid thrashing, a kernel may implement a *working set* model, which keeps a fixed number of pages of a process in memory at all times. Which pages (and the number of pages) that comprise this working set depends on the OS, but typically it is the pages accessed most recently. A kernel that wants to prepage a process also needs to have a working set defined for that process before the process's pages are swapped into memory.

*Virtual Memory*

Virtual memory is typically implemented via demand segmentation (fragmentation of processes from within, as discussed in a previous section) and/or demand paging (fragmentation of logical user memory as a whole) memory fragmentation techniques. When virtual memory is implemented via these "demand" techniques, it means that only the pages and/or segments that are currently in use are loaded into RAM.

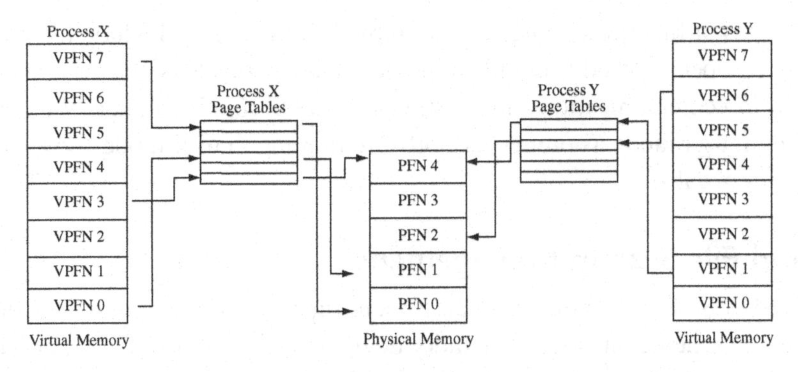


**Figure 9-38**
Virtual memory.[3]

As shown in Figure 9-38, in a virtual memory system, the OS generates *virtual* addresses based on the logical addresses, and maintains *tables* for the sets of logical addresses into virtual addresses conversions (on some processors table entries are cached into translation lookaside buffers (TLBs); see Chapters 4 and 5 for more on MMUs and TLBs). The OS (along with the hardware) then can end up managing more than one different address space for each process (the physical, logical, and virtual). In short, the software being managed by the OS views memory as one continuous memory space, whereas the kernel actually manages memory as several fragmented pieces which can be segmented and paged, segmented and unpaged, unsegmented and paged, or unsegmented and unpaged.

### 9.3.2 Kernel Memory Space

The kernel's memory space is the portion of memory in which the kernel code is located, some of which is accessed via system calls by higher-level software processes, and is where the CPU executes this code from. Code located in the kernel memory space includes required IPC mechanisms, such as those for message passing queues. Another example is when tasks are creating some type of fork/exec or spawn system calls. After the task creation system call, the OS gains control and creates the *Task Control Block (TCB)*, also referred to as a *Process Control Block (PCB)* in some OSs, within the kernel's memory space that contains OS control information and CPU context information for that particular task. Ultimately, what is managed in the kernel memory space, as opposed to the user space, is determined by the hardware, as well as the actual algorithms implemented within the OS kernel.

As previously mentioned, software running in user mode can only access anything running in kernel mode via *system calls*. System calls are the higher-level (user mode) interfaces to the kernel's subroutines (running in kernel mode). Parameters associated with system calls that need to be passed between the OS and the system caller running in user mode are then passed via registers, a stack, or in the main memory heap. The types of system calls typically fall under the types of functions being supported by the OS, so they include file systems management (i.e., opening/modifying files), process management (i.e., starting/stopping processes), and I/O communications. In short, where an OS running in kernel mode views what is running in user mode as processes, software running in user mode views and defines an OS by its system calls.

## 9.4 I/O and File System Management

Some embedded OSs provide memory management support for a temporary or permanent file system storage scheme on various memory devices, such as Flash, RAM, or hard disk. File systems are essentially a collection of files along with their management protocols (see Table 9-2). File system algorithms are middleware and/or application software that is *mounted* (installed) at some mount point (location) in the storage device.

In relation to file systems, a kernel typically provides file system management mechanisms for, at the very least:

- *Mapping* files onto secondary storage, Flash, or RAM (for instance).
- Supporting the primitives for manipulating files and directories:
  - *File Definitions and Attributes*: Naming Protocol, Types (executable, object, source, multimedia, etc.), Sizes, Access Protection (Read, Write, Execute, Append, Delete, etc.), Ownership, etc.

**Table 9-2: Middleware File System Standards**

| File System | Summary |
|---|---|
| FAT32 (File Allocation Table) | Where memory is divided into the smallest unit possible (called sectors). A group of sectors is called a cluster. An OS assigns a unique number to each cluster, and tracks which files use which clusters. FAT32 supports 32-bit addressing of clusters, as well as smaller cluster sizes than that of the FAT predecessors (FAT, FAT16, etc.) |
| NFS (Network File System) | Based on RPC (Remote Procedure Call) and XDR (Extended Data Representation), NFS was developed to allow external devices to mount a partition on a system as if it were in local memory. This allows for fast, seamless sharing of files across a network. |
| FFS (Flash File System) | Designed for Flash memory. |
| DosFS | Designed for real-time use of block devices (disks) and compatible with the MS-DOS file system. |
| RawFS | Provides a simple *raw file system* that essentially treats an entire disk as a single large file. |
| TapeFS | Designed for tape devices that do not use a standard file or directory structure on tape. Essentially treats the tape volume as a raw device in which the entire volume is a large file. |
| CdromFS | Allows applications to read data from CD-ROMs formatted according to the ISO 9660 standard file system. |

- *File Operations*: Create, Delete, Read, Write, Open, Close, etc.
- *File Access Methods*: Sequential, Direct, etc.
- Directory Access, Creation, and Deletion.

OSs vary in terms of the primitives used for manipulating files (naming, data structures, file types, attributes, operations, etc.), what memory devices files can be mapped to, and what file systems are supported. Most OSs use their standard I/O interface between the file system and the memory device drivers. This allows for one or more file systems to operate in conjunction with the OS.

I/O management in embedded OSs provides an additional abstraction layer (to higher-level software) away from the system's hardware and device drivers. An OS provides a uniform interface for I/O devices that perform a wide variety of functions via the available kernel system calls, providing protection to I/O devices since user processes can only access I/O via these system calls, and managing a fair and efficient I/O sharing scheme among the multiple processes. An OS also needs to manage synchronous and asynchronous communication coming from I/O to its processes—in essence be event-driven by responding to requests from both sides (the higher-level processes and low-level hardware)—and manage the data transfers. In order to accomplish these goals, an OS's I/O management scheme is typically made up of a generic device driver interface both to user processes and device drivers, as well as some type of buffer caching mechanism.

Device driver code controls a board's I/O hardware. In order to manage I/O, an OS may require all device driver code to contain a specific set of functions, such as startup, shutdown, enable, and disable. A kernel then manages I/O devices, and in some OSs file systems as well, as "black boxes" that are accessed by some set of generic APIs by higher-layer processes. OSs can vary widely in terms of what types of I/O APIs they provide to upper layers. For example, under Jbed, or any Java-based scheme, all resources (including I/O) are viewed and structured as objects. VxWorks, on the other hand, provides a communications mechanism, called *pipes*, for use with the VxWorks I/O subsystem. Under VxWorks, pipes are virtual I/O devices that include underlying message queue associated with that pipe. Via the pipe, I/O access is handled as either a stream of bytes (*block* access) or one byte at any given time (*character* access).

In some cases, I/O hardware may require the existence of OS buffers to manage data transmissions. Buffers can be necessary for I/O device management for a number of reasons. Mainly they are needed for the OS to be able to capture data transmitted via block access. The OS stores within buffers the stream of bytes being transmitted to and from an I/O device independent of whether one of its processes has initiated communication to the device. When performance is an issue, buffers are commonly stored in cache (when available), rather than in slower main memory.

## 9.5 OS Standards Example: POSIX (Portable Operating System Interface)

As introduced in Chapter 2, standards may greatly impact the design of a system component—and OSs are no different. One of the key standards implemented in off-the-shelf embedded OSs today is portable OS interface (POSIX). POSIX is based upon the IEEE *(1003.1-2001)* and The Open Group *(The Open Group Base Specifications Issue 6)* set of standards that define a standard OS interface and environment. POSIX provides OS-related standard APIs and definitions for process management, memory management, and I/O management functionality (see Table 9-3).

**Table 9-3: POSIX Functionality[14]**

| OS Subsystem | Function | Definition |
|---|---|---|
| Process Management | Threads | Functionality to support multiple flows of control within a process. These flows of control are called threads and they share their address space and most of the resources and attributes defined in the OS for the owner process. The specific functional areas included in threads support are:<br>• Thread management: the creation, control, and termination of multiple flows of control that share a common address space.<br>• Synchronization primitives optimized for tightly coupled operation of multiple control flows in a common, shared address space. |
| | Semaphores | A minimum synchronization primitive to serve as a basis for more complex synchronization mechanisms to be defined by the application program. |
| | Priority scheduling | A performance and determinism improvement facility to allow applications to determine the order in which threads that are ready to run are granted access to processor resources. |
| | Real-time signal extension | A determinism improvement facility to enable asynchronous signal notifications to an application to be queued without impacting compatibility with the existing signal functions. |
| | Timers | A mechanism that can notify a thread when the time as measured by a particular clock has reached or passed a specified value, or when a specified amount of time has passed. |
| | IPC | A functionality enhancement to add a high-performance, deterministic IPC facility for local communication. |

*(Continued)*

**Table 9-3: (Continued)**

| OS Subsystem | Function | Definition |
|---|---|---|
| Memory Management | Process memory locking | A performance improvement facility to bind application programs into the high-performance random access memory of a computer system. This avoids potential latencies introduced by the OS in storing parts of a program that were not recently referenced on secondary memory devices. |
| | Memory mapped files | A facility to allow applications to access files as part of the address space. |
| | Shared memory objects | An object that represents memory that can be mapped concurrently into the address space of more than one process. |
| I/O Management | Synchronionized I/O | A determinism and robustness improvement mechanism to enhance the data input and output mechanisms, so that an application can ensure that the data being manipulated is physically present on secondary mass storage devices. |
| | Asynchronous I/O | A functionality enhancement to allow an application process to queue data input and output commands with asynchronous notification of completion. |
| … | … | … |

How POSIX is translated into software is shown in Examples 16 and 17, examples in Linux and VxWorks of POSIX threads being created (note the identical interface to the POSIX thread create subroutine).

**Example 16: Linux POSIX Example[3]**

Creating a Linux POSIX thread:

```
if(pthread_create(&threadId, NULL, DEC threadwork, NULL)) {
printf("error");
...
}
```

Here, threadId is a parameter for receiving the thread ID. The second argument is a thread attribute argument that supports a number of scheduling options (in this case NULL indicates the default settings will be used). The third argument is the subroutine to be executed upon creation of the thread. The fourth argument is a pointer passed to the subroutine (pointing to memory reserved for the thread, anything required by the newly created thread to do its work, etc).

Example 17: VxWorks POSIX Example[4]
Creating a POSIX thread in VxWorks:

```
pthread_t tid;
int ret;

/* create the pthread with NULL attributes to designate default values */
ret = pthread_create(&threadId, NULL, entryFunction, entryArg);
....
```

Here, threadId is a parameter for receiving the thread ID. The second argument is a thread attribute argument that supports a number of scheduling options (in this case NULL indicates the default settings will be used). The third argument is the subroutine to be executed upon creation of the thread. The fourth argument is a pointer passed to the subroutine (pointing to memory reserved for the thread, anything required by the newly created thread to do its work, etc).

Essentially, the POSIX APIs allow for software that is written on one POSIX-compliant OS to be easily ported to another POSIX OS, since by definition the APIs for the various OS system calls must be identical and POSIX compliant. It is up to the individual OS vendors to determine how the internals of these functions are actually performed. This means that, given two different POSIX compliant OSs, both probably employ very different internal code for the same routines.

## 9.6 OS Performance Guidelines

The two subsystems of an OS that typically impact OS performance the most, and differentiate the performance of one OS from another, are the memory management scheme (specifically the process swapping model implemented) and the scheduler. The performance of one virtual memory-swapping algorithm over another can be compared by the number of page faults they produce, given the same set of memory references (i.e., the same number of page frames assigned per process for the exact same process on both OSs). One algorithm can be further tested for performance by providing it with a variety of different memory references and noting the number of page faults for various number of page frames per process configurations.

While the goal of a scheduling algorithm is to select processes to execute in a scheme that maximizes overall performance, the challenge OS scheduler's face is that there are a number of performance indicators. Furthermore, algorithms can have opposite effects on an indicator, even given the exact same processes. The main performance indicators for scheduling algorithms include:

- *Throughput*: number of processes being executed by the CPU at any given time. At the OS scheduling level, an algorithm that allows for a significant number of larger processes to be executed before smaller processes runs the risk of having a lower throughput. In an SPN (shortest process next) scheme, the throughput may even vary on the same system depending on the size of processes being executed at the moment.

- *Execution time*: average time it takes for a running process to execute (from start to finish). Here, the size of the process affects this indicator. However, at the scheduling level, an algorithm that allows for a process to be continually pre-empted allows for significantly longer execution times. In this case, given the same process, a comparison of a non-pre-emptable versus pre-emptable scheduler could result in two very different execution times.
- *Wait time*: total amount of time a process must wait to run. Again this depends on whether the scheduling algorithm allows for larger processes to be executed before slower processes. Given a significant number of larger processes executed (for whatever reason), any subsequent processes would have higher wait times. This indicator is also dependent on what criteria determine which process is selected to run in the first place—a process in one scheme may have a lower or higher wait time than if it is placed in a different scheduling scheme.

On a final note, while scheduling and memory management are the leading components impacting performance, to get a more accurate analysis of OS performance one must measure the impact of both types of algorithms in an OS, as well as factor in an OS's *response time* (essentially the time from when a user process makes the system call to when the OS starts processing the request). While no one factor alone determines how well an OS performs, OS performance in general can be *implicitly* estimated by how hardware resources in the system (the CPU, memory, and I/O devices) are utilized for the variety of processes. Given the right processes, the more time a resource spends executing code as opposed to sitting idle *can be* indicative of a more efficient OS.

## 9.7 Selecting the Right Embedded OS and BSPs

When selecting an embedded OS for an embedded design, the real questions to ask include:

- What is the master processor? Performance limitations? Memory footprint? Select an embedded OS that has been stably ported and supporting the hardware.
- What features are needed given cost, schedule, requirements, etc.? Is a kernel enough or more? How scalable should the embedded OS be?
- Does the class of the device require a special type of embedded OS along with certification?
- What is the budget? Embedded OSs can come as open-source, royalty-free, or royalty-based. What are the cost of the tools, the cost of the development license, and the royalty costs (i.e., per unit shipped)? Bad tools equals a *nightmare* for the development team.

- What are the skills of the team? What are developers experienced developing with? Don't skimp on the training, so, if training is required, is this computed this into cost of the embedded OS and the development schedule?
- How portable is the embedded OS? Is there a *BSP* compatible with the embedded OS and target hardware? The BSP is an optional component provided by the OS provider, the main purpose of which is simply to provide an abstraction layer between the OS and generic device drivers.

A BSP allows for an OS to be more easily ported to a new hardware environment, because it acts as an integration point in the system of hardware dependent and hardware independent source code. A BSP provides subroutines to upper layers of software that can customize the hardware and provide flexibility at compile time. Because these routines point to separately compiled device driver code from the rest of the system application software, BSPs provide runtime portability of generic device driver code. As shown in Figure 9-39, a BSP provides architecture-specific device driver configuration management and an API for the OS (or higher layers of software) to access generic device drivers. A BSP is also responsible for managing the initialization of the device driver (hardware) and OS in the system.

The device configuration management portion of a BSP involves architecture-specific device driver features, such as constraints of a processor's available addressing modes, endianess, and interrupts (connecting ISRs to interrupt vector table, disabling/enabling, control registers, etc.), and is designed to provide the most flexibility in porting generic device drivers to a new architecture-based board, with its differing endianess, interrupt scheme, and other architecture-specific features.

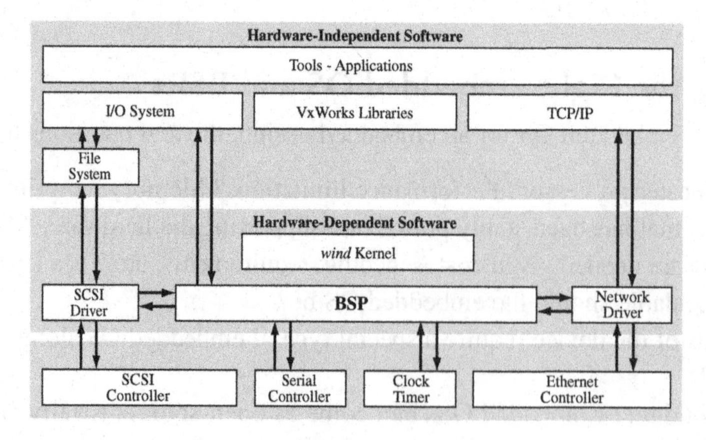


**Figure 9-39**
BSP within Embedded Systems Model.[4]

## 9.8 Summary

This chapter introduced the different types of embedded OSs, as well as the major components that make up most embedded OSs. While embedded OSs vary widely in their international design, what all embedded OSs have in common are the:

- Purpose
  - Partitioning tool
  - Provides abstraction layer for overlying code
  - Efficient and reliable management of system resources.
- Kernel Components
  - Process management
  - Memory management
  - IO system management.

Process management mechanisms, such as task implementation schemes, scheduling, synchronization, and scheduling, are what allow embedded OSs to provide the illusion of simultaneous multitasking over a single processor. The goals of an embedded OS are to balance between:

- Utilizing the system's resources (keeping the CPU, I/O, etc., as busy as possible).
- Task throughput and processing as many tasks as possible in a given amount of time.
- Fairness and ensuring that task starvation doesn't occur when trying to achieve a maximum task throughput.

This chapter also discussed the POSIX standard and its impact on the embedded OS market in terms of what function requirements are specified. The impact of OSs on system performance was discussed, specifically the importance of NOT underestimating the impact of embedded OS's internal design on performance. The key differentiators include:

- *Memory management scheme*: virtual memory swapping scheme and page faults.
- *Scheduling scheme*: throughput, execution time, and wait time.
- *Response time*: to make the context switch to a ready task and waiting time of task in ready queue.
- *Turnaround time*: how long a process takes to complete running.
- *Overhead*: the time and data needed to determine which tasks will run next.
- *Fairness*: what are the determining factors as to which processes get to run.

Finally, an abstraction layer that many embedded OSs supply, called a BSP, was introduced. The next chapter, Chapter 10, *Middleware and Application Software*, is the last of the software chapters, and discusses middleware and application software in terms of their impact on an embedded architecture.

## Chapter 9: Problems

1. [a] What is an operating system (OS)?
   [b] What does an OS do?
   [c] Draw a diagram showing where the OS fits in the Embedded Systems Model.
2. [a] What is a kernel?
   [b] Name and describe at least two functions of a kernel.
3. OSs typically fall under one of three models:
   A. Monolithic, layered, or microkernel.
   B. Monolithic, layered, or monolithic-modularized.
   C. Layered, client/server, or microkernel.
   D. Monolithic-modularized, client/server, or microkernel.
   E. None of the above.
4. [a] Match the types of OS model (see Q. 3) to Figures 9-40a–c.
   [b] Name a real-world OS that falls under each model.

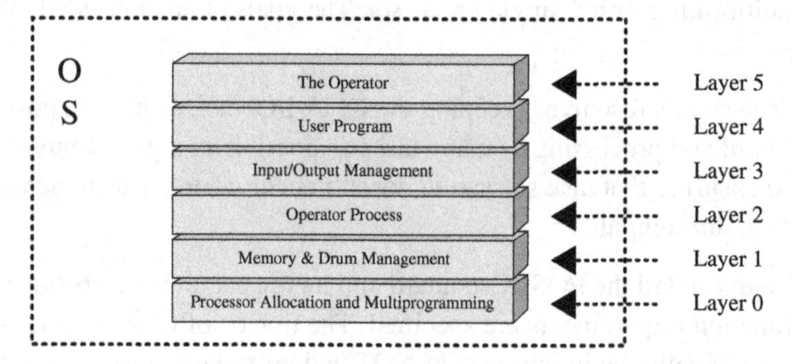


**Figure 9-40a**
OS block diagram 1.

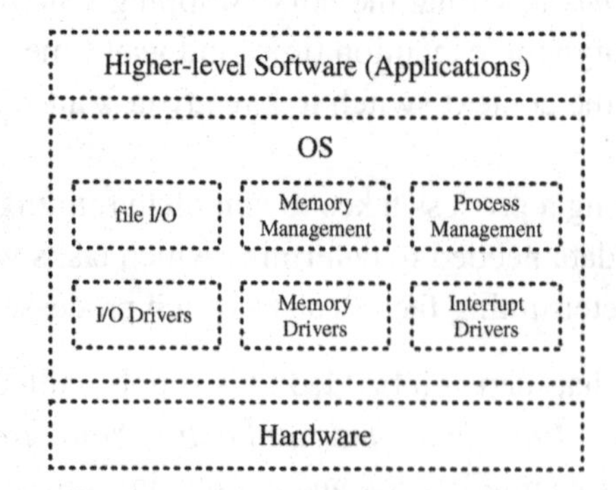


**Figure 9-40b**
OS block diagram 2.

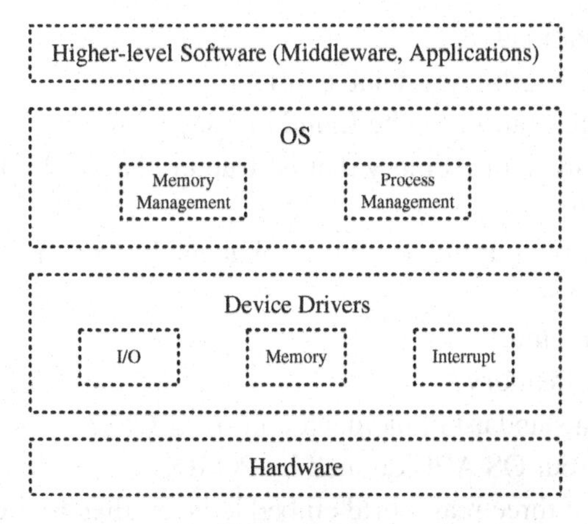


**Figure 9-40c**
OS block diagram 3.

5. [a] What is the difference between a process and a thread?
   [b] What is the difference between a process and a task?
6. [a] What are the most common schemes used to create tasks?
   [b] Give one example of an OS that uses each of the schemes.
7. [a] In general terms, what states can a task be in?
   [b] Give one example of an OS and its available states, including the state diagrams.
8. [a] What is the difference between pre-emptive and non-pre-emptive scheduling?
   [b] Give examples of OSs that implement pre-emptive and non-pre-emptive scheduling.
9. [a] What is a real-time operating system (RTOS)?
   [b] Give two examples of RTOSs.
10. [T/F] A RTOS does not contain a pre-emptive scheduler.
11. Name and describe the most common OS intertask communication and synchronization mechanisms.
12. [a] What are race conditions?
    [b] What are some techniques for resolving race conditions?
13. The OS intertask communication mechanism typically used for interrupt handling is:
    A. A message queue.
    B. A signal.
    C. A semaphore.
    D. All of the above.
    E. None of the above.
14. [a] What is the difference between processes running in kernel mode and those running in user mode?
    [b] Give an example of the type of code that would run in each mode.

15. [a] What is segmentation?
    [b] What are segment addresses made up of?
    [c] What type of information can be found in a segment?
16. [T/F] A stack is a segment of memory that is structured as a FIFO queue.
17. [a] What is paging?
    [b] Name and describe four OS algorithms that can be implemented to swap pages in and out of memory.
18. [a] What is virtual memory?
    [b] Why use virtual memory?
19. [a] Why is POSIX a standard implemented in some OSs?
    [b] List and define four OS APIs defined by POSIX.
    [c] Give examples of three real-world embedded OSs that are POSIX compliant.
20. [a] What are the two subsystems of an OS that most impact OS performance?
    [b] How do the differences in each impact performance?
21. [a] What is a BSP?
    [b] What type of elements are located within a BSP?
    [c] Give two examples of real-world embedded OSs that include a BSP.

## Endnotes

[1] Wind River Systems, VxWorks Programmer's Manual, p. 27.
[2] Esmertec, Jbed Programmer's Manual.
[3] *Linux Kernel 2.4 Internals*, T. Aivazian, 2002; *The Linux Kernel*, D. Rusling, 1999.
[4] Wind River Systems, VxWorks Programmer's Manual.
[5] Wind River Systems, VxWorks Programmer's Manual, pp. 21, 22.
[6] Esmertec, Jbed Programmer's Manual, pp. 21–36.
[7] Wind River Systems, VxWorks Programmer's Manual, p. 24.
[8] Wind River Systems, VxWorks Programmer's Manual, p. 23.
[9] *Conceptual Architecture of the Linux Kernel*, section 3, I. Bowman, 1998.
[10] Wind River Systems, VxWorks Programmer's Manual, p. 46.
[11] "The Executable and Linking Format (ELF)," M. Haungs, 1998.
[12] Sun Microsystems, java.sun.com website.
[13] http://support.microsoft.com/default.aspx?scid=kb%3Ben-us%3Bq121460
[14] www.pasc.org
[15] *Linux Kernel 2.4 Internals*, section 2.3, T. Aivazian, 2002; *The Linux Kernel*, chapter 15, D. Rusling, 1999.
[16] "Embedded Linux—Ready for RealTime," Whitepaper, p. 8, B. Weinberg, 2001.
[17] *Conceptual Architecture of the Linux Kernel*, section 3.2, I. Bowman, 1998.
[18] *The Linux Kernel*, section 4.1, D. Rusling, 1999.

CHAPTER 10

# Middleware and Application Software

**In This Chapter**

- Defining middleware
- Defining application software
- Introducing real-world networking and Java examples of middleware
- Introducing real-world networking and Java examples used in application software

The line between middleware and application software has been historically blurred. Furthermore, at this time, there is no formal consensus on how embedded systems middleware should be defined within the embedded system industry. Thus, until such time that there is a consensus, this chapter takes a more practical approach of introducing both middleware and application software together. The remaining sections of this chapter define middleware and application software concepts, and provide real-world pseudocode examples of middleware and application software.

## 10.1 What is Middleware?

In the most general terms, middleware software is any system software that is not the operating system (OS) kernel, device drivers, or application software. Middleware is software that has evolved to the point of being abstracted out of the application layer for a variety of reasons. One reason is that it may already be included as part of the off-the-shelf OS package. Other reasons to have removed software from the application into the middleware layer have been to allow reusability with other applications, to decrease development costs or time by purchasing it off-the-shelf through a third party vendor, or to simplify application code.

Remember, what determines if particular software component is "middleware" is where it resides within the embedded system's architecture and not because of its inherent purpose within the system, alone. Middleware is system software that typically either sits on the device drivers or on top of the OS and can sometimes be incorporated within the OS itself. Middleware acts as an abstraction layer that mediates between application software and underlying system software, such as an OS kernel or device driver software. Middleware is also software that can mediate and manage interactions between multiple applications.

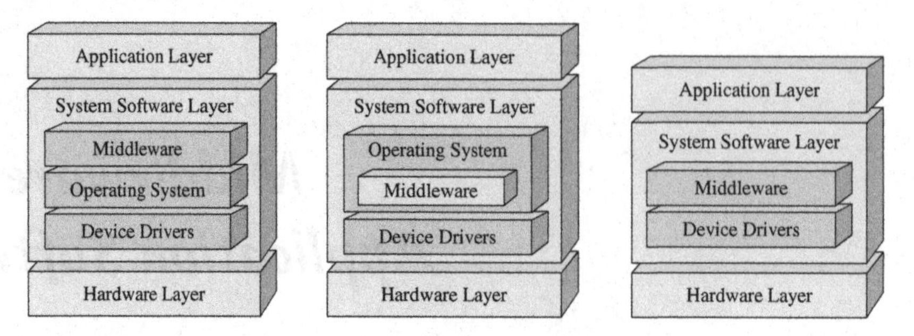


**Figure 10-1**
Middleware within the Embedded Systems Model.

These applications can be contained within the same embedded system or across multiple networked computer systems (see Figure 10-1).

The main reason why software teams incorporate different types of middleware is to achieve some combination of the following design requirements:

- *Adaptability*: enables overlying middleware and/or embedded applications to adapt to changes in availability of system resources.
- *Connectivity* and *Intercommunication*: provide overlying middleware and/or embedded applications the ability to transparently communicate with applications within other devices, via some user-friendly, standardized interface.
- *Flexibility* and *Scalability*: allow overlying middleware and/or embedded applications to be configurable and customizable in terms of functionality depending on application requirements, overall device requirements, underlying system software/hardware limitations, for example.
- *Portability*: allows overlying middleware and/or embedded applications to run on different embedded systems with different underlying system software and/or hardware layers.
- *Security*: ensures that the overlying middleware and/or embedded applications have authorized access to system resources.

There are many different types of middleware in the embedded systems arena, including message-oriented middleware (MOM), object request brokers (ORBs), remote procedure calls (RPCs), database/database access, and networking protocols above the device driver layer and below the application layers of the OSI (Open Systems Interconnection) model. However, within the scope of this book, all types of embedded systems middleware can be grouped into two general categories: core middleware and middleware that builds on these core components.

*Core middleware* software is more general purpose and the most commonly found type in embedded systems designs today that incorporates a middleware layer. Core middleware is also used as the foundation for more complex middleware software and can be further broken down into types, such as file systems, networking middleware, databases, and virtual machines to name a few. The reader will have a strong foundation to understanding, using and/or designing any middleware component successfully by understanding the different types of core middleware software.

More *complex middleware that builds on the core components* will vary widely from market to market and device to device, and generally falls under some combination of the following types:

- *Market-specific* complex middleware, meaning middleware that is unique to a particular family of embedded systems, such as a digital TV (DTV) standard-based software that sits on an OS or Java Virtual Machine (JVM).
- Complex *messaging* and *communication* middleware, such as:
  - Message oriented and distributed messaging, i.e., MOM, Message Queues, Java Messaging Service (JMS), Message Brokers, Simple Object Access Protocol (SOAP).
  - Distributed transaction, i.e., RPC, Remote Method Invocation (RMI), Distributed Component Object Model (DCOM), Distributed Computing Environment (DCE).
  - Transaction processing, i.e., Java Beans (TP) Monitor.
  - ORBs, i.e., Common Object Request Broker Object (CORBA), Data Access Object (DAO) Frameworks.
  - Authentication and security, i.e., Java Authentication and Authorization Support (JAAS).
  - Integration brokers.

A middleware element can be further categorized as *proprietary*, meaning it is closed software supported by a company that licenses it to others for use, or *open*, meaning it is standardized by some industry committee and can be implemented and/or licensed by any interested party.

More complex embedded systems usually have more than one middleware element, since it is unusual to find one technology that supports all specified application requirements. In this case, the individual middleware elements are typically selected based upon their interoperability with each other, so as to avoid later problems in integration. In some cases, integrated middleware packages of compatible middleware elements are available commercially, off-the-shelf, for use in embedded systems, such as the Sun embedded Java solutions, Microsoft's .NET Compact Framework, and CORBA from the Object Management Group (OMG), to name a few. Many embedded OS vendors also provide integrated middleware packages that run "out-of-the-box" with their respective OS and hardware platforms.

Section 10.3 of this chapter provides specific real-world examples of individual middleware networking elements, as well as integrated middleware Java packages.

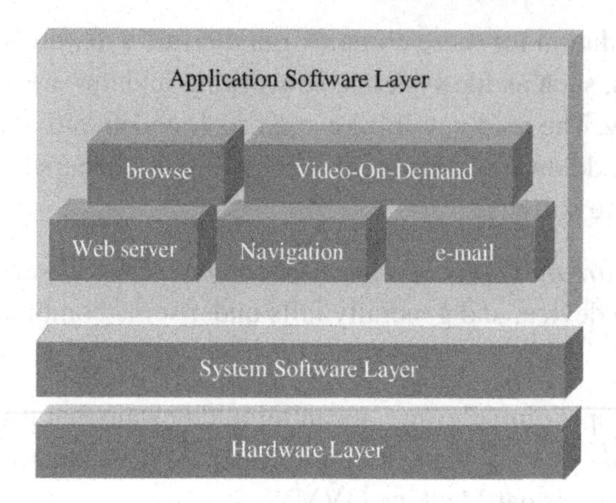


**Figure 10-2**
Application layer and Embedded Systems Model.

## 10.2 What Is an Application?

The final type of software in an embedded system is the *application* software. As shown in Figure 10-2, application software sits on top of the system software layer, and is dependent on, managed, and run by the system software. It is the software within the application layer that inherently defines what type of device an embedded system is, because the functionality of an application represents at the highest level the purpose of that embedded system and does *most* of the interaction with users or administrators of that device, if any exists. *(Note: I say* most *because features such as powering on or off the device when a user hits a button may trigger a device driver function directly for the power on/power off sequence, rather than bringing up an application—it depends on the programmer how that is handled.)*

Like embedded standards, embedded applications can be divided according to whether they are market specific (implemented in only a specific type of device, such as video-on-demand applications in an interactive DTV) or general-purpose (can be implemented across various types of devices, such as a browser).

Section 10.4 introduces real-world examples of types of application software and how they contribute to an embedded system's architecture.

## 10.3 Middleware Examples

One of the main strengths in using middleware is that it allows for the reduction of the complexity of the applications by centralizing software infrastructure that would traditionally be redundantly found in the application layer. However, in introducing middleware to a system, one introduces additional overhead, which can greatly impact scalability and performance. In short, middleware impacts the embedded system at all layers.

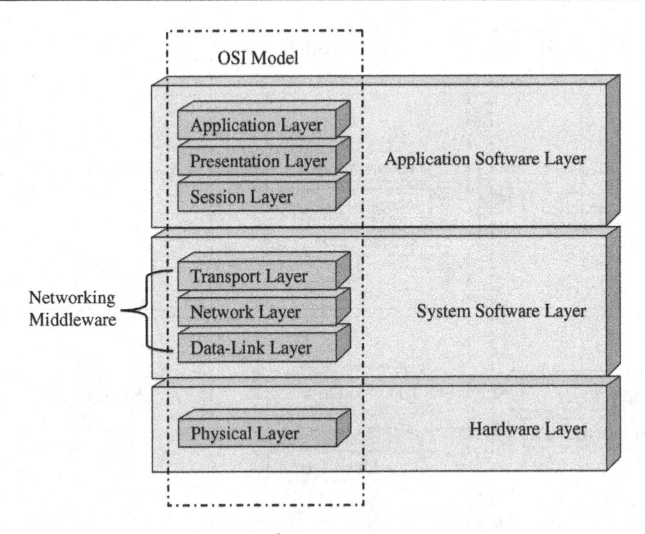


**Figure 10-3**
OSI model and middleware.

The goal of this section is not just about introducing some of the most common types of embedded systems middleware. It is also to show the reader the pattern behind different types of embedded middleware designs, and to help demonstrate a powerful approach to understanding and applying this knowledge to any embedded system's middleware component encountered in the future. Open-source and real-world examples of these types of middleware will be used when possible throughout this chapter to demystify the technical concepts. Examples of building real-world designs based on these types of middleware will be provided, and the challenges and risks to be aware of when utilizing middleware in embedded systems will also be addressed in this chapter.

### *10.3.1 Networking Middleware Driver Examples*

As discussed in Chapter 2, one of the simplest ways to understand the components needed to implement networking in an embedded device is to visualize networking components according to the OSI model and to relate that to the Embedded Systems Model. As shown in Figure 10-3, software that falls between the upper data-link and session layers can be considered networking middleware software components.

The examples given in this section, *User Datagram Protocol (UDP)* and *Internet Protocol (IP)* (shown in Figures 10-4a and b), are protocols that fall under the *TCP/IP (Transmission Control Protocol/Internet Protocol)* protocol stack and are typically implemented as middleware. As introduced in Chapter 2, this model is made up of four layers: the network access layer, Internet layer, transport layer, and the application layer. The TCP/IP application layer incorporates the functionality of the top three layers of the OSI model (the application,

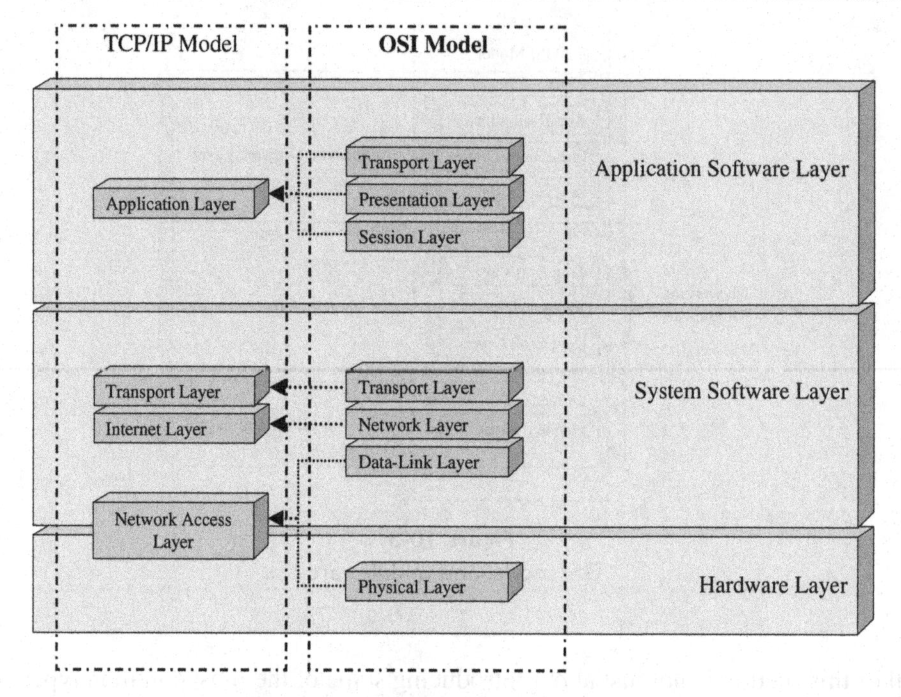


**Figure 10-4a**
TCP/IP, OSI models, and Embedded Systems Model block diagram.

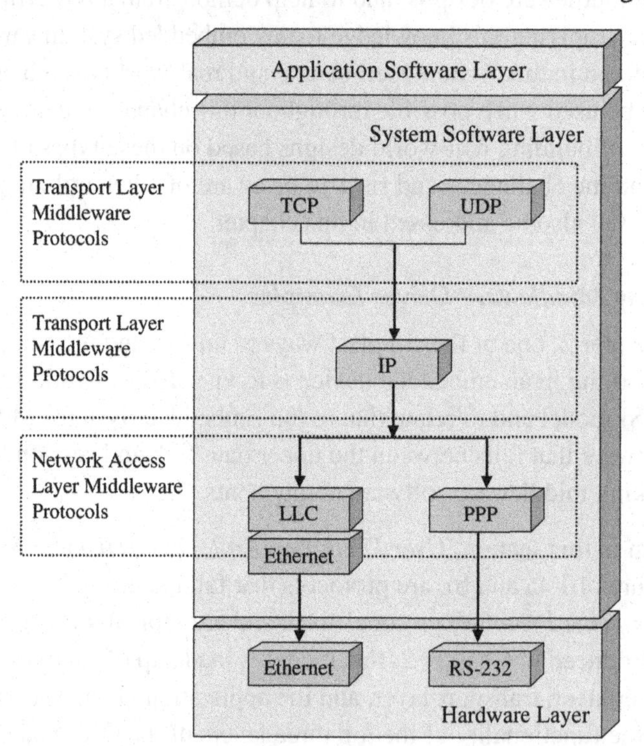


**Figure 10-4b**
TCP/IP model and protocols block diagram.

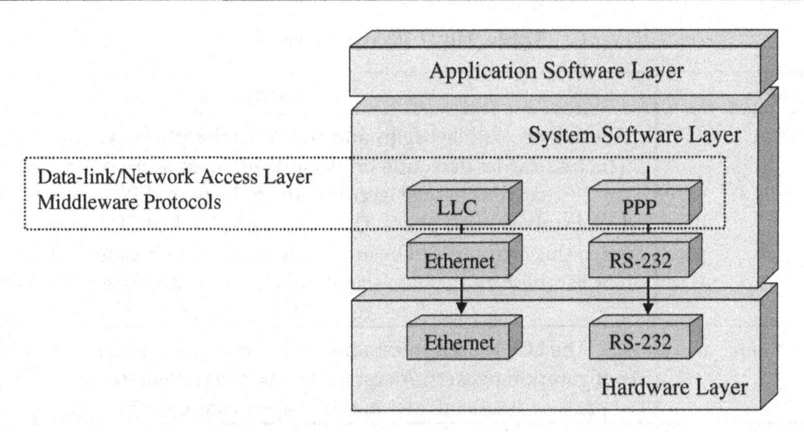


**Figure 10-5**
Data-link middleware.

presentation, and session layers) and the network access layer incorporates the layers of the OSI model (physical and data-link). The Internet layer corresponds to the network layer in the OSI model and the transport layers of both models are identical. This means that, in reference to TCP/IP, networking middleware falls under the transport, Internet, and upper portion of the network access layers (see Figure 10-4a).

*Network Access/Data-Link Layer Middleware Example: PPP (Point-to-Point Protocol)*

PPP (point-to-point protocol) is a common OSI data-link (or network access layer under the TCP/IP model) protocol that can encapsulate and transmit data to higher layer protocols, such as IP, over a physical serial transmission medium (see Figure 10-5). PPP provides support for both asynchronous (irregular interval) and synchronous (regular interval) serial communication.

PPP is responsible for processing data passing through it as frames. When receiving data from a lower layer protocol, for example, PPP reads the bit fields of these frames to ensure that entire frames are received, that these frames are error free, and that the frame is meant for this device (using the physical address retrieved from the networking hardware on the device), and to determine where this frame came from. If the data is meant for the device, then PPP strips all data-link layer headers from the frame, and the remaining data field, called a *datagram*, is passed up to a higher layer. These same header fields are appended to data coming down from upper layers by PPP for transmission outside the device.

In general, PPP software is defined via a combination of four submechanisms:

- *PPP encapsulation mechanism* (in RFC1661), such as the high-level data-link control (HDLC) framing in RFC1662 or the link control protocol (LCP) framing defined in RFC1661 to process (demultiplex, create, verify checksum, etc.).
- *Data-link protocol handshaking*, such as the LCP handshaking defined in RFC1661, responsible for establishing, configuring, and testing the data-link connection.

**Table 10-1: Phase Table[1]**

| Phase | Description |
|---|---|
| Link Dead | The link necessarily begins and ends with this phase. When an external event (such as carrier detection or network administrator configuration) indicates that the physical layer is ready to be used, PPP proceeds to the Link Establishment phase. During this phase, the LCP automaton (described later in this chapter) will be in the Initial or Starting states. The transition to the Link Establishment phase signals an Up event (discussed later in this chapter) to the LCP automaton. |
| Establish Link | The LCP is used to establish the connection through an exchange of configuration packets. An establish link phase is entered once a Configure-Ack packet (described later in this chapter) has been both sent and received. |
| Authentication | Authentication is an optional PPP mechanism. If it does take place, it typically does so soon after the establish link phase. |
| Network Layer Protocol | Once PPP has completed the establish or authentication phases, each network-layer protocol (such as IP, IPX, or AppleTalk) MUST be separately configured by the appropriate NCP. |
| Link Termination | PPP can terminate the link at any time. after which PPP should proceed to the Link Dead phase. |

- *Authentication protocols*, such as PAP (PPP authentication protocol) in RFC1334, used to manage security after the PPP link is established.
- *Network control protocols (NCPs)*, such as IPCP (Internet Protocol Control Protocol) in RFC1332, which establish and configure upper-layer protocol (OP, IPX, etc.) settings.

These submechanisms work together in the following manner: a PPP communication link, connecting both devices, can be in one of five possible phases at any given time, as shown in Table 10-1. The current phase of the communication link determines which mechanism (encapsulation, handshaking, authentication, etc.) is executed.

How these phases interact to configure, maintain, and terminate a point-to-point link is shown in Figure 10-6.

As defined by PPP layer 1 (i.e., RFC1662), data is encapsulated within the PPP frame, an example of which is shown in Figure 10-7.

The *flag* bytes mark the beginning and end of a frame, and are each set to 0×7E. The *address* byte is a HDLC broadcast address and is always set to 0xFF, since PPP does not assign individual device addresses. The *control* byte is an HDLC command for UI (unnumbered information) and is set to 0×03. The *protocol* field defines the protocol of the data within the information field (i.e., 0×0021 means the information field contains IP datagram, 0xC021 means the information field contains link control data, 0×8021 means the information field contains network control data; see Table 10-2). Finally, the *information* field contains the

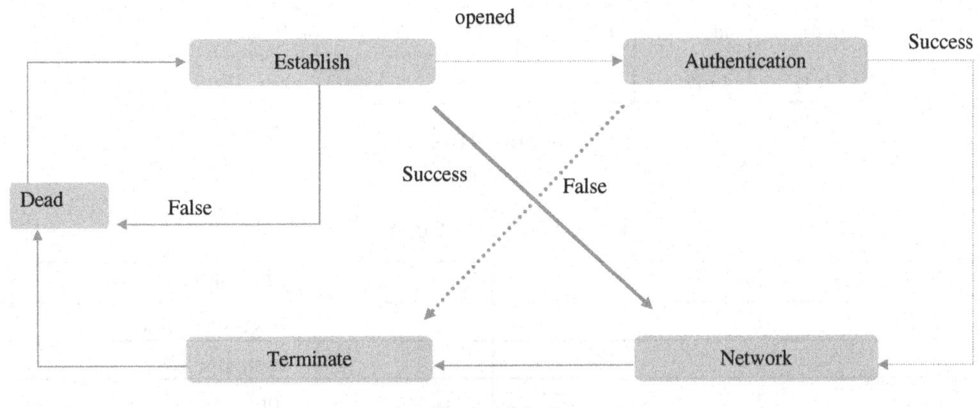


**Figure 10-6**
PPP phases.[1]

| Flag | Address | Control | Protocol | Information | FCS | Flag |
|---|---|---|---|---|---|---|
| 1 byte | 1 byte | 1 byte | 2 bytes | Variable | 2 bytes | 1 byte |

**Figure 10-7**
PPP HDLC-like frame.[1]

**Table 10-2: Protocol Information[1]**

| Value (hex) | Protocol Name |
|---|---|
| 0001 | Padding Protocol |
| 0003 to 001f | Reserved (transparency inefficient) |
| 007d | Reserved (Control Escape) |
| 00cf | Reserved (PPP NLPID) |
| 00ff | Reserved (compression inefficient) |
| 8001 to 801f | Unused |
| 807d | Unused |
| 80cf | Unused |
| 80ff | Unused |
| c021 | LCP |
| c023 | Password Authentication Protocol |
| c025 | Link Quality Report |
| c223 | Challenge Handshake Authentication Protocol |

data for higher-level protocols and the *FCS (frame check sequence)* field contains the frame's checksum value.

The data-link protocol may also define a frame format. An LCP frame, for example, is as shown in Figure 10-8.

| Code | Identifier | Length | Data [variable in size] | | |
|---|---|---|---|---|---|
| 1 byte | 1 byte | 2 bytes | Type | Length | Data |

**Figure 10-8**
LCP frame.[1]

**Table 10-3: LCP Codes[1]**

| Code | Definition |
|---|---|
| 1 | configure-request |
| 2 | configure-ack |
| 3 | configure-nak |
| 4 | configure-reject |
| 5 | terminate-request |
| 6 | terminate-ack |
| 7 | code-reject |
| 8 | protocol-reject |
| 9 | echo-request |
| 10 | echo-reply |
| 11 | discard-request |
| 12 | link quality report |

The *data* field contains the data intended for higher networking layers and is made up of information (type, length, and data). The *length* field specifies the size of the entire LCP frame. The *identifier* is used to match client and server requests and responses. Finally, the *code* field specifies the type of LCP packet (indicating the kind of action being taken); the possible codes are summarized in Table 10-3. Frames with codes 1–4 are called link configuration frames, 5 and 6 are link termination frames, and the rest are link management packets.

The LCP code of an incoming LCP datagram determines how the datagram is processed, as shown in the pseudocode example below.

```
...
if (LCPCode) {
     = CONFREQ: RCR(…); //see Table 10-3
end CONFREQ;
     = CONFACK: RCA(…); //see Table 10-3
end CONFACK;
     = CONFNAK or CONFREJ: RCN(…); //see Table 10-3
end LCPCode;
     = TERMREQ:
             event(RTR);
```

```
        end TERMREQ;
     = TERMACK:
        ...
     }
...
```

In order for two devices to be able to establish a PPP link, each must transmit a data-link protocol frame, such as LCP frames, to configure and test the data-link connection. As mentioned, LCP is one possible protocol that can be implemented for PPP, to handle PPP handshaking. After the LCP frames have been exchanged (and thereby a PPP link established), authentication can occur. It is at this point where authentication protocols, such as PPP Authentication Protocol (PAP), can be used to manage security, through password authentication and so forth. Finally, NCPs such as IPCP establish and configure upper-layer protocols in the network layer protocol settings, such as IP and IPX.

At any given time, a PPP connection on a device is in a particular *state*, as shown in Figure 10-9; the PPP states are outlined in Table 10-4.

*Events* (also shown in Figure 10-9) are what cause a PPP connection to transition from state to state. The LCP codes (from RFC1661) in Table 10-5 define the types of events that cause a PPP state transition.

As PPP connections transition from state to state, certain actions are taken stemming from these events, such as the transmission of packets and/or the starting or stopping of the Restart timer, as outlined in Table 10-6.

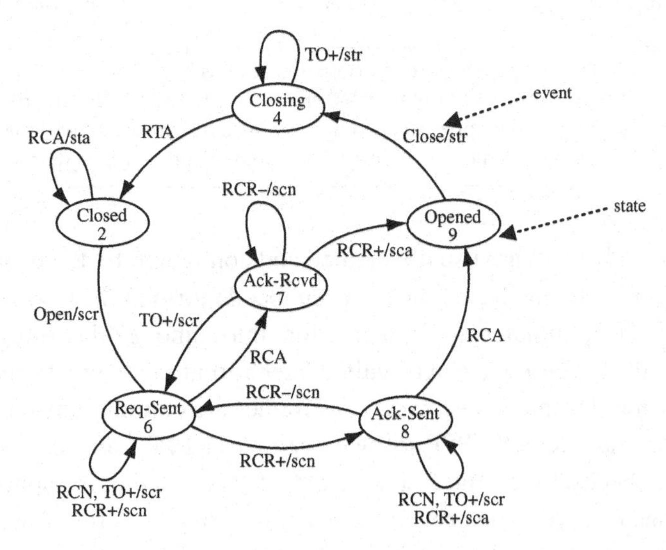


**Figure 10-9**
PPP connection states and events.[1]

**Table 10-4: PPP States[1]**

| States | Definition |
|---|---|
| Initial | PPP link is in the Initial state, the lower layer is unavailable (Down), and no Open event has occurred. The Restart timer is not running in the Initial state. |
| Starting | The Starting state is the Open counterpart to the Initial state. An administrative Open has been initiated, but the lower layer is still unavailable (Down). The Restart timer is not running in the Starting state. When the lower layer becomes available (Up), a Configure-Request is sent. |
| Stopped | The Stopped state is the Open counterpart to the Closed state. It is entered when the automaton is waiting for a Down event after the This-Layer-Finished action, or after sending a Terminate-Ack. The Restart timer is not running in the Stopped state. |
| Closed | In the Closed state, the link is available (Up), but no Open has occurred. The Restart timer is not running in the Closed state. Upon reception of Configure-Request packets, a Terminate-Ack is sent. Terminate-Acks are silently discarded to avoid creating a loop. |
| Stopping | The Stopping state is the Open counterpart to the Closing state. A Terminate-Request has been sent and the Restart timer is running, but a Terminate-Ack has not yet been received. |
| Closing | In the Closing state, an attempt is made to terminate the connection. A Terminate-Request has been sent and the Restart timer is running, but a Terminate-Ack has not yet been received. Upon reception of a Terminate-Ack, the Closed state is entered. Upon the expiration of the Restart timer, a new Terminate-Request is transmitted, and the Restart timer is restarted. After the Restart timer has expired Max-Terminate times, the Closed state is entered. |
| Request-Sent | In the Request-Sent state an attempt is made to configure the connection. A Configure-Request has been sent and the Restart timer is running, but a Configure-Ack has not yet been received nor has one been sent. |
| Ack-Received | In the Ack-Received state, a Configure-Request has been sent and a Configure-Ack has been received. The Restart timer is still running, since a Configure-Ack has not yet been sent. |
| Opened | In the Opened state, a Configure-Ack has been both sent and received. The Restart timer is not running. When entering the Opened state, the implementation SHOULD signal the upper layers that it is now Up. Conversely, when leaving the Opened state, the implementation SHOULD signal the upper layers that it is now Down. |

PPP states, actions, and events are usually created and configured by the platform-specific code at boot-time, some of which is shown in pseudocode form in the next several pages. A PPP connection is in an initial state upon creation; thus, among other things, the "initial" state routine is executed. This code can be called later at runtime to create and configure PPP, as well as respond to PPP runtime events (i.e., as frames are coming in from lower layers for processing). For example, after PPP software demixes a PPP frame coming in from a lower layer and the checksum routine determines the frame is valid, the appropriate field of the frame can then be used to determine what state a PPP connection is in and thus what associated software state, event, and/or action function needs to be executed. If the frame is to be passed to a higher layer protocol, then some mechanism is used to indicate to the higher layer protocol that there is data to receive (e.g., *IPReceive* for *IP).*

**Table 10-5: PPP Events[1]**

| Event Label | Event | Description |
|---|---|---|
| Up | Lower layer is Up | This event occurs when a lower layer indicates that it is ready to carry packets. |
| Down | Lower layer is Down | This event occurs when a lower layer indicates that it is no longer ready to carry packets. |
| Open | Administrative Open | This event indicates that the link is administratively available for traffic; that is, the network administrator (human or program) has indicated that the link is allowed to be Opened. When this event occurs, and the link is not in the Opened state, the automaton attempts to send configuration packets to the peer. |
| Close | Administrative Close | This event indicates that the link is not available for traffic; that is, the network administrator (human or program) has indicated that the link is not allowed to be Opened. When this event occurs, and the link is not in the Closed state, the automaton attempts to terminate the connection. Further attempts to reconfigure the link are denied until a new Open event occurs. |
| TO+ | Timeout with counter>0 | This event indicates the expiration of the Restart timer. |
| TO- | Timeout with counter expired | The Restart timer is used to time responses to Configure-Request and Terminate-Request packets. The TO+ event indicates that the Restart counter continues to be greater than zero, which triggers the corresponding Configure-Request or Terminate-Request packet to be retransmitted. The TO- event indicates that the Restart counter is not greater than zero and no more packets need to be retransmitted. |
| RCR+ | Receive configure request good | An implementation wishing to open a connection MUST transmit a Configure-Request. The Options field is filled with any desired changes to the link defaults. Configuration Options SHOULD NOT be included with default values. |
| RCR- | Receive configure request bad | |
| RCA | Receive configure ack | This event occurs when a valid Configure-Ack packet is received from the peer. The Configure-Ack packet is a positive response to a Configure-Request packet. An out of sequence or otherwise invalid packet is silently discarded. If every Configuration Option received in a Configure-Request is recognizable and all values are acceptable, then the implementation MUST transmit a Configure-Ack. The acknowledged Configuration Options MUST NOT be reordered or modified in any way. On reception of a Configure-Ack, the Identifier field MUST match that of the last transmitted Configure-Request. Additionally, the Configuration Options in a Configure-Ack MUST exactly match those of the last transmitted Configure-Request. Invalid packets are silently discarded. |
| RCN | Receive configure nak/rej | This event occurs when a valid Configure-Nak or Configure-Reject packet is received from the peer. The Configure-Nak and Configure-Reject packets are negative responses to a Configure-Request packet. An out of sequence or otherwise invalid packet is silently discarded. |

(*Continued*)

**Table 10-5: (Continued)**

| Event Label | Event | Description |
|---|---|---|
| RTR | Receive terminate request | This event occurs when a Terminate-Request packet is received. The Terminate-Request packet indicates the desire of the peer to close the connection. |
| RTA | Receive terminate ack | This event occurs when a Terminate-Ack packet is received from the peer. The Terminate-Ack packet is usually a response to a Terminate-Request packet. The Terminate-Ack packet may also indicate that the peer is in Closed or Stopped states and serves to resynchronize the link configuration. |
| RUC | Receive unknown code | This event occurs when an uninterpretable packet is received from the peer. A Code-Reject packet is sent in response. |
| RXJ+ | Receive code reject permitted or receive protocol reject | This event occurs when a Code-Reject or a Protocol-Reject packet is received from the peer. The RXJ+ event arises when the rejected value is acceptable, such as a Code-Reject of an extended code, or a Protocol-Reject of an NCP. These are within the scope of normal operation. The implementation MUST stop sending the offending packet type. The RXJ– event arises when the rejected value is catastrophic, such as a Code-Reject of Configure-Request or a Protocol-Reject of LCP! This event communicates an unrecoverable error that terminates the connection. |
| RXJ– | Receive code reject catastrophic or receive protocol reject | |
| RXR | Receive echo request, receive echo reply, or receive discard request | This event occurs when an Echo-Request, Echo-Reply, or Discard-Request packet is received from the peer. The Echo-Reply packet is a response to an Echo-Request packet. There is no reply to an Echo-Reply or Discard-Request packet. |

**Table 10-6: PPP Actions[1]**

| Action Label | Action | Definition |
|---|---|---|
| tlu | This layer up | This action indicates to the upper layers that the automaton is entering the Opened state. Typically, this action is used by the LCP to signal the Up event to an NCP, Authentication Protocol, or Link Quality Protocol, or MAY be used by an NCP to indicate that the link is available for its network layer traffic. |
| tld | This layer down | This action indicates to the upper layers that the automaton is leaving the Opened state. Typically, this action is used by the LCP to signal the Down event to an NCP, Authentication Protocol, or Link Quality Protocol, or MAY be used by an NCP to indicate that the link is no longer available for its network layer traffic. |
| tls | This layer started | This action indicates to the lower layers that the automaton is entering the Starting state, and the lower layer is needed for the link. The lower layer SHOULD respond with an Up event when the lower layer is available. This results of this action are highly implementation dependent. |

(*Continued*)

**Table 10-6: (Continued)**

| Action Label | Action | Definition |
|---|---|---|
| tlf | This layer finished | This action indicates to the lower layers that the automaton is entering the Initial, Closed, or Stopped states, and the lower layer is no longer needed for the link. The lower layer SHOULD respond with a Down event when the lower layer has terminated. Typically, this action MAY be used by the LCP to advance to the Link Dead phase or MAY be used by an NCP to indicate to the LCP that the link may terminate when there are no other NCPs open. This results of this action are highly implementation dependent. |
| irc | Initialize restart count | This action sets the Restart counter to the appropriate value (Max-Terminate or Max-Configure). The counter is decremented for each transmission, including the first. |
| zrc | Zero restart count | This action sets the Restart counter to zero. |
| scr | Send configure request | Configure-Request packet is transmitted. This indicates the desire to open a connection with a specified set of Configuration Options. The Restart timer is started when the Configure-Request packet is transmitted, to guard against packet loss. The Restart counter is decremented each time a Configure-Request is sent. |
| sca | Send configure ack | A Configure-Ack packet is transmitted. This acknowledges the reception of a Configure-Request packet with an acceptable set of Configuration Options. |
| scn | Send configure nak/rej | A Configure-Nak or Configure-Reject packet is transmitted, as appropriate. This negative response reports the reception of a Configure-Request packet with an unacceptable set of Configuration Options. Configure-Nak packets are used to refuse a Configuration Option value and to suggest a new, acceptable value. Configure-Reject packets are used to refuse all negotiation about a Configuration Option, typically because it is not recognized or implemented. The use of Configure-Nak versus Configure-Reject is more fully described in the chapter on LCP Packet Formats. |
| str | Send terminate request | A Terminate-Request packet is transmitted. This indicates the desire to close a connection. The Restart timer is started when the Terminate-Request packet is transmitted, to guard against packet loss. The Restart counter is decremented each time a Terminate-Request is sent. |
| sta | Send terminate ack | A Terminate-Ack packet is transmitted. This acknowledges the reception of a Terminate-Request packet or otherwise serves to synchronize the automatons. |
| scj | Send code reject | A Code-Reject packet is transmitted. This indicates the reception of an unknown type of packet. |
| ser | Send echo reply | An Echo-Reply packet is transmitted. This acknowledges the reception of an Echo-Request packet. |

*PPP (LCP) State Pseudocode*

- *Initial.* PPP link is in the Initial state, the lower layer is unavailable (Down), and no Open event has occurred. The Restart timer is not running in the Initial state.[1]

```
initial() {
  if (event) {
  = UP:
     transition(CLOSED); //transition to
     closed state
     end UP;

  = OPEN:
     tls(); //action
     transition(STARTING); //transition to
     starting state
     end OPEN;

  = CLOSE:
     end CLOSE; //no action or state
     transition
  = any other event:
     wrongEvent; //indicate that when PPP
                     in initial state
                   //no other event is
                    processed
  }
}
```

```
event (int event)
  {
  if (restarting() && (event = DOWN))
  return; //SKIP

  if (state) {
     = INITIAL:
       initial(); //call initial state routine
         end INITIAL;
     = STARTING:
         starting(); //call starting state
                    routine
         end STARTING;
     = CLOSED:
         closed(); //call closed state
                   routine
         end CLOSED;
     = STOPPED:
         stopped(); //call stopped state
                   routine
```

```
PPP(LCP) Action Pseudocode
   tlu () {
   ...
   event(UP); //UP event
triggered
   event(OPEN); //OPEN event
                triggered
   }
   tld() {
   ...

   event (DOWN); //DOWN event
                 triggered
   }
   tls() {

   ...
   event(OPEN); //OPEN event
                triggered
   }

   tls() {
   ...
   event(OPEN); //OPEN event
                triggered

   }

   tlf() {
   ...
   event(CLOSE); //close event
triggered
   }

irc(int event) {

   if (event = UP, DOWN, OPEN,
   CLOSE, RUC, RXJ+, RXJ-, or RXR)
   {
   restart counter = Max terminate;
   } else {
   restart counter = Max Configure;
   }
}
```

```
      end STOPPED;
   = CLOSING:
      closing(); //call closing state
      routine
      end CLOSING;
   = STOPPING:
      stopping(); //call stopping state
      routine
      end STOPPING;
   = REQSENT:
      reqsent(); //call reqsent state
      routine
      end REQSENT;
   = ACKRCVD:
      ackrcvd(); //call ackrcvd state
      routine
      end ACKRCVD;
   = ACKSENT:
      acksent(); //call acksent state
      routine
      end ACKSENT;
   = OPENED:
      opened(); //call opened state
      routine
      end OPENED;
   = any other state:
      wrongState; //any other state is
      considered invalid
   }
}
```

```
zrc(int time) {
   restart counter = 0;
   PPPTimer = time;
   }

   sca(…) {
   …
      PPPSendViaLCP (CONFACK);
   …
   }

   scn(…) {
   …
   if (refusing all Configuration
   Option negotiation) then{
   PPPSendViaLCP (CONFNAK);
   } else {
   PPPSendViaLCP (CONFREJ);
   }
   …
   }
   …
```

- *Starting*. The Starting state is the Open counterpart to the Initial state. An administrative Open has been initiated, but the lower layer is still unavailable (Down). The Restart timer is not running in the Starting state. When the lower layer becomes available (Up), a Configure-Request is sent.[1]

```
starting () {
   if (event) {
      = UP:
         irc(event); //action
         scr(true); //action
         transition(REQSENT); //transition to REQSENT state
         end UP;
      = OPEN:
         end OPEN; //no action or state transition
```

```
        = CLOSE:
            tlf(); //action
            transition(INITIAL); //transition to initial state
            end CLOSE;
    = any other event:
            wrongEvent++; //indicate that when PPP in starting state no other event
            is processed
        }
    }
```

- *Closed.* In the Closed state, the link is available (Up), but no Open has occurred. The Restart timer is not running in the Closed state. Upon reception of Configure-Request packets, a Terminate-Ack is sent. Terminate-Acks are silently discarded to avoid creating a loop.[1]

```
closed (){
    if (event) {
        = DOWN:
            transition(INITIAL); //transition to initial state
            end DOWN;
        = OPEN:
            irc(event); //action
            scr(true); //action
            transition(REQSENT); //transition to REQSENT state
            end OPEN;
        = RCRP, RCRN, RCA, RCN, or RTR:
            sta(…); //action
            end EVENT;
        = RTA, RXJP, RXR, CLOSE:
            end EVENT; //no action or state transition
        = RUC:
            scj(…); //action
            end RUC;
        = RXJN:
            tlf(); //action
            end RXJN;
        = any other event:
            wrongEvent; //indicate that when PPP in closed state no other event is
            processed
        }
    }
```

- *Stopped.* The Stopped state is the Open counterpart to the Closed state. It is entered when the automaton is waiting for a Down event after the This-Layer-Finished action, or after sending a Terminate-Ack. The Restart timer is not running in the Stopped state.[1]

```
stopped (){
   if (event) {
       = DOWN : tls(); //action
         transition(STARTING); //transition to starting state
         end DOWN;
       = OPEN : initializeLink(); //initialize variables
         end OPEN;
       = CLOSE : transition(CLOSED); //transition to closed state
         end CLOSE;
       = RCRP : irc(event); //action
         scr(true); //action
         sca(…); //action
         transition(ACKSENT); //transition to ACKSENT state
         end RCRP;
       = RCRN : irc(event); //action
         scr(true); //action
         scn(…); //action
         transition(REQSENT); //transition to REQSENT state
         end RCRN;
       = RCA, RCN or RTR : sta(…); //action
         end EVENT;
       = RTA, RXJP, or RXR:
         end EVENT;
       = RUC : scj(…); //action
         end RUC;
       = RXJN : tlf(); //action
         end RXJN;
       = any other event :
         wrongEvent; //indicate that when PPP in stopped state no other event is
         processed
       }
   }
```

- *Closing.* In the Closing state, an attempt is made to terminate the connection. A Terminate-Request has been sent and the Restart timer is running, but a Terminate-Ack has not yet been received. Upon reception of a Terminate-Ack, the Closed state is entered. Upon the expiration of the Restart timer, a new Terminate-Request is transmitted, and the Restart timer is restarted. After the Restart timer has expired Max-Terminate times, the Closed state is entered.[1]

```
closing (){
   if (event) {
       = DOWN : transition(INITIAL); //transition to initial state
          end DOWN;
       = OPEN : transition(STOPPING); //transition to stopping state
          initializeLink(); //initialize variables
          end OPEN;
       = TOP : str(…); //action
          initializePPPTimer; //initialize PPP Timer variable
          end TOP;
       = TON : tlf(); //action
          initializePPPTimer; //initialize PPP Timer variable
          transition(CLOSED); //transition to CLOSED state
          end TON;
       = RTR : sta(…); //action
          end RTR;
       = CLOSE, RCRP, RCRN, RCA, RCN, RXR, or RXJP:
          end EVENT; //no action or state transition
       = RTA : tlf(); //action
          transition(CLOSED); //transition to CLOSED state
          end RTA;
       = RUC : scj(…); //action
          end RUC;
       = RXJN : tlf(); //action
          end RXJN;
       = any other event:
          wrongEvent; //indicate that when PPP in closing state no other event is
          processed
       }
   }
```

- *Stopping*. The Stopping state is the Open counterpart to the Closing state. A Terminate-Request has been sent and the Restart timer is running, but a Terminate-Ack has not yet been received.[1]

```
stopping (){
   if (event) {
     = DOWN : tls(); //action
         transition(STARTING); //transition to STARTING state
         end DOWN;
     = OPEN : initializeLink(); //initialize variables
         end OPEN;
     = CLOSE : transition(CLOSING); //transition to CLOSE state
         end CLOSE;
     = TOP : str(…); //action
         initialize PPPTimer(); //initialize PPP timer
         end TOP;
```

```
    = TON : tlf(); //action
        initialize PPPTimer(); //initialize PPP timer
        transition(STOPPED); //transition to STOPPED state
        end TON;
    = RCRP, RCRN, RCA, RCN, RXJP, RXR : end EVENT; // no action or state transition
    = RTR : sta(…); //action
        end RTR;
    = RTA : tlf(); //action
        transition(STOPPED); //transition to STOPPED state
        end RTA;
    = RUC : scj(…); //action
        end RUC;
    = RXJN : tlf(); //action
        transition(STOPPED); //transition to STOPPED state
        end RXJN;
    = any other event : wrongEvent; //indicate that when PPP in stopping state no
                                      other event is
                                    //processed
    }
  }
```

- *Request-Sent.* In the Request-Sent state an attempt is made to configure the connection. A Configure-Request has been sent and the Restart timer is running, but a Configure-Ack has not yet been received nor has one been sent.[1]

```
reqsent (){
   if (event) {
     = DOWN : transition(STARTING); //transition to STARTING state
       end DOWN;
     = OPEN : transition(REQSENT); //transition to REQSENT state
       end OPEN;
     = CLOSE : irc(event); //action str(…); //action
       transition(CLOSING); //transition to closing state
       end CLOSE;
     = TOP : scr(false); //action
       initialize PPPTimer(); //initialize PPP timer
       end TOP;
     = TON, RTA, RXJP, or RXR : end EVENT; //no action or state transition
     = RCRP : sca(…); //action
       if (PAP = Server){
       tlu(); //action
       transition(OPENED); //transition to OPENED state
       } else { //client
       transition(ACKSENT); //transition to ACKSENT state
       }
       end RCRP;
```

```
      = RCRN : scn(…); //action
          end RCRN;
      = RCA : if (PAP = Server) {
          tlu(); //action
          transition(OPENED); //transition to OPENED state
      }
      else { //client
          irc(event); //action
          transition(ACKRCVD); //transition to ACKRCVD state
      }
          end RCA;
      = RCN : irc(event); //action scr(false); //action
          transition(REQSENT); //transition to REQSENT state
          end RCN;
      = RTR : sta(…); //action
          end RTR;
      = RUC : scj(…); //action
          break;
      = RXJN : tlf(); //action
          transition(STOPPED); //transition to STOPPED state
          end RXJN;
      = any other event : wrongEvent; //indicate that when PPP in reqsent state
                                                   no other event is
                                                 //processed
      }
}
```

- *Ack-Received.* In the Ack-Received state, a Configure-Request has been sent and a Configure-Ack has been received. The Restart timer is still running, since a Configure-Ack has not yet been sent.[1]

```
ackrcvd (){
   if (event) {
      = DOWN : transition(STARTING); //transition to STARTING state
          end DOWN;
      = OPEN, TON, or RXR: end EVENT; //no action or state transition
      = CLOSE : irc(event); //action
          str(…); //action
          transition(CLOSING); //transition to CLOSING state
          end CLOSE;
      = TOP : scr(false); //action
          transition(REQSENT); //transition to REQSENT state
          end TOP;
```

```
        = RCRP : sca(…); //action
            tlu(); //action
            transition(OPENED); //transition to OPENED state
            end RCRP;
        = RCRN : scn(…); //action
            end RCRN;
        = RCA or RCN : scr(false); //action
            transition(REQSENT); //transition to REQSENT state
            end EVENT;
        = RTR : sta(…); //action
            transition(REQSENT); //transition to REQSENT state
            end RTR;
        = RTA or RXJP : transition(REQSENT); //transition to REQSENT state
            end EVENT;
        = RUC : scj(…); //action
            end RUC;
        = RXJN : tlf(); //action
            transition(STOPPED); //event
            end RXJN;
        = any other event : wrongEvent; //indicate that when PPP in ackrcvd state
                                           no other event is
                                         //processed
        }
    }
```

- *Ack-Sent.* In the Ack-Sent state, a Configure-Request and a Configure-Ack have both been sent, but a Configure-Ack has not yet been received. The Restart timer is running, since a Configure-Ack has not yet been received.[1]

```
acksent (){
    if (event) {
        = DOWN : transition(STARTING);
            end DOWN;
        = OPEN, RTA, RXJP, TON, or RXR : end EVENT; //no action or state transition
        = CLOSE : irc(event); //action
            str(…); //action
            transition(CLOSING); //transition to CLOSING state
            end CLOSE;
        = TOP : scr(false); //action
            transition(ACKSENT); //transition to ACKSENT state
            end TOP;
        = RCRP : sca(…); //action
            end RCRP;
        = RCRN : scn(…); //action
            transition(REQSENT); //transition to REQSENT state
            end RCRN;
```

```
        = RCA : irc(event); //action
            tlu(); //action
            transition(OPENED); //transition to OPENED state
            end RCA;
        = RCN : irc(event); //action
            scr(false); //action
            transition(ACKSENT); //transition to ACKSENT state
            end RCN;
        = RTR : sta(…); //action
            transition(REQSENT); //transition to REQSENT state
            end RTR;
        = RUC : scj(…); //action
            end RUC;
        = RXJN : tlf(); //action
            transition(STOPPED); //transition to STOPPED state
            end RXJN;
        = any other event : wrongEvent; //indicate that when PPP in acksent state
        no other event is
                                                    //processed
        }
    }
```

- *Opened*. In the Opened state, a Configure-Ack has been both sent and received. The Restart timer is not running. When entering the Opened state, the implementation SHOULD signal the upper layers that it is now Up. Conversely, when leaving the Opened state, the implementation SHOULD signal the upper layers that it is now Down.[1]

```
opened (){
    if (event) {
        = DOWN :
            tld(); //action
            transition(STARTING); //transition to STARTING state
            end DOWN;
        = OPEN : initializeLink(); //initialize variables
            end OPEN;
        = CLOSE : tld(); //action
            irc(event); //action
            str(…); //action
            transition(CLOSING); //transition to CLOSING state
            end CLOSE;
        = RCRP : tld(); //action
            scr(true); //action
            sca(…); //action
            transition(ACKSENT); //transition to ACKSENT state
            end RCRP;
```

```
      = RCRN : tld(); //action
          scr(true); //action
          scn(…); //action
          transition(REQSENT); //transition to RCRN state
          end RCRN;
      = RCA : tld(); //action
          scr(true); //action
          transition(REQSENT); //transition to REQSENT state
          end RCA;
      = RCN : tld(); //action
          scr(true); //action
          transition(REQSENT); //transition to REQSENT state
          end RCN;
      = RTR : tld(); //action
          zrc(PPPTimeoutTime); //action
          sta(…); //action
          transition(STOPPING); // transition to STOPPING state
          end RTR;
      = RTA : tld(); //action
          scr(true); //action
          transition(REQSENT); // transition to REQSENT state
          end RTA;
      = RUC : scj(…); //action
          end RUC;
      = RXJP : end RXJP; //no action or state transition
      = RXJN : tld(); //action
          irc(event); //action
          str(…); //action
          transition(STOPPING); //transition to STOPPING state
          end RXJN;
      = RXR : ser(…); //action
          end RXR;
      = any other event : wrongEvent; //indicate that when PPP in opened state no other
      event is
                                      //processed
      }
  }
```

*Internet Layer Middleware Example: Internet Protocol (IP)*

The networking layer protocol called IP is based upon DARPA standard RFC791, and is mainly responsible for implementing addressing and fragmentation functionality (see Figure 10-10).

While the IP layer receives data as packets from upper layers and frames from lower layers, the IP layer actually views and processes data in the form of *datagrams*, whose format is shown in Figure 10-11.

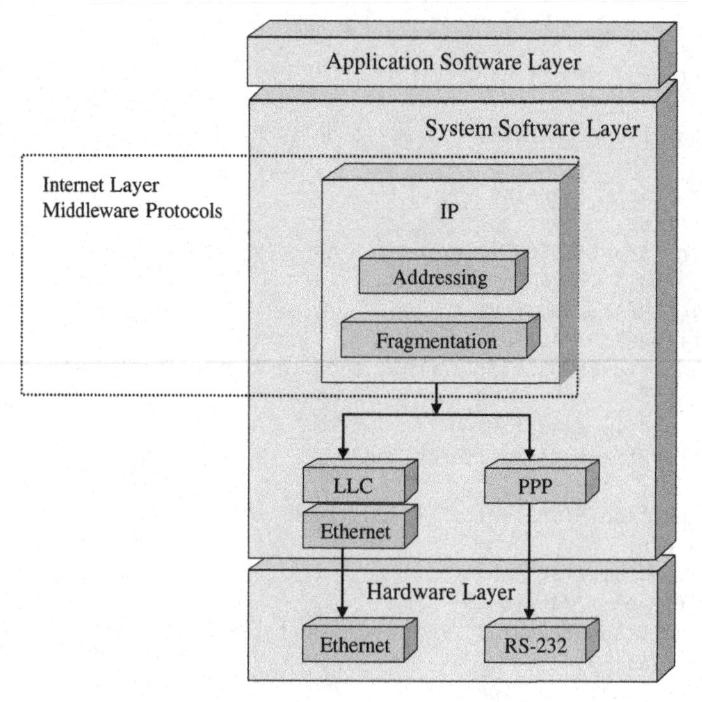


**Figure 10-10**
IP functionality.

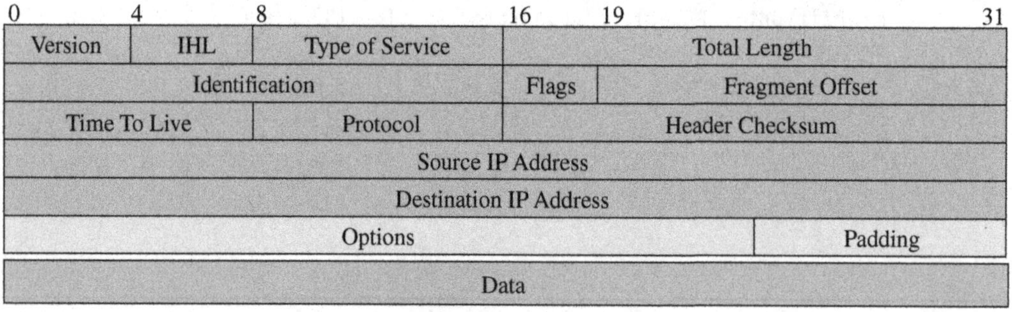

| 0 | 4 | 8 | 16 | 19 ... 31 |
|---|---|---|---|---|
| Version | IHL | Type of Service | Total Length | |
| Identification | | | Flags | Fragment Offset |
| Time To Live | | Protocol | Header Checksum | |
| Source IP Address | | | | |
| Destination IP Address | | | | |
| Options | | | | Padding |
| Data | | | | |

**Figure 10-11**
IP datagram.[2]

The entire IP datagram is what is received by IP from lower layers. The last field alone within the datagram, the data field, is the *packet* that is sent to upper layers after processing by IP. The remaining fields are stripped or appended, depending on the direction the data is going, to the data field after IP is finished processing. It is these fields that support IP addressing and fragmentation functionality.

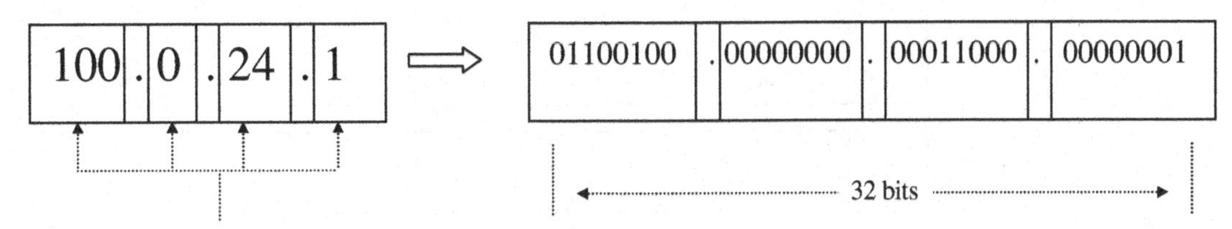


**Figure 10-12**
IP address.

**Table 10-7: IP Address Classes[2]**

| Class | IP Address Range | |
|---|---|---|
| A | **0**.0.0.0 | **127**.255.255.255 |
| B | **128**.0.0.0 | **191**.255.255.255 |
| C | **192**.0.0.0 | **223**.255.255.255 |
| D | **224**.0.0.0 | **239**.255.255.255 |
| E | **244**.0.0.0 | **255**.255.255.255 |

The source and destination IP address fields are the *networking addresses*, also commonly referred to as the *Internet* or *IP address*, processed by the IP layer. In fact, it is here that one of the main purposes of the IP layer, addressing, comes into play. IP addresses are 32 bits long, in "dotted-decimal notation," divided by "dots" into four octets (four 8-bit decimal numbers between the ranges of 0–255 for a total of 32 bits), as shown in Figure 10-12.

IP address are divided into groups, called *classes*, to allow for the ability of segments to all communicate without confusion under the umbrella of a larger network, such as the World Wide Web (WWW), or the Internet. As outlined in RFC791, these classes are organized into ranges of IP addresses, as shown in Table 10-7.

The classes (A, B, C, D, and E) are divided according to the value of the first octet in an IP address. As shown in Figure 10-13, if the highest order bit in the octet is a "0," then the IP address is a class "A" address. If the highest order bit is a "1," then the next bit is checked for a "0"; if it is, then it's a class "B" address, etc.

In classes A, B, and C, following the class bit or set of bits is the *network id*. The network id is unique to each segment or device connected to the Internet, and is assigned by Internet Network Information Center (InterNIC). The *host id* portion of an IP address is then left up to the administrators of the device or segment. Class D addresses are assigned for groups of networks or devices, called *host groups*, and can be assigned by the InterNIC or the IANA (Internet Assigned Numbers Authority). As noted in Figure 10-13, Class E addresses have been reserved for future use.

| Class | Bits | | | | Fields | |
|---|---|---|---|---|---|---|
| Class A | 0 | | | | NetID (7 bits) | Host Id (24 bits) |
| Class B | 1 | 0 | | | NetID (14 bits) | Host Id (16 bits) |
| Class C | 1 | 1 | 0 | | NetID (21 bits) | Host Id (8 bits) |
| Class D | 1 | 1 | 1 | 0 | Multicast Group Id (28 bits) | |
| Class E | 1 | 1 | 1 | 1 | Reserved (28 bits) | |

**Figure 10-13**
IP classes.[2]

### *IP Fragmentation Mechanism*

Fragmentation of an IP datagram is done for devices that can only process smaller amounts of networking data at any one time. The IP procedure for fragmenting and reassembling datagrams is a design that supports unpredictability in networking transmissions. This means that IP provides support for a variable number of datagrams containing fragments of data that arrive for reassembly in an arbitrary order and not necessarily the same order in which they were fragmented. Even fragments of differing datagrams can be handled. In the case of fragmentation, most of the fields in the first 20 bytes of a datagram, called the *header*, are used in the fragmentation and reassembling process.

The *version* field indicates the version of IP being transmitted (e.g., IPv6 is version 6). The *IHL (Internet Header Length)* field is the length of the IP datagram's header. The *total length* field is a 16-bit field in the header that specifies the actual length in octets of the entire datagram including the header, options, padding, and data. The implication behind the size of the total length field is that a datagram can be up to 65,536 ($2^{16}$) octets in size.

When fragmenting a datagram, the originating device splits a datagram "*N*" ways and copies the contents of the header of the original datagram into the all of the smaller datagram headers. The *Internet Identification (ID)* field is used to identify which fragments belong to which datagrams. Under the IP protocol, the data of a larger datagram must be divided into fragments, of which all but the last fragment must be some integral multiple of eight octet blocks (64 bits) in size.

The *fragment offset* field is a 13-bit field that indicates where in the entire datagram the fragment actually belongs. Data is fragmented into subunits of up to 8192 ($2^{13}$) fragments of eight octets (64 bits) each—which is consistent with the total length field being 65,536 octets

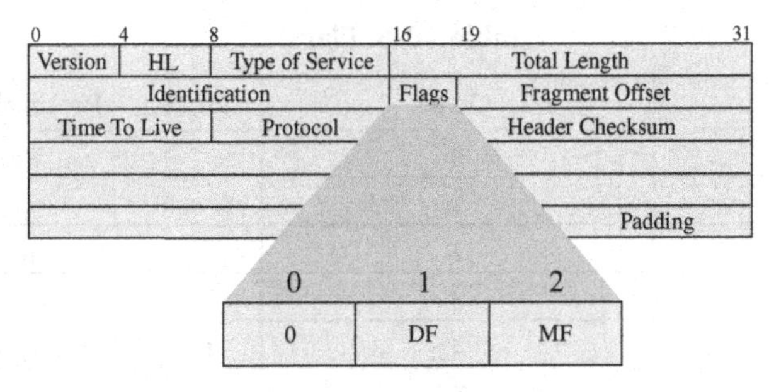


**Figure 10-14**
Flags.[2]

in size—dividing by 8 for eight octet groups = 8192. The fragment offset field for the first fragment would be "0," but for other fragments of the same datagram it would be equal to the total length (field) of that datagram fragment plus the number of eight octet blocks.

The flag fields (shown in Figure 10-14) indicate whether or not a datagram is a fragment of a larger piece. The *MF* (more fragments) flag of the *flag field* is set to indicate that the fragment is the last (the end piece) of the datagram. Of course, some systems don't have the capacity to reassemble fragmented datagrams. The *DF* (don't fragment) flag of the flag field indicates whether or not a device has the resources to assemble fragmented datagrams. It is used by one device's IP layer to inform another that it doesn't have the capacity to reassemble data fragments transmitted to it. Reassembly simply involves taking datagrams with the same ID, source address, destination address, and protocol fields, and using the fragment offset field and MF flags to determine where in the datagram the fragment belongs.

The remaining fields in an IP datagram are summarized as follows:

- *Time to live*: indicates the datagram's lifetime.
- *Checksum*: datagram integrity verification.
- *Options field*: provide for control functions needed or useful in some situations but unnecessary for the most common communications (i.e., provisions for timestamps, security, and special routing).
- *Type of service*: used to indicate the quality of the service desired. The type of service is an abstract or generalized set of parameters that characterizes the service choices provided in the networks that make up the Internet.
- *Padding*: Internet header padding is used to ensure that the Internet header ends on a 32-bit boundary. The padding is zero.
- *Protocol*: indicates the next level protocol used in the data portion of the Internet datagram. The values for various protocols are specified in "Assigned Numbers" RFC790 as shown in Table 10-8.

Following are pseudocode examples for sending and receiving processing routines for a datagram at the IP layer. Lower layer protocols (PPP, Ethernet, SLIP, etc.) call the "IPReceive" routine to

Table 10-8: Flags[2]

| Decimal | Octal | Protocol Numbers |
|---|---|---|
| 0 | 0 | Reserved |
| 1 | 1 | ICMP |
| 2 | 2 | Unassigned |
| 3 | 3 | Gateway-to-Gateway |
| 4 | 4 | CMCC Gateway Monitoring Message |
| 5 | 5 | ST |
| 6 | 6 | TCP |
| 7 | 7 | UCL |
| 8 | 10 | Unassigned |
| 9 | 11 | Secure |
| 10 | 12 | BBN RCC Monitoring |
| 11 | 13 | NVP |
| 12 | 14 | PUP |
| 13 | 15 | Pluribus |
| 14 | 16 | Telenet |
| 15 | 17 | XNET |
| 16 | 20 | Chaos |
| 17 | 21 | User Datagram |
| 18 | 22 | Multiplexing |
| 19 | 23 | DCN |
| 20 | 24 | TAC Monitoring |
| 21–62 | 25–76 | Unassigned |
| 63 | 77 | Any Local Network |
| 64 | 100 | SATNET and Backroom EXPAK |
| 65 | 101 | MIT Subnet Support |
| 66–68 | 102–104 | Unassigned |
| 69 | 105 | SATNET Monitoring |
| 70 | 106 | Unassigned |
| 71 | 107 | Internet Packet Core Utility |
| 72–75 | 110–113 | Unassigned |
| 76 | 114 | Backroom SATNET Monitoring |
| 77 | 115 | Unassigned |
| 78 | 116 | WIDEBAND Monitoring |
| 79 | 117 | WIDEBAND EXPAK |
| 80–254 | 120–376 | Unassigned |
| 255 | 377 | Reserved |

indicate to this layer to receive the datagram to disassemble, while higher layer protocols (such as TCP or UDP) call "IPSend" routine to transmit the datagram (for example).

```
ipReceive (datagram, …) {
        …
        parseDatagram(Version, InternetHeaderLength, TotalLength, Flags, …);
        …
        if (InternetHeaderLength "OR" TotalLength = OutOfBounds) OR
            (FragmentOffset = invalid) OR
            (Version = unsupported) then {
            … do not process as valid datagram …;
    } else {
        VerifyDatagramChecksum(HeaderChecksum…);
        if {HeaderChecksum = Valid) then
        …
        if (IPDestination=this Device) then {
        …
        if (Protocol Supported by Device) then {
            indicate/transmit to Protocol, data packet awaiting …;
            return;
            }
        …
            } else {
            … datagram not for this device processing …;
            } //end if-then-else Ipdestination …
        } else {
        … CHECKSUM INVALID for datagram processing …;
        } //end if-then-else headerchecksum…
    } //end if headerchecksum valid
    ICMP (error in processing datagram); //Internet Control Message Protocol used
     to indicate
     //datagram not processed successfully by this device
  } //end if-then-else (InternetHeaderLength …)
 }
    ipSend (packet, …) {
     …
     CreateDatagram(Packet, Version, InternetHeaderLength, TotalLength, Flags,
…)    sendDatagramToLowerLayer(Datagram);
     …
    }
```

*Transport Layer Middleware Example: User Datagram Protocol (UDP)*

The two most common transport layer protocols are TCP and UDP. One of the main differences between the two protocols is reliability. TCP is considered *reliable* because it requires acknowledgments from recipients of its packets. If it doesn't receive them, TCP then retransmits the unacknowledged data. UDP, on the other hand, is an *unreliable* transport layer

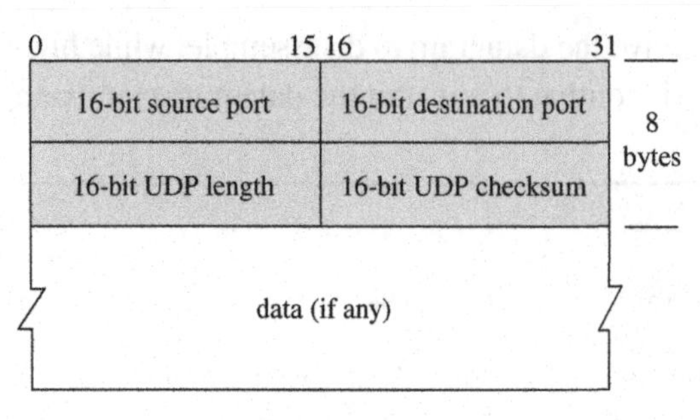


**Figure 10-15**
UDP diagram.[3]

protocol, because it never knows whether or not the recipient of its packets actually receives the data. In short, this example covers UDP, a simple, unreliable, datagram-oriented protocol based upon RFC768. The UDP packet is shown in Figure 10-15.

Transport layer protocols, such as UDP, sit on top of Internet layer protocols (such as IP), and are typically responsible for establishing and dissolving communication between two specific devices. This type of communication is referred to as *point-to-point* communication. Protocols at this layer allow for multiple higher-layer applications running on the device to connect point-to-point to other devices. While some transport layer protocols can also ensure reliable point-to-point data transmission, UDP is not one of them.

While the mechanisms for establishing communication on the server side can differ from those of the client device, both client and server mechanisms are based upon the transport layer *socket*. There are several types of sockets that a transport protocol can use, such as stream, datagram, raw, and sequenced packet, to name a few. UDP uses *datagram* sockets, a message-oriented socket handling data one message at a time (e.g., as opposed to a continuous stream of characters supported by a stream socket used by TCP). There is a socket on each end of a point-to-point communication channel, and every application on a device wanting to establish communication to another device does so by establishing a socket. Sockets are *bound* to specific ports on that device, where the port number determines the application incoming data is intended for. The two devices (client and server) then send and receive data via their sockets.

In general, on the server side a server application is running, listening to the socket, and waiting for a client to request a connection. The client essentially communicates to the server through its port (see Figure 10-16a). Ports are 16-bit unsigned integers, meaning each device has 65536 (0-65535) ports. Some ports are assigned to particular applications (FTP = ports 20–21, HTTP = port 80, etc.). UDP essentially includes the destination IP address and port number in the transmitted packet—there is no handshaking to verify the data is received in the correct order or even at all. The server determines if the received data is for one of its own applications by extracting the IP address and port number from the received packet.

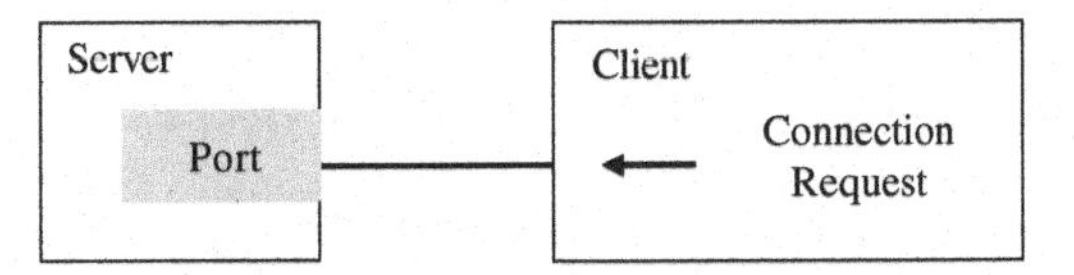


**Figure 10-16a**
Client connection request.

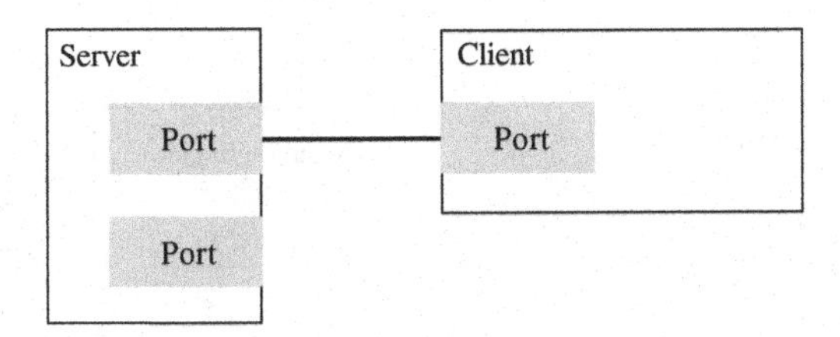


**Figure 10-16b**
Server connection established.

After the connection is successfully established, the client application establishes a socket for communication and the server then establishes a new socket to listen for incoming requests from other clients (see Figure 10-16b).

The pseudocode below demonstrates a sample UDP pseudocoded algorithm for processing an incoming datagram. In this example, if the socket for the received datagram is found, the datagram is sent up the stack (to the application layer), otherwise an error message is returned and the datagram discarded.

```
demuxDatagram(datagram) {
    ...
    verifyDatagramChecksum(datagram.Checksum);
    if (datagram.Length <= 1480 && datagram.Length >= 8) {
    ...
    if (datagram.Checksum VALID) then {
        findSocket(datagram, DestinationPort);
        if (socket FOUND) {
        sendDatagramToApp(destinationPort, datagram.Data); //send datagram to
        application return;
        } else {
        Icmp.send(datagram, socketNotFound); //indicate to Internet layer that
                                             //data will not reach intended
                                               application

        return;
        }
      }
    }
    discardInvalidDatagram();
  }
```

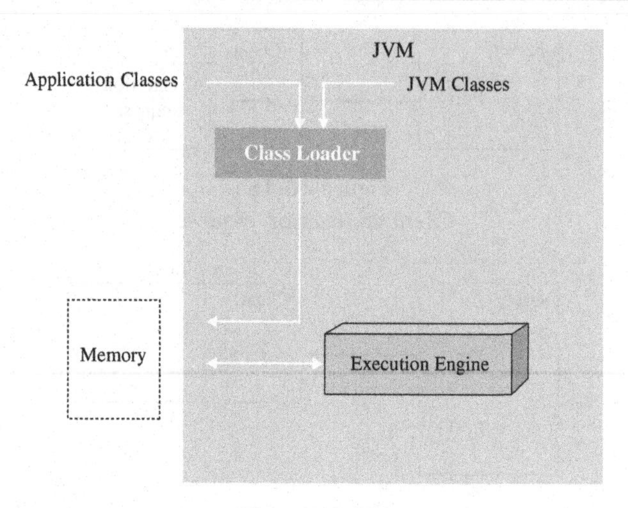


**Figure 10-17**
Internal JVM components.

*Embedded Java and Networking Middleware Example*

As introduced in Chapter 2, a *JVM* can be implemented within a system's middleware, and is made up of a class loader, execution engine, and Java API (application program interface) libraries (see Figure 10-17).

The type of applications in a Java-based design is dependent on the Java APIs provided by the JVM. The functionality provided by these APIs differs according to the Java specification adhered to, such as inclusion of the Real Time Core Specification from the J Consortium, Personal Java (pJava), Embedded Java, Java 2 Micro Edition (J2ME), and The Real Time Specification for Java from Sun Microsystems. Of these standards, pJava 1.1.8 and J2ME's Connected Device Configuration (CDC) standards are typically the standards implemented within larger embedded devices.

pJava 1.1.8 was the predecessor of J2ME CDC and in the long term may be replaced by CDC altogether. There is a pJava 1.2 specification from Sun, but as mentioned J2ME standards are intended to completely phase out the pJava standards in the embedded industry (by Sun). However, because there are JVMs on the market still supporting pJava 1.1.8, it will be used as a middleware example in this section to demonstrate what networking middleware functionality is implemented via the JVM.

The APIs provided by pJava 1.1.8 are shown in Figure 10-18. In the case of a pJava JVM implemented in the system software layer, these libraries would be included (along with the JVM's loading and execution units) as middleware components.

java.applet
java.awt
java.awt.datatransfer
java.awt.event
java.awt.image
java.beans
java.io
java.lang
java.lang.reflect
java.math
java.net
java.rmi
java.rmi.dgc
java.rmi.registry
java.rmi.server
java.security
java.security.acl
java.security.interfaces
java.sql
java.text
java.util
java.util.zip

**Figure 10-18**
pJava 1.1.8 APIs.[4]

In the pJava 1.1.8 specification, networking APIs are provided by the java.net package, shown in Figure 10-19.

The JVM provides an upper-transport layer API for remote interprocess communication via the *client/server model* (where the client requests data, etc., from the server). The APIs needed for client and servers are different, but the basis for establishing the network connection via Java is the *socket* (one at the client end and one at the server end). As shown in Figure 10-20, Java sockets use transport layer protocols of middleware networking components, such as TCP/IP discussed in the previous middleware example.

Of the several different types of sockets (raw, sequenced, stream, datagram, etc.), pJava 1.1.8 JVM provides datagram sockets, in which data messages are read in their entirety at one time, and stream sockets, where data is processed as a continuous stream of characters. JVM datagram sockets rely on the UDP transport layer protocol, while stream sockets use the TCP transport layer protocol. As shown in Figure 10-19, pJava 1.1.8 provides support for the client and server sockets, specifically one class for datagram sockets (called DatagramSocket, used for either client or server) and two classes for client stream sockets (Socket and MulticastSocket).

A socket is created within a higher-layer application via one of the socket constructor calls, in the DatagramSocket class for a datagram socket, in the Socket class for a stream socket, or

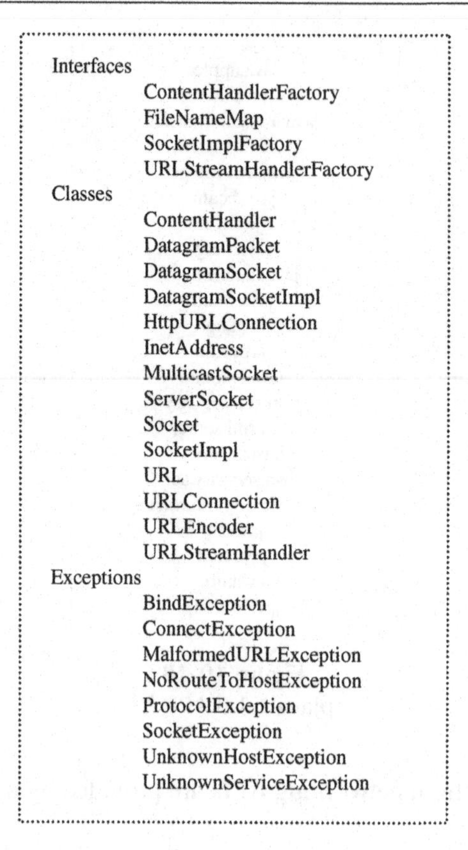
Interfaces
ContentHandlerFactory
FileNameMap
SocketImplFactory
URLStreamHandlerFactory
Classes
ContentHandler
DatagramPacket
DatagramSocket
DatagramSocketImpl
HttpURLConnection
InetAddress
MulticastSocket
ServerSocket
Socket
SocketImpl
URL
URLConnection
URLEncoder
URLStreamHandler
Exceptions
BindException
ConnectException
MalformedURLException
NoRouteToHostException
ProtocolException
SocketException
UnknownHostException
UnknownServiceException

**Figure 10-19**
java.net package.[4]

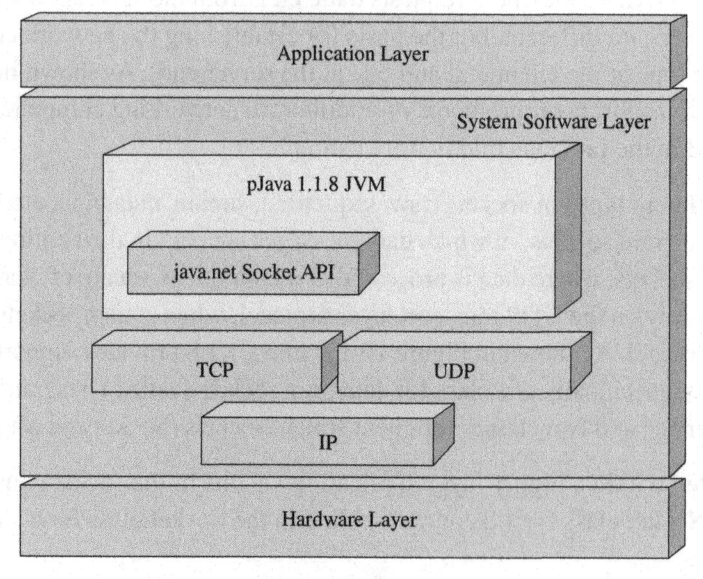


**Figure 10-20**
Sockets and the JVM.

**Socket Class Constructors**

**Socket()**
Creates an unconnected socket, with the system-default type of SocketImpl.
**Socket(InetAddress, int)**
Creates a stream socket and connects it to the specified port number at the specified IP address.
**Socket(InetAddress, int, boolean)**
Creates a socket and connects it to the specified port number at the specified IP address.
**Deprecated.**
**Socket(InetAddress, int, InetAddress, int)**
Creates a socket and connects it to the specified remote address on the specified remote port.
**Socket(SocketImpl)**
Creates an unconnected Socket with a user-specified SocketImpl.
**Socket(String, int)**
Creates a stream socket and connects it to the specified port number on the named host.
**Socket(String, int, boolean)**
Creates a stream socket and connects it to the specified port number on the named host. Deprecated.
**Socket(String, int, InetAddress, int)**
Creates a socket and connects it to the specified remote host on the specified remote port.

**MulticastSocket Class Constructors**

**MulticastSocket()**
Creates a multicast socket.
**MulticastSocket(int)**
Creates a multicast socket and binds it to a specific port.

**DatagramSocket Class Constructors**

**DatagramSocket()**
Constructs a datagram socket and binds it to any available port on the local host machine.
**DatagramSocket(int)**
Constructs a datagram socket and binds it to the specified port on the local host machine.
**DatagramSocket(int, InetAddress)**
Creates a datagram socket, bound to the specified local address.

**Figure 10-21**
Socket constructors in socket, multicast, and datagram classes.[4]

in the MulticastSocket class for a stream socket that will be multicast over a network (see Figure 10-21). As shown in the pseudocode example below of a Socket class constructor, within the pJava API, a stream socket is created, bound to a local port on the client device, and then connected to the address of the server.

```
Socket(InetAddress address, boolean stream)
{
X.create(stream); //create stream socket
X.bind(localAddress, localPort); //bind stream socket to port
If problem …
   X.close(); //close socket
else
   X.connect(address, port); //connect to server
}
```

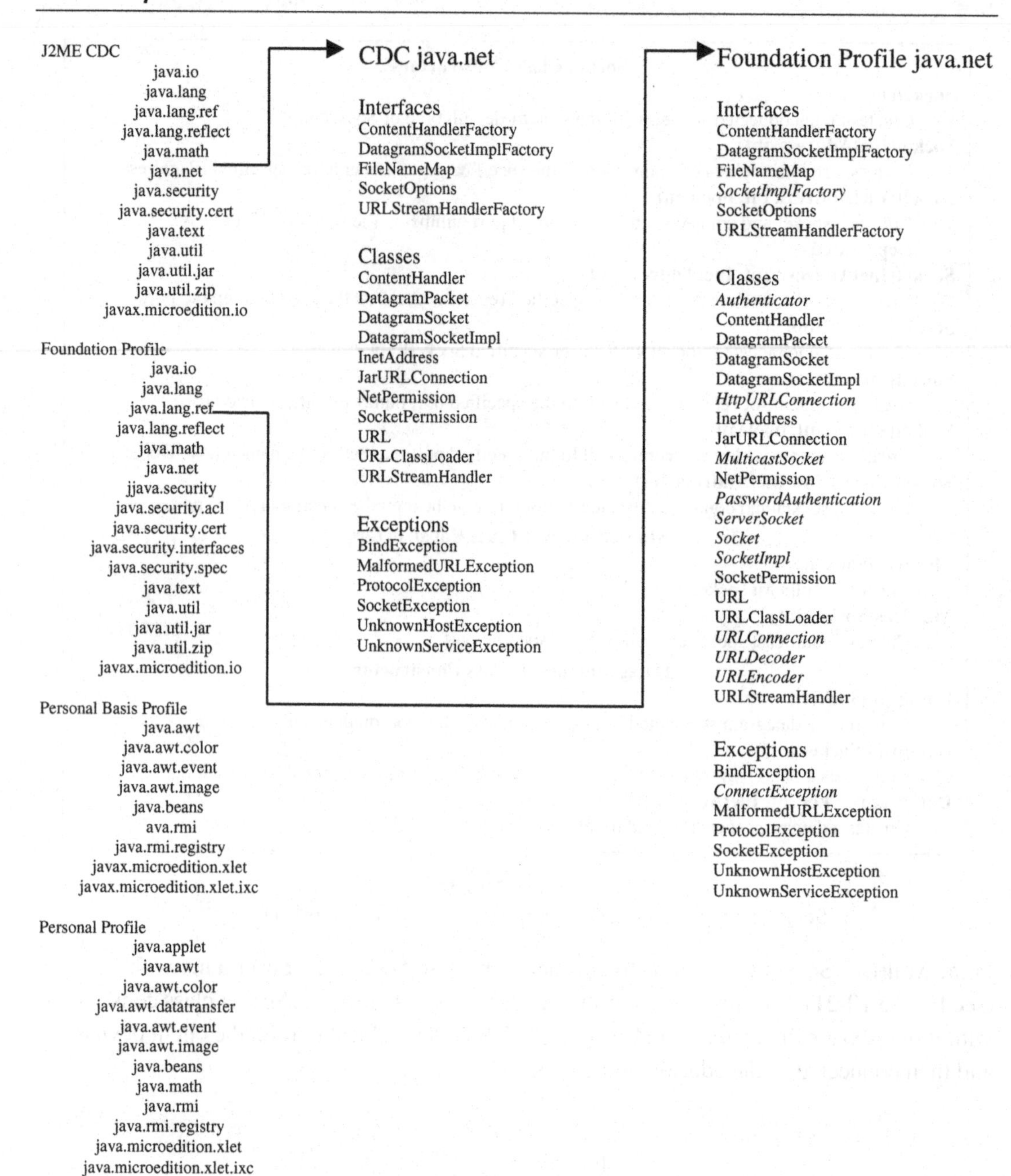


**Figure 10-22**
J2ME packages.[5]

In the J2ME set of standards, there are networking APIs provided by the packages within the CDC configuration and Foundation profile, as shown in Figure 10-22. In contrast to the pJava 1.1.8 APIs shown in Figure 10-18, J2ME CDC 1.0a APIs are a different set of libraries that would be included, along with the JVM's loading and execution units, as middleware components.

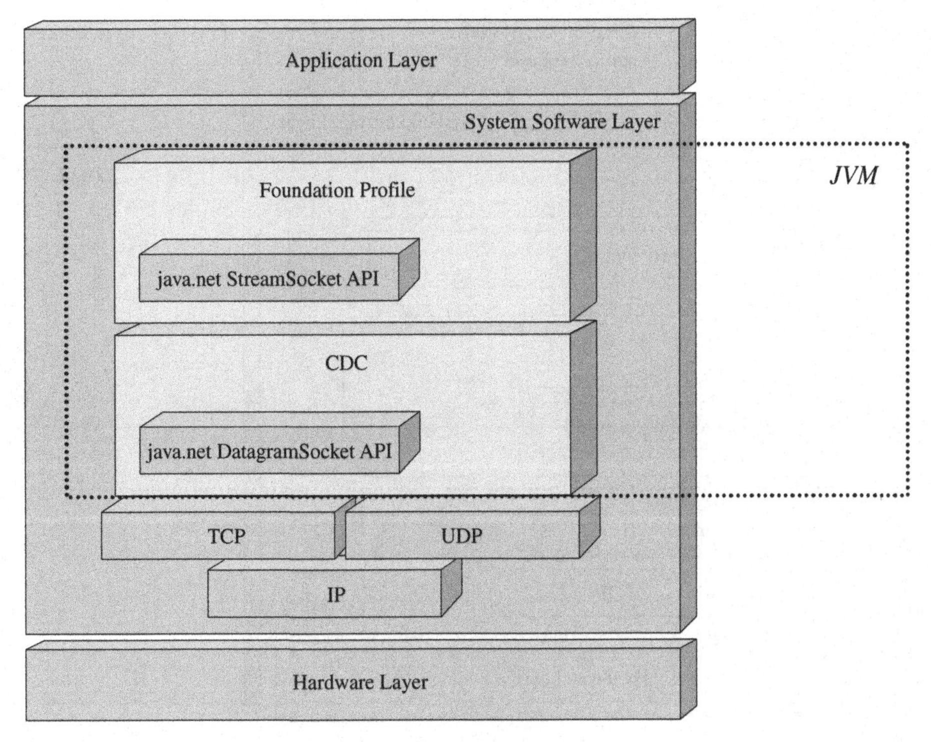


**Figure 10-23**
Sockets and the J2ME CDC-based JVM.

As shown in Figure 10-22, the CDC provides support for the client sockets. Specifically, there is one class for datagram sockets (called DatagramSocket and used for either client or server) under CDC. The Foundation Profile, which sits on top of CDC, provides three classes for stream sockets, two for client sockets (Socket and MulticastSocket) and one for server sockets (ServerSocket). A socket is created within a higher layer application via one of the socket constructor calls, in the DatagramSocket class for a client or server datagram socket, in the Socket class for a client stream socket, in the MulticastSocket class for a client stream socket that will be multicast over a network, or in the ServerSocket class for a server stream socket, for instance (see Figure 10-22). In short, along with the addition of a server (stream) socket API in J2ME, a device's middleware layer changes between pJava 1.1.8 and J2ME CDC implementations in that the same sockets available in pJava 1.1.8 are available in J2ME's network implementation, just in two different sub standards under J2ME as shown in Figure 10-23.

The J2ME connected limited device configuration (CLDC) and related profile standards are geared for smaller embedded systems by the Java community (see Figure 10-24).

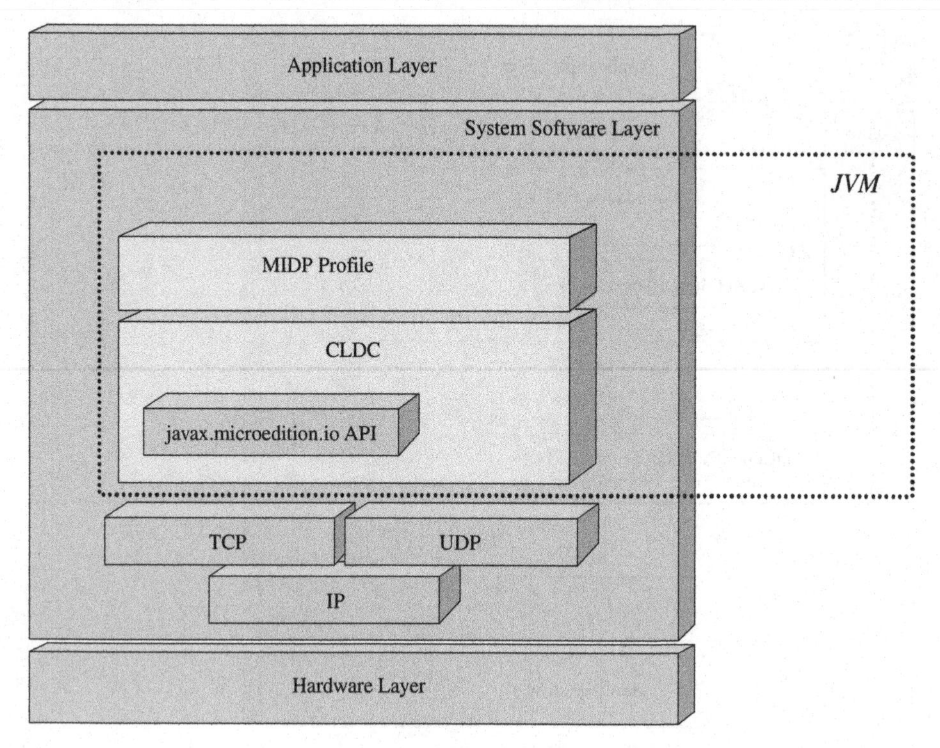


**Figure 10-24**
CLDC/MIDP stack and networking.

Continuing with networking as an example, the CLDC-based Java APIs (shown in Figure 10-25) provided by a CLDC-based JVM do not provide a .NET package, as do the larger JVM implementations (see Figure 10-25).

Under the CLDC implementation, a generic connection is provided that abstracts networking, and the actual implementation is left up to the device designers. The Generic Connection Framework (javax.microedition.io package) consists of one class and seven connection interfaces:

- *Connection*: closes the connection.
- *ContentConnection*: provides meta data info.
- *DatagramConnection*: create, send, and receive.
- *InputConnection*: opens input connections.
- *OutputConnection*: opens output connections.
- *StreamConnection*: combines Input and Output.
- *Stream ConnectionNotifier*: waits for connection.

The Connection class contains one method (Connector.open) that supports the file, socket, comm, datagram, and HTTP protocols, as shown in Figure 10-26.

```
J2ME CLDC 1.1
        java.io
        java.lang
        java.lang.ref
        java.util
        javax.microedition.io


J2ME MIDP 2.0
        java.lang
        java.util
        java.microedition.lcd.ui
        java.microedition.lcd.ui.game
        java.microedition.midlet
        java.microedition.rms
        java.microedition.io
        java.microedition.pki
        java.microedition.media
        java.microedition.media.control
```

**Figure 10-25**
J2ME CLDC APIs.[5]

```
Http Communication :
        -Connection hc = Connector.open ("http:/www.wirelessdevnet.com");

Stream-based socket communication :
        -Connection sc = Connector.open ("socket://localhost:9000");

Datagram-based socket communication:
        -Connection dc = Connector.open ("datagram://:9000);

Serial port communication :
        -Connection cc = Connector.open ("comm:0;baudrate=9000");
```

**Figure 10-26**
Connection class used.

## 10.4 Application Layer Software Examples

In some cases, applications can be based upon industry-accepted standards, as shown in Figure 10-27.

For example, for devices that need to be able to connect to other devices to transfer data or to have the remote device perform functions on command, an application networking protocol must be implemented in some form in the application layer. Networking protocols at the application level are independent of the software that they are implemented in, meaning an application-specific protocol can be implemented within a standalone application whose sole

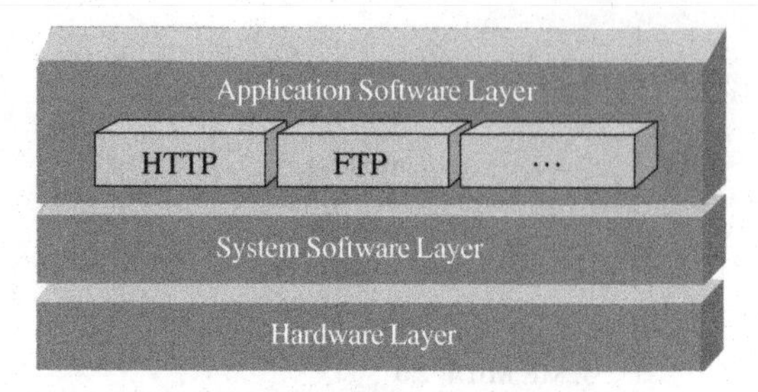


**Figure 10-27**
Application software and networking protocols.

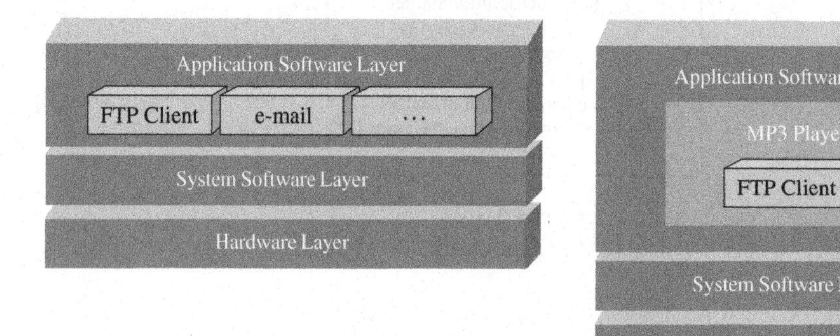


**Figure 10-28**
Application software and networking protocols.

function is the implementation of that protocol or can be implemented as a subunit of a larger application that provides many functions, as shown in Figure 10-28.

The next three examples demonstrate networking protocols implemented in both general-purpose and market-specific applications.

### *10.4.1 File Transfer Protocol (FTP) Client Application Example*

FTP is one of the simplest protocols used to securely exchange files over a network. FTP is based upon RFC959 and can be implemented as a standalone application, solely dedicated to transferring files between networked devices, or within applications such as browsers and MP3 applications. As shown in Figure 10-29, the FTP protocol defines the communication mechanisms between the device initiating the transfer, called the *FTP client* or *user-protocol interpreter* (user PI), and the device receiving the FTP connection, called the *FTP Server* or *FTP Site.*

Two types of connections can exist between FTP client and servers: the *control connection* in which commands are transmitted between the devices and the *data connection* in which the

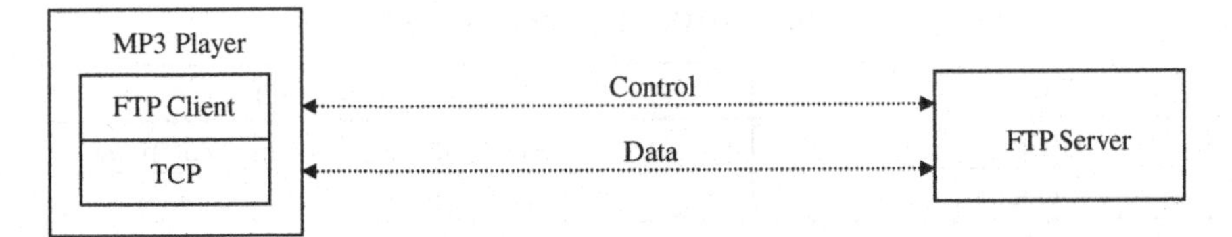


**Figure 10-29 FTP network**
E-mail can be transmitted to its ultimate destination directly, or via gateways and/or relay devices. The sender of an e-mail is referred to as the "SMTP client," whereas the recipient of an e-mail is the "SMTP server."

**Table 10-9: FTP Reply Codes[6]**

| Code | Definition |
|---|---|
| 110 | Restart marker reply |
| 120 | Service ready in "x" minutes |
| 125 | Data connection already open |
| 150 | File status ok |
| 200 | Command ok |
| 202 | Command not implemented |
| 211 | System help |
| ... | ... |

files are transmitted. An FTP session starts with the FTP client initiating a control connection by establishing a TCP connection to port 21 of the destination device. The FTP protocol requires its underlying transport protocol to be a reliable, ordered data stream channel, such as TCP (as shown in Figure 10-29). *(Note: The FTP connection mechanism is in part based upon RFC854, the Telnet (terminal emulation) protocol.)*

The FTP client, after transmitting its commands, waits for the FTP Site to respond with a *reply code* over the control connection; these codes are defined in RFC959 and shown in Table 10-9.

If the response from the FTP site is favorable, the FTP client then sends commands, like the ones shown in Table 10-10, that specify parameters for access control, such as username or password, and transfer criteria (data port, transfer mode, representation type, and file structure, etc.), as well as the transaction (store, retrieve, append, delete, etc.).

The following pseudocode demonstrates a possible initial FTP connection mechanism within an FTP client application in which access control commands are transmitted to an FTP site.

**Table 10-10: FTP Commands[6]**

| Code | Definition |
|---|---|
| USER | Username—access control command |
| PASS | Password—access control command |
| QUIT | Logout—access control command |
| PORT | Data port—transfer parameter command |
| TYPE | Representation type—transfer parameter command |
| MODE | Transfer mode—transfer parameter command |
| DELE | Delete—FTP service command |
| ... | ... |

```
FTP Client Pseudocode for access control commands USER and PASS
FTPConnect (string host, string login, string password) {
TCPSocket s=new TCPSocket(FTPServer, 21); //establishing a TCP connection to port
                                             21 of the
                                           //destination device
Timeout = 3 seconds;                       //timeout for establishing connection
                                             3 seconds
FTP Successful = FALSE;
Time = 0;
While (time<timeout) {
    read in REPLY;
    If response from recipient then {
    //login to FTP
        transmit to server ("USER "+login+"\r\n");
        transmit to server ("PASS "+password+"\r\n");
        read in REPLY;
        // reply 230 means user logged in, to proceed
        if REPLY not 230 {
                close TCP connection
                time = timeout;
        } else{
                time = timeout;
                FTP Successful = TRUE;
    }
        } else {
        time = time+1;
        } // end if-then-else response from recipient
     } // end while (time<timeout)
    }
```

In fact, as shown in the pseudocode below, the FTP client needs to provide mechanisms that support the user being able to transmit the different types of commands (shown in Table 10-10) available via FTP, as well as process any response codes received (shown in Table 10-9).

```
FTP Client Pseudocode
// "QUIT" access command routine
FTPQuit () {
    transmit to server ("QUIT");
    read in REPLY;
    // reply 221 means server closing control connection
    if REPLY not 221 {
    // error closing connection with server
    }
    close TCP connection
}
// FTP Service Command Routines
    // "DELE"
    FTPDelete (string filename) {
        transmit to server ("DELE"+filename);
        read in REPLY;
        // reply 250 means requested file action Ok
        if REPLY not 250 {
        // error deleting file
        }
    }
    // "RNFR"
    FTPRenameFile (string oldfilename, string newfilename) {
        transmit to server ("RNFR"+oldfilename);
        read in REPLY;
        // reply 350 means requested file action pending further information
        if REPLY not 350 {
        // error renaming file
    }
        transmit to server ("RNTO"+newfilename);
        read in REPLY;
        // reply 250 means requested file action Ok
        if REPLY not 250 {
        // error renaming file
        }
    }
...
```

The FTP Server initiates the data connection and any transfers according to the commands specified by the FTP client.

### 10.4.2 Simple Mail Transfer Protocol (SMTP) and E-Mail Example

SMTP is a simple application-layer ASCII protocol that is implemented within an *e-mail (electronic mail)* application for sending mail messages efficiently and reliably between devices (see Figure 10-30).

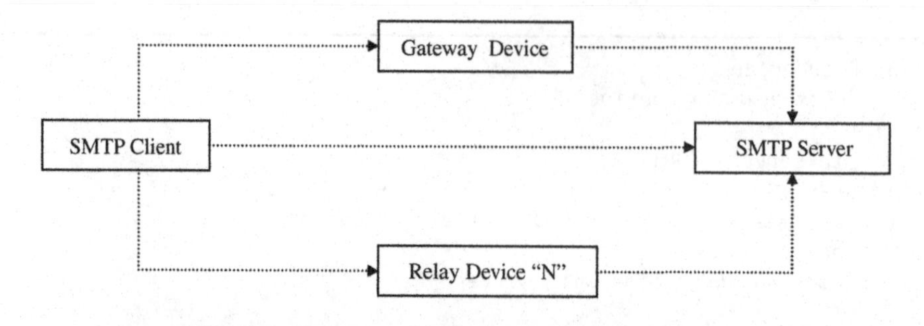


*E-mail can be transmitted to its ultimate destination directly, or via gateways and/or relay devices. The sender of an e-mail is referred to as the "SMTP client," whereas the recipient of an e-mail is the "SMTP server."*

**Figure 10-30**
SMTP network diagram.[7]

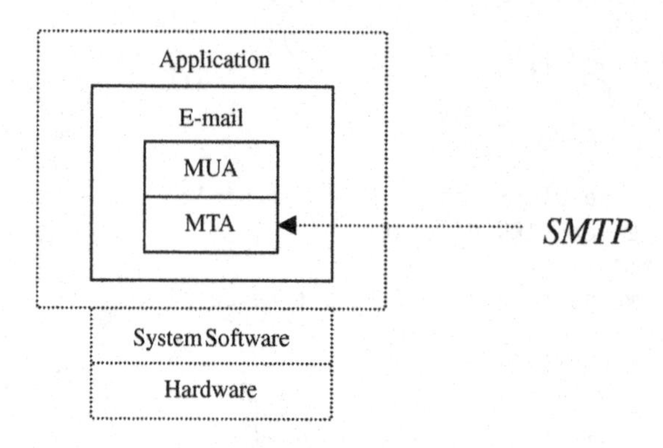


**Figure 10-31**
E-mail and the MUA and MTA components.[7]

SMTP was initially created in 1982 by ARPANET to replace file transfer protocols, which were too limited, used in e-mail systems of that time. The latest RFC ("Request for Comments") published in 2001, RFC2821, has become the *de facto* standard used in e-mail applications. According to RFC2821, e-mail applications are typically made up of two major components: the *Mail User Agent (MUA)*, which is the interface to the user that generates the e-mail, and the *Mail Transfer Agent (MTA)*, which handles the underlying SMTP exchange of e-mails. *(Note: as shown in Figure 10-31, in some systems the MUA and MTA can be separate, but layered in applications.)*

The SMTP protocol specifically defines two main mechanisms relative to sending e-mails:

- The e-mail message format.
- The transmission protocol.

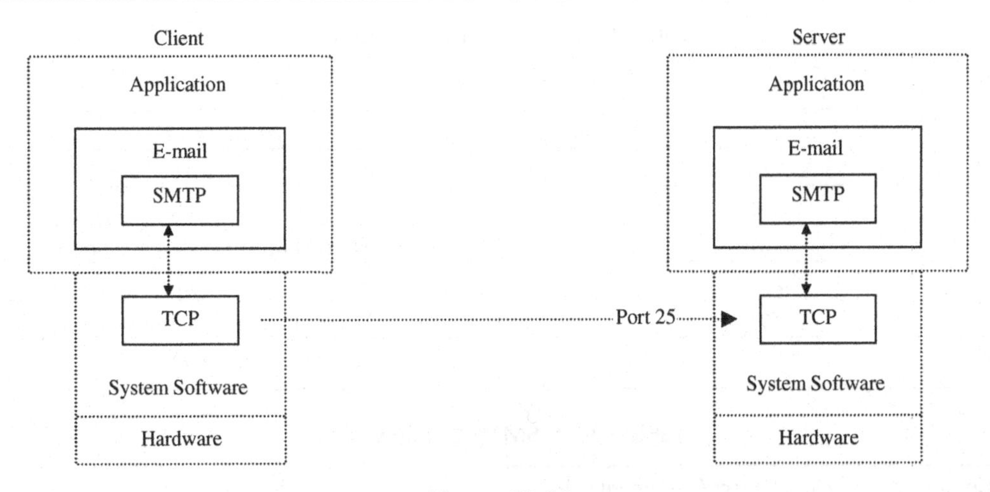


**Figure 10-32**
E-mail and TCP.

Because the sending of e-mail in SMTP is handled via *messages,* the format of an e-mail message is defined by the SMTP protocol. According to this protocol, an e-mail is made up of three parts: the *header*, the *body*, and the *envelope*. An e-mail's header, the format of which is defined in RFC2821, includes fields such as: Reply-To, Date, and From. The body is the actual content of the message being sent. According to RFC2821, the body is made up of NVT ASCII characters and is based upon RFC2045 *MIME specification* (the "Multipurpose Internet Mail Extensions Part One: Format of Internet Message Bodies"). RFC2821 also defines the contents of the envelope, which includes the addresses of the sender and recipients. After an e-mail is written and the "send" button is hit, the MTU adds a few additional headers, and then passes the content (the combination of the body and headers) to the MTA. The MTA also adds headers of its own, incorporates the envelope, and then begins the procedure of transmitting the e-mail to the MTA on another device.

The SMTP protocol requires its underlying transport protocol to be a reliable, ordered data stream channel, such as TCP (which is used in RFC2821; however, any reliable transport protocol is acceptable). As shown in Figure 10-32, when using TCP, SMTP on the client device starts the transmission mechanism by establishing a TCP connection to port 25 of the destination device.

The device transmitting the e-mail (the client) then waits for the receiver (the server) to respond with *reply code* 220 (a greeting). RFC2821 actually defines a list of reply codes, shown in Table 10-11, that a server must use when responding to a client.

Upon receiving the e-mail, the server starts by sending its identity, a fully qualified domain name of the server's host, and reply code to the client. If the reply code indicates the server won't accept the message, the client releases the connection. Also, if appropriate, an error report is sent

**Table 10-11: SMTP Reply Codes[7]**

| Code | Definition |
|---|---|
| 211 | System status |
| 214 | Help message |
| 220 | Service ready |
| 221 | Service closing transmission channel |
| 250 | Requested mail action completed |
| 251 | User not local, will forward |
| 354 | Start mail input ... |

**Table 10-12: SMTP Commands[7]**

| Command | Data Object (Argument) | Definition |
|---|---|---|
| HELO | Fully qualified domain name of the client host | How the client identifies itself |
| MAIL | Address of sender | Identifies originator of message |
| RCPT | Address of recipient | RECIPIENT—identifies who the e-mail is for |
| RSET | None | RESET—aborts the current mail transaction and causes both ends to reset. Any stored information about sender, recipients, or mail data is discarded |
| VRFY | A user or mailbox | VERIFY—lets the client ask the sender to verify a recipient's address, without sending mail to the recipient |

back to the client along with the undeliverable message. If the server responds with a reply code of 220, for instance, it means the server is willing to accept the e-mail. In this case, the client then informs the server who the sender and recipients are. If the recipient is at that device, the server gives the client one of the "go ahead and send e-mail" reply codes to the client.

The reply codes are a subset of SMTP mechanisms that are used in an e-mail transaction between devices. SMTP transmission protocol is based upon commands with arguments, referred to as mail objects that represent the data being transmitted, basically the envelope and content. There are specific commands to transmit specific types of data, such as: the MAIL command whose data object is the address of the sender (reverse-path), the RCPT command whose mail object is the forward-path (recipients address), and the DATA command whose mail object is the e-mail content (header and body) (see Table 10-12). The SMTP server responds to each command with a reply code.

SMTP defines different buffers that can be implemented on a server to include the various types of data, such as the "mail-data" buffer to hold the body of an e-mail, a "forward-path" buffer to hold the addresses of recipients, and "reverse-path" buffer to hold addresses of senders. This is because data objects that are transmitted can be held pending a confirmation by the sender that the "end of mail data" has been transmitted by the client device. This "end of mail data" confirmation (QUIT) is what finalizes a successful e-mail transaction. Finally, because TCP is a reliable byte stream protocol, checksums are usually not needed in an SMTP algorithm to verify the integrity of the data.

The following pseudocode is an example of SMTP pseudocode implemented in an e-mail application on a client device.

```
Email Application Task
    {
        Sender="xx@xx.com";
        Recipient="yy@yy.com";
        SMTPServer="smtpserver.xxxx.com"
        SENDER="tn@xemcoengineering.com",;
        RECIPIENT="cn@xansa.com";
        CONTENT="This is a simple e-mail sent by SMTP";
        SMTPSend("Hello!"); // a simple SMTP sample algorithm
        ...
    }
    SMTPSend (string Subject) {
        TCPSocket s = new TCPSocket(SMTPServer, 25); // establishing a TCP
                                                        connection to port 25 of the
                                                     // destination device
        Timeout = 3 seconds;                         // timeout for establishing
                                                        connection 3 seconds
        Transmission Successful = FALSE;
        Time = 0;
        While (time<timeout) {
        read in REPLY;
        If response from recipient then {
          if REPLY not 220 then {
                //not willing to accept e-mail
                close TCP connection
                time = timeout;
        } else {
                transmit to RECIPIENT ("HELO"+hostname);//client identifies itself
                read in REPLY;
        if REPLY not 250 then {
                //not mail action completed
                close TCP connection
                time = timeout;
        } else {
                transmit to RECIPIENT ("MAIL FROM:<"+SENDER+">");
                read in REPLY;
                if REPLY not 250 then {
                //not mail action completed
                close TCP connection
                time = timeout;
        } else {
```

```
            transmit to RECIPIENT ("RCPT TO:<"+RECEPIENT+">");
            read in REPLY;
            if REPLY not 250 then{
            //not mail action completed
            close TCP connection time = timeout;
    } else {
            transmit to RECIPIENT ("DATA");
            read in REPLY;
            if REPLY not 354 then{
            //not mail action completed
            close TCP connection
            time = timeout;
    } else {
            //transmit e-mail content according to STMP spec
            index = 0;
            while (index<length of content) {
            transmit CONTENT[index]; //transmit e-mail content character by
            character index = index+1;
    } //end while
            transmit to RECIPIENT ("."); // mail data is terminated by a line
            containing only
                    //a period
            read in REPLY;
    if REPLY not 250 then {
    //not mail action completed
    close TCP connection
    time = timeout;
    } else {
            transmit to RECIPIENT ("QUIT");
            read in REPLY;
            if REPLY not 221 then {
            //service not closing transmission channel
            close TCP connection
            time = timeout;
    } else {
            close TCP connection;
            transmission successful = TRUE;
            time = timeout;
            }//end if-then-else "." REPLY not 221
                    } //end if-then-else "." REPLY not 250
                } //end if-then-else REPLY not 354
            } // end if-then-else RCPT TO REPLY not 250
            } // end if-then-else MAIL FROM REPLY not 250
        } // end if-then-else HELO REPLY not 250
        } // end if-then-else REPLY not 220
  } else {
  time = time+1;
  } // end if-then-else response from recipient
  } // end while (time<timeout)
} // end STMPTask
```

### *10.4.3 Hypertext Transfer Protocol (HTTP) Client and Server Example*

Based upon several RFC standards and supported by the WWW Consortium, HTTP 1.1 is the most widely implemented application layer protocol, used to transmit all types of data over the Internet. Under the HTTP protocol, this data (referred to as a *resource*) is identifiable by its *URL (Uniform Resource Locator)*.

As with the other two networking examples, HTTP is a based upon the client/server model that requires its underlying transport protocol to be a reliable, ordered data stream channel, such as TCP. The HTTP transaction starts with the HTTP client opening a connection to an HTTP server by establishing a TCP connection to default port 80 (for example) of the server. The HTTP client then sends a *request message* for a particular resource to the HTTP server. The HTTP server responds by sending a *response message* to the HTTP client with its requested resource (if available). After the response message is sent, the server closes the connection.

The syntax of request and response messages both have headers that contain message attribute information that varies according to the message owner, and a body that contains optional data, where the header and body are separated by an empty line. As shown in Figure 10-33, they differ according to the first line of each message—where a request message contains the method (command made by client specifying the action the server needs to perform), the request-URL (address of resource requested), and version (of HTTP) in that order, and the first line of a response message contains the version (of HTTP), the status-code (response code to the client's method), and the status-phrase (readable equivalent of status-code).

Tables 10-13a and b list the various methods and reply codes that can be implemented in an HTTP Server.

The pseudocode example below demonstrates HTTP implemented in a simple web server.

| Request Message | Response Message |
| --- | --- |
| <method> <request-URL><version> | <version> <status-code><status-phrase> |
| <headers> | <headers> |
| "/r/n" [blank line] | "/r/n" [blank line] |
| <body> | <body> |

**Figure 10-33**
Request and response message formats.[8]

**Table 10-13a: HTTP Methods[8]**

| Method | Definition |
|---|---|
| DELETE | The DELETE method requests that the origin server delete the resource identified by the Request-URI. |
| GET | The GET method means retrieve whatever information (in the form of an entity) is identified by the Request-URI. If the Request-URI refers to a data-producing process, it is the produced data which shall be returned as the entity in the response and not the source text of the process, unless that text happens to be the output of the process. |
| HEAD | The HEAD method is identical to GET except that the server MUST NOT return a message body in the response. The meta information contained in the HTTP headers in response to a HEAD request SHOULD be identical to the information sent in response to a GET request. This method can be used for obtaining meta information about the entity implied by the request without transferring the entity body itself. This method is often used for testing hypertext links for validity, accessibility, and recent modification. |
| OPTIONS | The OPTIONS method represents a request for information about the communication options available on the request/response chain identified by the Request-URI. This method allows the client to determine the options and/or requirements associated with a resource, or the capabilities of a server, without implying a resource action or initiating a resource retrieval. |
| POST | The POST method is used to request that the destination server accept the entity enclosed in the request as a new subordinate of the resource identified by the Request-URI in the Request-Line. POST is designed to allow a uniform method to cover the following functions:<br>• annotation of existing resources;<br>• posting a message to a bulletin board, newsgroup, mailing list, or similar group of articles;<br>• providing a block of data, such as the result of submitting a form, to a data-handling process;<br>• extending a database through an append operation. |
| PUT | The PUT method requests that the enclosed entity be stored under the supplied Request-URI. If the Request-URI refers to an already existing resource, the enclosed entity SHOULD be considered as a modified version of the one residing on the origin server. If the Request-URI does not point to an existing resource, and that URI is capable of being defined as a new resource by the requesting user agent, the origin server can create the resource with that URI. |
| TRACE | The TRACE method is used to invoke a remote, application-layer loop-back of the request message. TRACE allows the client to see what is being received at the other end of the request chain and use that data for testing or diagnostic information. |

**Table 10-13b: HTTP Reply Codes[8]**

| Code | Definition |
|---|---|
| 200 | OK |
| 400 | Bad request |
| 404 | Not found |
| 501 | Not implemented |

```
HTTP Server Pseudocode Example
…
HTTPServerSocket (Port 80)
ParseRequestFromHTTPClient (Request, Method, Protocol);
If (Method not "GET", "POST", or "HEAD") then {
   Respond with reply code "501"; // not implemented
   Close HTTPConnection;
}
If (HTTP Version not "HTTP/1.0" or "HTTP/1.1") then {
   Respond with reply code "501"; // not implemented
Close HTTPConnection;
}
…
   ParseHeader, Path, QueryString(Request, Header, Path, Query);
       if (Length of Content>0) {
       if (Method="POST") {
       ParseContent(Request, ContentLength, Content);
       } else {
           // bad request - content but not post
           Respond with reply code to HTTPURLConnection "400";
           Close HTTPConnection;
       }
   }
   // dispatching servlet
   If (servlet(path) NOT found) {
       Respond with reply code "404"; // not found
       Close HTTPConnection
   } else {
   Respond with reply code "200"; // Ok
   …
   Transmit servlet to HTTPClient;
   Close HTTPConnection;
}
```

Tables 10-13a and b show the methods that can be transmitted by an HTTP client as illustrated in the following pseudocode example that demonstrates how HTTP can be implemented in a web client, such as a browser.

```
public void BrowserStart()
{
   Create "UrlHistory" File;
   Create URLDirectory;
   Draw Browser UI (layout, borders, colors, etc.)
   //load home page
   socket= new Socket(wwwserver, HTTP_PORT);
   SendRequest("GET"+filename+" HTTP/1.0\n");
       if(response.startsWith("HTTP/1.0 404 Not Found"))
```

```
        {
        ErrorFile.FileNotFound();
        } else
    {
        Render HTML Page (Display Page);
    }
    Loop Wait for user event
    {
        Respond to User Events;
        }
    }
```

### 10.4.4 Quick Note on Putting It All Together

After gaining an understanding of the overall networking requirements of the embedded system, it is then important to tune networking parameters at all layers of software to real-world performance needs accordingly. Even if the networking components are included as part of a middleware bundle purchased from an off-the-shelf embedded OS vendor, never assume that underlying system software is automatically configured by a vendor for a specific device's production-ready requirements.

For example, for design teams that use VxWorks, they have the option of purchasing an additional tightly networking stack with VxWorks. Networking components can be tuned to the requirements of the device via specific networking parameters provided by Wind River's VxWorks development environment (i.e., Tornado or Workbench) and source code.

In Table 10-14, as a specific example, tuning the TCP/IP middleware stack parameter TCP_MSS_DFLT, which is the TCP *Maximum Segment Size (MSS)*, requires analyzing both IP fragmentation and managing overhead.

Because TCP segments are repackaged into IP datagrams when data flows down the stack, size limitations of the IP datagrams must be taken into account otherwise fragmentation will occur at the IP layer if the TCP segment is too large. Meaning that if IP fragmentation is not analyzed properly and more than one datagram must be transmitted at the IP layer for the TCP segment data to be managed successfully, then system performance will be negatively impacted. To manage overhead properly with IP fragmentation concerns means recognizing that TCP and IP headers are not part of the data being transmitted. (These headers are transmitted along with the data). Otherwise, a decrease in MSS would reduce fragmentation, but could prove too inefficient due to the overhead if it is decreased too much.

Another specific example for tuning at the middleware layer for requirements and performance are the VxWorks TCP window sizes. TCP informs connections of how much data can be managed at any given time by utilizing socket window size parameters.

**Table 10-14: Tuning Parameters for Networking Components in VxWorks[9]**

| Networking Component | Parameter | Description | Value |
|---|---|---|---|
| TCP | TCP_CON_TIMEO_DFLT | Timeout intervals to connect (default 150 = 75 s) | 150 |
| | TCP_FLAGS_DFLT | Default value of the TCP flags | (TCP_DO_RFC1323) |
| | TCP_IDLE_TIMEO_DFLT | Seconds without data before dropping connection | 14,400 |
| | TCP_MAX_PROBE_DFLT | Number of probes before dropping connection (default 8) | 8 |
| | TCP_MSL_CFG | TCP maximum segment lifetime in seconds | 30 |
| | TCP_MSS_DFLT | Initial number of bytes for a segment (default 512) | 512 |
| | TCP_RAND_FUNC | A random function to use in tcp_init | (FUNCPTR)random |
| | TCP_RCV_SIZE_DFLT | Number of bytes for incoming TCP data (8192 by default) | 8192 |
| | TCP_REXMT_THLD_DFLT | Number of retransmit attempts before error (default 3) | 3 |
| | TCP_RND_TRIP_DFLT | Initial value for round-trip time (s) | 3 |
| | TCP_SND_SIZE_DFLT | Number of bytes for outgoing TCP data (8192 by default) | 8192 |
| UDP | UDP_FLAGS_DFLT | Optional UDP features: default enables checksums | (UDP_DO_CKSUM_ SND \| UDP_DO_ CKSUM_RCV) |
| | UDP_RCV_SIZE_DFLT | Number of bytes for incoming UDP data (default 41,600) | 41,600 |
| | UDP_SND_SIZE_DFLT | Number of bytes for outgoing UDP data (9216 by default) | 9216 |
| | IP_FLAGS_DFLT | Selects optional features of IP layer | (IP_DO_FORWARDING \| IP_DO_REDIRECT \| IP_DO_CHECKSUM_ SND \| IP_DO_ CHECKSUM_RCV) |
| IP | IP_FRAG_TTL_DFLT | Number of slow timeouts (2/s) | 60 |
| | IP_QLEN_DFLT | Number of packets stored by receiver | 50 |
| | IP_TTL_DFLT | Default TTL value for IP packets | 64 |
| | IP_MAX_UNITS | Maximum number of interfaces attached to IP layer | 4 |

For networking protocols that may require larger window sizes, such as satellite or ATM communication, the TCP socket receive and send buffer sizes can be tuned via TCP_RCV_SIZE_DFLT and TCP_SND_SIZE_DFLT parameters (see Table 10-14). What is typically recommended is that socket buffer sizes should be an even multiple of the MSS and three or more times the MSS value. Socket buffer sizes need to accommodate the Bandwidth (bytes/s)*Round Trip Time (s) in order to ensure meeting networking performance goals.

Whether learning about networking middleware and application software as covered in previous sections of this chapter, or another type of middleware (e.g., a database or file system), one should apply the same type of concepts as presented in this chapter to understand other types of middleware and overlying software and their impact on the architecture of the embedded design. The following sections provide other examples relative to the impact of middleware and application software on an embedded system's architecture.

*Impact of Other Types of Middleware and Application Software: Programming Languages*

Certain types of higher-level programming languages can impact the application layer architecture. One such language is Java; an application written in Java can integrate the JVM within it, shown in Figure 10-34. An example of this is the J2ME CLDC/MIDP JVM compiled into a .prc file (executable file on PalmOS) on a Palm Pilot personal data assistant (PDA) or a .exe file on a PocketPC platform. In short, a JVM can be implemented in hardware, as middleware either integrated within the OS or sitting on top of the OS, or as part of an application.

For better or worse, additional support is required relative to increased processing power and memory requirements when writing applications in a higher-level language that requires introducing an underlying JVM in the middleware layer or a JVM that must be integrated into the application itself. Introducing a JVM into an embedded device means planning for any additional hardware requirements and underlying system software by both the JVM and the overlying applications.

For example, several factors are impacted by the scheme a JVM utilizes in processing the overlying application byte code—whether it is an interpretation byte code scheme versus a just-in-time (JIT) compiler versus a way-ahead-of-time (WAT) compiler versus a dynamic adaptive compiler (DAC).

A JVM that utilizes some type of ahead-of-time (AOT) or WAT compilation can provide a big boost in performance when running on particular hardware, but this type of JVM may

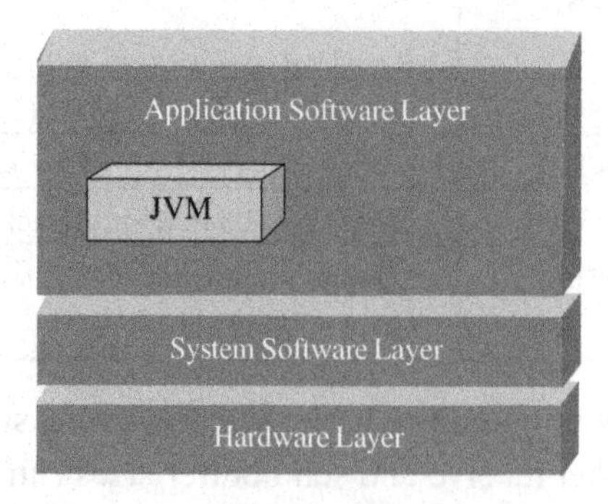


**Figure 10-34**
JVM compiled in application.

lack the ability to process dynamically downloaded Java byte code. If an AOT- or a WAT-based JVM does not support dynamic download capabilities and on-the-field, dynamic extensibility support is a non-negotiable requirement for the embedded system being designed, then it means needing to further investigate:

- The feasibility of deploying with another JVM based on a different byte code processing scheme that may run a bit slower then the faster JVM solution that lacks dynamic download and extensibility support. This also requires understanding any new requirements for more powerful underlying hardware in support of a different JVM implementation.
- The resources, costs, and time to implement this required functionality for the AOT/WAT-based JVM within the scope of the project.

Another example is investigating utilizing a JIT implementation of a JVM versus going with the JIT-based .NET Compact Framework solution of comparable performance within a particular embedded design. In addition to examining the available APIs provided by the JVM versus .NET Compact Framework embedded solutions for application requirements, it is important to consider the non-technical aspects of going with either particular solution. This means taking into consideration the availability of experienced programmers (e.g., Java versus C# programmers) when selecting between such alternative VM solutions. If there are no programmers available with the necessary skills for application development on that particular VM, factor in the costs and time involved in finding and hiring new resources, training current resources, etc.

Another case in which programming languages can impact the architecture of the application layer occurs when the application must process source code written in a scripting language, such as HTML (HyperText Markup Language, the language in which most of the files or pages found on the WWW are written) or Javascript (used to implement interactivity functionality in web pages, including images changing on-the-fly when hit with a cursor, forms interacting, calculations executed, etc.). As shown in Figure 10-35, a scripting language

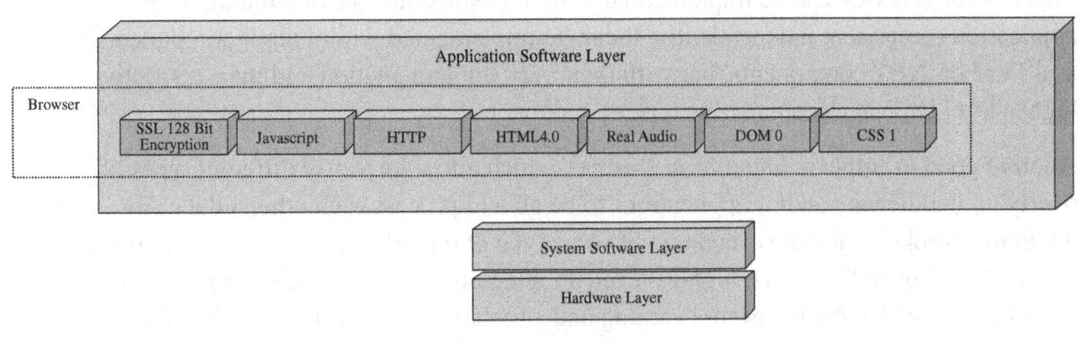


**Figure 10-35**
Browser and the Embedded Systems Model.

interpreter is then integrated into the application software (such as a web browser, an HTTP client itself, where multiple scripting language interpreters are integrated).

Finally, to ensure success in taking an embedded design that introduces the components and complexity of these type of programming languages within an embedded system's architecture, it requires programmers to plan carefully how overlying applications will be written. This means that that it is NOT the most elegant or the most brilliantly written application code that will ensure the success of the design, but the programmers that design applications in a manner that properly utilizes the underlying system software and hardware's powerful strengths and avoids weaknesses. For instance, a Java application that is written as a masterpiece by even the savviest programmer will not be worth much if, when it runs on the embedded system it was intended for, this application is so slow and/or devours so much of the embedded system's resources that the device simply cannot be shipped!

*Impact of Other Types of Middleware and Application Software: Complex Messaging, Communication, and Security*

A popular approach in embedded systems design to support complex, distributed application requirements within embedded systems is to overlay complex messaging, communication, and security middleware that build on core middleware.

For example, *Message Oriented and Distributed Messaging* provides *message* passing capabilities when some type of point-to-point and/or autonomous publish–subscribe messaging scheme is optimal. *Distributed Transaction* builds on other middleware (i.e., RPC builds on TCP and/or UDP depending on the scheme) and allows different types of communication across remote systems. *ORBs* build upon RPC, and allow for the creation of overlying middleware and/or application components that reside(s) as multiple objects, on the same embedded device and/or across more than one device.

*Authentication and Security* middleware provides code security, validation, and verification features, such as ensuring valid-type operations are performed (i.e., array bounds checking, type checks and conversions, checking for stack integrity (overflow) and memory safety). This type of software can be implemented within an embedded device middleware or application component independently. It can be implemented within other components such as a JVM or .NET components that with their very implementation and their respective higher-level languages contain this type of support.

Another good example is *Integration Brokers*, which allow for vastly different types of overlying middleware and/or applications to be able to process each other's data. An integration broker is not only made up of some type of underlying communication broker (e.g., a MOM or RPC). At the highest level, an integration broker is also composed of components that handle the event listening and generation that resides upon some type of core networking middleware. An integration broker's "TCP listener" component utilizes underlying TCP sockets, whereas a "file listener" component utilizes an underlying file system. An integration broker's transformer component handles any translation of data required as this data passes through the broker on its way to the destination (e.g., an "FTP adaptor" subcomponent that supports FTP or an "HTTP adaptor" subcomponent that supports HTTP).

## 10.5 Summary

In this chapter, middleware was defined as system software that mediates between application software and the kernel or device driver software, or software that mediates and serves different application software. Application software was defined as software that sits on top of system software and that essentially gives the device its character, differentiating it from other embedded devices, and application software is usually the software that directly interacts with a user of the device.

This chapter presented applications based on industry standards that would contribute to a device's architecture if implemented. This included the impact of networking functionality, as well as the impact of various high-level languages (specifically Java and scripting languages) on the middleware and application layer architecture. In short, the key to selecting the embedded middleware and overlying application software that best match the requirements of an embedded systems design *and* successfully taking this design to production within schedule and costs, includes:

- Determining if the middleware has been ported to support underlying system software and target the hardware's master CPU architecture in the first place. If not, it means calculating how much additional time, cost, and resources would be required to support within the design.
- Calculating additional processing power and memory requirements to support the middleware solution and overlying applications.
- Investigating the stability and reliability of the middleware implementation on real hardware and underlying system software.
- Planning around the availability of experienced developers.
- Evaluating development and debugging tool support.
- Checking up on the reputation of vendors.
- Ensuring access to solid technical support for the middleware implementation for developers.
- Writing the overlying applications properly.

Section IV puts all of these layers together, and discusses how to apply Sections I–III in designing, developing, and testing an embedded system.

## Chapter 10: Problems

1. What is middleware?
2. Which of Figures 10-36a–d is incorrect in terms of mapping middleware software into the Embedded Systems Model?
3. [a] What is the difference between general-purpose middleware and market-specific middleware?
   [b] List two real-world examples of each.
4. Where in the OSI model is networking middleware located?
5. [a] Draw the TCP/IP model layers relative to the OSI model.
   [b] Which layer would TCP fall under?
6. [T/F] RS-232-related software is middleware.
7. PPP manages data as:
   A. Frames.
   B. Datagrams.

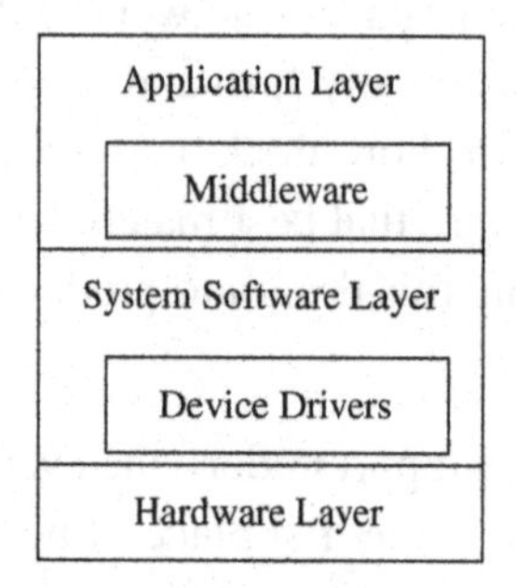


**Figure10-36a**
Example 1.

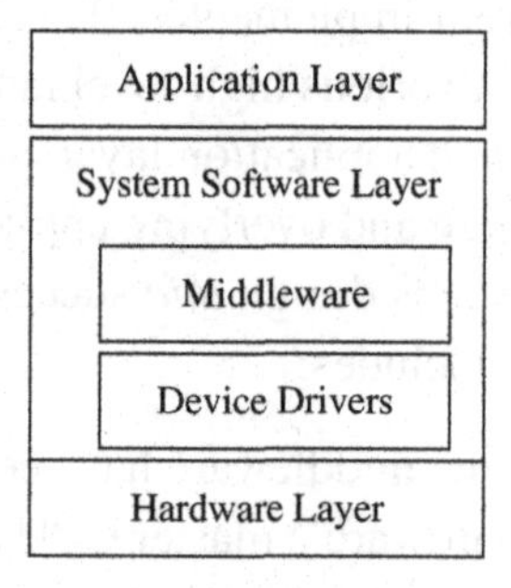


**Figure 10-36b**
Example 2.

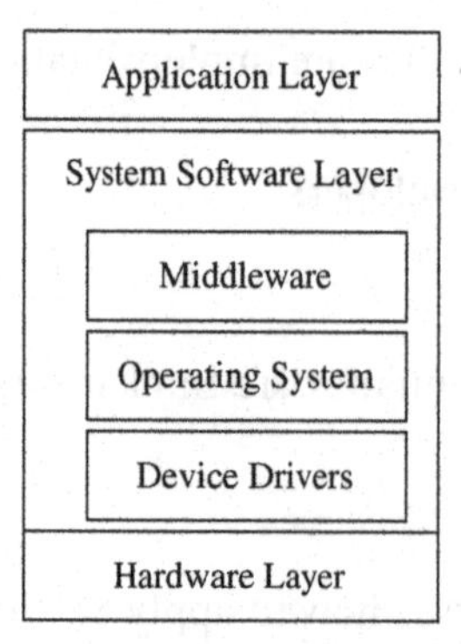


**Figure 10-36c**
Example 3.

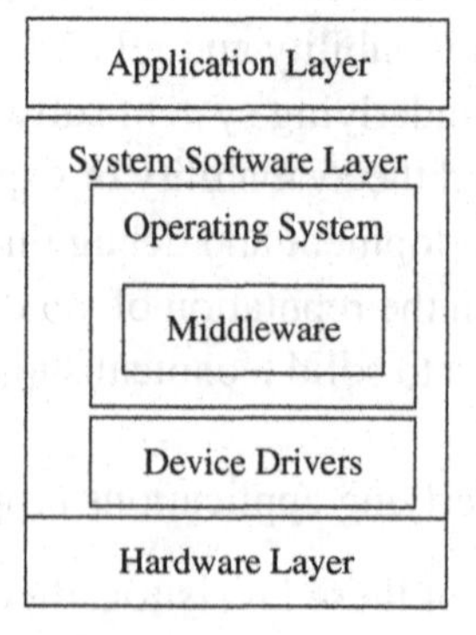


**Figure 10-36d**
Example 4.

C. Messages.
D. All of the above.
E. None of the above.

8. [a] Name and describe the four subcomponents that make up PPP software.
   [b] What RFCs are associated with each?
9. [a] What is the difference between a PPP state and a PPP event?
   [b] List and describe three examples of each.
10. [a] What is an IP address?
    [b] What networking protocol processes IP addresses?
11. What is the main difference between UDP and TCP?
12. [a] Name three embedded JVM standards that can be implemented in middleware.
    [b] What are the differences between the APIs of these standards?
    [c] List two real-world JVMs that support each of the standards.
13. [T/F] The .NET Compact Framework is implemented in the middleware layer of the Embedded Systems Model.
14. [a] What is application software?
    [b] Where in the Embedded Systems Model is application software typically located?
15. Name two examples of application layer protocols that can either be implemented as standalone applications whose sole function is that protocol or be implemented as a subcomponent of a larger multifunction application.
16. [a] What is the difference between an FTP client and an FTP server?
    [b] What type of embedded devices would implement each?
17. SMTP is a protocol that is typically implemented in:
    A. An e-mail application.
    B. A kernel.
    C. A BSP.
    D. Every application.
    E. None of the above.
18. [T/F] SMTP typically relies on TCP middleware to function.
19. [a] What is HTTP?
    [b] What types of applications would incorporate an HTTP client or server?
20. What type of programming languages would introduce a component at the application layer?

## Endnotes

[1] RFC1661: http://www.freesoft.org/CIE/RFC/1661/index.htm; RFC1334: http://www.freesoft.org/CIE/RFC/1334/index.htm; RFC1332: http://www.freesoft.org/CIE/RFC/1332/index.htm.
[2] RFC791: http://www.freesoft.org/CIE/RFC/791/index.htm.
[3] RFC768: http://www.freesoft.org/CIE/RFC/768/index.htm.

[4] Personal Java 1.1.8 API documentation, java.sun.com.
[5] Java 2 Micro Edition 1.0 API Documentation, java.sun.com.
[6] RFC959: http://www.freesoft.org/CIE/RFC/959/index.htm.
[7] RFC2821: http://www.freesoft.org/CIE/RFC/2821/index.htm.
[8] http://www.w3.org/Protocols/
[9] Wind River, VxWorks API Documentation and Project.

SECTION IV

# Putting It All Together: Design and Development

## Putting it all Together: Design and Development

The chapters of Sections II and III presented the fundamental technical details of the major embedded hardware and software elements that an engineer needs to be familiar with (at a minimum) in order to understand or create an architecture. As Chapter 1 indicated, Chapter 2, Section II, and Section III are all part of the first phase in designing an embedded system—defining the system (Stage 1: have a solid technical foundation).

Section IV continues to outline the remaining stages of designing an embedded system, covering the remaining five stages of defining a system (Chapter 11): Stage 2: understand the architecture business cycles of embedded models; Stage 3: define the architectural patterns and reference models; Stage 4: create the architectural structures; Stage 5: document the architecture; and Stage 6: analyze and evaluate the architecture. Chapter 12 outlines the remaining phases of designing an embedded system—implementing a system based upon the architecture, debugging and testing the system, and maintaining the system.

CHAPTER 11

# Defining the System—Creating the Architecture and Documenting the Design

**In This Chapter**

- Defining the stages of creating an embedded systems architecture
- Introducing the architecture business cycle and its effect on architecture
- Describing how to create and document an architecture
- Introducing how to evaluate and reverse engineer an architecture

This chapter is about giving the reader some practical processes and techniques that have proven useful over the years. This book has taken the systems, holistic approach to understanding embedded systems, because one of the *most powerful* methods of insuring success for a design team is to accept and address the reality that the successful engineering of an embedded systems product will be impacted by more than the pure technology, alone. This includes having the discipline in adhering to development processes and best practices, in order to avoid costly mistakes.

Defining the system and its architecture, if done correctly, is the phase of development that is the *most difficult* and the *most important* of the entire development cycle. Figure 11-1 shows the different phases of development as defined by the *Embedded System Design and Development Lifecycle Model*.[1]

This model indicates that the process of designing an embedded system and taking that design to market has four phases:

*Phase 1* *Creating the architecture*: the process of planning the design of the embedded system.

*Phase 2* *Implementing the architecture*: the process of developing the embedded system.

*Phase 3* *Testing the system*: the process of testing the embedded system for problems and then solving those problems.

*Phase 4* *Maintaining the system*: the process of deploying the embedded system into the field and providing technical support for users of that device for the duration of the device's lifetime.

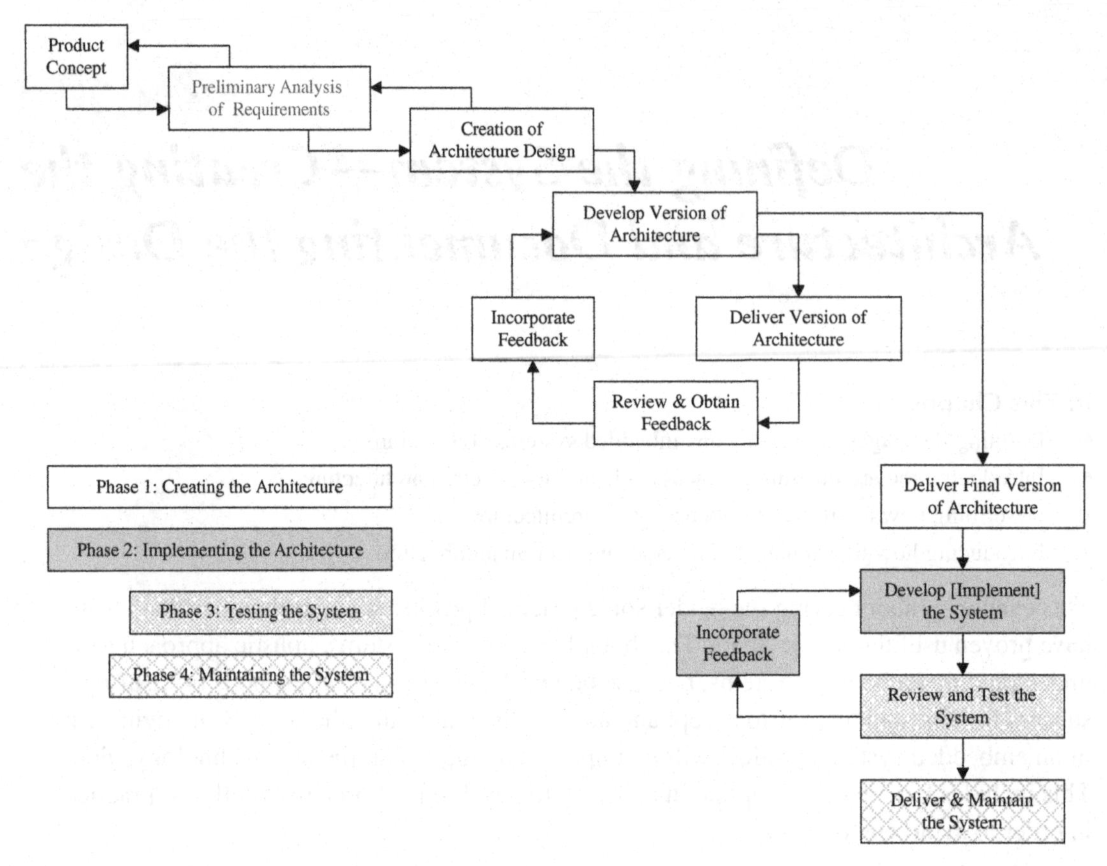


**Figure 11-1**
Embedded Systems Design and Development Lifecycle Model.[1]

This model also indicates that the most important time is spent in Phase 1, creating the architecture. At this phase of the process, no board is touched and no software is coded. It is about putting full attention, concentration, and investigative skills into gathering information about the device to be developed, understanding what options exist, and documenting those findings. If the right preparation is done in defining the system's architecture, determining requirements, understanding the risks, etc., then the remaining phases of development, testing, and maintaining the device will be simpler, faster, and cheaper. This, of course, assumes that the engineers responsible have the necessary skills.

In short, if Phase 1 is done correctly, then less time will be wasted on deciphering code that doesn't meet the system requirements or guessing what the designers' intentions were, which most often results in more bugs and more work. That is not to say that the design process is always smooth sailing. Information gathered can prove inaccurate, specifications can change, etc., but if the system designer is technically disciplined, prepared, and organized, new hurdles can be immediately recognized and resolved. This results in a development process that is much less stressful, with less time and money spent and wasted. Most importantly, the project will, from a technical standpoint, almost certainly end in success.

## 11.1 Creating an Embedded System Architecture

Several industry methodologies can be adopted in designing an embedded system's architecture, such as the Rational Unified Process (RUP), Attribute-Driven Design (ADD), the Object-Oriented Process (OOP), and the Model-Driven Architecture (MDA), to name a few. Within this book, I've taken a pragmatic approach by introducing a process for creating an architecture that combines and simplifies many of the key elements of these different methodologies. This process consists of six stages, where each stage builds upon the results of the previous stages. These stages are:

*Stage 1* Have a solid technical foundation.
*Stage 2* Understand the Architecture Business Cycles (ABCs) of embedded systems.
*Stage 3* Define the architectural patterns and reference models.
*Stage 4* Create the architectural structures.
*Stage 5* Document the architecture.
*Stage 6* Analyze and evaluate the architecture.

These six stages can serve as a basis for further study of one of the many, more complex, architectural design methodologies in the industry. However, if given a limited amount of time and resources to devote to full-time architectural methodology studies of the many industry schemes before having to begin the design of a real product, these six stages can be used directly as a simple model for creating an architecture. The remainder of this chapter will provide more details on the six stages.

**Author Note**

This book attempts to give a pragmatic process for creating an embedded systems architecture based upon some of the mechanisms that exist in the more complex industry approaches. I try to avoid using a lot of the specific terminology associated with these various methodologies because the same terms across different approaches can have different definitions as well as different terms possibly having identical meanings.

### *11.1.1 Stage 1: Have a Solid Technical Foundation*

In short, Stage 1 is about understanding the material presented in Chapters 2–10 of this book. Regardless of what portion of the embedded system an engineer or programmer will develop or work on, it is useful and practical to understand at a systems engineering level *all of the elements that can be implemented in an embedded system.* This includes the possible permutations of both the hardware and software as represented in the embedded systems

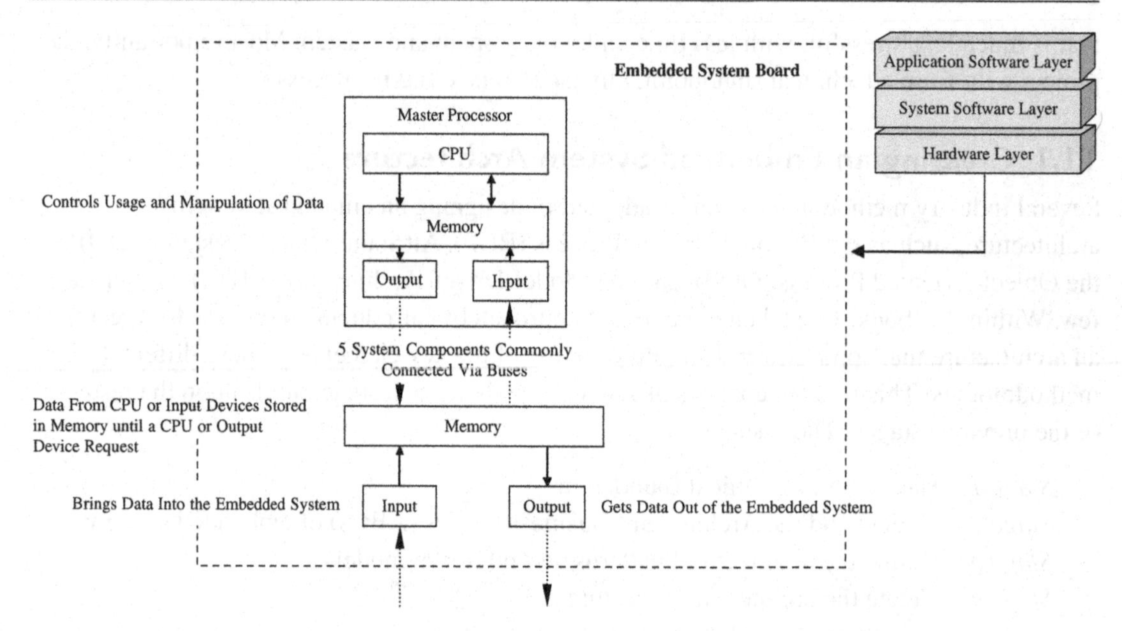


**Figure 11-2a**
von Neumann and Embedded Systems Model diagrams.

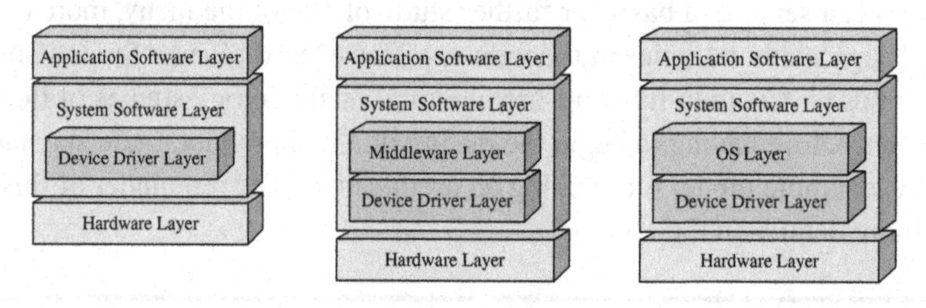


**Figure 11-2b**
System software layer and Embedded Systems Model diagrams.

model, such as the von Neumann model, reflecting the major components that can be found on an embedded board (see Figure 11-2a) or the possible complexities that can exist in the system software layer (see Figure 11-2b).

### 11.1.2 Stage 2: Understand the Architecture Business Cycles of Embedded Systems

The *Architecture Business Cycle (ABC)*[2] of an embedded device, shown in Figure 11-3, is the cycle of influences that impact the architecture of an embedded system and the influences that the embedded system in turn has on the environment in which it is built. These influences can be technical, business-oriented, political, or social. In short, the *ABCs of embedded systems* are the *many* different types of influences that generate the requirements of the system; the requirements in turn generate the architecture, the architecture then produces the

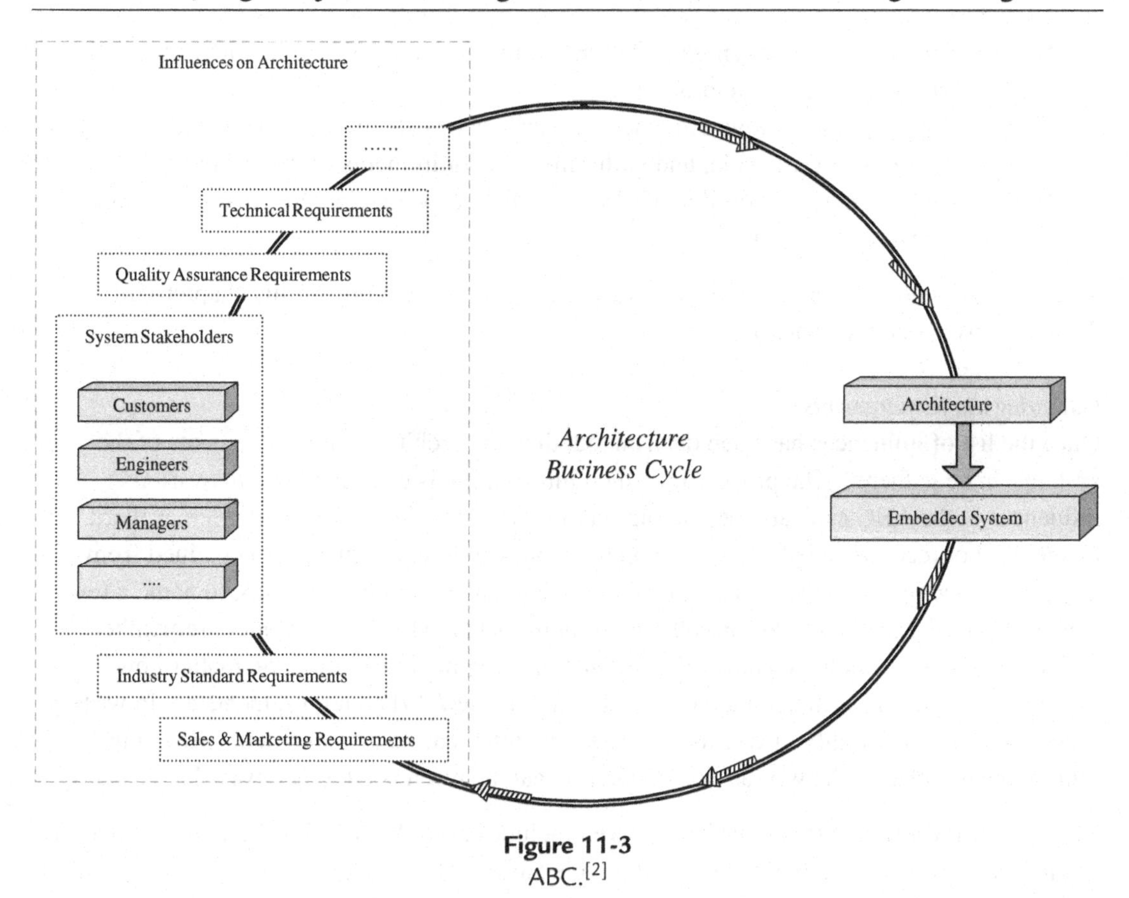


**Figure 11-3**
ABC.[2]

system, and the resulting system in turn provides requirements and capabilities back to the organization for future embedded designs.

What this model implies is that, for better or worse, architectures are not designed on technical requirements alone. For example, given the same type of embedded system, such as a cell phone or TV, with the exact same technical requirements designed by different design teams, the different architectures produced incorporate different processors, operating systems (OSs), and other elements. An engineer that recognizes this from the start will have much more success in creating the architecture for an embedded system. If the responsible architects of an embedded system identify, understand, and engage the various influences on a design at the start of the project, it is less likely that any of these influences will later demand design changes or delays after a lot of time, money, and effort has gone into developing the original architecture.

The steps of Stage 2 include:

*Step 1* Understanding that the ABC influences drive the requirements of an embedded system and that these influences are not limited to technical ones.

*Step 2* Specifically identifying all the ABC influences on the design, whether technical, business, political, and/or social.
*Step 3* Engaging the various influences as early as possible in the design and development lifecycle, and gathering the requirements of the system.
*Step 4* Determining the possible hardware and/or software elements that would meet the gathered requirements.

Steps 1 and 2 were introduced in detail above; the next few sections of this chapter will discuss steps 3 and 4 in more depth.

*Gathering the Requirements*

Once the list of influences has been determined, then the architectural requirements of the system can be gathered. The process by which information is obtained from the various influences in the ABC can vary, depending on the project, from the *informal*, such as word-of-mouth (not recommended), to *formal* methods in which requirements are obtained from finite-state machine models, formal specification languages, and/or scenarios, to name a few. Regardless of the method used for gathering requirements, what is important to remember is that information should be gathered *in writing* and any documentation, no matter how informal (even if it is written on a napkin), should be *saved*. When requirements are in writing, it decreases the probability of confusion or past communication disagreements involving requirements, because the written documentation can be referenced to resolve related issues.

The kind of information that must be gathered includes both the functional and non-functional requirements of the system. Because of the wide variety of embedded systems, it is difficult in this book to provide a list of functional requirements that could apply to all embedded systems. Non-functional requirements, on the other hand, can apply to a wide variety of embedded systems and will be used as real-world examples later in this chapter. Furthermore, from non-functional requirements certain functional requirements can be derived. This can be useful for those who have no specific functional requirements at the start of a project and only have a general concept of what the device to be designed should be able to do. Some of the most useful methods for deriving and understanding non-functional requirements are through outlining *general ABC features* and utilizing *prototypes*.

*General ABC features* are the characteristics of a device that the various "influence types" require. This means that the non-functional requirements of the device are based upon general ABC features. In fact, because most embedded systems commonly require some combination of general ABC features, they can be used as a starting point in defining and capturing system requirements for just about any embedded system. Some of the most common features acquired from various general ABC influences are shown in Table 11-1.

Another useful tool in understanding, capturing, and modeling system requirements is through utilizing a system *prototype*—a physically running model containing some

**Table 11-1: Examples of general ABC features**

| Influence | Feature | Description |
|---|---|---|
| Business (Sales, Marketing, Executive Management, etc.) | Salability | How the device will sell, will it sell, how many will it sell, etc. |
| | Time-to-market | When will the device be delivered with what technical features, etc. |
| | Costs | How much will the device cost to develop, how much can it sell for, is there overhead, how much for technical support once device is in field, etc. |
| | Device lifetime | How long the device will be on the market, how long will the device be functional in the field, etc. |
| | Target market | What type of device is it, who will buy it, etc. |
| | Schedule | When will it be in production, when will it be ready to be deployed to the marketplace, when does it have to be completed, etc. |
| | Capability | Specifying the list of features that the device needs to have for the target market, understanding what the device actually can do once it ships from production, are there any serious bugs in the shipping product, etc. |
| | Risks | Risks of lawsuits over device features or malfunctions, missing scheduled releases, not meeting customer expectations, etc. |
| Technical | Performance | Having the device appear to the user to be functioning fast enough, having the device do what it is supposed to be doing, throughput of processor, etc. |
| | User-friendliness | How easy it is to use, pleasant or exciting graphics, etc. |
| | Modifiability | How fast it can be modified for bug fixes or upgrades, how simple it is to modify, etc. |
| | Security | Is it safe from competitors, hackers, worms, viruses, and even idiot-proof, etc. |
| | Reliability | Does it crash or hang, how often does it crash or hang, what happens if it crashes or hangs, what can cause it to crash or hang, etc. |
| | Portability | How simple is it to run the applications on different hardware, on different system software, etc. |
| | Testability | How easily can the system be tested, what features can be tested, how can they be tested, are there any built in features to allow testing, etc. |
| | Availability | Will any of the commercial software or hardware implemented in the system be available when needed, when will they be available, what are the reputations of vendors, etc. |
| | Standards | See Industry below |
| | Schedule | See Business above |
| Industry | Standards | Industry standards (introduced in Chapter 2), which may be market specific (TV standards, medical device standards, etc.) or general purpose across different families of devices (programming language standards, networking standards, etc.) |
| Quality Assurance | Testability | See Technical above |
| | Availability | When is the system available to be tested |
| | Schedule | See Business above |
| | Features | See Business above. |
| | Quality assurance Standards | ISO9000, ISO9001, etc. (see Industry above.) |
| Customers | Cost | How much will the device cost, how much will it cost to repair or upgrade, etc. |
| | User-friendliness | See Technical above |
| | Performance | See Technical above |

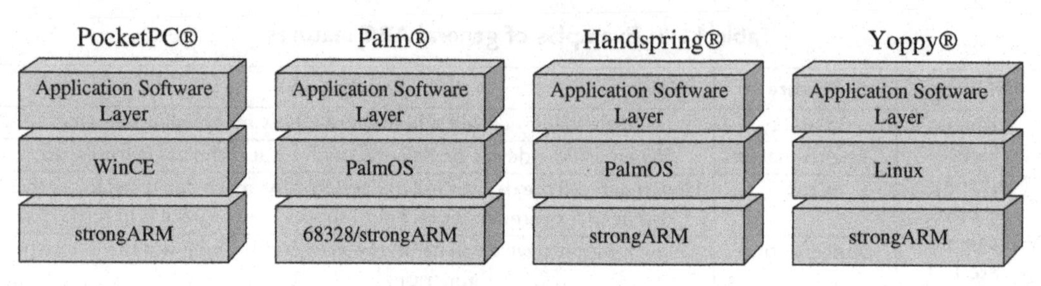


**Figure 11-4**
PDAs.

combination of system requirements. A prototype can be used to define hardware and software elements that could be implemented in a design, and indicate any risks involved in using these elements. Using a prototype in conjunction with the general ABC features allows you to accurately determine early in the project what hardware and software solutions would be the most feasible for the device to be designed.

A prototype can be developed from scratch or can be based upon a currently deployed product in the market, which could be any similar devices or more complex devices that incorporate the desirable functionality. Even devices in other markets may have the desired look and feel even if the applications are not what you are looking for. For example, if you want to design a wireless medical handheld device for doctors, consumer personal data assistants (PDAs) (shown in Figure 11-4) that have been successfully deployed into the market could be studied and adapted to support the requirements and system architecture of the medical device.

When comparing your product to similar designs already on the market, also notice when what was adopted in the design wasn't necessarily the best technical solution for that product. Remember, there could have been a multitude of non-technical reasons, from non-technical influences, for which a particular hardware or software component was implemented.

Some of the main reasons for looking at similar solutions is to save time and money by gathering ideas of what is feasible, what problems or limitations were associated with a particular solution, and, if technically the prototype is a reasonable match, understanding why that is the case. If there really is no device on the market that mirrors any of the requirements of your system, using an *off-the-shelf reference board* and/or *off-the-shelf system software* is another quick method to create your own prototype. Regardless of how the prototype is created, it is a useful tool in modeling and analyzing the design and behavior of a potential architecture.

*Deriving the Hardware and Software from the Requirements*

Understanding and applying the requirements to derive feasible hardware and/or software solutions for a particular design can be accomplished through:

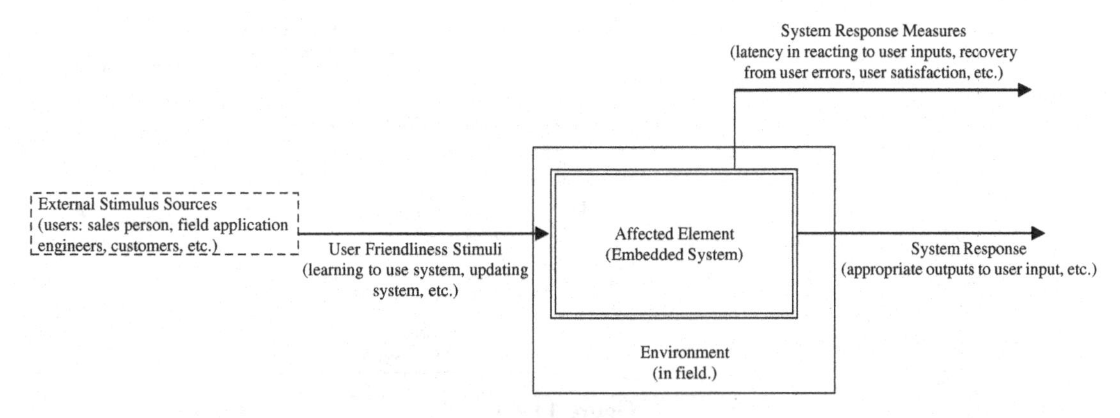


**Figure 11-5**
General ABC user-friendliness scenario.[2]

1. Defining a set of scenarios that outlines each of the requirements.
2. Outlining tactics for each of the scenarios that can be used to bring about the desired system response.
3. Using the tactics as the blueprint for what *functionality* is needed in the device and then deriving a list of specific hardware and/or software elements that contain this functionality.

As shown in Figure 11-5, outlining a scenario means defining:

- The external and internal *stimulus sources* that interact with the embedded system.
- The actions and events, or *stimuli*, that are caused by the stimulus sources.
- The *environment* that the embedded system is in when the stimulus takes place, such as in the field under normal stress, in the factory under high stress, outdoors exposed to extreme temperature, indoors, etc.
- The *elements* of the embedded system that could be affected by the stimulus, whether it is the entire system or the general hardware or software element within such as memory, the master processor, or data.
- The desired *system response* to the stimulus, which reflects one or more system requirements.
- How the system response can be *measured*, meaning how to prove the embedded system meets the requirements.

After outlining the various scenarios, the tactics can then be defined that bring about the desired system response. These tactics can be used to determine what type of functionality is needed in the device. These next several examples demonstrate how hardware and software components can be derived from non-functional requirements based upon performance, security, and testability of general ABC features.

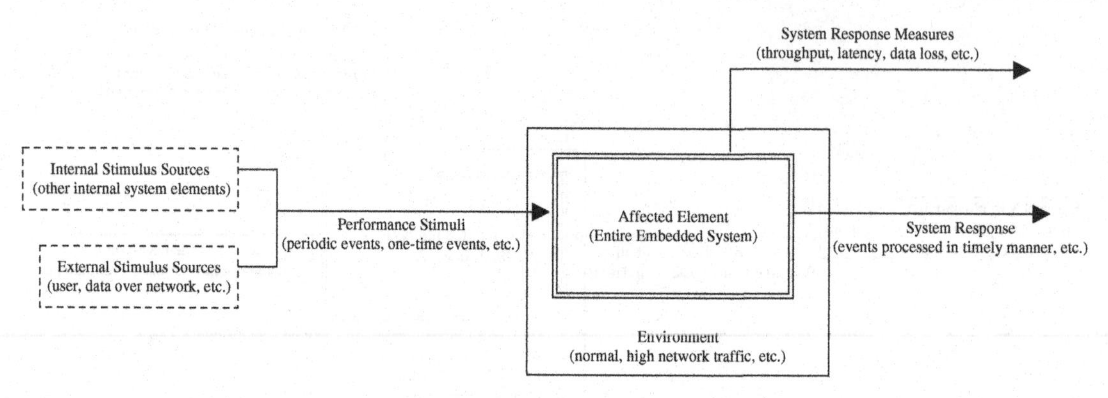


**Figure 11-6a**
General ABC performance scenario.[2]

*Example 1: Performance*

Figure 11-6a shows one possible scenario for a performance-based requirement. In this example, the *stimulus sources* that can impact performance are internal and/or external sources to the embedded system. These stimulus sources can generate one-time and/or periodic asynchronous *events*. According to this scenario, the *environment* in which these events take place occurs when there is normal to a high level of data for the embedded system to process. The stimulus sources generate events that impact the performance of the *entire embedded device*, even if it is only one or a few specific elements within the system that are directly manipulated by the events. This is because typically any performance bottlenecks within the system are perceived by the user as being a performance issue with the entire system.

In this scenario, a desirable system response is for the device to process and respond to events in a timely manner, a possible indicator that the system meets the desired performance requirements. To prove that the performance of the embedded system meets specific performance-based requirements, the system response can be measured and verified via throughput, latency, or data loss system response measures.

Given the performance scenario shown in Figure 11-6a, a method by which to bring about the desired system response is to control the *time period* in which the stimuli are processed and responses generated. In fact, by defining the specific variables that impact the time period, you can then define the tactics needed to control these variables. The tactics can then be used to define the specific elements within an architecture that would implement the functionality of the tactic in order to allow for the desired performance of the device.

For example, *response time*—a system response measure in this scenario—is impacted by the availability and utilization of resources within a device. If there is a lot of contention between multiple events that want access to the same resource, such as events having to block and

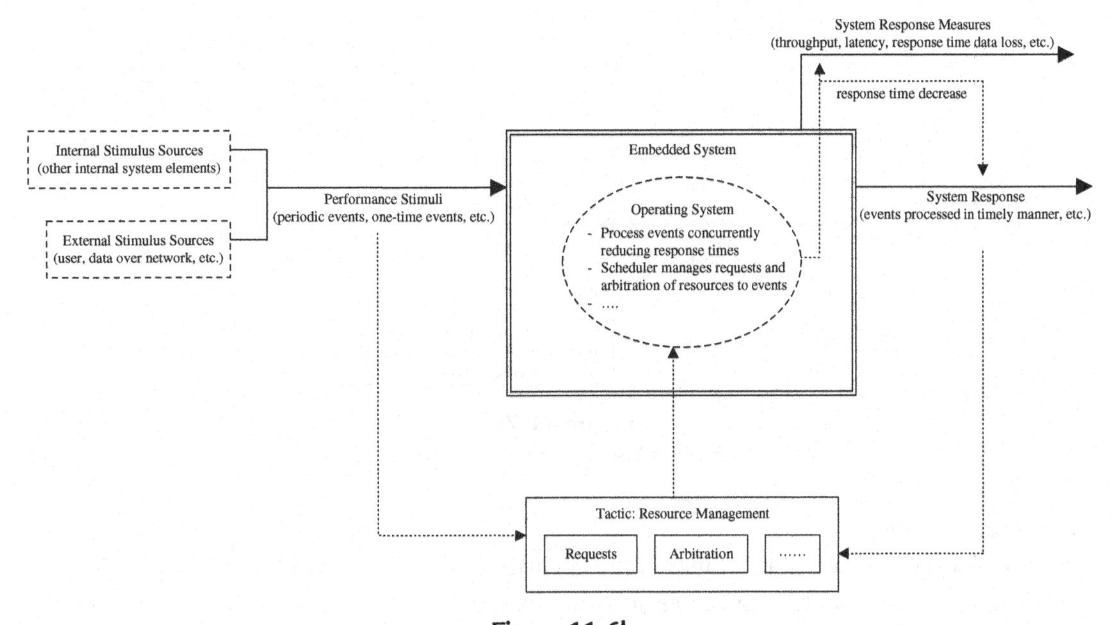


**Figure 11-6b**
Performance tactics and architectural elements.[2]

wait for other events to finish using a resource, the time waiting for the resource impacts the response time. Thus, a *resource management tactic* shown in Figure 11-6b that arbitrates and manages the requests of events allowing for fair and maximum utilization of resources could be used to decrease response time, and increase a system's performance.

A scheduler, such as that found within an OS, is an example of a specific software element that can provide resource management functionality. Thus, it is the OS with the desired scheduling algorithm that is derived for the architecture in this scenario example. In short, this example demonstrates that given the stimuli (events) and a desired system response (good performance), a tactic can be derived (resource management) to achieve the desired system response (good performance) measurable via a system response measure (response time). The functionality behind this tactic, resource management, can then be implemented via an OS through its scheduling and process management schemes.

**Author Note**

It is at this point where Stage 1 (Have a Solid Technical Foundation) is critical. In order to determine what software or hardware elements could support a tactic, one needs to be familiar with the hardware and software elements available to go into an embedded system and the functionality of these elements. Without this knowledge, the results of this stage could be disastrous to a project.

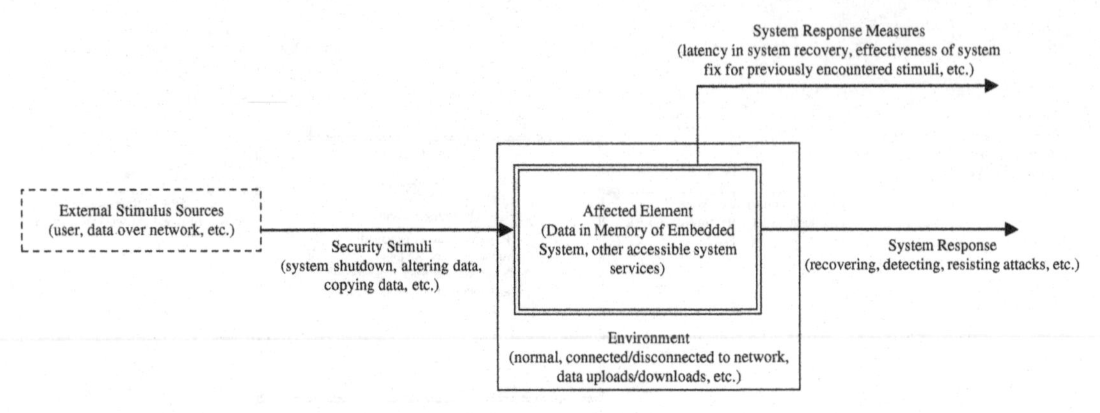


**Figure 11-7a**
General ABC security scenario.[2]

*Example 2: Security*
Figure 11-7a shows a possible scenario for a security-based requirement. In this example, the *stimulus sources* that can impact security are external, such as a hacker or a virus. These external sources can generate *events* that would access system resources, such as the contents of memory. According to this scenario, the *environment* in which these events can take place occurs when the embedded device is in the field connected to a network, doing uploads/ downloads of data. In this example, these stimulus sources generate events that impact the security of anything in *main memory* or any *system resource* accessible to the stimulus sources.

In this scenario, desirable system responses for the embedded device include defending against, recovering from, and resisting a system attack. The level and effectiveness of system security is measured in this example by such system response measures as determining how often security breaches (if any) occur, how long it takes the device to recover from a security breach, and its ability to detect and defend against future security attacks. Given the security scenario shown in Figure 11-7a, one method by which to manipulate an embedded system's response so that it can resist a system attack is to control the *access* external sources have to internal system resources.

To manipulate access to system resources, one could control the variables that impact system access through the authentication of external sources accessing the system and through limiting access to a system's resources to unharmful external sources. Thus, the *authorization* and *authentication* tactics shown in Figure 11-7b could be used to allow a device to track external sources accessing the device and then deny access to harmful external sources, thereby increasing system security.

Given a device resource impacted by a security breach (e.g., main memory), memory and process management schemes—such as those found within an OS, security application

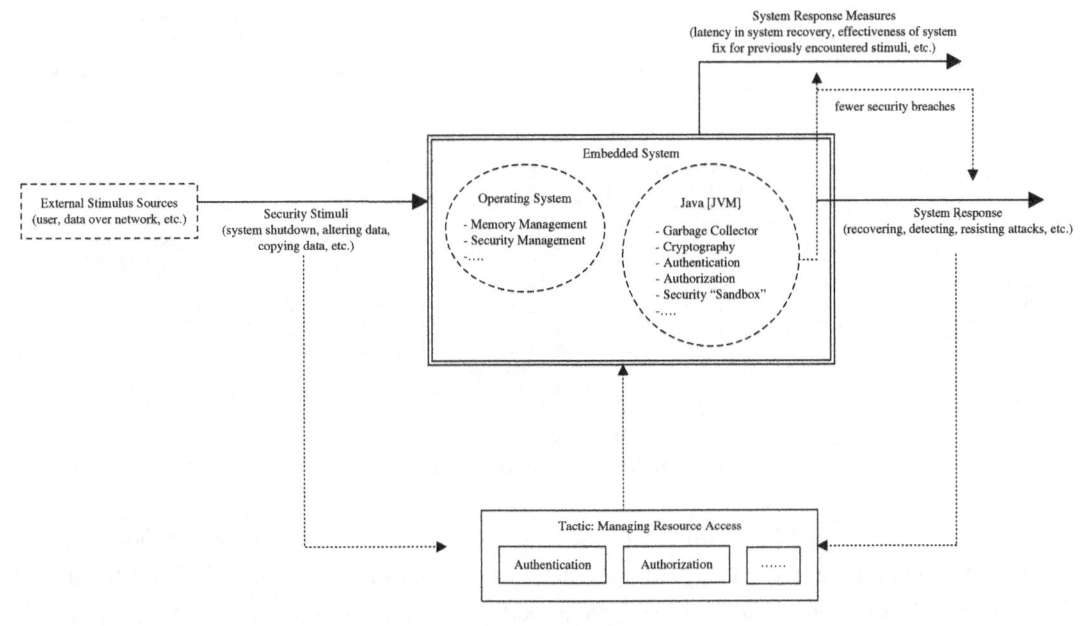


**Figure 11-7b**
Security tactics and architectural elements.[2]

program interfaces (APIs) and memory allocation/garbage collection schemes included when using certain higher-level programming languages such as Java, and network security protocols—are examples of software and hardware elements that can support managing access to memory resources. In short, this example shows that given the stimuli (attempts to access/delete/create unauthorized data) and a desired system response (detect, resist, recover from attacks), a tactic can be derived (managing access to resources) to achieve the desired system response (detect, resist, recover from attacks) measured via a system response measure (occurrences of security breaches).

*Example 3: Testability*

Figure 11-8a shows a possible scenario for a testability-based requirement. In this example, the *stimulus sources* that can impact testability are internal and external. These sources can generate *events* when hardware and software elements within the embedded system have been completed or updated, and are ready to be tested. According to the scenario in this example, the *environment* in which these events take place occurs when the device is in development, during manufacturing, or when it has been deployed to the field. The *affected elements* within the embedded system can be any individual hardware or software element, or the entire embedded device as a whole.

In this scenario, the desired system response is the easily controlled and observable responses to tests. The testability of the system is measured by such system response measures as the

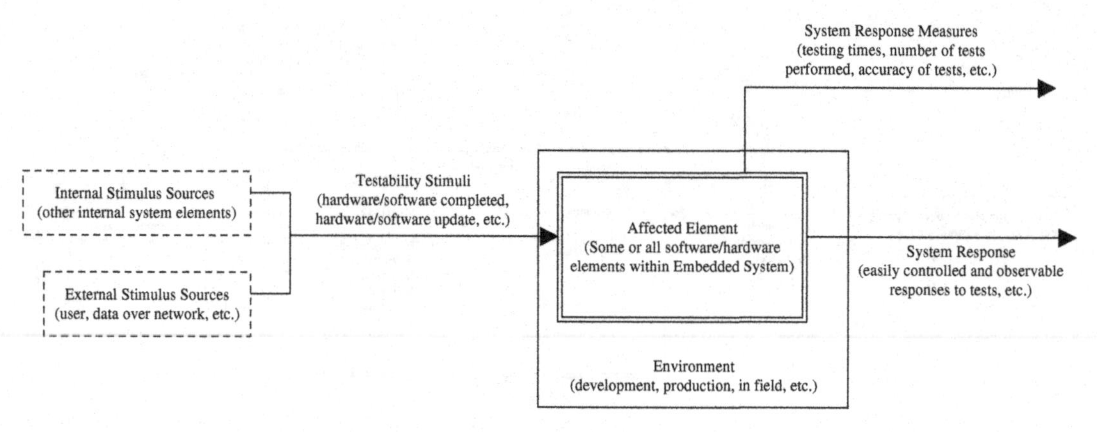


**Figure 11-8a**
General ABC testability scenario.[2]

number of tests performed, the accuracy of the tests, how long tests take to run, verifying data in registers or in main memory, and whether the actual results of the tests match the specifications. Given the testability scenario in Figure 11-8a, a method in which to manipulate an embedded system's response so that the responses to tests are controllable and observable is to provide accessibility for stimulus sources to the internal workings of the embedded system.

To provide accessibility into the internal workings of the system, one could control the variables that impact the desired system response, such as the ability to do runtime register and memory dumps to verify data. This means that the internal workings of the system have to be visible to and manipulable by stimulus sources to allow for requesting internal control and status information (states of variables, manipulating variables, memory usage, etc.), and receiving output based on the requests.

Thus, an internal monitoring tactic, as shown in Figure 11-8b, could be used to provide stimulus sources with the ability to monitor the internal workings of the system and allow this internal monitoring mechanism to accept inputs and provide outputs. This tactic increases the system's testability, since how testable a system is typically depends on how visible and accessible the internal workings of the system are.

Built-in monitors, such as those found on various processors or debugging software subroutines integrated into system software that can be called by a debugger on the development system to perform various tests, are examples of elements that can provide internal monitoring of a system. These hardware and software elements are examples of what could be derived from this scenario. In short, this example demonstrates that given the stimuli (element completed and ready to be tested) and a desired system response (easily test the element and observe results), a tactic can then be derived (internal monitoring of system) to achieve the desired system response (easily test the element and observe results) measurable via a system response measure (testing results, test times, testing accuracy, etc.).

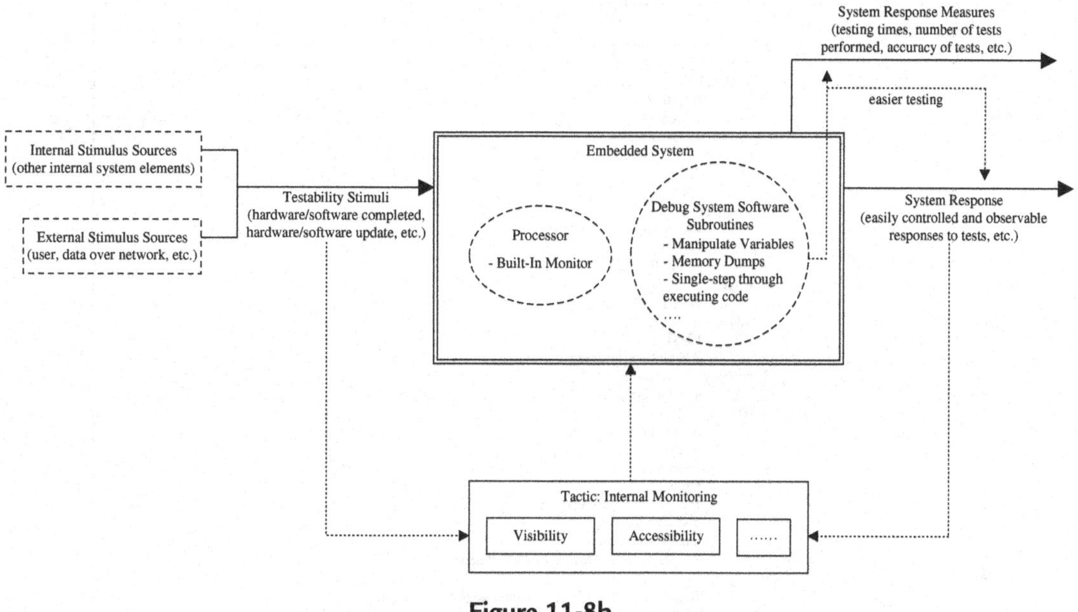


**Figure 11-8b**
General ABC testability scenario.[2]

> **Author Note**
>
> While these examples explicitly demonstrate how elements within an architecture can be derived from general requirements (via their scenarios and tactics), they also implicitly demonstrate that the tactics for one requirement may be counterproductive to another requirement. For example, the functionality that allows for security can impact performance or the functionality that allows for testing accessibility can impact the system's security. Also, note that:
>
> - A requirement can have multiple tactics.
> - A tactic isn't limited to one requirement.
> - The same tactic can be used across a wide variety of requirements.
>
> Keep these points in mind when defining and understanding the requirements of any system.

### 11.1.3 Stage 3: Define the Architectural Patterns and Reference Models

An architectural pattern (also referred to as a *architectural idiom* or *architectural style*) for a particular device is essentially a high-level *profile* of the embedded system. This profile is a description of the various types of software and hardware elements the device could consist of, the functions of these elements within the system, a topological layout of these elements (also known as a *reference model*), and the inter-relationships and external interfaces of

1. Application Layer
    1.1 Browser (Compliance based upon the type of web pages to render)
        1.1.1 International language support (Dutch, German, French, Italian and Spanish, etc.)
        1.1.2 Content type
            HTML 4.0
            Plain text
            HTTP 1.1
        1.1.3 Images
            GIF89a
            JPEG
        1.1.4 Scripting
            JavaScript 1.4
            DOM0
            DHTML
        1.1.5 Applets
            Java 1.1
        1.1.6 Styles
            CSS1
            Absolute positioning, z-index
        1.1.7 Security
            128 bit SSL v3
        1.1.8 UI
            Model Printing
            Scaling
            Panning
            Ciphers
            PNG image format support
            TV safe colors
            anti-aliased fonts
            2D Navigation (Arrow Key) Navigation
        1.1.9 Plug-Ins
            Real Audio Plug-in support and integration on Elate
            MP3 Plug-in support and integration on Elate
            Macromedia Flash Plug-in support and integration on Elate
            ICQ chat Plug-in support and integration on Elate
            Windows Media Player Plug-in support and integration
        1.1.10 Memory Requirement Estimate : 1.25 Mbyte for Browser
            16 Mbyte for rendering web pages
            3-4 Real Audio & MP3 Plug-In...
        1.1.11 System Software Requirement : TCP/IP (for an HTTP Client),...
    1.2 Email Client
        POP3
        IMAP4
        SMTP
        1.2.1 Memory Requirement Estimate : .25 Mbyte for Email Application
            8 Mbyte for managing Emails
        1.2.2 System Software Requirement : UDP-TCP/IP (for POP3, IMAP4, and SMTP),...
    1.3 Video-On-Demand Java Application
        1.3.1 Memory Requirement Estimate : 1 MB for application
            32 MB for running video,...
        1.3.2 System Software Requirement : JVM (to run Java), OS ported to Java, master processor supporting OS and JVM, TCP/IP (sending requests and receiving video),...
    1.4 ....

**Figure 11-9**
DTV-STB profile—application layer.

the various elements. Patterns are based upon the hardware and elements derived from the functional and non-functional requirements via prototypes, scenarios, and tactics.

Figure 11-9 shows an example of architectural pattern information. It starts from the top down in defining the elements for a digital TV set-top box (DTV-STB). This means that it starts with the types of applications that will run on the device, then outlining what system software and hardware these applications implicitly or explicitly require, any constraints in the system, etc.

The system profile can then be leveraged to come up with possible hardware and software reference models for the device that incorporates the relative elements. Figure 11-10 shows a possible reference model for the DTV-STB used as an example in Figure 11-9.

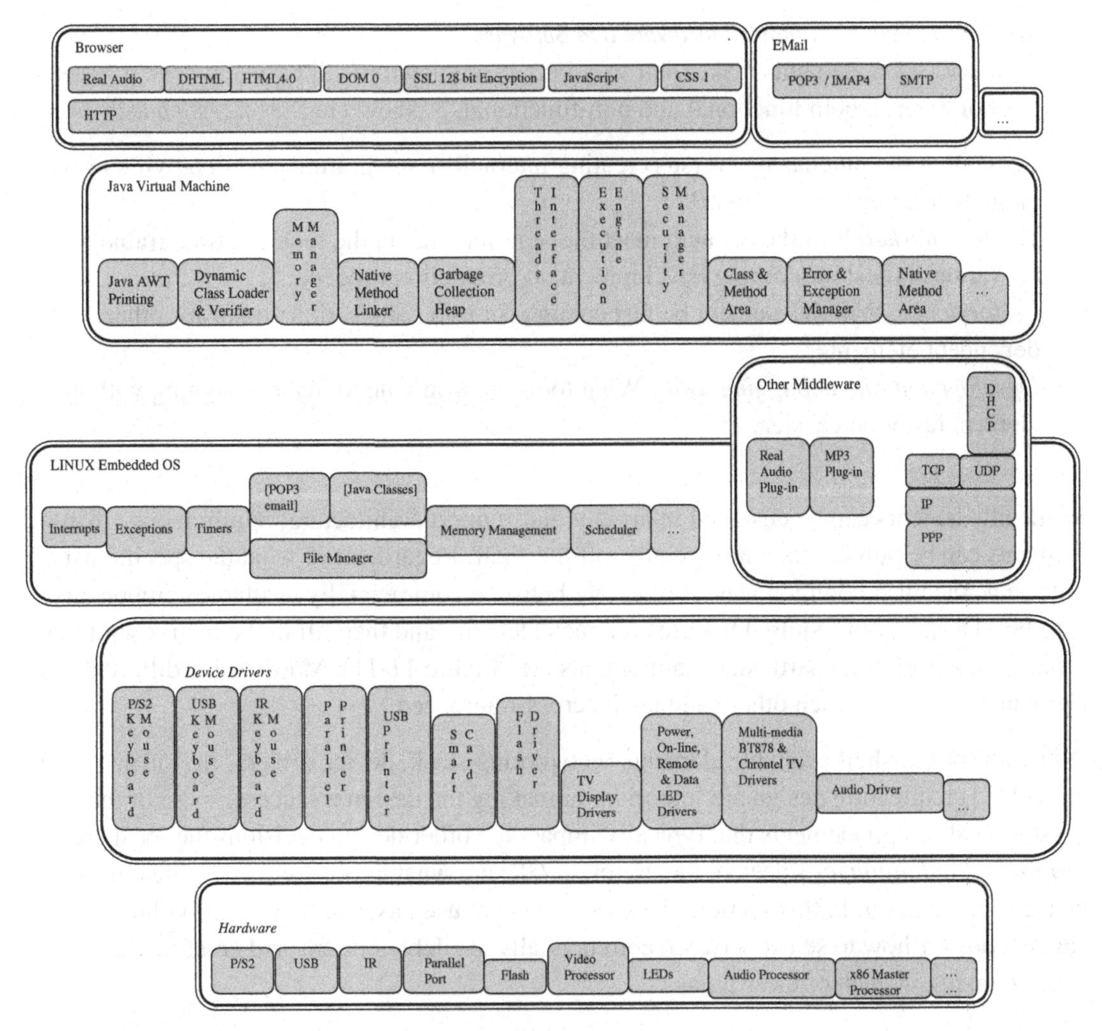


**Figure 11-10**
DTV-STB reference model.

## Author Recommendation

If the software requirements are known, map and decompose as much of the major software elements (OS, Java Virtual Machine (JVM), applications, networking, etc.) as possible before finalizing the hardware elements. This is because the hardware will limit (or enhance) what can be done via software and typically the less that is done in hardware, the cheaper the hardware. Then, while matching the software configuration with the possible master processor and board, come up with different models. This could include removing some functionality or implementing some software components within hardware, for example.

*Selecting Commercially Available Hardware and Software*

Regardless of what elements make their way into an architecture, all typically have to meet a basic set of criteria, both functional and non-functional, as shown in Stage 2, such as:

- *Cost.* Does the purchasing (versus creating internally), integrating, and deploying of the element meet cost restrictions?
- *Time-to-market.* Will the element meet the requirements in the required time frame (at various milestones during development, at production time, etc.)?
- *Performance.* Will the element be fast enough for user satisfaction and/or for other dependent elements?
- *Development and debugging tools.* What tools are available to make designing with the element faster and easier?
- …

While all elements can be designed internally that support architectural requirements, many elements can be purchased commercially, off the shelf. Regardless of what the specific list of criteria is, the most common way of selecting between commercially available components is to build a *matrix* of required features for each element, and then fill in the matrix with the products that fulfill the particular requirements (see Figure 11-11). Matrices for different elements that rely on each other can then be cross-referenced.

While all off-the-shelf available elements (networking stack, device drivers, etc.) in an embedded architecture design are important in making the design a success, some of the most critical design elements that typically impact the other design decisions the most are the *programming languages* selected, the use of an *OS*, and what *master processor* the embedded board is based upon. In this section, these elements are used as examples in providing suggestions for how to select between commercially available options and creating the relative matrix in each of these areas.

| | Requirement 1 | Requirement 2 | Requirement 3 | Requirement … | Requirement "N" |
|---|---|---|---|---|---|
| **Product 1** | YES<br>Features … | NO | NOT YET<br>Next Year | … | … |
| **Product 2** | YES<br>Features … | YES<br>Features … | YES<br>Features … | … | … |
| **Product 3** | NO | YES<br>Features … | NO | … | … |
| **Product 4** | YES<br>Features … | NOT YET<br>In 3 Months | NOT YET<br>In 6 Months | … | … |
| **Product …** | …. | …. | …. | …. | … |
| **Product "N"** | … | … | … | … | … |

**Figure 11-11**
Sample matrix.

*Example 4: Programming Language Selection*

All languages require a compiler to translate source into machine code for that processor, whether it's a 32-, 16-, 8-bit, etc., processor. With 4- and 8-bit based designs that contain kilobytes of memory (total read-only memory (ROM) and random access memory (RAM)), assembly language has traditionally been the language of choice. In systems with more powerful architectures, assembly is typically used for the low-level hardware manipulation or code that must run very fast. Writing code in assembly is always an option and, in fact, most embedded systems implement some assembly code. While assembly is fast, it is typically more difficult to program than higher-level languages; in addition, there is a different assembly language set to learn for each ISA.

C is typically the basis of more complex languages used in embedded systems, such as C++, Java, or Perl. In fact, C is often the language used in more complex embedded devices that have an OS, or that use more complex languages, such as a JVM or scripting languages. This is because OSs, JVMs, and scripting language interpreters, excluding those implemented in non-C applications, are usually written in C.

Using higher-level object-oriented languages, such as C++ or Java, in an embedded device is useful for larger embedded applications where modular code can simplify the design, development, testing, and maintenance of larger applications over procedural coding (of C, for instance). Also, C++ and Java introduce additional out-of-the-box mechanisms, such as security, exception handling, namespace, or type-safety, not traditionally found in the C language. In some cases, market standards may actually require the device to support a particular language (e.g., the Multimedia Home Platform (MHP) specification implemented in certain DTVs requires using Java).

For implementing embedded applications that are hardware and system software independent, Java, .Net languages (C#, Visual Basic, etc.), and scripting languages are the most common higher-level languages of choice. In order to be able to run applications written in any of these languages, a JVM (for Java), the .NET Compact Framework (for C#, Visual Basic, etc.), and an interpreter (for scripting languages, like JavaScript, HTML, or Perl) all need to be included within the architecture. Given the desire for hardware and system software independent applications, the correct APIs must be supported on the device in order for the applications to run. For example, Java applications targeted for larger, more complex devices are typically based upon the Personal Java (pJava) or Java 2 Micro Edition (J2ME) connected device configuration (CDC) APIs, whereas Java applications targeted for smaller, simpler systems would typically expect a J2ME connected limited device configuration (CLDC) supported implementation. Also, if the element is not implemented in hardware (i.e., JVM in a Java processor), then either it must be implemented in the system software stack, either as part of the OS, ported to the OS and to the master processor (as is the case with .NET to WinCE, or a JVM to VxWorks, Linux, etc., on x86, MIPS, strongARM, etc.), or it needs to be implemented as part of the application (HTML, JavaScript in a browser, a JVM in a .exe file on WinCE or .prc file on PalmOS, etc.). Also, note, as with any other significant software element introduced in the architecture, the minimal processing power and memory requirements have to be met in hardware in systems that contain these higher-level language elements in order for the code written in these languages to perform in a reasonable manner, if at all.

In short, what this example attempts to portray is that an embedded system can be based upon a number of different languages (e.g., assembly, C, Java, and HTML in an MHP-based DTV with a browser on an x86 board) or based upon only one language (e.g., assembly in a 21-inch analog TV based on an 8-bit TV microcontroller). As shown in Figure 11-12, the key is creating a matrix that outlines what available languages in the industry meet the functional and non-functional requirements of the device.

| | Real-Time | Fast Performance | MHP-Spec | ATVEF-Spec | Browser Application | ... |
|---|---|---|---|---|---|---|
| Assembly | YES | YES | NOT Required | NOT Required | NOT Required | ... |
| C | YES | YES Slower than assembly | NOT Required | NOT Required | NOT Required | ... |
| C++ | YES | YES Slower than C | NOT Required | NOT Required | NOT Required | ... |
| .NetCE (C#) | NO WinCE NOT RTOS | Depends on processor, slower than C on less powerful processors | NOT Required | NOT Required | NOT Required | ... |
| JVM (Java) | Depends on JVM's Garbage Collector and is OS ported to RTOS | Depends on JVM's byte code processing scheme (WAT almost as fast as C where interpretation requires more power-ful processor, i.e. 200+ MHz), slower than C on slower processors | YES | NOT Required | NOT Required | ... |
| HTML (Scripting) | Depends on what language written in, and the OS (an RTOS in C/assem-bly OK, .NetCE platform no, Java) depends on JVM | Slower because of the interpretation that needs to be done but depends on what language interpreter written in (see above cells of this column) | NOT Re-quired | YES | YES | ... |

**Figure 11-12**
Programming language matrix.

*Example 5: Selecting an OS*

The main questions surrounding the use of an OS within an embedded design are:

1. What type of systems typically use or require an OS?
2. Is an OS needed to fulfill system requirements?
3. What is needed to support an OS in a design?
4. How to select the OS that best fits the requirements?

Embedded devices based upon 32-bit processors (and up) typically have an OS, because these systems are usually more complex and have many more megabytes of code to manage than their 4-, 8-, and 16-bit based counterparts. Sometimes other elements within the architecture require an OS within the system, such as when a JVM or the .NET Compact Framework is implemented in the system software stack. In short, while any master CPU can support some type of kernel, the more complex the device, the more likely it is that a kernel will be used.

Whether an OS is needed or not depends on the requirements and complexity of the system. For example, if the ability to multitask, to schedule tasks in a particular scheme, to manage resources fairly, to manage virtual memory, and to manage a lot of application code is important, then using an OS will not only simplify the entire project, but may be critical to completing it. In order to be able to introduce an OS into any design, the overhead needs to be taken into account (as with any software element introduced), including processing power, memory, and cost. This also means that the OS needs to support the hardware (i.e., master processor).

Selecting an off-the-shelf OS, again, goes back to creating a matrix with requirements and the OS features. The features in this matrix could include:

- *Cost.* When purchasing an OS, many costs need to be taken into consideration. For example, if development tools come with an OS package, there is typically a fee for the tools (which could be a per team or per developer fee), as well as a license fee for the OS. Some OS companies charge a one-time license OS fee, whereas others charge an upfront fee along with royalties (a fee per device being manufactured).
- *Development and debugging tools.* This includes tools that are compatible with or are included in the OS package, such as technical support (a website, a support engineer, or Field Applications Engineer (FAE) to call for help), an IDE (integrated development environment), ICEs (in-circuit emulators), compilers, linkers, simulators, debuggers, etc.
- *Size.* This includes the footprint of the OS on ROM, as well as how much main memory (RAM) is needed when OS is loaded and running. Some OSs can be configured to fit into much less memory by allowing the developer to scale out unnecessary functionality.
- *Non-kernel related libraries.* Many OS vendors entice customers by including additional software with the OS package or optional packages available with the OS (device drivers, file systems, networking stack, etc.) that are integrated with the OS and (usually) ready to run out-of-the-box.

| | Tools | Portability | Non-kernel | Processor | Scheduling Scheme | … |
|---|---|---|---|---|---|---|
| vxWorks | Tornado IDE, SingleStep debugger, … | BSP | Device Drivers w/ BSP, graphics, networking, … | x86, MIPS, 68K, ARM, strongARM, PPC … | Hard Real-Time, Priority-based … | … |
| Linux | Depends on vendor for development IDE, gcc, … | Depends on vendor, some with no BSP | Device Drivers graphics, networking, … | Depends on vendor (x86, PPC, MIPS, …) | Depends on vendor, some are hard real-time, others soft-real time … | … |
| Jbed | Jbed IDE , Sun Java compiler, … | BSP | Device Drivers – the rest depends on JVM specification (graphics, networking, …) | PPC, ARM, … | EDF Hard Real Time Scheduling … | … |

**Figure 11-13**
OS matrix.

- *Standards support.* Various industries may have specific standards for software (such as an OS) that need to meet some safety or security regulations. In some cases, an OS may need to be formally certified. There are also general OS standards (i.e., POSIX) which many embedded OS vendors support.
- *Performance.* See Section 9.6 on OS performance guidelines.
- Of course, there are many other desirable features that can go into the matrix such as process management scheme, processor support, portability, and vendor reputation, but essentially it comes down to taking the time to create the matrix of desired features and OSs as shown in Figure 11-13.

*Example 6: Selecting a Processor*

The different ISA designs (application-specific, general purpose, instruction-level parallelism, etc.) are targeted for various types of devices. However, different processors that fall under different ISAs could be used for the same type of device, given the right components on the board and in the software stack. As the names imply, general purpose ISAs are used in a wide variety of devices and application-specific ISAs are targeted for specific types of devices or devices with specific requirements, where their purposes are typically implied by their names: TV microcontroller for TVs, Java processor for providing Java support, DSPs (digital signal processors) that repeatedly perform fixed computations on data, etc. As mentioned in Chapter 4, instruction-level parallelism processors are typically general purpose processors with better performance (because of their parallel instruction execution mechanisms). Furthermore, in general, 4-bit/8-bit architectures have been used for lower-end embedded systems, and 16-bit/32-bit architectures for higher-end, larger, and more expensive embedded systems.

| | Tools | Java-specific Features | OS Support | ... |
|---|---|---|---|---|
| **aJile aj100 Java Processor (Application Specific ISA)** | JEMBuilder, Charade debugger, | J2ME/CLDC JVM | NOT Needed... | |
| **Motorola PPC823 (General Purpose ISA)** | Tornado tools, Jbed Tools, Sun tools, Abatron BDM... | Implemented in software (Jbed, PERC, CEE-J, ...) | Coming Soon — Linux, vxWorks, Jbed, Nucleus Plus, OSE, ... | ... |
| **Hitachi Camelot Superscaler SoC (Instruction Level Parallel ISA)** | Tornado Tools, QNX Tools, JTAG, ... | Coming Soon — Implemented in software (IBM, OTI, Sun VMs..) | Coming Soon — QNX, vxWorks, WinCE, Linux, .... | ... |

**Figure 11-14**
Processor matrix.

As with any other design decision, selecting a processor means selecting it based on the requirements as well as identifying its impact on the remainder of the system, including software elements, as well as other hardware elements. This is especially important with hardware, since it is the hardware that impacts what enhancements or constraints are implementable in software. What this means is creating the matrix of requirements and processors, and cross-referencing this matrix with the other matrices reflecting the requirements of other system components. Some of the most common features that are considered when selecting a processor include the cost of the processor, power consumption, development and debugging tools, OS support, processor/reference board availability and lifecycle, reputation of vendor, and technical support and documentation (data sheets, manuals, etc.). Figure 11-14 shows a sample matrix for master processor selection for implementing a Java-based system.

### *11.1.4 Stage 4: Create the Architectural Structures*

After Stages 1–3 have been completed, the architecture can be created. This is done by decomposing the entire embedded system into hardware and/or software elements, and then further decomposing the elements that need breaking down. These decompositions are represented as some combination of various types of structures (see Table 1-1 in Chapter 1 for examples of structure types). The patterns defined under Stage 3 that most satisfy the system requirements (the most complete, the most accurate, the most buildable, highest conceptual integrity, etc.) should be used as the foundations for the architectural structures.

It is ultimately left up to the architects of the system to decide which structures to select and how many to implement. While different industry methodologies have different recommendations, a technique favored by some of the most popular methodologies, including the RUP, ADD, and others, is the "4 + 1" model shown in Figure 11-15.

The "4 + 1" model states that a system architect should create *five* concurrent structures per architecture at the very least and each structure should represent a different viewpoint of the system. What is literally meant by "4 + 1" is that *four* of the structures are responsible for

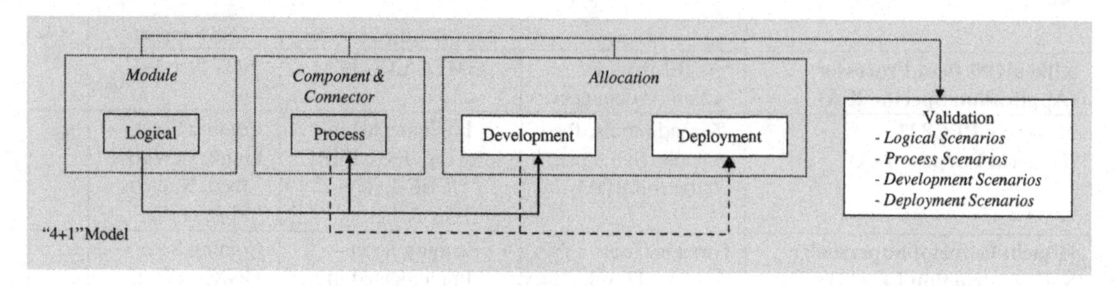


**Figure 11-15**
"4 + 1" model.[2]

capturing the various requirements of the system. The *fifth* structure is used to validate the other four, ensuring that there are no contentions between the structures and that all structures describe the exact same embedded device, from their various viewpoints.

What is specifically shown in Figure 11-15 is that the four base structures of the "4 + 1" model should fall under the *module*, *component & connector*, and *allocation* structural types. Within each of these structural family types, this model after being adapted for use in embedded systems by this author specifically recommends:

*Structure 1* One *logical* structure, which is a modular structure of the key functional hardware and software elements in the system, such as objects for an object-oriented based structure, or a processor, OS, etc. A logical modular structure is recommended because it demonstrates how the key functional hardware and software requirements of the system are being met by displaying the elements that meet the requirements, as well as their inter-relationships. This information can then be leveraged to build the actual system through these functional elements, outlining which functional elements need to be integrated with which other functional elements, as well as outlining what functional requirements are needed by various elements within the system to execute successfully.

*Structure 2* One *process* structure, which is the component and connector structure reflecting the concurrency and synchronization of processes in a system that would contain an OS. A process structure is recommended because it demonstrates how non-functional requirements, such as performance, system integrity, and resource availability, are being met by an OS. This structure provides a snapshot of the system from an OS process standpoint, outlining the processes in the system, the scheduling mechanism, the resource management mechanisms, etc.

Two *allocation* structures:

*Structure 3* A *development* structure describing how the hardware and software map into the development environment. A development structure is recommended because it provides support for non-functional requirements related to the buildability of the hardware and software. This includes information on any constraints of the development environment like the IDE, debuggers, and compilers; complexity of the programming language(s) to be used; and other requirements. It demonstrates this buildability through the mapping of the hardware and software to the development environment.

*Structure 4* A *deployment/physical* structure showing how software maps into the hardware. A *deployment/physical* structure is recommended because, like the process structure, it demonstrates how non-functional requirements such as hardware resource availability, processor throughput, performance, and reliability of the hardware are met by demonstrating how all the software within the device maps to the hardware. This essentially defines the hardware requirements based on the software requirements. This includes the processors executing code/data (processing power), memory storing the code/data, buses transmitting code/data, etc.

As seen from the definitions of these structures, this model assumes that (1) the system has software (development, deployment, and process structures) and (2) the embedded device will include some type of OS (process structure). Essentially, modular structures apply universally regardless of what software components are in the system, or even if there is no software, as is the case with some older embedded system designs. With embedded designs that don't require an OS, then other component and connector structures, such as system resource structures like memory or input/output (I/O), can be substituted that represent some of the functionality that is typically found in an OS, such as memory management or I/O management. As with embedded systems with no OS, for embedded devices with no software, hardware-oriented structures can be substituted for the software-oriented structures.

The "+1" structure, or fifth structure, is the mapping of a subset of the most important scenarios and their tactics that exist in the other four structures. This ensures that the various elements of the four structures are not in conflict with each other, thus validating the entire architectural design. Again, keep in mind that these specific structures are *recommended* for particular types of embedded systems, not *required* by the "4 + 1" model. Furthermore, implementing five structures, as opposed to implementing fewer or more structures, is also a recommendation. These structures can be altered to include additional information reflecting the requirements. Additional structures can be added if they are needed to accurately reflect a view of the system not captured by any of the other structures created. The important thing that the model is trying to relay regarding the number of structures is that it is very difficult to reflect all the information about the system in only one type of structure.

Finally, the arrows to and from the four primary structures shown in Figure 11-15 of the "4 + 1" module represent the fact that, while the various structures are different perspectives of the same embedded system, they are not *independent* of each other. This means that at least one element of a structure is represented as a similar element or some different manifestation in another structure, and it is the sum of all of these structures that makes up the architecture of the embedded system.

**Author Note**

While the "4 + 1" model was originally created to address the creation of a software architecture, it is adaptable and applicable to embedded systems architectural hardware and software design as a whole. In short, the purpose of this model is to act as a tool to determine how to select the right structures and how many to select. Essentially, the same fundamentals of the "4 + 1" model concerning the number, types, and purposes of the various structures can be applied to embedded systems architecture and design regardless of how structural elements are chosen to be represented by the architect or how strictly an architect chooses to abide by the various methodology notations (i.e., symbols representing various architectural elements within a structure) and styles (object oriented, hierarchical layers, etc.).

Many architectural structures and patterns have been defined in various architecture books (do your research), but some useful books include *Software Architecture in Practice* (L. Bass, P. Clements, and R. Kazman, Addison-Wesley, 2003), *A System of Patterns: Pattern-Oriented Software Architecture* (F. Buschmann, R. Meunier, H. Rohnert, P. Sommerlad, and M. Stal, Wiley, 1996), and *Real-Time Design Patterns: Robust Scalable Architecture for Real-Time Systems* (B. P. Douglass, Addison-Wesley, 2003). These can all be applied to embedded systems design.

### 11.1.5 Stage 5: Document the Architecture

Documenting the architecture means documenting all of the structures—the elements, their functions in the system, and their inter-relationships—in a coherent manner. How an architecture is actually documented ultimately depends on the standard practices decided by the team and/or management. A variety of industry methodologies provide recommendations and guidelines for writing architectural specifications. These popular guidelines can be summarized in the following three steps:

*Step 1* *A document outlining the entire architecture.* This step involves creating a table of contents that outlines the information and documentation available for the architecture, such as: an overview of the embedded system, the actual requirements supported by the architecture, the definitions of the various structures, the inter-relationships between the structures, outlining the documentation representing the various structures, and how these documents are laid out (modeling techniques, notation, semantics, etc.).

*Step 2* *A document for each structure.* This document should indicate which requirements are being supported by the structure and how these requirements are being supported by the design, as well as any relative constraints, issues, or open items. This document should also contain graphical and non-graphical (tabular, text, etc.) representation of each of the various elements within the structure. For instance, a graphical representation of the structural elements and relationships would include an index containing a textual summary of the various elements, their behavior, their interfaces, and their relationships to other structural elements. It is also recommended that this document or a related subdocument outline any interfaces or protocols used to communicate to devices outside the embedded system from the point-of-view of the structure. While in embedded systems there is no one template for documenting the various structures and related information, there are popular industry techniques for modeling the various structural related information. Some of the most common include the *Universal Modeling Language (UML)* by the Object Management Group (OMG) that defines the notations and semantics for creating state charts and sequence diagrams that model the behavior of structural elements, and the ADD that, among other templates, provides a template for writing the interface information. Figure 11-16a–c shows examples of templates that can be used to document the information of an architectural design.

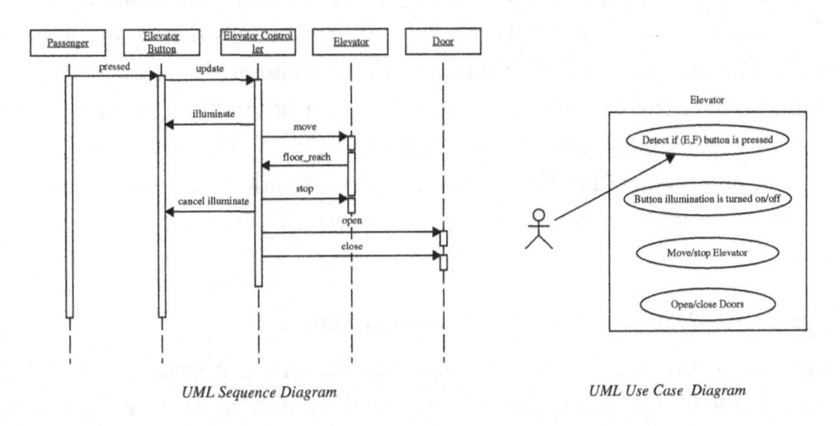


**Figure 11-16a**
UML diagrams.[3]

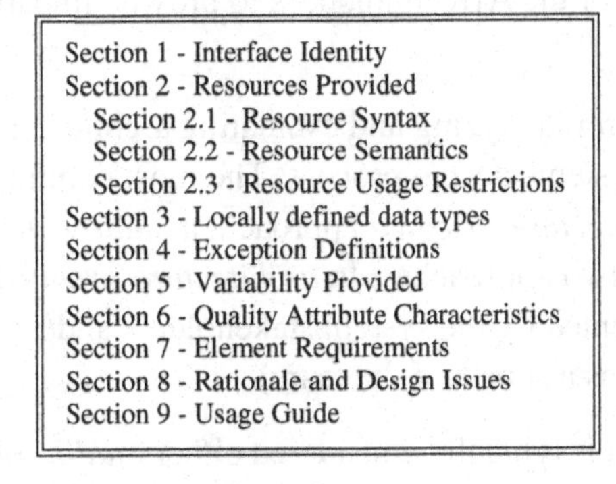

Section 1 - Interface Identity
Section 2 - Resources Provided
  Section 2.1 - Resource Syntax
  Section 2.2 - Resource Semantics
  Section 2.3 - Resource Usage Restrictions
Section 3 - Locally defined data types
Section 4 - Exception Definitions
Section 5 - Variability Provided
Section 6 - Quality Attribute Characteristics
Section 7 - Element Requirements
Section 8 - Rationale and Design Issues
Section 9 - Usage Guide

**Figure 11-16b**
ADD interface template.[2]

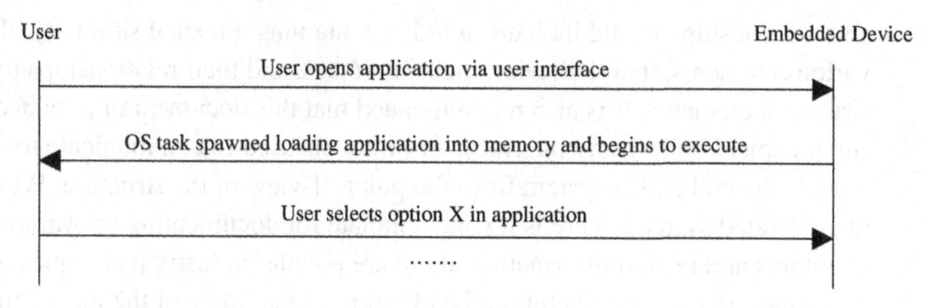


**Figure 11-16c**
(Rough and informal) sequence diagram.

*Step 3* *An architecture glossary*. This document lists and defines all of the technical terms used in all of the architectural documentation.

Regardless of whether the architecture documentation is made up of text and informal diagrams or is based upon a precise UML template, the documentation should reflect the various readers' points-of-view, not only the writer's. This means that it should be useful and unambiguous regardless of whether the reader is a beginner, non-technical, or very technical (it should have high-level "use case" models outlining the various users and how the system can be used, sequence diagrams, state charts, etc.). Also, the various architectural documentation should include the different kinds of information the various readers (stakeholders) need in order to do their analysis and provide feedback.

### 11.1.6 Stage 6: Analyze and Evaluate the Architecture

While there are many purposes for reviewing an architecture, primarily it is to determine if an architecture meets the requirements, and to evaluate the potential risks, and possible failure, of a design long before it has been built. When evaluating an architecture, *who* reviews the architecture must be established, as well as the process of *how* evaluations should occur. In terms of the "who," outside of the architects and stakeholders, the evaluation team should include an engineer outside of the ABC influences to provide impartial perspectives on the design.

There are many techniques for analyzing and evaluating architectures that can be adapted and used in an embedded systems design process. The most common of these approaches typically fall under an *architecture-oriented* approach, a *quality attribute-based* approach, or some *combination* of these two approaches. In *architecture-oriented* approaches, scenarios to be evaluated are implemented by the system stakeholders and/or an evaluation team (with stakeholder representatives as a subset of the team).

A *quality attribute* approach is typically considered either *qualitative* or *quantitative*. Under the qualitative-analysis approach, different architectures with the same *quality attributes*

**Table 11-2: Architecture analysis approaches[2]**

| Methodology | Description |
|---|---|
| Architecture Level Prediction of Software Maintenance (ALPSM) | Maintainability evaluated via scenarios. |
| Architecture Tradeoff Analysis Method (ATAM) | Quality attribute (quantitative) approach that defines problem areas and technical implications of an architecture via question and measurement techniques. Can be used for evaluating many quality attributes. |
| Cost Benefit Analysis Method (CBAM) | Extension of ATAM for defining the economical implications of an architecture. |
| ISO/IEC 9126-1–4 | Architecture analysis standards using internal and external metric models for evaluation (relevant to the functionality, reliability, usability, efficiency, maintainability and portability of devices). |
| Rate Monotonic Analysis (RMA) | Approach which evaluates the real-time behavior of a design. |
| Scenario-Based Architecture Analysis Method (SAAM) | Modifiability evaluated through scenarios defined by stakeholders (an architecture-oriented approach). |
| SAAM Founded on Complex Scenarios (SAAMCS) | SAAM extension–flexibility evaluated via scenarios defined by stakeholders (an architecture-oriented approach). |
| Extending SAAM by Integration in the Domain (ESAAMI) | SAAM extension–modifiability evaluated via scenarios defined by stakeholders (an architecture-oriented approach). |
| Scenario-Based Architecture Reengineering (SBAR) | Variety of quality attributes evaluated via mathematical modeling, scenarios, simulators, objective reasoning (depends on attribute). |
| Software Architecture Analysis Method for Evolution and Reusability (SAAMER) | Evaluation of evolution and reusability via scenarios. |
| Software Architecture Evaluation Model (SAEM) | A quality model that evaluates via different metrics depending on GQM (Goal Question Metrics) technique. |

(also known as features of a system that non-functional requirements are based upon) are compared by an architect and/or by an evaluation team depending on the specific approach. Quantitative-analysis techniques are measurement-based, meaning particular quality attributes of an architecture and associated information are analyzed, as well as models associated with the quality attributes and related information being built. These models, along with associated characteristics, can then be used to determine the best approach to building the system. There is a wide variety of both quality attribute-based and architecture-oriented approaches, some of which are summarized in Table 11-2.

As seen in Table 11-2, some of these approaches analyze only certain types of requirements, whereas others are meant to analyze a wider variety of quality attributes and scenarios. In order for the evaluation to be considered successful, it is important that (1) the members of the evaluation team understand the architecture, such as the patterns and structures, (2) these members understand how the architecture meets the requirements, and (3) everyone on the team agree that the architecture meets the requirements. This can be accomplished via mechanisms introduced in these various analytic and evaluation approaches (see Table 11-2 for examples), general steps including:

*Step 1* Members of the evaluation team obtain copies of the architecture documentation from the responsible architect(s) and it is explained to the various team members the evaluation process, as well as the architecture information within the documentation to be evaluated.

*Step 2* A list of the architectural approaches and patterns is compiled based upon feedback from the members of the evaluation team after they have analyzed the documentation.

*Step 3* The architect(s) and evaluation team members agree upon the exact scenarios derived from the requirements of the system (the team responding with their own inputs of the architect's scenarios: changes, additions, deletions, etc.), and the priorities of the various scenarios are agreed upon in terms of both importance and difficulty of implementation.

*Step 4* The (agreed upon) more difficult and important scenarios are those on which the evaluation team spends the most evaluation time because these scenarios introduce the greatest risks.

*Step 5* Results the evaluation team should include (at the very least) are the (1) uniformly agreed upon list of requirements/scenarios, (2) benefits (i.e., the return-on-investment (ROI) or the ratio of benefit to cost), (3) risks, (4) strengths, (5) problems, and (6) any of the recommended changes to the evaluated architectural design.

## 11.2 Summary

This chapter introduced a simple process for creating an embedded systems architecture that included six major stages: have a solid technical base (Stage 1), understand the ABC of embedded systems (Stage 2), define the architectural patterns and reference models (Stage 3), create the architectural structures (Stage 4), document the architecture (Stage 5), and analyze and evaluate the architecture (Stage 6). In short, this process uses some of the most useful mechanisms of the various popular industry architectural approaches. The reader can use these mechanisms as a starting point for understanding the variety of approaches, as well as for creating an embedded system architecture based upon this simplified, pragmatic methodology.

The next and final chapter in this text, Chapter 12, *The Final Phases of Embedded Design: Implementation and Testing*, discusses the remaining phases of embedded system design: the implementation of the architecture, the testing of the design, and the maintainability issues of a design after deployment.

## Chapter 11: Problems

1. Draw and describe the four phases of the Embedded System Design and Development Lifecycle Model.
2. [a] Of the four phases, which phase is considered the most difficult and important?
   [b] Why?
3. What are the six stages in creating an architecture?
4. [a] What are the ABCs of embedded systems?
   [b] Draw and describe the cycle.
5. List and define the four steps of Stage 2 of creating the architecture.
6. Name four types of influences on the design process of an embedded system.
7. Which method is least recommended for gathering information from ABC influences?
   A. Finite-state machine models.
   B. Scenarios.
   C. By phone.
   D. In an e-mail.
   E. All of the above.
8. Name and describe four examples of general ABC features from five different influences.
9. [a] What is a prototype?
   [b] How can a prototype be useful?
10. What is the difference between a scenario and a tactic?
11. In Figure 11-17, list and define the major components of a scenario.

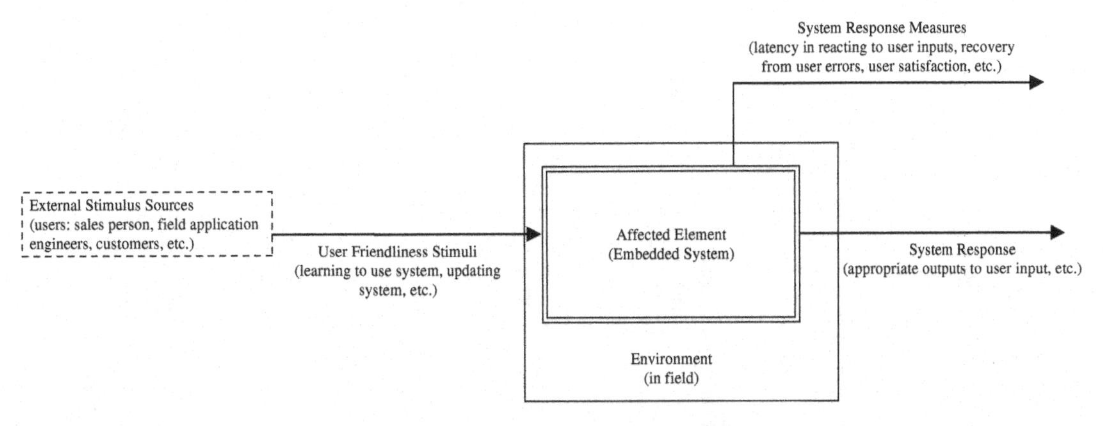


**Figure 11-17**
General ABC user-friendliness scenario.[2]

12. [T/F] A requirement can have multiple tactics.
13. What is the difference between an architectural pattern and a reference model?
14. [a] What is the "4 + 1" model?
    [b] Why is it useful?
    [c] List and define structures that correspond to the "4 + 1" model.

15. [a] What is the process for documenting an architecture?
    [b] How can a particular structure be documented?
16. [a] List and define two common approaches for analyzing and evaluating an architecture.
    [b] Give at least five real-world examples of either.
17. What is the difference between a qualitative and quantitative quality attribute approach?
18. What are the five steps introduced in the text as a method by which to review an architecture?

## Endnotes

[1] The Embedded Systems Design and Development Lifecycle Model is specifically derived from the Software Engineering Institute (SEI)'s Evolutionary Delivery Lifecycle Model and the Software Development Stages Model.

[2] Based on the software architectural brainchildren of the SEI, read *Software Architecture in Practice*, L. Bass, P. Clements, and R. Kazman, Addison-Wesley, 2nd edn, 2003, or go to www.sei.cmu.edu for more information. White papers: "A Survey on Software Architecture Analysis Methods," L. Dobrica and E. Niemela, *IEEE Transactions on Software Engineering*, 28(7), 638–653, 2002; "The 4 + 1 View Model of Architecture," P. Kruchten, *IEEE Software*, 12(6), 42–50, 1995.

[3] http://mini.net/cetus/oo_uml.html#oo_uml_examples

CHAPTER 12

# The Final Phases of Embedded Design: Implementation and Testing

**In This Chapter**

- Defining the key aspects of implementing an embedded systems architecture
- Introducing quality assurance methodologies
- Discussing the maintenance of an embedded system after deployment
- Conclusion of book

## 12.1 Implementing the Design

Having the explicit architecture documentation helps the engineers and programmers on the development team to implement an embedded system that conforms to the requirements. Throughout this book, real-world suggestions have been made for implementing various components of a design that meet these requirements. In addition to understanding these components and recommendations, it is important to understand what *development tools* are available that aid in the implementation of an embedded system. The development and integration of an embedded system's various hardware and software components are made possible through development tools that provide everything from loading software into the hardware to providing complete control over the various system components.

Embedded systems aren't typically developed on one system alone (e.g., the hardware board of the embedded system), but usually require *at least* one other computer system connected to the embedded platform to manage development of that platform. In short, a development environment is typically made up of a *target* (the embedded system being designed) and a *host* (a PC, Sparc Station, or some other computer system where the code is actually developed). The target and host are connected by some transmission medium, whether serial, Ethernet, or other method. Many other tools, such as utility tools to burn EPROMs (erasable programmable read-only memory (ROM)s) or debugging tools, can be used within the development environment in conjunction with the host and target. (See Figure 12-1.)

The key development tools in embedded design can be located on the host or on the target, or can exist standalone. These tools typically fall under one of three categories: *utility*, *translation*, and *debugging* tools. *Utility tools* are general tools that aid in software or

Embedded Systems Architecture.

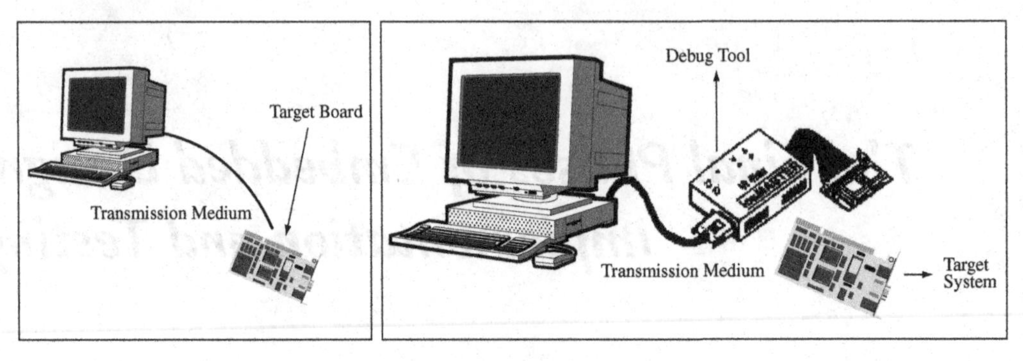


**Figure 12-1**
Development environments.

hardware development, such as editors (for writing source code), VCS (Version Control Software) that manages software files, and ROM burners that allow software to be put onto ROMs. *Translation tools* convert code a developer intends for the target into a form the target can execute, and *debugging tools* can be used to track down and correct bugs in the system. Development tools of all types are as critical to a project as the architecture design, because without the right tools, implementing and debugging the system would be very difficult, if not impossible. In short, an embedded developer needs a solid software tool box to help ensure success. This means asking:

- Will the tool help write better source code, faster?
- Who is actually using what tool?
- Why and how is the tool being used?

**Real-World Advice**

***The Embedded Tools Market***

The embedded tools market is a small, fragmented market, with many different vendors supporting some subset of the available embedded CPUs, operating systems (OSs), Java Virtual Machines (JVMs), etc. No matter how large the vendor, there is not yet a "one-stop shop" where all tools for most of the same type of components can be purchased. Essentially there are many different distributions from many different tool vendors, each supporting their own set of variants or supporting similar sets of variants. Responsible system architects need to do their research and evaluate available tools *before* finalizing their architecture design to ensure that both the right tools are available for developing the system, and the tools are of the necessary quality. Waiting several months for a tool to be ported to your architecture, or for a bug fix from the vendor after development has started, is not a good situation to be in.

*Based on the article "The Trouble with the Embedded Tools Market," J. Ganssle, Embedded Systems Programming, April 2004.*

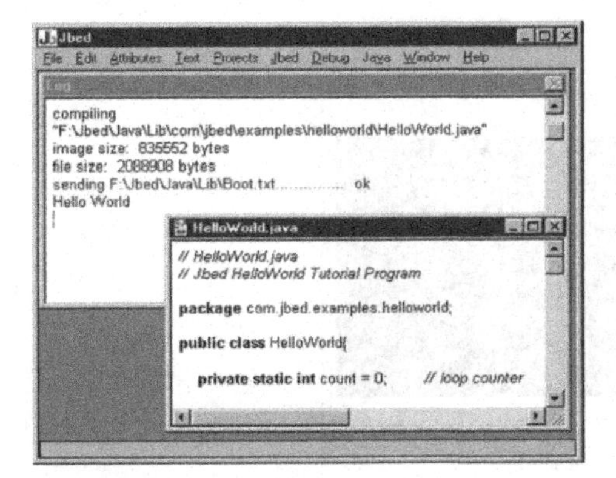


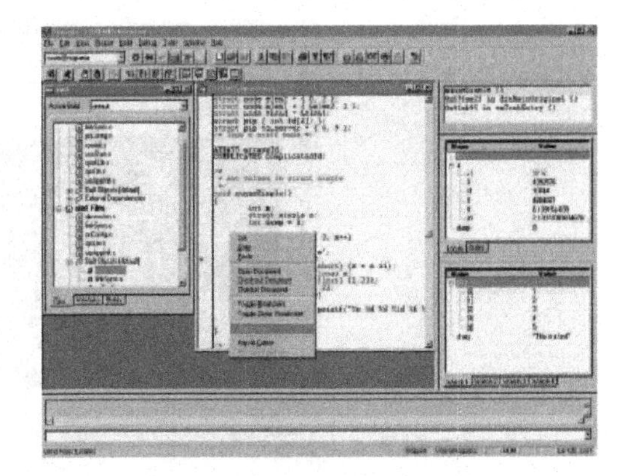

**Figure 12-2**
IDEs.[1]

## 12.1.1 The Main Software Utility Tool: Writing Code in an Editor or Integrated Development Environment (IDE)

Source code is typically written with a tool such as a standard ASCII text editor, or an *IDE* located on the host (development) platform, as shown in Figure 12-2. An IDE is a collection of tools, including an ASCII text editor, integrated into one application user interface. While any ASCII text editor can be used to write any type of code, independent of language and platform, an IDE is specific to the platform and is typically provided by the IDE's vendor, a hardware manufacturer (in a starter kit that bundles the hardware board with tools such as an IDE or text editor), OS vendor, or language vendor (Java, C, etc.).

## 12.1.2 Computer-Aided Design (CAD) and the Hardware

*CAD* tools are commonly used by hardware engineers to simulate circuits at the electrical level in order to study a circuit's behavior under various conditions before they actually build the circuit.

Figure 12-3a is a snapshot of a popular standard circuit simulator, called PSpice. This circuit simulation software is a variation of another circuit simulator that was originally developed at University of California, Berkeley, called SPICE (Simulation Program with Integrated Circuit Emphasis). PSpice is the PC version of SPICE and is an example of a simulator that can do several types of circuit analysis, such as non-linear transient, non-linear DC, linear AC, noise, and distortion, to name a few. As shown in Figure 12-3b, circuits created in this simulator can be made up of a variety of active and/or passive elements. Many commercially available electrical circuit simulator tools are generally similar to PSpice in terms of their overall purpose, and mainly differ in what analysis can be done, what circuit components can be simulated, or the look and feel of the user interface of the tool.

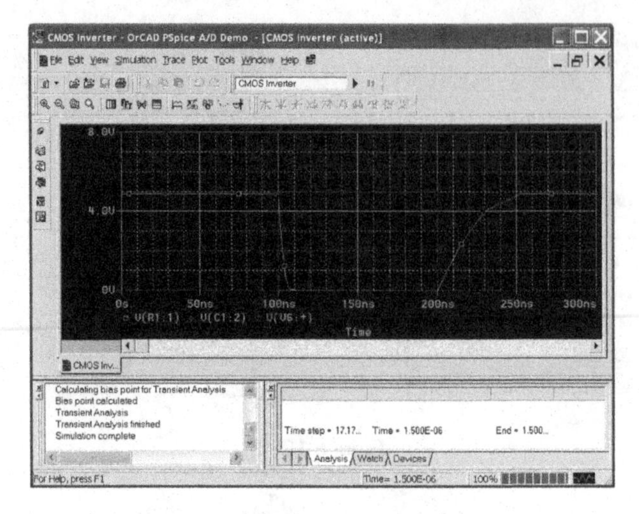


**Figure 12-3a**
PSpice CAD simulation sample.[2]

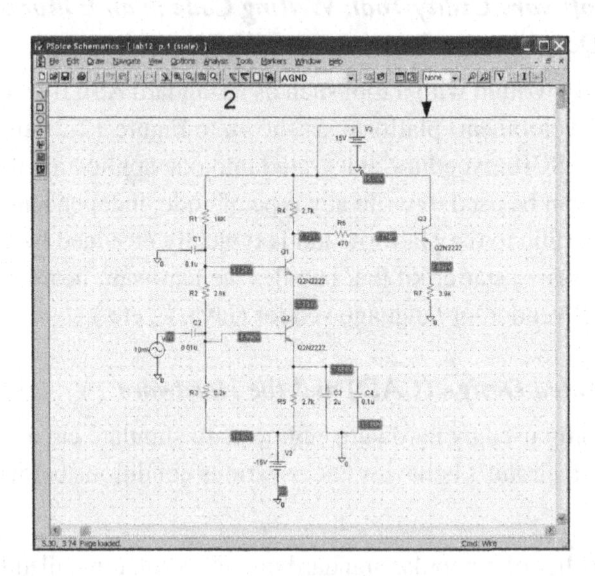


**Figure 12-3b**
PSpice CAD circuit sample.[2]

Because of the importance of and costs associated with designing hardware, there are many industry techniques in which CAD tools are utilized to simulate a circuit. Given a complex set of circuits in a processor or on a board, it is very difficult, if not impossible, to perform a simulation on the whole design, so a hierarchy of *simulators* and *models* is typically used. In fact, the use of models is one of the most critical factors in hardware design, regardless of the efficiency or accuracy of the simulator.

At the highest level, a behavioral model of the entire circuit is created for both analog and digital circuits, and is used to study the behavior of the entire circuit. This behavioral model can be created with a CAD tool that offers this feature or can be written in a standard programming language. Then depending on the type and the makeup of the circuit, additional models are created down to the individual active and passive components of the circuit, as well as for any environmental dependencies (e.g., temperature ) that the circuit may have.

Aside from using some particular method for writing the circuit equations for a specific simulator, such as the tableau approach or modified nodal method, there are simulating techniques for handling complex circuits that include one or some combination of:[1]

- Dividing more complex circuits into smaller circuits and then combining the results.
- Utilizing special characteristics of certain types of circuits.
- Utilizing vector-high-speed and/or parallel computers.

### *12.1.3 Translation Tools—Preprocessors, Interpreters, Compilers, and Linkers*

Translating code was first introduced in Chapter 2, along with a brief introduction to some of the tools used in translating code, including preprocessors, interpreters, compilers, and linkers. As a review, after the source code has been written, it needs to be translated into machine code, since machine code is the only language the hardware can directly execute. All other languages need development tools that generate the corresponding machine code the hardware will understand. This mechanism usually includes one or some combination of *preprocessing*, *translation*, and/or *interpretation* machine code generation techniques. These mechanisms are implemented within a wide variety of translating development tools.

Preprocessing is an optional step that occurs before either the translation or interpretation of source code, and whose functionality is commonly implemented by a *preprocessor*. The preprocessor's role is to organize and restructure the source code to make translation or interpretation of this code easier. The preprocessor can be a separate entity, or can be integrated within the translation or interpretation unit.

Many languages convert source code, either directly or after having been preprocessed, to target code through the use of a *compiler*—a program which generates some target language (machine code, Java byte code, etc.) from the source language (assembly, C, Java, etc.) (see Figure 12-4).

A compiler typically translates all of the source code to a target code at one time. As is usually the case in embedded systems, most compilers are located on the programmer's host machine and generate target code for hardware platforms that differ from the platform the compiler is actually running on. These compilers are commonly referred to as *cross-compilers*. In the case of assembly, an assembly compiler is a specialized cross-compiler referred to as an *assembler* and will always generate machine code. Other high-level language

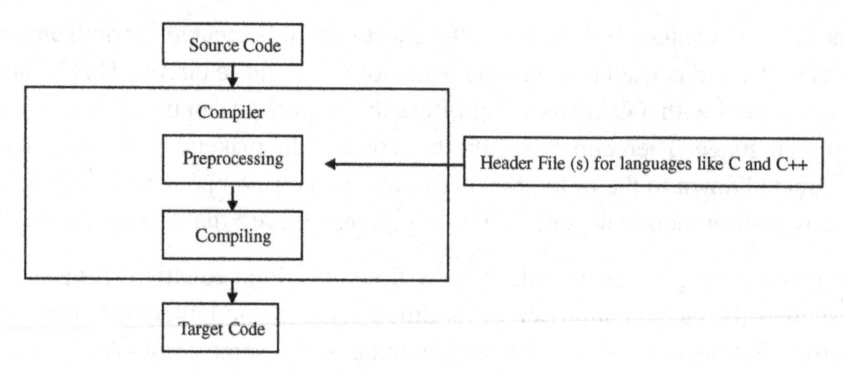


**Figure 12-4**
Compilation diagram.

compilers are commonly referred to by the language name plus "compiler" (e.g., Java compiler, C compiler). High-level language compilers can vary widely in terms of what is generated. Some generate machine code, while others generate other high-level languages, which then require what is produced to be run through at least one more compiler. Still other compilers generate assembly code, which then must be run through an assembler.

After all the compilation on the programmer's host machine is completed, the remaining target code file is commonly referred to as an *object file* and can contain anything from machine code to Java byte code, depending on the programming language used. As shown in Figure 12-5, a *linker* integrates this object file with any other required system libraries, creating what is commonly referred to as an *executable* binary file, either directly onto the board's memory or ready to be transferred to the target embedded system's memory by a *loader.*

One of the fundamental strengths of a translation process is based upon the concept of software placement (also referred to as *object placement*), the ability to divide the software into modules and relocate these modules of code and data anywhere in memory. This is an especially useful feature in embedded systems, because:

1. Embedded designs can contain several different types of physical memory.
2. They typically have a limited amount of memory compared to other types of computer systems.
3. Memory can typically become very fragmented and defragmentation functionality is not available out-of-the-box or is too expensive.
4. Certain types of embedded software may need to be executed from a particular memory location.

This software placement capability can be supported by the master processor, which supplies specialized instructions that can be used to generate "position-independent code," or it could be separated by the software translation tools alone. In either case, this capability depends on whether the assembler/compiler can process only absolute addresses, where the starting

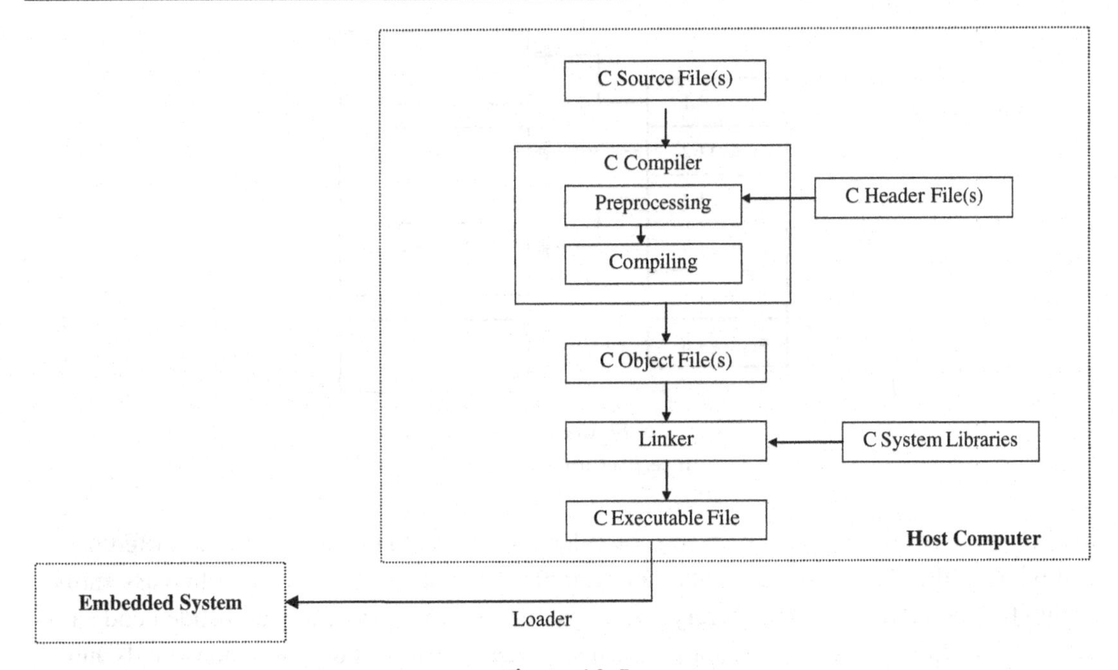


**Figure 12-5**
C example compilation/linking steps and object file results.

address is fixed by software before the assembly processes code, or whether it supports a relative addressing scheme in which the starting address of code can be specified later and where module code is processed relative to the start of the module. Where a compiler/assembler produces relocatable modules or process instruction formats, and may do some translation of relative to physical (absolute) addresses, for example, the remaining translation of relative addresses into physical addresses, essentially the software placement, is done by the linker.

While the IDE, preprocessors, compilers, linkers, etc. reside on the host development system, some languages, such as Java and scripting languages, have compilers or *interpreters* located on the target. An interpreter generates (interprets) machine code one source code line at a time from source code or target code generated by an intermediate compiler on the host system (see Figure 12-6).

An embedded developer can make a big impact in terms of selecting translation tools for a project by understanding how the compiler works and, if there are options, by selecting the strongest possible compiler. This is because the compiler, in large part, determines the size of the *executable* code by how well it translates the code.

This not only means selecting a compiler based on support of the master processor, particular system software, and the remaining toolset (a compiler can be acquired separately, as part of a starter kit from a hardware vendor, and/or integrated within an IDE); it also means selecting a compiler based upon a feature set that optimizes the code's simplicity, speed, and size. These

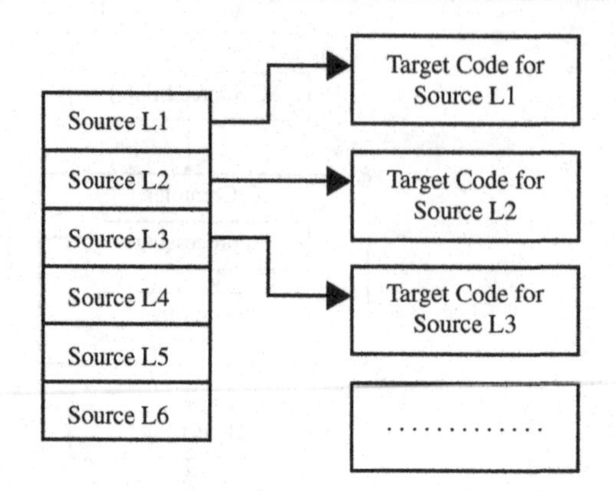


**Figure 12-6**
Interpretation diagram.

features may, of course, differ between compilers of different languages, or even different compilers of the same language, but as an example would include allowing in-line assembly within the source and standard library functions that make programming embedded code a little easier. Optimizing the code for performance means that the compiler understands and makes use of the various features of a particular *Instruction Set Architecture (ISA)*, such as math operations, the register set; knowing the various types of on-chip ROM and random access memory (RAM), the number of clock cycles for various types of accesses, etc. By understanding how the compiler translates the code, a developer can recognize what support is provided by the compiler, and learn, for example, how to program in the higher-level language supported by the compiler in an efficient manner ("*compiler-friendly* code") and when to code in a lower-level, faster language, such as assembly.

**Real-World Advice**

*The Ideal Embedded Compiler*

Embedded systems have unique requirements and constraints atypical of the non-embedded world of PCs and larger systems. In many ways, features and techniques implemented in many embedded compiler designs evolved from the designs of non-embedded compilers. These compilers work fine for non-embedded system development, but don't address the different requirements of embedded systems development, such as limited speed and space. One of the main reasons that assembly is still so prevalent in embedded devices that use higher-level languages is that developers have no *visibility* into what the compiler is doing with the higher-level code. Many embedded compilers provide no information about how the code is generated. Thus, developers have no basis to make programming decisions when using a higher-level language to improve on size and performance. Compiler features that would address some of the needs, such as the size and speed requirements unique to embedded systems design, include:

- A compiler listing file that tags each line of code with estimates of expected execution times, an expected range of execution time, or some type of formula by which to do the calculation (gathered from the target-specific information of the other tools integrated with the compiler).
- A compiler tool that allows the developer to see a line of code in its compiled form and tag any potential problem areas.
- Providing information on size of the code via a precise size map, along with a browser that allows the programmer to see how much memory is being used by particular subroutines.

Keep these useful features in mind when designing or shopping for an embedded compiler.

*Based on the article "Compilers Don't Address Real-Time Concerns," J. Ganssle, Embedded Systems Programming, March 1999.*

### 12.1.4 Debugging Tools

Aside from creating the architecture, debugging code is probably the most difficult task of the development cycle. Debugging is primarily the task of locating and fixing errors within the system. This task is made simpler when the programmer is familiar with the various types of debugging tools available and how they can be used (the type of information shown in Table 12-1).

As seen from some of the descriptions in Table 12-1, debugging tools reside and interconnect in some combination of standalone devices, on the host, and/or on the target board.

**A Quick Comment on Measuring System Performance with Benchmarks**

Aside from debugging tools, once the board is up and running, benchmarks are software programs that are commonly used to measure the performance (latency, efficiency, etc.) of individual features within an embedded system, such as the master processor, the OS, or the JVM. In the case of an OS, for example, performance is measured by how efficiently the master processor is utilized by the scheduling scheme of the OS. The scheduler needs to assign the appropriate time quantum—the time a process gets access to the CPU—to a process, because if the time quantum is too small, thrashing occurs.

The main goal of a benchmark application is to represent a real workload to the system. There are many benchmarking applications available. These include EEMBC (Embedded Microprocessor Benchmark Consortium) benchmarks, the industry standard for evaluating the capabilities of embedded processors, compilers, and Java; Whetstone, which simulates arithmetic-intensive science applications; and Dhrystone, which simulates systems programming applications, used to derive MIPS introduced in Section II. The drawbacks of benchmarks are that they may not be very realistic or reproducible in a real-world design that involves more than one feature of a system. Thus, it is typically much better to use real embedded programs that will be deployed on the system to determine not only the performance of the software, but the overall system performance.

In short, when interpreting benchmarks, ensure you understand exactly what software was run and what the benchmarks did or did not measure.

**Table 12-1: Debug Tools**

| Tool Type | Debugging Tools | Descriptions | Examples of Uses and Drawbacks |
|---|---|---|---|
| Hardware | In-circuit emulator (ICE) | Active device replaces microprocessor in system | • typically most expensive debug solution, but has a lot of debugging capabilities<br>• can operate at the full speed of the processor (depends on ICE) and to the rest of the system it is the microprocessor<br>• allows visibility and modifiability of internal memory, registers, variables, etc., real-time<br>• similar to debuggers, allows setting breakpoints, single stepping, etc.<br>• usually has overlay memory to simulate ROM<br>• processor dependent<br>... |
| | ROM emulator | Active tool replaces ROM with cables connected to dual port RAM within ROM emulator, simulates ROM. It is an intermediate hardware device connected to the target via some cable (i.e., BDM) and connected to the host via another port | • allows modification of contents in ROM (unlike a debugger)<br>• can set breakpoints in ROM code and view ROM code real-time<br>• usually doesn't support on-chip ROM, custom ASICs, etc.<br>• can be integrated with debuggers<br>... |
| | Background debug mode (BDM) | BDM hardware on board (port and integrated debug monitor into master CPU) and debugger on host, connected via a serial cable to BDM port. The connector on cable to BDM port, commonly referred to as wiggler. BDM debugging sometimes referred to as On-Chip Debugging (OCD) | • usually cheaper than ICE, but not as flexible as ICE<br>• observe software execution unobtrusively in real-time<br>• can set breakpoints to stop software execution<br>• allows reading and writing to registers, RAM, I/O ports, etc.<br>• processor/target dependent, Motorola proprietary debug interface<br>... |
| | IEEE 1149.1 Joint Test Action Group (JTAG) | JTAG-compliant hardware on board | • similar to BDM, but not proprietary to specific architecture (is an open standard)<br>... |
| | IEEE-ISTO Nexus 5001 | Options of JTAG port, Nexus-compliant port, or both, several layers of compliance (depending on complexity of master processor, engineering choice, etc.) | • offers scalable debug functions depending on level of compliance of hardware<br>... |

| | | | |
|---|---|---|---|
| | Oscilloscope | Passive analog device that graphs voltage (on vertical axis) versus time (on horizontal axis), detecting the exact voltage at a given time | • monitor up to two signals simultaneously<br>• can set a trigger to capture voltage given specific conditions<br>• used as voltmeter (though a more expensive one)<br>• can verify circuit is working by seeing signal over bus or I/O ports<br>• capture changes in a signal on I/O port to verify segments of software are running, calculate timing from one signal change to next, etc.<br>• processor independent<br>... |
| | Logic analyzer | Passive device that captures and tracks multiple signals simultaneously and can graph them | • can be expensive<br>• typically can only track two voltages (VCC and ground); signals in-between are graphed as either one or the other<br>• can store data (whereas only storage oscilloscopes can store captured data)<br>• two main operating modes (timing, state) to allow triggers on changes of states of signal (i.e., high-to-low or low-to-high)<br>• capture changes in a signal on I/O port to verify segments of software are running, calculate timing from one signal change to next, etc. (timing mode)<br>• can be triggered to capture data from a clock event off the target or an internal logic analyzer clock<br>• can trigger if processor accesses off-limits section of memory, writes invalid data to memory, or accesses a particular type of instruction (state mode)<br>• some will show assembly code, but usually cannot set break point and single-step through code using analyzer<br>• logic analyzer can only access data transmitted externally to and from processor, not the internal memory, registers, etc.<br>• processor independent and allows view of system executing in real time with very little intrusion<br>... |

*(Continued)*

**Table 12-1: (Continued)**

| Tool Type | Debugging Tools | Descriptions | Examples of Uses and Drawbacks |
|---|---|---|---|
| | Voltmeter | Measures voltage difference between two points on circuit | • to measure for particular voltage values<br>• to determine if circuit has any power at all<br>• cheaper than other hardware tools<br>... |
| | Ohmmeter | Measures resistance between two points on circuit | • cheaper than other hardware tools<br>• to measure changes in current/voltage in terms of resistance (Ohm's law: $V = IR$)<br>... |
| | Multimeter | Measures both voltage and resistance | • same as volt and ohm meters<br>... |
| *Software* | Debugger | Functional debugging tool | Depends on the debugger — in general:<br>• loading/single-stepping/tracing code on target<br>• implementing breakpoints to stop software execution<br>• implementing conditional breakpoints to stop if particular condition is met during execution<br>• can modify contents of RAM, typically cannot modify contents of ROM<br>... |
| | Profiler | Collects the timing history of selected variables, registers, etc. | • capture time-dependent (when) behavior of executing software<br>• capture execution pattern (where) of executing software<br>... |
| | Monitor | Debugging interface similar to ICE, with debug software running on target and host. Part of monitor resides in ROM of target board (commonly called debug agent or target agent), and a debugging kernel on the host. Software on host and target typically communicate via serial or Ethernet (depends on what is available on target). | • similar to print statement but faster, less intrusive, works better for soft real-time deadlines, but not for hard real-time<br>• similar functionality to debugger (breakpoints, dumping registers and memory, etc.)<br>• embedded OSs can include monitor for particular architectures<br>... |

<table>
<tr><td></td><td>Instruction set simulator</td><td>Runs on host and simulates master processor and memory (executable binary loaded into simulator as it would be loaded onto target) and mimics the hardware</td><td>• typically does not run at exact same speed of real target, but can estimate response and throughput times by taking into consideration the differences between host and target speeds<br>• verify assembly code is bug free<br>• usually doesn't simulate other hardware that may exist on target, but can allow testing of built-in processor components<br>• can simulate interrupt behavior<br>• capture variable, memory and register values<br>• more easily port code developed on simulator to target hardware<br>• will not precisely simulate the behavior of the actual hardware in real-time<br>• typically better suited for testing algorithms rather than reaction to events external to an architecture or board (waveforms and such need to be simulated via software)<br>• typically cheaper than investing in real hardware and tools<br>...</td></tr>
<tr><td rowspan="2">Manual</td><td colspan="3">Readily available, free or cheaper than other solutions, effective, simpler to use but usually more highly intrusive than other types of tools; not enough control over event selection, isolation, or repeatability. Difficult to debug real-time system if manual method takes too long to execute.</td></tr>
<tr><td>Print statements</td><td>Functional debugging tool, printing statements inserted into code that print variable information, location in code information, etc.</td><td>• to see output of variables, register values, etc. while the code is running<br>• to verify segment of code is being executed<br>• can significantly slow down execution time<br>• can cause missed deadlines in real-time system.<br>...</td></tr>
</table>

(*Continued*)

**Table 12-1: (Continued)**

| Tool Type | Debugging Tools | Descriptions | Examples of Uses and Drawbacks |
|---|---|---|---|
| | Dumps | Functional debugging tool that dumps data into some type of storage structure at runtime | • same as print statements but allows faster execution time in replacing several print statements (especially if there is a filter identifying what specific types of information to dump or what conditions need to be met to dump data into the structure)<br>• see contents of memory at runtime to determine if any stack/heap over-runs<br>… |
| | Counters/timers | Performance and efficiency debugging tool in which counters or timers reset and incremented at various points of code | • collect general execution timing information by working off system clock or counting bus cycles, etc.<br>• some intrusiveness<br>… |
| | Fast display | Functional debugging tool in which LEDs are toggled or simple LCD displays present some data | • similar to print statement but faster, less intrusive, working well for real-time deadlines<br>• allows confirmation that specific parts of code are running<br>… |
| | Output ports | Performance, efficiency, and functional debugging tool in which output port toggled at various points in software | • with an oscilloscope or logic analyzer, can measure when port is toggled and get execution times between toggles of port<br>• same as above but can see on oscilloscope that code is being executed in first place<br>• in multitasking/multithreaded system assign different ports to each thread/task to study behavior<br>… |

Some of these tools are active debugging tools and are intrusive to the running of the embedded system, while other debug tools passively capture the operation of the system with no intrusion as the system is running. Debugging an embedded system usually requires a combination of these tools in order to address all of the different types of problems that can arise during the development process.

**Real-World Advice**

*The Cheapest Way To Debug*

Even with all the available tools, developers should still try to reduce debugging time and costs, because (1) the cost of bugs increases the closer to production and deployment time the schedule gets, and (2) the cost of a bug is logarithmic (it can increase 10-fold when discovered by a customer versus if it had been found during development of the device). Some of the most effective means of reducing debug time and cost include:

- Not developing too quickly and sloppily. The cheapest and fastest way to debug is to *not insert any bugs in the first place.* Fast and sloppy development actually delays the schedule with the amount of time spent on debugging mistakes.
- *System inspections.* This includes hardware and software inspections throughout the development process that ensures that developers are designing according to the architecture specifications, and any other standards required of the engineers. Code or hardware that doesn't meet standards will have to be "debugged" later if system inspections aren't used to flush them out quickly and cheaply (relative to the time spent debugging and fixing all that much more hardware and code later).
- *Don't use faulty hardware or badly written code.* A component is typically ready to be redesigned when the responsible engineer is fearful of making any changes to the offending component.
- *Track the bugs* in a general text file or using one of the many bug tracking off-the-shelf software tools. If components (hardware or software) are continually causing problems, it may be time to redesign that component.
- *Don't skimp on the debugging tools.* One good (albeit more expensive) debugging tool that would cut debug time is worth more than a dozen cheaper tools that, without a lot of time and headaches, can barely track down the type of bugs encountered in the process of designing an embedded system.

And finally what I (the author of this book) believe is one of the best methods by which to reduce debug times and costs: *read the documentation* provided by the vendor and/or responsible engineers first, before trying to run or modify anything. I have heard many, many excuses over the years—from "I didn't know what to read" to "Is there documentation?"—as to why an engineer hasn't read any of the documentation. These same engineers have spent hours, if not days, on individual problems with configuring the hardware or getting a piece of software running correctly. I know that if these engineers had read the documentation in the first place, the problem would have been resolved in seconds or minutes—or might not have occurred at all.

If you are overwhelmed with documentation and don't know what to read first, anything titled along the lines of "Getting Started...," "Booting up the system...," or "README" are good indicators of a place to begin. ☺ Moreover, take the time to read *all* of the documentation provided with any hardware or software to become familiar with what type of information is there, in case it's needed later.

*Based on the article "Firmware Basics for the Boss," Jack Ganssle, Embedded Systems Programming, February 2004.*

### 12.1.5 System Boot-Up

With the development tools ready to go and either a reference board or development board connected to the development host, it is time to start up the system and see what happens. System boot-up means that some type of powerON or reset source, such as an internal/external hard reset (generated by a check-stop error, the software watchdog, a loss of lock by the PLL, debugger, etc.) or an internal/external soft reset (generated by a debugger, application code, etc.), has occurred. When power is applied to an embedded board (because of a reset), start-up code, also referred to as *boot code, bootloader, bootstrap* code, or *BIOS (basic input/output system)* depending on the architecture, in the system's ROM is loaded and executed by the master processor. Some embedded (master) architectures have an internal program counter that is automatically configured with an address in ROM in which the start of the boot-up code (or table) is located, while others are hardware wired to start executing at a specific location in memory.

Boot code differs in length and functionality depending on where in the development cycle the board is, as well as the components of the actual platform that need initialization. The same (minimal) general functions are performed by boot code across the various platforms, which are basically initializing the hardware, which includes disabling interrupts, initializing buses, setting the master and slave processors in a specific state, and initializing memory. This first hardware initialization portion of boot-up code is essentially the executing of the initialization device drivers, as discussed in Chapter 8. How initialization is actually done (i.e., the order in which drivers are executed) is typically outlined by the master architecture documentation or in documentation provided by the manufacturers of the board. After the hardware initialization sequence, executed via initialization device drivers, the remaining system software, if any, is then initialized. This additional code may exist in ROM, for a system that is being shipped out of the factory, or loaded from an external host platform (see the callout box with bootcodeExample).

```
bootcodeExample ()
{
...
       // Serial Port Initialization Device Driver
       initializeRS232(UART, BAUDRATE, DATA_BITS, STOP_BITS, PARITY);
```

```
        // Initialize Networking Device Driver
        initializeEthernet(IPAddress, Subnet, GatewayIP, ServerIP);
        //check for host development system for down loaded file of rest of code to RAM
        // through Ethernet
        // start executing rest of code (define memory map, load OS, etc.)
...
}
```

*MPC823-Based Board Booting Example*

The MPC823 processor contains a reset controller that is responsible for responding to all reset sources. The actions taken by the reset controller differ depending on the source of a reset event, but in general the process includes reconfiguring the hardware and then sampling the data pins or using an internal default constant to determine the initial reset values of system components.

The data pins sample represents initial configuration (setup) parameters as shown in Figure 12-7c.

The Embedded Planet RPXLite Board assumes that on-board ROM (Flash) contains the bootloader monitor/program, called PlanetCore, originally created by Embedded Planet. The PowerPC processor and on-board memory start up in a default configuration set via hardware (CS0 is an output pin that can be configured to be the global chip select for the boot device, HRESET/SRESET, data pins, etc.), and has no dedicated accessible PC register.

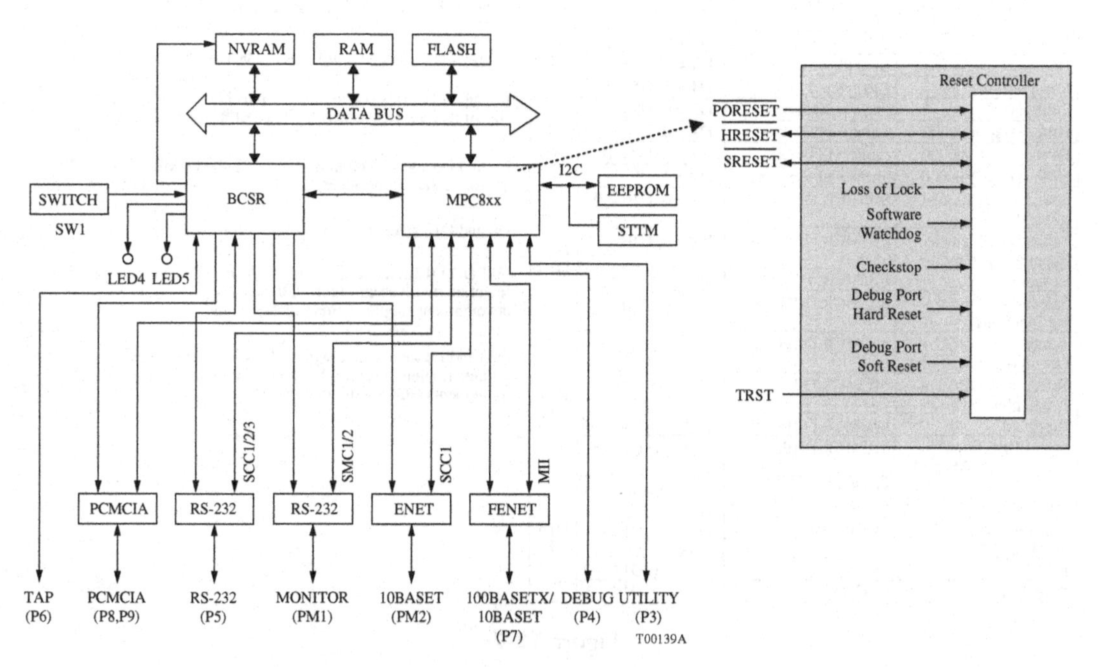


**Figure 12-7a**

Interpretation diagram.[3] © *2004 Freescale Semiconductor, Inc. Used by permission.*

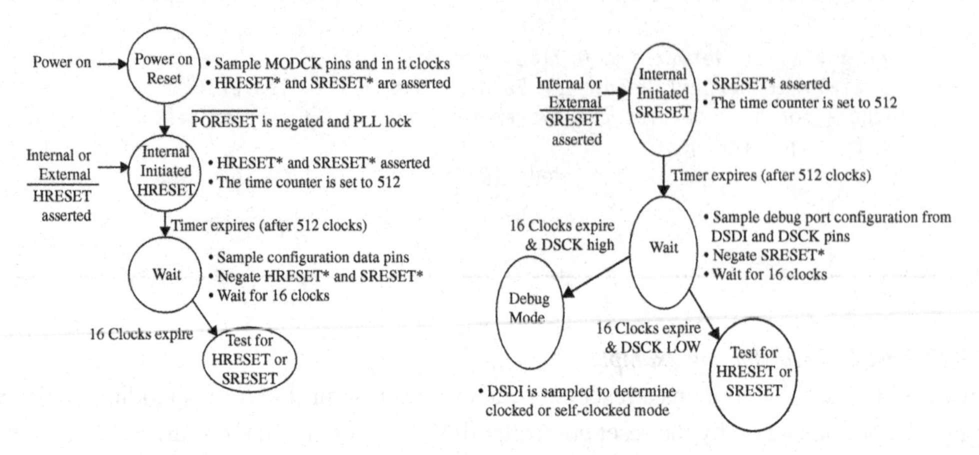


**Figure 12-7b**
Interpretation diagram.[3]

| If.. | | then.. | |
|---|---|---|---|
| No external arbitration | SIUMCR.EARB = 0 | D0 = 0 | D0 |
| External arbitration | SIUMCR.EARB = 1 | D0 = 1 | |
| EVT at 0 | MSR.IP = 0 | D1 = 1 | D1 |
| EVT at 0xFFF00000 | MSR.IP = 1 | D1 = 0 | |
| Do not activate memory controller | BR0.V = 0 | D3 = 1 | D3 |
| Enable CS0 | BR0.V = 1 | D3 = 0 | |
| Boot port size is 32 | BR0.PS = 00 | D4 = 0, D5 = 0 | D4 D5 |
| Boot port size is 8 | BR0.PS = 01 | D4 = 0, D5 = 1 | |
| Boot port size is 16 | BR0.PS = 10 | D4 = 1, D5 = 0 | |
| Reserved | BR0.PS = 11 | D4 = 1, D5 = 1 | |
| DPR at 0 | immr = 0000xxxx | D7 = 0, D8 = 0 | D7 D8 |
| DPR at 0×00F00000 | immr = 00F0xxxx | D7 = 0, D8 = 1 | |
| DPR at 0×FF000000 | immr = FF00xxxx | D7 = 1, D8 = 0 | |
| DPR at 0×FFF00000 | immr = FFF0xxxx | D7 = 1, D8 = 1 | |
| Select PCMCIA functions, Port B | SIUMCR.DBGC = 0 | D9 = 0, D10 = 0 | D9 D10 |
| Select Development Support functions | SIUMCR.DBGC = 1 | D9 = 0, D10 = 1 | |
| Reserved | SIUMCR, DBGC = 2 | D9 = 1, D10 = 0 | |
| Select program tracking functions | SIUMCR.DBGC = 3 | D9 = 1, D10 = 1 | |
| Select as in DBGC + Dev. Supp. comm pins | SIUMCR.DBPC = 0 | D11 = 0, D12 = 0 | D11 D12 |
| Select as in DBGC + JTAG pins | SIUMCR.DBPC = 1 | D11 = 0, D12 = 1 | |
| Reserved | SIUMCR.DBPC = 2 | D11 = 1, D12 = 0 | |
| Select Dev. Supp. comm and JTAG pins | SIUMCR.DBPC = 3 | D11 = 1, D12 = 1 | |
| CLKOUT is GCLK2 divided by 1 | SCCR.EBDF = 0 | D13 = 0, D14 = 0 | D13 D14 |
| CLKOUT is GCLK2 divided by 2 | SCCR.EBDF = 1 | D13 = 0, D14 = 1 | |
| Reserved | SCCR.EBDF = 2 | D13 = 1, D14 = 0 | |
| Reserved | SCCR.EBDF = 3 | D13 = 1, D14 = 1 | |

D0 specifies whether external arbitration or the internal arbiter is to be used.

D1 controls the initial location of the exception vector table, and the IP bit in the machine state register is set accordingly.

D3 specifies whether Chip Select 0 is active on reset.

If Chip Select 0 is active on reset pins D4 and D5 specify the port size of theboot ROM, with a choice of 8, 16, or 32 bits.

D7 and D8 specify the initial value for the IMMR registers. There are four different possible locations for the internal memory map.

D9 and D10 select the configuration for the debug pins.

D11 and D12 select the configuration of the debug port pins. This selection involves configuring these pins either as JTAG pins, or development support communication pins.

D13 and D14 determine which clock scheme is in use; one clock scheme implements GCLK2 divided by one, and the second implements GCLK2 divided by two.

**Figure 12-7c**
Interpretation diagram.[3]

| Chip Select | Port Size | Function/Address | Comment |
|---|---|---|---|
| CS0# | ×32 | FLASH (×32)<br>FFFF FFFF minus actual FLASH size | Reset vector at IP = 1:<br>0000 0100<br>Vector set at IP = 1 in hardware<br>BRO set at FFFF minus Flash size<br>2, 4, 8, or 16 Mbytes |
| CS1# | ×32 | SDRAM (×32)<br>0000 | 16, 32, or 64 Mbytes |
| CS2# | | Expansion Header<br>UUUU | Routed to expansion receptacle |
| CS3# | ×32 | Control and status Registers<br>FA40 | Byte and/or word accessible |
| CS4# | ×8 | NVRAM/RTC or SRAM/RTC<br>FA00 | OK, 32K, 128K, or 512 Kbytes<br>Also available at Expansion Receptacle |
| CS5# | | Expansion Header<br>UUUU | Routed to expansion receptacle |
| CS6# | ×16<br>or<br>U | PCMCIA Slot B Chip Select Even Bytes<br>or<br>Chip Select 6 to I/O Header<br>UUUU | OP2 in MPC850 PCMCIA control register selects mode:<br>L = PCMCIA Slot B enabled<br>H = CS6# to expansion header enabled |
| CS7# | ×16<br>or<br>U | PCMCIA Slot B Chip Select Odd Bytes<br>or<br>Chip Select 7 to I/O Header<br>UUUU | OP2 in MPC850 PCMCIA control register selects mode:<br>L = PCMCIA Slot B enabled<br>H = CS7# to I/O expansion header enabled |
| IMMR | ×32 | Value at rest = FF00 0000,<br>then set to FA20 0000 | |

**Figure 12-7d**

Interpretation diagram.[3]

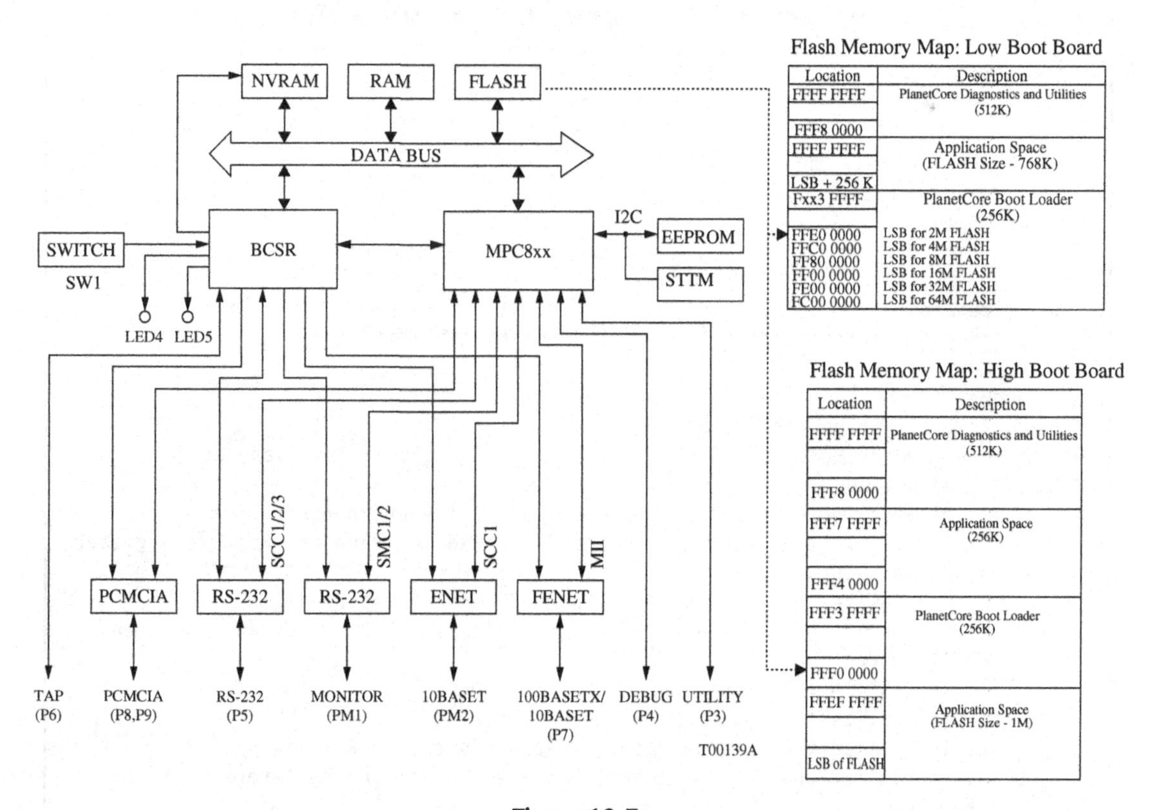


**Figure 12-7e**

Interpretation diagram.[3] *© 2004 Freescale Semiconductor, Inc. Used by permission.*

The default configuration executed via hardware includes the configuration of only one bank of memory, whose base address is determined by D7 and D8, where 00 = 0x00000000, 01 = x00F000000, 10 = 0xFF000000, and 11 = 0xFFF00000. This bank is in some type of ROM (i.e., Flash) and is where the boot code resides. After the board has been powered on, the PowerPC processor executes the boot code in this memory bank to complete the initialization and configuration sequence. In fact, all the MPC8xx processor series (not just the MPC823) require either a high or low boot, depending on the specific board and revision, meaning PlanetCore is located either at the high end of Flash or at the low end of Flash. PlanetCore starts at virtual address 0xFFF00000 if it is located at the high end of Flash. On the other hand, PlanetCore is in the first sectors of the Flash (i.e., located at virtual address 0xFC000000 for 64 MB of Flash) if it is located at the low end of Flash.

On this MPC823-based board, after the hardware initialization sequence initializing the processor, the CPU begins executing the PlanetCore bootloader code. As shown in the following callout box, all of the hardware specific to the MPC823 architecture as well as specific to the board is initialized (serial, networking, etc.) via this type of boot code.

```
/****************************************************************
 * c_entry
 * Description :
 * -------------
 *
 * First C-function
 *
 * Return values :
 * ---------------
 *
 * Never returns
****************************************************************/
int c_entry(void){
BootLoader
- Board initialization for custom BSP
Initializing the MPC823, itself (not board initialization), involves about 24
steps, which includes :
        1. Disable the data cache to prevent a machine check error from occurring.
        2. Initialize the Machine State Register and the Save and Restore Register
        1 with a value of 0x1002.
        3. Initialize the Instruction Support Control Register, ICTRL,
        modifying it so that the core is not serialized (which has an impact on
        performance).
        4. Initialize the Debug Enable Register, DER.
        5. Initialize the Interrupt Cause Register, ICR.
        6. Initialize the Internal Memory Map Register, IMMR.
        7. Initialize the Memory Controller Base and Options registers as
        required.
```

```
        8. Initialize the Memory Periodic Timer Pre-scalar Register, MPTPR.
        9. Initialize the Machine Mode Registers, MAMR and MBMR.
       10. Initialize the SIU Module Configuration Register, SIUMCR. Note that
           this step configures many of the pins shown on the right hand side of
           the main pin diagram in the User Manual.
       11. Initialize the System Protection Register, SYPCR. This register
           contains settings for the bus monitor and the software watchdog.
       12. Initialize the Time Base Control and Status Register, TBSCR.
       13. Initialize the Real Time Clock Status and Control Register, RTCSC.
       14. Initialize the Periodic Interrupt Timer Register, PISCR.
       15. Initialize the UPM RAM arrays using the Memory Command Register and
           the Memory Data Register. We also discuss this routine in the chapter
           regarding the memory controller.
       16. Initialize the PLL Low Power and Reset Control Register, PLPRCR.
       17. Is not required, although many programmers implement this step. This
           step moves the ROM vector table to the RAM vector table.
       18. Changes the location of the vector table. The example shows this
           procedure by getting the Machine State Register, setting or clearing
           the IP bit, and writing the Machine State Register back again.
       19. Disable the instruction cache.
       20. Unlock the instruction cache.
       21. Invalidate the instruction cache.
       22. Unlock the data cache.
       23. Verify whether the cache was enabled, and if so, flush it.
       24. Invalidate the data cache.
- Initialization of all components: processor, clocks, EEPROM, I2C, serial,
Ethernet 10/100, chip selects, UPM machine, DRAM initialization, PCMCIA (Type I
and II), SPI, UART, video encoder, LCD, audio, touch screen, IR, ...
Flash Burner
Diagnostics and Utilities
- Test DRAM
- Command line interface
}
[3]
```

### *MIPS32-Based Booting Example*

The Ampro Encore M3 Au1500 based board assumes that on-board ROM (i.e., Flash) contains the bootloader monitor/program, called YAMON, originally created by MIPS Technologies. Where this boot ROM is mapped on the Au1500 is based upon the requirements of the MIPS architecture itself, which specifies that upon a reset, a MIPS processor must fetch the Reset exception vector from address 0xBFC00000. Basically, when a cold boot occurs on an MIPS32-based processor, a reset exception occurs which performs a full reset "hardware" initialization sequence that (in general) puts the processor in a state of executing instructions from unmapped, uncached memory, initializes registers (such as

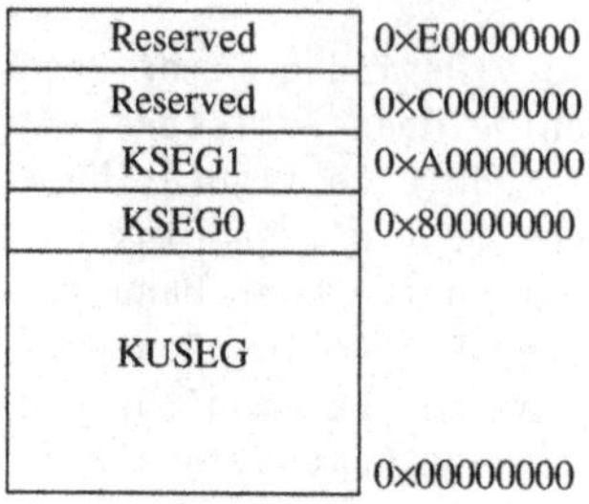

| | |
|---|---|
| Reserved | 0×E0000000 |
| Reserved | 0×C0000000 |
| KSEG1 | 0×A0000000 |
| KSEGO | 0×80000000 |
| KUSEG | 0×00000000 |

**Figure 12-8a**
Interpretation diagram.[4]

Rando, Wired, Config, and Status) for reset, and then loads the PC with 0xBFC0_0000, the Reset Exception Vector.

0xBFC0_0000 is a virtual address, not a physical address. All addresses under the MIPS32 architecture are virtual addresses, meaning the actual physical memory address on the board is translated when processing, such as instruction fetches and data loading and storing. The upper bits of the virtual address define the different regions in the memory map; for example:

- KUSEG (2 GB virtual memory from 0x0000000 to 0x7FFFFFFF).
- KSEG0 (512 MB virtual memory from 0x8000000 to 9FFFFFFF), which is a direct map to physical addresses and inherently cacheable.
- KSEG1 (512 MB virtual memory from 0xA000000 to BFFFFFFF), which is a direct map to physical addresses and inherently non-cacheable.

This means virtual addresses (KSEG0) 0x80000000 (a cacheable view of physical memory) and (KSEG1) 0xA00000000 (a non-cacheable view of physical memory) both map directly onto physical address 0x00000000. The MIPS32 Reset Exception Vector (0xBFC0_0000) is located in the last 4 MB of the KSEG1 region of memory—a non-cacheable region that can execute even if other board components are not yet initialized. This means that a physical address of 0x1FC00000 is generated for the first instruction fetch from the boot ROM. Basically, the programmer puts the start of the boot code (i.e., YAMON) at 0x1FC0_0000, which is the value the PC is set to start executing upon power-on, and effectively could occupy the entire 4 M of space (0x1FC00000 thru 0x1FFFFFFF) or more. (See Figures 12-8a–c.)

The physical address 0×1FC000000 is fixed for the Reset Exception Vector (the start of YAMON) on the MIPS32, regardless of how much physical memory is actually on the board—meaning the Flash chip on the board has to be integrated so that it correlates to this physical address. In the case of the Ampro Encore M3 Board, there is 2 MB of Flash memory on the board.

On this MIPS32-based board, after the hardware initialization sequence initializing the processor, the CPU begins executing the code located at the address within the PC register.

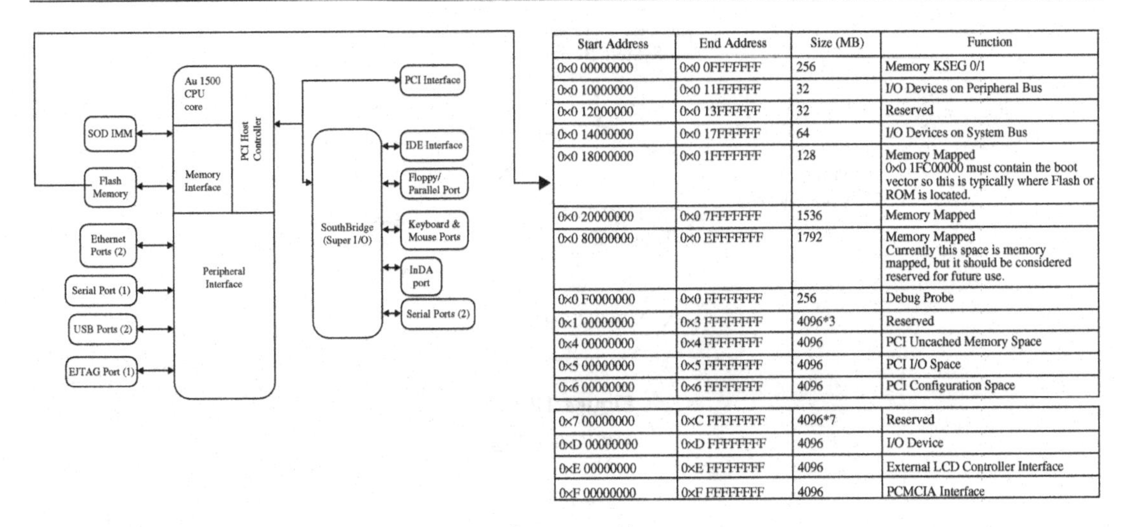


| Start Address | End Address | Size (MB) | Function |
|---|---|---|---|
| 0x0 00000000 | 0x0 0FFFFFFF | 256 | Memory KSEG 0/1 |
| 0x0 10000000 | 0x0 11FFFFFF | 32 | I/O Devices on Peripheral Bus |
| 0x0 12000000 | 0x0 13FFFFFF | 32 | Reserved |
| 0x0 14000000 | 0x0 17FFFFFF | 64 | I/O Devices on System Bus |
| 0x0 18000000 | 0x0 1FFFFFFF | 128 | Memory Mapped<br>0x0 1FC00000 must contain the boot vector so this is typically where Flash or ROM is located. |
| 0x0 20000000 | 0x0 7FFFFFFF | 1536 | Memory Mapped |
| 0x0 80000000 | 0x0 EFFFFFFF | 1792 | Memory Mapped<br>Currently this space is memory mapped, but it should be considered reserved for future use. |
| 0x0 F0000000 | 0x0 FFFFFFFF | 256 | Debug Probe |
| 0x1 00000000 | 0x3 FFFFFFFF | 4096*3 | Reserved |
| 0x4 00000000 | 0x4 FFFFFFFF | 4096 | PCI Uncached Memory Space |
| 0x5 00000000 | 0x5 FFFFFFFF | 4096 | PCI I/O Space |
| 0x6 00000000 | 0x6 FFFFFFFF | 4096 | PCI Configuration Space |
| 0x7 00000000 | 0xC FFFFFFFF | 4096*7 | Reserved |
| 0xD 00000000 | 0xD FFFFFFFF | 4096 | I/O Device |
| 0xE 00000000 | 0xE FFFFFFFF | 4096 | External LCD Controller Interface |
| 0xF 00000000 | 0xF FFFFFFFF | 4096 | PCMCIA Interface |

**Figure 12-8b**
Interpretation diagram.[4]

| Sys Address | Flash Address | Sectors | Description |
|---|---|---|---|
| bfC00000 – bfC03FFF | 00000000 – 00003FFF | 0 | Reset Image (16kB) |
| bfC04000 – bcC05FFF | 00004000 – 00005FFF | 1 | Boot Line (8kB) |
| bfC06000 – bfC07FFF | 00006000 – 00007FFF | 2 | Parameter Flash (8kB) |
| bfC08000 – bfC0FFFF | 00008000 – 0000FFFF | 3 | User NVRAM (32kB) |
| bfC10000 – bfC8FFFF | 00010000 – 0008FFFF | 4–11 | YAMON Little Endian (512kB) |
| bfC90000 – bfD0FFFF | 00090000 – 0010FFFF | 12–19 | YAMON Big Endian (512kB) |
| bfD10000 – bfDEFFFF | 00110000 – 001EFFFF | 20–33 | System Flash (896kB) |
| bfDF0000 – bfDFFFFF | 001F0000 – 001FFFFF | 34 | Environmental Flash (64kB) |

**Figure 12-8c**
Interpretation diagram.[4]

In this case it is the YAMON bootloader code that has been ported to the Ampro Encore M3 Board. All of the hardware specific to the MIPS32 architecture (i.e., initialization of interrupts) as well as specific to the board (serial, networking, etc.) is initialized via the YAMON program, which is proprietary software available from MIPS.

*System Booting with an OS*

Typically, 32-bit architectures include a more complex system software stack that includes an OS and, depending on the OS, may also include a *Board Support Package (BSP)*. Regardless of where the additional bootstrapping code comes from, if the system includes an OS, then it is the OS (with its BSP if there is one) that is initialized and loaded. While the boot sequence

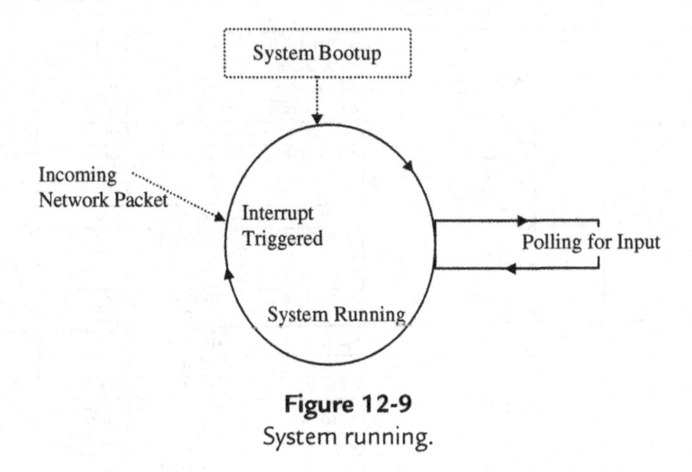


**Figure 12-9**
System running.

of a particular OS may vary, all architectures essentially execute the same steps when initializing and loading different embedded OSs.

For example, the boot sequence for a Linux kernel on an *x86 architecture* occurs via a BIOS that is responsible for searching for, loading and executing the Linux kernel, which is the central part of the Linux OS that controls all other programs. This is basically the "init" parent process that is started (executed) next. Code within the init process takes care of setting up the remainder of the system, such as forking tasks to manage networking/serial port etc. The VxWorks *Real-Time Operating System (RTOS)* boot sequence on most architectures, on the other hand, occurs via a VxWorks boot ROM that performs the architecture and board-specific initialization, and then starts the multitasking kernel with a user booting task as its only activity.

After the completion of the start-up sequence, an embedded system then typically enters an infinite loop, waiting for events that trigger interrupts, or actions triggered after some components are polled (see Figure 12-9).

## 12.2 Quality Assurance and Testing of the Design

Among the goals of testing and assuring the quality of a system are finding bugs within a design and tracking whether the bugs are fixed. Quality assurance and testing is similar to debugging, discussed earlier in this chapter, except that the goals of debugging are to actually fix discovered bugs. Another main difference between debugging and testing the system is that debugging typically occurs when the developer encounters a problem in trying to complete a portion of the design, and then typically *tests-to-pass* the bug fix (meaning tests only to ensure the system minimally works under normal circumstances). With testing, on the other hand, bugs are discovered as a result of trying to break the system, including both testing-to-pass and *testing-to-fail*, where weaknesses in the system are probed.

As soon as embedded hardware is available with device drivers to bring up the board, it is important to have an integration, verification, and test strategy from day 1. From the start of the project, plan the different types of testing. What guarantees problems later is when developers do not verify or unit test these available components. Not testing proactively and adequately is one of the most common mistakes within embedded design teams. It is irrelevant if a particular component is home-grown or a BSP, OS, and/or middleware software from an external vendor, for example. Never assume that because a particular off-the-shelf component has come from an expensive external vendor that it is bug-free and production ready. More importantly, do not assume that any third-party component that comes out-of-the-box is tuned or tested to a specific project's embedded system's requirements until team members see it running and have verified it with their own two eyes. In fact, many embedded hardware and software vendors deliver their components to their customers with manuals and development tools, because the expectation is that their customers will tune and test their component to meet the requirements of a particular embedded device design.

It does not matter "how" testing gets done (e.g., whether it is an individual responsible engineer that is assigned the role of test engineer or via a formal test group) as long as the testing gets done.

Ensuring the quality of source code is visible along with feature feedback from day 1, requires executing a disciplined test strategy as soon as any software is running on the system. Verify and test all components—including any prototypes that will later be used as the final design's foundation. Then, fix the hardware and/or source code defects as they are found. This is because having an unstable system with unreliable hardware and/or code is "worse" than having no system. Track all hardware and software defects, including measure defect rate in specific components as they are found. Highly problematic components are more expensive to debug than to replace, so it is critical to monitor these defects to ensure knowing when these unstable components are in a state that they must be replaced or rewritten.

Under testing, bugs usually stem from either the system not adhering to the architectural specifications (i.e., behaving in a way it shouldn't according to documentation, not behaving in a way it should according to the documentation, behaving in a way not mentioned in documentation) or the inability to test the system. The types of bugs encountered in testing depend on the type of testing being done. In general, testing techniques fall under one of four models: *static black-box* testing, *static white-box* testing, *dynamic black-box* testing, or *dynamic white-box* testing (see the matrix in Figure 12-9). Black-box testing occurs with a tester that has no visibility into the internal workings of the system (no schematics, no source code, etc.). Black-box testing is based on general product requirements documentation, as opposed to white-box testing (also referred to *clear-box* or *glass-box* testing) in which the tester has access to source code, schematics, etc. Static testing is done while the system is not running, whereas dynamic testing is done when the system is running.

Within each of the models (shown in Figure 12-10), testing can be further broken down to include, for example:

- *Unit/module testing*: incremental testing of individual elements within the system.
- *Compatibility testing*: testing that the element doesn't cause problems with other elements in the system.

| | Black Box Testing | White Box Testing |
|---|---|---|
| Static Testing | Testing the product specifications by:<br>1. looking for high-level fundamental problems, oversights, omissions (pretending to be customer, research existing guidelines/standards, review and test similar software, etc.).<br>2. low-level specification testing by insuring completeness, accuracy, preciseness, consistency, relevance, feasibility, etc. | Process of methodically reviewing hardware and code for bugs without executing it. |
| Dynamic Testing | Requires definition of what software and hardware does, includes:<br>• *data testing*, which is checking info of user inputs and outputs<br>• *boundary condition testing*, which is testing situations at edge of planned operational limits of software<br>• *internal boundary testing*, which is testing powers-of-two, ASCII table<br>• *input testing*, which is testing null, invalid data<br>• *state testing*, which is testing modes and transitions between modes software is in with state variables<br>e.g., race conditions, repetition testing (main reason is to discover memory leaks), stress (starving software = low memory, slow cpu, slow network), load (feed software = connect many peripherals, process large amount of data, web server has many clients accessing it, etc.). | Testing running system while looking at code, schematics, etc.<br>Directly testing low-level and high-level based on detailed operational knowledge, accessing variables and memory dumps. Looking for data reference errors, data declaration errors, computation errors, comparison errors, control flow errors, subroutine parameter errors, I/O errors, etc. |

**Figure 12-10**
Testing model matrix.[5]

- *Integration testing*: incremental testing of integrated elements.
- *System testing*: testing the entire embedded system with all elements integrated.
- *Regression testing*: rerunning previously passed tests after system modification.
- *Manufacturing testing*: testing to ensure that manufacturing of the system didn't introduce bugs.

From these types of tests, an effective set of test cases can be derived that verify that an element and/or system meets the architectural specifications, as well as validate that the element and/or system meets the actual requirements, which may or may not have been reflected correctly or at all in the documentation. Specific testing methodologies and templates for test cases, as well as the entire testing process, have been defined in several popular industry quality assurance and testing standards, including ISO9000 Quality Assurance standards, Capability Maturity Model (CMM), and the ANSI/IEEE 829 Preparation, Running, and Completion of Testing standards.

Once the test cases have been completed and the tests are run, how the results are handled can vary depending on the organization, but typically vary between *informal*, where information is exchanged without any specific process being followed, and *formal* design reviews, or peer reviews where fellow developers exchange elements to test, walkthroughs where the responsible engineer formally walks through the schematics and source code, inspections where someone other than the responsible engineer does the walkthrough, etc.

To be the most effective, code inspections should be incorporated into the test strategy from the start. Code inspections also need to do more than look for "pretty" code, meaning ensuring that programming language "best practices" have been followed and actively look for bugs. It is cheaper and faster to do stringent code inspections as soon as source code compiles on target hardware, before testing formally begins.[7]

In order for code inspections to be the most effective and most efficient, ensure that the right "type" of team members are doing the actual code inspections. For instance, to code inspect device drivers that manage various hardware components on the target, ensure developers with knowledge and understanding of the "hardware" or even the actual hardware engineers are part of the inspection team. Generally, code inspection teams should have some combination of the following type of roles:[7]

1. *Moderator*: responsible for managing the code inspection process and meeting(s).
2. *Reader*: reads out-loud the source code and relative specifications for the operational investigation. Should not be the developer that created the source code being inspected.
3. *Recorder*: fills in code inspection check list report and documents any agreed upon open items.
4. *Author*: helps explain source to code inspection team, to discuss errors found, and future rework that needs to be done. Should be the developer of the source code being inspected.

As shown in Table 12-2, the code inspection process should also include some type of checklist of what is being inspected and the location of documented results.

Finally, as with debugging, there are a wide variety of automation and testing tools and techniques that can aid in the speed, efficiency, and accuracy of testing various elements. These include load tools, stress tools, interference injectors, noise generators, analysis tools, macro recording and playback, and programmed macro, including tools listed in Table 12-1.

**Table 12-2: Example of Code Inspection Checklist for "C" Source**

| Parameter/Function Name | Number of Errors | | Error Type |
|---|---|---|---|
| | Major | Minor | |
| | | | *Code does not meet firmware standards* |
| | | | Header block |
| | | | Naming consistency |
| | | | Comments |
| | | | Code layout and elements |
| | | | *Recommended Coding practices* |
| | | | Auto-generated code not manually edited |
| | | | Don't use magic numbers hard coded in the source code, i.e., place constant numerical values directly into the source |
| | | | Avoid using global variables |
| | | | Initialize all defined variables |
| | | | Function size and complexity unreasonable |
| | | | Unclear expression of ideas in the code |
| | | | Poor encapsulation |
| | | | Data types do not match |
| | | | Poor logic—won't function as needed |
| | | | Exceptions and error conditions not caught (e.g., return codes from malloc())? |
| | | | Switch statement without a default case (if only a subset of the possible conditions used)? |
| | | | Incorrect syntax, such as improper use of ==, =, &&, &, etc. |
| | | | Non-re-entrant code in dangerous places |
| ... | ... | .... | Other ... |

**Real-World Advice**

*The Potential Legal Ramifications (in the United States) of NOT Testing*

The US laws of product liabilities are considered very strict, and it is recommended that those responsible for the quality assurance and testing of systems receive training in products liability law in order to recognize when to use the law to ensure that a critical bug is fixed and when to recognize that a bug could pose a serious legal liability for the organization.

The general areas of law under which a consumer can sue for product problems are:

- Breach of contract (i.e., if bug fixes stated in contract are not forthcoming in timely manner).
- Breach of warranty and implied warranty (i.e., delivering system without promised features).
- Strict and negligence liability for personal injury or damage to property (i.e., bug causes injury or death to user).
- Malpractice (i.e., customer purchases defective product).
- Misrepresentation and fraud (i.e., product released and sold that doesn't meet advertised claims, whether intentionally or unintentionally).

Remember, these laws apply whether your "product" is embedded consulting services, embedded tools, an actual embedded device, or software/hardware that can be integrated into a device.

*Based on the chapter "Legal Consequences of Defective Software," in Testing Computer Software, C. Kaner, J. Falk, and H. Q. Nguyen, 2nd edn, Wiley, 1999.*

## 12.3 Conclusion: Maintaining the Embedded System and Beyond

This chapter introduced some key requirements behind implementing an embedded system design, such as understanding utility, translation, and debugging development tools. These tools include IDE and CAD tools, as well as interpreters, compilers, and linkers. A wide range of debugging tools useful for both debugging and testing embedded designs were discussed, from hardware ICEs, ROM emulators, and oscilloscopes to software debuggers, profilers, and monitors, just to name a few. This chapter also discussed what can be expected when booting up a new board, providing a few real-world examples of system bootcode.

Finally, even after an embedded device has been deployed, there are responsibilities that typically need to be met, such as user training, technical support, providing technical updates, and bug fixes. In the case of user training, for example, architecture documentation can be leveraged relatively quickly as a basis for technical, user, and training manuals. Architecture documentation can also be used to assess the impact involved in introducing updates (new features, bug fixes, etc.) to the product while it is in the field, mitigating the risks of costly recalls or crashes, or on-site visits by Field Application Engineers (FAEs) that could be required at the customer site. Contrary to popular belief, the responsibilities of the engineering team last throughout the lifecycle of the device and do *not* end when the embedded system has been deployed to the field.

To ensure success in embedded systems design, it is important to be familiar with the phases of designing embedded systems, especially the importance of first creating an architecture. This requires that all engineers and programmers, regardless of their specific responsibilities and tasks, have a strong technical foundation by understanding at the systems level all the major components that can go into any embedded system's design. This means that hardware engineers understand the software and software engineers understand the hardware at the systems level, at the very least. It is also important that the responsible designers adopt, or come up with, an agreed-upon methodology to implement and test the system, and then have the discipline to follow through on the required processes.

It is the hope of the author that you have appreciated the architectural approach of this book and found it a useful tool as a comprehensive introduction to the world of embedded systems design. There are unique requirements and constraints related to designing an embedded system, such as those dictated by cost and performance. Creating an architecture addresses these requirements very early in a project, allowing a design team to mitigate risks. For this reason alone, the architecture of an embedded device will continue to be one of the most critical elements of any embedded system project.

## Chapter 12: Problems

1. What is the difference between a host and a target?
2. What high-level categories do development tools typically fall under?
3. [T/F] An IDE is used on the target to interface with the host system.
4. What is CAD?
5. In addition to CAD, what other techniques are used to design complex circuits?
6. [a] What is a preprocessor?
   [b] Provide a real-world example of how a preprocessor is used in relation to a programming language.
7. [T/F] A compiler can reside on a host or a target, depending on the language.
8. What are some features that differentiate compiling needs in embedded systems versus in other types of computer systems?
9. [a] What is an object file?
   [b] What is the difference between a loader and a linker?
10. [a] What is an interpreter?
    [b] Name three real-world languages that require an interpreter.
11. An interpreter resides on:
    A. The host.
    B. The target and the host.

   C. In an IDE.
   D. A and C only.
   E. None of the above.
12. [a] What is debugging?
    [b] What are the main types of debugging tools?
    [c] List and describe four real-world examples of each type of debugging tool.
13. What are five of the cheapest techniques to use in debugging?
14. Boot code is:
    A. Hardware that powers on the board.
    B. Software that shuts down the board.
    C. Software that starts-up the board.
    D. All of the above.
    E. None of the above.
15. What is the difference between debugging and testing?
16. [a] List and define the four models under which testing techniques fall.
    [b] Within each of these models, what are five types of testing that can occur?
17. [T/F] Testing-to-pass is testing to ensure that system minimally works under normal circumstances.
18. What is the difference between testing-to-pass and testing-to-fail?
19. Name and describe four general areas of law under which a customer can sue for product problems.
20. [T/F] Once the embedded system enters the manufacturing process, the design and development team's job is done.

## Endnotes

[1] *The Electrical Engineering Handbook*, chapter 27, R. C. Dorf, IEEE Computer Society Press, 2nd edn, 1998.
[2] "Short Tutorial on PSpice," B. Rison, <http://www.ee.nmt.edu/~rison/ee321_fall02/Tutorial.html>
[3] Embedded Planet, RPXLite Board Documentation; Freescale, PowerPC MPC823 User's Manual.
[4] Ampro, Encore M3 Au150 Documentation.
[5] *Software Testing*, R. Patton, Sams, 2nd edn, 2005.
[6] "A Guide to Code Inspections," J. Ganssle, <http://www.ganssle.com/inspections.htm>; "Code Inspection Process," Wind River Services.
[7] "A Boss's Quick-Start to Firmware Engineering," J. Ganssle, <http://www.ganssle.com/articles/abossguidepi.htm>.

APPENDIX A

# Projects and Exercises

The projects found in this Appendix are designed to complement an embedded systems architecture class and corresponding laboratory course. They are intended to help the reader become familiar with embedded systems concepts, hardware, software, and development and diagnostic tools that may be available to the student. Because of the variations between labs at different institutions, these exercises encourage students to independently investigate, acquire, and work with different platforms as an engineer in the real world would do. Ultimately, the goal of these exercises is to allow the reader to be able to:

- Investigate, understand, and articulate the primary characteristics of any embedded system through creating an architecture, including any design criteria and constraints.
- Investigate, understand, and articulate embedded system software and hardware standards, and their importance.
- Understand, articulate, and implement the overall design process of an embedded system as defined by the Embedded Systems Design and Development Lifecycle Model.
- Understand the development and debugging tools and environment when designing an embedded system.
- Search out and gather information from various sources, such as the Internet, professional magazines, or embedded conferences.
- Learn how to work in an engineering team environment, including understanding the importance and benefits of teamwork, as well as the potential problems.

When doing the projects, the following guidelines are recommended:[1]

1. Create project reports for each project, containing information similar to that shown in the following sample project report.
2. Original references should always be read, understood, and cited.
3. All technical work used, whether published, unpublished, proprietary, open-source, etc., should always be credited to the source ("engineer so-and-so says this is how the field should be configured," "section of code was modified from … to …," etc.).

[1]Recommended project guidelines are based on guidelines for projects from the Specification and Modeling of Reactive Real-Time Systems course by Professor Edward A. Lee at the University of California at Berkeley.

4. It is not recommended that hardware or software be built from scratch for projects in this lab book. This material is intended for a systems engineering course in which projects are not evaluated on the basis of how much effort is put into a specific element, but rather on how effective the entire architecture of the embedded system is. To be able to design the most interesting and exciting of projects given limitations in time and budget, it is not recommended to "reinvent the wheel" if it isn't absolutely necessary. Use the Internet to find commercial hardware and software (evaluations, open source, etc.), and leverage any relevant work already engaged in.

**Sample Project Report**

Student Name:

Project Number and Title:

Contribution to the lab: %______ Description of Contribution:

List any team members and their contributions to the lab:

Name 1: ________________ %______ Description of Contribution:

Name 2: ________________ %______ Description of Contribution:

Name 3: ________________ %______ Description of Contribution:

.

.

.

Name "N": ________________ %______ Description of Contribution:

Project Summary:

What were the steps taken to implement the project:

What where the results of the project (attach any generated output):

Comments/Suggestions:

Attach any documentation (schematics, users manuals, etc.) and software used in this project.

5. These projects reflect a serious effort to go beyond the material in the text, requiring readers to obtain additional information from the Internet, journals, or books, for example. This is how the reader will most likely be expected to function on the job. There typically is no hand-holding provided to professionals. In fact, the most successful engineers are typically the ones who can learn and function quickly and independently. The sooner readers are comfortable with developing and maintaining the latest technical skills on their own, the more successful they are likely to be.
6. Learn to use the relevant Integrated Development Environments (IDEs), languages, hardware, etc., at least to the level of proficiency required to complete the project. Get the compiler, simulator, or design environment and install it, read all the documentation, and run it.
7. Students are encouraged to work in a team (of two or more), as this is typically how things are done in the real world. Working in a team is also an excellent way to produce a more ambitious and exciting project in a given amount of time, with each team member being assigned a portion of the project.

**Real-World Advice**

It is recommend that labs used for this course be stocked with as many different boards and system software elements as possible, to encourage students to gain the hands-on experience of figuring out and bringing up different hardware and software as quickly as possible (meaning in hours or minutes, not days and weeks). With this type of lab, students can learn the skills of a seasoned professional, and can become comfortable with the endless permutations of hardware and software elements available to embedded developers. Students need to discover that this vast array of possibilities is nothing to be intimidated by as long as they pay attention to the details and learn to work with elements they have never had hands-on experience with before. For example, they need to learn to check for cables needing to be swapped or plugged in, jumpers on a board set to desired configurations, boot code configured, and especially to search for and read the documentation.

In these projects, I encourage students to experience many different elements, such as different boards with different master CPUs and multiple operating systems (OSs), because I feel that in many cases the expectations placed on students today relative to what they need to succeed after graduation are far too low. I have heard arguments that the reason not to have multiple architectures in the lab, for instance, is because it is too complicated for students to figure out the different boards in a reasonable time. In today's electronic gadget world where many students at a much earlier age have mastered games, cell phones, DVD players, PDAs, etc., *it isn't expecting too much to ask them to apply those skills to learning embedded systems engineering. Learning the types of skills introduced in these labs will save students years of having to learn it in the real world. This is especially crucial because, with pressing international competition, many students can no longer afford those years of hand-holding if they want to compete for engineering jobs after graduation.*

# 1 Section I Projects

## 1.1 Project 1: The Product Concept

Using Table A-1, select a product concept and use the Internet to find at least four commercially available products that fall under that product category. Using available on-line documentation, compile a document that outlines their main features, their similarities, and their differences. Create a product concept for your device based on this information.

## 1.2 Project 2: Modeling the Design of the Product Concept

Outline the developmental process of the project based on the product concept of Project 1 using:

A. The big-bang model
B. The code-and-fix model
C. The waterfall model
D. The spiral model
E. The Embedded Systems Design and Development Lifecycle Model

**Table A-1: Examples of embedded systems and their markets[1]**

| Market | Embedded Device |
|---|---|
| Automotive | Ignition system |
| | Engine control |
| | Brake system (i.e., antilock braking system) |
| Consumer electronics | Digital and analog televisions |
| | Set-top boxes (DVDs, VCRs, cable boxes, etc.) |
| | Personal data assistants (PDAs) |
| | Kitchen appliances (refrigerators, toasters, microwave ovens) |
| | Automobiles |
| | Toys/games |
| | Telephones/cell phones/pagers |
| | Cameras |
| | Global Positioning Systems (GPS) |
| Industrial control | Robotics and control systems (manufacturing) |
| Medical | Infusion pumps |
| | Dialysis machines |
| | Prosthetic devices |
| | Cardiac monitors |
| Networking | Routers |
| | Hubs |
| | Gateways |
| Office automation | Fax machines |
| | Photocopiers |
| | Printers |
| | Monitors |
| | Scanners |

Draw each model from product conception to product completion, describing the strengths and drawbacks of each model. What would be required when using each model for the project to succeed? What factors could cause failure?

## 1.3 Project 3: The Embedded Systems Model and the Product Concept

Given the provided documentation on the four commercially available products of Project 1, outline what specific hardware and/or software elements are specified (if none, then find products that do provide more information in their documentation). Draw the Embedded Systems Model for each product and show where in this model each of these elements would fall.

## 1.4 Project 4: the Product Concept and Recent Developments

Given the list of technical magazines in Table A-2, or any other relevant magazine (there are many more not shown in Table A-2), select and summarize 10 articles from at least five different magazines released this month that impact features of your product concept.

**Table A-2: Examples of technical magazines**

| Magazine | Website |
|---|---|
| *C/C++ Users Journal* | http://www.cuj.com/ |
| *C++ Report* | http://www.creport.com/ |
| *Circuit Cellar* | http://www.circellar.com/ |
| *CompactPCI Systems* | http://www.picmgeu.org/magazine/CPCI_magazine.htm |
| *Compliance Engineering (CE)* | www.ce-mag.com |
| *Dedicated Systems Magazine* | http://www.realtime-magazine.com/magazine/magazine.htm |
| *Design News* | http://www.designnews.com/index.asp?cfd=1 |
| *Dr. Dobb's Journal* | http://www.ddj.com/ |
| *Dr. Dobb's Embedded Systems* | http://www.ddjembedded.com/resources/articles/2001/0112g/0112g.htm |
| *EE Product News* | http://www.eepn.com/ |
| *EDN Asia* | http://www.ednasia.com/ |
| *EDN Australia* | http://www.electronicsnews.com.au/ |
| *EDN China* | http://www.ednchina.com/Cstmf/BCsy/index.asp |
| *EDN Japan* | http://www.ednjapan.com/ |
| *EDN Korea* | http://www.ednkorea.com/ |
| *EDN Magazine—Europe* | http://www.reed-electronics.com/ednmag/ |
| *EDN Taiwan* | http://www.edntaiwan.com/ |
| *EE Times Asia Edition* | http://www.eetasia.com/ |
| *EE Times China Edition* | http://www.eetchina.com/ |
| *EE Times France* | http://www.eetimes.fr/ |
| *EE Times Germany* | http://www.eetimes.de/ |
| *EE Times Korea* | http://www.eetkorea.com/ |
| *EE Times North America* | http://www.eet.com/ |
| *EE Times Taiwan* | http://www.eettaiwan.com/ |
| *EE Times UK* | http://www.eetuk.com/ |
| *Electronic Design* | http://www.elecdesign.com/Index.cfm?Ad=1 |

(*Continued*)

Table A-2: (Continued)

| Magazine | Website |
|---|---|
| *Elektor France* | http://www.elektor.presse.fr/ |
| *Elektor Germany* | http://www.elektor.de/ |
| *Elektor Netherlands* | http://www.elektuur.nl/ |
| *Elektor UK* | http://www.elektor-electronics.co.uk/ |
| *Electronics Express Europe* | http://www.electronics-express.com/ |
| *Electronics Supply and Manufacturing* | http://www.my-esm.com/ |
| *Embedded Linux Journal* | http://www.linuxjournal.com/ |
| *Embedded Systems Engineering* | http://www.esemagazine.co.uk/ |
| *Embedded Systems Europe* | http://www.embedded.com/europe |
| *Embedded Systems Programming—North America* | http://www.embedded.com/ |
| *European Medical Device Manufacturer* | http://www.devicelink.com/emdm/ |
| *Evaluation Engineering* | http://www.evaluationengineering.com/ |
| *Handheld Computing Magazine* | http://www.hhcmag.com/ |
| *Hispanic Engineer* | http://www.hispanicengineer.com/artman/publish/index.shtml |
| *IEEE Spectrum* | http://www.spectrum.ieee.org/ |
| *Java Developers Journal* | http://sys-con.com/java/ |
| *Java Pro* | http://www.ftponline.com/javapro/ |
| *Linux Journal* | http://www.linuxjournal.com/ |
| *Linux Magazine* | http://www.linux-mag.com/ |
| *Medical Electronics Manufacturing* | http://http://www.medicalelectronicsdesign.com/ |
| *Design and Development of Medical Electronic Products* | http://www.devicelink.com/mem/index.html |
| *Microwaves & RF* | http://www.mwrf.com/ |
| *Microwave Engineering Europe* | http://www.kcsinternational.com/microwave%20engineering%20europe.html |
| *Military and Aerospace Electronics* | http://mae.pennnet.com/home.cfm |
| *MSDN Magazine* | http://msdn.microsoft.com/msdnmag/ |
| *PC/104 Embedded Solutions* | http://www.pc104online.com/ |
| *Pen Computing Magazine* | http://www.pencomputing.com/ |
| *PocketPC Magazine* | http://pocketpcmag.com/ |
| *Portable Design* | http://pd.pennnet.com/home.cfm |
| *Practical Electronics* | http://www.epemag.wimborne.co.uk/ |
| *RTC Magazine* | http://www.rtcmagazine.com/ |
| *Silicon Chip* | http://www.siliconchip.com.au/ |
| *TRONIX* | http://www.tronix-mag.com/ |
| *US Black Engineering* | http://www.blackengineer.com/artman/publish/index.shtml |
| *VMEBus Systems* | http://www.vmebus-systems.com/ |
| *Wired* | http://www.wired.com/wired/ |
| *Wireless Systems Design* | http://www.wsdmag.com/ |

### 1.5 Project 5: Finding Market-Specific Standards

Using Table A-1, select three products each under a different market, or three products assigned by the instructor, and use the Internet, journals, books, the list of magazines in Table A-2, or your own magazines, to find at least six recent market-specific standards that apply to each of the selected types of embedded systems. At least two of the market-specific standards should be competing standards. Compile three documents (one for each product) that outline what requirements each of these standards impose on the respective embedded system.

### 1.6 Project 6: Finding General-Purpose Standards

Given the products and market-specific standards in Project 1, find at least six recent general-purpose standards that apply to the selected type of embedded system. At least two of the general-purpose standards should be competing standards. Compile three documents, one for each product selected in Project 1, that outline what requirements each of these general-purpose standards impose on the respective embedded system.

### 1.7 Project 7: Standards and the Embedded Systems Model

Compile three documents, one for each product selected under Project 1, that map into the Embedded Systems Model what hardware and/or software elements each of the standards of Projects 1 and 2 define. There can be more than one model reflecting competing standards.

## 2 Section II Projects

### 2.1 Project 1: Hardware Documentation

This project is based upon one of the most efficient ways of learning how to read or draw a hardware diagram, the Traister and Lisk method,[2] which involves:

*Step 1* Learning the basic symbols that can make up the type of diagram, such as timing or schematic symbols. To aid in the learning of these symbols, rotate between this step and Steps 2 and/or 3.

*Step 2* Reading as many diagrams as possible, until reading them becomes boring (in that case rotate between this step and Steps 1 and/or 3) or comfortable (so there is no longer the need to look up every other symbol while reading).

*Step 3* Drawing a diagram to practice simulating what has been read, again until it either becomes boring (which means rotating back through Steps 1 and/or 2) or comfortable.

Therefore, this project is made up of three exercises, each exercise reflecting a step within the Traister and Lisk method.[2]

*Exercise 1: The Symbols, Conventions, and Rules of Schematic Diagrams*

Generate a report that lists three different standards (organizational, regional, and/or international) that define symbols, conventions, and rules for schematic diagrams. Use the Internet, books, journals, etc., to gather this information.

*Exercise 2: Reading Schematic Diagrams*

Using schematics available from the professor or acquired on the Internet (e.g., search on "schematic diagrams") and approved by your professor, select three schematic diagrams based on different boards, and write a report identifying the symbols, conventions, and rules of these schematic diagrams.

*Exercise 3: Drawing Schematic Diagrams*

For this project, either use a program provided in your labs or find and download an evaluation copy of a program that allows for drawing schematic diagrams. There are several of them, so do a web search (e.g., search on "drawing schematic diagrams" or "schematic diagram software"). It may be necessary to evaluate two or three programs before finding one that is stable and has the symbols and features you need.

Because this isn't a class on building electronic circuits from scratch, this exercise is designed to enable the reader, whether hardware engineer or programmer, to become comfortable with finding, evaluating, and using schematic diagram software and drawing circuits. Being able to document the hardware is an important part of the process of creating an architecture.

Using the schematic application, draw the schematics and create your own files of these schematics from Exercise 2, using the schematic diagram software.

## 2.2 Project 2: Simulating a Circuit

Because of the importance and costs associated with designing hardware, computer-aided design (CAD) tools are commonly used by hardware engineers to simulate circuits at the electrical level in order to study a circuit's behavior under various conditions before they actually build the circuit. There are many commercially available electrical circuit simulator tools that are generally similar in terms of their overall purpose, but differ in what analysis can be done, what circuit components can be simulated, and the look and feel of the user interface of the tool, for example.

There are many industry techniques in which CAD tools are used to simulate a circuit, and circuits created in this simulator can be made up of a variety of active and passive elements. This project is about being able to find, evaluate, and use a CAD tool to simulate a simple circuit. Unless otherwise specified by the professor, you won't be asked to perform a simulation on an entire complex set of circuits in a processor or on a board; it is so difficult, if not impossible, to perform a simulation on an entire design that a hierarchy of simulators and models are typically used.

For this project, either use a CAD tool provided in your labs, or find and download an evaluation copy of a program that allows for simulating circuits. There are several of them, so do a web search (search on the program "PSpice" or "circuit simulator"). It may be necessary to evaluate two or three programs before finding one that is stable and has the symbols and features you need.

Read the documentation provided with the CAD tool to understand how to create and simulate a simple circuit. This includes understanding how to input your circuit, what types of output files are generated, any semantic rules that must be followed, the symbols that can be used to build the circuit, and how to actually simulate the circuit that you have created and analyze the output. Some CAD tools come with tutorials, while others have many on-line

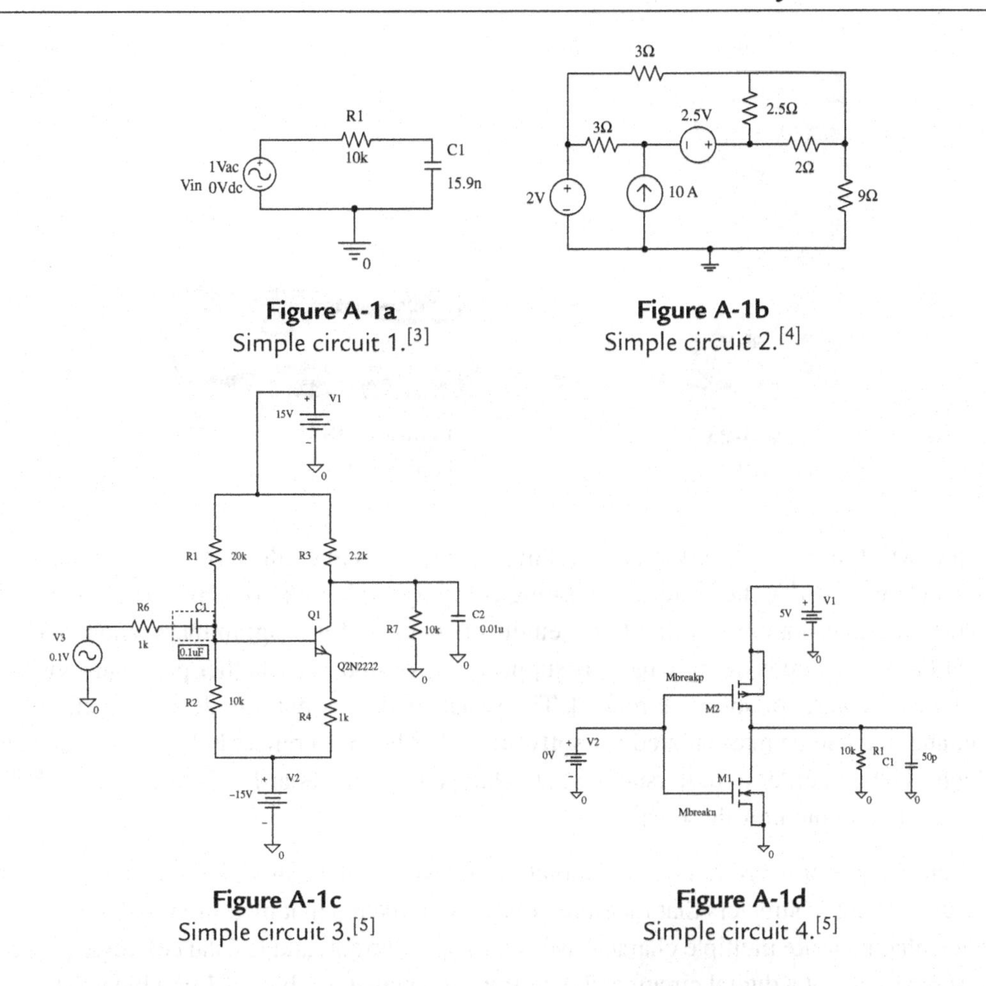


**Figure A-1a**
Simple circuit 1.[3]

**Figure A-1b**
Simple circuit 2.[4]

**Figure A-1c**
Simple circuit 3.[5]

**Figure A-1d**
Simple circuit 4.[5]

tutorials from various sources available (e.g., do a web search for "PSpice tutorial" if using PSpice—some tutorials will provide a link for obtaining a free evaluation version of the tool). Input and simulate four simple circuits provided by your professor or the simple circuits shown in Figures A-1a–d.

## 2.3 Project 3: Working with a Real Board

**Warning**

To avoid injury or damage to a circuit, and to accurately measure signals, carefully follow all instructor and device directions about wearing a grounding strap and using lab equipment. For example, when connecting probes to a board, be careful not to short adjacent wires or pins, which could result in damage to the board. Also, when measuring signals with a device in which one probe is attached to ground, because of noise produced from other circuit elements, it is recommended that you connect to a ground pin that is in close proximity to the signal being measured.

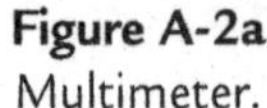

**Figure A-2a**
Multimeter.

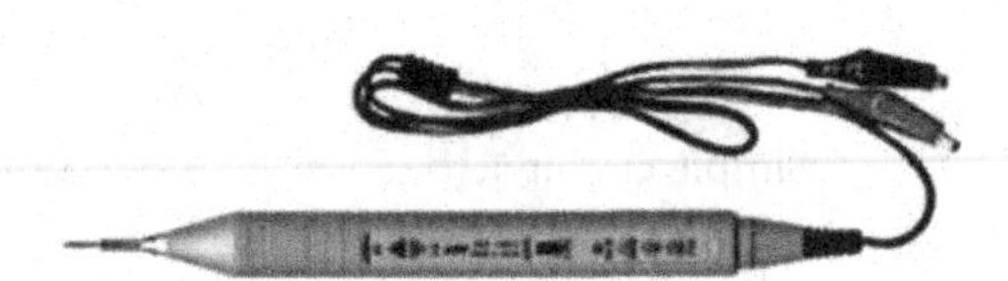

**Figure A-2b**
Logic probe.

This project will familiarize you with the techniques and tools that allow you to understand a circuit's behavior, verify the system is working correctly, and track down problems on a board. This project is not only for hardware engineers but also for programmers who need to verify that their software is doing what it is supposed to, or find out whether problems with the board are hardware- or software-related. The variables that essentially describe a circuit's behavior, and that can be manipulated via software or hardware to reflect behavior, are current and voltage. Thus, in order to understand what is happening on a board, it is necessary to be able to measure and monitor these variables.

Many different types of measurement and monitoring devices can be used, including ammeters that measure current, voltmeters that measure voltage, ohmmeters that measure resistance, multimeters that measure multiple characteristics (voltage, current, and resistance), logic probes that measure voltage of a digital circuit and determine if a signal is a binary 1 or a binary 0, and oscilloscopes that can graph voltage signals, to name a few. In general, many measurement devices, such as voltmeters, ohmmeters, and ammeters, measure board characteristics using two probes—one for positive terminal (the red probe) and one for negative terminal (the black probe). Measurements are taken at various points in the circuit by inserting the metal tips of these probes into the board. Figures A-2a and b are examples of two such measurement devices along with their probes, and Figure A-2c shows how these probes can be inserted into a circuit.

The next several exercises involve the student becoming comfortable using these tools and interacting with a board. Because different labs contain different sets of tools with different types of boards and circuitry, the exercises in this project will serve as an outline that, along with instructor directions, can be used to complete the project.

*Exercise 1: Using an Ammeter with an Embedded Board*

An ammeter measures the current flowing through a circuit. This is accomplished by hooking up an ammeter's probes in series, meaning the probes have to be hooked up to the board so that the current flowing through the section of the circuit you want to measure also flows through

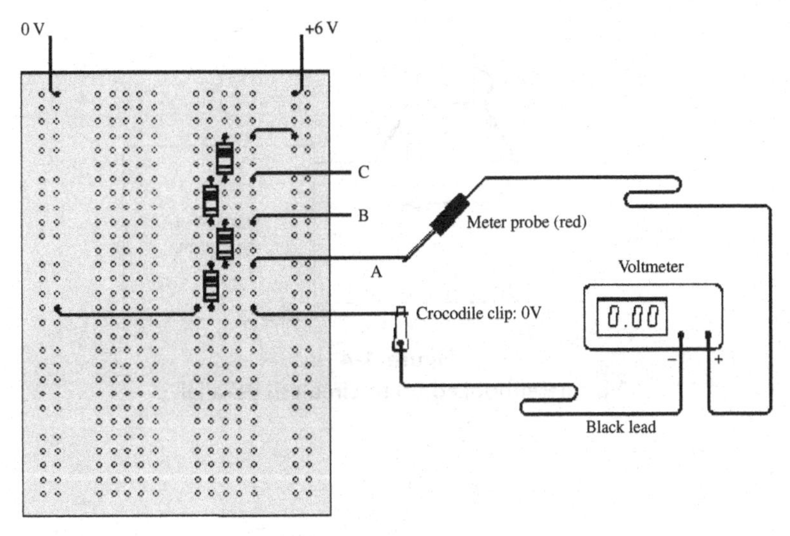


**Figure A-2c**
Inserting probes into circuit.

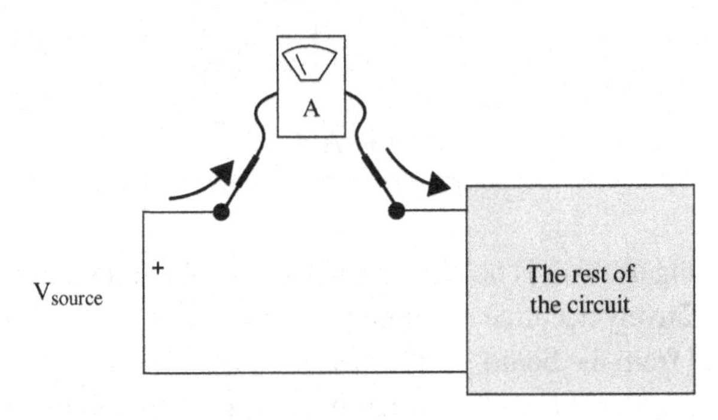


**Figure A-3**
Ammeter hooked up to circuit in series.

the ammeter (see Figure A-3). Use the ammeter to measure current using the instructions and circuits outlined to you by the instructor on the hardware available in your lab.

*Exercise 2: Using a Voltmeter with an Embedded Board*

The voltage between two locations on a circuit can be measured across a voltmeter. As shown in Figure A-4, a voltmeter is connected to the circuit in parallel, meaning that the voltmeter needs to be connected so that the voltage across it is identical to the voltage across the section of the circuit being measured. Use the voltmeter to measure voltage using the instructions and circuits outlined to you by the instructor on the hardware available in your lab.

*Exercise 3: Using an Ohmmeter with an Embedded Board*

The resistance of a board element is measured using an ohmmeter and can also be used to check for short or open circuits. As shown in Figure A-5, an ohmmeter measures a circuit element's resistance with the power disconnected.

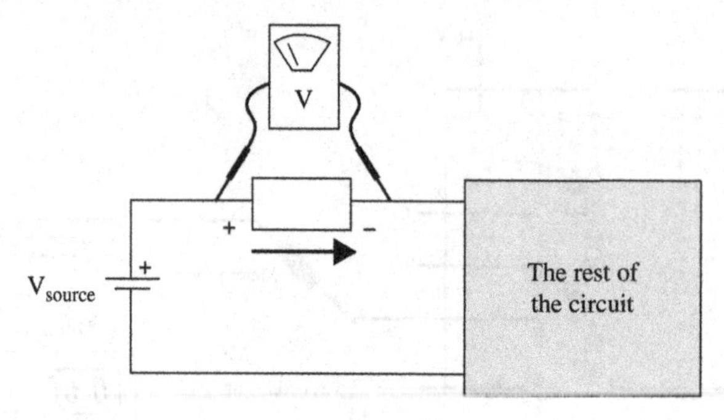


**Figure A-4**
Voltmeter hooked up to circuit in parallel.

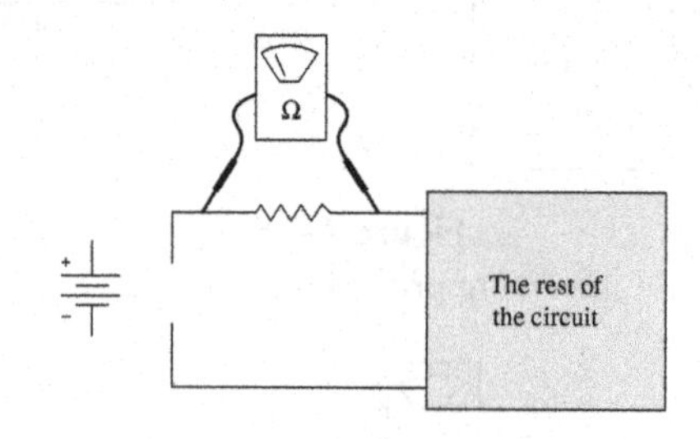


**Figure A-5**
Ohmmeter measuring circuit elements with no power.

In a simple circuit, as long as there is no electron flow (due to a voltage source) then the measurement can be relatively accurate. In more complex circuits it is recommended that the element be removed from the board, so the measurement won't be adversely affected by other circuit elements. Use the ohmmeter to measure resistance of circuit elements using the instructions and circuits outlined to you by the instructor on the hardware available in your lab.

*Exercise 4: Using a Logic Probe with an Embedded Board*

Because the data an embedded board processes is digital by nature, typically voltage levels are used to indicate a binary 1 or binary 0. What these levels are depends on the circuit, such as +5, –3, or –12 V for a binary one and 0, +3, or +12 V for a binary zero. A logic probe is a device that measures the voltage in a digital circuit, and indicates whether the signal is a binary one or binary zero.

*Note: Sometimes a logic analyzer is referred to as a logic probe, but a logic analyzer is a different, more complex type of measurement tool.*

In general, logic probes have two probes (e.g., a black probe connected to ground and a red probe connected to the circuit at a voltage source indicated by the instructor for your particular circuit and logic probe), as well as an additional probe with a metal tip that is connected to the section of the circuit to be measured. Use the logic probe to measure for

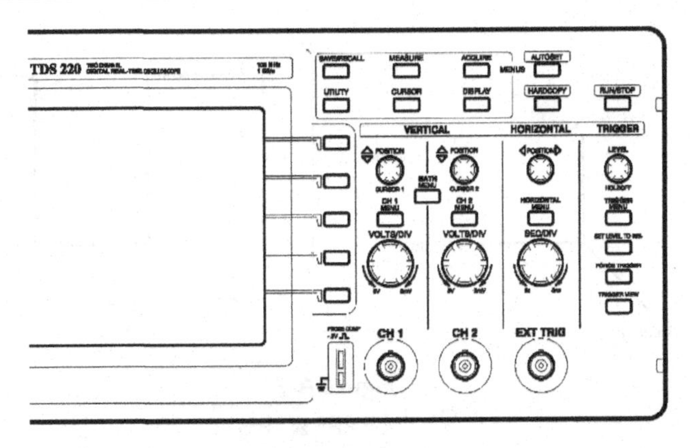


**Figure A-6**
Typical oscilloscope front panel.[6] *Courtesy Tektronix, Inc.*

logical 0 and logical 1 data using the instructions and digital circuits outlined to you by the instructor on the hardware available in your lab.

*Exercise 5: Using an Oscilloscope with an Embedded Board*

One of the most indispensable tools for designing and debugging an embedded system is an oscilloscope—a measurement device that displays electrical signals in the form of graphs. All of the switches and input ports needed to graph signals are typically located on the front panel of an oscilloscope. As seen in Figure A-6, there is a screen on the front of an oscilloscope, as well as a control and connector section. There are several types of oscilloscopes, so read the documentation provided and become familiar with the oscilloscope in your own lab. This includes understanding your oscilloscope's features and limitations, as well as how it connects to a circuit. This is because how an oscilloscope interacts with a circuit will impact any measurements taken.

As shown in Figure A-7, an oscilloscope typically displays a three-dimensional graph within its screen—a horizontal *X*-axis typically representing time when the input to the X-axis is connected to a clock, a vertical *Y*-axis typically representing voltage when the input to the *Y*-axis is connected to measure voltage, and a *Z*-axis representing the intensity of the signal via the brightness of the display. This graph can relay several types of key information about a signal, such as signal frequency, voltage relative to time, and distortions/noise in signals. It can be used to distinguish between AC and DC portions of signals, and to verify software by monitoring an I/O port being toggled in software, to determine if a software routine is executing and how often. These are just a few applications of this versatile and powerful device.

Learning to use an oscilloscope is essentially a six-step process:

*Step 1* Learn the general oscilloscope terminology. This means reading the documentation provided with the oscilloscope or going on-line and looking for

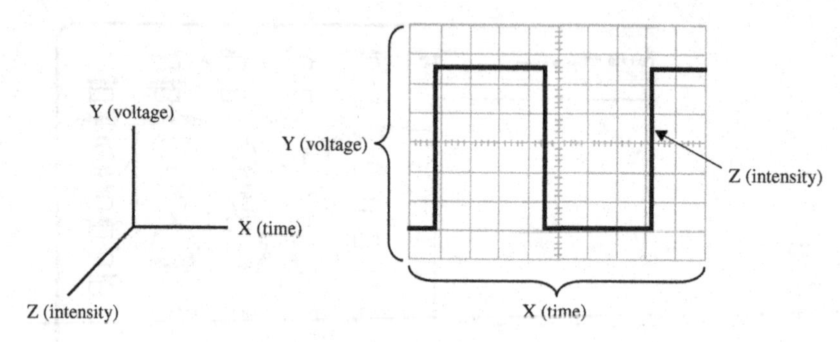


**Figure A-7**
Oscilloscope graph.[7]

an oscilloscope tutorial to tell you the types of waveforms that can be observed by your particular scope, how to measure these waveforms, and what features differentiate one oscilloscope from another. The latter is especially important, because different types of oscilloscopes are better suited for certain types of circuits.

*Step 2* Grounding. This means learning how to ground yourself and the oscilloscope correctly. This step is very important for your safety and for protecting the circuit being measured.

*Step 3* Learn the controls of your oscilloscope and how to set them. Again, read the documentation provided by the manufacturer of the oscilloscope, buy a book, or go on the web and understand what all the controls are, what they do, and how to set them.

*Step 4* Using the probes. Learn how to connect the probes to the oscilloscope as well as to the board, without shorting anything and be able to take accurate measurements.

*Step 5* Calibrating the scope. Most oscilloscopes have a waveform to allow the user to calibrate the oscilloscope after the setup of the probe is complete, to insure that the oscilloscope has the correct electrical properties when you are taking measurements.

*Step 6* Learn how to take the actual measurements. While some digital oscilloscopes will take measurements automatically, it is useful to learn how to take measurements manually to allow you to be able to work with a variety of oscilloscopes and to verify automatically generated results. This means learning how to read the graphs generated on the display, including any grid markings and what these markings represent.

In short, the key to learning to use an oscilloscope is to PRACTICE! So, use an oscilloscope to measure various signals outlined to you by the instructor on the hardware available in your lab.

## 3 Section III Projects

Software developers often get hardware from board vendors or internal hardware designers with schematics that may or may not be accurate, and with these hardware documents they are then responsible for figuring out how to get the hardware up and running. It is usually easier to start with an off-the-shelf board similar to the hardware that will be in the final product and to get the major system software components running on the off-the-shelf board. Even if there is only a simulator, this allows the software engineers to familiarize themselves with the development environment and system software components that will be used during the design and development process of the embedded system.

These projects were designed to reflect what usually happens on the job, where those responsible for embedded software development have to adapt to new development environments and new system software elements on various types of hardware and/or hardware simulation environments as they move from project to project.

### *3.1 Project 1: Introduction to IDEs*

An IDE is commonly provided as the software development environment for designing the software of a particular embedded system.

After software has been designed with an IDE, it is then downloaded onto the target board to run. Sometimes an IDE is provided with an emulator that plugs into an embedded board or sometimes this is provided by an OS vendor. Whatever the situation, it is important to understand what is provided with this IDE (compilers, linkers, debuggers, etc.) and what additional tools can be integrated with it.

In this project, the reader will become familiar with whatever IDE is available for the project, and will learn how to write, compile, and download the software to be run and debugged on a simulator or real target board.

Either execute the tutorial provided in the vendor's IDE manual or instructor for an IDE available in your project, or go to a commercial embedded OS vendor (there are at least 100 of them), such as Wind River for VxWorks/Linux, Mentor Graphics for Nucleus Plus, or Microsoft for WinCE, and get an evaluation copy of their IDE for their OS platform running on a simulator or available hardware in your project. You can also search the Internet for free, open-source OS IDE packages.

### *3.2 Project 2: Using an Embedded OS*

In this project you will need to work with two different OSs, each with their relative IDEs, either on a target board or using simulators. If the IDE used in Project 1 doesn't support a multitasking embedded OS, first repeat Project 1 with an IDE specifically for an embedded

OS until you are comfortable working with the IDE. If the IDE in Project 1 is an OS vendor IDE then this can count as one of the OSs. Implement Exercises 1–3 on both OSs.

*Exercise 1: Multitasking and Intertask Synchronization*

This exercise will provide an introduction in the basic concepts of process management in a multitasking OS. Create five tasks that will operate on shared memory, concurrently calling some function that increments a shared variable, for example.

In order to implement this correctly, only one task should be allowed to update the shared variable at a time, meaning mutual exclusion mechanisms need to be used. To provide mutual exclusion to the critical region (i.e., the increment function), you will need to use whatever synchronization mechanisms are available in the OS (i.e., semaphores).

*Exercise 2: Producer/Consumer*

In this exercise, you will implement a popular problem in the OS arena, reflected in the producer/consumer problem. This problem concerns multiple concurrent tasks modifying a bounded buffer.

Create two concurrent tasks, a consumer task and a producer task. The consumer task randomly removes data from the bounded buffer as long as the buffer is not empty. The producer task randomly adds data to the bounded buffer as long as the buffer is not full. The key here is to insure that memory is managed correctly by the tasks, keeping in mind the restrictions on the size of memory in embedded devices (i.e., memory is allocated on an as-needed basis and deallocated when no longer in use). Also, use task synchronization mechanisms to insure that the buffer boundaries are respected.

Have these tasks generate output reflecting what is produced or consumed, as well as outline in your project report what guarantees there are that there will be no contention between the two tasks.

*Exercise 3: Dining Philosophers*

A very popular concurrency problem which many designers using an embedded OS face is reflected in the dining philosophers problem. In dining philosophers, five philosophers are seated at a round table, with food in the center of the table. As shown in Figure A-8, each philosopher has a fork to the left and right of them, for a total of five forks.

Each philosopher requires both the fork on the left and the fork on the right in order to eat, thus requiring that each philosopher share their forks with their neighbors. Based on the assumption that no philosopher is ever done eating, the problem is that is if all philosophers grab their right forks, then everyone is waiting for the left fork, and vice-versa. Furthermore, if philosophers don't put down their forks, a neighboring philosopher starves.

The purpose of this problem is to use the OS and create a scheme in which philosophers can get both forks to eat and no philosopher starves. Create five tasks, one for each philosopher,

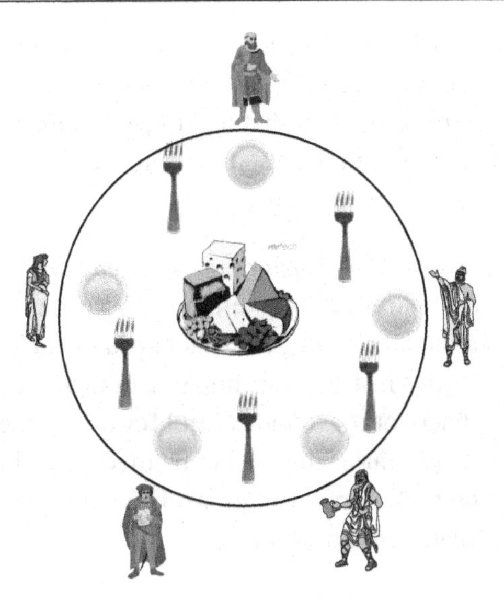

**Figure A-8**
Dining philosophers.

as well as any other functions that may be required (GetForkA, GetForkB, PutForkA, PutForkB, etc.). Use the OS task synchronization mechanisms to insure that there are no deadlocks and that critical sections of code (i.e., a fork) are respected by other task "philosophers." Have your application output the number of philosophers eating after any fork acquisition.

### 3.3 Project 3: Middleware and the Java Virtual Machine (JVM)

The purpose of this project is to gain experience using embedded middleware software, specifically a JVM.

*Exercise 1: Learning Java*
This exercise is for readers who have no Java experience. In the real world, even the most knowledgeable experts may have to quickly learn and evaluate a new language for possible use in a design. Thus, in this exercise, go online to http://www.oracle.com/technetwork/java/index.html and run the Java Tutorial or purchase a book on Java.

*Exercise 2: Two Different Embedded JVM Standards*
For this exercise, download evaluations of two different JVMs that support two different embedded Java standards (a pJava and J2ME CLDC/MIDP JVMS, a J2ME CDC and J2ME CLDC JVMS, a pJava and a J2ME/CDC/Personal Profile JVMS, etc.) and write three Java applications assigned by your instructor or implement the Java code samples provided at http://www.oracle.com/technetwork/java/index.html for one of the standards to run on both JVMs. It is not necessary to have the JVMs ported to an embedded OS in this exercise, as

long as the JVMs are embedded JVMs. This means that you can use embedded JVMs ported to a Windows PC or UNIX station directly from Sun Microsystems, for example. Report the differences in libraries that were missing or that had to be renamed (if any) in the application to get the application to run on both implementations. How did your results compare to the marketed notion that Java is platform independent?

*Exercise 3: Two Different Embedded JVMs Supporting the Same Standard*

For this exercise, download evaluations of two different JVMs that support the same standard (i.e., Sun's pJava implementation and Tao's Elate/Intent pJava implementation) and write three Java applications assigned by your instructor or implement three java code samples provided at http://www.oracle.com/technetwork/java/index.html for one of the standards to run on both JVMs. Report the differences in libraries that had to be modified (if any) in the application to get the application to run on both implementations. How did your results compare to the marketed notion that Java is platform-independent?

## 4 Section IV Projects

In the projects in this section, the student or team of students will apply the technical foundations provided in the text, along with the methodology for creating an embedded systems architecture, to take a product concept and develop a prototype of it. Specifically, this will be the implementation of Phase 1 of the Embedded System Design and Development Lifecycle Model[8] shown in Figure A-9. Phase 1 of this model is Creating the Architecture, which is the process of planning the design of the embedded system. The first four steps of this process—product concept, preliminary analysis of requirements, creation of architecture design, and developing versions of the architecture—are reflected in the projects of this section.

As stated at the beginning of this Appendix, it is recommended that the students work in teams, especially for these projects, for the real-world experience needed in how to work in a team environment, as well as because many of the activities in a project can be executed in parallel by different team members.

### 4.1 Project 1: Product Concept and Preliminary Analysis of Requirements

The key to a successful project is the planning and preparation. This means starting with defining the project goals, objectives, scope, timetable, and the roles and responsibilities of the various team members.

In this project, create a project plan that outlines what it is exactly that you want to build, with what features, in what time frame, and with what resources.

### 4.2 Project 2: Creating an Architecture

As discussed in the main text, several industry methodologies can be adopted in designing an embedded system's architecture, such as the Rational Unified Process (RUP),

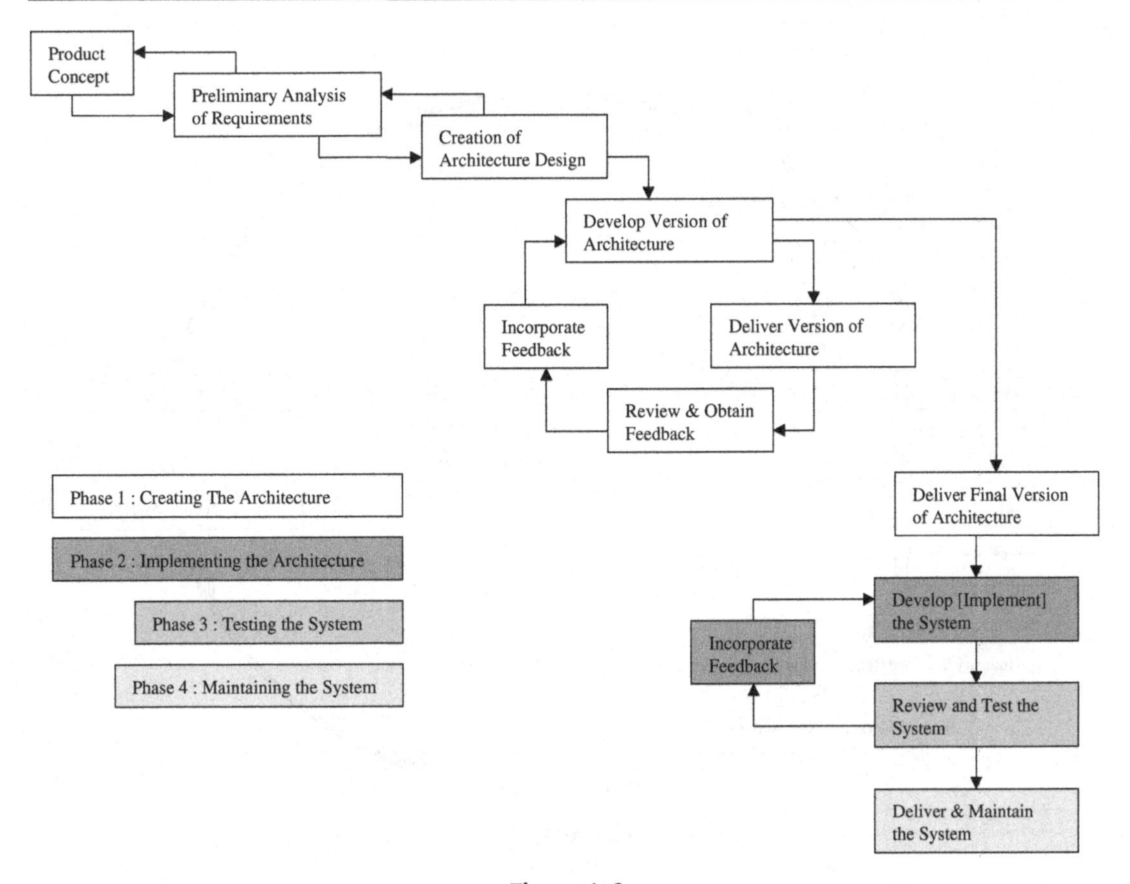


**Figure A-9**
Embedded Systems Design and Development Lifecycle Model.[8]

Attribute-Driven Design (ADD), and the Object-Oriented Process (OOP), to name just a few. However, if given a limited amount of time and resources as in most lab courses, this project will provide exercises based on the pragmatic approach I have introduced in the main text for creating an architecture. This process is a six-stage process that combines and simplifies many of the key elements of these popular industry different methodologies. In review, these stages are:

*Stage 1* Have a solid technical base.
*Stage 2* Understand the Architecture Business Cycle (ABC) of embedded systems.
*Stage 3* Define the architectural patterns and reference models.
*Stage 4* Create the architectural structures.
*Stage 5* Document the architecture.
*Stage 6* Analyze and evaluate the architecture.

It is assumed that the reader has by now gained a solid technical base from the material studies in previous chapters and understands how to further independently learn new skills

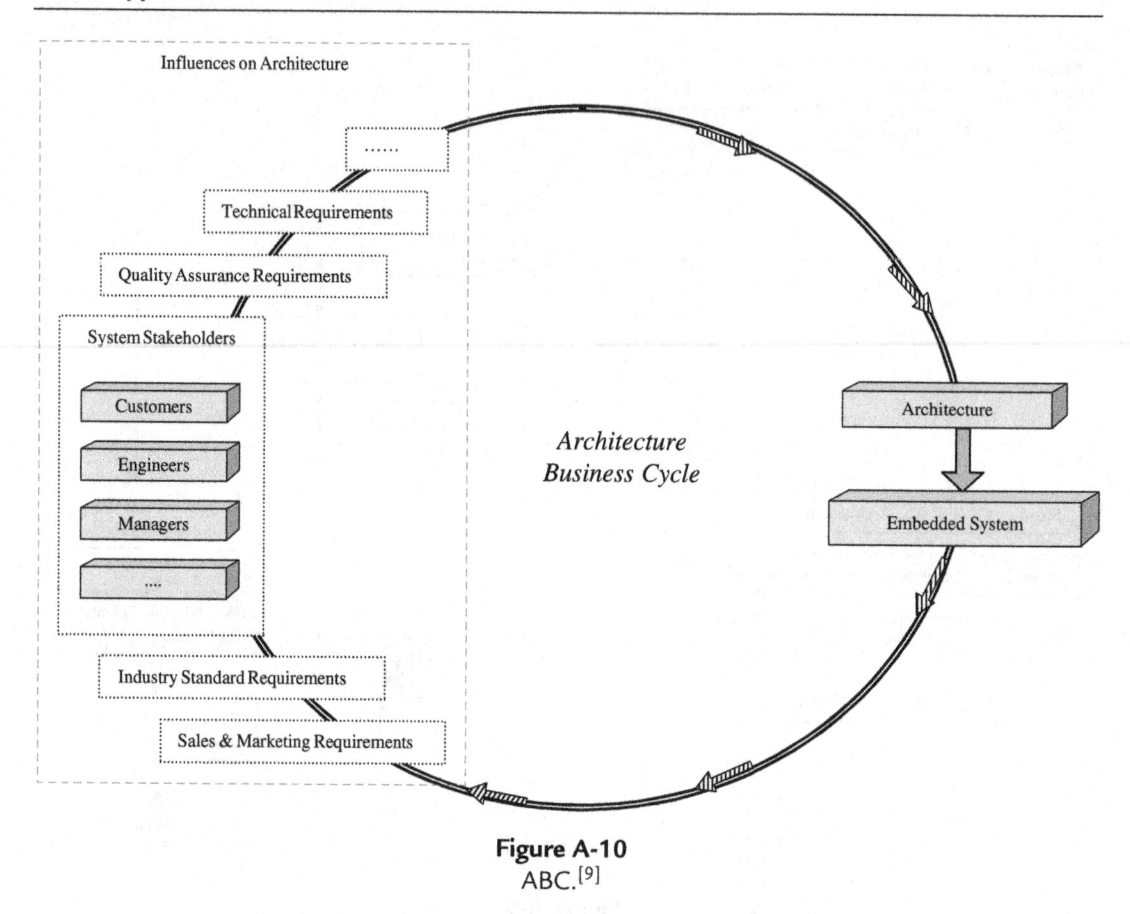


**Figure A-10**
ABC.[9]

if necessary for implementing a project. Thus, the exercises in this project will reflect Stages 2–6, defining the ABC (Exercise 1), defining architectural patterns and reference models (Exercise 2), creating and documenting the architectural structures (Exercise 3), and analyzing and evaluating the architecture (Exercise 4).

*Exercise 1: Defining the ABC*

In this exercise, create a requirements specification identifying all the stakeholders (i.e., the professor, yourself, your team mates), and gather both the functional and non-functional requirements for your product concept. Use the process outlined in Chapter 11 of the main text, including how to derive hardware and/or software functionality that may be necessary to include in your design based on these requirements. (See Figure A-10.)

*Exercise 2: Defining the Architectural Patterns and Reference Models*

Again using Chapter 11 of the main text as a reference, create a product high-level systems reference specification outlining at least two possible patterns that your team can implement in a final prototype, and include a reference model of the various possible hardware and software components in each pattern.

*Exercise 3: Creating and Documenting the Architectural Structures*

In this exercise, the patterns defined in the high-level specification in Exercise 3 will be turned into an actual architecture. Using the Internet, recommendations in the main text, or any other relevant sources, define which architectural structures will be used to represent your design. Then, either using an industry methodology (such as Universal Modeling Language (UML)) or your own team's scheme, document the structures for each pattern.

*Exercise 4: Analyzing and Evaluating the Architecture*

In this exercise, the team will review the architectures and select which would be most feasible to implement in a final prototype. Using the guidelines under Section 11.1.6 on "Stage 6" in the main text, identify and document the risks associated with the design, and highlight what the critical success factors are.

Review the architecture with the instructor and prepare necessary action items to integrate the required measures to mitigate the risks back into the architecture design, as well as any other changes to requirements. Repeat the other exercises of this project as necessary until the architecture has been finalized.

### 4.3 Project 3: Finalizing the Prototype

Given the architecture created in Project 2, implement a prototype using the tools available to you in your lab or from other sources. The team lead is responsible for partitioning the various responsibilities to the team members and insuring that each team member has what they need to do their work. The team lead is also responsible for gathering the day-to-day reports from team members on progress and/or missed milestones, as well as reports reflecting any meetings of the team. At the end of this project, you should have the prototype itself, the architecture documentation, and the reports reflecting the process of implementing the architecture into the finalized prototype.

## Endnotes

[1] *Embedded System Building Blocks*, p. 61, J. J. Labrosse, Cmp Books, 1995; *Embedded Microcomputer Systems*, p. 3, J. W. Valvano, CL Engineering, 2nd edn, 2006.

[2] *Beginner's Guide to Reading Schematics*, p. 49, R. J. Traister and A. L. Lisk, TAB Books, 2nd edn, 1991.

[3] http://tuttle.merc.iastate.edu/ee333/spice/pspicetutorial/basics/pspicebasics.htm

[4] http://www.ee.olemiss.edu/atef/engr360/tutorial/ex1.html

[5] http://www.ee.nmt.edu/~rison/ee321_fall02/Tutorial.html

[6] Tektronix, "Digital Storage Oscilloscopes TDS1002, TDS1012, TDS2002, TDS2012, TDS, TDS2022, TDS2024" datasheet, 2014.

[7] "Oscilloscope Tutorial," Hitesh Tewari.

[8] The Embedded Systems Design and Development Lifecycle Model is specifically derived from the Software Engineering Institute (SEI)'s Evolutionary Delivery Lifecycle Model and the Software Development Stages Model.

[9] Based on the software architectural brainchildren of the Software Engineering Institute (SEI); *Software Architecture in Practice*, L. Bass, P. Clements, and R. Kazman, Addison-Wesley, 2nd edn, or go to www.sei.cmu.edu for more information, 2003.

APPENDIX B

# Schematic Symbols

These symbols are a subset of industry-accepted schematic symbols representing electronic elements on schematic diagrams. Note that symbols for the same electronic device can differ internationally, as well as depending on what standards are being adhered to by a particular organization (NEMA, IEEE, JEDEC, ANSI, IEC, DoD, etc.). If there are any unfamiliar symbols within a schematic, it is always best to ask the engineer responsible for drafting the schematic.

| | | |
|---|---|---|
| AC Voltage Source |  | Voltage source that generates AC (alternating current). Because an AC voltage source can come from a variety of different components (outlet, oscillator, signal generator, etc.), the type of AC source is typically stated somewhere on the schematic. |
| Antenna<br>Balanced<br>General<br>Loop [Shielded]<br>Loop [Unshielded]<br>Unbalanced |   | A transducer made up of conductive material (wires, metal rod, etc.) used to transmit and receive wireless signals (radio waves, IR, etc.). |
| Attenuator<br>Fixed<br>Variable |   | Commonly used for a variety of purposes including to extend the dynamic range of certain devices (power meters, amplifiers, etc.), reduce signal levels, match circuits, and to balance out unequal signal levels in transmission lines, just to name a few. |
| Battery/DC Cell | 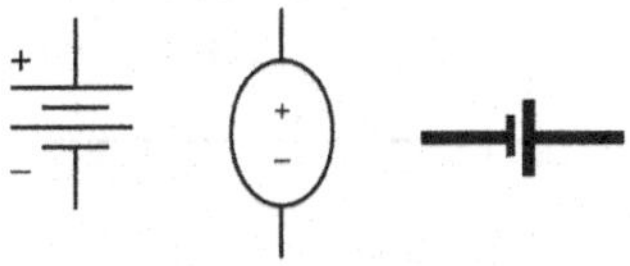 + − | Voltage source that creates voltage through a chemical reaction in a battery. |

| | | |
|---|---|---|
| Buffer [Amplifier] | 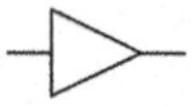 | An electrical device that is used to provide compatibility between two signals (i.e., interfacing the output of a CMOS to the input of a TTL). |
| Capacitors<br>Non-Polarized General<br>Feedthrough<br>Non-Polarized/ Bipolar Fixed<br>Polarized Fixed [Electrolytic]<br>Variable Single<br>Split-Stator |   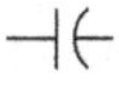   + | A passive electrical element that stores electric charge in a circuit.<br>The feedthrough capacitor is uniquely constructed to provide lower parallel inductance, better decoupling capability for all high d*I*/d*t* environments, significant noise reduction in digital circuits, EMI suppression, broadband I/O filtering, VCC power line conditioning in comparison to other types of capacitors.<br>A non-polar/bipoal fixed capacitor has no "implicit" polarity, thus can be connected in any way into a circuit.<br>A fixed polarized has an "explicit" polarity, thus there is only one way to connect it into a circuit.<br>A variable capacitor has capacitance that can be varied on-the-fly.<br>The split-stator capacitor is a variable capacitor used to preserve balance in a circuit. |
| Cathode<br>Cold<br>Directly Heated<br>Indirectly Heated |   | (1) The positively charged pole (terminal) of a voltage source. (2) The negatively charged electrode of a device (e.g., diode) that acts as an electron source. |
| Cavity Resonator | 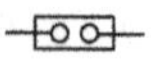 | A component that contains and maintains an oscillating electromagnetic field. |
| Circuit Breaker [Single Pole] |  | An electrical component that ensures that a current load doesn't get too large by shutting down the circuit when its overheat sensor senses there is too much current. |

| Name | Symbol | Description |
|---|---|---|
| Coaxial Cable | | A type of cabling made up of two layers of physical wire: one center wire and one grounded wire shielding. Coaxial cables also include two layers of insulation: one between the wire shielding and center wire, and one layer above the wire shielding. The shielding allows for a decrease in interference (electrical, radiofrequency, etc.). |
| Connector<br>Female<br><br>Male | 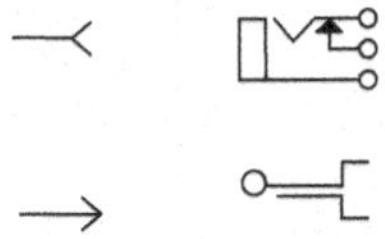 | An electrical component that interconnects different types of subsystems. |
| Crystal | 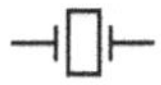 | An electrical component that determines an oscillator's frequency. A crystal is typically made up of two metal plates separated by quartz, with two terminals attached to each plate. The quartz within a crystal vibrates when current is applied to the terminals and it is this frequency that impacts the frequency at which the oscillator operates. |
| Delay Line | | An electrical component that delays the transmission of a signal. |
| Diode | | Two-terminal semiconductor device that allows current flow in one direction and blocks current flow in the opposite direction. |
| Diode | C A | Diodes made of Si or Ge are more common and typically cheaper. |
| Light-Emitting Diode (LED) |  C A | All diodes emit light; LEDs are made from special semiconductive material, which optimizes the light. |
| Photodiode/ Photosensitive | 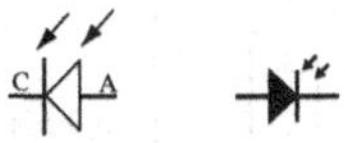 C A | The photodiode optimizes the fact that diodes are light sensitive, i.e., solar cells that convert light into electrical energy. |
| Zener | C A | The Zener diode is designed with a specific reverse breakdown voltage that causes a specific amount of resistance when blocking current flow. |

| | | |
|---|---|---|
| Flip-Flop | | Flip-flops are sequential circuits that are called such because they function by alternating (flip-flopping) between two output states (0 and 1) depending on the input. |
| RS | 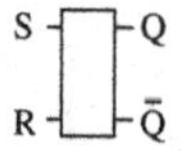 S, R, Q, $\bar{Q}$ | The RS flip-flop alternates between the two output lines (Q and Q NOT) depending on the R and S inputs. |
| JK | 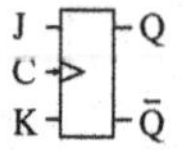 J, C, K, Q, $\bar{Q}$ | The JK flip-flop alternates between the two output lines (Q and Q NOT) depending on the J and K inputs, as well as the clock signal (C). |
| D | 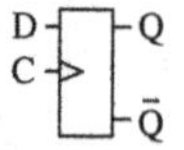 D, C, Q, $\bar{Q}$ | The D flip-flop alternates between the two output lines (Q and Q NOT) depending on the D input, as well as the clock signal (C). |
| Fuse | 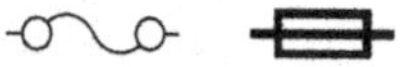 | An electrical component that protects a circuit from too much current by breaking the circuit when a high enough current passes through it. |
| Gates | Standard NEMA ANSI | A more complex type of electronic switching circuit designed to perform logical binary operations. |
| AND | 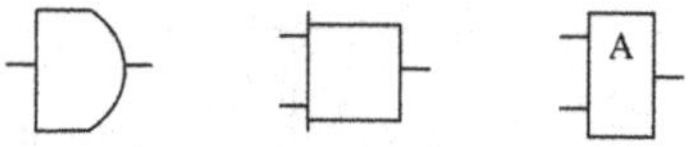 A | An AND gate's output is 1 when both inputs are 1. |
| OR |  OR | An OR gate's output is 1 if either of the inputs is 1. |
| NOT/Inverter | | An electrical device that inverts (i.e., a HIGH to a LOW or vice-versa) a logical level input. |
| NAND | 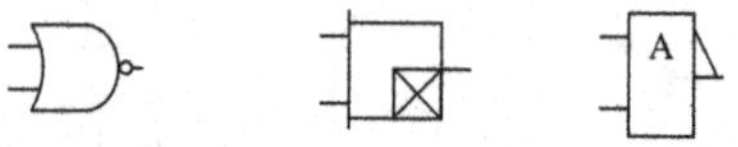 A | A NAND gate's output is 0 when both inputs are 1. |
| NOR | 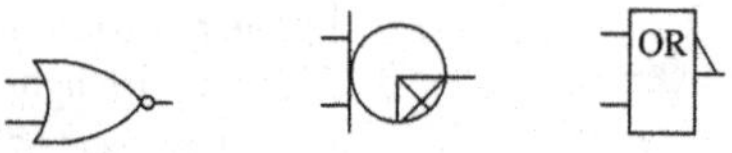 OR | A NOR gate's output is 0 if either of the inputs is 1. |
| XOR | 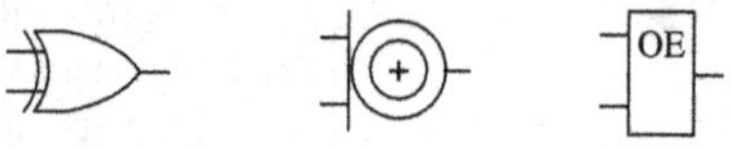 + OE | A XOR gate's output is 1 (or On, or High, etc.) if only one input (but not both) is 1. |
| Ground | | |
| Circuit |  | |
| Earth |  | An arbitrary point for "0" potential voltage that a circuit is connected to. |
| Special |  | |

| Component | Symbol | Description |
|---|---|---|
| Inductor [Coil] | | An electrical component made up of coiled wire surrounding some type of core (air, iron, etc.). When a current is applied to a conductor, energy is stored in the magnetic field surrounding the coil allowing for an energy storing and filtering effect. |
| Air Core |  | |
| Iron Core | 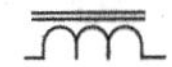 | |
| Tapped |  | |
| Variable |  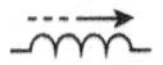 | |
| Integrated Circuit (IC) Generic | 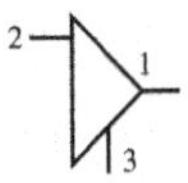 2 1 3 | An electrical device made up of several other discrete electrical active elements, passive elements, and devices (transistors, resistors, etc.)—all fabricated and interconnected on a continuous substrate (chip). |
| Jack | | An electrical device designed to accept a plug. |
| Coaxial | 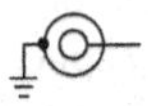 | |
| Two-Conductor |  | |
| Three-Conductor |  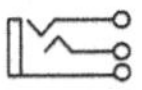 | |
| Phono | 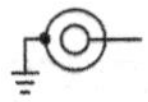 | |
| Lamp | | An electrical device that produces light. |
| Incandescent |    | An incandescent lamp produces light via heat. |
| Neon | 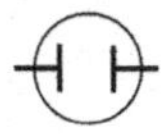  | A neon lamp produces light via neon gas. |
| Xenon Flash | 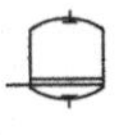 | Xenon flash lamps produces large flashes of bright white light via some combination that includes high voltage, electrodes, and gas. |
| Loudspeaker | 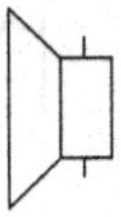 | A type of transducer that converts variations of electrical current into sound waves. |

| Meter | ± ± ± M | A measurement device that measures some form of electrical energy. |
|---|---|---|
| Ammeter | A | An ammeter is a meter that measures current in a circuit. |
| Galvanometer | G | A galvanometer is a meter that measures small amounts of current in a circuit. |
| Voltmeter | V | A voltmeter is a meter that measures voltage. |
| Wattmeter | W P | A wattmeter is a meter that measures power. |
| Microphone | | A type of transducer that converts sound waves into electrical current. |
| Condenser Microphone | | A condenser microphone uses changes in capacitance in proportion to changes in sound waves to produce its conversions. |
| Dynamic | | A dynamic microphone uses a coil that vibrates to sound waves and a magnetic field to generate a voltage that varies in proportion to sound variations. |
| Electret | | An electret microphone is dynamic and uses a small transistor amplifier. |
| ECM Microphone | | |
| Plug | | |
| Two-Conductor | | |
| Three-Conductor | | Electrical components used to connect one subsystem into the jack of another subsystem. |
| Phono/RCA | | |
| Rectifier | | |
| Semiconductor | | |
| Silicon-Controlled [thyristor] | A G C; A G P N P N C | A four-layer PNPN (three P–N junction) device that functions as a cross between a diode and a transistor. |
| Tube-Type | | |

| | | |
|---|---|---|
| Relay | | An electromagnetic switch. |
| Double Pole Double Throw |  | A DPDT relay contains two contacts that can be toggled both ways (on and off). |
| Double Pole Single Throw | 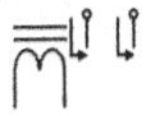 | An DPST relay contains two contacts that can only be switched on or off. |
| Single Pole Double Throw | 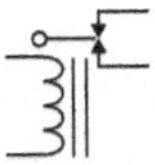 | An SPDT relay contains one contact that can be toggled both ways (on and off). |
| Single Pole Single Throw | 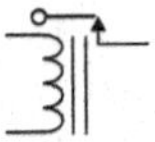 | An SPST relay contains one set of contacts and can only be switched one way (on or off). |
| Resistor | | Used to limit current in a circuit. |
| Fixed | 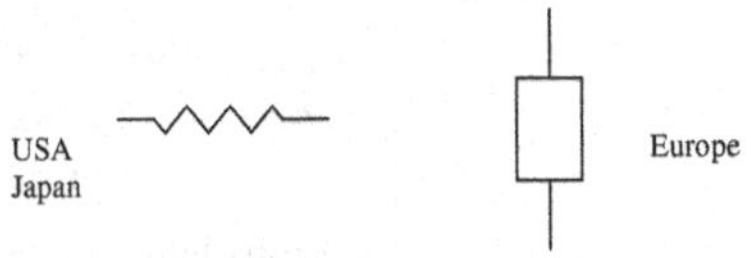 USA Japan Europe | Fixed resistors have resistance value set at manufacturing. |
| Variable/ Potentiometers | 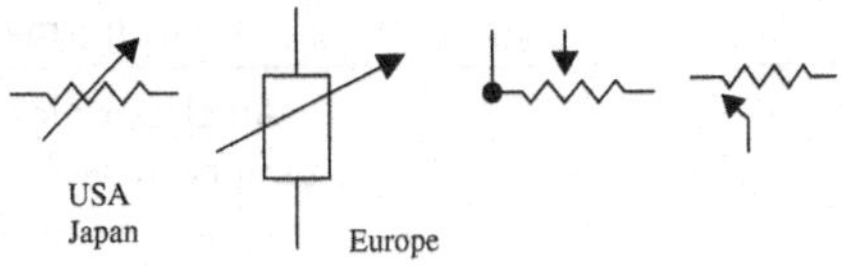 USA Japan Europe | Variable resistors have a dial that allows a change in resistance values on-the-fly. |
| Rheostat | 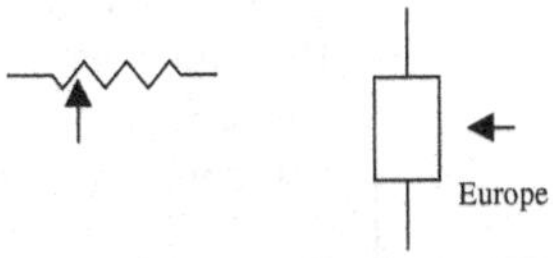 Europe | Potentiometer variable-resistance control, similar to potentiometer, but with three discrete areas of control. The part of the circuit connected off the arrow can be varied in resistance to the two circuit points connected to the other two leads. |
| Photosensitive/ Photoresistor |  | Photosensitive resistors have resistance that changes on-the-fly depending on the amount of light photo resistors are exposed to. |
| Thermally Sensitive/ Thermistor |  T | Thermistors have resistance changes on-the-fly depending on the temperature the thermistor is exposed to (typically resistance decreases as temperature increases). |

| | | |
|---|---|---|
| Switch | | An electrical device used to turn an electrical current flow on or off. |
| Single Pole Single Throw | | An SPST switch contains one set of contacts that can only be switched on or off (one way). |
| Single Pole Double Throw | 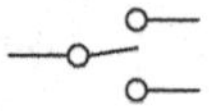 | An SPDT switch contains one contact that can be toggled on and off (both ways). |
| Double Pole Single Throw | 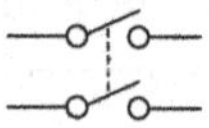 | An DPST switch contains two contacts that can only be switched one way (on or off) |
| Double Pole Double Throw | 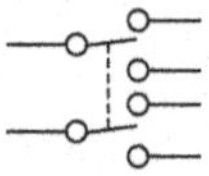 | A DPDT contains two contacts that can be toggled on and off (both ways). |
| Normally Closed Push Button |  | A normally closed push button switch is a switch in the form of a button that is normally closed. |
| Normally Open Push Button | 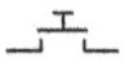 | A normally open push button switch is a switch in the form of a button that is normally open. |
| Thermocouple | | An electronic circuit that relays temperature differences via current flowing through two wires joined at either end. Each wire is made of different materials with one junction of the connected wires at the stable lower temperature, while the other junction is connected at the temperature to measured. |
| Transformer | | A type of inductor that can increase or decrease the voltage of an AC signal. |
| Air Core | 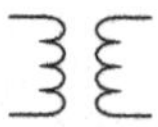 | |
| Iron Core |  | |
| Tapped Primary | 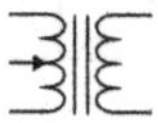 | |
| Tapped Secondary | 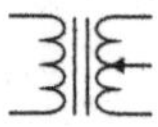 | |

| | | |
|---|---|---|
| Transistor | | Three-terminal semiconductor device that provides current amplification, as well as can act as a switch. |
| Bipolar/BJT (Bipolar Junction Transistor) | 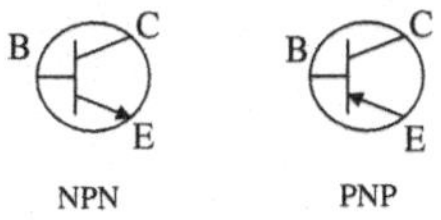 B C E NPN; B C E PNP | A bipolar transistor is made of alternating P-type and N-type semiconductive material (meaning both positive and negative charges used to conduct—hence the name "bipolar." |
| Junction FET (Field-Effect Transistor) | 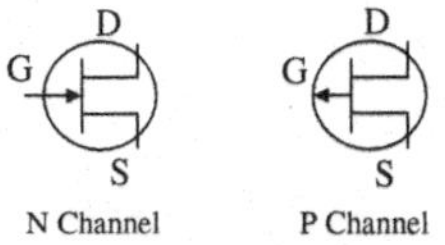 G D S N Channel; G D S P Channel | A junction FET is also made up of both N-type and P-type material, however, unipolar, involving only positive or negative charges to conduct. Gate voltage applied across P-N junction. |
| MOSFET (Metal Oxide Semiconductor FET) | 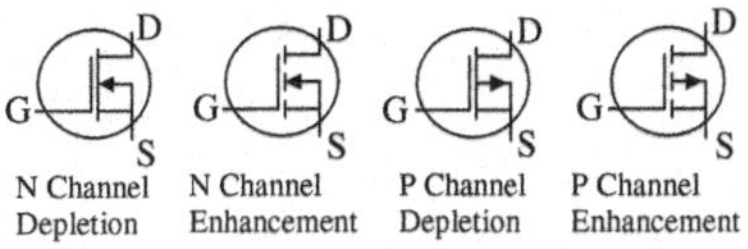 G D S N Channel Depletion; G D S N Channel Enhancement; G D S P Channel Depletion; G D S P Channel Enhancement | A MOSFET is similar to junction FET, except gate voltage applied across insulator. |
| Photosensitive (Phototransistor) |  B C E | A photosensitive transistor is a bipolar transistor designed to leverage a transistor's sensitivity to light. |
| Wire | | |
| Wire | | Wires are conductors that carry signals between the other components on a board. |
| Wires Crossing and Connected |  | The wires crossing and connected symbol represents two connected wires. |
| Wires Crossing and Unconnected | 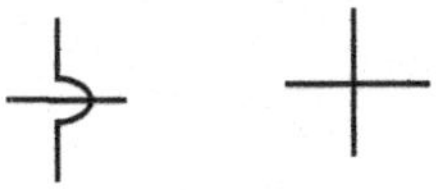 | The wires crossing and unconnected symbol represents two wires crossing on the board but not connected. |

APPENDIX C

# Acronyms and Abbreviations

## A

| | |
|---|---|
| AC | Alternating Current |
| ACK | Acknowledge |
| A/D | Analog-to-Digital |
| ADC | Analog-to-Digital Converter |
| ALU | Arithmetic Logic Unit |
| AM | Amplitude Modulation |
| AMP | Ampere |
| ANSI | American National Standards Institute |
| AOT | Ahead-of-Time |
| API | Application Programming Interface |
| ARIB-BML | Association of Radio Industries and Business of Japan |
| AS | Address Strobe |
| ASCII | American Standard Code for Information Interchange |
| ASIC | Application Specific Integrated Circuit |
| ATM | Asynchronous Transfer Mode, Automated Teller Machine |
| ATSC | Advanced Television Standards Committee |
| ATVEF | Advanced Television Enhancement Forum |

## B

| | |
|---|---|
| BDM | Background Debug Mode |
| BER | Bit Error Rate |
| BIOS | Basic Input/Output System |
| BML | Broadcast Markup Language |
| BOM | Bill of Materials |
| bps | Bits per Second |
| BSP | Board Support Package |
| BSS | "Block Started by Symbol," "Block Storage Segment," "Blank Storage Space," ... |

## C

| | |
|---|---|
| CAD | Computer-Aided Design |
| CAN | Controller Area Network |
| CAS | Column Address Select |
| CASE | Computer-Aided Software Engineering |
| CBIC | Cell-Based IC or Cell-Based ASIC |
| CDC | Connected Device Configuration |
| CEA | Consumer Electronics Association |
| CEN | European Committee for Standardization |
| CISC | Complex Instruction Set Computer |
| CLDC | Connected Limited Device Configuration |
| CMOS | Complementary Metal Oxide Silicon |
| COFF | Common Object File Format |
| CPLD | Complex Programmable Logic Device |
| CPU | Central Processing Unit |
| CRT | Cathode Ray Tube |
| CTS | Clear-to-Send |

## D

| | |
|---|---|
| DAC | Digital-to-Analog Converter |
| DAG | Data Address Generator |
| DASE | Digital TV Applications Software Environment |
| DAVIC | Digital Audio Visual Council |
| dB | Decibel |
| DC | Direct Current |
| D-Cache | Data Cache |
| DCE | Data Communications Equipment |
| Demux | Demultiplexor |
| DHCP | Dynamic Host Configuration Protocol |
| DIMM | Dual Inline Memory Module |
| DIP | Dual Inline Package |
| DMA | Direct Memory Access |
| DNS | Domain Name Server, Domain Name System, Domain Name Service |
| DoD | Department of Defense |

| | |
|---|---|
| DPRAM | Dual Port RAM |
| DRAM | Dynamic Random Access Memory |
| DSL | Digital Subscriber Line |
| DSP | Digital Signal Processor |
| DTE | Data Terminal Equipment |
| DTVIA | Digital Television Industrial Alliance of China |
| DVB | Digital Video Broadcasting |

## E

| | |
|---|---|
| EDA | Electronic Design Automation |
| EDF | Earliest Deadline First |
| EDO RAM | Extended Data Out Random Access Memory |
| EEMBC | Embedded Microprocessor Benchmarking Consortium |
| EEPROM | Electrically Erasable Programmable Read-Only Memory |
| EIA | Electronic Industries Alliance |
| ELF | Extensible Linker Format |
| EMI | Electromagnetic Interference |
| EPROM | Erasable Programmable Read-Only Memory |
| ESD | Electrostatic Discharge |
| EU | European Union |

## F

| | |
|---|---|
| FAT | File Allocation Table |
| FCFS | First Come First Served |
| FDA | Food and Drug Administration (USA) |
| FDMA | Frequency Division Multiple Access |
| FET | Field-Effect Transistor |
| FFS | Flash File System |
| FIFO | First In First Out |
| FM | Frequency Modulation |
| FPGA | Field Programmable Gate Array |
| FPU | Floating-Point Unit |
| FSM | Finite-State Machine |
| FTP | File Transfer Protocol |

## G

| | |
|---|---|
| GB | Gigabyte |
| Gbit | Gigabit |
| GCC | GNU C Compiler |
| GDB | GNU Debugger |
| GHz | Gigahertz |
| GND | Ground |
| GPS | Global Positioning System |
| GUI | Graphical User Interface |

## H

| | |
|---|---|
| HAVi | Home Audio/Video Interoperability |
| HDL | Hardware Description Language |
| HL7 | Health Level Seven |
| HLDA | Hold Acknowledge |
| HLL | High-level Language |
| HTML | HyperText Markup Language |
| HTTP | HyperText Transport Protocol |
| Hz | Hertz |

## I

| | |
|---|---|
| IC | Integrated Circuit |
| $I^2C$ | Inter-Integrated Circuit Bus |
| I-Cache | Instruction Cache |
| ICE | In-Circuit Emulator |
| ICMP | Internet Control Message Protocol |
| IDE | Integrated Development Environment |
| IEC | International Engineering Consortium |
| IEEE | Institute of Electrical and Electronics Engineers |
| IETF | Internet Engineering Task Force |
| IGMP | Internet Group Management Protocol |
| INT | Interrupt |
| I/O | Input/Output |
| IP | Internet Protocol |

| | |
|---|---|
| IPC | Interprocess Communication |
| IR | Infrared |
| IRQ | Interrupt ReQuest |
| ISA | Instruction Set Architecture |
| ISA Bus | Industry Standard Architecture Bus |
| ISO | International Standards Organization |
| ISP | In-System Programming |
| ISR | Interrupt Service Routine |
| ISS | Instruction Set Simulator |
| ITU | International Telecommunication Union |

## J

| | |
|---|---|
| JEDEC | JEDEC Solid State Technology Association, formerly known as the Joint Electron Devices Engineering Council |
| JIT | Just-in-Time |
| J2ME | Java 2 MicroEdition |
| JTAG | Joint Test Access Group |
| JVM | Java Virtual Machine |

## K

| | |
|---|---|
| kB | Kilobyte |
| kbit | Kilobit |
| kbps | Kilobits per Second |
| kHz | kilohertz |
| KVM | K Virtual Machine |

## L

| | |
|---|---|
| LA | Logic Analyzer |
| LAN | Local Area Network |
| LCD | Liquid Crystal Display |
| LED | Light-Emitting Diode |
| LIFO | Last In First Out |
| LSb | Least Significant Bit |
| LSB | Least Significant Byte |
| LSI | Large-Scale Integration |

## M

| | |
|---|---|
| mΩ | Milliohm |
| MΩ | Megaohm |
| MAN | Metropolitan Area Network |
| MCU | Microcontroller |
| MHP | Multimedia Home Platform |
| MIDP | Mobile Information Device Profile |
| MIPS | Millions of Instructions per Second, Microprocessor without Interlocked Pipeline Stages |
| MMU | Memory Management Unit |
| MOSFET | Metal Oxide Silicon Field-Effect Transistor |
| MPSD | Modular Port Scan Device |
| MPU | Microprocessor |
| MSb | Most Significant Bit |
| MSB | Most Significant Byte |
| MSI | Medium-Scale Integration |
| MTU | Maximum Transfer Unit |
| MUTEX | Mutual Exclusion |

## N

| | |
|---|---|
| ns | Nanosecond |
| NAK | NotAcKnowledged |
| NAT | Network Address Translation |
| NCCLS | National Committee For Clinical Laboratory Standards |
| NEMA | National Electrical Manufacturers Association |
| NFS | Network File System |
| NIST | National Institute of Standards and Technology |
| NMI | Non-Maskable Interrupt |
| NTSC | National Television Standards Committee |
| NVRAM | Non-Volatile Random Access Memory |

## O

| | |
|---|---|
| OCAP | Open Cable Application Forum |
| OCD | On-Chip Debugging |
| OEM | Original Equipment Manufacturer |

| | |
|---|---|
| OO | Object Oriented |
| OOP | Object Oriented Programming |
| OS | Operating System |
| OSGi | Open Systems Gateway Initiative |
| OSI | Open Systems Interconnection |
| OTP | One-Time Programmable |

# P

| | |
|---|---|
| PAL | Programmable Array Logic, Phase Alternating Line |
| PAN | Personal Area Network |
| PC | Personal Computer |
| PCB | Printed Circuit Board, Process Control Block |
| PCI | Peripheral Component Interconnect |
| PCP | Priority Ceiling Protocol |
| PDA | Personal Data Assistant |
| PDU | Protocol Data Unit |
| PE | Presentation Engine, Processing Element |
| PID | Proportional Integral Derivative |
| PIO | Parallel Input/Output |
| PIP | Priority Inheritance Protocol, Picture-In-Picture |
| PLC | Programmable Logic Controller, Program Location Counter |
| PLD | Programmable Logic Device |
| PLL | Phase Locked Loop |
| POSIX | Portable Operating System Interface X |
| POTS | Plain Old Telephone Service |
| PPC | PowerPC |
| PPM | Parts Per Million |
| PPP | Point-to-Point Portocol |
| PROM | Programmable Read-Only Memory |
| PSK | Phase Shift Keying |
| PSTN | Public Switched Telephone Network |
| PTE | Process Table Entry |
| PWM | Pulse Width Modulation |

# Q

| | |
|---|---|
| QA | Quality Assurance |

# R

| | |
|---|---|
| RAM | Random Access Memory |
| RARP | Reverse Address Resolution Protocol |
| RAS | Row Address Select |
| RF | Radiofrequency |
| RFC | Request for Comments |
| RFI | Radiofrequency Interference |
| RISC | Reduced Instruction Set Computer |
| RMA | Rate Monotonic Algorithm |
| RMS | Root Mean Square, Rate Monotonic Scheduling |
| ROM | Read-Only Memory |
| RPM | Revolutions per Minute |
| RPU | Reconfigurable Processing Unit |
| RTC | Real-Time Clock |
| RTOS | Real-Time Operating System |
| RTS | Request to Send |
| RTSJ | Real Time Specification for Java |
| R/W | Read/Write |

# S

| | |
|---|---|
| SBC | Single-Board Computer |
| SCC | Serial Communications Controller |
| SECAM | Système Électronique pour Couleur avec Mémoire |
| SEI | Software Engineering Institute |
| SIMM | Single Inline Memory Module |
| SIO | Serial Input/Output |
| SLD | Source Level Debugger |
| SLIP | Serial Line Internet Protocol |
| SMPTE | Society of Motion Picture and Television Engineers |
| SMT | Surface Mount |

| | |
|---|---|
| SNAP | Scalable Node Address Protocol |
| SNR | Signal-to-Noise Ratio |
| SoC | System – On-Chip |
| SOIC | Small Outline Integrated Circuit |
| SPDT | Single Pole Double Throw |
| SPI | Serial Peripheral Interface |
| SPST | Single Pole Single Throw |
| SRAM | Static Random Access Memory |
| SSB | Single Sideband Modulation |
| SSI | Small-Scale Integration |

## T

| | |
|---|---|
| TC | Technical Committee |
| TCB | Task Control Block |
| TCP | Transmission Control Protocol |
| TDM | Time Division Multiplexing |
| TDMA | Time Division Multiple Access |
| TFTP | Trivial File Transfer Protocol |
| TLB | Translation Lookaside Buffer |
| TTL | Transistor-Transistor Logic |

## U

| | |
|---|---|
| UART | Universal Asynchronous Receiver Transmitter |
| UDM | Universal Design Methodology |
| UDP | User Datagram Protocol |
| ULSI | Ultra-Large-Scale Integration |
| UML | Universal Modeling Language |
| UPS | Uninterruptible Power Supply |
| USA | United States of America |
| USART | Universal Synchronous Asynchronous Receiver Transmitter |
| USB | Universal Serial Bus |
| UTP | Untwisted Pair |

## V

| | |
|---|---|
| VHDL | Very-High-Speed Integrated Circuit Hardware Design Language |
| VLIW | Very-Long Instruction Word |
| VLSI | Very-Large-Scale Integration |
| VME | Versa Module Eurocard |
| VoIP | Voice Over Internet Protocol |
| VPN | Virtual Private Network |

## W

| | |
|---|---|
| WAN | Wide Area Network |
| WAT | Way-Ahead-of-Time |
| WDT | Watchdog Timer |
| WLAN | Wireless Local Area Network |
| WML | Wireless Markup Language |
| WOM | Write-Only Memory |

## X

| | |
|---|---|
| XCVR | Transceiver |
| XHTML | eXtensible HyperText Markup Language |
| XML | eXtensible Markup Language |

APPENDIX D

# Glossary

## A

**Absolute Memory Address** The physical address of a specific memory cell.

**Accumulator** A special processor register used in arithmetic and logical operations to store an operand used in the operation, as well as the results of the operation.

**Acknowledge (ACK)** A signal used in bus and network "handshaking" protocols as an acknowledgment of data reception from another component on the bus (on an embedded board for bus handshaking) or from another embedded system via some networking transmission medium (for network handshaking).

**Active High** Where a logic value of "1" is a higher voltage than a logic value of "0" in a circuit.

**Active Low** Where a logic value of "0" is a higher voltage than a logic value of "1" in a circuit.

**Actuator** A device used for converting electrical signals into physical actions, commonly found in flow-control valves, motors, pumps, switches, relays, and meters.

**Adder** A hardware component that can be found in a processor's CPU that adds two numbers.

**Address Bus** An address bus carries the addresses (of a memory location, or of particular status/control registers) between board components. An address bus can connect processors to memory, as well as processors to each other.

**Ahead-of-Time Compiler (AOT)** See *Way-Ahead-of-Time (WAT) Compiler*.

**Alternating Current (AC)** An electric current whose voltage source changes polarity of its terminals over time, causing the current to change direction with every polarity change.

**Ammeter** A measurement device that measures the electrical current in a circuit.

**Ampere** The standard unit for measuring electrical current, defined as the charge per unit time (meaning the number of coulombs that pass a particular point per second).

**Amplifier** A device that magnifies a signal. There are many types of amplifiers (log, linear, differential, etc.), all differing according to what how they modify the input signal.

**Amplitude** A signal's size. For an AC signal it can be measured via the high point of an AC wave from the equilibrium point (center) to the wave's highest peak or by performing the RMS (root mean square) mathematical scheme, which is by (1) finding the square of the waveform function, (2) averaging the value of the result of step (1) over time, and (3) taking the square root of the results of step (2). For a DC signal, it is its voltage level.

**Amplitude Modulation (AM)** The transmission of data signals via modifying (modulating) the amplitude of a waveform to reflect the data (i.e., a "1" bit being a wave of some amplitude, and a "0" bit being a wave with a different amplitude).

**Analog** Data signals represented as a continuous stream of values.

**Analog-to-Digital Converter (ADC)** A device that converts analog signals to digital signals.

**AND Gate** A gate whose output is 1 when both inputs are 1.

**Anion** A negative ion, meaning an atom that gains electrons.

**Anode** (1) The negatively charged pole (terminal) of a voltage source. (2) The positively charged electrode of a device (i.e., diode), which accepts electrons (allowing a current to flow through the device).

**Antenna** A transducer made up of conductive material (wires, metal rod, etc.) used to transmit and receive wireless signals (radio waves, IR, etc.).

**Antialiased Fonts** Fonts in which a pixel color is the average of the colors of surrounding pixels. It is a commonly used technique in digital televisions for evening (smoothing) displayed graphical data.

**Application Layer** The layer within various models (OSI, TCP/IP, Embedded Systems Model, etc.), which contains the application software of an embedded device.

**Application Programming Interface (API)** A set of subroutine calls that provide an interface to some type of component (usually software) within an embedded device (OS APIs, Java APIs, MHP APIs, etc.).

**Application Specific Integrated Circuit (ASIC)** An application-specific ISA-based IC that is customized for a particular type of embedded system or in support of a particular application within an embedded system. There are mainly full-custom, semi-custom, or programmable types of ASICs. PLDs and FPGAs are popular examples of (programmable) ASICs.

**Architecture** See *Instruction Set Architecture.*

**Arithmetic Logic Unit (ALU)** The component within a processor's CPU, which executes logical and mathematical operations.

**Aspect Ratio** A ratio of width to height (in memory the number of bits per address to the total number of memory addresses, the size or resolution of a display, etc.).

**Assembler** A compiler that translates assembly language into machine code.

**Astable Multivibrator** A sequential circuit in which there is no state it can hold stable in.

**Asynchronous** A signal or event that is independent of, unrelated to, and uncoordinated with a clock signal.

**Attenuator** A device that reduces (attenuates) a signal (the opposite of what an amplifier does).

**Autovectoring** The process of managing interrupts via priority levels rather than relying on an external vector source.

## B

**Background Debug Mode (BDM)** Components used in debugging an embedded system. BDM components include BDM hardware on the board (a BDM port and an integrated debug monitor in the master CPU), and debugger on the host (connected via a serial cable to BDM port). BDM debugging is sometimes referred to as On-Chip Debugging (OCD).

**Bandwidth** On any given transmission medium, bus, or circuitry: the frequency range of an analog signal (in hertz, the number of cycles of change per second) or digital signal (in bps, the number of bits per second) traveling through it (as in the case of a bus or transmission medium) or being processed by it (as in the case of a processor).

**Basic Input/Output System (BIOS)** Originally the boot up firmware on x86-based PCs, now available for many off-the-shelf embedded x86-based boards and a variety of embedded OSs.

**Battery** A voltage source where voltage is created through a chemical reaction within it. A battery is made up of two metals submerged within a chemical solution, called an electrolyte, that is in liquid (wet cell) or paste (dry cell) form. Basically, the two metals respond with different ionic state after they are exposed to the electrolyte. Wet cells are used in automobiles (car batteries) and dry cells are used in many different types of portable embedded systems (radios, toys, etc.).

**Baud Rate** The total number of bits per some unit of time (kbits/s, Mbits/s, etc.) that can be transmitted over some serial transmission link.

**Bias** An offset (such as voltage or current) applied to a circuit or electrical element to modify the behavior of the circuit or element.

**Big-Endian** A method of formatting data in which the lowest-order byte (or bits) is stored in the highest byte (or bits). For example, if the highest-order bits are from left to right in descending order in a particular 8-bit ISA, big-endian mode in this ISA would mean that bit 0 of the data would be stored from left to right in ascending order (the value of "B3h/10110011b" would be stored as "11001101b"). In a 32-bit ISA, for instance, where the highest-order bytes are stored from left to right in descending order, big-endian mode in this ISA would mean that byte 0 of the data is stored from left to right in the word in ascending order (i.e., the value of "B3A0FF11h" would be stored as "11FFA0B3h").

**Binary** A base-2 number system used in computer systems, meaning the only two symbols are a "0" or "1." These symbols are used in a variety of combinations to represent all data.

**Bit Error Rate (BER)** The rate at which a serial communication stream loses and/or transfers incorrect data bits.

**Bit Rate** The (number of actual data bits transmitted/total number of bits that can be transmitted) * the baud rate of the communications channel.

**Black-Box Testing** Testing that occurs with a tester that has no visibility into the internal workings of the system (no schematics, no source code, etc.) and is basing testing on general product requirements documentation.

**Block Started by Symbol (BSS)** BSS is several different things depending on the context and who is asked, including "Block Started by Symbol," "Block Storage Segment," and "Blank Storage Space." The term "BSS" originated from the 1960s and while not everyone agrees on what the BSS acronym stands for, it is generally agreed upon that BSS is a statically allocated memory space containing the source code's uninitialized variables (data).

**Board Support Package (BSP)** A software provided by many embedded off-the-shelf OS vendors that allows their OSs to be ported more easily over various boards and architectures. BSPs contain the board and architecture-specific libraries required by the OS, and allow for the device drivers to be integrated more easily for use by the OS through BSP APIs.

**Bootloader** Firmware in an embedded system that initializes the system's hardware and system software components.

**Breakpoint** A debugging mechanism (hardware or software), which stops the CPU from executing code.

**Bridge** A component on an embedded board that interconnects and interfaces two different buses.

**Bus** A collection of wires that interconnect components on an embedded board.

**Byte** A byte is defined as being some 8-bit value.

**Byte Code** Byte (8-bit) size opcodes that have been created as a result of high-level source code (such as Java or C#) being compiled by a compiler (a Java or some Intermediate Language (IL) compiler) on a host development machine. It is byte code that is translated by a Virtual Machine (VM), such as the Java Virtual Machine (JVM) or a .NETCE Compact Framework virtual machine.

**Byte Order** How data bits and/or bytes are represented and stored in a particular component of a computer system.

## C

**Cache** Very fast memory that holds copies of a subset of main memory, to allow for faster CPU access to data and instructions typically stored in main memory.

**Capacitor** Used to store electrostatic energy, a capacitor is basically made up of conductors (two parallel metal plates), separated by an insulator (a dielectric such as air, ceramic, polyester, mica, etc.). The energy, itself, is stored in an electric field created between the two plates given the right environment.

**Cathode** (1) The positively charged pole (terminal) of a voltage source. (2) The negatively charged electrode of a device (diode) that acts as an electron source.

**Cation** A positive ion, meaning an atom that has lost electrons.

**Cavity Resonator** A component that contains and maintains an oscillating electromagnetic field.

**Central Processing Unit (CPU)** (1) The master/main processor on the board. (2) The processing unit within a processor that is responsible for executing the indefinite cycle of fetching, decoding, and executing instructions while the processor has power.

**Checksum** A numerical value calculated from some set of data to verify the integrity of that data, commonly used for data transmitted via a network.

**Chip** See *Integrated Circuit (IC).*

**Circuit** A closed system of electronic components in which a current can flow.

**Circuit Breaker** An electrical component that insures that a current load doesn't get too large by shutting down the circuit when its overheat sensor detects there is too much current.

**Class** Used in object-oriented schemes and languages to create objects, a class is a prototype (type description) that is made up of some combination of interfaces, functions (methods), and variables.

**Clear-Box Testing** See *White-Box Testing.*

**Clock** An oscillator that generates signals resulting in some type of waveform. Most embedded boards include a digital clock that generates a square waveform.

**Coaxial Cable** A type of cabling made up of two layers of physical wire, one center wire and one grounded wire shielding. Coaxial cables also include two layers of insulation, one between the wire shielding and center wire, and one layer above the wire shielding. The shielding allows for a decrease in interference (electrical, radiofrequency, etc.).

**Compiler** A software tool that translates source code into assembly code, an intermediary language opcode, or into a processor's machine code directly.

**Complex Instruction Set Computer (CISC)** A general-purpose ISA which typically is made up of many, more-complex operations and instructions than other general-purpose ISAs.

**Computer-Aided Design (CAD) Tools** Tools used to create technical drawings and documentation of the hardware, such as schematic diagrams.

**Computer-Aided Software Engineering (CASE) Tools** Design and development tools that aid in creating an architecture and implementing a system, such as UML tools and code generators.

**Conductor** A material that has fewer impediments to an electric current (meaning it more easily loses/gains valence electrons) allowing for an electrical current to flow more easily through it than through other types of materials. Conductors typically have three or fewer valence electrons.

**Connector** An electrical component that interconnects different types of subsystems.

**Context** The current state of some component within the system (registers, variables, flags, etc.).

**Context Switch** The process in which a system component (interrupts, an OS task, etc.) switches from one state to another.

**Coprocessor** A slave processor that supports that master CPU by providing additional functionality, and that has the same ISA as the master processor.

**Coulomb** In electronics, the charge of one electron is too small to be of practical use, so in electronics, the unit for measuring a charge is called a coulomb (named after Charles Coulomb who founded Coulomb's law), and is equal to that of $6.28 \times 10^{18}$ electrons.

**Critical Section** A set of instructions that are flagged to be executed without interruption.

**Cross-Compiler** A compiler that generates machine code for hardware platforms that differs from the hardware platform the compiler is actually residing and running on.

**Crystal** An electrical component that determines an oscillator's frequency. A crystal is typically made up of two metal plates separated by quartz, with two terminals attached to each plate. The quartz within a crystal vibrates when current is applied to the terminals, and it is this frequency that impacts the frequency at which the oscillator operates.

**Current** A directed flow of moving electrons.

## D

**Daisy Chain** A type of digital circuit in which components are connected in series (in a "chain-like" structure) and where signals pass through each of the components down through the entire chain. Components at the top of the chain essentially can impact (slow down, block, etc.) a signal being received by components further down in the chain.

**Datagram** What the networking data received and processed by the networking layer of the OSI model or corresponding layer in other networking models (the Internet layer in the TCP/IP model) is called.

**Data Communications Equipment (DCE)** The device that the DTE wants to serially communicate with, such as an I/O device connected to the embedded board.

**Data Terminal Equipment (DTE)** The initiator of a serial communication, such as a PC or embedded board.

**Deadlock** An undesired result related to the use of an operating system, in which a set of tasks are blocked, awaiting an event to unblock that is controlled by one of the tasks in the blocked set.

**Debugger** A software tool used to test for, track down, and fix bugs.

**Decimal** A base-10 number system, meaning there are 10 symbols (0–9), used in a variety of combinations to represent data.

**Decoder** A circuit or software that translates encoded data into the original format of the data.

**Delay Line** An electrical component that delays the transmission of a signal.

**Demodulation** Extracting data from a signal that was modified upon transmission to include a carrier signal and the added transmitted data signal.

**Demultiplexor (Demux)** A circuit which connects one input to more than one output, where the value of the input determines which output is selected.

**Device Driver** Software that directly interfaces with and controls hardware.

**Dhrystone** A benchmarking application which simulates generic systems programming applications on processors, used to derive the MIPS (millions of instructions per second) value of a processor.

**Die** The portion of an integrated circuit that is made of silicon, that can either be enclosed in some type of packaging or connected directly to a board.

**Dielectric** An insulative layer of material found in some electrical components, such as capacitors.

**Diode** A two-terminal semiconductor device that allows current flow in one direction and blocks current that flows in the opposite direction.

**Differentiator** A circuit that calculates a mathematical (calculus) derivative output based on a given input.

**Digital** A signal that is expressed as some combination of one of two states – a "0" or "1."

**Digital Signal Processor (DSP)** A type of processor that implements a datapath ISA and is typically used for repeatedly performing fixed computations on different sets of data.

**Digital Subscriber Line (DSL)** A broadband networking protocol that allows for the direct digital transmission of data over twisted pair wired (POTS) mediums.

**Digital-to-Analog Converter (DAC)** A device that converts digital signals to analog signals.

**Direct Current (DC)** Current that flows constantly in the same direction in a circuit. DC current is defined by two variables: *polarity* (the direction of the circuit) and *magnitude* (the amount of current).

**Direct Memory Access (DMA)** A scheme in which data is exchanged between I/O and memory components on a board with minimal interference from and use of the master processor.

**Disassembler** Software that reverse-compiles the code, meaning machine language is translated into assembly language.

**Domain Name Service (DNS)** An OSI model session layer networking protocol that converts domain names into internet (network layer) addresses.

**Dual Inline Memory Module (DIMM)** A type of packaging which memory ICs can come in, specifically a minimodule (PCB) that can hold several ICs. A DIMM has protruding pins from one side (both on the front and back) of the module that connect into a main embedded motherboard, and where opposing pins (on the front and back of the DIMM) are each independent contacts.

**Dual Inline Package (DIP)** A type of packaging that encloses a memory IC, made up of ceramic or plastic material, with pins protruding from two opposing sides of the package.

**Dual Port Random Access Memory (DPRAM)** RAM that can connect to two buses allowing for two different components to access this memory simultaneously.

**Dynamic Host Configuration Protocol (DHCP)** A networking layer networking protocol that provides a framework for passing configuration information to hosts on a TCP/IP-based network.

**Dynamic Random Access Memory (DRAM)** RAM whose memory cells are circuits with capacitors that hold a charge in place (the charges or lack thereof reflecting the data).

## E

**Earliest Deadline First (EDF)** A real-time, preemptive OS scheduling scheme in which tasks are scheduled according to their deadline, duration, and frequency.

**Effective Address** The memory address generated by the software. This is the address that is then translated into the physical address of the actual hardware.

**Electrically Erasable Programmable Read-Only Memory (EEPROM)** A type of ROM which can be erased and reprogrammed more than once, the number of times of erasure and reuse depending on the EEPROM. The contents of EEPROM can be written and erased "in bytes" without using any special devices. This means the EEPROM can stay on its residing board and the user can connect to the board interface to access and modify EEPROM.

**Electricity** Energy generated by the flow of electrons through a conductor.

**Electron** A negatively charged subatomic particle.

**Emitter** One of three terminals of a bipolar transistor.

**Encoder** A device that encodes (translates) a set of data into another set of data.

**Endianness** See *Byte Order.*

**Energy** The amount of work performed that can be measured in units of joules (J) or watts * time.

**Erasable Programmable Read-Only Memory (EPROM)** A type of ROM that can be erased more than one time using other devices that output intense short wavelength, UV light into the EPROM package's built-in transparent window.

**Ethernet** One of the most common LAN protocols, implemented at physical and data-link layers of the OSI model.

**Extended Data Out Random Access Memory (EDO RAM)** A type of RAM commonly used as main and/or video memory; it is a faster type of RAM that can send a block of data and fetch the next block of data simultaneously.

## F

**Farad** The unit of measurement in which capacitance is measured.

**Field Programmable Gate Array (FPGA)** A type of programmable ASIC implementing the application specific ISA model.

**Firmware** Any software stored on ROM.

**Flash Memory** A CMOS-based faster and cheaper variation of EEPROM. Flash can be written and erased in blocks or sectors (a group of bytes). Flash can also be erased electrically, while still residing in the embedded device.

**Flip-Flop** One of the most commonly used types of latches in processors and memory circuitry. Flip-flops are sequential circuits that are called such because they function by alternating (flip-flopping) between both states (0 and 1) and the output is then switched (e.g., from 0 to 1 or from 1 to 0). There are several types of flip-flops, but all essentially fall under either the asynchronous or synchronous categories.

**Fuse** An electrical component that protects a circuit from too much current by breaking the circuit when a high enough current passes through it. Fuses can also be used in some types of ROMs as the mechanism to store data.

## G

**Galvanometer** A measurement device that measures smaller amounts of current in a circuit.

**Garbage Collector** A language-related mechanism that is responsible for deallocating unused memory at runtime.

**Gate** A more complex type of electronic switching circuit designed to perform logical binary operations, such as AND, OR, NOT, NOR, NAND, XOR.

**Glass-Box Testing** See *White-Box testing.*

**Ground** In a circuit, the negative reference point for all signals.

## H

**Half Duplex** An I/O communications scheme in which a data stream can be transmitted and received in either direction, but in only one direction at any one time.

**Handshaking** The process in which protocols are adhered to by components on a board or devices over a network that want to initiate and/or terminate communication.

**Hard Real Time** Describes a situation in which timing deadlines are always met.

**Hardware** All of the physical components of an embedded system.

**Harvard Architecture** A variation of the von Neumann model of computer systems, which differs from von Neumann in that it defines separate memory spaces for data and instructions.

**Heap** A portion of memory used by software for dynamic allocation of memory space.

**Heat Sink** A component on a board that extracts and dissipates heat generated by other board components.

**Henry** The unit of measurement for inductance.

**Hertz** The unit of measurement for frequency in terms of cycles per second.

**High-Level Language** A programming language that is semantically further away from machine language, more resembles human language, and is typically independent of the hardware.

**Hit Rate** A cache memory term indicating how often desired data is located in cache relative to the total number of times cache is searched for data.

**Host** The computer system used by embedded developers to design and develop embedded software; it can be connected to the embedded device and/or other intermediary devices for downloading and debugging the embedded system.

**Hysteresis** The amount of delay in a device's response to some change in input.

## I

**In-Circuit Emulator (ICE)** A device used in the development and debugging of an embedded system which emulates the master processor on an embedded board.

**Inductance** The storage of electrical energy within a magnetic field.

**Inductor** An electrical component made up of coiled wire surrounding some type of core (air, iron, etc.). When a current is applied to a conductor, energy is stored in the magnetic field surrounding the coil allowing for an energy storing and filtering effect.

**Infrared (IR)** Light in the THz (1000 GHz, $2 \times 10^{11} - 2 \times 10^{14}$ Hz) range of frequencies.

**Instruction Set Architecture (ISA)** The features that are built into an architecture's instruction set, including the types of operations, types of operands, and addressing modes, to name a few.

**Insulator** A type of component or material which impedes the movement of an electric current.

**Integrated Circuit (IC)** An electrical device made up of several other discrete electrical active elements, passive elements, and devices (transistors, resistors, etc.)—all fabricated and interconnected on a continuous substrate (chip).

**Interpreter** A mechanism that translates higher-level source code into machine code, one line or one byte code at a time.

**Interrupt** An asynchronous electrical signal.

**Interrupt Handler** The software that handles (processes) the interrupt and is executed after the context switch from the main instruction stream as a response to the interrupt.

**Interrupt Service Routine (ISR)** See *Interrupt Handler.*

**Interrupt Vector** An address of an interrupt handler.

**Inverter** A NOT gate that inverts a logical level input, such as from HIGH to a LOW or vice-versa.

## J

**Jack** An electrical device designed to accept a plug. There are many type of jacks, including coaxial, two-plug, three-plug, and phono, just to name a few.

**Joint Test Access Group (JTAG)** A serial port standard that defines an external interface to ICs for debugging and testing.

**Just-in-Time (JIT) Compiler** A higher-level language compiler that translates code via interpretation in the first pass, and then compiles into machine code that same code to be executed for additional passes.

## K

**Kernel** The component within all operating systems that contains the main functionality of the OS, such as process management, memory management, and I/O system management.

## L

**Lamp** An electrical device that produces light. There are many types of lamps used on different types of embedded devices, including neon (via neon gas), incandescent (producing light via heat), and xenon flash lamps (via some combination that includes high voltage, gas, and electrodes), to name a few.

**Large-Scale Integration (LSI)** A reference to the number of electronic components in an IC. An LSI chip is an IC containing 3000–100,000 electronic components per chip.

**Latch** A bistable multivibrator that has signals from its output fed back into its inputs, and can hold stable at only one of two possible output states: 0 or 1. Latches come in several different subtypes, including S–R, Gated S–R, and D.

**Latency** The length of elapsed time it takes to respond to some event.

**Least Significant Bit (LSb)** The bit furthest to the right of any binary version of a number.

**Least Significant Byte (LSB)** The 8 bits furthest to the right of any binary version of a number (e.g., the two digits furthest to the right of any hexadecimal version of a number larger than a byte).

**Light-Emitting Diode (LED)** Diodes that are designed to emit visible or IR light when in forward bias in a circuit.

**Lightweight Process** See *Thread.*

**Linker** A software development tool used to convert object files into executable files.

**Little-Endian** Data represented or stored in such a way that the LSB and/or the LSb is stored in the lowest memory address.

**Loader** A software tool that relocates developed software into some location in memory.

**Local Area Network (LAN)** A network in which all devices are within close proximity to each other, such as in the same building or room.

**Logical Memory** Physical memory as referenced from the software's point-of-view, as a one-dimensional array. The most basic unit of logical memory is the byte. Logical memory is made up of all the physical memory (registers, ROM, and RAM) in the entire embedded system.

**Loudspeaker** See *Speaker.*

**Low-Level Language** A programming language which more closely resembles machine language. Unlike high-level languages, low-level languages are hardware dependent, meaning there is typically a unique instruction set for processors with different architectures.

## M

**MAC Address** The networking address located on networking hardware. MAC addresses are internationally unique due to the management of allocation of the upper 24 bits of these addresses by the IEEE organization.

**Machine Language** A basic language consisting of 1 s and 0 s that hardware components within an embedded system directly transmit, store, and/or execute.

**Medium-Scale Integration (MSI)** A reference to the number of electronic components in an IC. An MSI chip is an IC containing 100–3000 electronic components per chip.

**Memory Cell** Physical memory circuit that can store one bit of memory.

**Memory Management Unit (MMU)** A circuit used to translate logical addresses into physical addresses (memory mapping), as well as handling memory security, controlling cache, handling bus arbitration between the CPU and memory, and generating appropriate exceptions.

**Meter** A measurement device that measures some form of electrical energy, such as voltage, current, or power.

**Microcontroller** Processors that have most of the system memory and peripherals integrated on the chip.

**Microphone** A type of transducer that converts sound waves into electrical current. There are many types of microphones used on embedded boards, including condenser microphones that use changes in capacitance in proportion to changes in sound waves to produce conversions, and dynamic microphones that use a coil that vibrates to sound waves, and a magnetic field to generate a voltage that varies in proportion to sound variations, to name a few.

**Microprocessor** Processors that contain a minimal set of integrated memory and I/O peripherals.

**Most Significant Bit (MSb)** The bit furthest to the left of any binary version of a number.

**Most Significant Byte (MSB)** The 8 bits furthest to the left of any binary version of a number (e.g., the two digits furthest to the left of any hexadecimal version of a number larger than a byte).

**Multitasking** The execution of multiple tasks in parallel.

**Multivibrator** A type of sequential logical circuit designed so that one or more of its outputs are fed back as input.

## N

**NAND Gate** A gate whose output is 0 when both inputs are 1.

**Noise** Any unwanted signal alteration from an input source or any part of the input signal generated from something other then a sensor.

**Non-Volatile Memory (NVM)** Memory that contains data or instructions that remain even when there is no power in the system.

**NOR Gate** A gate whose output is 0 if either of the inputs is 1.

**NOT Gate** See *Inverter.*

## O

**On-Chip Debugging (OCD)** Refers to debugging schemes in which debugging capabilities are built into the board and master processor.

**One-Time Programmable (OTP)** A type of ROM that can only be programmed (permanently) one time outside the manufacturing factory, using a ROM burner. OTPs are based upon bipolar transistors, in which the ROM burner burned out fuses of cells to program them to "1" using high voltage/current pulses.

**Operating System (OS)** A set of software libraries that serve two main purposes in an embedded system: providing an abstraction layer for software on top of the OS to be less dependent on hardware (making the development of middleware and applications that sit on top of the OS easier), and managing the various system hardware and software resources to ensure the entire system operates efficiently and reliably.

**OR Gate** A gate whose output is 1 if either of the inputs is 1.

## P

**Packet** A unit to describe some set of data being transmitted over a network at one time.

**Parallel Port** An I/O channel that can transmit or receive multiple bits simultaneously.

**Plug** An electrical component used to connect one subsystem into the jack of another subsystem. There are many types of plugs, such as two-conductor, three-conductor, and phono/RCA.

**Polling** Repeatedly reading a mechanism (such as a register, flag, or port) to determine if some event has occurred.

**Printed Circuit Board (PCB)** Thin sheets of fiberglass which all the electronics within the circuit sit on. The electric path of the circuit is printed in copper, which carries the electrical signals between the various components connected on the board.

**Process** A creation of the OS that encapsulates all the information that is involved in the execution of a program, such as a stack, PC, the source code and data.

## R

**Random Access Memory (RAM)** Volatile memory in which any location within it can be accessed directly (randomly, rather than sequentially from some starting point), and whose content can be changed more than once (the number depending on the hardware).

**Read-Only Memory (ROM)** A type of non-volatile memory that can be used to store data on an embedded system permanently.

**Rectifier** An electronic component that allows current to flow in only one direction.

**Reduced Instruction Set Computer (RISC)** An ISA that usually defines simpler operations made up of fewer instructions.

**Register** A combination of various flip-flops that can be used to temporarily store data or delay signals.

**Relay** An electromagnetic switch. There are many types of relays, including the DPDT (Double Pole Double Throw) relay which contains two contacts that can be toggled both ways (on and off), a DPST (Double Pole Single Throw) relay which contains two contacts that can only be switched on or off, a SPDT (Single Pole Double Throw) relay which contains one contact that can be toggled both ways (on and off), and an SPST

(Single Pole Single Throw) relay which contains one set of contacts and can only be switched one way (on or off).

**Real-Time Operating System (RTOS)** An OS in which tasks meet their deadlines, and related execution times are predictable (deterministic).

**Resistor** An electronic device made up of conductive materials that have had their conductivity altered in some fashion in order to allow for an increase in resistance.

**Romizer** A device used to write data to EPROMs.

## S

**Scheduler** A mechanism within the OS that is responsible for determining the order and the duration of tasks to run on the CPU.

**Semaphore** A mechanism within the OS which can be used to lock access to shared memory (mutual exclusion), as well as can be used to coordinate running processes with outside events (synchronization).

**Semiconductor** Material or electrical component whose base elements have a conductive nature that can be altered by introducing other elements into their structure, meaning it has the ability to behave both as a conductor (conducting part of the time) and as an insulator (blocking current part of the time).

**Serial Port** An I/O channel that can transmit or receive one bit at any given time.

**Speaker** A type of transducer that converts variations of electrical current into sound waves.

**Switch** An electrical device used to turn an electrical current flow on or off.

## T

**Target** The embedded system platform, connected to the host, being developed.

**Task** See *Process*.

**Thermistor** A resistor with a resistance changes on-the-fly depending on the temperature the thermistor is exposed to. A thermistor's resistor typically decreases as temperature increases.

**Thermocouple** An electronic circuit that relays temperature differences via current flowing through two wires joined at either end. Each wire is made of different materials with one junction of the connected wires at the stable lower temperature, while the other junctional connection is at the temperature to be measured.

**Thread** A sequential execution stream within a task. Threads are created within the context of a task, meaning a thread is bound to a task. Depending on the OS, a task can also own one or more threads. Unlike tasks, threads of a task share the same resources, such as working directories, files, I/O devices, global data, address space, and program code.

**Throughput** The amount of work completed in a given period of time.

**Tolerance** Represents at any one time how much more or less precise the parameters of an electrical component are based on its actual labeled parameter value. The actual values should not exceed plus or minus ("±") the labeled tolerance.

**Transceiver** A physical device which receives and transmits data bits over a networking transmission medium.

**Transducer** An electrical device that transforms one type of energy into another type of energy.

**Transformer** A type of inductor that can increase or decrease the voltage of an AC signal.

**Transistor** Some combination of P-type and N-type semiconductor material, typically with three terminals connecting to one type of each material. Depending on the type of transistor, they can be used for a variety of purposes, such as current amplifiers (amplification), in oscillators (oscillation), in high-speed integrated circuits, and/or in switching circuits (DIP switches and push buttons commonly found on off-the-shelf reference boards).

**Translation Lookaside Buffer (TLB)** A portion of cache used by an MMU for allocating buffers that store address translations.

**Trap** Software and internal hardware interrupts that are raised by some internal event to the master processor.

**Truth Table** A table that outlines the possible input(s) of a logic circuit or Boolean equation, and the relative output(s) to the input(s).

**Twisted Pair** A pair of tightly interwrapped wires used for digital and analog data transmission.

## U

**Ultra-Large-Scale Integration (ULSI)** A reference to the number of electronic components in an IC. A ULSI chip is an IC containing over 1,000,000 electronic components per chip.

**Universal Asynchronous Receiver Transmitter (UART)** A serial interface that supports asynchronous serial transmission.

**Universal Synchronous Asynchronous Receiver Transmitter (USART)** A serial interface that supports both synchronous and asynchronous serial transmission.

**Untwisted Pair (UTP)** A pair of parallel wires used for digital and analog data transmission.

## V

**Very-Large-Scale Integration (VLSI)** A reference to the number of electronic components in an IC. A VLSI chip is an IC containing 100,000–1,000,000 electronic components per chip.

**Virtual Address** A memory location based upon a logical address that allows for the expansion of the physical memory space.

**Voltage Divider** An electrical circuit made up of a few or more resistors that can decrease the input voltage of a signal.

**Voltmeter** A measurement device that measures voltage.

## W

**Wattmeter** A measurement device that measures power.

**Way-Ahead-of-Time (WAT) Compiler** A compiler that translates higher-level code directly into machine code.

**White-Box Testing** Testing that occurs with a tester that has visibility into the system's inter-workings, such as having access to source code and schematics information.

**Wire** A component made up of conductive material that carries signals between components on a board (i.e., bus wires) or between devices (i.e., wired transmission mediums).

## X

**XOR Gate** A gate whose output is 1 (or on, or high) if only one input (but not both) is 1.

# Index

*Note:* Page numbers followed by "*f*", "*t*" and "*b*" refers to figures, tables and Boxes respectively.

## J



## L

## P

Printed in the United States
By Bookmasters